Research Design and Statistical Analysis

This fully updated fourth edition of *Research Design and Statistical Analysis* provides comprehensive coverage of the design principles and statistical concepts necessary to make sense of real data. The guiding philosophy is to provide a strong conceptual foundation so that readers can generalize to new situations they encounter in their research, including new developments in data analysis.

Key features include:

- Emphasis on basic concepts such as sampling distributions, design efficiency, and expected mean squares, relating the research designs and data analyses to the statistical models that underlie the analyses.
- Detailed instructions on performing analysis using both R and SPSS.
- Pedagogical exercises mapped to key topic areas to support students as they review their understanding and strive to reach their higher learning goals.

Incorporating the analyses of both experimental and observational data, and with coverage that is broad and deep enough to serve a two-semester sequence, this textbook is suitable for researchers, graduate students, and advanced undergraduates in psychology, education, and other behavioral, social, and health sciences.

The book is supported by a robust set of digital resources, including data files and exercises from the book in an Excel format for easy import into R or SPSS; R scripts for running example analysis and generating figures; and a solutions manual.

Caren M. Rotello is Professor Emerita at the University of Massachusetts Amherst. She received her PhD in psychology from Stanford University.

Jerome L. Myers is Professor Emeritus at the University of Massachusetts Amherst. He received his PhD in psychology from the University of Wisconsin.

Arnold D. Well is Professor Emeritus at the University of Massachusetts Amherst. He received his PhD in experimental psychology from the University of Oregon.

Robert F. Lorch, Jr. is Professor Emeritus at the University of Kentucky. He received his PhD in psychology from the University of Massachusetts Amherst.

Research Design and Statistical Analysis

Fourth Edition

Caren M. Rotello, Jerome L. Myers, Arnold D. Well and Robert F. Lorch, Jr.

Routledge
Taylor & Francis Group
NEW YORK AND LONDON

Designed cover image: carloscastilla via Getty Images

Fourth edition published 2025
by Routledge
605 Third Avenue, New York, NY 10158

and by Routledge
4 Park Square, Milton Park, Abingdon, Oxon, OX14 4RN

Routledge is an imprint of the Taylor & Francis Group, an informa business

First edition published by Lawrence Erlbaum Associates 1995
Third edition published by Routledge 2010

Library of Congress Cataloging-in-Publication Data
Names: Rotello, Caren M., author. | Myers, Jerome L., author. | Well, A. (Arnold), author. | Lorch, Robert Frederick, 1952– author.
Title: Research design and statistical analysis / Caren M. Rotello, Jerome L. Myers, Arnold D. Well and Robert F. Lorch, Jr.
Description: Fourth edition. | New York, NY : Routledge, 2025. | Includes bibliographical references and index.
Identifiers: LCCN 2024029026 (print) | LCCN 2024029027 (ebook) | ISBN 9781032592107 (hardback) | ISBN 9781032897288 (paperback) | ISBN 9781003453550 (ebook)
Subjects: LCSH: Experimental design. | Mathematical statistics.
Classification: LCC QA279 .M933 2025 (print) | LCC QA279 (ebook) | DDC 519.5—dc23/eng/20240724
LC record available at https://lccn.loc.gov/2024029026
LC ebook record available at https://lccn.loc.gov/2024029027

ISBN: 978-1-032-59210-7 (hbk)
ISBN: 978-1-032-89728-8 (pbk)
ISBN: 978-1-003-45355-0 (ebk)

DOI: 10.4324/9781003453550

Typeset in Sabon
by Apex CoVantage, LLC

Access the Support Material: www.routledge.com/9781032897288

Contents

Preface

Caren M. Rotello, Jerome L. Myers, Arnold D. Well, and Robert F. Lorch, Jr.

Like the previous editions, this fourth edition of *Research Design and Statistical Analysis* is intended as a resource for researchers and a textbook for graduate and advanced undergraduate students. The guiding philosophy of the book is to provide a strong conceptual foundation so that readers can generalize concepts to new situations they will encounter in their research, including new developments in data analysis and more advanced methods that are beyond the scope of this book. Toward this end, we continue to emphasize basic concepts such as sampling distributions, design efficiency, and expected mean squares, and we relate the research designs and data analyses to the statistical models that underlie the analyses. We discuss the advantages and disadvantages of various designs and analyses. We pay particular attention to the assumptions involved, the consequences of violating the assumptions, and alternative analyses if assumptions are seriously violated.

As in previous editions, an important goal is to provide coverage that is broad and deep enough so that the book can serve as a textbook for a two-semester sequence. Such sequences are common; typically, one semester focuses on experimental design and the analysis of data from such experiments, and the other semester focuses on observational studies and regression analyses of the data. Incorporating the analyses of both experimental and observational data within a single textbook provides continuity of concepts and notation in the typical two-semester sequence and facilitates developing relationships between analysis of variance and regression analysis. At the same time, it provides a resource that should be helpful to researchers in many different areas, whether analyzing experimental or observational data.

Content Overview

Also like the previous editions, this edition can be viewed as consisting of four parts:

1. Data exploration and basic concepts such as sampling distributions, elementary probability, principles of hypothesis testing, measures of effect size, properties of estimators, and confidence intervals on both differences among means and on standardized effect sizes.
2. Between-subject designs; these are designs with one or more factors in which each subject provides a single score. Key elements in the coverage are the statistical models underlying the analysis of variance for these designs, the role of expected mean squares in justifying hypothesis tests and in estimating effects of variables, the interpretation of interactions, and procedures for testing contrasts and for controlling Type 1 error rates for such tests.

3. Extension of these analyses to repeated-measures designs; these are designs in which subjects contribute several scores. We discuss nesting and counterbalancing of variables in research designs, present quasi-*F* ratios that provide approximate tests of hypotheses and consider the advantages and disadvantages of different repeated-measures and mixed designs.
4. The fourth section provides a comprehensive introduction to correlation and regression, with the goal of developing a general framework for analysis that incorporates both categorical and quantitative variables. The basic ideas of regression are developed first for one predictor, and then extended to multiple regression. The expanded section on multiple regression discusses both its usefulness as a tool for prediction and its role in developing explanatory models. Throughout, there is an emphasis on interpretation and on identifying common errors in interpretation and usage.

As in the third edition, each of these major sections ends with an Integrated Analysis chapter that brings together the major ideas and themes of the section in the context of a data analysis problem.

New to This Edition

Although the fourth edition shares the overall goals of the previous editions, there are many modifications and additions. These include (1) revisions of the chapters from the third edition with a focus on improving accessibility for students; (2) demonstrations and examples using R, a free software environment for statistical computing and highly flexible graphics; (3) new "using software" sections to provide more detailed support for users of both SPSS and R; (4) exercise solutions in both R and SPSS; and (5) new exercises. In addition to the modifications of the text, there is a substantial amount of support material available at www.routledge.com/9781032897288. This includes the following: (1) All of the data files used in the text and for most of the exercises in Excel format, for easy import into SPSS or R; (2) R scripts for running all of the example analyses and for generating most of the figures in each chapter; and (3) a solutions manual and the text's figures and tables for instructors only.

Use of Statistical Software

We assume that readers will have access to statistical software of some kind. Although we have used R for most of our examples, the analyses we illustrate are available in most statistical software. The new "using software" sections provide detailed instructions on how to use both R and SPSS to perform the analyses introduced in the immediately preceding sections.

Why R?

You may wonder why R should be the software to choose for data analysis and graphical display. After all, there are many commercial statistical software packages such as SAS, SPSS, Systat, and Minitab. However, the ever-growing population of R users suggests that

R has many appealing features, many of which provide significant advantages over other products. Some of these features are as follows:

- R is free, an important consideration for most students, instructors, and researchers.
- R runs on several platforms: Windows, MacOS, and Linux. R is fast, accurate, and flexible, and in many instances faster and more versatile than the commercial statistical packages.
- R has many functions that make it easy to transform or reorganize data, or to quickly summarize, manipulate, or combine data sets. We recommend taking the time to learn *dplyr* (Wickham, Francois, Henry, Muller, & Vaughan, 2023), a powerful R package for data transformation.
- R's graphing functions are exceptional, providing total control over font size, color, the position of the legend, the types of symbols, the organization of multiple panels, and many other aspects of a plot. We recommend taking the time to learn *ggplot2* (Wickham, 2016), a powerful R package for data visualization.
- R is open-source, which allows for many contributors. This in turn means that someone is likely to have developed a function to do what you seek to do. It also means that the latest statistical methods are often available to users of R before they are available in commercial packages. Furthermore,
    - if you can't find an appropriate function, you can write your own.
- Help is readily available from many sources: R's help function, manuals provided by the R developers, responses on the internet to others who may have had a similar question, and at dozens of websites containing explanations and lecture notes by many researchers and instructors.

We note that all analyses were performed using R 4.3.3 running within the RStudio 2023.12.1 environment. R and RStudio are available for free download at https://posit.co/download/rstudio desktop/ and are frequently updated. A set of manuals for R is available at https://cran.r-project.org/manuals.html#R-admin.

What About SPSS Users?

For instructors and students more comfortable with the pull-down menus available in SPSS, we continue to provide examples and instructions for doing so. We note that all analyses were performed using SPSS 29.0, and that future versions of the software may provide slightly different output, other options for analysis, or somewhat different syntax.

*G*Power*

We have used G*Power 3.1 for power analyses in many of the chapters and in some exercises. This very versatile software provides *a priori* analyses for many designs and analyses, as well as figures showing the central and noncentral distributions for the test and parameters under consideration. G*Power 3.1 can be freely downloaded from the website (https://psychologie.hhu.de/arbeitsgruppen/allgemeine-psychologie-und-arbeitspsychologie/gpower). Readers should register there to download the software and to be notified of any further updates. Excellent discussions of the use of G*Power have been written by Faul, Erdfelder, Lang, and Buchner (2007) and Faul, Erdfelder, Buchner, and Lang (2009).

Exercises

As in previous editions, each chapter ends with a set of exercises. New to this edition, we have labeled each exercise with a primary purpose. Answers to odd-numbered exercises are provided at the back of the book; all answers, and the R scripts used to generate them, are available in the password-protected Instructor's Solution Manual available at the book's website. There are more than 30 exercises in the four Integrated Analysis chapters to serve as a further review of the material in the preceding chapters.

Support Material

For the fourth edition, a variety of support materials are available at www.routledge.com/9781032897288. These include the following.

Data Files

Many data sets can be accessed, including all data sets used in analyses presented in the chapters, so that these analyses can be re-created in either R or SPSS. Also, there are additional data sets used in the exercises that are included in the support materials. Some of the exercises require data sets that are specific to R users; these are available for download and use within R from packages we will specify using squiggle brackets (e.g., {ggplot2}). All data files on the website are available in Excel format, to make them easily accessible for both R and SPSS users. Some of these data sets have been included to provide instructors with an additional source of classroom illustrations and exercises. For example, we may have used one of the tables in the book to illustrate analysis of variance, but the file can also be used to illustrate tests of contrasts that could follow the omnibus analysis. A listing of the data files and descriptions of them are available in the support material.

Teaching Tools

There is information for instructors only. Specifically, there is a solutions manual for all the exercises in the book and electronic files of the figures in the book.

Errata

Despite our best intentions, some errors may have crept into the book. We will maintain an up-to-date listing of corrections.

Acknowledgments

We wish to express our gratitude to J. Michael Royer for permission to use the data from his 1999 study; to Jennifer Wiley and James F. Voss for permission to use the *Wiley–Voss* data; and to Ira Ockene for permission to use the *Seasons* data. The *Seasons* research was supported by National Institutes of Health, National Heart, Lung, and Blood Institute Grant HL52745 awarded to University of Massachusetts Medical School, Worcester, Massachusetts. We are also grateful to the American Statistical Association, the Biometric Society, and the Biometrika Trustees for their permission to reproduce statistical tables.

Special thanks go to our late colleague, Alexander Pollatsek, and the many graduate assistants who, over the years, have contributed to our thinking about the teaching of statistics, and to the staff at Routledge who made this edition possible.

Finally, we are indebted to our spouses, Vincent M. Rotello, Nancy A. Myers, Susan Well, and Elizabeth Lorch, for their encouragement and patience during this long process.

Part 1

Foundations of Research Design and Data Analysis

Chapter 1

Planning the Research

1.1 Overview

There are three essential stages in carrying out an effective research project. In the first stage, the research is planned: objectives are stated; decisions are made about the treatments to be included, the measures to be obtained, and the type and number of participants; the research design is determined; and possible patterns of results and their implications are contemplated. This stage is reflected primarily in the Method section of the final research report. In the second stage, the data are collected and analyzed; descriptive statistics are calculated; population parameters are estimated; and inferential tests are performed to determine whether any obtained effects are larger than those we would expect to occur due to chance. The outcomes of the analyses are presented in words, tables, and graphs in the Results section of the final report. The final stage is the interpretation of the results, which typically is presented in the Discussion section of the research report: What do the results tell us about the answers to the questions we initially asked? Answering the questions that motivated the research is, of course, our ultimate goal; however, correct conclusions about what our results mean are totally dependent upon the first two stages. If the study is designed in such a way that factors other than the independent variable may have influenced the results, we may be led to incorrect conclusions. Even if the design of the study is sound, our statistical tests may fail to reveal effects that are present in the population from which we have sampled if our measures are very variable, or if we collect too few observations. Finally, despite a sound design and adequate procedures, data analyses that violate assumptions underlying the statistical procedures may lead us to incorrect inferences. We may think of the planning and analysis stages as providing input to the inferential stage and, as an adage states, "garbage in, garbage out."

In this chapter, we focus on the initial stage of planning the research. We will present an overview of the decisions that confront the researcher at the onset of a new project, and of the factors that should influence those decisions. In subsequent chapters, we will return to many of the issues raised in this chapter, explicitly linking the decisions made in the planning stage to aspects of the data analysis. Indeed, the decisions made in planning the research are the major factors influencing the results of statistical analyses, and subsequently influencing the conclusions that are drawn.

Chapter 1 is organized by the major decisions that must be made in planning a study:

- *The independent variable.* What is the question being asked in the research study? The question must be translated into an independent variable to be manipulated in an experiment or a predictor variable to be measured in an observational study.

DOI: 10.4324/9781003453550-2

- *The dependent variable.* What measure or measures should we use? In any study, there is a choice of measures. Different measures will provide different information and have different psychometric properties. Some measures may be better windows than others on the phenomenon we are studying. Some measures may be more sensitive than others to variation in the behaviors that are of interest. Some measures may be more reliable than others as indicators of some aspect of ability or performance. How should we balance these considerations in deciding among alternative measures?
- *The participant population.* Who is the target of the research question? Are we interested in healthy, elderly adults? Do we wish to study brain development in rats? And how should observations be sampled from the relevant population? Shall we sample from a diverse population, perhaps varying widely in attributes such as age, socioeconomic status, and ethnicity? Or should our sampling process be more tightly constrained? What are the implications of our decisions about sampling for conclusions that we hope to be able to make?
- *Nuisance variables.* In addition to decisions about the independent and dependent variables, the researcher must carefully consider the possible influences of other variables in the research study. What other variables may plausibly influence the dependent variable? These other variables are potentially a nuisance in two very important respects. First, if they are not taken into account, they may confound the independent variable. In that case, it may be impossible to determine whether a difference in the mean scores across various conditions is due to an effect of the independent variable or to effects of the nuisance variables that are correlated with the independent variable. Second, even if steps are taken to make certain that the independent variable is not confounded with other variables, nuisance variables contribute random variability to the data. This "error variance" is a concern because it can prevent us from detecting effects of the independent variable.
- *The research design.* In some circumstances, the research question demands an observational study; in other circumstances, the research question can be addressed by an experiment; in all circumstances, there are many options in designing the final study. The choice among the options must be informed by all the questions considered to this point: What is the independent variable? What is the population to be studied? What measure has been chosen? What are the potential nuisance variables, and how – and how much – are they expected to influence the dependent variable?
- *The statistical analyses.* The statistical analyses should be planned before collecting the data for two reasons. First, planning the analyses has the healthy effect of forcing the researcher to specify the questions to be addressed by the data. Second, it enables the researcher to be sure that, given the planned research design, the targeted questions can be answered by a statistical analysis.

The decisions made in the planning stage are interrelated; therefore, any sequencing of those decisions is a bit arbitrary and may be somewhat misleading. Nevertheless, for expository purposes, we will sequence the six categories of considerations by the order in which the decisions are typically confronted by the researcher.

1.2 The Independent Variable

Research begins with a question. What is the best way to teach the logic of a simple experiment to fourth-graders? Do people's attitudes toward abortion affect the likelihood that they will vote in a presidential election? Which dosage schedule minimizes the side effects of a drug? The answer to the question begins with the identification of a variable to be manipulated in an experiment or observed in a natural setting. In an experiment, we call that variable an "independent variable"; in an observational setting, we often refer to the variable as a "predictor variable"; for the present, we will use the term "independent variable" to apply to both situations. The translation of a research question into a relevant independent variable is guided, in part, by the researcher's assumptions about plausible answers to the question. In the question about science education, for example, the choice of teaching interventions might be based on current theory about teaching science. In the drug example, the selection of different dosages for comparison will likely be based on existing knowledge of the drug.

Beyond the general process of translating a question into a relevant independent variable, it is useful to distinguish between quantitative and qualitative variables. *Quantitative variables* are defined by the *amount* of a variable. The research question about drug dosage provides an example of a quantitative variable. The conditions of an experiment to relate drug dosage to the occurrence of side effects will compare conditions that differ solely in the amount of drug administered to participants. *Qualitative variables* involve the *type* of treatment. The research question about science teaching provides an example of a qualitative variable. The relevant experiment will compare two or more teaching interventions to determine which is the most effective. In this example, the different teaching methods would constitute the levels of the independent variable.

1.2.1 Quantitative Independent Variables

In the case of a quantitative independent variable, the experimenter is generally interested in questions relating to the form of the function relating the independent variable, X, to the dependent variable, Y. What is the rate of change in the incidence of side effects as drug dosage is increased? At what point on the dose effect function are side effects at a minimum? Does the function that relates side effects to drug dosage differ for two alternative drug treatments? Given the focus on the form of the function, the levels of the independent variable should be chosen to cover a wide enough range to detect any behavioral change within that range. Further, the number and spacing of levels across that range should be sufficient to clearly define the shape of the function, including its maximum or minimum (e.g., the drug dosage that minimizes side effects).

Theoretical considerations may also dictate the choice of levels of a quantitative independent variable. Suppose the investigator is trying to decide between two theories, one of which predicts a step function of change in Y as a function of X and another which predicts a gradual change. In this case, examination of a broad range of levels may be sacrificed to concentrate on more levels within a narrow range, presumably permitting a clearer view of the shape of the function within that range.

1.2.2 Qualitative Independent Variables

In studies investigating a quantitative independent variable, the specific numerical levels are of less interest in themselves than in the information they provide about a function relating the independent and dependent variables. In our drug dosage example, it may not be critical whether the levels of drug dosage are 10, 20, and 40 mg or 15, 25, and 50 mg. In contrast, a qualitative independent variable is one whose levels have typically been chosen specifically because they are of direct interest to the researcher. In our science teaching example, the teaching interventions to be compared in such an experiment will be specifically chosen. The choice may be based on theory, or on previous research findings, or on observations of behavior in naturally occurring circumstances. For example, two teaching interventions might be constructed to represent opposing teaching philosophies. Alternatively, an educational researcher may observe that a novel method of teaching science has produced impressive scores relative to national norms, so it is decided to compare the novel method with a more conventional method in a controlled experiment.

1.3 The Dependent Variable

In addition to the independent variable, a researcher must also select a dependent variable, or measure of performance. The choice of the dependent variable should be based on several considerations. First, the dependent variable should be a *valid* measure of the behavior of interest. By valid, we mean that the measure should reflect the behavior of interest in the research. It is often a straightforward matter to select a valid measure. For example, if a researcher wishes to study how the organization of information in a text influences recall of the text, the dependent variable will be recall. If the incidence of side effects of a drug is of interest, the researcher will probably have a list of side effects to check and will simply count the number of instances of side effects for each participant in each condition. However, there are many situations where the selection of a valid measure is less straightforward. If a researcher wants to study how the organization of information in a text influences comprehension of the text, the use of recall as a measure of comprehension is less convincing because recall is influenced by factors other than comprehension. A researcher wishing to assess teaching effectiveness might use student ratings of the overall quality of instruction. However, if by "effectiveness" we mean how well students learned in a course, student ratings may not be a good measure of their learning.

The choice of a valid measure should be guided by the goals of the research and what the researcher knows about the domain. This includes both theory and the empirical literature. For example, if a clinical researcher hypothesizes specific factors that underlie depressive behavior, scales designed to measure those factors may be included in a study of depression. Similarly, if an experimenter has a theory that predicts effects on response time as well as accuracy in studies of memory, then both types of measures should be recorded. Further, as a measure is used by many researchers over time, a knowledge base develops that should inform future use of the measure. For example, recall measures were common in early studies of text comprehension, but fell into disfavor as it was discovered that recall performance is influenced by memory factors that have little to do with what most would define as text comprehension.

Practical constraints are frequently an important influence on choice of measures. A personality researcher may feel that the ideal way to measure some clinical state is in an intensive interview, but if the research goal requires a large number of participants, a paper-and-pencil scale may be used because it is easily administered to many individuals and can also be scored by machine.

Within the constraints imposed by the researcher's goals and by practical limits of time in administration and scoring, statistical considerations are relevant to selection of the dependent variable. All other factors being equal, the most *reliable* measure – the one that is least variable within conditions – should be chosen. In some instances, high reliability is easily attained. For example, response times may be consistently accurately recorded. However, in other cases, reliability is less easily assured. For example, reliability is a major consideration in research in which the dependent variable is an attitude scale or other measurement instrument. Many articles and books have been written on the topic of measurement, and we cannot do justice to the issues here. Interested readers should consult some of the many sources available (e.g., Downing & Haladnya, 2006; Martin & Bateson, 2007; Nunnally & Berstein, 1994).

The *sensitivity* of the measure is also an important concern when selecting a dependent variable. All else being equal, a measure that is capable of detecting small differences in performance is preferable to a measure that can detect only gross differences. In our example of a study of drug side effects, the ultimate side effect would be death of a participant; however, if that was the only category measured, the resulting failure to record less drastic side effects would seriously compromise our ability to identify an optimal dosage to minimize side effects. Perhaps less obviously, measures that are relatively sensitive across many ranges of behavior will not be able to reveal differences among experimental conditions if performance is close to some maximum or minimum value. This would be the case, for example, if two methods of teaching are evaluated by scores on a test so easy that the scores are high regardless of the instructional method. In this example, simply increasing the difficulty of the test should improve its sensitivity.

Another factor that may affect the outcome of the statistical analysis is the *distribution of the dependent variable*. Many common statistical procedures rest on the assumption that the data are normally distributed; that is, that the distribution curve has, at least approximately, a bell shape. This is but one of several assumptions that underlie various methods for estimating and testing effects. We will have much to say throughout this book about the assumptions that underlie the methods we discuss – what these assumptions are, the consequences when they are violated, and the alternative methods appropriate when the consequences may invalidate our conclusions.

Finally, we do not want to leave the impression that there is a single "best" dependent variable in a given experiment. Different measures provide different information, so it is often advisable to use multiple measures of behavior in the same experiment. The result will be a richer understanding of the phenomenon of interest.

1.4 The Participant Population

With the independent variable identified and the dependent variable(s) selected, who will participate in the study? The answer to this question has immediate implications for our ability to generalize our research conclusions. Several considerations shape the choice of participants in most research.

Practical considerations often play a major role in defining the participant population. In clinical research, for example, financial and time limitations may force a researcher to draw a sample of participants from the local Veterans Administration hospital. In social and cognitive research, the easy accessibility of students in introductory psychology courses makes them a nearly irresistible source of study participants. These practical considerations are understandable, but it is important to recognize that the constraints on sampling that they represent serve to define the participant population. The sampling process, in turn, has implications for how our research conclusions may be generalized. In some situations, practical constraints on sampling may not seriously constrain conclusions. If the study involves basic sensory or perceptual processes, the results from a sample of college students can probably be generalized to individuals of a similar age who do not attend college. On the other hand, if a study of problem-solving strategies is conducted on students from a highly selective college, the conclusions may not be applicable to individuals who do not attend college or, for that matter, to students at less selective colleges.

Beyond the reality of practical limitations, the purpose of the research should be an important factor in the sampling of participants. If the investigator wants to investigate the cognitive abilities of older adults, then that is the population to be represented in the study. This seems obvious, but matters usually are more complicated. For example, what range of older ages will be included in the study? The actual sample of participants may be a function of several considerations beyond the very general goal.

Theoretical considerations may influence the choice of participants. Suppose the investigator who is studying cognitive functioning of older adults hypothesizes that speed and accuracy in memory and problem-solving tasks are a direct function of the degree to which people continue to engage in intellectual pursuits. Of course, this depends on the definition of intellectual activities, but the researcher might administer a questionnaire to ensure that he or she had a range of individuals with respect to whether they participated in activities such as doing puzzles, reading and discussing those readings, or learning new things such as a foreign language.

The participants' previous histories will be an important constraint on the sample. If the investigator wants older individuals with no obvious cognitive impairments, individuals who may be suffering from such impairments should be excluded from the study. If part of the research goal is to generalize to a population of older individuals representing a broad spectrum of experience, some attempt should be made to include people from various social and economic classes, ethnic backgrounds, and work experiences.

Although perhaps less obvious than the preceding considerations, the control of error variance is an issue in selecting research participants. The investigator may want to include a wide range of attributes in the study so that there is a firm statistical basis for generalizing to a broad population. However, the more diversified the sample, the more variable the data will be and, therefore, the less clear will be the effects of independent variables on the behavior of interest. For example, if we want to evaluate some method of improving memory in a sample of older adults, we are likely to have a better estimate of the effect of the method in a sample whose members are similar in attributes such as age, educational experience, and socioeconomic status than in one with wide variation in these attributes.

The tradeoff in sampling from a more narrowly defined population is that, although the data are less variable, inferences about individuals outside the range studied are more speculative. For example, we may assume that if a method of improving memory is less effective for the older members of our sample, it will be even less effective with individuals older than those in our sample. However, that is a hypothesis that is only indirectly supported by our data. Ideally, investigators should think about the population to which they wish to generalize, consider the implications for the possible variability of the data, and then come to a decision about the attributes to be controlled in recruiting participants. Often, we can compensate for a heterogeneous sample by having a larger sample (see Section 1.6), by the choice of the research design (Chapter 12), and by statistical means (Chapters 12 and 24).

1.5 Nuisance Variables

Many of the decisions facing the researcher reflect the fact that scores are influenced by variables other than the independent variables of interest. Such variables have been labeled "irrelevant" (to the researcher's interest; Myers, 1979) or "nuisance" variables (Kirk,

1995). Participants in research differ in ability, prior experience, attitudes, age, gender, and many other attributes that may influence their responses in a study. Even a single individual may make different responses to the same stimulus at different points in time because of factors such as fatigue or practice, a change in calibration of equipment, a change in room temperature, or variation in items on a test or rating instrument.

The presence of nuisance variables is a threat to the success of a study in two potential ways. First, care must be taken to ensure that nuisance variables are not confounded with the independent variable; otherwise, differences among experimental conditions may not be unambiguously attributed to the independent variable. Second, the presence of nuisance variables may make it more difficult to decide whether the independent variable has had an effect or whether the effect is due to chance.

1.5.1 Confounding

Suppose we observe that students taught science by a lab-based curriculum show higher achievement than students taught science by a text-based curriculum. Can we be sure that the difference in achievement is due to the difference in the curricula? Are the students taught by the lab-based approach more likely to come from an environment that provides more parental support or more resources (e.g., lab equipment)? Are the teachers using the lab-based curriculum better trained in science or more experienced? These are just some of the potential *confounds* that may compromise our hypothetical comparison of curricula. This is but one illustration of how differences between the means of various conditions may be attributed to the effects of the independent variable when those differences are in part, or entirely, due to effects of one or more nuisance variables. Such incorrect attribution of effects may occur whenever variables that are not of primary interest are related to the independent variable.

Issues of public policy provide an important example because they often involve separating the contributions of many variables from those of the variable of interest. For example, a relevant policy issue is whether school choice, an option that enables children to attend charter schools rather than their public school, results in more effective education of those children. However, differences in academic performance between students who choose charter schools and those who remain in public schools cannot simply be attributed to differences in the schools. Children who choose charter schools may have more motivated or wealthier parents than those public school students who do not opt out of the public school system.

Many other examples of possible confounds due to nuisance variables could be presented. Relations among variables cause problems of interpretation for true experiments, for observational studies, and for correlational studies. Campbell and Stanley (1963) present an excellent discussion of several classes of nuisance variables, together with consideration of the advantages and disadvantages of relevant research designs. Although primarily concerned with educational research, their discussion is relevant to any area.

1.5.2 Error Variance

Error variance consists of those chance fluctuations in scores that are attributable to the effects of nuisance variables. Consider Figure 1.1. The two points on the x-axis represent two levels of an independent variable, perhaps two methods of teaching science. In that

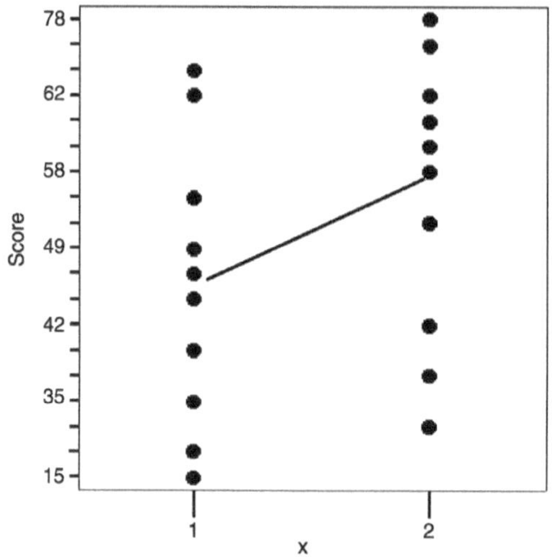

Figure 1.1 Two data sets with a line connecting mean scores.

example, the individual points on the graph represent the scores of individual students on a science achievement test, and the points connected by the line would represent the average scores of individuals in the two groups. The average score at Method 2 is higher than that at Method 1, suggesting that the second method is superior. However, error variance weakens that conclusion. Note that several scores at Method 2 are lower than some of the scores at Method 1, even lower than the mean score at Method 1. Can we be sure that Method 2 represents the better method? Or might the difference in mean performance be due to chance differences between the groups in average ability or motivation, or some other factor? Note that if there were no variability within a condition, all the points would fall at the means, and the advantage of Method 2 would be clear. Unfortunately, error variance is always present, albeit to varying degrees, in behavioral research.

1.6 Research Design

In an ideal world, a well-articulated research question would be rewarded by definitive empirical results. But an ideal world would not include nuisance variables, so even well-articulated research questions are sometimes answered in a misleading way or not at all. The study of research design and inferential statistics deals to a large extent with the problems posed by nuisance variables. We have already encountered the two main problems: nuisance variables can introduce confounds that make it difficult or impossible to know the true effect of our independent variables; and nuisance variables produce error variance that hinders a researcher's ability to discern the effects of independent variables. In this section, we elaborate these two problems and discuss some strategies for coping with them.

1.6.1 Observation Versus Experimentation

Not all research questions may be approached in the same way. The nature of the approach taken by a researcher has important implications for the researcher's ability to cope with potential confounds and with error variance.

Some questions readily lend themselves to experimentation. Our example of determining the relationship between drug dosage and the incidence of side effects is a good case in point. Typically, an animal model would be used to investigate this question in a laboratory setting. The experimenter has precise control over the administration of drug dosages and extensive control over the conditions under which the drug is administered (e.g., animal weight, feeding and watering, time of day). As we will see, the control associated with experiments makes experimentation eminently suited to dealing with the threats of confounds and error variance.

However, many important questions necessitate a loss of experimental control. For example, if we want to know the effects of a traumatic event on depression, a true experiment is not possible because manipulating the independent variable would be unethical. Instead, a researcher might collect relevant information through interviews or questionnaires administered to individuals who have undergone some trauma. In such circumstances, a researcher cannot prevent nuisance variables from confounding variables of interest. For example, if the researcher is specifically interested in the effects of a traumatic combat experience on depression, a comparison of individuals who have experienced combat trauma with individuals who have not experienced combat trauma could easily be compromised by other differences between the groups. Some of the potential confounds might be avoided by careful selection of a comparison group (e.g., a comparison group comprised of soldiers not involved in combat would be preferable to a comparison group comprised of college students). Still, it is often impossible to identify comparison groups that are matched in all respects to the group of interest. Similarly, the loss of control inherent to observational studies is frequently associated with increased error variance.

Finally, there are many examples of questions that may be approached through either experimentation or observation. Our example of comparing science curricula is relevant. Many educational researchers would approach this question with observational methods, comparing the achievement levels of schools that use a lab-based curriculum with those of schools using a text-based curriculum. One argument for adopting this approach is that an experimental approach disrupts the natural educational environment, with the consequence that the findings of an experiment will ultimately not generalize to the classroom. Another argument is that the barriers to a true experimental approach are so great as to render an experiment impractical. However, many educational researchers counter that an experimental approach can be situated in the classroom without seriously compromising the natural educational environment. Although conceding that the barriers to such an approach are significant, they argue that the effort to surmount the barriers is rewarded by a clearer understanding of what instructional methods are effective and why. There are valid points on both sides of the argument, and there is a place for both observational and experimental research. Thus, it is important to have tools to address the effects of nuisance variables in both types of research.

1.6.2 Dealing With Threats to Valid Inferences

To draw a valid conclusion about the effect of an independent variable, a researcher must be able to rule out the possibility that nuisance variables could explain the apparent effect of the independent variable.

Control of Nuisance Variables by Random Assignment

The most effective means for ruling out nuisance variables as an alternative explanation of an observed effect is to guarantee that the independent variable is not confounded with any relevant nuisance variables. This is possible only in a true experiment. When the independent variable is manipulated, we have control over which individuals are assigned to which treatments. In such cases, the potential effects of many nuisance variables can be countered by *random assignment* of individuals to conditions. For example, in a study comparing effects of several instructional methods on problem solving, we would not want to select participants in such a way as to increase the likelihood that more motivated individuals would be assigned to a particular method. However, that might happen if the first volunteers, possibly the more motivated individuals, were assigned to one method. One solution is to randomly assign participants to conditions, using a method that ensures that each participant has an equal chance of being assigned to each instructional method. Randomization does not ensure that the various instructional groups are perfectly matched on those nuisance variables that might influence learning. However, randomization does ensure that over many replications of the experiment, no treatment will have an advantage. In any one experiment, one condition might have an advantage (e.g., include more motivated participants) but statistical analyses that are based on the assumption of randomization take these chance differences into account.

Randomization does not apply only to the assignment of participants to conditions. In many studies, the participant is tested on several different items in each condition. For example, a researcher might be interested in whether there is a difference in ease of understanding sentences as a function of whether they are in the active or passive voice. Accordingly, the researcher might create an active and a passive version of each member of a set of sentences. The time each participant needed to read each sentence would be recorded; half of the sentences would be randomly selected for presentation in the active voice, and the other half of the sentences would be in the passive voice. A different random assignment could be used for each participant. In addition, the order of presentation of the sentences could be randomly sequenced for each participant. Random assignment of sentences to the active and passive voices addresses the possibility that some sentences are inherently more difficult to read by providing an equal chance of their being presented in either condition to each participant. Similarly, random sequencing of the presentation of the sentences addresses concerns about the influences on reading time of fatigue, boredom, practice, or other time-related effects that influence reading time.

Randomization guards against attributing effects to an independent variable when other variables are responsible for the effects. However, it is not a cure-all. For example, we might wish to study the effects on depression of a newly developed pill. We could have interviewers rate the state of depression before and after treatment. However, just the knowledge of being treated might influence the participants' responses. One solution would be to have a control condition, a second group of depressed patients who are given a placebo (e.g., a sugar pill). Participants would be randomly assigned to either the experimental or the control condition. However, if the interviewer is aware of the condition of each individual being interviewed, this could (presumably unintentionally) bias the results. The response to this concern is to run what is termed a double-blind experiment, in which neither interviewer nor participant is aware of the condition. Kirk (1995, pp. 22–24) discusses several other methods that supplement randomization.

Measurement of Nuisance Variables

In the absence of experimental control, a researcher can never be completely certain that a nuisance variable might not be responsible, at least in part, for an apparent effect of an independent variable. The researcher's best strategy is to identify and measure relevant nuisance variables and then hope to demonstrate that those variables cannot account for the observed results. For example, consider a study that attempts to determine the effect of combat trauma on depression. A comparison of soldiers who have with those who have not experienced trauma might be compromised by, say, differences in the average education levels of the two groups. Better educated soldiers may be more likely to receive assignments that make them less likely to experience combat. If education level is recorded and included in the analyses of the data, it may be possible to demonstrate that it is unrelated to depression and therefore cannot explain an observed relation between combat trauma and depression. Alternatively, education level might be related to depression, but not in a way that can explain an observed relation between trauma and depression. Of course, it is possible that the nature of the relationship between education and trauma make it impossible to clearly discern the nature of the relation between trauma and depression. Measurement is the best a researcher can do in this situation, but it does not substitute for the ability to exert control over variables.

1.6.3 Dealing With Error Variance

If there is an effect of our independent variable in the population, we hope that the effect will be clear in our sample. Unfortunately, that doesn't always happen. The reason is error variance. An analogy is useful here. Think of error variance as noise and the effect of the independent variable as a signal embedded in that noise. If the signal is sufficiently strong relative to the noise, it will be detected; however, if the signal is too weak, it will be missed. A well-designed and conducted study will establish circumstances in which the signal-to-noise ratio is high and thus the probability of detecting an effect of the independent variable is high. This detection probability is the concept of statistical power: *power* is the probability of detecting an effect of an independent variable in an experiment when an effect exists in the population. We seek to conduct studies with high power; our primary means to accomplish this goal is to minimize error variance. Several strategies are available.

Control by Uniform Conditions

Random assignment of participants or items to treatments provides protection against wrongly attributing effects to the independent variable. However, randomization does not reduce error variance. One way to reduce error variance is to hold the conditions of the study constant except for variation in the levels of the independent variables. For example, all participants should receive the same instructions in the same environment, and all animal participants should receive the same handling, feeding, and housing. To the extent that sources of variance other than the independent variable are held constant, error variance will be reduced and effects of independent variables will be more easily detected.

It is possible and desirable to seek uniformity of conditions in observational studies, as well as experiments. A good example is found in a study by Räkkönen, Matthews, Flory, Owens, and Gump (1999). They studied the relation between ambulatory blood pressure (BP) and personality characteristics, taking BP measurements from all participants in the same time period on the same three days of the week, two work days and one nonwork day. In this way, they reduced possible effects of the time at which measurements were taken. Similarly, researchers have the option of selecting the participants of observational studies in such a way as to match participants on attributes (e.g., age and socioeconomic status) that may influence the behavior of interest (e.g., learning in a study comparing teaching interventions), but which are not of interest to the researcher.

Control by Design

The design of the research is a plan for data collection. We can minimize the effects of error variance by choosing a design that permits us to calculate the contribution of one or more nuisance variables. In the data analysis, that error variance is subtracted from the total variability in the data set. Many of the design variations also enable the researcher to equate treatment conditions for one or more nuisance variables, thus ensuring that those variables are not responsible for any differences that are found between conditions.

One procedure that is often used in experiments is *blocking*, sometimes also referred to as *stratification*, or *matching*. Typically, we divide the pool of participants into blocks based on some variable whose effects are not of primary interest to us, such as gender identity or ability level. Then we randomly assign participants within each block to the different levels of the independent variable. As an example of the blocking design, suppose we wish to carry out a study of memory in which participants are taught to use one of three possible mnemonic strategies. We might measure the memory span of participants before the start of the research. We could then randomly distribute the three participants with the highest memory span score among the three conditions, then do the same with the three next highest scoring participants, and so on, until all three conditions are filled. In this design, the groups are roughly matched on initial memory ability, a variable that might influence their memory of materials in the experiment. Thus, any possible differences between means of the strategy groups cannot be attributed to initial memory span. As an added result of the matching, a statistical analysis that treats the level of initial memory span as a second independent variable can now remove the variability due to that variable. The blocking design is said to be more *efficient* than a simpler design that randomly assigns participants to conditions without regard to memory span. (Chapter 9 presents the analysis of data when a blocking design has been used.).

Blocking can also be used in some studies that are not true experiments. If we wish to study attitudes in different racial groups about some issue, we can attempt to recruit participants of different races who are matched on variables that might influence responses, such as age, income, and education.

For some independent variables, even greater efficiency can be achieved if we test the same participant at each level of the independent variable. This variant of blocking is known as a *repeated-measures design*. A common example occurs in studies of forgetting in which measures are obtained at various times after the material was studied. The advantage of repeated-measures designs is that participants can be treated as another independent variable in the statistical analysis, thus removing from the data variability due to differences among participants. Ideally, the remaining error variance will be due to errors of measurement caused by random fluctuations over time in factors such as the participant's attentiveness, the difficulty of the stimuli, and the temperature in the room. Thus, the repeated-measures design is a particularly effective way to minimize error variance, although it is not without its drawbacks. The major drawbacks of the design are the possibility of systematic biases due to time-related effects (e.g., practice and fatigue) and the possibility of *carry-over effects* (i.e., influences of participation in earlier conditions on performance in subsequent conditions). (A full treatment of this design is presented in Chapters 13–16.)

Control by Measurement

Often, blocking is not practical. Morrow and Young (1997) studied the effects of exposure to literature on the reading scores of third-graders. Although reading scores were obtained before the start of the school year (pretest scores), the composition of the third-grade classes was established by school administrators prior to the study. Therefore, the blocking design was not a possibility. However, the pretest score could still be used to reduce error variance by removing that portion of the variability in the dependent variable (i.e., the posttest score) that was predictable from the pretest score. This adjustment, called *analysis of covariance*, is thus a statistical means to reduce error variance, as opposed to the design approach of blocking. (Analysis of covariance is discussed in Chapters 12 and 24.)

Usually, the greater efficiency that comes with more complicated designs and analyses has a cost. For example, additional information is required for both blocking and the analysis of covariance. Furthermore, the appropriate statistical analysis associated with more efficient approaches is usually based on more assumptions about the nature of the data. In view of this, a major theme of this book is that there are many possible designs and analyses, and many considerations in choosing among them. We would like to select our design and method of data analysis with the goals of minimizing potential confounds and reducing error variance as much as possible. However, our decisions in these matters may be constrained by the resources that are available and by the assumptions that must be made about the data. Ideally, the researcher should be aware of the pros and cons of the different designs and analyses and the tradeoffs that must be considered in making the best choice.

Control by Sample Size

The final strategy for dealing with error variance is through the choice of sample size. The size of the sample does not actually affect the individual error variance in the design; however, it does influence the amount of error in estimates of the effects of the independent variables. As sample size increases, estimates of effects become more precise; as a consequence, the power to detect the effects improves. Thus, in general, increasing sample size increases statistical power in any design. The relationship between sample size and power is well defined for a given design, so we can – and should – use that relationship in planning how many observations to make in any experiment or observational study. Sometimes, we may find that the sample size required to achieve a certain level of power is impractical. However, it is better to know that before we begin a study so that we can modify our plans by revising our research design or choosing other, less variable, measures.

1.7 Statistical Analyses

Statistical analyses are performed after the data are collected, of course. However, the analyses should be planned *before* the data are collected, as part of the process of evaluating the adequacy of the research design. Indeed, some researchers argue that the analyses should be recorded (even "preregistered") before the data are collected to reduce the number of choices available to the analyst once the data have been viewed, because those choices can inflate the probability of reaching erroneous conclusions. Gelman and Loken (2014) referred to this as the "garden of forking paths"; it has also been termed *p-hacking* (Simmons, Nelson, & Simonsohn, 2011). The analysis of data broadly consists of two phases:

(1) an *exploratory* or *descriptive phase*, in which measures of central tendency (e.g., means, medians), variability, and shape of distributions should be calculated and graphed; and (2) an *inferential phase*, in which population parameters are estimated and hypotheses about them are tested.

Too often, researchers neglect the first, exploratory phase. However, the descriptive statistics *are* the results of an experiment and they therefore deserve particular attention. That attention begins during the planning phase. The research plans should include decisions about what descriptive statistics will be calculated and what plots of the data will be useful to view. The researcher should imagine, for example, possible patterns of means across conditions, and what different patterns would reveal about the behavior being studied. We will present the reasons for, and examples of, exploratory analyses in the next chapter. For now, we note that such analyses may suggest additional hypotheses and may also inform us about potential problems for the statistical analyses to be carried out in the next phase.

The goals of the research will often dictate very specific hypotheses or questions. For example, if there are several levels of the independent variable, are there comparisons between certain levels that are of particular interest? Or, if the independent variable is quantitative, is it of interest to examine certain features of the function relating the dependent and independent variables, such as whether it is best represented by a straight line or a curve? These rather specific sorts of questions should be considered before the data are collected for two reasons. First, considering specific questions before the data are collected ensures that the design allows us to answer those questions. Second, inferences based on tests of hypotheses formulated before the results are viewed have less likelihood of yielding erroneous conclusions than tests of hypotheses suggested by the observed results.

The power of the statistical test should be an important consideration in planning the research. Estimates of effect size and error variance obtained from related research or pilot studies, together with a desired level of power to detect effects of a certain minimum size, should determine the amount of data collected. Just how this is done is considered in several chapters in this book, and computer software is provided on the website for the book to carry out the calculations.

Finally, it is useful to specify the planned inferential tests in detail to make certain that the experimental design has been thoroughly evaluated and all potential sources of variability have been identified. This practice will help guard against oversights in planning the design. For example, an educational researcher planning an experiment to compare two methods of teaching science might assign two classes of students at random to a lab-based teaching method and two other classes to a text-based method. However, this plan is flawed because it neglects an important potential influence on performance; namely, individual classes may differ in performance regardless of their assignment to different teaching methods. An implication of this observation is that the research plan should include many more than four classes so that variability due to classes may be adequately evaluated.

1.8 Generalizing Conclusions

A final planning consideration is to identify issues concerning the generalizability of any findings to emerge from the research. Campbell and Stanley (1963) distinguished between internal and external validity of inferences. *Internal validity* refers to the question of whether observed effects can validly be attributed to the independent variable. *External validity* refers to the extent to which the inferences that are drawn can be generalized to other conditions or populations. In much of this chapter, we have been concerned with internal validity; in particular, we have discussed the role of randomization and research design in protecting against potential confounds. We touched briefly on the participant of external validity when we cautioned that researchers should consider the population to which their inferences apply. For example, if a treatment of depression is successful in a study with all female participants, or in one with individuals in a certain age range, we should not conclude that the treatment would also reduce depression in males, or in older participants, without some firm basis in theory or data. Perhaps the most common setting in which we need to be careful about generalizing beyond the population sampled is the academic, in which the participants are usually college students within a limited range of age and intelligence, or the children of faculty and students.

Similar issues arise with respect to generalizing conclusions based on a particular method or measure. For example, recall and recognition scores have been shown to reflect very different memory processes. And different results have been obtained in studies of reading in which reading times are calculated as differences between button presses that bring on successive words in a text and studies in which reading times are recorded by tracking eye movements.

Studies in which the independent variable is quantitative confront the researcher with other problems of generalization. There are practical limits to the levels that can be included. Conclusions about the shape of the function relating dependent and independent variables require interpolation between, and extrapolation beyond, the levels observed.

Still another common problem for external validity involves generalization to a population of items. When emotional states or attitudes are inferred from responses to a set of pictures, we must consider whether the results are specific to those particular pictures or have more general implications. Similar inferential issues confront researchers in the area of language comprehension, whose stimuli are sentences, or words that are usually constrained as to length or familiarity.

One approach to the issue of external validity is to systematically vary the populations sampled or the measures obtained. This can be done within a study or over a series of studies. For example, a study of the effects of some treatment might include gender of the participants as a second variable. Or, as has often been done, both recall and recognition measures might be recorded in studies of memory. The inclusion of variables that enable tests of the generalizability of results should be a consideration in planning the research. Nevertheless, it is important to understand that results can never be generalized on every dimension. As Campbell and Stanley (1963) have pointed out, the very act of conducting an experiment raises the question of whether the results would apply in a less controlled, real-world setting. Whereas we can provide a logical basis for arguing that certain methods, such as randomization, increase internal validity, we can only argue for external validity by extrapolating beyond the methods and results of the current study. We should take great care before making such extrapolations.

1.9 Summary

Careful planning of a study may save much wasted effort. Such plans should have several goals:

- The research plan should include steps to ensure that results attributed to the independent variable are not due to other uncontrolled variables. In experiments, random assignment is the key procedure for guarding against systematic biases due to nuisance variables.

In less controlled research environments, the researcher must identify nuisance variables of interest and measure them so that their influences may be assessed.

- Error variance should be controlled as much as possible to ensure maximal power to detect effects of the independent variable. Several strategies address control of error variance, including ensuring uniformity of conditions, selecting reliable dependent variables, choice of research design, and measurement of nuisance variables for purposes of statistical adjustment.
- Decisions about the number of observations to collect in a study should be informed by *a priori* power considerations based on realistic estimates of effect size.
- Statistical analyses, including descriptive statistics, should be planned as part of the process of planning the research design.
- Issues pertaining to the bases for generalizing conclusions should be considered during the planning phase.

Exercises

1.1 [Identifying design and sample characteristics] A researcher requested volunteers for a study comparing several methods to reduce weight. Participants were told that if they were willing to be in the study, they would be assigned randomly to one of three methods. Thirty individuals agreed to this condition and participated in the study.

a) Is this an experiment or an observational study?
b) Is the sample random? If so, characterize the likely population.
c) Describe and discuss an alternative research design.

1.2 [Identifying populations and sources of variance] A study of computer-assisted learning of arithmetic in third-grade students was carried out in a private school in a wealthy suburb of a major city.

a) Characterize the population that this sample represents. Consider whether the results permit generalizations about computer-assisted instruction (CAI) for the broad population of third-grade students. Present your reasoning.
b) This study was done by assigning one class to CAI and one to a traditional method. Discuss some potential sources of error variance in this design.

1.3 [Identifying limits of observational study] Investigators who conducted an observational study reported that children who spent considerable time in day care were more likely than other children to exhibit aggressive behavior in kindergarten (Stolberg, 2001). Although this suggests that placement in day care may cause aggressive behavior – either because of the day-care environment or because of the time away from parents – other factors may be involved.

a) What factors other than time spent in day care might affect aggressive behavior in the study cited by Stolberg?
b) If you were carrying out such an observational study, what might be done to attempt to understand the effects upon aggression of factors other than day care?
c) An alternative approach to the effects of day care upon aggressive behavior would be to conduct an experiment. How would you do this and what are the pros and cons of this approach?

1.4 [What constitutes evidence for a hypothesis?] It is well known that the incidence of lung cancer in individuals who smoke cigarettes is higher than in the general population.

a) Is this evidence that smoking causes lung cancer?
b) If you were a researcher investigating this question, what further lines of evidence would you seek?

1.5 [Connecting data to hypotheses] In the *Seasons* study (the data are in the *Seasons* file in the *Seasons* folder on the website for this book), we found that the average depression score was higher for participants identified as male with only a high school education than for those with at least some college education. Discuss the implications of this finding. Consider whether the data demonstrate that providing a college education will reduce depression.

1.6 [Understanding variables] In a 20-year study of cigarette smoking and lung cancer, researchers recorded the incidence of lung cancer in a random sample of smokers and nonsmokers, none of whom had cancer at the start of the study.

a) What are the independent and dependent variables?
b) For each, state whether the variable is discrete or continuous.
c) What variables other than these might be recorded in such a study? Which of these are discrete or continuous?

Chapter 2

Describing the Data

2.1 Overview

Too often, students – and even established researchers – begin data analysis by calculating a few statistics, usually means or correlations, and then immediately perform statistical tests. This common approach to data analysis puts the cart before the horse by emphasizing inferential statistics over descriptive statistics. In fact, a thorough description of the data is critical to both a complete understanding of the data and appropriate application of inferential tests. Data analyses should begin with graphing of the distribution of observations and calculation of several descriptive statistics. Such exploration of the data serves several purposes.

First, a data set is a researcher's window on the population under study. Inferential tests perform the important function of helping researchers to determine which trends in the data reflect corresponding characteristics of the population, but inferential tests will only be applied to those aspects of the data that are considered by the researcher in the first place. For example, consider a clinical researcher studying depression who observes that the pattern of means on the Beck Depression Inventory appears to vary as a function of marital status. If information is not also gathered about the distribution of scores for each marital category, the researcher is likely to overlook the possibility that a few extreme scores are responsible for the pattern of means. As Hoaglin, Mosteller, and Tukey (1983, p. 2) have said: "Exploratory data analysis emphasizes flexible searching for clues and evidence, whereas confirmatory data analysis stresses evaluating the available evidence." We must look for the evidence before we can evaluate it appropriately.

Second, in many instances, description is the goal of the study. Consider the example of a school district superintendent who wishes to evaluate the scores on a standardized reading test. The superintendent has several questions she would like to answer. How should the overall level of performance of students in the district be characterized relative to national norms? Are most students performing near the average? Or is there considerable variability in reading level? If there is variability, are there students whose scores clearly indicate the need for remediation? Are there factors that seem related to variation in performance, such as class size or teacher experience?

Yet a third reason for a thorough descriptive analysis is that careful examination of the data may serve to generate hypotheses that the investigator had not had in mind when the study was designed. For example, our hypothetical district superintendent may find an unexpected difference between the means of students who identify as male and female. She may then ask whether this is a difference that would hold for similarly instructed students

DOI: 10.4324/9781003453550-3

with the same prior experiences, or whether it is due to a chance fluctuation in performance of the two groups. Or the superintendent may note that the average score in her district is below that for the state as a whole. She may ask whether this is a chance result or one that will repeat itself with future generations of students.

Finally, description of the data is important to ensure that assumptions underlying subsequent, inferential statistics will be valid for the data being analyzed. Questions about group differences (e.g., boys and girls), or the difference in performance between a group (e.g., the school district) and some standard (e.g., scores for the state as a whole), are questions of statistical inference. Such questions require further analyses, and those analyses rest on certain assumptions about the shape and variability of the distribution of data. Therefore, we first must explore our data to assess whether assumptions underlying planned inferential analyses are being met. If not, we may consider modifications of our planned analyses. Among many possibilities, such modifications may include deleting extreme scores, transforming data, or using statistical procedures other than those originally planned.

In this chapter, we present the tools for the necessary description and exploration of data. The major objectives of this chapter are to:

- *Present some ways of graphing the data that provide different views of the distribution of scores.* We graph the data in different ways because different graphs provide different sorts of information. Some data plots provide only a general sense of the distribution, whereas others provide a more detailed look and enable us more readily to address questions such as whether there are scores that are markedly different from most of the others.
- *Present several summary statistics to reflect key properties of distributions.* Specifically, we will present measures of the average, variability, and shape of the distribution of scores. In the case of measures of average and variability, alternative statistics will be presented because they provide somewhat different information about a distribution. In addition, different indices of the same general characteristic have different properties that make them more or less desirable, depending upon the situation. Thus, in addition to the mean and standard deviation, we will consider the median and interquartile range.
- *Introduce ways of characterizing relationships between two variables.* We will present examples of scatter diagrams, which graphically present such relationships. We will also present and illustrate two ways of summarizing the relationship between two variables: the correlation coefficient and linear regression. The correlation coefficient is a numerical measure of the direction and strength of a relationship, while linear regression tells us how much one variable changes as a function of changes in the other variable.

Good statistical practice should begin with the description and visualization of the data, using graphs and summary statistics such as those presented in this chapter.

2.2 Graphical Summaries of the Data

The first step in describing a data set is to get an overall view of the data by graphing the distribution of observations. An understanding of the distribution of scores provides fundamental information, as well as a context for interpreting specific properties of the distribution, such as its mean and standard deviation.

Table 2.1 The *Royer* second-grade addition scores

93	82	94	90	82	84	100	85	76	74
95	50	88	100	89	69	94	94	89	100
50	83	95	72	47	100	100	94		

In a study of basic arithmetic skills, Royer, Tronsky, and Chan (1999) collected both accuracy and response time scores from students in grades 1 to 8 for addition, subtraction, and multiplication.[1] Table 2.1 presents the percent correct addition score for 28 students in second grade. What do we do with these numbers? Even with only 28 values, it can be hard to get a sense of how the students are doing. For example, it isn't easy to see whether all of the children are doing well learning simple addition, or if there are particular individuals who are lagging behind or who need more challenge. Let's use some different types of graphs to visualize the data. Many of these graphs are available in the *Explore* module of SPSS or from the *Graphs* menu. Similarly, basic plots can be generated in Excel. We will focus on plotting using R because doing so is easy and the results are nicely customizable.[2]

2.2.1 *Histograms*

One of the simplest graphical summaries of data is a histogram. Histograms plot the frequency of groups of scores. Figure 2.1 presents a histogram for the data set of Table 2.1. In this graph, the label on the left hand y-axis (the ordinate) shows the number of scores represented by each bar, that is, the frequency. Histograms sometimes show the proportion of scores in each bin, rather than the frequency, which changes the y-axis but not the distribution of data. The x-axis (the abscissa) presents the addition scores, which have been divided into intervals, called bins, of five points each (e.g., 96–100, 91–95, etc.). The width of the bins or, equivalently, the total number of bins, can be adjusted to best capture the shape of the data. The goal in constructing a graph of a distribution is to summarize the important characteristics of the distribution. A summary involves data reduction and a corresponding loss of information; however, an accurate summary must retain any important information. Thus, researchers must exercise judgment in deciding on an appropriate degree of resolution (i.e., bin size) for a graph.

Once the histogram is constructed, important characteristics of the distribution should now be more evident than they were from the list of scores in Table 2.1. It is clear in Figure 2.1 that most students performed very well. Indeed, the greatest number of scores appears in the bin showing 96%–100% correct. However, it is also evident that a few students performed very poorly, and there is a gap of at least 10 percentage points between the three lowest scores and the lowest of the remaining 25 scores. This gap is typical of many real data sets, as is the obvious asymmetry in the distribution. Micceri (1989) examined 440 distributions of achievement scores and other psychometric data and noted the prevalence of such departures from the classic bell shape as asymmetry (skew) and "lumpiness" (more than one mode, the most frequently observed value). Similarly, after analyzing many data distributions based on standard chemical analyses of blood samples, Hill and Dixon (1982) concluded that their real-life data distributions were "asymmetric, lumpy, and have relatively few unique values" (p. 393). We raise this point because the inferential procedures most commonly encountered in journal articles rest on strong assumptions about the shape

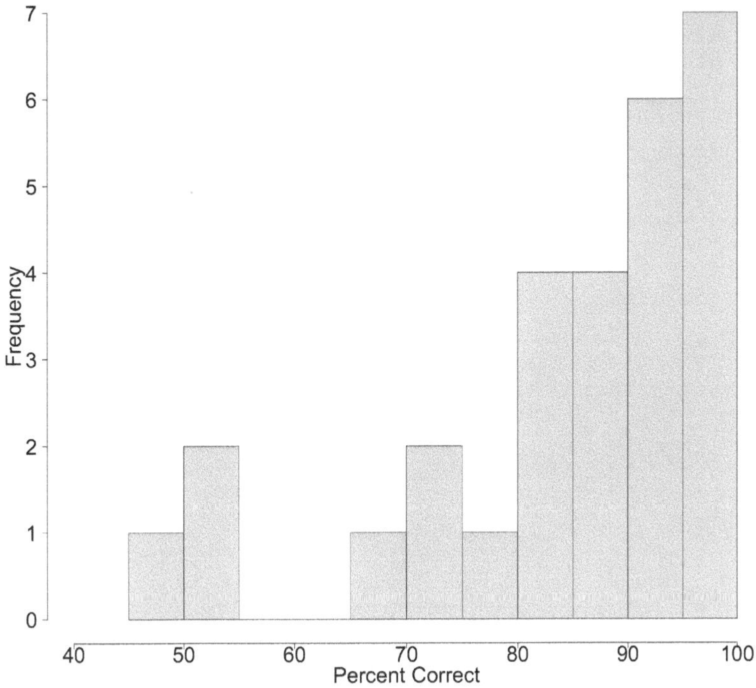

Figure 2.1 Histogram of the data in Table 2.1.

of the distribution of data in the population. These assumptions are often not met, and it is therefore important to understand the consequences of the mismatch between assumptions and the distribution of data. We will consider those consequences and possible alternative analyses later in this book. For now, we again emphasize that exploring the data is a first step in evaluating the validity of assumptions that play a role in the next, inferential, stage of analysis.

2.2.2 Box Plots

Histograms provide only one way to look at our data. A very different view is provided by box plots, which emphasize important characteristics of the data while sacrificing much of the detail available in a histogram. Figure 2.2 presents a box plot of the same data depicted in the histogram in Figure 2.1. The box plot quickly provides information about the main characteristics of the distribution of scores: (1) the median, the value at the center of the distribution; (2) the variability, or spread, of the scores; (3) the degree of skewness, or asymmetry, in the distribution; and (4) extreme scores, or outliers. We will consider each of these aspects of the box plot, but we first provide an overview of its elements.

Creating the Box

The construction of a box plot begins by placing the scores of the data set in order by rank, from lowest to highest. Once this is done, we can draw the box depicted in Figure 2.2. The

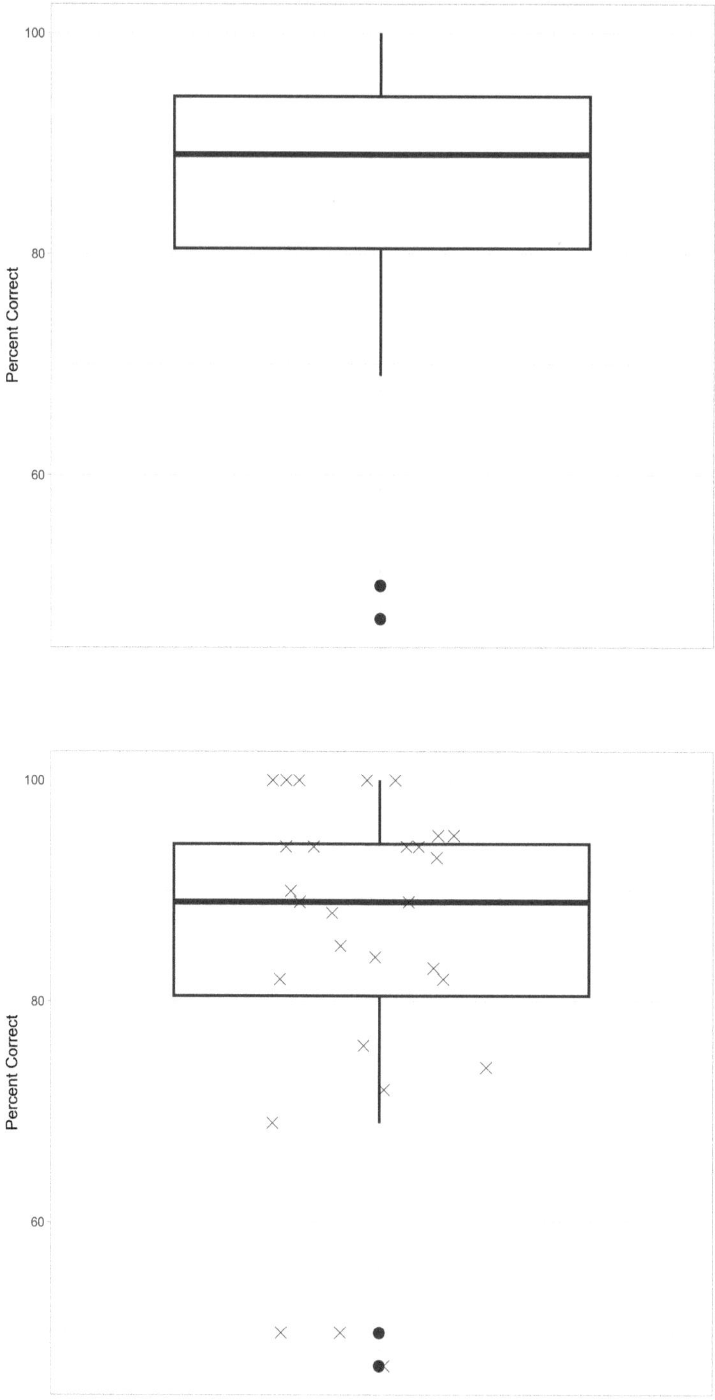

Figure 2.2 Box plots of the data in Table 2.1.

high end corresponds to the score that exceeds 75% of the scores; this is the third quartile. The lower end corresponds to the score that exceeds 25% of the scores; this is the first quartile. Thus, 50% of the data lies between the limits indicated by the top and bottom of the box.

There are several ways to calculate the quartiles. The definition usually found in introductory statistics texts is often awkward, requiring linear interpolation. Somewhat simpler to calculate are the *hinges* (Tukey, 1977); these are used by both the *boxplot* function in R, and in SPSS, for the top and bottom of the box. As an example of their calculation, consider the *Royer* data of Table 2.1. Then take the following steps.

1. Find the location, or *depth*, of the median; $d_m = (n + 1)/2$. With 28 scores, $d_m = 29/2 = 14.5$. When d_m has a fractional value – that is, when n is an even number – drop the fraction. We use brackets to represent the integer; that is, $[d_m] = 14$.
2. Find the depth of the lower and upper hinges, d_{lh} and d_{uh}. These are simply the medians of the lower and of the upper 14 scores. The depth of the lower hinge is

$$\begin{aligned} d_{LH} &= ([d_M]+1)/2 \\ &= (14+1)/2 = 7.5 \end{aligned} \tag{2.1a}$$

and the depth of the upper hinge is

$$\begin{aligned} d_{UH} &= (n+1) - d_{LH} \\ &= (28+1) - 7.5 = 21.5 \end{aligned} \tag{2.1b}$$

The depth of 7.5 for the lower hinge means that the lower hinge is the score midway between the seventh score (76) and the eighth score (82), or 79. The depth of the upper hinge is 21.5, which is midway between the seventh and eighth scores from the top; the corresponding score is 94.5 in the *Royer* data. The length of the box in Figure 2.2 is therefore 94.5 - 79, or 15.5. This distance between the hinges is one measure of variability and is usually referred to as the *H-spread*, to signify the spread of scores between the hinges. The box-plot statistics can also be expressed in percentiles. In that case, quartiles are computed by linear interpolation and the values of the first and third quartiles are 77.5 and 94.75.

The Median

The horizontal line between the top and bottom of the box represents the median. The median is the middle score in an ordered set of scores. In the data of Table 2.1, in which the number of scores, n, is an even number, the median falls halfway between the middle two scores. Those scores are in the 14th and 15th positions when the scores are ordered from lowest to highest. Both of the scores are 89, and so the median has that value. If the two scores had differed, the median would have been computed as the mean of the two values.

The median is useful in an initial exploration of data because it is not affected by extreme scores. Unlike the mean, whose value changes if any individual score is changed, the median is considered a *resistant* or *robust statistic* because it is unaffected by changes in the value of any single score. For example, if we replaced the 47 in the data set of Table 2.1 with a score of 67, the median would be unchanged, but the mean would be increased. Because of this property of the median, there are circumstances in which it provides a more representative index of the location of a distribution than does the mean. For example, if the salaries of corporate officers are included in reporting mean salaries at a company, we may have a distorted sense of how well everyone is paid because the officers' salaries will greatly increase the mean.

The Whiskers

The lines extending vertically from the box are often called whiskers, and they extend to the lowest and highest scores that are not outliers. Note that the median is closer to the top of the box than to the bottom and that the top whisker is shorter than the bottom one. This reflects the same straggling left tail that we saw reflected in the histogram of Figure 2.1. In other words, both figures illustrate that half of the data fall in a relatively narrow range between 89 and 100, whereas the remaining half are spread over a wider range of scores; the distribution is asymmetric or *skewed.*

Outliers

The circles in the box plot represent outliers, the two scores of 50 and the score of 47 in Table 2.1. Notice that only two circles are shown, even though there are three outliers. That is simply because the two points reflecting the scores of 50 appear in exactly the same location, superimposed on one another. The steps to define an outlier are as follows:

1. Calculate the *H*-spread. This is 94.5 – 79, or 15.5, in the present example.
2. Find 1.5 times the *H*-spread; this is (1.5)(15.5) = 23.25.
3. Subtract the preceding result (23.25) from the bottom limit of the box and add it to the top limit. In the current example, these values are 79 – 23.25 = 55.75 and 94.5 + 23.25 = 117.75.
4. An outlier is any score beyond those limits. Therefore, the scores of 47 and 50 are outliers. Extreme outliers are scores that are more than three times the *H*-spread above or below the hinges. SPSS displays outliers with a circle and extreme outliers with an asterisk. In R's *boxplot* function in the {graphics} package, the *range* parameter is the multiplier on the *H*-spread. It has a default value of 1.5 (as in step 2 of defining an outlier) and can be set to 3.0 to identify extreme outliers.

Outliers are important to detect because they often correspond to cases that merit special attention. In the current example, they draw attention to three students who are performing much worse than any other student in the group of second-grade males. These students may require remedial aid, or there may be other causes of their poor performance that require investigation.

Outliers are also important to detect in studies where the goal is to draw inferences about the population from which the sample has been drawn. Many statistical tests involve calculating the sample mean, which, as we have already noted, is quite sensitive to extreme scores. Thus, it is important to be aware of outliers when interpreting patterns of means. In addition, many statistical tests assume a normal distribution of scores. Inferences based on such tests can be suspect when the distribution is asymmetric, which often occurs when there are extreme scores in the sample. In the current example, our estimate of the population mean is the sample mean, 84.61. If we eliminate the three outlying scores, the mean is now estimated to be 88.88, close to the sample median. Furthermore, the distribution is far more symmetric in shape when the outliers are eliminated. Thus, our interest in both the individuals who had the outlying scores and the effects of outliers on inferences drawn from the sample dictates a need to identify outliers.

Integrating the Components

Our box plot provides a sense of several important aspects of the data. The median of 89 tells us that at least half of the second-grade students have a good grasp of addition. We also can see from the position of the lower hinge that approximately 75% of the students have scored in the high 70s or above. We can also see the straggling lower tail, usually referred to as *negative skew* or *left skew*; the top 25% of the students have scores between 95 and 100 whereas the lowest 25% fall between 47 and 79, reminding us that there are several students who have had problems with the test.

One caution with boxplots is that they can sometimes obscure important aspects of the data. We noticed this already: there are only two points displayed for the three outliers in the data. Another common example occurs when the data are bimodal, with two clusters of scores. In that case, the central box may hide the score clusters. This limitation can be overcome in R by including the individual scores in the boxplot ("jittered" slightly so that they can be seen individually), as in the lower panel of Figure 2.2. Notice that plotting the individual observations reveals all three outliers in these data, not just the two suggested by the boxplot alone.

2.2.3 A Graphic Check on Normality

Because many commonly used statistical procedures assume that the data were sampled from a normally distributed population, it is helpful to have ways of looking at possible violations of this assumption. Specific characteristics of a distribution imply nonnormality. For example, because the normal distribution is symmetric, *skewness* (asymmetry) in a data plot indicates a distribution that is not normal. Another indication is a nonzero value of *kurtosis* (roughly, more, or fewer, scores in the tails than we would expect in a normal distribution). However, a more direct indication is available in many computing packages. These programs rank order the scores and then plot the standardized scores (z scores; see Section 2.4 for further discussion) against the actual scores. Figure 2.3 presents two such plots. Using R's *qqnorm* function in the {stats} package, we first plotted multiplication response times in the *Royer* data for students in the fifth through eighth grades. If data are normally distributed, the points will fall on a straight line. This is clearly not the case for the multiplication response times (*RT*). However, if we transform the multiplication times into response speeds by taking the reciprocals of the response times, we obtain the second plot shown in Figure 2.3. The multiplication speeds, except for a few scores at each extreme, do fall quite close to the line, indicating that the response speeds are more nearly normally distributed than the response times. Implicit in this illustration is a possible strategy for dealing with situations where we desire normally distributed observations, but the data do not conform to the distribution; namely, we can sometimes transform the data. We will have more to say about transformations, as well as other distributional forms, in later chapters.

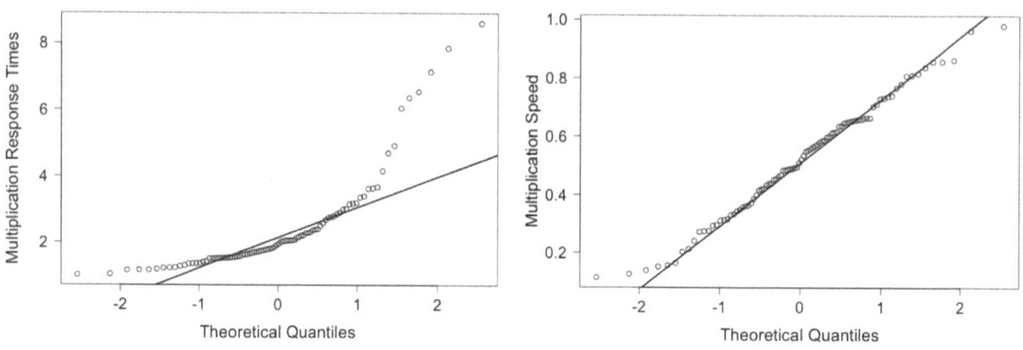

Figure 2.3 Normal probability (*Q–Q*) plots of multiplication response times and speeds.

We might ask whether departures from the straight line that we observe are large enough to affect subsequent inferential analyses. We will have more to say about this and its implication for data analysis in Chapter 6, in which we discuss the *t* test and its assumptions.

2.3 Numerical Summaries of the Data

A graph provides an immediate picture of the overall shape of a distribution, the variability in the observations, and whether there are outliers in the data. Additionally, box plots provide specific information about the location of the distribution (i.e., the median); box plots provide a specific measure of variability, too (i.e., the *H*-spread). Numerical estimates of specific characteristics of a distribution are very useful in helping us to understand a data set. Further, some measures are prominent in inferential procedures, most notably the mean and standard deviation. Thus, in this section, we focus on the mean and measures of variability that complement the mean; specifically, the standard deviation, variance, and standard error of the mean.

Most statistical packages readily provide several useful descriptive statistics for a data set. For example, the data of Table 2.1 were submitted to the Explore module in SPSS to generate the statistics shown in Table 2.2. Similarly, the R function *fivenum* in the {stats} package provides a five number summary of the data: the minimum, lower hinge, median, upper hinge, and maximum. R has many functions for summarizing data, such as mean and median, that can be applied to vectors of data individually or to several variables simultaneously. For example, the *summary* function in the {base} package provides descriptive statistics on each variable in a data frame. We will also make use of the *summarise* function in {dplyr} to calculate specific summary statistics.

2.3.1 *Measures of Location: Mean, Median, and Mode*

When we talk about the location of the data, we are referring to its *central tendency*: what values are typical, or near the middle of the data set? We have already met one measure of location, the *median*, which is the value above which, and below which, half of the data fall. The median value may not be represented in the data. For example, the median of $Y = [1, 3, 3, 9, 10, 11]$ is 6, the average of the two middle scores, 3 and 9. The *mode* describes the

Table 2.2 Summary statistics for the data of Table 2.1 from SPSS

		Statistic	*Std. error*
Mean		84.61	2.89
95% Confidence interval for mean	Lower bound	78.68	
	Upper bound	90.54	
Median		89.00	
Variance		234.025	
Std. deviation		15.298	
Minimum		47	
Maximum		100	
Interquartile range		17.25	
Skewness		−1.326	.441
Kurtosis		1.124	.858

location of the data by identifying its most frequently observed value. For the second grade addition scores in Table 2.1, the modal score is 100. Bimodal data sets have two such peaks of frequently observed scores.

The familiar arithmetic *mean*, symbolized by $\bar{Y}$ (Y bar), is just the sum of all scores divided by the number of scores. Expressing this as an equation, we have

$$\bar{Y} = \sum_{i=1}^{n} Y_i \,/\, n \tag{2.2}$$

where Σ represents the mathematical operation of summation,[3] Y_i represent the score for individual i, and n represents the number of scores in a sample. For example, the mean of Y = [1, 2, 3, 5, 9, 10, 12] is 42/7 = 6.

The widespread use of the mean reflects two advantages it has over other measures of location. First, it has certain properties that are desirable when estimating the mean of the population from which the sample was drawn. In Chapter 5, we will discuss the role of the sample mean as an estimator of the population mean. Second, unlike the median or mode, the mean can be manipulated algebraically. The algebraic properties of the mean can be proven by the rules of summation in Appendix A. They can be demonstrated by using the preceding set of numbers, or any other numbers you choose.

1. Adding a constant, k, to every score in the sample results in the mean being increased by the amount k; that is, $\Sigma(Y + k)/n = \bar{Y} + k$. For example, if we add 10 to each of the values in the preceding set of 12 numbers, the mean increases from 6 to 16.
2. Multiplying every score in the sample by a constant, k, results in the mean being multiplied by k; that is, $\Sigma(kY)/n = k\bar{Y}$. For example, multiplying each of the scores in the example by 2 increases the mean from 6 to 12.
3. Perhaps most importantly, means can be combined across subgroups in the data by weighting the separate means by their respective sample sizes. The result is the overall mean for the data set. Note that you cannot meaningfully combine other measures of location, such as medians or modes.

Let's demonstrate how means are combined, using the data in Table 2.3, which represent hypothetical depression scores from four different clinics. It is tempting to add the four means and divide by 4 to get the mean of the combined data sets. However, because the four sample sizes vary, this won't do. The mean for Clinic C should carry more weight and that for Clinic B less weight in combining the means because of their relative sample sizes. The correct approach requires us to obtain the sum of all the scores in all four data sets and then divide by *N*, the sum of the *n*s. We obtain the sum of scores for each clinic by multiplying the clinic mean by the number of scores for that clinic. Summing these four sums, and dividing the grand total by *N*, the total number of scores, we have the grand mean of all the scores:

$$\bar{Y} = [(26 \times 17.5) + (17 \times 18.3) + (31 \times 19.2) + (24 \times 22.6)] / 98 = 19.426$$

We might have slightly rewritten this:

$$\bar{Y} = \left(\frac{26}{98}\right)(17.5) + \left(\frac{17}{98}\right)(18.3) + \left(\frac{31}{98}\right)(19.2) + \left(\frac{24}{98}\right)(22.6)$$

This equation suggests that the mean can be represented as a sum of weighted values, where the weights are proportions or probabilities. The weight for Clinic A is 26/98 because 26 of the 98 depression scores come from Clinic A. To take a somewhat different example, consider a student who has spent two semesters in College A and compiles a 3.2 average. She then transfers to College B, where she earns a 3.8 average for the next three semesters. The student's overall grade point average (GPA) for the five semesters is calculated as in the preceding example of the clinic means. The overall GPA is a weighted mean where the weights are 2/5 and 3/5:

$$\bar{Y} = (2/5)(3.2) + (3/5)(3.8) = 3.56$$

In general, the preceding calculations may be represented by

$$\bar{Y} = \Sigma p(y) \cdot y \tag{2.3}$$

Equation 2.3 is the formula for a ***weighted mean***. It indicates that each distinct value of Y is to be multiplied by its weight, $p(y)$, the proportion of all scores with that value. All of these products are then added together. Note that the usual expression for the arithmetic mean, $\bar{Y} = \Sigma Y/n$, is a special case of the preceding formula for the weighted mean; here each of the n scores in the sample is given a weight of $1/n$.

Table 2.3 Example of means based on different sample sizes

	Clinics			
	A	*B*	*C*	*D*
Mean	17.5	18.3	19.2	22.6
n	26	17	31	24

Two other properties of the mean, proven in Appendix A, may help convey the sense in which it reflects the central tendency of a data set:

1. The mean is the balance point of the data in the sense that the sum of the deviations about the mean is zero; that is, $\Sigma(Y - \bar{Y}) = 0$.
2. The sum of squared deviations of scores from the mean is smaller than the sum of squared differences taken from any point other than the mean; that is, $\Sigma[Y - (\bar{Y} + k)]^2$ has its smallest value when $k = 0$.

Many of the advantages of the mean as a descriptive statistic are attributable to the fact that every value in a set of scores is incorporated into the calculation of the mean. However, the mean has a potential weakness that derives from this same fact. Namely, a few extreme scores can bias the value of the mean and may result in the mean taking on a value that does not typify the location of the data. We gave an example earlier in this chapter where a single outlier had a substantial effect on the value of the mean. In such a situation, the median is the better description of location because its value is affected only by score ranks, not by their individual values. Because the median changes little when values of individual scores are changed, we say that the median is a *resistant* or *robust statistic*. Resistance is a desirable property in a statistic, but it is not a property of the mean. We will find that this fact has several implications for our analyses of means when we study inferential tests.

2.3.2 Measures of Variability: Variance, Standard Deviation, and Standard Error

One common way to measure the spread or variability of a set of scores is to compute their *variance*. The sample variance, S^2, is computed as: $S^2 = \Sigma(Y - \bar{Y})^2/n$. Consider the information summarized by this formula. First, note that the calculation is based on a measure of the discrepancy of individual scores from the average score in the sample, $(Y - \bar{Y})$. This makes the variance a nice complement to the mean. Also, like the mean, the variance incorporates the value of each individual score and, thus, the variance possesses some of the same statistical properties we observed in the mean. Second, the squaring of the deviation scores is necessitated by the fact that $\Sigma(Y - \bar{Y}) = 0$, as we observed in Section 2.3.1. Finally, the squared deviations are summed and divided by n, so the final value represents *the average squared deviation of scores about the mean.*

Although the formula defining the sample variance involves division by n, we will divide by $n - 1$, as most statistical packages do. The divisor $n - 1$ is used instead of n because it results in a better estimate of the population variance, a fact that will be explained in Chapter 5. We denote this revised definition by s^2, rather than S^2, to indicate that we are dividing by $n - 1$. Thus, our formula for s^2 is:

$$s^2 = \Sigma(Y - \bar{Y})^2 / (n - 1) \tag{2.4}$$

Generally, s^2 will be computed with the aid of a calculator or statistical package; however, we illustrate its calculation here with a simple data set to make sure that the formula is understood. For the following set of seven scores, Y = [1, 2, 3, 5, 9, 10, and 12], the sum of the scores is 42, and, therefore, $\bar{Y} = 42/7 = 6$. The deviations from the mean are $(Y - \bar{Y})$ = −5, −4, −3, −1, 3, 4, and 6. Squaring these deviations, $(Y - \bar{Y})^2$ = 25, 16, 9, 1, 9, 16, and 36.

Summing the squared deviations, we have $\Sigma(Y - \bar{Y})^2 = 112$. Then, $s^2 = 112/6 = 18.667$. In words, the variance of the scores is 18.667 squared units.

As a descriptive measure, the variance is rather awkward because it expresses the variance on a different scale (i.e., squared units) than the data being described. Because of this, the preferred measure of spread for descriptive purposes is the *standard deviation*, which is simply the square root of the variance. In the preceding example, $s = \sqrt{18.667} = 4.320$.

Two properties of the standard deviation should be noted:

1. *If* $Y' = Y + k$, $s_{y'} = s_y$. When a constant is added to all scores in a distribution, the standard deviation is unchanged. Each score is increased (or decreased) by the same amount so the spread of scores is unchanged. The range and the *H*-spread are also unchanged when a constant is added, and it is a desirable property of any measure of variability.
2. *If* $Y' = kY$, $s_{y'} = |k| s_y$, where $|k|$ indicates the absolute value of k. When each score is multiplied by a constant, k, the standard deviation of the new scores is the absolute value of k times the old standard deviation. Thus, the standard deviation of the transformed distribution is changed by multiplication or division.

These properties are proven in Appendix A.

Although the standard deviation is less intuitive than other measures of variability (e.g., *H*-spread), it has two important advantages. First, the standard deviation is important in drawing inferences about populations from samples. It is a component of formulas for many significance tests, for procedures for estimating population parameters and the bounds on them, and for measures of relations among variables. Second, the standard deviation (and the variance) can be manipulated arithmetically in ways that other measures cannot. For example, if we know the standard deviations, means, and sample sizes of two sets of scores, we can calculate the standard deviation of the combined data set without access to the individual scores. This relation between the variability within groups of scores and the variability of the total set plays an important role in data analysis. Both properties of the standard deviation will prove important throughout this book.

The main drawback of the standard deviation is that, like the mean, it can be greatly influenced by a single outlying score. Recall that for $Y = [1, 2, 3, 5, 9, 10, \text{and } 12]$, $\bar{Y} = 6$ and $s = 4.320$. Suppose we add one more score. If that score is 8, a value within the range of the scores, the new mean and standard deviation are 6.25 and 4.062, a fairly small change. However, if the additional score is 20, we now have $\bar{Y} = 7.75$ and $s = 6.364$. The standard deviation has increased by almost 50% with the addition of one extreme score. In contrast, the *H*-spread is resistant to extreme scores and is often a more useful measure for describing the variability in a data set. We again emphasize that there is no single best measure of variability (or of location or shape), but that there is a choice, and that different measures may prove useful for different purposes and may sometimes supplement each other.

2.3.3 *The Standard Error of the Mean* (SEM)

Ordinarily, we view the data set as a sample from some population. If we are studying the arithmetic scores of second-grade students, we may wish to draw conclusions about the performance of other second-graders who were not in our study but who share many of the attributes of our sample that might affect performance, such as teacher experience, curricula, and class size. The standard error of a statistic provides some sense of how much the statistic might vary if other samples were taken from the same population. A familiar example is the *SEM*. This statistic tells us how much error there is in the sample mean as an estimate of the mean of the population from which the sample was drawn. The *SEM* can be best understood if we assume that many random samples of size n are drawn from the same population and the mean is calculated each time. If we record all of the values of the sample mean we observe and tabulate the proportion of times each value occurs, we have constructed the *sampling distribution of the mean* for samples of size n. The *SEM* that is calculated from a single sample is an estimate of the standard deviation of the sampling distribution of the mean. If the *SEM* is small, the one sample mean we have is likely to be a good estimate of the population mean because the small *SEM* suggests that the mean will not vary greatly across samples, and therefore any one sample mean will be close to the population mean. We will have considerably more to say about the *SEM* and its role in drawing inferences in later chapters. At this point, we introduced it because of its close relation to the standard deviation and because it provides an index of the variability of the sample mean.

The *SEM* is a simple function of the standard deviation:

$$SEM = s / \sqrt{n} \tag{2.5}$$

For example, in Table 2.2 the standard deviation is 15.298. Dividing by the square root of 28 (the n based on Table 2.1), we have 15.298/5.292, or 2.89, the value given in the Std. error column of Table 2.2. Note an immediate and important implication of Equation 2.5: The error in the sample mean as an estimate of the population mean depends on both the variability in the sample and the size of the sample. Specifically, as the variability in the sample increases, the variability in the sample mean also increases. However, as the size of the sample increases, the variability in the sample mean decreases.

2.3.4 Displaying Means and Standard Errors

A graph of the means for various conditions often provides a quick comparison of those conditions. The graph is more useful still when accompanied by a visual representation of variability, such as s, the *SEM*, or the confidence interval bounds. The selection of a procedure for graphing the data should depend upon the nature of the independent variable. Although graphics programs will provide a choice, the American Psychological Association's *Publication Manual* (2001) recommends that "*Bar graphs* are used when the independent variable is categorical" and "*Line graphs* are used to show the relation between two quantitative variables" (p. 178). We believe this is good advice.

When the independent variable consists of categories that differ in type rather than in amount, we should make it clear that the shape of a function relating the independent and dependent variables is not a meaningful concept; thus, the choice of a bar graph. Figure 2.4 presents mean depression scores[4] from the *Seasons* data set (found on the *Data Files* page on the book's website) as a function of marital status (Married or Divorced) and employment status; the numbers on the x-axis are the researchers' codes: 1 = employed full time; 2 = employed part time; 3 = no occupation. At least in this sample, depression scores are, on average, higher for divorced individuals than for those who are married, regardless of their employment status. Additionally, full-time employment is associated with lower depression levels, on average, than either part-time employment or no occupation. The vertical lines at the top of each bar show +/− one *SEM*. Note that the *SEM* bars indicate

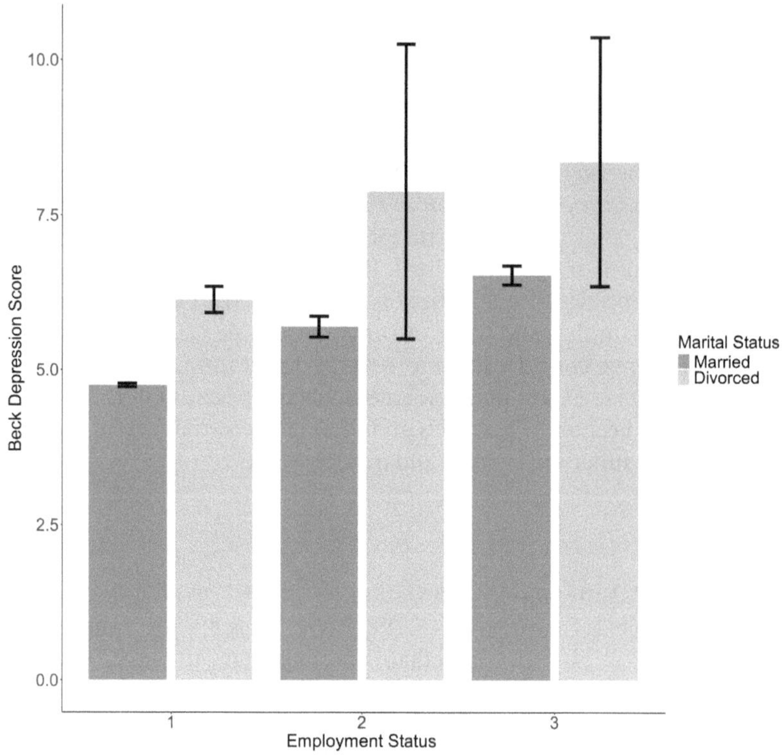

Figure 2.4 Bar graph of mean depression scores as a function of employment and marital status, with *SEMs*.

that the data are much more variable for divorced individuals who are employed part time. However there are only two individuals represented in that bar, so the data must be interpreted with caution.

When the independent variable consists of categories that do differ in amount, it is meaningful to illustrate the shape of a function relating the independent and dependent variables; thus, a line graph is a good choice. Figure 2.5 presents mean times in seconds to do subtraction problems (*subrt*) from the *Royer* data set as a function of grade. Because grade level is quantitative, it is useful to plot the data as line curves, providing a sense of the shape of the functional relation. We can see in Figure 2.5 that response times decrease as a function of grade and level off around the sixth grade. Variability also decreases with grade as indicated by the general decrease in the length of the *SEM* bars.

Software capable of bar and line plots usually offers several options such as the choice of placing different plots in one panel or in separate panels, or choosing the error bars to represent standard deviations, standard errors, or confidence intervals. The best advice is to become thoroughly familiar with the software being used, and then to think carefully about which options will enable you to best communicate the points you believe are important. R offers tremendous flexibility and control over the visual display of your data; the {ggplot2} package is especially powerful and worth the time investment to learn.

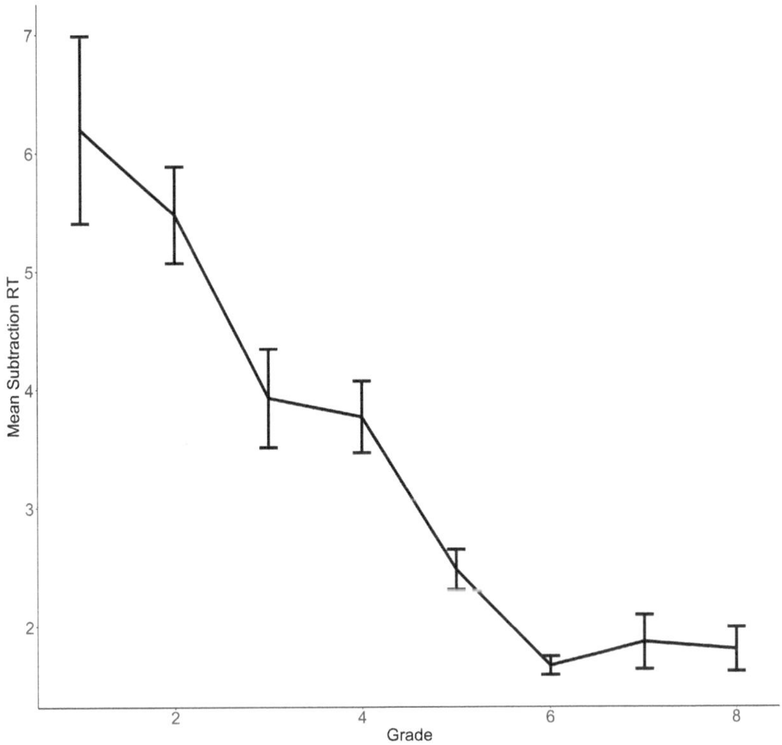

Figure 2.5 Line graph of mean subtraction response times (*RT*) as a function of grade, with *SEMs*.

2.3.5 *The Interquartile Range* (IQR)

One other measure of spread deserves mention here. The value of the *IQR*, like the closely related *H*-spread of the box plot, provides a measure of spread that is more resistant to extreme scores than is the standard deviation. Like the *H*-spread measure, *IQR* is a resistant statistic because it ignores scores in the tails of a distribution. The value of the *IQR* is the difference between the 75th and 25th percentiles; it is the range of the middle 50% of the data. Calculating percentiles usually requires interpolation between points, and there are several ways to do that (Hyndman & Fan, 1996). In R, we can use the *quantile* function in {stats} to obtain the 25% and 75% percentiles for the *Royer* data in Table 2.1, 80.50 and 94.25, and then use subtraction to find the *IQR* = 94.25 – 80.50 = 13.75. Alternatively, the *IQR* {stats} function provides the *IQR* in one step.[5]

2.4 Standardized (z) Scores

An individual's score, by itself, is not very informative. If a student earns a score of 70 on a math test, is that good or bad? If told that a score of 70 was at the 90th percentile – that it had been exceeded by only 10% of all scores – we would probably feel that the performance was quite good. On the other hand, we would be considerably less impressed if the same raw score fell at the 40th percentile. Although information about percentile values tells us where a score ranks relative to other scores, it provides no information about distances among scores. For example, a score at the 90th percentile could be many points higher than the average or only a little higher; it depends on how variable the scores are. Standardized, or z, scores tell us more about where the score is located within the distribution, measured in units of the variability of the data. Specifically, a z score tells us how many standard deviation units the score is above or below the mean. Given a distribution of scores with mean, $\bar{Y}$, and standard deviation, s, the z score corresponding to a score Y is calculated as.

$$z_Y = (Y - \bar{Y}) / s \tag{2.6}$$

For example, if the mean is 75 and the standard deviation is 15, for a score of 90, we would have $z_{90} = (90 - 75)/15 = 1.0$; thus, this score is one standard deviation above the mean. In R, the *scale* function in the {base} package computes z scores for a set of data.

Standardizing a group of scores changes the scale to one of standard deviation units, permitting comparisons with scores that were originally on a different scale. Nevertheless, there are aspects of the original distribution that remain unchanged. The following are two things that remain constant:

1. *An individual's* z *score has the same percentile rank as that individual's original score.* This is because subtracting a constant, $\bar{Y}$, from every score does not change the rank order of the scores, nor is the order changed by dividing all scores by a constant, s.
2. *The shape of the distribution of* z *scores is the same as that of the original data.* Subtraction of $\bar{Y}$ shifts the original distribution and division by s squeezes the points closer together, but shape information is preserved. If the original distribution was symmetric, the distribution of z scores will also be symmetric. Likewise, if the original distribution was skewed, the distribution of z scores will be, too.

As we shall see in Chapter 5, z scores are used in drawing inferences when scores can reasonably be assumed to be normally distributed. However, the preceding point should make clear that z *scores are not necessarily (or even usually) normally distributed.* Their distribution depends on the distribution of the scores prior to the z transformation.

Two other characteristics of z scores should be noted:

1. *The mean (and therefore also the sum) of a set of* z *scores is zero.* We stated earlier that when a constant is subtracted from every score, the mean is also changed by that constant. In the case of z score, $\bar{Y}$ is subtracted from each of the Y values. Therefore, the mean of the z scores is $\bar{Y} - \bar{Y}$, or 0.
2. *The variance of a group of* z *scores is 1.0 and, therefore, so is the standard deviation.* Because the average z score is zero, we need only square each member of the group of z scores, sum the squared values, and divide by $n - 1$ to obtain the variance, S_z^2. Doing so,

$$\begin{aligned} s_z^2 &= \sum \frac{(Y - \bar{Y})^2}{s^2} / (n-1) \\ &= \sum \frac{(Y - \bar{Y})^2}{n-1} / s^2 \\ &= s^2 / s^2 = 1 \end{aligned}$$

Standardized scores can be useful in our understanding of a data set because they provide a way of comparing performances on different scales. For example, the mean and standard deviation of the subtraction accuracy scores in the *Royer* data set were 88.840 and 11.457, respectively; the corresponding values for multiplication accuracy scores were 87.437 and 13.996. Even though a subtraction score of 70 is numerically higher than a multiplication score of 65, it reflects slightly worse performance relative to the other scores in the distribution. For subtraction, $z_{70} = -1.64$ (that is, the score of 70 is 1.64 standard deviations below the mean of the subtraction scores), whereas for multiplication, $z_{65} = -1.60$.

Standardized scores play other roles as well. In Chapter 5, we will discuss how z scores provide percentile information when the population distribution has an approximate bell shape. Also, in Chapter 18, we will consider the correlation coefficient as an average product of z scores.

2.5 Measures of the Shape of a Distribution: Skewness and Kurtosis

We have considered several measures of location (mean, median, and mode) and variability (variance, standard deviation, standard error, and *IQR*) that summarize important properties of a distribution. In addition to these measures, it is often also useful to obtain measures of the shape of a distribution. Measures of shape permit a more precise description of a data set than measures of location and variability alone. They also provide another way to assess the validity of the assumption of normality. This is important because when a data distribution departs markedly from normality, reliance on the mean and standard deviation may lead to incorrect inferences. For example, although the mean and median of a symmetric population distribution are identical, the sample mean is usually considered the better estimate of the location of the population distribution. However, if the population distribution has long straggling tails, the sample median is more likely to have a value close to that of the center of the population distribution (Rosenberger & Gasko, 1983), and therefore would be a better estimate of the location of the population. If the distribution is skewed (asymmetric), a transformation, or a test based on ranks, may be helpful (see Chapter 6 for a discussion of these alternatives). The data plots in Section 2.2 provide a first step in assessing the shape of the distribution, but the measures in this section will provide summary numbers, aiding in the evaluation of shape.

There is yet a third reason for our interest in measures of shape. An important stage in understanding the processes that underlie behavior is the construction of mathematical or computational models, models precise enough to predict the behavior in question. Comparing predicted and observed measures of the shape of the data distribution provides additional tests of such models.

Two aspects of shape have received the most attention from statisticians: (1) the degree of *skewness*, or departure from symmetry; and (2) *kurtosis*, or the proportion of data in the extreme tails and peak of the distribution. Indices of these two attributes of shape can be obtained from various computer packages, although there are frequently small differences in the details of the calculation that can result in slightly different values, especially for smaller samples.

2.5.1 *Skewness*

Skewness statistics are designed to reflect departures from symmetry. The standard definition of skewness is the average cubed deviation of scores from the mean divided by the cubed standard deviation. Because the mean and standard deviation are influenced by outliers in the data, the dependence of skewness on cubed standard deviations means outliers have an exaggerated effect. In R, the *skewness* function of the {moments} package provides this value. When skew is 0, the distribution is symmetric about the mean. For the data in Table 2.1, skewness equals −1.25.[6] The negative value indicates that the left tail of the data is longer than the right, consistent with the fact that the median has a higher value than the mean. We call data with a long left tail "left skewed." Positive skewness values indicate a long or heavy right tail, and we describe those data as "right skewed."

Skewness is a statistic calculated from a sample, analogous to a sample mean, and so its value varies from sample to sample. The standard error of skewness depends only on the size of the sample, N: $SE_{\text{skew}} = \sqrt{\frac{6N(N-1)}{(N-2)(N+1)(N+3)}}$. For the data in Table 2.1, SE_{skew} = 0.44. A rough indication of substantial skew is to compare the absolute value of skewness to twice the SE_{skew}. If skew exceeds twice the SE_{skew}, the observed asymmetry may reflect asymmetry in the population rather than chance variability in the sample. Use caution, however, as this assessment is error prone (Wright & Herrington, 2011).

2.5.2 *Kurtosis*

Kurtosis values also reflect departures from the normal distribution and are generally sensitive to the height of the peak and to the tail weight of a distribution. Conceptually, leptokurtosis takes mass from the normal distribution and moves it to the tails and the peak in such a way that the variance is not changed; the result looks "peaky" with long tails. Platykurtic distributions also hold variance constant but move mass from the peak and tails of the normal distribution to the shoulders, resulting in a flat looking shape ("Student" [Gosset] 1927 said they look like a platypus). Some examples are shown in Figure 2.6, following DeCarlo (1997).

Both panels of Figure 2.6 show a normal distribution with a mean of 0. The histogram on the left shows data sampled from a *t* distribution with 5 degrees of freedom. We will learn more about the *t* distribution in Chapter 6. For now, you only need to know that it is symmetric and that its variance is 5/3. The normal distribution in that panel has been scaled to also have a variance of 5/3. You can see the higher peak, skinnier shoulders, and – if you squint – the heavy tails of the leptokurtosis in the *t* distribution. In the right panel, we see a histogram of data sampled from a rectangular distribution, which has a flat peak, heavy shoulders, and light tails compared to a normal distribution with the same variance. The rectangular distribution is clearly platykurtic.

The standard definition of kurtosis is analogous to skew, except that we use the 4th power rather than cubing the deviations of scores from the mean in the numerator, and we square the variance in the denominator. As in skewness, outliers have an outsized influence on kurtosis. For a normal distribution, kurtosis = 3, so some statistics packages report kurtosis after having subtracted 3; this value of kurtosis is called $g2$. The *kurtosis* function of the {moments} package in R provides the untransformed values: for the sample from the *t* distribution, kurtosis = 7.20; for the rectangular sample, it is 1.80 (and so $g2$ = 4.20 and −1.20, respectively).

Like the standard error of skew, the standard error of kurtosis depends only on the size of the sample: $SE_{\text{kurtosis}} = \sqrt{\frac{4\left(N^2-1\right)SE_{skew}^2}{(N-3)(N+5)}}$. For the data in Table 2.1, SE_{kurtosis} = 0.86. If

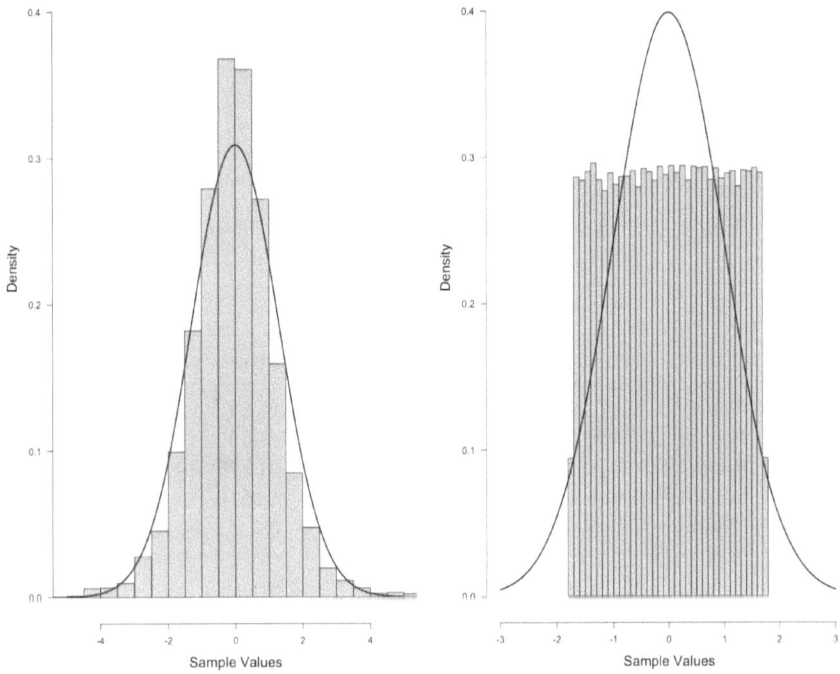

Figure 2.6 Histograms showing leptokurtosis in $t(5)$ distribution (left panel) and platykurtosis in uniform distribution (right panel), both compared to a standard normal distribution (solid curves).

the transformed value of kurtosis exceeds twice its *SE*, the observed shape may reflect the shape of the population rather than chance variability in the sample. Similarly to the interpretation of skewness, however, this eyeball assessment is overly simplistic and prone to error (Wright & Herrington, 2011).

Heavy-tailed distributions – those with positive kurtosis, measured as *g*2 – are of particular interest because inferences based on the assumption of normality have often been shown to be affected by this departure from normality. With such data, increasing the sample size or removing extreme scores ("trimming" the top and bottom 5% of scores) may improve the accuracy of our inferences. However, kurtosis is sensitive to more than just the tails of the distribution and therefore interpretation is often difficult. Good discussions of kurtosis are provided in several sources (e.g., Balanda & MacGillivray, 1988; DeCarlo, 1997; Wright & Harrington, 2011). These and other articles and chapters (e.g., Hogg, 1974; Rosenberger & Gasko, 1983) also suggest alternative measures of tail weight that are often more resistant to outlying points.

2.6 Comparing Two Data Sets

Our focus up until now has been on examining the characteristics of single data sets. However, a comparison of graphs and statistics for two or more data sets can often be even more useful, suggesting hypotheses about the effects of variables and providing an initial assessment of assumptions underlying subsequent statistical tests. In this section, we review what we have done previously, but now in the context of comparisons of data sets. We look at two examples.

Table 2.4 Data (Y) to be compared with the data of Table 2.1

	31	32	79	83	83	85	85	85	87	87
Y	87	89	89	89	89	89	89	90	90	91
	91	91	91	92	92	93	95	95		

Table 2.5 Comparison of statistics for the *Royer* data of Table 2.1 with those for the *Y* data of Table 2.4

	Royer		*Y*	
	Statistic	*Std. error*	*Statistic*	*Std. error*
Mean	84.61	2.89	84.61	2.92
Median	89.00		89.00	
Variance	234.025		238.099	
Std. deviation	15.298		15.530	
Minimum	47		31	
Maximum	100		95	
Range	53		64	
Interquartile range	17		6	
Skewness	−1.25	.441	−3.02	.441
Kurtosis	3.73	.858	10.82	.858

2.6.1 *Example: The Royer Data Revisited*

A particularly clear example of how just comparing measures of location and variability can obscure potentially important differences between conditions is provided by comparing the distribution of the data in Table 2.4 with that in Table 2.1. The data in Table 2.4 are artificial and we have used the letter *Y* to label the dependent variable. Table 2.5 presents statistics that provide a comparison of the two data sets. In most respects the statistics based on Tables 2.1 and 2.4 are either identical or very similar. The means are both 84.61 and the medians are both 89. The variances are very similar for the two data sets. These statistics would lead us to believe that the two distributions also are very similar. However, the *Y* data have much larger skewness and kurtosis values, indicating some shape differences in the distributions. Replotting the *Royer* data, together with plots of the *Y* data, provides more information about the difference between the two distributions.

Figure 2.7 presents box plots of the two data sets. A comparison of the plots reveals that despite being equal in both means and medians, the *Y* distribution contains two scores much lower than any in the *Royer* data. Furthermore, despite the near equality of the variances in Table 2.5, the *Y* data are actually less variable when the outliers are excluded from the two data sets. If the two data sets represented two methods of teaching arithmetic, we might prefer that associated with the *Y* data; if outliers are excluded from both sets of data, the mean is higher for the *Y* data, and scores fall within a narrower range.

2.6.2 *Example: Age Differences in Depression Scores*

In the preceding section, we contrived a comparison between a real data set, the *Royer* addition scores, and one we created to make the point that the summary statistics usually

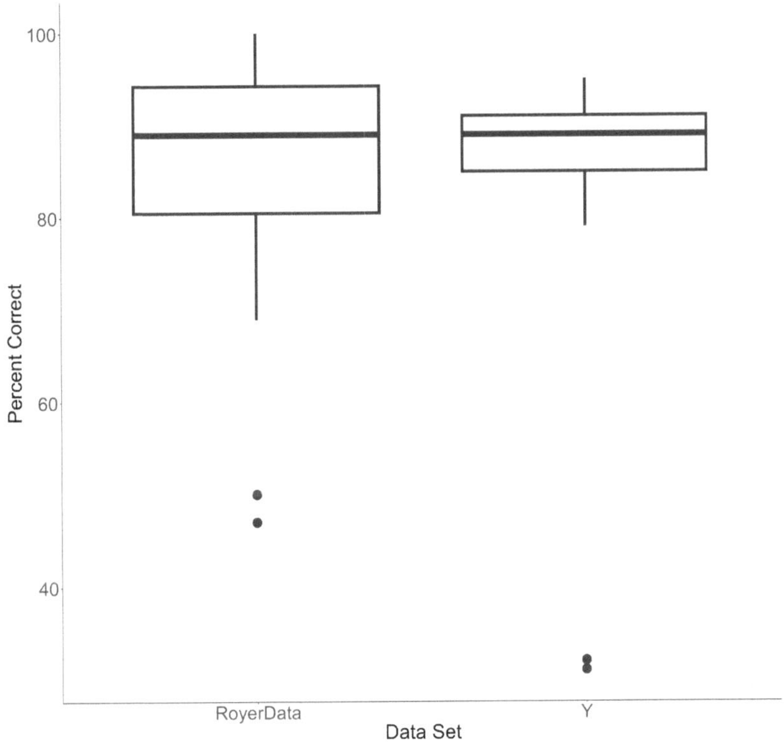

Figure 2.7 Box plots of the data in Tables 2.1 and 2.4.

examined (e.g., mean, variance) can be misleading, or at least fail to provide a complete understanding of the data. Real data can make the same point. In examining Beck Depression scores for the winter season for males of various ages, we found a difference between the mean of the youngest group (< 40 years, mean = 6.599) and that of a group between 50 and 59 years old (mean = 4.613). However, the medians were identical, 4.500.

Plotting the winter Beck Depression data for the two groups is a first step in clarifying why the means are further apart than the medians. Figure 2.8 presents box plots for the two groups. As in all plots of Beck Depression scores, most of the scores are at the low (normal) end of the scale for both age groups. We say that the distributions are skewed to the right because of the straggling right-hand (upper) tails. The explanation for the difference in means is readily apparent. The younger group has several individuals with scores above 18; these are depicted as outliers in the box plot for that group, some of which are extreme. In contrast, no score is an outlier in the older group. Although the medians are identical, the greater number of extremely high scores in the younger group has moved that mean higher than the mean of the older group. Just why there are more extremely depressed males in the under-40 group is not clear. However, the point for the present is that only by comparing the distributions of the two age groups do we achieve a full understanding of the nature of the difference between the two groups. Consideration of just the means and/or medians is misleading.

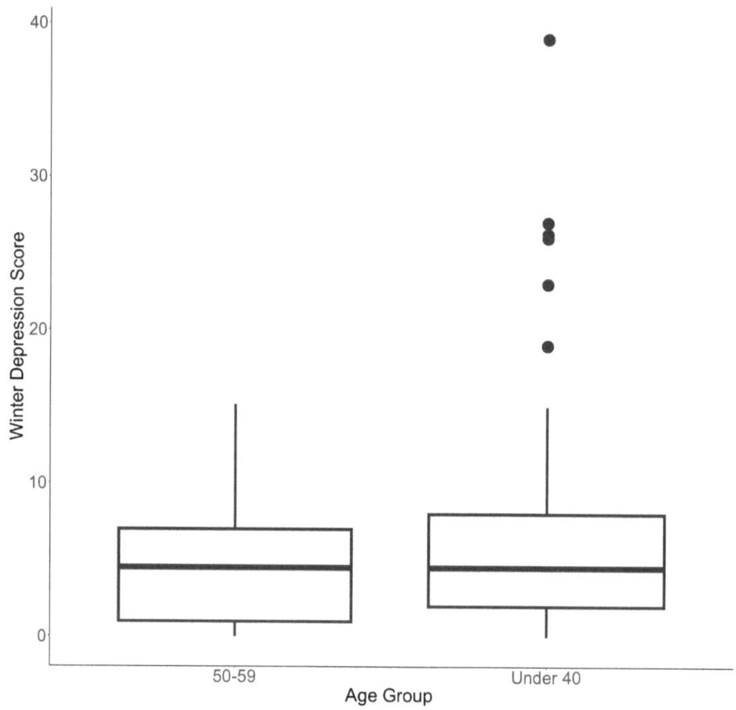

Figure 2.8 Box plots of winter depression scores for two age groups of men.

2.7 Relationships Among Quantitative Variables

Thus far, we have considered a few ways to graph and summarize distributions of single variables. However, we are often interested in how two or more variables are *related* to one another. For example, we may wish to know how cholesterol level changes with age, or whether mathematics skills are related to verbal skills in children. Because variability is always present, when variables are related, they are usually not perfectly related. Tall fathers tend to have tall sons, but because of a host of factors, the tallest fathers do not always have the tallest sons.

Variables may be related in ways that vary in type and degree, so we need ways to graphically represent these relationships and statistics to characterize them. In this section, we very briefly introduce:

- *Scatterplots* as a way of graphically exploring the relationship between two variables.
- The *correlation coefficient* as an index of the extent to which the relationship is linear.
- *Regression* as a way of generating the linear equation that best predicts one variable from another.

2.7.1 *Scatterplots*

Consider how subtraction and multiplication performance might be related in elementary school children. We might expect that children who are better at subtraction will also be

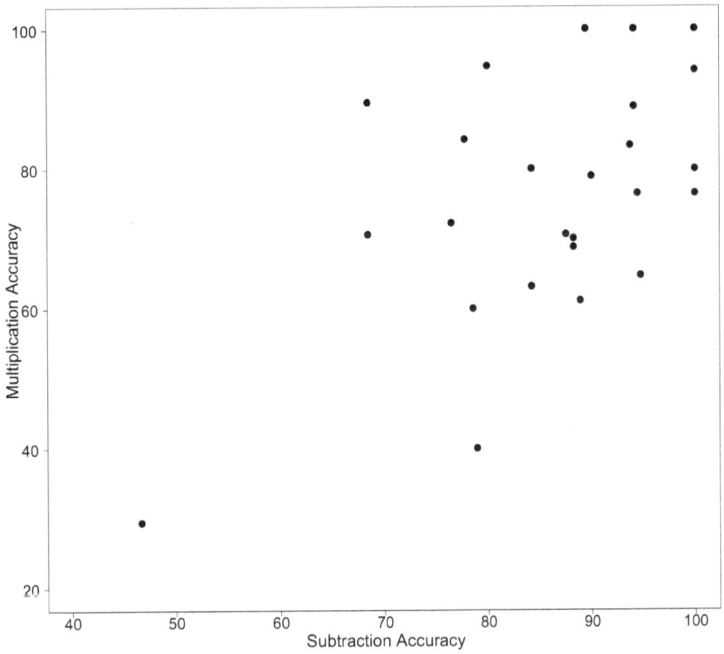

Figure 2.9 Scatterplot of multiplication and subtraction accuracy for 28 third-grade children.

better at multiplication. Perhaps the best way of determining whether there is a relationship between the two variables is to use a *scatterplot*, a plot in which each data point has coordinates that represent the scores for one of the students. The scatterplot for the 28 third-graders for whom we have both multiplication accuracy (percent correct) and subtraction accuracy scores in the *Royer* data set is presented in Figure 2.9. The scatterplot in Figure 2.10 shows the mean time to answer multiplication problems plotted against accuracy.

In both scatterplots, we see some organization to the distribution of data points, although it is imperfect. In Figure 2.9, there is a tendency for higher multiplication scores to go together with higher subtraction scores; when this happens, we say there is a *positive relationship* between the two variables. In Figure 2.10, there is a tendency for children who are more accurate to take less time to answer; this is an example of a *negative* relationship. In both scatterplots, there is a good deal of variability in the distribution of scores. Although we might imagine a line being drawn through the scatterplot to summarize the general tendency in the distribution, there is a lot of "scatter" of points about that line. Next, we would like to discuss some statistics that characterize the relationship.

2.7.2 *Correlation*

Correlation is an index of strength of linear relationship between two variables. The equation for a straight line is given by

$$Y = b_0 + b_1 X \tag{2.7}$$

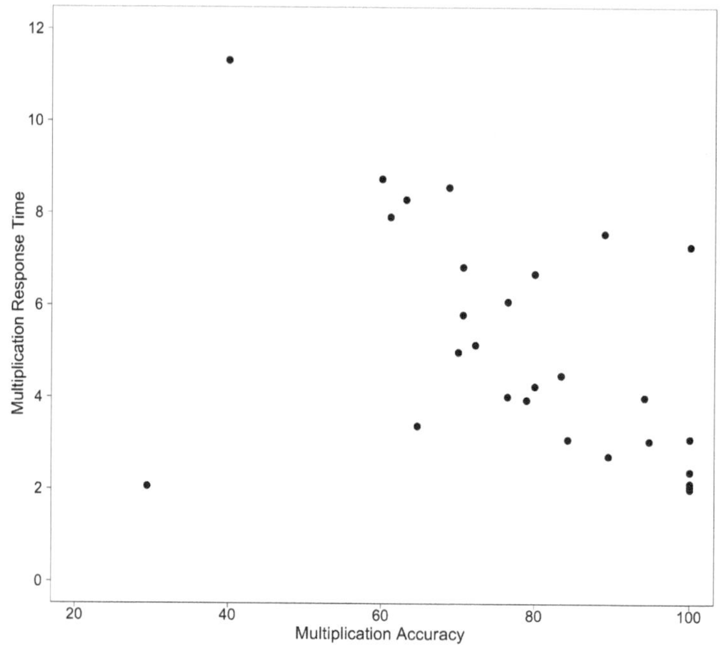

Figure 2.10 Scatterplot of multiplication response time and accuracy for 28 third-grade children.

All data points (*X*, *Y*) that satisfy this equation fall on a straight line. The constant b_1 is the *slope* of the line – the amount by which *Y* changes when *X* is increased by one unit. The slope is *positive* if *Y* increases as *X* increases; it is *negative* if *Y* decreases as *X* increases. The constant b_0 is the value of *Y* when *X* = 0.

The *correlation coefficient*, *r*, is a number between 1 and –1 that provides a measure of the extent to which the relationship between *Y* and *X* is described by a straight line, and whether the straight line has a positive or negative slope. If *r* is +1, all the data points fall exactly on a straight line with a positive slope; if *r* is –1, all data points fall on a straight line with a negative slope; if *r* is 0, there is no overall tendency for there to be a linear relationship. Intermediate values of *r* are used as measures of the strength of the linear relationship. All other things being equal, scatterplots with higher *r*s will display tighter clustering of data points around a straight line.

A useful formula for the correlation coefficient is given by

$$r_{XY} = \frac{1}{N-1}\sum_{i=1}^{N} z_{X_i} z_{Y_i} \tag{2.8}$$

Although there are many expressions for the correlation coefficient, they are all equivalent to Equation 2.8 and will give the same result within rounding error. Values of *r* are readily obtained from any statistical package: the *cor* {stats} function in R works well. When the correlation is computed for the multiplication and subtraction accuracy data for third-graders in the *Royer* data set, *r* = 0.59. For accuracy and response time on multiplication tests,

$r = -0.49$. Both values suggest strongly linear relationships, although we will see later that some nonlinear relationships can yield similar r values. Because the correlation coefficient is inherently a z score measure, it should be interpreted with caution. We discuss interpretation of correlation in great detail in Chapter 18.

2.7.3 *Linear Regression*

Although the correlation coefficient is an index of how well a bivariate distribution is fit by a straight line, it does not describe the line. In Chapter 18, we show that the straight line that best predicts Y from X has its slope and intercept given by

$$b_1 = r\frac{s_Y}{s_X} \tag{2.9a}$$

$$\text{and} \quad b_0 = \bar{Y} - b_1\bar{X} \tag{2.9b}$$

For example, the regression equation that predicts multiplication accuracy from subtraction accuracy for third-graders in the *Royer* data set is

$$\textit{predicted multiplication accuracy} = 1.631 + 0.883 \times \textit{subtraction accuracy}$$

Therefore, we would predict that a student with subtraction accuracy of 90 would have a multiplication accuracy of 81, rounding to the nearest integer. Every 10 point increase in subtraction accuracy is associated with, on average, an 8.8 point increase in multiplication accuracy. The regression equation is useful for prediction and it provides a description of the relationship between multiplication and subtraction accuracy.

There is much more to be said about correlation and regression, and more generally about measures relating two or more variables. Chapters 18–20 expand upon this very brief introduction to bivariate relationships, and Chapters 21–25 continue our discussion by considering more than two variables and nonlinear relationships.

2.8 Summary

In this chapter, we presented various ways of looking at data. The process of understanding our results begins with exploration of the data. Exploration should go beyond computing means and standard deviations and correlation coefficients. Therefore, we considered ways of graphing the distribution of observations (e.g., histograms, box plots, scatterplots), of calculating measures of location (e.g., median, mean) and variability (e.g., standard deviation, *IQR*), and of calculating measures of shape (skewness, kurtosis). Data description and exploration serves several important purposes:

- A thorough description is essential to understanding the data.
- Exploration provides a check on assumptions underlying subsequent statistical tests, such as the assumption of a normal distribution of scores.
- Data exploration can be a source of hypotheses. A thorough description of a data set often leads to discovery: Unexpected differences between conditions may be noticed and pursued. Or details of a distribution of scores may produce a deeper understanding of the nature of the effect of a manipulation.

More extensive discussions of ways of plotting data and of resistant statistics may be found in many sources. In particular, we recommend the three volumes edited by Hoaglin et al. (1983; Hoaglin, Mosteller, & Tukey, 1985, 1991). They provide a clear presentation of many topics beyond the scope of this book, as well as further discussion and examples of topics we have introduced. Other suggestions of possible ways of plotting data, and references to useful sources, may be found in the report of the American Psychological Association's Task Force on Statistical Inference (Wilkinson & Task Force, 1999). There are also powerful tools for data manipulation and plotting available in many packages within R. Of particular note are {dplyr} for data transformation and {ggplot2} for data visualization.

Exercises

2.1 [Calculating summary statistics] We have scores for 16 individuals on a measure of problem-solving ability: $Y = 21, 40, 34, 34, 16, 37, 21, 38, 32, 11, 34, 38, 26, 27, 33, 47$. Without using statistical software, find (a) the mean, (b) the median, (c) $(\Sigma_i Y_i)^2$, (d) $\Sigma_i Y_i^2$, (e) the standard deviation, and (f) the upper and lower hinges for these data. (g) Check your answers using R or another software package, then compute skew and kurtosis.

2.2 [Effect of data transformation on summary statistics]

a) Transform the scores in Exercise 2.1 to a new scale so that they have a mean of 100 and a standard deviation of 15.
b) What will the new values of the median and hinges be?

2.3 [Understanding means and variances] Given the five scores 37, 53, 77, 30, and 28,

a) What sixth score must be added so that all six scores together have a mean of 47?
b) What sixth score should be added so that the set of six scores has the smallest possible variance?

2.4 [Assessing normality] Following are several sets of scores in ranked order. For each data set, is there any indication that it does not come from a normal distribution? Explain, using graphical evidence to support your conclusion.

a) $X = 10, 16, 50, 50, 50, 55, 55, 55, 57, 61, 61, 62, 63, 72, 73, 75, 83, 85, 107, 114$.
b) $Y = 15, 25, 26, 37, 37, 39, 45, 45, 48, 49, 49, 52, 53, 61, 61, 63, 68, 70, 72, 76$.
c) $Z = 99, 10, 12, 14, 14, 15, 16, 16, 16, 17, 18, 24, 28, 31, 32, 32, 35, 47, 59$.
d) Find the mean and median of data sets (b) and (c). How do the results relate to your conclusion based on the graphic evidence?

2.5 [Interpreting equations with data] We have the following data set:

$X = 6, 5, 7, 1, 11$
$Y = 7, 11, 14, 21, 9$

Calculate (a) $\Sigma_{i=1}^{5}(X_i + Y_i)$ (b) $\Sigma_{i=1}^{5} X_i^2$; (c) $\left(\sum_{i=1}^{5} X_i\right)^2$; (d) $\Sigma_{i=1}^{5} X_i Y_i$;
(e) $\Sigma_{i=1}^{5}(X_i + 5Y_i^2 + 27)$; (f) the variance of X, of Y, and of X + Y.

2.6 [Interpreting statistical notation with data] We have five subjects tested in conditions C_1–C_3:

Subject	C_1	C_2	C_3
1	7	11	3
2	31	15	12
3	16	40	5
4	21	42	19
5	35	45	4

Given that Y_{ij} is the *i*th score in the *j*th column, find $(a)\bar{Y}_{.1}$; $(b)\bar{Y}_{2}.$; $(c)\bar{Y}..$; $(d)\sum_{i=1}^{5}\sum_{j=1}^{3}Y_{ij}^2$.

2.7 [Practice calculating statistics with data] In Exercise 2.6, find (a) $\sum_i^5(\bar{Y}_{i.}-\bar{Y}_{..})^2$; (b) $\sum_{j=1}^3(\bar{Y}_{.j}-\bar{Y}_{..})^2$; (c); (c) $\sum_{j=1}^3\sum_{i=1}^5(Y_{ij}-\bar{Y}_{..})^2$.

2.8 [Exploring and comparing data] This problem uses the *EX2_8* data set at the website. Using any software that you have, explore the distributions and compare them by graphing one or more of the following: histograms, box plots, or violin plots. In addition, base your discussion on descriptive statistics (include the median, as well as measures of skewness and kurtosis) and probability (*Q–Q*) plots. Summarize what you have learned about the distributions of *X*, *Y*, and *Z*. Be sure to refer to such concepts as the location, spread, and shapes of the distributions and whether they seem to be normally distributed. Compare X, Y, and *Z* in terms of spread and shape.

2.9 [Effect of standardization on distributions] Suppose we standardize *X*, *Y*, *Z* (that is, convert to standard or *z* scores) in the *EX2_8* data set. How would this affect the characteristics of the distributions? Do the standardization to check your answer.

2.10 [Exploring real data] The *Sayhlth* data set on the *Seasons* page of the book's website includes self-ratings from 1 (excellent health) to 4 (fair); only three participants rated themselves in poor health, and they were excluded from this file. The file also includes *Beck_D* (depression) scores for each season. It is reasonable to suspect that individuals who feel less healthy will be more depressed.

a) Selecting any statistics and graphs you believe may help, evaluate this hypothesis.
b) Discuss any trends over seasons. Are there differences in trends as a function of self-ratings of health?

2.11 [Choosing how to display data] The *Seasons* study was carried out to see whether there was variation in the physical or psychological characteristics of the sampled population over seasons.

a) Using the data in the *Seasons* file on the book's website, plot *beck_a* (anxiety; *beck_a1*,. . ., *beck_a4*) means as a function of seasons for each age group (*agegrp*). Use either a bar or line graph. Which do you think is preferable? Why?
b) Discuss the graph in part (a), noting any effects of age, seasons, or both.
c) Box or violin plots reveal a skewed distribution with many outliers. Do you think the pattern of outliers contributed to your conclusions in part (b)?

2.12 [Comparing data on different scales] Scores for 50 students on two tests may be found in the *EX2_12* file. One student received a score of 41 on Test 1 and a 51 on Test 2. She was delighted with the improvement.

a) Should she have been? Explain.
b) What is the minimum score on Test 2 that would be an improvement for this student?
c) Graph the data for each test and describe the distributions.
d) Plot the data to show the relationship between scores on the two tests and provide a statistic summarizing that relationship.
e) If a student had a score of 40 on Test 1, what score would you predict for Test 2?

2.13 [Exploring real data] The file *mlb salaries 22_86_07* at the website contains major league baseball player average salaries by team and league for the years 1986, 2007, and 2022.

a) Describe the distribution of the team payrolls in 2022. Support any conclusions by citing statistics and graphs.
b) Describe the distribution of team payrolls in 2007, separately for each league. What differences or similarities do you see between the leagues?
c) What is the relationship between salaries in 2007 and 2022? Include comparisons of measures of location, shape, and spread, as well as graphs showing the relation, and a measure of the relation, between the two sets of payrolls. Does the relationship between relative payroll sizes in the two years depend on which league we look at?
d) Evaluate how the increase in the teams' payroll from 1986 to 2022 compares with inflation, which was 185% over that time period.
e) How are team rank and average salary related in 2022? Use both numerical and graphical summaries.

Notes

1 The complete data set can be downloaded by following the link from the home page of the book's website to *Data Files*, and then to *Royer*; also see the *Royer Readme* file there.
2 The R scripts for most figures are available at the website for the book. Throughout, we will use squiggle brackets to denote R packages and italics to denote specific functions within those packages. Note that R is a rich and constantly evolving platform; there are many ways to create a graph or carry out an analysis. Feel free to try out different functions in other packages.
3 The presentation of formulas for some of these statistics involves the use of a summation sign, a capital Greek sigma, Σ. Although our use of notation is intended to be easily understood, readers may find it helpful to refer to Appendix A, which reviews the algebra of summation and derives several properties of statistics that we state in this chapter.
4 The means are averages of the four seasonal depression scores and were calculated only for those subjects who had been tested in all four sessions.
5 These percentiles and *IQR* differ slightly from the values reported by SPSS in Table 2.2. Use the type = 6 option in *quantile* or *IQR* to interpolate using the same calculational method as SPSS.
6 There are several ways to calculate skewness (e.g., Hillebrand, 1986, pp. 43–45). Different functions may also yield slightly different values of skew depending on whether they use a sample or population standard deviation. These differences are largest for small sample sizes. In particular, SPSS reports skewness = −1.33 for the data in Table 2.1.

Chapter 3

Basic Concepts in Probability

3.1 Overview

In Chapter 2, we discussed various descriptive statistics and their uses. These statistics summarize important aspects of a sample of data and form the basis for making inferences about the population from which the sample was obtained. However, because sample statistics vary across samples, our inferences may be incorrect. A sample is not a miniature version of the population, so the mean or variance of a sample is not likely to be the same as the mean or variance of the population from which the sample was selected. This is true for any other statistic we might calculate. How then can we justify using statistics from a single sample to draw inferences about the population? As we'll see, we use *probability models*, which describe the possible outcomes in an experiment and their probabilities, to tell us how likely our observed sample statistic is.

Let's begin by considering an example. Imagine that we want to test whether extrasensory perception (ESP) exists. We decide to conduct an experiment to test the ability of Rachel, who claims to have ESP. An experimenter is seated in a room with a pack consisting of four different cards. On each of 20 trials, the experimenter shuffles the cards well, then randomly picks one, looks at it, and tries to transmit its contents mentally. Rachel sits in a room in a different building, is familiar with the pack of cards, and knows when each trial of the experiment is to occur. For each trial, she tries to "perceive" and then record the card chosen by the experimenter.

How well must Rachel perform before we decide that the evidence is sufficiently strong to support her claim to have ESP? We first note that if Rachel is guessing randomly, the most likely outcome is that she will be correct on 25% of the trials. However, we must recognize that Rachel may do better than 25% correct even if she guesses. Before we can conclude that Rachel is not simply guessing, we want convincing evidence that her performance cannot easily be attributed to guessing alone. How much better than 25% correct must Rachel do before we are convinced? Twenty correct responses (i.e., 100% correct) would be impressive support for Rachel's claim of ESP because such a high level of performance is extremely unlikely to result if she was just guessing. But what about 19 correct in 20 trials? Or 12? Or 8? Those outcomes still seem very unlikely given guessing, but how much evidence against guessing do they provide? To answer this question, we must be able to calculate the probability of any possible outcome under the assumption that Rachel is guessing. We should also note that we are concerned with the entire population of potential responses that characterizes Rachel's ESP ability, not just the 20 responses we happen to collect in the experiment. If we ran the experiment several times, it would almost certainly turn out somewhat differently each time.

DOI: 10.4324/9781003453550-4

This example makes the important point that drawing inferences in an objective manner requires that we find the probabilities of different possible outcomes. Suppose that Rachel makes eight correct responses. Is this strong enough evidence to reject the hypothesis that Rachel is randomly guessing on each trial? Later, we will show that there is a better than 10% chance that Rachel would make eight or more correct responses even if she randomly guessed on each trial. This is very useful information on which to base a decision. We might decide that a 10% chance of being wrong is too high, and therefore conclude that obtaining eight correct responses is not sufficiently strong evidence to reject the guessing hypothesis.

The subsequent chapters of this book are concerned with procedures for drawing inferences about characteristics of populations from samples of observations. The ability to assign probabilities to possible outcomes is central to statistical inference. Therefore, some background in probability is essential to understanding the logic and assumptions underlying statistical tests, and to interpreting the tests. In this chapter, we provide a basic introduction to probability. The organization of topics is as follows:

- *Basic concepts for analyzing event structure.* The ability to compute probabilities of events depends, in part, on being able to analyze how simple events (e.g., specific sequences of correct and incorrect guesses) comprise more complex events (e.g., total number of correct guesses in a 20-trial experiment).
- *Computing probabilities.* Once we have the tools for analyzing the structure of events, we will introduce a definition of the probability of an event and some basic rules of probability. We will use the rules and our analysis of the structure of events to develop two important laws – the additive and multiplicative laws – that will be very useful in computing probabilities of complex events. We will also discuss concepts such as independence and conditional probability that are important for understanding many statistical tests.
- *Probability distributions.* The first two sections of the chapter provide the foundation for the concept of a distribution of a random variable, or probability distribution.
- *Connecting probability theory to real-world experiments.* The final section will provide a transition from the theoretical constructs of probability theory to the real world of experiments and data. In this section, probability distributions will be presented as hypothetical distributions that may be used to evaluate empirical results.

3.2 Basic Concepts for Analyzing the Structure of Events

3.2.1 *Simple Experiments and Elementary Events*

Suppose we have a class of 100 students, 60 psychology majors and 40 linguistics majors. The instructor, a kindly statistician, gives no grades lower than C. The number of students in each major receiving each grade is presented in Table 3.1.

Suppose further that the major of each student, along with their grade, is written on a separate slip of paper and the 100 slips are placed in a box. We can determine the probability that a slip of paper randomly selected from the box has a particular major and/or grade written on it. We can use this *simple experiment* to introduce some basic ideas about probability.

Table 3.1 Distribution of grades for a hypothetical statistics class

	Grade			
Major	A	B	C	Total
Linguistics	12	24	4	40
Psychology	15	36	9	60
Total	27	60	13	100

A simple experiment is a well-defined process that leads to a single outcome (Hays, 1981). All probabilities are defined with respect to an experiment. Randomly drawing a slip of paper from the box and observing the grade and major listed on the slip is an example of a simple experiment. Another example is recording whether each of Rachel's 20 responses is correct or wrong.

The possible outcomes of a simple experiment are called *elementary events*, and the complete set of elementary events is called the *sample space* for the experiment. In the statistics class example, the sample space consists of the 100 combinations of major and grade across all the students in the class. In the ESP example, the sample space consists of every possible sequence of correct and incorrect guesses across 20 trials.

3.2.2 *Combining Elementary Events Into Event Classes*

We are often not as interested in the probability of a particular elementary event as we are in the probability of meaningful collections of elementary events that are called *event classes* or, simply, *events*. For example, a student in the statistics class might be interested in the probability of getting a grade of A or B. In the ESP experiment, we would likely be interested in a summary of performance (e.g., proportion of correct responses) across all of the trials in the experiment.

Events can be combined in important ways. Events may be "joined" such that an elementary event may be simultaneously classified into two (or more) event classes. In our statistics class example, a person may be both a linguistics major *and* have earned an A in the class. In our ESP experiment, Rachel might pick the correct card on every trial of the experiment; that is, she may be correct on the first and second and third and . . . and twentieth trial. Thus, we can talk about *joint events* that are comprised of the *intersections* of two or more events. We will use the conjunction "and" to refer to the joining of two or more events, but we will understand the term to refer to the logical operation of intersection.

Events may also "unite" to form *compound events*. In contrast to joint events such as "*Linguistics and A*," we can have compound events (or unions of events) such as "*Linguistics or A*." The *union* of two events occurs if an elementary event may be classified as **either or both** of the events. In our statistics class experiment, the union of *linguistics* with *A* occurs if we observe "*linguistics*" or "*A*" or "*linguistics* and *A*" (i.e., a linguistics major who earned a grade of A). In a simplified version of our ESP example, the outcome of at least one correct guess in *two* trials is comprised of the events: "correct on the first trial only" *or* "correct on the second trial only" *or* "correct on both trials" (i.e., correct on trial 1 and on trial 2). We will use the disjunction "or" to refer to the union of two or more events but understand the term to refer to the logical operation of union (i.e., either or both).

3.2.3 *Characterizations of Sets of Events*

Event classes may be related to one another in different ways. One important kind of relationship is *complementarity*. The complement of event *A* is the collection of elementary events that do not qualify as instances of event *A*. The complement of *A* is expressed as "*not-A*" or $\tilde{A}$. In our statistics example, the complement of *psychology major* is *linguistics major*; the complement of the *C* is "*A* or *B*." One useful application of the concept of complementarity is that it is sometimes easier to approach the analysis of a problem by redefining it in terms of the complement of the event specified in the problem. For example, if asked to compute the probability that Rachel will make at least one correct guess in five trials of our ESP experiment by guessing, we could approach the calculation by computing the probability of one correct, the probability of two correct, the probability of three correct, and so on. But a simpler approach would be to realize that getting at least one correct is equivalent to *not* making errors on all five trials; this realization would suggest an approach based on computing the probability of zero correct in five trials.

Another kind of relationship that characterizes some sets of events is *mutual exclusion*. Two events are mutually exclusive if they cannot jointly occur. The events *A*, *B*, and *C* are mutually exclusive, as are *correct* and *incorrect*. If you get an A on an exam, you do not get a B. In contrast, *A* and *psychology major* are not mutually exclusive; it is possible that a psychology student will get a grade of A. We will find that it is easier to compute probabilities of mutually exclusive events than those involving events that are not mutually exclusive.

In part because of the tractability of probability calculations involving mutually exclusive events, it is generally desirable to define the outcomes for an experiment in terms of event classes that are *mutually exclusive* and *exhaustive* (i.e., account for all possible elementary events in a sample space). A set of mutually exclusive and exhaustive events partitions the sample space. That is, it accounts for every possible event, and it does so such that each possible outcome is unambiguously classified into a single event class. For example, the set of events "*A*, *B*, and *C*" partitions the sample space for the statistics class; so does the set "*psychology* and *linguistics*." Similarly, the six cells of Table 3.1, corresponding to the six combinations of major and grade, also partition the sample space. As a final example, any event and its complement partition a sample space.

Another important possible relationship between events is that of *independence*. A formal definition of independence must be deferred until probability calculations are introduced. An informal definition is that two events are independent if the occurrence of one event provides no information about the likelihood of occurrence of the other event. That is, independent events are unrelated events. In our ESP example, it is probably reasonable to assume that the trial-to-trial outcomes are independent of one another. If Rachel is randomly guessing, knowing that she made a correct response on the first trial does not help us to predict her performance on any other trial. Because the way in which the terms "independence" and "mutual exclusion" are commonly used in everyday language, some students confuse the two concepts. However, the two terms have very different meanings when we talk about probability: If two events, E_1 and E_2, are mutually exclusive (e.g., correct, incorrect), then knowing that E_1 occurs tells us that E_2 does not occur (i.e., their joint probability is zero). However, if E_1 and E_2 are independent, then the probability of their joint occurrence is greater than zero, except in the trivial case where the probability of one of the events is zero. Thus, mutually exclusive events are not independent. As we will see, the concept of independence is often a major assumption of statistical tests.

3.2.4 *Combining Events in the ESP Experiment*

To illustrate the potential usefulness of some of the concepts introduced to this point, suppose that Rachel does not have ESP and randomly guesses on every trial. If so, we would expect her to be correct 25% of the time in the long run because there are four equally likely choices on each trial. In evaluating the strength of the evidence against guessing, we want to be able to translate our hypothesis that she is guessing on each trial into the probability that she makes a specific number of correct responses in a given number of trials. In order to do this, we need to consider all of the possible sequences of correct and incorrect guesses that might occur. A useful conceptual tool for this purpose is a tree diagram like the one in Figure 3.1. Each branch of the diagram in Figure 3.1 represents a distinct combination of possible outcomes for a four-trial experiment.

Suppose that we want to compute the probability that Rachel makes exactly one correct guess in a four-trial experiment (i.e., 25% correct). To do this, we must analyze the various ways in which individual trial outcomes may produce one correct result in four trials. Looking at Figure 3.1, we can distinguish four distinct sequences of trial outcomes that correspond to one correct guess: Rachel can be correct on the first trial, but wrong on trials 2 through 4 (call this "sequence 1"), or she can be correct on only the second trial (sequence 2), or only on the third trial (sequence 3), or only on the fourth trial (sequence 4). Each of these sequences represents a joint event; e.g., (correct on trial 1) and (error on trial 2)

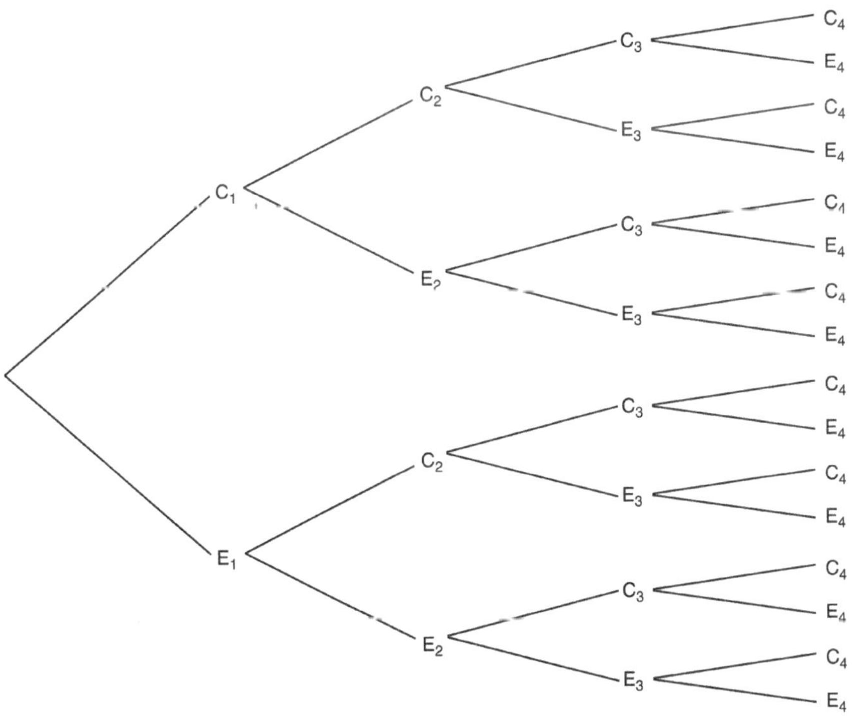

Figure 3.1 Tree diagram for four trials with two possible responses on each trial (*C* = correct, *E* = error). The numerical subscripts represent trial number.

and (error on trial 3) and (error on trial 4). We do not care on which trial the correct response occurs – the outcome of one correct response occurs if sequence 1 or sequence 2 or sequence 3 or sequence 4 is observed. Further, the four sequences are mutually exclusive events. Thus, the response of one correct in four trials can be represented as the union of four, mutually exclusive joint events (i.e., the four sequences of trial outcomes). This precise specification of the relationship between individual trial outcomes and one possible experimental outcome is necessary if we are to translate a hypothesis about the probability of a correct response on any trial into a calculation of the probability (p) of a given experimental outcome.

3.3 Computing Probabilities

In probability theory, probabilities are numbers assigned to the events in a sample space. We will denote the probability of an event, A, by $p(A)$. The assignment of probabilities to events follows some simple rules.

3.3.1 *Some Basic Rules of Probability*

The sample space, S, consists of all the elementary events, and the probabilities of the elementary events must sum to 1. This must be the case because the elementary events are, by definition, mutually exclusive and exhaustive. In the simplified ESP experiment example, the set of events (0, 1, 2, 3, and 4 correct responses) are mutually exclusive and exhaustive, so their probabilities must also sum to 1. In general,

$$p(S)=1 \tag{3.1}$$

It should be evident from the definition of probability in Equation 3.1 that a probability must have a value within the range from 0 to 1. More precisely,

$$0 \leq p(A) \leq 1 \tag{3.2}$$

The sample space can always be partitioned into two mutually exclusive and exhaustive sets of elementary events; call these A and $\tilde{A}$ ("not-A," which is called the "complement of A"). Given that $p(S) = 1$, it follows that $p(A \text{ or } \tilde{A}) = p(A) + p(\tilde{A}) = 1$, and therefore that

$$p(\tilde{A}) = 1 - p(A) \tag{3.3}$$

For example, if event A is "zero correct responses" in the four-trial ESP experiment, then $\tilde{A}$ is "one or more correct responses." Given that $p(S) = 1$, p(one or more correct) must equal 1 − p(zero correct).

Finally, it follows from the preceding rules that if we define a set of events that are mutually exclusive and exhaust the possible events in a sample space, then the sum of their probabilities will equal 1. That is, if the set of events A, B, C,. . . constitute a partition of S, then

$$p(A \text{ or } B \text{ or } C \text{ or} \ldots) = p(A) + p(B) + p(C) + \ldots = 1 \tag{3.4}$$

3.3.2 Assigning Probabilities to Marginal Events

We have some simple rules that define properties of probabilities, but how do we assign probabilities to actual events in a meaningful way? In the case where all of the elementary events in a sample space are equally likely, the probability of event A is simply the ratio of the number of elementary events in A to the total number of events, $n(A)$, in the sample space, $n(S)$. That is, the probability of an event A is its relative frequency in the sample space:

$$p(A) = \frac{n(A)}{n(S)} \tag{3.5}$$

To be concrete, let's return to Table 3.1 and the example of grades in a statistics class. Suppose we want to find the probability that a slip of paper randomly selected from the box will correspond to a person who got a B; $p(B)$ is simply the number of slips of paper with a B divided by the total number of slips of paper:

$$p(B) = \frac{60}{100} = .60$$

In the preceding example, we computed the probability of a particular grade without regard to the major of the student. In Table 3.1, the numbers of students with As, Bs, and Cs and the numbers of psychology majors and linguistics majors are presented in the margins of Table 3.1. Because of this, the probability of observing a particular value of some variable (e.g., grade) without respect to any other variable (e.g., major) is referred to as a *marginal probability*.

3.3.3 Joint Probabilities

The probability of obtaining a particular *combination* of events is referred to as a *joint probability*. For example, $p(A$ and *psychology*), which is read as "the probability of A and *psychology*," is the probability of the joint event *<A, psychology>*; that is, the probability of selecting a slip of paper with both "A" and "psychology major" written on it. If the probabilities of the elementary events are equal, $p(A$ and *psychology*) can be obtained by computing the relative frequency of the joint event in the sample space:

$$p(A \text{ and } psychology) = \frac{n(A \text{ and } psychology)}{n(S)} \tag{3.6}$$

where $n(A$ and *psychology*) is the number of elementary events that belong to *both* the events A and *psychology*. For the data of Table 3.1, $p(A$ and *psychology*) = .15, because 15 of the 100 slips of paper correspond to grades of A obtained by psychology majors. Similarly, if the events B and *linguistics* correspond to "getting a grade of B" and "being a linguistics major," respectively, $p(B$ and *linguistics*) = 24/100 = .24. Note that $p(A)$ must always be at least as large as $p(A$ and *psychology*) because the event A will always contain at least as many elementary events as the joint event *<A, psychology>*. These ideas may be clarified by reconsidering Table 3.1. Each column represents the event of a letter grade and has two nonoverlapping parts. For example, the column representing the event <*A*> consists of the joint events <*A, psychology*> and <*A, linguistics*>. Note that *n*(*A*) = *n*(*A* and *psychology*) + *n*(*A* and *linguistics*), and it follows from Equations 3.5 and 3.6 that *p*(*A*) = *p*(*A* and *psychology*) + *p*(*A* and *linguistics*). Also note that *p*(*A* and *psychology*) = *p*(*psychology* and *A*).

3.3.4 Probabilities of Unions of Events

The *union* of two events consists of all the elementary events belonging to either or both of them. The elementary events forming the union of events *A* and *psychology* are the following cells of Table 3.1: <*A, psychology*>, <*A, linguistics*>, <*B, psychology*>, and <*C, psychology*>. The expression *p*(*A* or *psychology*) refers to the probability of obtaining an elementary event belonging to either *A* or *psychology*; that is, falling into any of the four cells just noted. Therefore,

$$p(A \text{ or } psychology) = \frac{n(A \text{ or } psychology)}{n(S)} = \frac{12 + 15 + 36 + 9}{100} = .72 \tag{3.7}$$

because 72 of the 100 elementary events belong either to *A* or to *psychology* or to both. Note that *n*(*A* or *psychology*) does not generally equal *n*(*A*) + *n*(*psychology*). As should be clear from Table 3.1, this sum counts twice the 15 elementary events that belong to both *A* and *psychology*. Verify for yourself that *p*(*A* or *psychology*) = *p*(*A*) + *p*(*psychology*) – *p*(*A* and *psychology*). Also verify that *p*(*A* or *linguistics*) = 55/100 = .55. In general, if E_1 and E_2 are two events of interest,

$$p(E_1 \text{ or } E_2) = p(E_1) + p(E_2) - p(E_1 \text{ and } E_2) \tag{3.8}$$

We subtract $p(E_1$ and $E_2)$ because E_1 and E_2 are not mutually exclusive. If they were, the probability of their intersection would be zero and Equation 3.8 would still be true.

3.3.5 Conditional Probabilities

We are often interested in computing the probabilities of events when considering only a subset of events in a sample space. For example, we may be interested in the probability of obtaining a grade of A if only the psychology majors are considered. This probability is called a *conditional probability* because it is the probability of *A given the condition that psychology major occurs*. It is represented by *p*(*A* | *psychology major*), and is read as "the probability of *A* given *psychology major*." There are 60 slips of paper labeled "psychology major" and 15 of them correspond to grades of A. Therefore, *p*(*A* | *psychology*) = 15/60 = .25. More generally, *p*(*A* | *psychology major*) is the proportion of all elementary events in *psychology* that also belong to *A*; that is,

$$p(A \mid psychology) = \frac{n(A \text{ and } psychology)}{n(psychology)} = \frac{p(A \text{ and } psychology)}{p(psychology)} \tag{3.9}$$

Verify that for the current example, *p*(*B* | *psychology major*) = 36/60 = .60; *p*(*psychology major* | *A*)= 15/27 = .56; and *p*(*A* | *B*) = 0/60 = 0.

Two important things about conditional probabilities should be noted. First, people have a tendency to confuse conditional probabilities with joint probabilities. Look carefully at Equations 3.6 and 3.9. The conditional probability *p*(*A* | *psychology major*) is the probability of selecting a slip of paper that is labeled "A" if the selection is made from only the 60 slips labeled "psychology." The joint probability *p*(*A* and *psychology*) is the probability of selecting a slip labeled both "A" and "psychology" if selection is randomly made from *all* 100 slips of paper. A conditional probability must be at least as large as – and is generally larger than – the corresponding joint probability because the set from which we sample is a subset of the entire sample. For example, when we calculate *p*(*A* | *psychology*), we divide by only the number of psychology majors, a number less than the size of the total sample. Bear in mind, however, that although joint and conditional probabilities are not the same, they are related, as can be seen in Equation 3.9.

The second thing to note about conditional probabilities is that for any two events, *A* and *psychology major*, there are two *opposite* conditional probabilities, *p*(*A* | *psychology major*) and *p*(*psychology major* | *A*). These two conditional probabilities will generally not have the same values; in our current example, *p*(*A* | *psychology major*) = 15/60 = .25 and *p*(*psychology major* | *A*) = 15/27 = .56. As this example illustrates, the denominators are based on different subsets of the entire sample, and these often will have different numbers of elementary events. As we will see, conditional probabilities are very useful in characterizing relationships between events.

3.3.6 Mutually Exclusive and Independent Events

Two events E_1 and E_2 are *mutually exclusive* if they are incompatible; that is, if an elementary event belongs to E_1, it cannot belong to E_2. It follows that if E_1 and E_2 are mutually exclusive, $p(E_1 \text{ and } E_2) = 0$, $p(E_1 \mid E_2) = 0$, and $p(E_2 \mid E_1) = 0$. In our current example (Table 3.1), *p*(*A* and *B*) = 0, because a student who receives a grade of A in the course did not receive a B. Earlier, we encountered the partitioning rule (Equation 3.4), in which we summed the probabilities of the elementary events in a sample space. It is because any partitioning is based on mutually exclusive events that we summed the probabilities of events in Equation 3.4; it is because the set of events in any partitioning is also *exhaustive* that the sum equals 1.

We introduced the concept of independence in discussing concepts for analyzing the structure of events. At that time, we said that two events are independent if the occurrence of one event provides no information about the likelihood of occurrence of the other event. Now that we have introduced the concept of conditional probabilities, we can state the concept of independence more precisely: Two events E_1 and E_2 are *independent* if $p(E_1 \mid E_2) = p(E_1)$; that is, if the probability of event E_1 is the same whether or not event E_2 occurs.

We may wish to ask questions such as "Is getting a grade of A independent of the major of the student?" This is another way of asking whether the probability of getting an A is the same for linguistics majors and psychology majors. If there is independence, *p*(*A* | *psychology*) = *p*(*A* | *linguistics*) = *p*(*A*). Returning to Table 3.1, *p*(*A* | *psychology*) = 15/60 = .25, *p*(*A* | *linguistics*) = 12/40 = .30, and *p*(*A*) = .27. Clearly, for this data set, getting an A is not independent of being a psychology or linguistics major because *p*(*A* | *psychology*) is not equal to *p*(*A* | *linguistics*); therefore, <*A*, *psychology major*> and <*A*, *linguistics major*> are pairs of events that are not independent. On the other hand, *p*(*B* | *psychology*) = *p*(*B* | *linguistics*) = *p*(*B*), so that, for this data set, getting a grade of B is independent of the student's major. For both psychology and linguistics students, the probability of getting a B is.60.

We may also wish to ask more general questions such as "Are the variables of grade and major independent of each other?" In order for the answer to be "yes," each of the six pairs of events formed by combining levels of major and grade (specifically, <*A, linguistics*>, <*A, psychology*>, <*B, linguistics*>, <*B, psychology*>, <*C, linguistics*>, and <*C, psychology*>) would have to be independent. The variables major and grade are not independent of each other in this example because, as we have already shown, <*A, psychology*> and <*A, linguistics*> are pairs of events that are not independent.

Two important points about independence should be noted. First, if E_1 and E_2 are two independent events, $p(E_1 \text{ and } E_2) = p(E_1) \times p(E_2)$. To see why this is so, consider the definition of conditional probability given by Equation 3.9:

$$p(E_1 \mid E_2) = p(E_1 \text{ and } E_2) / p(E_2)$$

Multiplying both sides of this equation by $p(E_2)$ yields

$$p(E_1 \mid E_2) \times p(E_2) = p(E_1 \text{ and } E_2)$$

But we know that if E_1 and E_2 are independent, $p(E_1 \mid E_2) = p(E_1)$. Replacing $p(E_1 \mid E_2)$ by $p(E_1)$ in the last equation, we have, *if* E_1 *and* E_2 are *independent events,*

$$p(E_1 \text{ and } E_2) = p(E_1) \times p(E_2)$$

The second important point is that if events E_1 and E_2 are mutually exclusive, they cannot be independent. By definition, if E_1 occurs, then E_2 cannot occur if the two events are mutually exclusive. Therefore, if E_1 and E_2 are mutually exclusive, their joint probability and both conditional probabilities must be zero; that is

$$p(E_1 \text{ and } E_2) = 0, p(E_1 \mid E_2) = 0, \text{ and } p(E_2 \mid E_1) = 0$$

In contrast, if E_1 and E_2 are independent and their probabilities are both greater than zero, then their joint probability is greater than zero.

3.3.7 The Additive Law for Compound Events

We are now ready to integrate much of what we have developed to this point into two very useful probability laws. The first law is the *additive law for compound events*. We will often apply this law in the special case in which we have two or more mutually exclusive events, so we begin with that situation.

If E_1 and E_2 are mutually exclusive events,

$$p(E_1 \text{ or } E_2) = p(E_1) + p(E_2) \tag{3.10}$$

This can be extended to any number of mutually exclusive events:

$$p(E_1 \text{ or } E_2 \text{ or} \ldots \text{ or } E_n) = p(E_1) + p(E_2) + \ldots + p(E_n) \tag{3.11}$$

In English, the probability of the union of two or more mutually exclusive events is the sum of their individual probabilities.

Of course, events are not always mutually exclusive. As we explained in Section 3.2.4, if events E_1 and E_2 are not mutually exclusive, then the probability of their joint occurrence is (usually) greater than zero and must be taken into account, so that

$$p(E_1 \text{ or } E_2) = p(E_1) + p(E_2) - p(E_1 \text{ and } E_2) \tag{3.12}$$

For example, in Table 3.1, $p(A$ or *psychology major*$) = p(A) + p($*psychology*$) - p(A$ and *psychology*$) = .27 + .60 - .15 = .72$, exactly as we calculated by summing event probabilities using Equation 3.7.

3.3.8 The Multiplication Law for Joint Events

As we did for the additive law, we start with a special case. If E_1 and E_2 are two independent events,

$$p(E_1 \text{ and } E_2) = p(E_1) \times p(E_2) \tag{3.13}$$

The law can be extended to any number of independent events, $E_1, E_2, \ldots, E_n$:

$$p(E_1 \text{ and } E_2 \text{ and} \ldots \text{ and } E_n) = p(E_1) \times p(E_2) \times \ldots \times p(E_n) \tag{3.14}$$

In English, if two or more events are independent, then the probability of their joint occurrence is equal to the product of their individual probabilities.

Note that Equations 3.13 and 3.14 do not hold if the events are not independent. For example, *A* and *psychology major* are not independent and $p(A$ and *psychology*$) = .15$, but $p(A)p($*psychology*$) = (.27)(.60) = .162$. However, the multiplication rule can be extended to events that are not independent. In this case,

$$p(E_1 \text{ and } E_2) = p(E_1) \times p(E_2 \mid E_1) = p(E_2) \times p(E_1 \mid E_2) \tag{3.15}$$

Equation 3.15 follows directly from the definition of conditional probability, $p(E_1 \mid E_2) = p(E_1 \text{ and } E_2)/p(E_2)$. Multiplying both sides of this last equation by $p(E_2)$ yields $p(E_2) \times p(E_1 \mid E_2) = p(E_1 \text{ and } E_2)$. For example, applying Equation 3.15 to the data of Table 3.1, we can see that $p(A$ and *psychology*$) = p($*psychology*$)p(A \mid$ *psychology*$) = (.60)(15/60) = .15$.

Although the multiplication and addition rules are quite simple, people often mix them up, possibly because the statement of the multiplication rule uses the word "and" and the statement of the addition rule does not. It should be emphasized that the multiplication rule tells us how to calculate $p(E_1 \text{ and } E_2)$, the probability of the joint occurrence of E_1 and E_2. The addition rule tells us how to calculate $p(E_1 \text{ or } E_2)$, the probability that E_1 *or* E_2 occurs. This union of E_1 and E_2 (i.e., E_1 *or* E_2) includes the joint event $\langle E_1 \text{ and } E_2 \rangle$ but it also includes occurrences of E_1 without E_2 and of E_2 without E_1.

3.3.9 Bayes' Rule

There is one more probability rule that is useful to know. In Section 3.3.5, we noted that $p(E_1 \mid E_2)$ and $p(E_2 \mid E_1)$ are not the same thing and that they usually have different values.

For any two events, E_1 and E_2, there is an important relationship between $p(E_1 \mid E_2)$ and $p(E_2 \mid E_1)$. From Equation 3.15, we know that

$$p(E_1 \mid E_2)p(E_2) = p(E_2 \mid E_1)p(E_1)$$

We can divide both sides by $p(E_2)$ to obtain Bayes' rule:

$$p(E_1 \mid E_2) = \frac{p(E_2 \mid E_1)p(E_1)}{p(E_2)} \tag{3.16}$$

Earlier, we computed the probability of a psychology major earning an A in a statistics class, using the data in Table 3.1: $p(A \mid psychology\ major) = 15/60 = .25$. Using the same data, we computed $p(psychology\ major \mid A) = 15/27 = .56$. Now, let's use Bayes' rule to find the conditional probability of a psychology major earning an A:

$$p(A \mid psychology) = \frac{p(psychology \mid A)p(A)}{p(psychology)} = \frac{\left(\frac{15}{27}\right)\left(\frac{27}{100}\right)}{\left(\frac{60}{100}\right)} = 0.25$$

Bayes' rule allows us to calculate probabilities we care about – like the probability that we have a disease if we receive a positive test result – from probabilities we know: the base rate of the disease in the population and the probability of a positive test result for individuals who do and do not have the disease (see Exercise 3.5). Bayes' rule also provides the foundation for Bayesian statistics, an approach that allows the researcher to decide how much belief to have in a hypothesis after observing some data, $p(\text{hypothesis} \mid \text{data})$, given a set of prior beliefs about the world, $p(\text{hypothesis})$, and a probability model, $p(\text{data} \mid \text{hypothesis})$.

Table 3.2 summarizes much of what has been presented in Section 3.3. It includes important definitions and the rules embodied in Equations 3.10–3.16.

Table 3.2 Some probability definitions and rules

Some probability definitions	
Probability of event *A*	$p(A) = n(A)/n(S)$
Probability of the joint events *A* and *B*	$p(A \text{ and } B) = n(A \text{ and } B)/n(S)$
Probability of the union of events *A* and *B*	$p(A \text{ or } B) = n(A \text{ or } B)/n(S)$
Conditional probability of *A* given *B*	$p(A \mid B) = p(A \text{ and } B)/p(B) = n(A \text{ and } B)/n(B)$
Bayes' rule	$p(A \mid B) = \frac{p(B \mid A)p(A)}{p(B)}$
Some probability rules	
The addition rule for unions of events	$p(A \text{ or } B) = p(A) + p(B) - p(A \text{ and } B)$
Special case of the addition rule if the events are mutually exclusive	$p(A \text{ or } B) = p(A) + p(B)$
The multiplication rule for joint events	$p(A \text{ and } B \text{ and } C) = p(A)\ p(B \mid A)\ p(C \mid A \text{ and } B)$
Special case of the multiplication rule for independent events	$p(A \text{ and } B \text{ and } C) = p(A)\ p(B)\ p(C)$

3.3.10 *Computing Probabilities of Events in the ESP Experiment*

Let's apply what we have learned about probability to calculate the probability of one correct guess in a four-trial ESP experiment. In Section 3.2.4, we established that there were four sequences of trial outcomes corresponding to the experimental outcome of exactly one correct guess in four trials. If C and E are used to denote guesses that are "correct" and in "error," and we use subscripts to denote trials, then the four sequences of interest are: $<C_1\ E_2\ E_3\ E_4>$, $<E_1\ C_2\ E_3\ E_4>$, $<E_1\ E_2\ C_3\ E_4>$, and $<E_1\ E_2\ E_3\ C_4>$. Each of these sequences represents a joint event, and it is reasonable to assume that trial outcomes are independent. If we hypothesize that Rachel guesses on every trial, then $p(C) = .25$ and $p(E) = 1 - p(C) = .75$. Applying the multiplicative law, the probability of each sequence is $(.25)(.75)(.75)(.75) = .1055$. However, we are interested not in the probability of a specific sequence, but in the probability of any one of the sequences occurring. That is, we wish to compute:

$$p\left[\left\langle C_1E_2E_3E_4\right\rangle \text{ or } \left\langle E_1C_2E_3E_4\right\rangle \text{ or } \left\langle E_1E_2C_3E_4\right\rangle \text{ or } \left\langle E_1E_2E_3C_4\right\rangle\right]$$

The sequences are mutually exclusive and we want the probability of their union, so we may apply the additive law and sum their probabilities. Thus, the probability of exactly one correct guess in four trials if Rachel is guessing is $(.1055 + .1055 + .1055 + .1055) = .422$. In English, if Rachel is guessing from among four equally likely alternatives, then the probability is .422 that she will be correct exactly once in the four-trial experiment.

3.4 Probability Distributions

We are not interested in just one possible outcome of an experiment. It is possible that Rachel will get 0, 1, 2, 3, or 4 items correct in our four-trial ESP experiment. The probabilities of each of these possible outcomes based on a *statistical model* – a set of assumptions about how the responses are generated that can be used to calculate the probabilities is referred to as a *probability distribution* or a *probability model*.

3.4.1 *Random Sampling, Random Variables, and Distributions of Random Variables*

In any experiment, we consider our observations to be a sample from a much larger, possibly infinite, population of observations. We attempt to obtain a sample that is representative of the population from which the observations are drawn. In calculating probabilities, we often assume *random sampling*. In our ESP example, the set of trials in the experiment is a sample from an infinite number of trials that characterizes Rachel's ability to generate correct responses. Random sampling means that each of these potential trials with its accompanying response has an equal opportunity of being sampled.

Although the goal of random sampling is to obtain a representative set of observations from a population, samples will differ from one another and thus are not perfect miniatures of the population from which they are drawn. Put another way, sample statistics are *variables* because they can take on different values for different samples. When the numerical value of a variable is determined by a chance event, that variable is called a *random variable*. Therefore, sample statistics as well as the outcomes of chance events, such as throwing a pair of dice or guessing the identity of a card (scored 1 if correct and 0 if wrong), are random variables.

3.4.2 *Discrete Random Variables*

In many situations, a random variable may take only a limited number of values within its range. In our ESP experiment, if the random variable of interest is the number of correct responses in 20 trials, the possible values are the 21 integers from 0 to 20. Alternatively, we might characterize the results of the ESP experiment in terms of the *proportion* of correct responses, in which case the possible values of our random variable would be each of the integers divided by 20. In either case, there is a relatively small number of possible values so that the variable is discrete. This contrasts with other variables such as the amount of time taken to make a response. Time to respond is a continuous variable that can take on an infinite number of possible values if measured precisely enough.

We will encounter many situations involving discrete variables that are of interest. We have already given a great deal of attention to one – the number (or proportion) of correct guesses in an ESP experiment. Probability calculations for discrete random variables are often simple. With just a basic knowledge of probability, we were able to compute the probability that Rachel makes exactly one correct guess in a four-trial ESP experiment by randomly guessing. We could similarly calculate the probability of each other potential outcome of the four-trial experiment and we will do so in Chapter 4. The graph of the resulting probability distribution is given in Figure 4.1. In general, a probability distribution associates a probability with every possible value of a random variable.

3.4.3 *Continuous Random Variables*

Imagine that we are interested in the distribution of times it takes healthy adults in the 20- to 40-year age range to complete a 1-mile run. The range of times will be finite (i.e., presumably between 3.75 and, say, 25 minutes). However, if we could measure time to any degree of precision, then any value between the extremes might be observed. Unlike our example of correct guesses in an ESP experiment, the random variable of time to complete a mile run is a *continuous random variable.*

Arguably, we deal only with discrete random variables in practice because there are always limits to the precision of our measurements. However, continuous functions often provide excellent approximations to distributions of discrete random variables, so they will turn out to be very useful to us in many inferential applications. Therefore, it is important to develop a basic understanding of the properties of distributions of continuous random variables.

In general, the probability of occurrence of any *exact* value of a continuous random variable is essentially zero. For this reason, we do not talk about the probability of a value of a continuous random variable. Instead, we refer to the *probability density* at a particular value of a continuous random variable. We will symbolize the probability density of a variable Y at value a by

$$f(a) = \text{probability density of } Y \text{ at } a$$

Graphically, the probability density at a particular value of a continuous random variable is the height of the probability distribution at that value.

Although an exact value of a continuous random variable has a probability of zero, it is meaningful to talk about the probability of observing a value within an *interval*. Graphically, the probability of observing a value within some interval is the area under the curve for the distribution of that random variable. This is how we will typically work with probability computations when dealing with distributions of continuous random variables. For example, if we know that IQ scores in a population are normally distributed with a certain mean and standard deviation, we can find the probability that a person randomly selected from the population has an IQ of, say, 130 or greater, by using the characteristics of the normal distribution.

3.5 Connecting Probability Theory to Data

Probability distributions are theoretical entities, but they have very useful applications. We have already hinted frequently at some of those applications. At least three closely related applications of probability distributions to data interpretation can be identified. First, we have already seen that we can, in some circumstances, calculate the probabilities of various experimental outcomes given an explicit model of performance. In our ESP example, we calculated the probability that Rachel would make exactly one correct guess in a four-trial experiment given the assumption that she was guessing randomly among four, equally likely alternatives on each trial. We indicated that we could also compute the probability of any other possible outcome of the ESP experiment. Thus, a basic knowledge of probability allows a precise statement of the predictions of a specific hypothesis for the possible outcomes of an experiment.

A second application of a probability distribution is related to the first. Namely, a probability distribution like the one that we generated for our four-trial ESP experiment provides us with two clear expectations regarding the experimental outcome. One is that the probability distribution identifies the most likely outcome of an experiment. In our ESP example, that outcome is one correct in four trials, or 25% correct, if Rachel is guessing. The other expectation is that there will be *sampling error* in any experiment. Although the most likely outcome of the experiment is one correct response, that outcome has only a .422 probability of occurring; there is a .578 probability of observing a different outcome. The very fact that there is a *distribution* of possible outcomes embodies the reality of sampling error in an experiment.

Finally, the ability to generate a probability distribution representing the potential outcomes of an experiment and their corresponding probabilities provides us with critical information for evaluating the actual outcome of an experiment. In our ESP example, if Rachel makes only one correct guess in four trials, we will doubt that she has ESP because that outcome is quite likely under the hypothesis that she is simply guessing. On the other hand, if she is correct on all four trials, we will doubt that her performance is due to guessing because the probability of being correct on all four trials is very low (i.e., .004). Of course, we may remain skeptical that Rachel has ESP and so we will look for other explanations of her high accuracy.

Although we can see the relevance of probability theory and probability distributions to interpreting empirical observations in scientific studies, we must make an additional link between probability theory and the process of interpreting data. We have defined probability as the relative frequency of an event in a sample space. If we select a piece of paper with a grade and college major written on it from the distribution given in Table 3.1, all the elementary events are specified, so computing the probability of an event is simple. However, in most real experiments, we do not have such comprehensive knowledge of the characteristics of the sample space, so we are unable to compute the precise probability of each possible event. Indeed, the reason for conducting studies is to learn about the characteristics of populations of interest. We sample observations from populations and use characteristics of samples to estimate corresponding characteristics of populations. We define the probability of an event in these circumstances as its relative frequency of occurrence in the long run.

The definition of probability as the long-run relative frequency of an event is obviously related to our initial definition of probability as the relative frequency of an event in a sample space. However, defining probability in terms of long-run relative frequency has the advantage of suggesting a way to empirically estimate the probability of an event; namely, by computing the proportion of times the event occurs in some large number of observations. For example, we can compute the proportion of times that Rachel responds correctly in our ESP experiment and use that as our best indication of the level of Rachel's ESP ability. By "long run," we mean the relative frequency of occurrence of an event in an infinite number of observations. Because we can never make an infinite number of observations, we must assume that any calculation we base on a finite sample of observations is an imperfect estimate. However, *if we make* N *independent observations under the exact same conditions, then the observed relative frequency of occurrence of an event,* Y, *approaches the probability of the event,* p(Y), *as* N *approaches infinity.* This statement is usually referred to as Bernoulli's theorem. It tells us that the relative frequency of some event Y in a sample of observations may be used to estimate the probability of Y, and that the accuracy of the estimate will increase with the number of observations. This provides the connection we need to relate the probability to the application of analyzing and interpreting empirical observations.

3.6 Summary

Our goal in this chapter was to present the rudiments of probability theory in preparation for developing basic concepts in statistical inference in Chapters 4 and 5. Our major points were the following:

- Observations in behavioral experiments are probabilistic, so we must be able to compute the probabilities of events under different assumptions about the population if we are to have a basis for interpreting experimental outcomes.
- The starting point for computing probabilities is the ability to analyze the structure of events; that is, to determine how elementary events are combined to form the event classes. Such an analysis involves both determining how events may be combined (i.e., joint events and compound events) and how events are related (i.e., mutual exclusion, independence).
- The structure of events has implications for computing the probabilities of those events from a knowledge of the probabilities of the elementary events comprising them. We will have particular use for the *additive* and *multiplicative laws.*
- An important application of probability theory is the derivation of probability distributions that provide fundamental information to guide processes of statistical inference.
- Finally, defining the probability of an event in terms of its long-run relative frequency provides a connection between probability theory and empirical observations that provides the basis for data analysis. We develop that connection further in Chapter 4.

Exercises

3.1 [Basic probability calculations] Suppose an experiment similar to the one described in the chapter is designed to test for the existence of ESP. The major difference is that now there are five cards in the pack instead of four. Suppose that the subject, *S*, has no ESP and randomly picks one of the five cards on each trial. Assuming independence, what is the probability that *S* is (a) correct on the first trial? (b) correct on each of the first three trials? (c) correct on the second trial but wrong on the first and third? (d) correct on exactly one of the first three trials? (e) correct on at least one of the first three trials? (f) correct on exactly two of the first three trials? (g) correct for the first time on the fifth trial?

3.2 [Conditional and joint probabilities] Suppose a certain trait is associated with eye color. Three hundred randomly selected individuals are studied with the following results:

Trait	*Eye color*		
	Blue	*Brown*	*Other*
Yes	70	30	20
No	20	110	50

Suppose a person is chosen at random from the 300 in the study.

a) For each of the following pairs of events, indicate whether they are exhaustive, whether they are mutually exclusive, and whether they are independent: (i) "Yes" and "No"; (ii) "Blue" and "Brown"; (iii) "Yes" and "Brown."
b) Find: (i) $p(\text{Blue} \mid \text{Yes})$; (ii) $p(\text{Yes} \mid \text{Blue})$; (iii) $p(\text{Yes or Blue})$; (iv) $p(\text{Yes and Blue})$.
Suppose two people are chosen at random from the 300.
c) What is the probability that the first person has the trait and has brown eyes?
d) What is the probability that both people have the trait and have brown eyes if they are selected *with replacement* (that is, after the first person is selected, he or she is returned to the pool and may be selected again)?
e) What is the probability that both people have the trait and have brown eyes if they are selected *without replacement* (that is, once the person has been selected on the first choice, he or she is no longer available to be chosen)?

3.3 [Reasoning with probabilities] Two individuals, A and B, are throwing darts at a target. The probability of hitting the target on a given trial is .7 for A and .6 for B (i.e., $p(A) = .7$ and $p(B) = .6$). Assuming that the results of these throws are *independent* of one another, answer the following: If A and B each take a single throw at the target, what is the probability that

a) both A and B hit the target?
b) neither A nor B hit the target?
c) A hits the target and B misses it?
d) A misses the target and B hits it?
e) the target is hit at least once?

If A and B each throw two darts at the target, what is the probability that

f) the target is hit at least once?
g) the target is missed at least once?

3.4 [Conditional and joint probabilities] The staff at FiveThirtyEight.com rated a set of Super Bowl ads on several dimensions, including whether they were funny or not. Here are their results, with the ads grouped by product category:

Rating	*Product Type*						
	Beer	*Financial Trading Platform*	*Football*	*Snack Food*	*Soft Drink*	*Vehicle*	*TOTAL*
Funny	78	12	2	24	26	26	168
Not Funny	25	1	9	1	20	20	76
TOTAL	103	13	11	25	46	46	244

Suppose a Super Bowl ad is selected at random.

a) Find: (i) $p(\text{Vehicle} \mid \text{Funny})$; (ii) $p(\text{Funny} \mid \text{Vehicle})$; (iii) $p(\text{Funny} \ \& \ \text{Vehicle})$; (iv) $p(\text{Funny or Vehicle})$; (v) $p(\text{Funny})$
b) For each of the following pairs of events, determine whether they are exhaustive, whether they are mutually exclusive, and whether they are independent: (i) "Vehicle" and "Beer"; (ii) "Beer" and "Not Funny"; (iii) "Funny" and "Not Funny"

3.5 [Bayes' Rule] The following demonstrates why it is hard to screen populations for the presence of low-incidence diseases: enzyme-linked immunosorbent assays (ELISA) are used to screen donated blood for the presence of the HIV virus. The test detects antibodies, substances that the body produces when the virus is present. But the test is not completely accurate. It can be wrong in two ways: (1) by giving a positive result when there are no antibodies (false positive) or (2) by giving a negative result when there are antibodies present (false negative). When antibodies are present, ELISA gives a positive result with probability about .997 and a negative result (false negative) with probability about .003. When antibodies are not present, ELISA gives a positive result (false positive) with a probability of about .015 and a negative result with probability .985. That is, $p(\text{correct positive}) = p(\text{positive} \mid \text{HIV}) = .997$; $p(\text{false negative}) = p(\text{negative} \mid \text{HIV}) = .003$; $p(\text{false positive}) = p(\text{positive} \mid \text{no HIV}) = .015$; $p(\text{correct negative}) = p(\text{negative} \mid \text{no HIV}) = .985$. Suppose 100,000 blood samples are obtained from a population for which the incidence of HIV infection is 1.0%; that is, $p(\text{HIV}) = .01$.

a) Using the information given earlier, fill in the cells in the following 2 × 2 table.

Test results	*HIV*	*No HIV*	*Total*
Positive			
Negative			
Total			

b) Given that a randomly chosen sample tests positive, what is the probability that the individual is infected?
c) Given that a randomly chosen sample tests negative, what is the probability that the individual is not infected?

3.6 [Reasoning with probabilities] We are often able to use key words such as "and," "or," and "given" to decide among probability rules. In the following problem, we use more everyday language. So read carefully and for each part think about whether the wording dictates marginal, joint, or conditional probability.

Suppose that a survey of 200 people in a college town has yielded the following data on attitudes toward liberalizing laws on the sale of marijuana. Suppose a person is selected at random.

Attitude	*Younger (Age 18–40)*		*Older (Age 41–65)*		
	Student	*Non-student*	*Student*	*Non-student*	*Row total*
For	70	10	2	15	97
Against	5	30	1	60	96
No opinion	5	0	2	0	7
Col. total	80	40	5	75	200

a) What is the probability that someone is *for* if that person is under 30 years old?
b) What is the probability that a randomly selected individual is in the older group with no opinion?
c) What is the probability that a student in the younger group has no opinion?
d) What is the probability that a student has no opinion?
e) What is the probability that someone with no opinion is in the younger group?

3.7 [Conditional probabilities] A study reported in the local newspapers indicated that a psychological test has been developed with the goal of predicting whether elderly people are at high risk of developing dementia in the near future. For healthy people at age 79, the probability of developing dementia within the next 4 years is approximately .20. In the study, a group of healthy 79-year-olds was given the test. For those who went on to develop dementia within the next 4 years, the probability of a positive test at age 79 was found to be .17; that is, $p(\text{positive} \mid \text{dementia}) = .17$. For those who did not develop dementia within the next 4 years, the probability of a positive test was .008; that is, $p(\text{positive} \mid \text{no dementia}) = .008$.

a) What is $p(\text{negative} \mid \text{dementia})$?
b) What is $p(\text{negative} \mid \text{no dementia})$?

From the data given earlier, find the predictive accuracy of the test. That is, find the probability that a 79-year-old who takes the test will develop dementia within the next 4 years

c) if the test result is positive?
d) if the test result is negative?

3.8 [Real life probability] Many states have, as part of their lottery offerings, a game in which you can choose to bet by choosing any four digits. The winning number is a four-digit number chosen at random with replacement (i.e., the same digit can be chosen more than once).

a) What is the probability that your chosen number will exactly match the winning number?
b) What is the probability that the digits in your chosen number will match those in the winning number without regard to order?

3.9 [Probability models] Given a pair of fair dice (i.e., for each die, the probability of getting any of the outcomes 1, 2, 3, 4, 5, or 6 is 1/6), if you roll the dice after shaking them well,

a) What is the probability of getting a 12?
b) What is the probability of getting a 7?

3.10 [Probability models] Generate the probability distribution for the outcome of a roll of a pair of dice. What is the probability

a) of getting an outcome of 8 or greater?
b) of rolling an outcome that is even?
c) of rolling better (more) than 8 or less than 4?
d) of rolling an 8 or higher, *or* an outcome that is even?

3.11 [Working with real data] In R, load the "diamonds" data set from the {ggplot2} package. Diamond colors are rated D–J with "D" being the best; their cuts are rated from Fair to Ideal, with "Ideal" being the best.

a) Make a 2 × 2 table of the number of diamonds in this data set that have a better or less desirable color ("F" or better; "G" or worse), and better or worse cut ("Very Good" or better; "Good" or worse).
b) Use that table to determine whether better color is independent of cut.

Chapter 4

Developing the Fundamentals of Hypothesis Testing Using the Binomial Distribution

4.1 Overview

Our goal in conducting research is to use our data as a window on the population from which they are sampled. To make this generalization, we must have a way to distinguish trends in the data that reflect corresponding patterns in the population from trends in the data that may be specific to the sample. The procedures for making this critical distinction are collectively known as *inferential statistics*. Most of the subsequent chapters of this book are concerned with two very general inferential procedures: Testing specific hypotheses about population parameters and estimating the values of population parameters. The focus of the current chapter is on developing the logic of *hypothesis testing*. Parameter estimation will be discussed in Chapter 5.

In Chapter 3, we established that probability theory is central to the decision-making process of inferential statistics. To elaborate that role and to facilitate presentation of some of the rather abstract concepts underlying hypothesis testing, we will continue to use our ESP experiment as an example. We again have Rachel choosing among four different cards in a 20-trial experiment. Because we are skeptics, we are unwilling to concede that Rachel has ESP unless guessing can be ruled out as a reasonable explanation of her performance in our experiment. If Rachel simply randomly guesses on each trial, the probability of a correct response is 1/4 or .25. Therefore, a finding of five correct responses on the 20 trials is consistent with chance performance and would provide no evidence whatsoever for ESP. But suppose that 7/20 or .35 of Rachel's responses were correct. Is this result different enough from five correct to convince us that she performed better than what could reasonably be expected by guessing? If not, how many correct responses would it take to convince us? 10/20? 15/20? 20/20? How are we to select a criterion level of performance to decide that Rachel is not simply guessing?

We must develop a conceptual framework to answer this question, and the first sections of this chapter provide that foundation. Suppose that Rachel achieved a .35 success rate in the sample of 20 trials (i.e., seven correct choices). In addressing the issue of ESP, we are concerned with the entire population of potential responses that characterizes Rachel's ESP ability, not just the 20 responses we happened to collect in the experiment. If we ran the experiment again, it would almost certainly turn out somewhat differently. This is the fundamental obstacle to making correct inferences from a sample to a population: Because our observations are subject to *sampling error*, any conclusion we make about the population may be wrong. Therefore, we must be able to assess the likelihood of making an error if

DOI: 10.4324/9781003453550-5

we decide based on our experimental results to reject the hypothesis that Rachel is simply guessing. The major topics of this chapter are as follows:

- *What are the requirements for a test of a hypothesis?* We develop the tools for stating a hypothesis and using data to assess the hypothesis.
- *The binomial distribution as a sampling distribution.* We will use probability rules from Chapter 3 to derive a specific theoretical probability distribution called the *binomial distribution.* The binomial serves as the *sampling distribution* for the ESP experiment and others like it, and the concept of a sampling distribution is fundamental to the process of making statistical inferences.
- *Hypothesis testing.* Hypothesis testing uses the concept of a sampling distribution, along with a decision rule, to decide whether the sample of data we collected provides sufficient evidence to reject some hypothesis about the population.
- *Statistical power.* The power of an experiment is the probability that we reject a specific *null hypothesis* when it is appropriate to do so. Several factors that influence power will be described.
- *Assumptions.* Statistical procedures are always based on assumptions that may or may not be valid. When the assumptions are violated, we must understand the potential consequences for our interpretation of the data.

4.2 What Do We Need to Know to Test a Hypothesis?

Not all hypotheses that a researcher might generate are amenable to the procedures of hypothesis testing. We need two things to directly test a hypothesis. First, we must be able to specify a "point hypothesis." A *point hypothesis* is an exact statement about the state of the world. For example, the hypothesis that Rachel is guessing is a point hypothesis because it can be expressed by a precise probability statement. In our four-choice procedure, the guessing hypothesis can be stated as $p(\text{correct on any trial}) = 0.25$.

Second, we must be able to use the point hypothesis to derive the "sampling distribution" for the outcomes of interest in an experiment. A *sampling distribution* is a probability distribution that assigns a probability to every possible outcome that might be observed (i.e., is "sampled") in an experiment, given our point hypothesis. A sampling distribution may be thought of as the predictions of a hypothesis. As such, a sampling distribution provides the objective information needed to evaluate the tenability of a hypothesis. For example, the guessing hypothesis predicts that a result of four correct guesses in a four-trial ESP experiment has a probability of only $(1/4)^4$ or .0039. This probability information provides an objective basis for deciding that if Rachel is correct on all four trials, she almost surely is not guessing in the experiment.

Let's develop these ideas with our ESP example. To keep things manageable, we will start by considering a four-trial experiment. We will measure performance in the experiment by counting the number of correct responses across the four trials; we will call this measure Y. Given that each response is either correct (C) or in error (E), Y can now take on only five possible values (0–4 correct).

We want to determine the probability of each possible experimental outcome under the hypothesis that Rachel does not have ESP so she is randomly guessing on each trial. Under this hypothesis, $p(C) = .25$ because she is guessing among four equally likely alternatives. This hypothesis refers to performance on individual trials, but our interest centers on the

Table 4.1 Possible patterns of correct (*C*) and error (*E*) responses for four trials; the subscripts denote the trial numbers

Pattern	*Number correct (Y)*
$<E_1\ E_2\ E_3\ E_4>$	0
$<E_1\ E_2\ E_3\ C_4>$	1
$<E_1\ E_2\ C_3\ E_4>$	1
$<E_1\ C_2\ E_3\ E_4>$	1
$<C_1\ E_2\ E_3\ E_4>$	1
$<E_1\ E_2\ C_3\ C_4>$	2
$<E_1\ C_2\ E_3\ C_4>$	2
$<E_1\ C_2\ C_3\ E_4>$	2
$<C_1\ E_2\ E_3\ C_4>$	2
$<C_1\ E_2\ C_3\ E_4>$	2
$<C_1\ C_2\ E_3\ E_4>$	2
$<C_1\ C_2\ C_3\ E_4>$	3
$<C_1\ C_2\ E_3\ C_4>$	3
$<C_1\ E_2\ C_3\ C_4>$	3
$<E_1\ C_2\ C_3\ C_4>$	3
$<C_1\ C_2\ C_3\ C_4>$	4

summary measure of performance in the experiment, Y, which we will use to evaluate the hypothesis. Therefore, we must translate our hypothesis about trial performance into corresponding probabilities for the different possible values of Y.

The first step in deriving the probability distribution for Y is to identify all the ways in which trial outcomes may combine to form the experimental outcomes. With four trials that can each have two outcomes, there are 2^4 or 16 possible distinct patterns of correct and error responses, each associated with a possible value of Y. In Chapter 3, we found that a *tree diagram* is a useful way to enumerate all possible sequences of trial outcomes. Figure 3.1 diagrammed the sequences for our four-trial experiment and Table 4.1 summarizes those patterns. The 16 possible sequences of trial outcomes may be considered elementary events that partition the sample space for the experiment.

Having specified the structure of the sample space for the experiment, we want to use our analysis to compute the probability of each possible value of Y. The probability distribution for Y is obtained by using a *statistical model*, which is a set of assumptions about how the responses are generated that can be used to calculate the probabilities. In the current example, a desirable model would be one that captures the essential features of what we mean by random guessing and allows us to calculate the probability of each possible value of Y. We employ the following model for the simplified ESP experiment:

1. The probability of a correct response, π, is .25 on each trial.[1]
2. The responses are *independent* of one another; that is, the probability of a correct response on any trial does not depend on the outcomes of any other trials.

The first assumption is based on the fact that Rachel chooses from among four equally likely card values. The independence assumption seems reasonable because each card is replaced in the pack after each trial, the cards are well shuffled, and the selection of each card is done at random. The result on each trial should not be influenced by the outcome of any previous trial.

Given these assumptions, we can apply the multiplicative law to compute the probability of each sequence of trial outcomes because each sequence corresponds to a joint event. We assume trial outcomes are independent, so the probability of any sequence may be found by multiplying probabilities. The probability of the sequence $<C_1\ C_2\ C_3\ C_4> = (.25)(.25)(.25)(.25) = .0039$. The probability of the sequence $<E_1\ E_2\ E_3\ E_4> = (.75)(.75)(.75)(.75) = .3164$. The probability of each single sequence corresponding to one correct and three errors is $(.25)(.75)(.75)(.75) = .1055$. However, there are four distinct sequences corresponding to the outcome of one correct in four trials, so the probability of observing one correct is greater than .1055. The four sequences are mutually exclusive and we are interested in any one of them occurring, so the additive law may be used to compute the probability of the compound event. Summing up the probabilities of the four possible sequences gives us a value of .4220 for the probability of one correct in four trials. By the same approach, we can compute that the probability of two correct in four trials is .2109, and the probability of three correct in four trials is .0469.

We have now translated our hypothesis that Rachel is guessing into a distribution that represents the probability of each possible outcome of our four-trial experiment if Rachel does, indeed, guess on each trial. That probability distribution is summarized graphically in Figure 4.1.

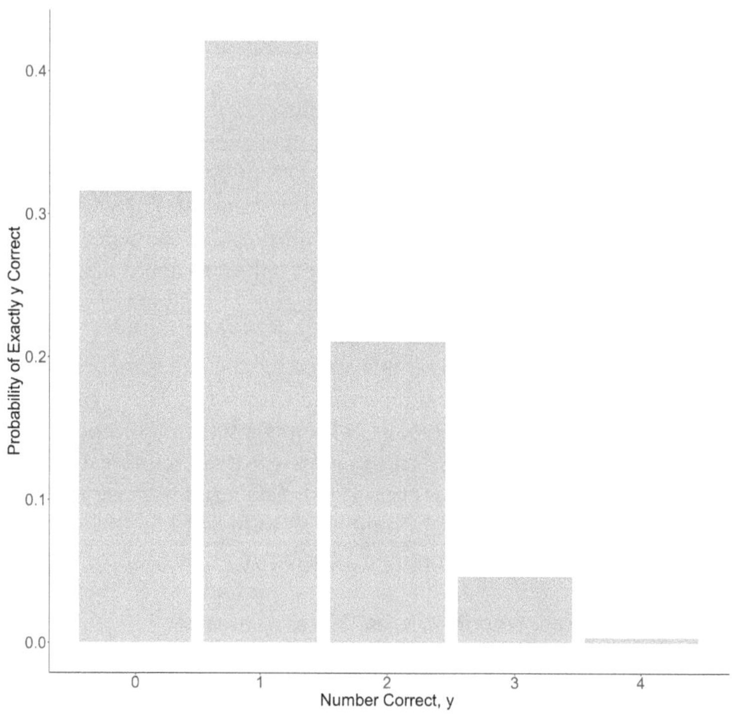

Figure 4.1 Theoretical probability distribution for the number of correct responses, Y, with n = 4 and $p(C)$ =.25.

Keep in mind why the probability distribution presented in Figure 4.1 is useful: to decide whether the responses were produced by guessing, we need some idea of what to expect if they were, in fact, produced by guessing. The probability distribution we generated is a *theoretical* probability distribution because it was generated with a statistical model. Here, our statistical model is a theory about the patterns of responses that would be produced by guessing. If the assumptions we made are valid and if we performed many random replications of the ESP experiment, then in .3164 of the experiments there would be zero correct responses; in .4220 of the experiments there would be exactly one correct response; and so on. In short, the proportions of experiments yielding various values of Y would match the theoretical probability values in Figure 4.1.

The idea of repeating an experiment many times and obtaining the value of some statistic from each experiment is basic to the inferential procedures described throughout this book. The idea is important enough to summarize the general steps:

1. A model is formulated.
2. On the basis of this model, we obtain the theoretical distribution of a statistic of interest given many repetitions of the experiment. This is the *sampling distribution* of the statistic.
3. The sampling distribution is then used together with the collected data to draw inferences about the population.

How can we use the theoretical sampling distribution (Figure 4.1) to evaluate the hypothesis that Rachel is guessing on each trial (i.e., $\pi = .25$)? We are interested in knowing whether Rachel actually has ESP, which would mean that the true probability of her making a correct response is greater than .25. Therefore, a finding that Rachel is correct more than 25% of the time would be evidence supporting the ESP hypothesis over the guessing hypothesis. But how much better than 25% correct must Rachel be before we conclude that she is not just guessing? One possible decision rule is to conclude that the population probability is greater than .25 if two or more responses in the sample (i.e., at least 50%) are correct. However, this is not a very good rule because from Figure 4.1 we see that it would not be surprising to get at least two correct responses simply by guessing. The probability of two or three or four correct responses if $\pi = .25$ is $(.2109 + .0469 + .0039) = .2617$. That is, even if Rachel is just guessing, there is greater than one chance in four that she would answer correctly at least twice in four trials. A better rule would be to require that all four responses be correct before concluding that something besides random guessing was going on. Four correct responses could also occur just by random guessing, but this is much less likely. According to our guessing model, the probability of correct responses on all four trials is only .0039. Therefore, if all four responses are correct, either the subject is guessing and an unlikely event has occurred, or the value of π in the population of responses is greater than .25.

We can now summarize the rationale underlying hypothesis testing. We begin by assuming that chance, and chance alone, determines outcomes in an experiment. That hypothesis, called the *null hypothesis*, is used to generate the sampling distribution of possible outcomes. Some outcomes will be quite unlikely if the chance-based model is valid. If one of these unlikely outcomes is obtained in the experiment, then we conclude that the model is not valid and that something other than chance is involved. That "something else" is the *alternative hypothesis*, which is always mutually exclusive with the null hypothesis. In the ESP example, either Rachel does not have ESP and merely guesses (null hypothesis,

H_0: $\pi = 0.25$) or she does have ESP and uses it to more accurately predict the cards drawn (alternative hypothesis, H_A: $\pi > 0.25$).

We have developed a simple example of an experiment for the purposes of sketching the role of a sampling distribution in the process of evaluating a specific hypothesis. The example that we used – a four-trial experiment in which Rachel guessed among four equally likely alternatives – is one specific example from a much more general class of experiments. If we had chosen to do a 20-trial experiment and changed the experimental procedure to require Rachel to guess among 10 alternatives, it should be obvious that we would need to specify a new sampling distribution to describe the possible outcomes for this more ambitious experiment. However, the approach to deriving the appropriate sampling distribution would be the same as that illustrated for our simpler experiment. We turn now to the task of generalizing our approach.

4.3 The Binomial Distribution

Figure 4.1 presents the probability distribution for the number of correct responses when there are four trials ($n = 4$) and the probability of a correct response on each trial is .25 ($\pi = .25$). For different values of n and π we could always find the desired probabilities by drawing the appropriate tree diagram and applying the multiplicative and additive laws. A tree diagram is very helpful for understanding the probability calculation; however, using a tree diagram quickly becomes very tedious as the sample size, n, increases.[2] It is helpful to have a general formula for the probability distribution given any combination of n and π. Specifically, we are interested in developing a formula that will give us a way to derive the sampling distribution for any experiment with the following three characteristics:

1. On each trial there are exactly *two* possible outcomes; examples might be correct/error, success/failure, head/tail, or live/die. The two outcomes possible on each trial might be referred to as A and $\tilde{A}$ with probabilities π and $1 - \pi$, respectively.
2. The value of π is constant over trials.
3. Trials are independent. The probability of an outcome of any trial does not depend on the outcome of any other trial.

4.3.1 The Binomial Function

We want a formula for calculating the probability of y responses as a function of n and π. We denote the binomial probability function by $p(y; n, \pi)$ to indicate that it is the probability of obtaining y responses of type A when there are n trials with $p(A) = \pi$ on each trial.

In determining the probability of each possible outcome, y, of a four-trial ESP experiment, we first identified each of the 16 possible sequences in the experiment.[3] We found that the probability of any given sequence could be computed using the multiplication rule for independent events. Suppose we wish to find the probability of obtaining exactly three A responses in four trials. For example, the probability of the sequence <*A*, *A*, *A*, *Ã*> would be $(\pi)(\pi)(\pi)(1 - \pi)$ or $\pi^3(1 - \pi)$.

Further, we observed that there were multiple distinct sequences for most values of y. For example, there are four combinations with three As and one $\tilde{A}$, so that $p(3; 4, \pi)$: (<*Ã*, *A*, *A*, *A*> *or* <*A*, *Ã*, *A*, *A*> *or* <*A*, *A*, *Ã*, *A*> *or* <*A*, *A*, *A*, *Ã*>). Because the four combinations are mutually exclusive, the additive law for mutually exclusive events is applicable:

The probability of three A responses and one $\tilde{A}$ response in any order is the sum of the probabilities of these four combinations, or $4\pi^3(1 - \pi)$. In general, to find the probability of having y A responses, we calculate the probability of a combination having exactly y A responses and then multiply by the number of ways that outcome can occur.

The approach can be generalized to any value of n. If we can assume the trials are independent, the probability of any one specific combination of y A responses and $(n - y)$ $\tilde{A}$ responses is $\pi^y(1 - \pi)^{n-y}$. The probability of exactly y A responses and $(n - y)$ $\tilde{A}$ responses is

$$p(y;n,\pi) = k\pi^y(1-\pi)^{n-y} \tag{4.1}$$

where k is the number of combinations consisting of exactly y A and $(n - y)$ $\tilde{A}$ responses.

We just about have our binomial function; all we need is a formula for k, the number of ways in which y A and $(n - y)$ $\tilde{A}$ responses can be combined. This number of combinations can be shown to be (see Appendix 4.1):

$$k = \binom{n}{y} = \frac{n!}{y!(n-y)!} \tag{4.2}$$

where $n! = (n)(n - 1)(n - 2)\ldots(3)(2)(1)$ and $0! = 1$. Note that

$$\binom{n}{y} \text{ and } \binom{n}{n-y}$$

have the same value. Substituting $y = 0, 1, 2, 3$, and 4 in turn into Equation 4.2, verify that k takes on the values 1, 4, 6, 4, and 1, respectively; these are the numbers of combinations that appear in Table 4.1. Replacing k in Equation 4.1 with the expression on the right side of Equation 4.2 yields the binomial probability function:

$$p(y;n,\pi) = \binom{n}{y}\pi^y(1-\pi)^{n-y} = \frac{n!}{y!(n-y)!}\pi^y(1-\pi)^{n-y} \tag{4.3}$$

Equation 4.3 describes a family of distributions, with a different probability distribution for each distinct combination of values of π and n. Table C.1 in Appendix C presents specific binomial distributions for selected values of π and n. We will restrict our examples to situations that are described by the distributions in Table C.1 to save ourselves the trouble of computing binomial distributions from scratch. In Section 4.3.3, we will show how software can be used for these calculations.

Now we can calculate the probabilities of various outcomes in the ESP experiment if Rachel was generating responses by random guessing. If $n = 20$ and $\pi = .25$, we can find the probability of, say, seven or more correct responses. In Table C.1, we first find the section of rows that corresponds to $n = 20$, and then we use the column corresponding to $p = .25$. Reading from the table, we see that $p(Y = 0) = .0032$, $p(Y = 1) = .0211$, $p(Y = 2) = .0669$, and so on. These are the same values that would be obtained using Equation 4.3. We can find the probability of seven or more responses in several ways. One way is to find the probabilities for 7, 8, 9,. . ., 20 correct and add these 14 numbers together. The other is to find the probabilities of 0, 1, 2,. . ., 6 correct, add these seven numbers together, and then subtract the sum from 1. The probability of getting 0–6 correct responses is .0032 + .0211

+ .0669 + .1339 + .1897 + .2023 + .1686 = .7857. So the probability of seven or more correct responses must be 1 – .7852 = .2143.

Putting this all together, we see that a result of seven correct responses in 20 tries is not very convincing evidence that Rachel was doing anything other than randomly guessing on each trial because there would be better than two chances in 10 of getting at least seven correct just by guessing. However, if Rachel got, say, 13 correct responses in 20 trials, we could pretty much rule out the random guessing model, because we can see from Table C.1 that there is only a probability of about .0002 (two chances in 10,000) of getting 13 or more correct responses by random guessing.

4.3.2 *What Happens When* n *and* π *Change?*

Figure 4.2 presents several binomial distributions for various values of n and π. For easier comparisons across different values of n, the x-axes show the proportion of A responses in the sample, $P = Y/n$, rather than Y. For example, when $n = 10$ and $\pi = .5$, we expect to

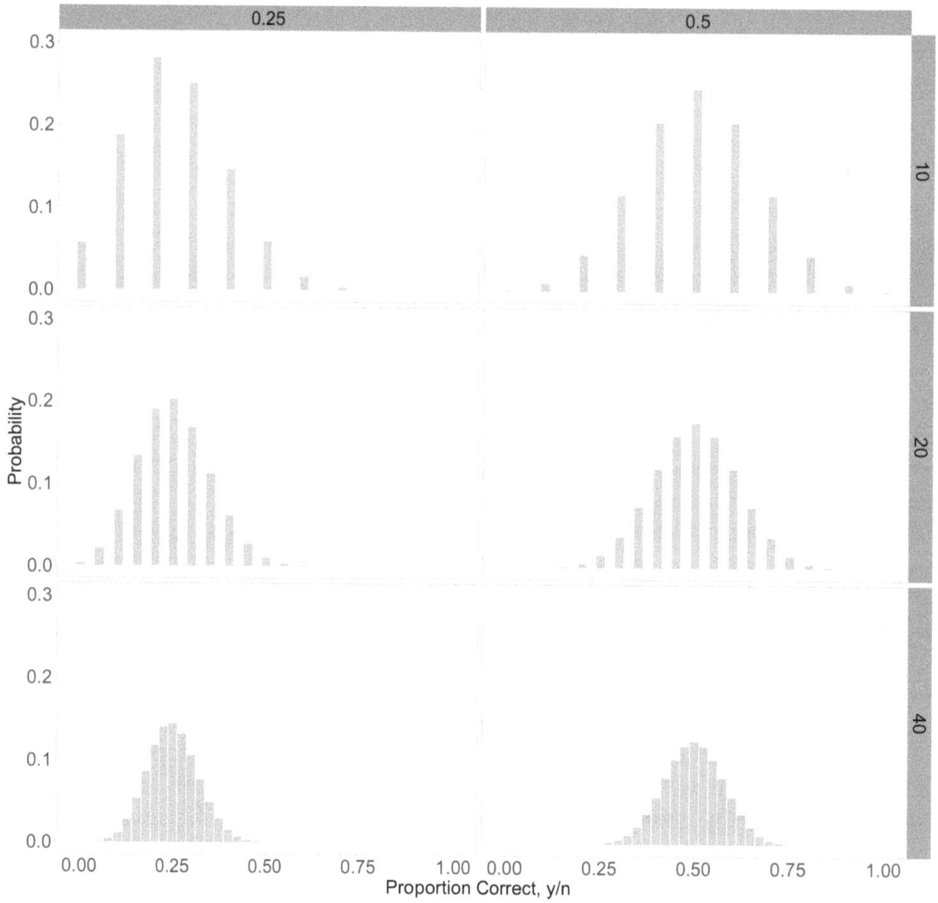

Figure 4.2 Binomial distributions as a function of sample size and probability.

observe 40% correct responding (four A responses in 10 trials) with probability .2051. In the long run (i.e., if the experiment were repeated many times), the proportion of experiments with four A and six $\tilde{A}$ responses should equal .2051, if the binomial model is correct. Several points should be noted about these distributions. First, when $\pi = .5$, the distributions are more symmetric than for values of π closer to 0 or to 1. Second, when $\pi = .25$ and skewness (asymmetry) is present, as n increases, the distribution becomes more symmetric about the value of $P = Y/n$ that equals π. Third, the distributions appear more continuous in form as n increases. These observations are important because if n is sufficiently large, particularly when π is close to .5, the binomial distribution looks much like the normal distribution. In that case, a normal distribution can be used to get binomial probabilities with considerably easier calculations.

A fourth point to note about Figure 4.2 is that the probability of getting a value of Y/n close to π increases as n increases. This illustrates Bernoulli's theorem in action. For example, consider the probability that Y/n lies in the range from .4 to .6; that is, $p(.4 \leq Y/n \leq .6)$ when π is actually .5. When n is 10, $p(.4 \leq Y/n \leq .6)$ is the probability that $Y = 4$, 5, or 6, which is .2051 + .2461 + .2051, or .6563. When n is 20, $p(.4 \leq Y/n \leq .6)$ is the probability that $Y = 8$, 9, 10, 11, or 12, which is .1201 + .1602 + .1762 + .1602 + .1201, or .7368. When n is 40, the probability is .8461. This point is very important; it means that as n grows larger, the proportion of A responses observed in a single experiment is more likely to be close to the proportion of A responses in the population. We prefer larger data sets to smaller ones not because of some deeply ingrained work ethic; we do so because larger data sets are more likely to produce sample statistics that are closer to the corresponding population parameters. Statistics that have a higher probability of being within any fixed range of the population parameter as n increases are called *consistent* estimators of the parameter. We will have more to say about this desirable property of sample statistics, and about other properties important to estimating population parameters, in Chapter 5.

4.3.3 *Using Software to Compute the Binomial Function*

In Section 4.3.1 we calculated the probability of exactly Y successes out of n independent trials, each with probability of success equal to π (see Equation 4.3). We showed how Table C.1 presents specific binomial distributions for selected values of π and n. Those values, as well as their sums, can also be obtained with distribution calculators in statistical software such as R. For example, in R, the probability of observing exactly Y correct responses out of n trials, each of which has a probability π of occurring, is calculated by the *dbinom* function in the base {stats} package. The function has three input parameters, one each for Y (called x), n (called *size*), and π (called *prob*). For example, *dbinom*(x = 1, size = 4, prob = .05) returns 0.171475, which, when rounded to four decimal places, you can verify as the second entry in Table C.1; Figure 4.3 shows this example.

```
> dbinom(x = 1, size = 4, prob = .05)
[1] 0.171475
> |
```

Figure 4.3 Example binomial calculation in R.

We are often interested in knowing the probability of Y or fewer, or Y or more, successes out of n trials. For example, in Section 4.3.1, we used Table C.1 to find $P(Y < 7 \mid n = 20, \pi = 0.25)$ by summing $P(Y = 0) + \ldots + P(Y = 6) = 0.7852$. Again, distributional calculators make these computations easier. In R, the *pbinom* function in the {stats} package calculates the probability of Y or fewer success: *pbinom*(x = 6, size = 20, prob = .25) = 0.7858. If we are interested in the probability of more than Y successes, rather than Y or fewer, we can calculate 1 – *pbinom*(6, 20, .25) = .2142. Alternatively, a change of default option will compute the probabilities above a particular value of Y: $P(Y > 6)$: *pbinom*(6, 20, .25, lower.tail = FALSE) = 0.2142.

4.4 Hypothesis Testing

4.4.1 What Is a Hypothesis Test?

Having derived the binomial distribution as a family of sampling distributions that characterizes a general domain of experiments, we move on to formally develop a general procedure for testing hypotheses.

A hypothesis test consists of a set of procedures to determine whether a sample of collected data provides sufficient evidence to reject some hypothesis of no effect in the population, often called the *null hypothesis* (H_0). In the ESP example, the null hypothesis is that Rachel is guessing or, more formally, that the true probability of a correct guess is .25. We represent this hypothesis as

$$H_0 : \pi = .25$$

In contrast to the null hypothesis is the *alternative hypothesis* (H_1), which is motivated by the claim that Rachel has ESP and therefore the probability of a correct response is higher than chance. We represent this hypothesis as

$$H_1 : \pi > .25$$

In our 20-trial experiment, the experimenter randomly chooses among four cards on each trial and Rachel attempts to identify each card without seeing it. We consider the sample of 20 responses to be a random sample selected from the hypothetical population of responses that characterizes Rachel's ESP ability. Our estimate of the probability of a correct response in the population is the proportion of correct responses in the sample, $P = Y/n$. We can use our estimate as evidence to evaluate the null hypothesis because we can generate the probabilities of possible outcomes assuming that the null hypothesis is true. If we assume that the null hypothesis is true and that trials are independent, Y should have a binomial distribution. Letting $n = 20$ and $\pi = .25$ (the value if H_0 is true) in Equation 4.3, we can generate the values of $p(Y = y)$ found in the $\pi = .25$ column of Table 4.2.

Next, we determine whether our sample of data provides sufficient evidence to reject H_0 in favor of H_1. There are two closely related but slightly different procedures for doing this, the "p-value" and the "rejection region" approach. Box 4.1 summarizes the steps following the p-value approach; Box 4.2 summarizes the steps following the rejection region approach. We illustrate each approach for our 20-trial ESP experiment.

Table 4.2 The binomial distribution for n = 20 and π =.25, .35, and .50

Number correct	$p(Y = y)$		
y	π = .25	π = .35	π = .50
0	.0032	.0002	.0000
1	.0211	.0020	.0000
2	.0669	.0100	.0002
3	.1339	.0323	.0011
4	.1897	.0738	.0046
5	.2023	.1272	.0148
6	.1686	.1712	.0370
7	.1124	.1844	.0739
8	.0609	.1614	.1201
9	.0271	.1158	.1602
10	.0099	.0686	.1762
11	.0030	.0336	.1602
12	.0008	.0136	.1201
13	.0002	.0045	.0739
14	.0000	.0012	.0370
15	.0000	.0003	.0148
16	.0000	.0000	.0046
17	.0000	.0000	.0011
18	.0000	.0000	.0002
19	.0000	.0000	.0000
20	.0000	.0000	.0000

Box 4.1 Steps for Testing Hypotheses Using the *p*-Value Approach

1. State the null and alternative hypotheses, H_0 and H_1.
2. Decide on the *test statistic* that will be used to assess the evidence against H_0.
3. Determine what *sampling distribution* the test statistic should have if H_0 is true.
4. Decide on the *significance level* that will be used as the criterion for deciding whether to reject the null hypothesis.
5. Compute the p-value for the experimental result. For a one-sided alternative hypothesis, the p-value is calculated from only one tail (e.g., the upper tail if H_1: test statistic > the value in H_0; the lower tail if H_1: test statistic < the value in H_0). For a two-sided alternative hypothesis (e.g., H_1: test statistic ≠ the value stated in H_0), the p-value is calculated by summing areas in each tail of the sampling distribution of H_0.
6. Reject H_0 in favor of H_1 if $p \leq \alpha$. If we reject H_0, we say that our result is "statistically significant at level α." If $p > \alpha$, we say that we have failed to reject H_0 or that we have insufficient evidence to reject H_0.

Box 4.2 Steps for Testing Hypotheses Using the Rejection Region (or Critical Region) Approach

Steps 1–4 are the same as for the *p*-value approach of Box 4.1.

5. Determine the *rejection region* or *critical region* for the hypothesis test.
6. Reject the H_0 in favor of H_1 if the value of the test statistic falls in the rejection region.

We have already done much of the work required for a hypothesis test; namely, we have completed the first three steps in the process. To summarize:

1. Our hypotheses are H_0: $\pi = .25$, H_1: $\pi > .25$.
2. Our test statistic is the number of correct guesses, Y.
3. Assuming that the probability of a correct response remains constant over the 20 trials and that the trial outcomes are binary (e.g., correct or incorrect) and independent, our test statistic, Y, will follow a binomial distribution with $n = 20$ and $\pi = .25$, as presented in Table 4.2.

Next, according to both approaches, we must decide on the criterion for how unlikely a result must be according to the null hypothesis before we are willing to reject the null hypothesis; the criterion is called the "significance level" of the test. In theory, the significance level, α, may be set at different values, and the value chosen for a particular situation may depend on a variety of considerations. For example, in a pilot experiment, a researcher may select a relatively lenient criterion, perhaps $\alpha = .10$. In contrast, if the efficacy of some new drug treatment is being tested against an established drug that has already been proven effective, a researcher might decide on a stringent criterion before rejecting the old treatment in favor of the new, perhaps $\alpha = .001$. In practice, only relatively small values of α are considered; generally, $\alpha = .05$ or less.

The remaining steps are as follows:

4. We will select a significance level of $\alpha = .05$.
5. It is at this point that the two hypothesis-testing approaches diverge. According to the *p*-value approach in Box 4.1, we next conduct our experiment and convert our observed result, say $Y = 7$, into its corresponding *p*-value. The *p*-value is the probability of observing a result at least as extreme as seven, assuming that the null hypothesis is true. This corresponds to the conditional probability $p(Y \geq 7 \mid \pi = .25)$. To compute this probability, refer to the sampling distribution that assumes H_0 is true and identify the tail(s) of the distribution that are consistent with the predictions of H_1, then sum the probability of the observed value of Y with the probabilities of all the values of Y that are more extreme. Using the $\pi = .25$ column of Table 4.2, we see $p(Y \geq 7 \mid \pi = .25)$ equals .1124 + .0609 + 0271 + .0099 + .0030 + .0008 + .0002 + 0 = .2143.
6. Comparing the *p*-value to α, we fail to reject H_0 because the *p*-value of .2143 is not less than or equal to $\alpha = .05$.

Because statistical packages display *p*-values in their outputs, it is probably most natural to think about hypothesis testing in terms of *p*-values. However, the alternative approach of framing hypothesis tests in terms of rejection regions makes it easier to think about important concepts such as statistical power (see Section 4.5). Thus, we illustrate the approach summarized in Box 4.2 for the same example.

To execute the alternative approach of determining the rejection region for a hypothesis test, we must first identify the types of outcomes that would favor the alternative hypothesis because we never reject the null unless the alternative provides a better account of the data. In our example, the alternative hypothesis claims that $\pi > .25$, so it predicts that Rachel will make more correct choices than expected by chance; thus, we will consider rejecting the null hypothesis only if Rachel makes lots of correct responses. Our goal is to specify the range of number of correct responses that (a) would be consistent with the alternative hypothesis, but (b) very unlikely under the null hypothesis. Using the sampling distribution specified by the null hypothesis (i.e., the $\pi = .25$ column of Table 4.2), we sum the probabilities beginning at the bottom of the column (i.e., 20 correct, 19 correct, etc.). We cumulate the probabilities at the upper tail of the distribution until the cumulative probability is as close to α as possible, without exceeding it. In our example, if H_0 is true, the probability that Y has a value of nine or greater is $p(Y \geq 9) = .0271 + .0099 + .0030 + .0008 + .0002 + 0 = .0410$. Therefore, our decision rule for this experiment is to reject H_0 if nine or more responses are correct. We call the part of the distribution for which $Y \geq 9$ the *rejection region* because we will reject H_0 if the test statistic, Y, falls in this region; the value of Y that demarcates the rejection region is called the *critical value* of the test statistic. Given the result of $Y = 7$ in our sample, we fail to reject H_0 because this value of Y does not fall in the rejection region. In this example, the rejection region consists of the uppermost .041 of the distribution, so, in effect, we have specified that $\alpha = .041$. Because Y is a discrete variable, we cannot find a value of Y that cuts off exactly .05 of the distribution.

Several important points should be noted about the logic of hypothesis testing. First, a statistically significant result means that a value of the test statistic has occurred that is very unlikely if H_0 is true. However, "unlikely" is not the same as "impossible," and we may sometimes reject a null hypothesis even when it is true. Such incorrect rejections of the null hypothesis are called *Type 1 errors*; these are false positives. The significance level α is the probability of such errors; that is, $\alpha = p(\text{reject } H_0 \mid H_0 \text{ true})$, the conditional probability of rejecting H_0 given that H_0 is true. A useful way to conceptualize this is that if the responses were generated by random guessing (H_0 true), and if we were to replicate the experiment many times, we can expect to obtain $Y \geq 9$ in .041 of these experiments. If we obtain Y greater than or equal to 9 when H_0 is true, we will incorrectly reject H_0. By setting α at a particular level, we specify the level of risk of a Type 1 error that we are willing to tolerate.

The *p*-value is the probability of the observed data given that the null hypothesis is true, or $p(\text{observed data} \mid H_0 \text{ true})$. Two types of errors are often made in interpreting *p*-values. First, it is common to confuse this conditional probability with its opposite, $p(H_0 \text{ true} \mid \text{observed data})$, the probability that the null hypothesis is true given the results of the experiment. This is not correct. It would be nice if a significance test gave us the probability that H_0 was true, but it does not.

The second error that researchers often make is to interpret *p*-values as measures of effect size. For instance, if a test of an effect produces a smaller *p*-value in Experiment A than in Experiment B, it is sometimes asserted that the effect is larger or more important in Experiment A. This is also not correct: *p*-values convey useful information, but they depend on sample size and variability as well as effect size, so direct comparisons of *p*-values are rarely useful. This is one reason why we will present ways to find confidence intervals and estimate effect sizes in the following chapters. Indeed, we will have a great deal to say about both.

4.4.2 One- and Two-Tailed Tests

The test we just discussed is an example of a *one-tailed* or *directional* test. Because we were interested in whether performance was *better* than chance, the rejection region consisted of only the largest possible values of Y (the upper tail of the binomial distribution). Not all hypothesis tests are directional. There are many situations in which a departure from the null hypothesis in either direction would be of interest. For example, in the case of Royer's data on arithmetic skills (see Chapter 2), we might wish to know whether there is a significant difference in performance for addition and subtraction; if there were, it might influence the way in which these skills were taught. To test for a difference, we could assign a plus to each student who had a higher addition than subtraction score and a minus to each student who had a higher subtraction score. Then we could ask whether the probability of a plus (or a minus) was significantly *different from* .5. A result in either direction (i.e., addition better than subtraction or subtraction better than addition) would be of interest. As another example, University of Massachusetts Medical School researchers collected data on seasonal variation in clinical states such as depression and anxiety. Comparing depression scores in winter and summer, we might assign a plus if the winter score was higher, and a minus if it was lower. In both of these examples, H_0 would be

$$H_0 : \pi = .5$$

The alternative hypothesis would be

$$H_1 : \pi \neq .5$$

In this case, H_1 is nondirectional; that is, we will reject H_0 if we find strong evidence that π is greater than *or* is less than .5. We may want α to be .05, but now the rejection region is split in two – half in the upper tail and half in the lower tail of the distribution. Suppose n is again 20. Turning to the column labeled $\pi = .50$ in Table 4.2, assuming equal weight is given to both directions, and that α is to be close to .05, H_0 would be rejected if $Y \leq 5$ *or* $Y \geq 15$. This is usually referred to as a *two-tailed* or *nondirectional* test. Note that if we use these rejection regions, the actual probability of a Type 1 error is .021 + .021 = .042. If we found $Y = 16$, we would reject H_0 because the statistic falls in the rejection region. What is the *p*-value? Using Table 4.2, we find $p(Y \geq 16 \mid \pi = .50) = .0046 + .0011 + .0002 + 0 = .0059$. But because H_1 is nondirectional, we must also consider equally extreme results in the lower tail of the distribution, so that the *p*-value is equal to $p(Y \geq 16 \mid \pi = .50) + p(Y \leq 4 \mid \pi = .50) = 2 \times .0059 = .0118$. The procedure we have just outlined (using the binomial distribution to test the hypothesis H_0: $\pi = .50$) is called the *sign test*.

4.4.3 *Using Software to Identify Rejection Regions*

We can use software to identify the rejection region or regions under the null hypothesis. Doing so has several advantages, including the speed and accuracy of the calculations, as well as a much larger range of possible parameters (Y, n, π) than are summarized in Table C.1. In R, the *qbinom* function in the {stats} package computes critical values of the binomial distribution. The input parameters are the *size* of the sample (n), the *prob*ability of success (π), and a value related to α. By default, the function returns critical values for the lower tail of the distribution. We consider three cases:

1. Directional alternative hypothesis with lower tail critical value (e.g., H_1: $\pi < .5$). In this case, *qbinom* provides the critical value directly. For example, if n = 8 and $\alpha = .05$, *qbinom*(.05, size = 8, prob = .5) returns the value 2. This means we can reject H_0 when $Y < 2$ or, equivalently, when $Y \leq 1$.
2. Directional alternative hypothesis with upper tail critical value (e.g., H_1: $\pi > .5$). Here, we can use the option *lower.tail* = FALSE to computer upper-tail probabilities: *qbinom*(.05, size = 8, prob = .5, lower.tail = FALSE) = 6, meaning we can reject the null when $Y > 6$ or, equivalently, when $Y \geq 7$.
3. Two-tailed alternative hypothesis (e.g., H_1: $\pi \neq .5$). In this case, we divide α by 2, placing half of the probability in each tail. Then we apply *qbinom* twice, once for the lower tail and once for the upper tail rejection region, following the same guidance as for the directional tests. For example, with $n = 8$ and $\alpha = .05$, we use $\alpha/2$ to identify the lower critical value: *qbinom*(.025, size = 8, prob = .5) = 1, meaning we can reject when $Y < 1$ or, equivalently, when $Y = 0$. For the upper critical value, *qbinom*(.025, size = 8, prob = .5, lower.tail = FALSE) = 7, meaning that we can reject the null when $Y > 7$ or, equivalently, when $Y = 8$.

4.5 The Power of a Statistical Test

4.5.1 *Errors in Decision Making*

In deciding to reject or not to reject a null hypothesis, we can make two types of errors. If the null hypothesis is true, rejecting it is a *false positive* that we call a *Type 1 error*. As we have just discussed, the probability of a Type 1 error is α, the significance level. This means that the Type 1 error rate is under the direct control of the researcher. If the null hypothesis is false, we can make a different kind of error by failing to reject it; this is a *false negative*. Failure to reject a null hypothesis when it is false is called a *Type 2 error*, and its probability, p(fail to reject H_0 | H_0 is false), is referred to as β (Greek letter beta). The complementary conditional probability, that is, p(reject H_0 | H_0 is false), is called the *power* of the test. When H_0 is false, only two decisions are possible: fail to reject it (with probability β) or reject it (with probability $1 - \beta$). Therefore, power = $1 - \beta$.

The following table may help to clarify the meanings of α, β, and power:

Truth in the Population	*Decision*	
	Reject H_0	*Fail to reject H_0*
H_0 True	α = Type 1 error rate	$1 - \alpha$
H_0 False	power = $1 - \beta$	β = Type 2 error rate

The rows represent two mutually exclusive events: H_0 is either true or false. Given either of these events, the researcher may make one of two mutually exclusive decisions: reject or fail to reject H_0. The cell probabilities are conditional probabilities representing the probability of the decision given the event. Because one of the two decisions must be made, the probabilities in each row sum to 1.

4.5.2 *Computing Power*

Power is a conditional probability: It is the probability of rejecting the null hypothesis given that the null hypothesis is wrong. Thus, the first step in computing power is to determine the conditions under which the null hypothesis will be rejected (see Box 4.2). Once the rejection region for H_0 is found, we must compute the probability of observing a result in the rejection region assuming that H_0 is wrong and that some specific alternative hypothesis, H_A, is correct. It is important to be clear about the distinction between H_1 and H_A. H_1 is the class of alternative hypotheses that determines where the rejection region is placed (right or left tail, or in both tails). In contrast, H_A is a *specific alternative hypothesis*; power is calculated for the test of H_0 assuming H_A to be true. Of course, the true value of the population parameter is never known. However, if some value of the parameter is assumed, the power of the test against that alternative can be calculated.

Let's illustrate the calculation of power for our ESP experiment. Suppose we wish to test H_0: $\pi = .25$ against the alternative hypothesis H_1: $\pi > .25$. Further, let's assume that Rachel has rather weak ESP abilities; more specifically, we will assume that her abilities will permit her to correctly choose the identity of a card 35% of the time. In this example, this is our *specific alternative hypothesis*,

$$H_A : \pi = .35$$

The H_0 and H_A specify different sampling distributions for our experiment, as illustrated in Figure 4.4. Our example assumes that H_A correctly describes the probabilities of different possible outcomes for the experiment, but the extensive overlap between the two distributions means that the two hypotheses will not be easy to discriminate empirically. Thus, we can expect that power will be low.

The general steps in computing power are the same for all statistical tests. These steps are summarized in Box 4.3. Applying this procedure to our example:

1. Calculate the theoretical sampling distribution of Y assuming H_0 to be true. In this example, the distribution is presented in the .25 column of Table 4.2.
2. Determine the rejection region. In this example ($n = 20$, $\alpha = .05$, H_1: $\pi > .25$), the rejection region is $Y \geq 9$. As you can see in Figure 4.4, because the outcomes are discrete, we chose the largest possible α that does not exceed 0.05; in this case, $\alpha = 0.0409$.
3. Calculate the probability distribution of Y assuming H_A is true. In this example, π in Equation 4.3 is replaced by .35. The results are presented in the $\pi = .35$ column of Table 4.2.
4. Sum the probabilities for $Y \geq 9$ (the rejection region) in the .35 column. This sum is the conditional probability $p(Y \geq 9 \mid H_A = .35)$ and is the power of the test of H_0 against H_A. In this case, power = .1158 + .0686 + .0336 + .0136 + .0045 + .0012 + .0003 = .2376.

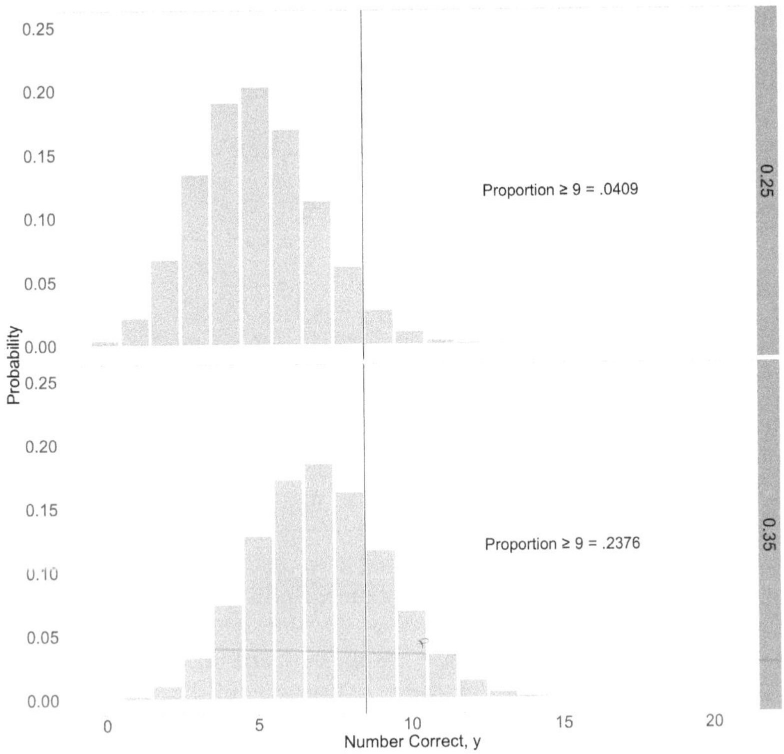

Figure 4.4 Binomial distributions for Y for $\pi = .25$ and $\pi = .35$ with $n = 20$, and the areas under each distribution where $Y \geq 9$.

Box 4.3 Steps in Computing the Power of a Test

1. Determine the theoretical sampling distribution of Y assuming H_0 to be true.
2. Determine the rejection region.
3. Assume that the null hypothesis is incorrect and that some specific alternative hypothesis, H_A, is correct.
4. Compute the probability of a result in the rejection region using the sampling distribution specified by the alternative hypothesis. The resulting value is the conditional probability of observing an outcome in the rejection region given that H_A is true. This is the *power* of the test.

These calculations mean that a test of H_0: $\pi = .25$ has .238 probability of yielding a significant result if π is actually .35. If π is actually .35, the experimental data will lead to an erroneous conclusion with probability $\beta = 1 - .238$, or .762. As anticipated, our experiment does not have a very good chance of discriminating H_0 from the assumed true state of the world, H_A.

Power is one of the most difficult concepts for students to grasp – probably because so many different things must be kept in mind. We need to consider two sampling distributions (those specified by H_0 and H_A), as well as the information provided by H_1 and α. To find the rejection region for the test, we start with the sampling distribution that assumes H_0 is true. We then need α to tell us how big the rejection region is, and we need H_1 to tell us which tail to consider, or whether both tails must be evaluated. Power is just the proportion of the H_A sampling distribution that falls in the rejection region.

4.5.3 *Factors Affecting Power*

We have just seen that many factors are relevant to the calculation of the power of an experiment: The sampling distributions specified by H_0 and H_A as well as the directionality of the alternative hypothesis specified by H_1. Some of those factors are affected by other experimental details. For example, the sampling distributions of H_0 and of H_A are affected by the sample size, *n*. As we saw in Figure 4.2, as the sample size increases, the observed sample proportion, Y/n, is closer to the true value, π, and extreme sample proportions are less likely. Less variability in the sampling distributions mean that H_0 and H_A overlap by a smaller amount, making them easier to distinguish. In this section, we show how effect size, sample size, α-level, and directionality of H_1 influence power.

Let's consider the null hypothesis that $\pi = .5$. With a directional alternative, H_1: $\pi < .5$, and a sample size of 15, we can use the methods of Section 4.5.2 to calculate the power of our study. We need to define the rejection region for a particular alpha-level, say $\alpha = .06$. Using Table C.1, $P(Y \le 4 \mid n = 15, \pi = .5) = .0000 + .0005 + .0032 + .0139 + .0417 = .0593$, so we reject the null hypothesis when $Y \le 4$. Next, we need a specific alternative hypothesis, say H_A: $\pi = .25$. Then, we calculate $P(Y \le 4 \mid n = 15, \pi = .25)$, the part of the sampling distribution for H_A that falls in the rejection region. Turning to Table C.1, we calculate $P(Y \le 4 \mid n = 15, \pi = .25) = .0134 + .0668 + .1559 + .2252 + .2252 = .6865$, which is the power for this set of assumptions. We can plot that value on a graph with possible values of H_A on the x-axis and power on the y-axis. In the upper panel of Figure 4.5, that point – (.25, .6865) – is marked with a dark dot. If we do the same calculations for other possible specific alternative hypotheses, assuming that H_A: $\pi = .05$ or .10 or .85, for example, we will find the power for those cases. In fact, those calculated power values, as well as others, are shown as the solid curve in the upper panel of Figure 4.5; this curve is called a *power curve*. This power curve shows that power is higher when our H_A is further from H_0 in the direction of H_1 (i.e., small values of π) and it drops to 0 for possible effect sizes, or assumptions about H_A, that are in the opposite direction from H_1 (i.e., large values of π).

The other curves in the upper panel of Figure 4.5 show the power of a one-tailed test under different assumptions about sample size and alpha-level. These power functions have been plotted for four different conditions: (1) $n = 20$, $\alpha = .20$; (2) $n = 20$, $\alpha = .06$; (3) $n = 15$, $\alpha = .06$; and (4) $n = 15$, $\alpha = .20$. The lower panel of Figure 4.5 shows power curves for the same four sets of assumptions about sample size and alpha-level, this time assuming a two-tailed test. To generate these curves, alpha was divided equally between the two tails, and two rejection regions were defined, one for each tail. The part of the sampling distribution of H_A that falls in each rejection region was then summed; this is the power for a two-tailed test for that specific combination of assumptions about α, H_A, and sample size.

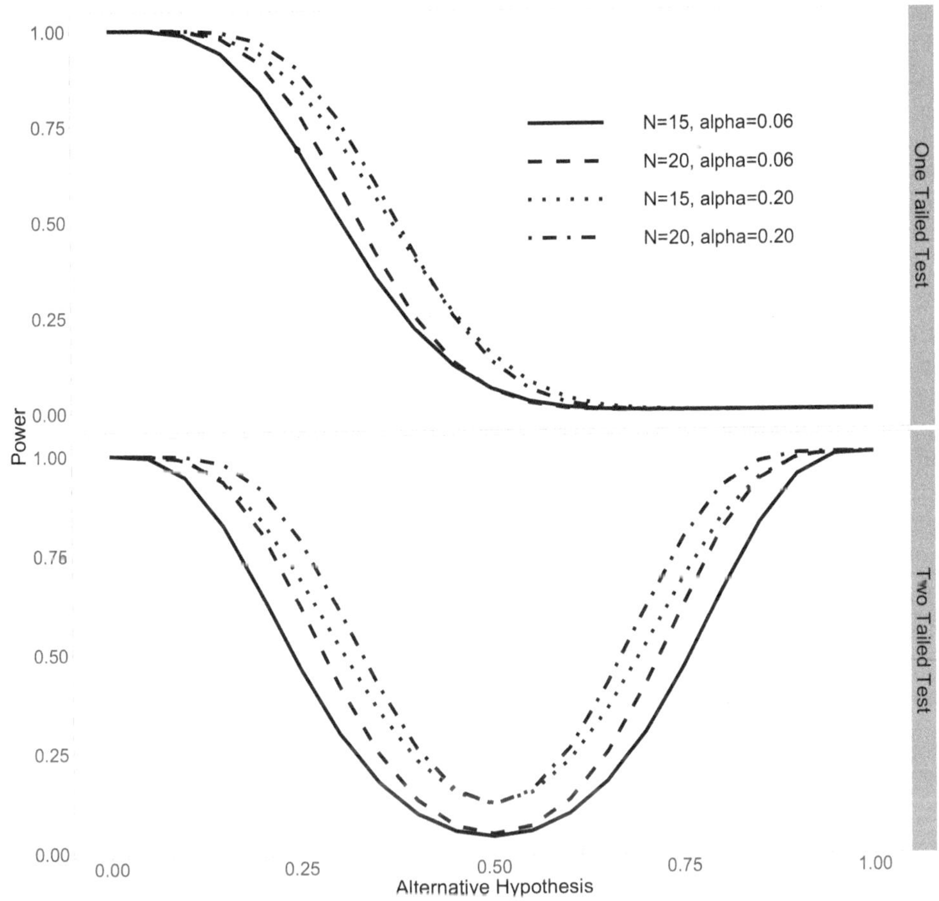

Figure 4.5 Power functions based on the binomial distribution.

Four points should be noted that are typical of power functions for all statistical tests:

1. *Power increases as the effect size increases.* The larger the difference between the sampling distribution of the null hypothesis and that of the specific alternative hypothesis – that is, the less those distributions overlap – the more likely it is that Y will fall in the rejection region. This is illustrated in both panels of Figure 4.5 by the fact that the curves rise as the value of the specific alternative hypothesis diverge from 0.5, the null hypothesis value of π. In the upper panel of Figure 4.5, the same consequence of effect size is observed exclusively for smaller values of the specific alternative because the rejection region is defined by H_1: ($\pi < .5$).
2. *Power is affected by whether H_1 is directional.* Compare the curves in the upper panel of Figure 4.5 to the left-hand side of the corresponding curves in the lower panel. Notice that the curves in the upper panel begin to rise sooner, and rise faster, than the curves in the lower panel. This illustrates that a one-tailed test has more power than the corresponding

two-tailed test whenever the specific alternative is less than .5. This is because the rejection region is concentrated in that tail of the distribution for a one-tailed test, whereas it is divided in half and distributed over both tails for a two-tailed test. On the other hand, the one-tailed test has virtually no power against specific alternatives of the form $\pi > .5$, whereas the two-tailed test does have power to reject H_0 against these alternatives. In short, a one-tailed test has more power than a two-tailed test if – and only if – the direction of the hypothesis is correct.

3. *Power increases as α increases*. This is because an increase in α means the size of the rejection region becomes larger. For example, when $n = 20$ and the alternative is one-tailed, the rejection region increases from $Y \leq 6$ to $Y \leq 7$ as α increases from .06 to .20. Because power is also calculated for this larger set of *Y* values, it too becomes larger. The relationship between power and α is evident in both panels of Figure 4.5: The power curves for $\alpha = .20$ fall above those for $\alpha = .06$.
4. *Power increases as* n *increases*. Comparing the two curves for $\alpha = .06$ (or for $\alpha = .20$) in either panel, you can see that the $n = 20$ rise more quickly than the $n = 15$. This demonstrates the effect of sample size on power and it follows from our discussion of the fact that Y/n is a consistent estimator of π. Recall that the property of consistency (see Section 4.3.3 and Figure 4.2) means that as *n* increases, Y/n is more likely to be close to the true value of the parameter. Therefore, if H_0 is false, it is more likely that the data will demonstrate that as *n* increases.

4.5.4 The Importance of A Priori Power Calculations

Power is important for researchers to understand and calculate *a priori* (in advance of collecting the data). Doing so allows an informed choice among possible statistical tests of the hypothesis and of various experimental design options: All else being equal, select the design and the test that afford the greatest power for a particular sample size. Additionally, a researcher who calculates *a priori* power can assess the consequence of potential violations of the assumptions of the statistical test. Finally, and most importantly, the relationships among power, effect size, and sample size can be used to decide how much data should be collected. For example, suppose we want power of at least .80 to reject H_0 with a one-tailed test if π is at least .6. We can derive power functions for various values of *n* like those depicted in Figure 4.5. The *n* we want for our study is the one that gives rise to a power function in which power (the y-axis value) is at least .80 when the specific alternative hypothesis value (π) is .6.

The last point deserves further comment. The null hypothesis is almost always false. If we collect enough data, we are likely to obtain statistically significant results. Whether the results will be of practical importance or theoretical significance is another matter. The effect may be trivially small or in a direction that makes no sense in terms of any theory or practical concern. Therefore, it makes good sense before we collect data to ask the following questions: *What is the smallest size effect that would be of interest* and *what power do we want to detect (i.e., find significance for) such an effect?* The answer to these two questions will be major factors in determining the sample size for our research. Sometimes the required *n* will be impractically large and we will face the challenge of redesigning the research so that a smaller sample will achieve the desired power against the specific alternative hypothesis we had in mind.

Finally, low powered studies should be avoided. The obvious reason is that low powered designs have a low probability of detecting a real effect of interest; that's just a restatement of the definition of power. A less obvious, and more insidious, reason to avoid low powered designs is that, when they do conclude that the null hypothesis can be rejected, they typically will dramatically overestimate the true effect size. This overestimation occurs because, in underpowered designs, only large effect sizes allow the rejection of the null hypothesis, and as we saw in Figure 4.2, smaller samples are more likely to produce sample statistics that are far from the true value. Said differently, when the sample size is small, sampling error is large and only very large effects in the sample provide sufficient evidence to conclude there is an effect in the population. As the sample size increases, sampling error decreases and smaller effect sizes in the sample will be sufficient to demonstrate an effect in the population.

4.5.5 Post Hoc Power Calculations Are Not Useful

In contrast to *a priori* power calculations, which should always be run, *post hoc* power calculations (run after the data have been collected) do not provide useful information and should be avoided. The desire to calculate *post hoc* power tends to occur in two different situations, both involving a failure to reject the null hypothesis. In the first case, the experimenter hopes to claim that their observed effect is "real" and was "only" nonsignificant because the experiment had low power. This argument is often advanced when the effect is in the expected direction, the observed p-value is close to α, and no definitive replication data are available. In the second type of situation, the experimenter hopes to claim that the null hypothesis is true because the experiment had high power to detect a real effect of some reasonable size, and yet it failed to do so. These are both flawed arguments.

Researchers hoping to claim their nonsignificant result is "real" calculate post hoc power in their experiment using the observed effect size as the best available estimate of the population effect size. This approach appears reasonable, yet it creates a circular logic: Effect sizes large enough to yield high power would have resulted in a significant test, and effect sizes that do not produce significant tests will also yield relatively low power estimates. In fact, post hoc power, the observed effect size, and the p-value are all interrelated: A low p-value is associated with high post hoc power, and vice versa (Hoenig & Heisey, 2001). In short, computing post hoc power using the observed effect size does not provide any information that is independent of the p-value.

Researchers hoping to declare the null hypothesis to be true are making a logical error. The logic of a hypothesis test is that the null hypothesis can be used to generate a sampling distribution that assigns probabilities to outcomes; we reject the null when we observe an outcome that is very unlikely ($p \leq \alpha$), given that sampling distribution. In other words, testing a hypothesis provides information about the probability of the data, given the null hypothesis, $p(\text{observed data} \mid H_0 \text{ true})$. In contrast, the experimenter is trying to make a claim about a different value, $p(H_0 \text{ true} \mid \text{observed data})$. This logical error is akin to being told that "if it rains, the ground is wet," observing wet grass and concluding that it must have rained, thus failing to consider alternative causes such as sprinklers. Researchers interested in learning about alternatives to the logic of hypothesis testing should consider courses on Bayesian statistics, which are beyond the scope of this text.

4.5.6 *Using Software to Compute Power*

In Box 4.3, we showed the steps involved in calculating power in a binomial setting. Those steps included determining the rejection region of the sampling distribution assuming H_0 is true and computing the probability of observing an outcome in the rejection region within a specific alternative hypothesis, H_A. The tools presented in Sections 4.3.4 and 4.4.3 allow us to use distributional calculators for these steps in computing power.

We can use R to identify the rejection region under the null hypothesis, as described in Section 4.4.3. For example, assume that H_0: $\pi = .25$, H_1: $\pi > .25$, H_A: $\pi = .35$, and $n = 20$. We start by using the *qbinom* function to identify the one-tailed rejection region of the sampling distribution assuming the null hypothesis is true: The *qbinom*(.95, size = 20, prob = .25) function returns the value of Y, 8, above which only 5% of the sampling distribution falls when $n = 20$ and $\pi = .35$. Using that value of Y and the specific H_A: $\pi = .35$, the tools described in Section 4.3.4 allow us to compute the probability of observing that many successes (or more) assuming the specific alternative hypothesis is true: *pbinom*(x = 8, size = 20, prob = .35, lower.tail = FALSE) = .2376. This means that our power for a sample of size 20, $p(y \geq 9 \mid \pi = .35, n = 20)$, is equal to .2376. Of course, this equals the power we computed in Section 4.5.2 using Table C.1.

Another helpful tool designed specifically for computing power is a freely available, convenient, and flexible program called G*Power 3.1 (Faul et al., 2007).[4] To calculate post hoc power for a sample of 20, with $y \geq 9$ and H_A: $\pi = .35$, open G*Power 3.1 and select *Exact* as the *Test family*, then select *Proportions: Difference from Constant (one sample case)* as the *Statistical test*. Then, for *Type of power analysis*, select *Post hoc: Compute achieved power – given α, sample size, and effect size*. Here, the constant proportion is .25; the effect size, *g*, is the difference between H_A and H_0, or .1 = .35 – .25; and $\alpha = .05$ with a one-tailed test. Insert those values and then click the *Calculate* button in the lower right of the dialog box. The results are displayed in Figure 4.6. We see that power = 0.2376, the same as we calculated previously. Also notice that the critical value of Y is shown as the "critical N = 9," defining the same rejection region we calculated previously.

The figure at the top of the results box in Figure 4.6 deserves some explanation: The distribution on the left is the binomial for $n = 20$ and $\pi = .25$ and has a mean of $n\pi = (20)(.25) = 5.0$. The distribution on the right is for $\pi = .35$ and has a mean of $(20)(.35) = 7.0$. The critical N of 9 informs us that if Rachel gets 9 or more correct in 20 trials, we can reject the null hypothesis at $\alpha = .05$.

A more important advantage of G*Power is that we can use it easily to calculate the sample size we need to achieve a certain minimum *a priori* power level. Let's see how many trials we would need Rachel to complete in the ESP experiment for power of .8. Within the *Exact* test family, select *Proportions: Difference from Constant (one sample case)* and the *A priori: Compute required sample size – given α, power, and effect size* power analysis. Enter the parameters as before: one-tailed test with effect size 0.1, constant proportion = .25, $\alpha = .05$, and desired power = .8, and click *Calculate*. Doing so reveals we need a total of 129 trials, a much larger study than we have conducted so far.

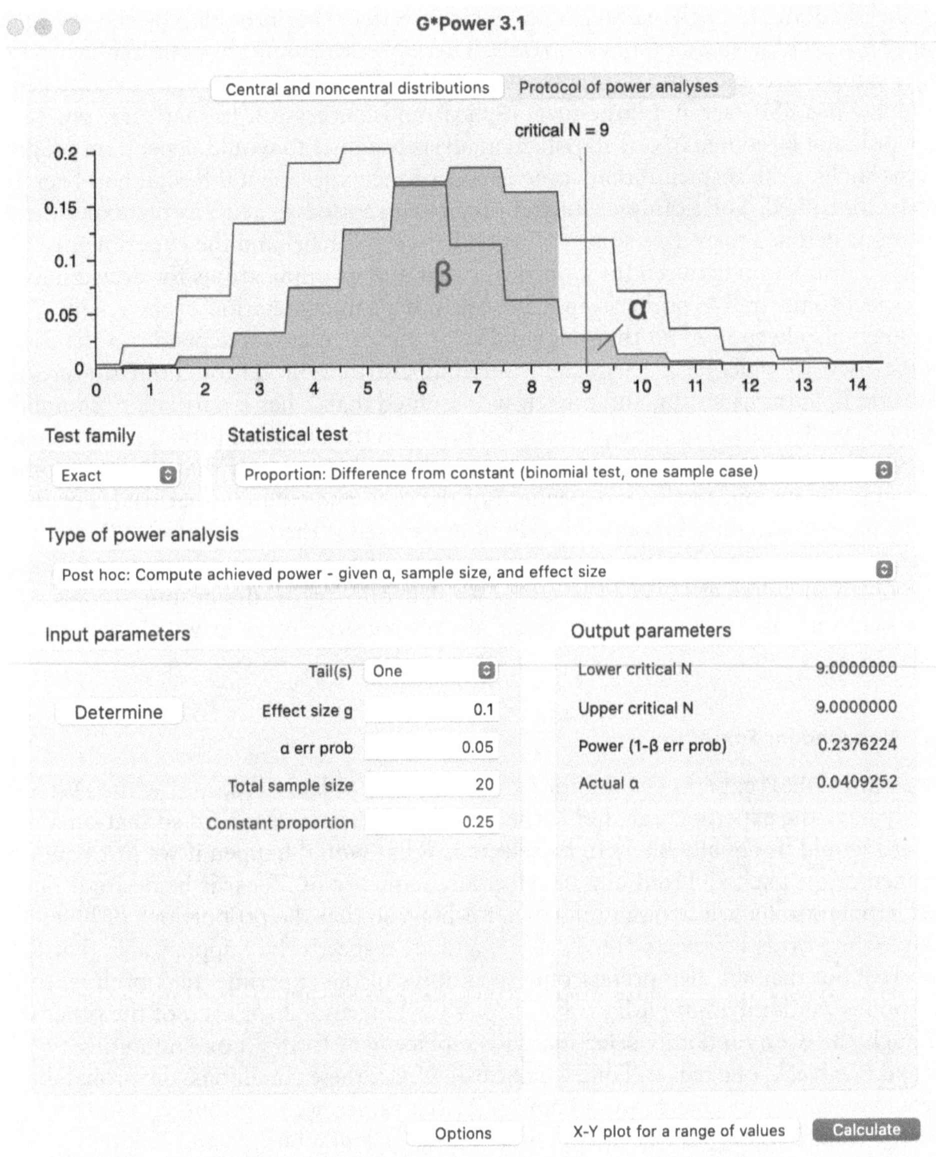

Figure 4.6 G*Power 3.1 output for post hoc power given H_0: π =.25, H_1 π >.5, H_A: π =.35, and n = 20.

4.6 When Assumptions Fail

Suppose the significance test on the ESP data allows us to reject the null hypothesis that H_0: $\pi = .25$ at $\alpha = .05$. Do we conclude that Rachel has ESP? Not necessarily. For one thing, we may have made a Type 1 error. In our example, a Type 1 error could occur because Rachel guessed an unusually high number of cards correctly, simply by chance. For a controversial effect such as ESP, we likely want stronger evidence than that provided by the $\alpha = .05$ significance level with a small number of trials. Therefore, we might insist on replication with a larger sample size and a much smaller value for α. Of course, we might remain skeptical that Rachel has ESP even if a large-scale replication is successful. In that case, our conclusions would not be rooted in our statistical analysis but instead would depend on pragmatic concerns such as the suspicion that some aspect of the experimental design had been inadequately controlled. For example, Rachel may have spotted a "tell" in the experimenter's behavior, or perhaps there was some collusion between Rachel and the experimenter. There is always a distinction between the empirical result and its implications for decision making with respect to the null hypothesis and the researcher's interpretation of the result.

We must also keep in mind that the validity of our calculations depends on the assumptions we used to generate a particular sampling distribution. Throughout the preceding sections on hypothesis testing and power, we assumed that Y has a particular binomial distribution. The derivation of the equation that provides the binomial probabilities (Equation 4.3) rests upon the assumption that π is constant across trials and that the probability of an outcome on any one trial is independent of the outcome on any other trial. For the ESP experiment, the particular binomial distribution we used is the one with $n = 20$ and $\pi = .25$ (for random guessing with 20 trials and four cards). If these assumptions do not hold, we cannot take our calculations at face value. We generally try to design our studies so that the assumptions are satisfied, but we must always consider ways in which they might be violated.

4.6.1 *Nonrandom Sampling*

Suppose in the ESP experiment the four cards are blue, black, red, and white. We would normally have the experimenter shuffle the pack before selecting a card so that on any trial each card would be equally likely to be selected. What would happen if we just handed the experimenter the pack and told him to select any sequence of 20 cards he desired? Suppose he just happens to have a strong preference for blue, so that the proportions of blue, black, red, and white cards he selects are .7, .1, .1, and .1, respectively. Suppose also that Rachel has no ESP but that she also prefers blue (or knows of the experimenter's preference). Say she responds randomly, but with probability .4 for blue and .2 for each of the other colors (as though she were randomly selecting with replacement from a box containing two blue cards and one black, one red, and one white card). Under these conditions, the probability that she would be correct by chance would not be .25 but rather $p(\text{C}) = p(\text{Blue}_{\text{guess}} \text{ and Blue}_{\text{selected}}) + p(\text{Black}_{\text{guess}} \text{ and Black}_{\text{selected}}) + p(\text{Red}_{\text{guess}} \text{ and Red}_{\text{selected}}) + p(\text{White}_{\text{guess}} \text{ and White}_{\text{selected}}) = (.4)(.7) + (.2)(.1) + (.2)(.1) + (.2)(.1) = .28 + .02 + .02 + .02 = .34$. This may seem like an extreme example; however, even more subtle deviations from the binomial assumptions can undermine our calculations.

It is important to distinguish between general knowledge about preferences and psychic ability. For example, given k options and no obvious reason to choose one option over the others, you might think that people would choose each option with probability $1/k$ – but they usually do not. Conduct your own "experiment": If you tell a group of people that you are thinking of a number between 1 and 10 and that you want them to guess that number and write it down, the integers 1–10 will not be chosen equally often. Usually, the extremes 1 and 10 will be avoided and more people will guess "7" than any other number. People do have preferences, and the design of ESP research must take this into account. If the cards were well shuffled before each selection, so that the experimenter could not select blue cards over the others, there would be no preference for Rachel to capitalize on. We should also point out that although here we have been talking about the ESP example, response preferences must be considered in other kinds of research as well.

4.6.2 *Violation of the Independence Assumption*

The assumption of independence can also be problematic for many kinds of social science research. Different measures or repeated observations taken from the same participant will usually be correlated. Also, whenever responses are obtained from members of the same discussion group, school class, or litter of animals, the responses obtained are likely to be correlated. Social, environmental, and biological factors will tend to affect the members of such units in a similar way. Violations of the independence assumption will frequently result in a Type 1 error rate very different from the error rate assumed by the experimenter. Some assumptions can be violated with minor consequences, but the independence assumption is often quite critical. Moreover, this assumption plays some role in all statistical test procedures. Here, the binomial test will be used to illustrate the consequences of violating the independence assumption, but the implications are much more general.

Consider a study in which 10 pairs of participants discuss a topic. After the discussion, each of the 20 individual participants casts a "yes" or "no" vote on the issue under consideration. Suppose that previous research has established that votes are evenly divided between the two positions when there is no discussion. However, theoretical principles lead the researcher to believe that "yes" votes will be more frequent than "no" votes following discussion. Thus, the null hypothesis is H_0: $\pi(\text{yes}) = .5$ and the alternative hypothesis is H_1: $\pi(\text{yes}) > .5$. If the significance level α is set equal to .06, the binomial table indicates that H_0 should be rejected if the observed number of "yes" responses is 14 or more.

There is a problem with this procedure: The two individuals in each discussion pair may have influenced each other and their responses may not be independent. To clarify the consequences of this dependency for the binomial test, let's consider an extreme example. Suppose the joint probabilities of votes for the two members of each pair (M_1 and M_2) were the following:

		M_1		
		Yes	No	
M_2	Yes	.50	0	.50
	No	0	.50	.50
		.50	.50	

In this case, the dependence within pairs is complete: The conditional probability of a "yes" vote is 1 when the partner votes "yes" and 0 when the partner votes "no." Note, however, that the null hypothesis is true; $\pi(\text{yes}) = .5$.

Recall that the researcher had sampled 10 pairs from this population and, based on the binomial distribution table, had decided to reject H_0 if there were 14 or more "yes" votes from the 20 individuals. Unknown to the researcher, the two members of each pair vote the same way. Therefore, the probability of 14 or more "yes" votes is really the probability that

7, 8, 9, or 10 pairs vote "yes." There are only 10 independent events; they are the pair (not the individual) votes. If this violation of the independence assumption occurs, the probability of a Type 1 error is not the .058 computed by the researcher; rather it is the probability that $Y = 7, 8, 9$, or 10 when $n = 10$ and $p(\text{yes}) = .5$. Using Table C.1, that probability can be shown to be .172. This means that the actual Type 1 error rate is about three times the size of the rate assumed by the researcher. Most researchers would feel that a Type 1 error rate of .172 was unacceptably high.

Although complete dependence between the members of the pair is improbable in a real experiment, some dependence is often likely. Consequently, the distortion in Type 1 error rate will be less than in our example, but there will be distortion. Frequently, the true error rate will be intolerably high. The opposite result occurs when responses are negatively related. For example, suppose that the null hypothesis is false, but in a high proportion of pairs, the partners agree to split their votes. In cases such as this, power will be greatly reduced. Thus, depending on the nature of the dependency, either Type 1 or Type 2 error rates will be increased. Positive dependencies are far more likely and, therefore, the greatest danger is an increased rate of rejection of true null hypotheses.

Independence is only one assumption that plays a role in many statistical tests. In general, the consequences of failures of assumptions are not simple and have to be thought through in each research situation. Many factors affect error rates. We have already discussed the effects of both the magnitude and direction of failures of the independence assumption. A third factor is the particular assumption that is violated. Some assumptions, despite being used in the derivation of the test statistic, are less critical, so their violation has little effect on error rates. A fourth factor is sample size; certain (but not all) assumptions are less critical when there are many observations. Appendix 4.2 provides an example of the interaction of assumptions and sample size.

In summary, every inferential procedure involves a statistical distribution and the derivation of that distribution rests on certain assumptions. If the assumptions are met, then the inferential logic we've developed will hold true. If the assumptions are not met, then an apparently significant test statistic may reflect that violation rather than a real effect that justifies rejecting the null hypothesis. The consequences of violating these assumptions will vary depending on the factors noted earlier. Throughout this book, we will emphasize the statistical model underlying each inferential procedure, detailing the conditions that cause assumptions to be violated, the results of such violations, and alternative analyses that remedy the situation when the violations are severe enough to make the proposed analysis untrustworthy.

4.7 Summary

Our goal in Chapter 4 was to introduce the basic concepts of statistical inference in the context of a relatively simple probability distribution, the binomial.

- The conceptual framework for statistical inference is based on the idea of taking a *sample* from a *population* and using the information in the sample to make certain claims about the population. The process of inferring population characteristics from sample characteristics involves sampling error and is, therefore, probabilistic in nature.

- We applied our knowledge of probability to derive the probability of each possible outcome of an ESP experiment according to the hypothesis that the subject was guessing on each trial. The resulting hypothetical probability distribution, a *sampling distribution*, is fundamental to the process of statistical inference because it provides the probabilities upon which we base our statistical decision making.
- The nature of the sampling distribution that is relevant to a given experiment depends on certain assumptions we make about the population. For our ESP experiment and related types of experiments, the *binomial distribution* is the appropriate sampling distribution. We therefore used the binomial distribution to illustrate the role of sampling distributions for an important procedure in inferential statistics, *hypothesis testing*.
- We developed the logic of hypothesis testing and identified the two types of errors we may commit when we conduct a significance test: We may reject H_0 when it is true (*Type 1 error*) or fail to reject H_0 when it is false (*Type 2 error*).
- We also developed an important topic in hypothesis testing, *power*. Statistical power is the probability of rejecting H_0 when it is false. Given H_0, H_1, and α, we can calculate the power to test H_0 against a specific alternative hypothesis, H_A. Power is a particularly important consideration during the planning stage of research. Power considerations inform the choices of experimental design and statistical test, and should be used to determine the sample size.
- In each section, the computed examples relied on Table C.1, which lists $P(Y = y \mid n, \pi)$ for select values of n and π. In practice, it is faster and more accurate to use software for these calculations, which also has the advantage of allowing a wide range of input parameters. For these reasons, we showed how to use R to calculate binomial probabilities, identify rejection regions, and compute power. We also used G*Power 3.1 to compute power and choose a sample size.
- Finally, we discussed the fact that any statistical procedure is based on certain assumptions associated with the derivation of the sampling distribution on which the procedure is based. If these assumptions are not valid, any decisions based on our calculations may also not be valid.

Appendix 4.1

Understanding the Combinatorial Formula

Consider five individuals who are running for positions on the city council; the two top vote getters will be elected. First consider all the possible assignments of individuals to ranks where the ranks are the position in the final vote. There are five possibilities for the first position in the vote count, four possibilities for the second position (e.g., A could be followed by B, C, D, or E). The total number of sequences is (5)(4)(3)(2)(1) or 5!. In general there are $n!$ sequences of n objects.

Suppose the question is: How many outcomes can this election have? Here, by "outcome" we mean patterns of election and non-election. For example, A and B might be elected and C, D, and E fail to be elected. Notice that the order of finish within each of the two classes (elected and non-elected) is irrelevant. The following sequences are all equivalent in that they constitute the same outcome: A and B elected, and C, D, and E not elected:

A,B/C,D,E	B,A/C,D,E
A,B/C,E,D	B,A/C,E,D
A,B/D,C,E	B,A/D,C,E
A,B/D,E,C	B,A/D,E,C
A,B/E,C,D	B,A/E,C,D
A,B/E,D,C	B,A/E,D,C

Note that the two (2!) possible sequences of A and B, paired with the six (3!) possible combinations of C, D, and E, correspond to one *combination* (A and B elected; C, D, and E not elected). In general, $r!(n - r)!$ sequences will correspond to a single combination when n items are split into one class with r items and one with $n - r$ items. Therefore, the number of combinations is $n!/r!(n - r)!$. In our example, the number of ways the election can turn out is

$$\binom{5}{3} = \binom{5}{2} = \frac{5!}{2!3!} = \frac{120}{(2)(6)} = 10$$

In general, the number of different ways of selecting r items from n items is

$$\binom{n}{r} = \frac{n!}{r!(n-r)!}$$

Read this as "n choose r."

Appendix 4.2

Sample Size and Violations of the Independence Assumption

Although violations of assumptions can often lead to erroneous inferences, the consequences can sometimes (though not always) be minimized by using large samples. The violation of the independence assumption in calculating probabilities provides a nice illustration of this point. Consider an urn containing five red and five black balls. We draw a marble three times from the urn. If we assume that the marble is replaced and the urn is thoroughly shaken after each draw, so that we have independence, according to the multiplication rule, the probability of drawing three red marbles is $p(R_1 \text{ and } R_2 \text{ and } R_3) = p(R)p(R)p(R) = (5/10)^3 = .125$. However, suppose that the drawn marble is not replaced each time. This violates our assumption of independence. We can see this by the following analysis: If a red ball is drawn on trial 1, the probability of drawing a second red ball is now 4/9, whereas if a black ball is drawn on the first draw, the probability of a red on the second draw is 5/9. In fact, the probability of drawing a red (or black) ball on any trial depends on the sequence of preceding draws. So, although our assumption of independence leads us to conclude that $p(R_1 \text{ and } R_2 \text{ and } R_3) = .125$, the true probability is $p(R_1 \text{ and } R_2 \text{ and } R_3) = p(R_1)\, p(R_2 \mid R_1)\, p(R_3 \mid R_1 \text{ and } R_2) = (5/10)(4/9)(3/8) = .063$, roughly half the inferred probability.

Now suppose the urn consists of 50 red and 50 black balls. Our assumption of independence leads us to the same probability of drawing three red balls in a row, .125. This time, however, the true probability if we select without replacement is $p(R_1 \text{ and } R_2 \text{ and } R_3) = (50/100)(49/99)(48/98) = .121$, and the true and inferred probabilities are quite close; that is, the violation of the independence assumption did not lead to a very large error.

There are two implications of our examples that extend beyond simple probability calculations and violations of the independence assumption. First, violations may lead to wildly incorrect conclusions, as the urn with 10 marbles attests. Second, the consequences of violations of assumptions may be less damaging when the sample size is large. Neither of these statements will be true for every inferential procedure, but they are often true and therefore worth bearing in mind.

Exercises

4.1 [Interpreting α and power] Assume that in a particular research area 30% of the null hypotheses tested are true. Suppose a very large number of experiments are conducted, each with $\alpha = .05$ and power = .80.

a) What proportion of true null hypotheses will be rejected?
b) What proportion of false null hypotheses will not be rejected?
c) What proportion of nonrejected null hypotheses will actually be true?
d) What proportion of all null hypotheses will be rejected?

4.2 [Logic of hypothesis tests] For each of the following, state the null and alternative hypotheses:

a) The recovery rate for a disease is known to be 25%. A new drug is tried with a sample of people who have the disease to determine whether the probability of recovering is increased.
b) An experiment such as that described in Exercise 3.1 is conducted to assess the evidence for the existence of ESP.
c) In the ESP experiment of Exercise 3.1, a proponent of ESP (Claire Voyant?) claims that she will be successful on greater than 60% of the trials.

4.3 [Calculations with binomial distribution] Use the binomial table (Appendix Table C.1) to find the rejection region in each of the following cases (π is the population probability), and then use your answers to confirm your understanding of the *binom* family of distribution functions in R or in SPSS:

Case	H_0	H_1	*n*	*α*
(a)	$\pi = .25$	$\pi > .25$	20	.01
(b)	$\pi = .25$	$\pi > .25$	5	.01
(c)	$\pi = .25$	$\pi < .25$	20	.05
(d)	$\pi = .5$	$\pi \neq .5$	20	.01

4.4 [Interpreting *p*-values] In an experiment, data are collected such that when a hypothesis test is conducted, the null hypothesis is rejected with $p = .003$.

a) Can you conclude that H_0 is true with probability .003? Why or why not?
b) Can you conclude that H_1 is true with probability .997? Why or why not?

4.5 [Logic of hypothesis tests] In each of the following, (i) state the null and alternative hypotheses, (ii) state n, and (iii) state the appropriate rejection region, assuming $\alpha = .05$.

a) An important quality in clinical psychologists is empathy, the ability to perceive others as they perceive themselves. In a simplified version of one investigation of empathy, five first-year graduate students were asked to rate a target individual on a particular trait, as they believed the individual would rate herself. A four-point scale was used. The question of interest was whether the raters would do better than chance.

b) In a study of group problem solving, the investigator uses the solution rate for individuals in a previous study to predict that 40% of three-person groups will reach the correct solution. Fifteen groups are run in the study. The question of interest is whether the theory is correct.

4.6 [Sign test, sample size determination] Suppose a sign test is to be done with H_0: $\pi = .50$, H_1: $\pi < .50$, $n = 20$, $\alpha = .05$. Suppose further that the number of successes, y, is 7.

a) What is the rejection region?
b) What do we conclude?
c) What is the power of the test if π is actually .35?
d) If π is actually .35, how many cases do we need to have power = .90?
e) If π is actually .35, how many cases do we need to have power = .90 for a two-tailed test?

4.7 [Binomial hypothesis testing] The data set *EX4_7* contains data for 20 cases on two variables, Y_1 and Y_2. Each variable is dichotomous; Y_1 has seven values of 1 and thirteen of 0 whereas Y_2 has five values of 1 and fifteen of 0. Test the hypothesis H_0: $\pi = .5$ at $\alpha = .05$ for both Y_1 and Y_2. In R, use the *binom.test* function in the {stats} package. In SPSS, you would start by selecting *Nonparametric Tests* from the *Analyze* menu, then *Legacy Dialogs, Binomial.* Add both Y1 and Y2 to the *Test Variable List*, and insert .50 for the *Test Proportion*, and then click on *OK*.

4.8 [Sign test] Ten students take a course to improve reasoning skills. Before the course, they took a pretest designed to measure reasoning ability and after the course they took a posttest of equal difficulty. The results for the 10 students are as follows:

Student	*1*	*2*	*3*	*4*	*5*	*6*	*7*	*8*	*9*	*10*
Pretest Score	25	27	28	31	29	30	32	21	25	20
Posttest Score	28	29	33	36	32	34	31	18	32	25

The instructors of the course want to decide whether performance on the posttest is significantly better than performance on the pretest by looking at the signs of the difference scores, on the reasoning that if the course had no effect whatsoever, each student would be equally likely to get a + or –.

a) State H_0 and H_1.
b) Perform a sign test on these data (here, use $\alpha = .06$) and report your conclusion.

4.9 [Binomial test and power] A researcher studying memory performs an experiment that compares two strategies for remembering pairs of words. Twelve students are each given several sets of word pairs to learn. They learn half the sets by rote memorization and the other half by using imagery. The order of conditions is counterbalanced appropriately. It is found that nine students do better with the imagery strategy and three do better with rote memorization.

a) Using the binomial distribution, test the null hypothesis that both strategies are equally effective using $\alpha = .05$. Write down the appropriate null and alternative hypotheses and describe the steps you take in testing the null hypothesis. What is the result of the significance test?
b) What is the power of this test if the probability of doing better using the imagery strategy is actually .9 in the population (so that the probability of doing better using the rote strategy would be .1)?

4.10 [Power] Reconsider the study of empathy described in Exercise 4.5, part (a).

a) If the true probability of an empathetic response is .5, what is the power of the significance test in your answer to the earlier question?
b) How many subjects would be required to get power = .80 for the test of the null hypothesis at $\alpha = .05$?
c) What is meant by "true probability" here?

Notes

1 We will use π (the Greek letter pi) to represent the proportion of correct responses in the population from which the sample was selected and p to represent the proportion of correct responses in the sample. We typically use Greek letters to represent population parameters and Latin letters to represent sample statistics, to reduce confusion between them. Do not confuse p, the proportion in the sample, with $p(A)$, the probability of an outcome.

2 With four trials each with two possible outcomes, the tree diagram consists of $2^4 = 16$ branches. With 20 trials, there would be $2^{20} = 1{,}048{,}576$ branches.

3 Note that the trial outcomes generally need not come from the same individual. For example, each A or $\tilde{A}$ could represent a single success or failure by n different participants on a single trial. Then each pattern would represent a possible set of outcomes for the n subjects and π would be the proportion of correct responses in the population from which the sample of subjects was selected. From now on, we will use the more general term *combination* to refer to a pattern of A and $\tilde{A}$ responses.

4 Versions of G*Power 3.1 are available at https://www.psychologie.hhu.de/arbeitsgruppen/allgemeine-psychologie-und-arbeitspsychologie/gpower

Chapter 5

Further Development of the Foundations of Statistical Inference

5.1 Overview

In Chapter 2, we explored a subset of data in the study in which Royer and his colleagues collected accuracy scores and response times on simple arithmetic skills from students in first to eighth grades. Usually, this is a preliminary step in addressing questions about the data of a population of students represented by our sample. Among the many questions we might ask of the data we collect are as follows:

- What is the average addition score of the population of first-graders?
- What are reasonable bounds on our estimate of the population average?

The answers to questions such as the first one involve *point estimates*; we use sample data to estimate a point, a single parameter value, in the population of scores. The answer to questions such as the second one involves *interval estimates*; we use sample data to estimate an interval within which we believe the population parameter falls.

There are many other questions we could ask of these data. For example,

- Does the average score of the sampled population differ from some standard set by the government?
- Does the average score differ from that of some other population of first-graders, perhaps first-grade students taught by a different method?

These last two questions imply tests of the null hypothesis, tests that are based on the logic presented in Chapter 4. Most researchers are primarily concerned with these questions. However, point and interval estimates of population parameters should take precedence. For example, rather than immediately asking whether there is a statistically significant difference between two means, we should first ask what the size of the observed difference is and place some bounds on our estimate. Testing for significance is equivalent to asking whether there is an effect, but our understanding of the behavior in question will proceed more rapidly if we ask what the magnitude of the effect is, a question that encompasses the possibility that there is little or no effect.

Both estimates and significance tests are influenced by the variability in our data. The variability in the population distribution is reflected in the variability of the distribution of the sample data and in the distribution of sample statistics. The greater the population variability, the more a sample statistic, such as the sample mean or variance, will vary from

DOI: 10.4324/9781003453550-6

sample to sample; that is, the greater will be the variance of the *sampling distribution* of the statistic. As a result of sampling variability, no statistic from a single sample will exactly match the parameter it estimates, and inferences about the parameter may be in error. In this chapter, we consider the relation between the sampling distribution of a statistic and both estimation and hypothesis testing.

The major topics of this chapter are organized as follows:

- *The use of sample statistics to estimate population parameters.* Familiar statistics such as the sample mean and standard deviation are useful estimators because they provide important information about the "typical" score and variability of scores in a population. However, other sample statistics can be used to estimate the average score or the variability in a population, such as the median and the *IQR*. We will see that the choice among the estimators of a parameter rests on comparisons of characteristics of the sampling distributions of the sample statistics.
- *The sampling distribution of the sample mean* plays a critical role in many inferential procedures. This is because the central limit theorem assures us that the sampling distribution of the mean is approximately normal in shape under a wide range of circumstances. For this reason, we can use the normal distribution as a basis for probability computations underlying inferential procedures concerning means of populations.
- *The normal distribution may be used to construct interval estimates and hypothesis tests concerning population means.* The construction and interpretation of interval estimates is introduced in this chapter. In addition, we present the use of the normal distribution to test hypotheses about means and to do power calculations. The discussion of hypothesis testing and power calculations serves, in part, to emphasize that the logic of those procedures is the same regardless of the application. Relations among interval estimates, hypothesis tests, and power calculations are also discussed.
- *The validity of the assumptions* underlying our inferential procedures affects the validity of the inferences we draw from our data. Using confidence intervals and hypothesis tests based on the normal distribution, we discuss these assumptions and the consequences of violating them.

In summary, in this chapter we will discuss sampling distributions and their role in estimating population parameters and in testing hypotheses about those parameters. We introduce the normal distribution to exemplify these concepts and procedures, in part because the normal distribution plays several important roles in inferential procedures that we will study in subsequent chapters of this text.

5.2 Using Sample Statistics to Estimate Population Parameters

5.2.1 *Populations, Samples, and Sampling Distributions*

Usually, a researcher has a sample and wants to draw inferences about a population. As we discussed in Chapter 4, there is a basic problem in drawing inferences from the sample to the population: The values of the sample statistics are not identical to those of the population parameters in which we are interested. For example, the sample mean will vary over independent replications of a study, as will all other statistics we can compute from a sample. Two important questions this variability raises are as follows:

- How is a particular statistic distributed over repeated samples?
- Can more than one statistic be used to estimate a particular population parameter? If so, how do we decide between these possible estimates of the parameter?

These are some of the questions we will consider in Section 5.2. The answers require an understanding of sampling distributions, the topic we turn to next.

5.2.2 *What Is a Sampling Distribution?*

The concept of a sampling distribution was introduced in Chapter 4, where we developed the binomial as a sampling distribution for frequency (or proportions). However, the concept of a sampling distribution is much more general than the context in which it was introduced in Chapter 4. It is implicit in statistical inference. For example, consider the following marketing study. Fifty individuals are randomly sampled from some well-defined population and asked to try a new brand of breakfast cereal. The number of grams of the cereal they consume is recorded as a measure of enjoyment. We might wish to test whether the mean of the sampled population is different from the standard serving size of 40 grams. The mean weight of cereal consumed by our participants is 60 grams. If the sample mean changed little from one sample to another, this value would probably provide strong evidence against the hypothesis that the population mean equaled 40 grams, and thus would suggest that the new cereal is worth marketing further. On the other hand, if the sample mean was quite variable over samples, then a sample value of 60 grams could well have occurred even when the population mean was 40 grams.

It is useful to picture many random replications of the 50-participant sampling experiment, with each replication giving rise to a value of the mean, $\bar{Y}$. If we tabulate all the possible values of $\bar{Y}$ and their associated relative frequencies of occurrence, the result is called the *sampling distribution of the mean* for samples of size 50. Knowing the properties of this sampling distribution will help us evaluate inferences made from a single sampled value of $\bar{Y}$. If we know that the sampling distribution has little variability, we have considerable confidence that our one estimate is close to the population parameter; conversely, we are less satisfied with an estimate when the variability of the sampling distribution is high. Furthermore, as we shall see, if we have knowledge of the shape of the sampling distribution – e.g., that the population of scores is normally distributed – we can draw various inferences about the parameter. This application of sampling distributions to statistical inference was introduced in Chapter 4, and we return to it soon to develop procedures for drawing inferences about population means. However, before doing so, we will consider another important use of sampling distributions of statistics; namely, to study properties of statistics as estimators of population parameters.

5.2.3 *Criteria for Estimators*

We have established that researchers usually want to draw inferences about population parameters from their sample data. For example, they want to know the mean of the population, or estimate some measure of variability (e.g., the standard deviation). In fact, there are often multiple ways to estimate the value of a given population parameter. For example, in a distribution that is unimodal and symmetric, the sample mean, median, and mode all estimate the population mean. Given this, how should we choose among the options? What criteria should be used when selecting the statistic to provide a point estimate of a population parameter?

In our marketing example, we focused on the sampling distribution of the mean. However, *every* statistic has a sampling distribution, because each time a new sample is drawn from a population, the sample statistic is based on a new set of values. Just as we can imagine a distribution of sample means, there are sampling distributions of medians, modes, variances, standard deviations, and so on. And like any distribution of a random variable, any sampling distribution has a characteristic shape, a mean, a variance, and other properties. The criteria we will require for good point estimates relate to several properties of the estimator's sampling distribution: The mean and variance of that distribution and how rapidly the mean of the sampling distribution approaches the parameter value as sample size increases. Before discussing these properties, it will prove useful to briefly consider the concept of an expected value.

5.2.4 Expected Values and Desirable Properties of Estimators

The *expected value of a random variable* is the mean of the variable's theoretical distribution. In the case of an individual score, Y, its expected value is the population mean; that is, $E(Y) = \mu$. The theoretical variance of Y is $\sigma^2 = E(Y - \mu)^2$; that is, the population variance is the average squared deviation of scores about the population mean. With respect to the issue of criteria for selecting estimators, we are interested in the mean and variance of the sampling distribution of potential estimators of population parameters. (For more information about expected values, and important proofs involving them, see Appendix B.) In general, we may think in terms of a population parameter that we will represent by the Greek letter θ (theta); this could be the population mean, variance, a measure of skewness, or any other measure that in theory might be a function of the random variable of interest. There are many possible estimators of θ; we will use the symbol $\hat{\theta}$ (theta-hat) to represent an estimator. We will briefly consider three criteria for choosing an estimator: *unbiasedness*, *consistency*, and *relative efficiency*.

Unbiasedness

An estimator, $\hat{\theta}$, is an unbiased estimator of the population parameter, θ, if the mean of its sampling distribution equals the parameter being estimated. That is, if

$$E(\hat{\theta}) = \theta \tag{5.1}$$

In Appendix B, we show that the sample mean, $\bar{Y}$, is an unbiased estimate of the population mean, μ. That seems intuitively reasonable. However, just because a statistic and a parameter share the same name does not guarantee that the statistic is an unbiased estimator of the parameter. An example of a biased estimator is S^2 as an estimator of the population variance, σ^2, where $S^2 = \Sigma(Y_i - \bar{Y})^2/n$. In Appendix B, we show that

$$E\left(s^2\right) = \sigma^2 \tag{5.2}$$

where $s^2 = \Sigma(Y_i - \bar{Y})^2/(n - 1)$. Therefore, because $S^2 = [(n - 1)/n)]s^2$,

$$E\left(S^2\right) = \left(\frac{n-1}{n}\right)\sigma^2 \tag{5.3}$$

so $E(S^2) < \sigma^2$. Thus, S^2 is a biased estimator of the population variance (although the bias decreases as n increases), but s^2 is an unbiased estimate. Because S^2 tends to underestimate σ^2, we will follow the usual practice of calculating s^2 rather than S^2 for the sample variance.

Consistency

Again, let $\hat{\theta}$ be some estimator of θ. $\hat{\theta}$ is a *consistent estimator* of θ if its value is more likely to be arbitrarily close to θ as n increases.[1] A familiar example of a consistent estimator is the sample mean of n independently distributed scores. From Appendix 5.1, $\sigma^2_{\bar{Y}} = \sigma^2_Y/n$; therefore, the sampling variability of $\bar{Y}$ about μ must decrease as n increases. Because inferences based on consistent estimators are more likely to be correct as sample size increases, consistency is an important property of an estimator.

Relative Efficiency

The third major criterion for selecting an estimator is the variance of the sampling distribution of that estimator, particularly when compared to the variance of an alternative estimator. The less variable the sampling distribution of the estimator is, the more likely it is that any single estimate will have a value close to that of the population parameter. As you might expect, the relative efficiency of two estimators, $\hat{\theta}_1$ and $\hat{\theta}_2$, can be expressed as a ratio:

$$RE_{1\ to\ 2} = \frac{E\left(\hat{\theta}_2 - \theta\right)^2}{E\left(\hat{\theta}_1 - \theta\right)^2} \tag{5.4}$$

is a measure of the efficiency of the estimator in the denominator, $\hat{\theta}_1$, relative to that in the numerator, $\hat{\theta}_2$.

5.2.5 *Which Estimator?*

Most of the estimation and hypothesis-testing procedures presented in statistics texts, and in published journal articles, make use of the sample mean, and the unbiased variance estimate, s^2. If the population from which the data are drawn has a normal distribution, these statistics will be efficient relative to their competitors. Consequently, estimates based upon them are more likely to be close to the true value of the parameter being estimated, and hypothesis tests are more likely to lead to correct inferences. But what if the population distribution is not normal? We will address this question by considering the relative efficiencies of several estimators of μ for different population distributions.

Rosenberger and Gasko (1983) derived the variances of various estimators for several population distributions and sample sizes. Table 5.1 presents their results for $n = 20$ for the mean, the median, and the 10% trimmed mean; this last statistic is obtained by rank ordering the scores in the sample, discarding the highest and lowest 10% (the top and bottom

Table 5.1 Variances and relative efficiencies (*RE*) of three estimators of the population mean for $n = 20$, from Rosenberger and Gasko (1983)

	Normal distribution		*Mixed-normal distribution*	
Statistic	*Variance*	*RE*	*Variance*	*RE*
Mean	.050	1.000	.298	1.000
Median	.073	.685	.079	3.772
Trimmed Mean	.053	.943	.061	4.885

Note: For the mixed-normal, one observation comes from a normal distribution with $\sigma = 10$.

two scores for $n = 20$), and then calculating the arithmetic mean of the remaining scores. The theoretical variances of these statistics when the distribution of the random variable is normal are presented in the second column of Table 5.1. The third column contains the efficiencies (*RE*) of each estimator relative to the mean; these are obtained by taking ratios of the variances, as in Equation 5.4. When the population of scores is normal, the mean is the more efficient statistic and therefore the best estimator of the population mean. However, the situation is quite different if we make one change in our sampling procedure. Suppose 19 of the 20 scores in each sample are drawn from the population having $\mu = 0$ and $\sigma = 1$; however, one score is drawn at random from a population for which $\mu = 0$ and $\sigma = 10$. This second population looks much like the first except that extreme scores are more likely. Think of the extreme scores as coming from those rare individuals who pay no attention to the task instructions. Such scores might contribute to the variance, increasing the proportion of very small and very large scores. Variances of both statistics and efficiencies relative to the mean are presented in the fourth and fifth columns of Table 5.1. The interesting result here is that the variances of the sampling distributions of both the trimmed mean and the median are markedly less than that of the mean, and thus their relative efficiencies are greater. In fact, Rosenberger and Gasko found that, for the seven distributions they studied, the mean was the most efficient estimator only when samples were drawn from a normally distributed population. In the other six distributions they investigated, some other estimator (not necessarily the median) was at least slightly more efficient, and sometimes much more so.

Contrary to intuition and popular usage, the sample mean is not always the best estimator of the population mean. Other estimators may be more efficient when the population is skewed, or is symmetric but with long tails (i.e., high variability), or when there are a few outlying scores, as in the example of Table 5.1. This happens because the sampling variance of the mean is increased much more than that of the median by the inclusion of even a few extreme scores. Micceri's (1989) review of many real data sets suggests that normality is the exception. In many cases, inferential procedures that do not rest upon the sample mean may provide more powerful tests of hypotheses. Several procedures based on assigning ranks to scores, and analyzing those ranks, are presented in this book; these *nonparametric procedures* will be useful in fairly simple designs. Another possible approach is implicit in the results presented in Table 5.1. This involves *trimming data* from the tails of the data set, thus reducing the effect of extreme scores (Hogg, Fisher, & Randles, 1975).

5.2.6 *Summary*

To sum up the developments of this section, unbiasedness, consistency, and efficiency are desirable properties in the statistics we use in drawing inferences. However, in some situations, there may be a tradeoff between bias and efficiency; a biased estimator may have less sampling variability than an unbiased one. Unless the bias is very extreme, and no correction for bias can be found, the more efficient estimator is to be preferred. It is more important to have an estimate that is more likely to be close to the parameter being estimated than to have one that, if many such estimates were obtained, would be correct on the average. The prevalent use of inferential procedures based on $\bar{Y}$ and $\sigma_{\bar{Y}}$, as we will see throughout the remaining chapters, reflects the fact that these statistics are known to be both unbiased and efficient when the data are normally distributed. However, there will be situations in which we will encounter distributions for which other statistics will be more efficient. Exploring data using the sorts of graphs and tabular outputs illustrated in Chapter 2 will help the researcher in determining when alternative estimators and approaches to inference should be considered.

5.3 The Sampling Distribution of the Sample Mean

Despite the observation that the sample mean is not always the best choice of estimator, it is conventionally the estimator of choice in procedures designed for making inferences about population means. The reason for this is that we can specify the sampling distribution of the mean for a wide range of circumstances. The theoretical key to this fact is the central limit theorem.

5.3.1 *The Central Limit Theorem*

The central limit theorem states:

> *If a sample is large enough, the sampling distribution of its mean will be approximately normal, regardless of the shape of the underlying population.*

This theorem about the large-sample shape of the sampling distribution is important because it establishes that the normal distribution may be used to compute probabilities of values of sample means under a wide range of circumstances. This will be true even when samples are drawn from a population that is not normally distributed. Probability statements, such as "the Type 1 error rate is .05" will be at least approximately correct. As one example, the normal distribution may be used to perform significance tests when dealing with a population that is binomially distributed. This is because the proportion of successes, y/n, is also a mean: if we assign a score of 1 to a success and 0 to a failure, y (the number of successes) is a sum of these zeros and ones, and a sum divided by the number of observations is a mean.

How rapidly the sampling distribution approaches normality as the sample size, n, increases will depend on the shape of the population. A sample size of 30 is usually "large enough" to ensure that the normal closely approximates the shape of the sampling distribution; however, a larger sample size may be necessary if the population is skewed, or is lumpy (i.e., has few values and several modes). We saw an example of the influence of population shape on the sampling distribution in Figure 4.2 of the preceding chapter: When the population is symmetrically distributed ($\pi = .5$), the sampling distribution is approximately normal for samples of size 20, whereas there is still some slight asymmetry in the sampling distribution for $n = 40$ when the population parameter, π, is .25.

One other point about the central limit theorem should be noted. It applies not only to means, but also to all linear combinations. A *linear combination* is a sum of scores, each multiplied by some number, or weight, w. That is,

$$\textit{linear combination} = w_1Y_1 + w_2Y_2 + \ldots + w_jY_j + \ldots + w_nY_n \tag{5.5}$$

We saw an example of a linear combination in Chapter 2, in which the means of four different clinics of different sizes were combined to obtain an overall, or grand, mean. The weights in that example were the n_j/N, the proportion of scores in the jth clinic divided by the total number of scores.

Finally, we note that there is a tendency to overgeneralize in applying the central limit theorem. Although it holds for all linear combinations, it does not hold for nonlinear combinations of variables such as sums or averages of scores raised to a power. For example, the sampling distribution of the variance is not normal because it involves sums of squared deviations. A second exception is the correlation coefficient, which was defined in Chapter 2 and is discussed further in Chapter 17.

5.3.2 The Mean and Variance of the Sampling Distribution of the Mean

Knowing that the sampling distribution of the mean is normal in shape is just part of the information needed to completely specify the sampling distribution. The normal distribution has two parameters, its mean and variance; therefore, we must know the values of the mean and variance of the relevant sampling distribution. Fortunately, even though we never observe the sampling distribution of a statistic, we can derive its properties without drawing even one sample from the population. This point may be clearer if we consider an example.

Assume that we toss a single die. As usual, the die has six sides, each with a different one of the values from 1 to 6. If this experiment is carried out many times and the resulting number is recorded each time, we have the distribution displayed in the upper panel of Figure 5.1; in the long run, each possible integer from 1 to 6 will occur on 1/6 of the trials (i.e., with probability 0.167), assuming the trial outcomes are independent and that each value has probability 1/6 of occurring. This is the population distribution.

Now let us change the experiment slightly so that a trial consists of tossing two dice. If we record the mean of the two numbers that come up on each of many trials and – still assuming independence and equal probability of the six values for each die – the sampling distribution of the trial mean will be that depicted in the lower panel of Figure 5.1. The distribution now has a definite peak. For example, the sample mean is more likely to equal 3.5 than 1 or 6. The reason for this follows from the multiplication rule for independent events. The mean will equal 1 only if both dice on a trial result in a 1, an event that occurs with probability $p = 1/6 \times 1/6 = 0.028$. On the other hand, the mean will equal 3.5 if one die shows a 1 and the other shows a 6, or if either die has a 2 while the other shows a 5, or if the result is a 3 and a 4. Therefore, there are six outcomes which can yield a sum of 7, or a mean of 3.5. Each outcome has a probability of 1/36, so the probability

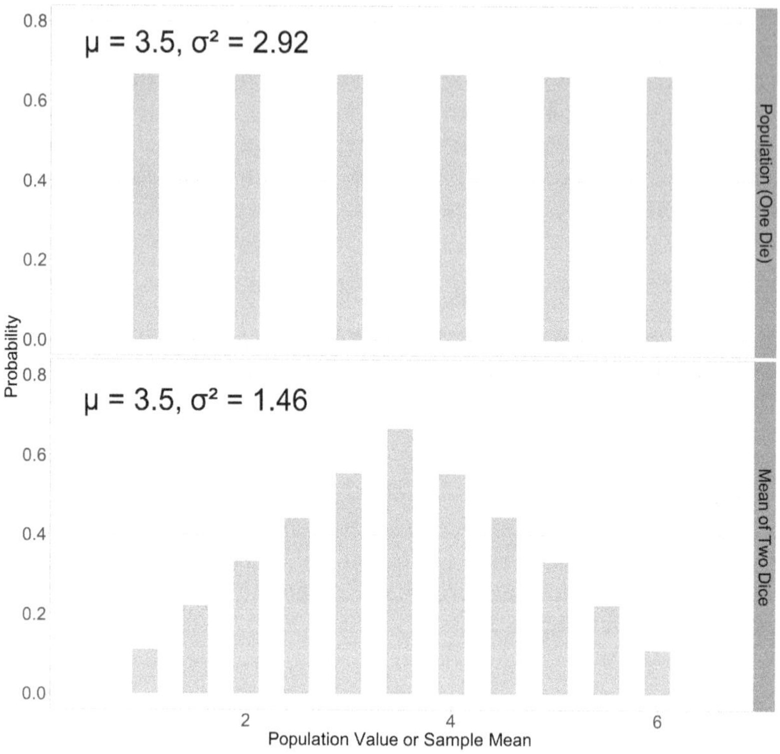

Figure 5.1 Population distribution for number of pips on a die (upper panel) and the sampling distribution of the mean number of pips displayed when n = 2 (lower panel).

of a mean of 3.5 is 6 × 1/36 = 0.167. If we were to further increase the number of dice tossed in each replication of the experiment, the resulting sampling distribution of the mean would be more closely approximated by the normal distribution, as expected under the central limit theorem.

Figure 5.1 includes values of the population mean and variance (upper panel), and the mean and variance of the sampling distribution when two dice are thrown (lower panel). Note that the mean of the sampling distribution of the mean, $\mu_{\bar{Y}}$, is identical to the mean of the population, μ_Y; both are 3.5. Also note that when there are two scores in the sample, the variance of the sample means, $\sigma^2_{\bar{Y}}$, is 1/2 the population variance; in Figure 5.1, the population variance, σ^2, is 2.92 and the variance of the sampling distribution, $\sigma^2_{\bar{Y}}$, is 2.92/2, or 1.46. In general, if the scores are independently distributed, $\sigma^2_{\bar{Y}} = \sigma^2_Y/n$, where n is the sample size. If we constructed the sampling distribution of means for samples of size 10, its variance would be .292. This relation between the variance of the sampling distribution and that of the population holds regardless of the shape of the population, *provided the scores are independently distributed.* Appendix 5.1 contains a proof that $\sigma^2_{\bar{Y}} = \sigma^2_Y/n$, as well as more general results for linear combinations, including cases in which scores are not independently distributed.

In summary,

- The average of many sample means will be the same as the population mean.
- The sample-to-sample variability of the sample mean will be less when n is large because the variance of the sample mean equals the population variance divided by the sample size.

Therefore, the value of a sample mean is more likely to be close to the value of the population mean it estimates as sample size increases. This makes sense; the larger the sample, the more likely it is to resemble other samples from the same population and the closer its mean will be to those of other samples and to the population mean.

5.3.3 Summary

We have determined that sample means are unbiased and consistent estimators of population means. Under some conditions – but not all – sample means are also relatively efficient estimators. Most importantly, the central limit theorem tells us that the sampling distribution of the sample mean is approximately normally distributed, with $E(\bar{Y}) = \mu$ and $\sigma^2_{\bar{Y}} = \sigma^2_Y/n$, if the sample size is sufficiently large. Thus, we can use the normal distribution as a basis for making inferences about population means. To do this, we need to know something about the normal distribution.

5.4 The Normal Distribution

5.4.1 Why Is the Normal Distribution Important?

The justification we have presented for using the normal distribution as the basis for making inferences about means is sufficient to establish the importance of the normal distribution in statistics. However, there are other important roles played by the normal distribution that are worth considering before we examine the properties of the normal distribution.

A second reason the normal distribution is important is that the derivations of several important theoretical distributions rest upon the assumption that observations are sampled from a normally distributed population. Specifically, the derivations of the χ^2, t, and F distributions all rest upon that assumption. In fact, many random variables have an approximately normal distribution. Consideration of almost any measurement taken on an individual, such as their height or weight, may clarify why this is so. The person's height might be represented as

$$Y = \mu + \varepsilon \tag{5.6}$$

where Y is the measured height, μ is the mean of the population of heights, and ε (epsilon) is a sum of "errors," positive and negative deviations from the population mean due to many random factors that affect the height being assessed. Such factors would include age, childhood nutrition, country of origin, posture, and precision of the measurement tool, among other variables. The central limit theorem states that the sum of many such effects will be normally distributed. Therefore, if ε can be viewed as a sum of many independent random effects like the ones we have indicated, it (and therefore Y) will tend to be normally distributed.

We do not want to give the impression that the normality assumption applies in all circumstances. Although the normal distribution is a reasonable approximation to the distribution of many variables, many others are not normally distributed. We cited published reviews of data sets in Chapter 2 and presented examples from the *Royer* and *Seasons* data sets that make this point. In later chapters, we will consider the consequences of nonnormality and alternatives to those classical statistical procedures that rest on the assumption of normality. Nevertheless, because of its central role in statistical inference, we will devote the remainder of this chapter to considering the normal distribution. The normal distribution also merits our consideration because it provides a relatively simple context within which to continue our development of inferential procedures such as interval estimation and hypothesis testing.

5.4.2 The Normal Distribution's Probability Density Function and z Scores

The normal distribution is characterized by its density function:

$$f(y) = \frac{1}{\sigma\sqrt{2\pi}} e^{-(y-\mu)^2/2\sigma^2} \tag{5.7}$$

where μ and σ are the mean and standard deviation of the population and π and e are mathematical constants. The random variable Y can take on any value between $-\infty$ and $+\infty$, and the curve is symmetric about its mean, μ.

Infinitely many normal distributions are possible, one for each combination of mean and variance. However, inferences based on these normal distributions are made possible by the fact that all possible normal distributions are related to a single distribution. This *standardized normal distribution* is obtained by subtracting the distribution mean from each score and dividing the difference by the distribution standard deviation; specifically, it is the distribution of the z score

$$z = (Y - \mu)/\sigma \tag{5.8}$$

As we showed in Chapter 2, the mean of the distribution of z scores is zero and its standard deviation is one. This is true of any complete set of z scores. In addition, if the variable Y is normally distributed, the corresponding distribution of z scores also will be normal. In this case, the variable z is often referred to as a *standardized normal deviate*.

Standardization provides information about the relative position of an individual score. For example, assume a normally distributed population of scores with $\mu = 500$ and $\sigma = 15$. We can write this assumption like this: $Y \sim N(500, 15)$, where the tilde (~) sign means "is distributed as" and the N indicates a normal distribution with the mean and standard deviation given. A value Y of 525 would correspond to a z score of 1.67; $z = (525 - 500)/15 = 1.67$, meaning that a raw score of 525 is 1.67 standard deviation units above the population mean of 500. Turning to Appendix Table C.2, we find that $F(z) = .9525$ when z is 1.67. $F(z)$ is the proportion of standardized scores less than z in a normally distributed population of such scores. In this example, we may conclude that the score of 525 exceeds 95.25% of the population. Of course, this conclusion may not be valid if our values of μ and σ are incorrect or if *Y* is not normally distributed.

Equation 5.8 defined a *z* score as $(Y - \mu)/\sigma$. In fact, this is just a special case of a general formula for a *z* score. Instead of *Y*, we could have any observed quantity; examples would be the sample mean, the difference between two sample means, or some other statistic. Call this *V* for *observed variable*. To transform *V* into a *z* score, subtract the expected value of *V*, and then divide the difference by the standard deviation of the sampling distribution of *V*. Thus, a general formula for *z* is

$$z_V = [V - \mathrm{E}(V)] / \sigma_V \tag{5.9}$$

The variable z_v will be normally distributed if – *and only if* – *V* is normally distributed. In that case, we can assess the probability that *V* exceeds some specified value by referring to Appendix Table C.2 to obtain $1 - F(z) = \alpha$.

5.4.3 Using Software to Calculate z Scores and Normal Distribution Probabilities

Using normal distributions tables, like Table C.2, is straightforward; we can obtain α, the probability that *V* exceeds a specified value. Of course, we can use the same table to find the *z* score corresponding to a particular probability. For accuracy and speed, we can also use software to calculate these same values. In Section 4.3.4, we used distributional calculators in R to compute the binomial function; here, we do the same type of computation for the normal distribution.

In R, the distribution functions are part of the base {stats} package. The *pnorm* function takes a *z* score as input and returns the probability that a score is less than that *z*. For example, *pnorm*(1.67, mean = 0, sd = 1) returns 0.9525, meaning that 95.25% of *z* scores are less than 1.67. The default assumption of *pnorm* is to calculate probabilities using the standard normal distribution, *N*(0, 1), so we also obtain 0.9525 using the simpler command *pnorm*(1.67). We can also skip the step of calculating the *z* score itself if we provide the mean and standard deviation of the population, like so: *pnorm*(525, mean = 500, sd =15) = 0.9525 because the *z* score for *x* = 525 is 1.67. To find the probability of a score greater than 1.67, we can either use 1 – *pnorm*(1.67) or we can change an option to compute probabilities above *z*, *pnorm*(1.67, lower.tail = FALSE); either command will return 0.0475.

In SPSS, we begin by creating a variable, *x*, with value 1.67. Then, from the *Transform* and *Compute Variable* pull-down menus, choose the *Cdf.Normal* function within the *CDF & Noncentral CDF* function group. Click the "up arrow" to move the function to the *Numeric Expression* box, where it will appear as CDF.NORMAL(?,?,?). The order of these inputs is shown in a box below the number pad onscreen: First the *z* score or variable value, then the mean and standard deviation. Insert the variable *x*, then type 0 and 1, and add a name for the *Target Variable*, which will hold the result. Click the OK button to find the output is 0.9525. You may need to adjust the number of decimal places using the *Variable View* tab in the *Data* window. To find the probability of a score above 1.67, simply compute 1 – CDF.NORMAL(1.67, 0, 1), which will return 0.0475.

5.5 Inferences About Population Means

In this section, we apply concepts developed in Chapter 4 to inferences based on the normal distribution. The application of normal probabilities assumes knowledge of the population variance. Using the data we have collected, we can only estimate that variance, a fact that introduces some error into our calculations. Although more accurate results require using the t distribution, we will wait until the next chapter to develop that topic. For now, we will continue to focus on the normal distribution to further illustrate and extend ideas introduced in Chapter 4.

One of the measures available to us in the *Seasons* data set is the seasonal total cholesterol score (*TC*1, . . ., *TC*4). We calculated the average over the four seasons for those participants who were measured in all four seasons (some participants missed at least one of the four sessions); the data are in the file, *TC_Data*; the link is on the *Seasons* page on the website. We used the sample data to estimate the mean *TC* of a subpopulation – namely males 50 and older (Agegrps 3 and 4 in the file). These data are in the file *TC_Data over 50*. Keeping in mind that doctors frequently recommend that *TC* should be at 200 or less, we wanted to know whether the mean in the subpopulation differed from this recommended maximum level. In what follows, we analyze the *TC* data of the 117 males in *Agegrps* 3 and 4 to estimate the mean of a population of such individuals and to test the null hypothesis that the population mean equals 200.

5.5.1 A First Look at the Data

As we emphasized in Chapter 2, we should explore our data before conducting any inferential procedures. We begin our exploration of the *TC* data with Figure 5.2, which presents a

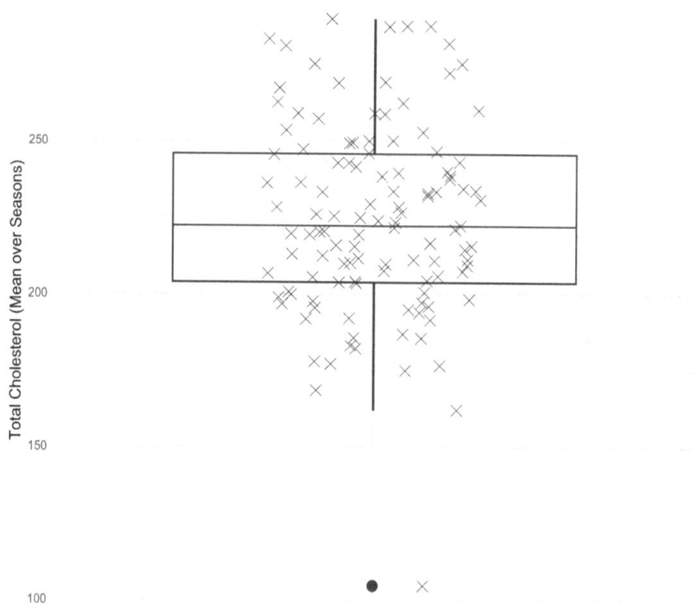

Figure 5.2 Box plot of total cholesterol scores of men age 50 and older.

box plot of the data overlaid with individual scores. Several aspects of the plot are of interest. First, because the median approximately bisects the box and the whiskers are of about equal length, it appears that the distribution is symmetric. Second, the median is clearly above the recommended maximum *TC* level of 200. Furthermore, the lower hinge is also above 200, meaning that at least 75% of the participants have *TC* scores above the recommended maximum. There is some good news, however. None of the participants have a *TC* level as high as 300, a value that would clearly signal a high-risk patient. Nevertheless, the box plot warns us that cholesterol level may be a problem for many of the patients. One other point should be noted about the box plot. There is one extreme outlier, a *TC* score close to 100. This is such an unusual score that it might be wise to recheck this patient's cholesterol level. One possibility that should be considered is that the score represents a data entry error.

The *Q–Q* plot in Figure 5.3 shows the expected *z* scores (on the x-axis) under the assumption that *TC* is normally distributed. The great majority of scores fall on or very near the straight line; the one clear exception is the outlier noted previously. We can confirm that the data are approximately normally distributed by observing a nonsignificant result of the Shapiro–Wilk test in R (using the *shapiro.test* function in the {stats} package), which can be thought of as the square of the correlation in the *Q–Q* plot. Although these test results provide confirmation of normality, we should be aware that there are occasions where they

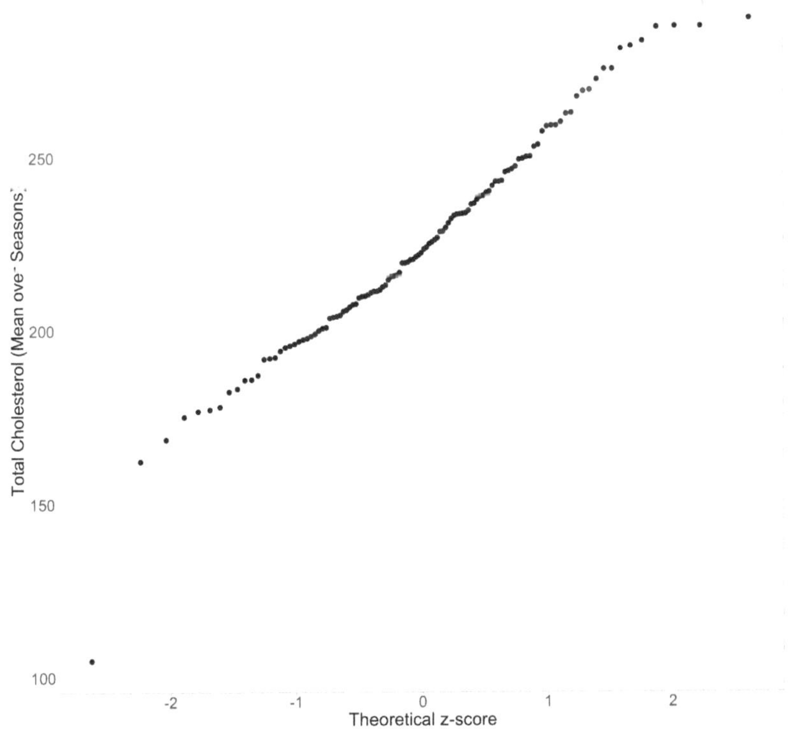

Figure 5.3 *Q–Q* plot of the total cholesterol scores of men age 50 and older.

will be misleading. When samples are large, even small departures from normality may yield significant results; when samples are small (≤ 30), the Shapiro–Wilk test has low power (Razali & Wah, 2011). Small but statistically significant deviations from the expected normal distribution may have little effect on subsequent hypothesis tests that assume normality. In view of this, we are concerned only when the *Q–Q* plot signals marked departures from normality or the presence of outliers that require our attention.

Given the findings of our data exploration, it seems reasonable to assume that the population from which the *TC* scores were sampled has a normal distribution. Therefore, our subsequent analyses will be based on the standard normal distribution. We were troubled by the one outlier, so we performed the analyses to be described with and without that case. We found little difference in our two sets of results and therefore will report the analyses based on the entire sample of 117 scores. The values we need for this example are $n = 117$, $\bar{Y} = 224.684$, and $s = 31.302$. We are now ready to estimate an interval containing the population mean, and test whether that mean differs significantly from the theoretical value of 200.

5.5.2 A Confidence Interval for μ

In Chapter 4, we introduced the inferential procedure of hypothesis testing. In this section, we introduce a new inferential procedure called *interval estimation*. Interval estimation answers the question, "What is the plausible range of values of θ?" where θ is a population parameter. Thus, interval estimation is a very general and useful procedure that is almost invariably more meaningful than the binary outcome of a significance test (i.e., reject or fail to reject the null hypothesis).

Although various population parameters may be estimated, we are currently interested in estimating the mean of a population, μ. Let's begin by illustrating the procedure for our current example. The sample mean, $\bar{Y} = 224.684$, provides a point estimate of μ, the population mean *TC* score for men older than 50 years. However, the sample mean might be close to the parameter, or it might reflect considerable error. To have a sense of the accuracy of such estimates, we calculate an interval estimate or *confidence interval* (CI), a pair of numbers which provide reasonable bounds for the parameter being estimated.

The procedure for constructing a CI for μ is based on what we have learned about the sampling distribution of the sample mean; namely, that the distribution of the sample mean is normal with an expected value equal to μ and variance equal to the population variance divided by the sample size, *n*. Imagine drawing many samples of 117 *TC* scores each from a population of *TC* scores. Further assume that the mean of each sample is converted into a standardized (z) score by subtracting the mean of the sampling distribution and then dividing the difference by the *standard error* (*SE*) of the mean. Suppose we want a 95% CI. If the scores are independently sampled from a normal population, Appendix Table C.2 tells us that 95% of the sampled z scores will lie between −1.96 and 1.96. That is,

$$p\left(-1.96 \leq \frac{\bar{Y}-\mu}{\sigma_{\bar{Y}}} \leq 1.96\right) = .95 \tag{5.10}$$

To obtain the bounds on μ, consider each inequality separately. First, consider

$$-1.96 \leq (\bar{Y}-\mu)/\sigma_{\bar{Y}}$$

Solving for μ, we have the upper bound:

$$\mu \leq \bar{Y} + 1.96\sigma_{\bar{Y}}$$

Similarly, we can solve for the lower bound. From the inequality

$$(\bar{Y} - \mu) / \sigma_{\bar{Y}} \leq 1.96$$

we arrive at

$$\bar{Y} - 1.96\sigma_{\bar{Y}} \leq \mu$$

Putting it all together, we have

$$p\left(\bar{Y} - 1.96\sigma_{\bar{Y}} \leq \mu \leq \bar{Y} + 1.96\sigma_{\bar{Y}}\right) = .95 \tag{5.11}$$

Recall that $\bar{Y} = 224.684$ and $s = 31.302$. Dividing s by the square root of n, $\sqrt{117}$, we have 2.894, an estimate of $\sigma_{\bar{Y}}$. Because n is large and the statistic s is a consistent estimate of σ, we expect s to be very close in value to σ and feel justified in using it in our calculations.[7] Substituting into Equation 5.11, we find that the lower and upper bounds of the 95% CI for μ are

$$CI = \bar{Y} \pm 1.96\sigma_{\bar{Y}} = 224.684 \pm 5.672 = 219.012, 230.356$$

In words, we can be 95% confident that, for men over 50, the population mean *TC* falls somewhere between 219.01 and 230.36. Consistent with our earlier look at plots of the data, we see that the lower limit of the confidence interval, 219, is above the recommended maximum *TC* of 200, but we are reasonably confident that the population mean is not dangerously high for the population of over-50 men from which we have sampled.

What exactly do these numerical limits mean? In what sense do we have 95% confidence that the population mean is contained within them? We may not say that "the *probability* is .95 that μ lies between 219 and 230" because either μ is in this interval or it is not. Furthermore, if we were to draw another sample from the same population of *TC* scores, the mean would change, giving rise to different limits. What 95% confidence means is that *if we draw many samples and find the 95% confidence interval for each sample, in the long run, 95% of these intervals will contain* μ. Therefore, our *confidence* is .95, or 95%, that the actual value of μ falls within the confidence interval we calculated.

The narrower the interval, the better our estimate of μ. Returning to Equation 5.11, we can see that the interval width depends on the *SE*; the smaller the variability of the sample mean, the smaller will be the distance between the two limits. Recalling that $\sigma_{\bar{Y}} = \sigma / \sqrt{n}$, it follows that the width of the interval decreases with increased sample size and with decreased variability. Therefore, we can increase the precision of our estimate by doing whatever we can to reduce error variance and by collecting as many observations as is practical.

A third factor, not immediately obvious in Equation 5.11, also affects the width of the confidence interval: The confidence level we choose. The higher our confidence level, the wider the interval must be. Conceptually, this is akin to fishing with a large net versus a small one: We are more likely to catch a fish with a large net (high confidence), but we learn much less about where the fish prefer to swim (low precision). Concretely, let's calculate a 90% CI rather than a 95% CI. Doing so requires a different critical value of z. Turning to Appendix Table C.2, we see that when the level of confidence is set at .90 (i.e., 5% in each tail of the distribution), the critical z score is 1.645. Replacing 1.96 by 1.645 in Equation 5.10 or 5.11, the new limits are 219.92 and 229.44; we have less confidence but a slightly narrower interval. This is the tradeoff between confidence level and interval width.

5.5.3 *A Test of the Null Hypothesis*

We originally asked whether the mean of the sampled population of *TC* scores differed from a value of 200. The confidence interval limits we calculated suggests that the answer is "yes." We reason as follows: First, we have 95% confidence that the computed interval, which has the limits 219 and 230, contains the population mean. Second, that interval does not contain 200. Thus, we conclude with 95% confidence that the population mean *TC* score differs from 200.

Many researchers skip calculating the confidence interval and, instead, directly test whether the population mean equals the theoretical value. We believe this is a mistake because it addresses the question, Is the mean 200? rather than the more interesting question, What is the mean? Nevertheless, the practice of hypothesis testing is widespread. Furthermore, a presentation of the test permits us to again address related concepts such as Type 1 and Type 2 errors. For these reasons, we will use the standardized normal distribution to test whether μ differs from 200. In Section 5.7, we will present a more detailed discussion of the relation between the confidence interval and the significance test.

When we introduced the logic of hypothesis testing in Chapter 4, we emphasized that the logic of the procedure is the same for any situation (see Boxes 4.1 and 4.2). Thus, the first step is to establish a null and an alternative hypothesis; these are again designated H_0 and H_1 respectively. Of course, our hypotheses in Chapter 4 concerned the probability of an event, whereas we are now concerned with the value of a population mean. Letting μ_{TC} represent the mean of the population of *TC* scores, the null hypothesis is that $\mu = 200$ and is stated as

$$H_0 : \mu_{TC} = 200$$

and the alternative hypothesis is

$$H_1 : \mu_{TC} \neq 200$$

Once these two hypotheses have been formulated, we need a test statistic whose value will enable us to decide between them. We will estimate the population mean with the mean *TC* from our sample, and then convert the sample mean to a z score. The z statistic will serve as our test statistic. Because the sample mean is normally distributed for a large sample, the sampling distribution for our test statistic, z, is also normally distributed.

Recall the general form of the z statistic (Equation 5.9): $z = [V - E(V)]/\sigma_V$. To test whether the mean *TC* score is significantly different from 200, we replace V by $\bar{Y}_{TC}$, $E(V)$ by the population mean specified by H_0 (μ_{hyp}), and σ_v by the *SE*.[3] Consequently, we have

$$z = \frac{\bar{Y}_{TC} - \mu_{hyp}}{\sigma / \sqrt{n}} \tag{5.12}$$

Substituting the values presented earlier,

$$z = \frac{224.684 - 200}{2.894} = 8.53$$

This z score informs us that the observed mean is more than eight standard deviation units above the hypothesized mean of 200.

Now that we have a numerical value for our test statistic, 8.53, we can use it to decide between H_0 and H_1. We do this by determining those values of z that would lead to rejection of H_0 in favor of H_1. Such values constitute the rejection region, the set of possible values of z that are consistent with H_1 and very improbable if H_0 is assumed to be true. Indeed, those values are so improbable in the sampling distribution for H_0 that their occurrence leads us to reject H_0. An arbitrarily chosen value, α (alpha), defines exactly how unlikely "so unlikely" is. Traditionally, researchers have set α at .05 or .01. We want very strong evidence against H_0 before we reject it.

Once we have decided on a value of α, we can establish a rejection region for our study (see Box 4.2). Because our alternative hypothesis is nondirectional, we will consider as evidence against the null hypothesis both sample means that are well below and sample means that are well above the hypothesized population mean of 200. Thus, we need to determine the values of z that cut off the lower 2.5% and the upper 2.5% of the normal distribution. Turning to Appendix Table C.2, we find that 1.96 is exceeded by .025 of the standardized normal curve; because the curve is symmetric, .025 of the area also lies below −1.96. Thus, if the null hypothesis is true, there is only a 5% probability of obtaining a value greater than 1.96 or less than −1.96. Equivalently, we reject H_0 if the absolute value of z, $|z|$, is greater than 1.96. Obviously, the z we calculated is much larger than 1.96 and therefore we reject H_0. The obtained value of the test statistic, 8.53, falls well into the rejection region, so we can reject the null hypothesis and conclude that the mean *TC* level of adult males over 50 is greater than the recommended maximum of 200.

Recall from Chapter 4 that there is an alternative approach that we may choose for testing hypotheses based on computing the p-value of the observed result (see Box 4.1). The p-value is the probability that the value of the test statistic would be at least as extreme as we actually obtained, if the null hypothesis was true. We reject the null hypothesis if $p \leq \alpha$. Here, the p is $p(z > 8.53) + p(z < -8.53)$; to three decimal places, $p = .000$.

5.5.4 One-Tailed vs Two-Tailed Tests

The two-tailed rejection region was selected for our hypothesis test because our alternative hypothesis was nondirectional; that is, we tested whether cholesterol scores were different from 200 in either direction (lower or higher than 200). However, because low cholesterol scores are good, our interest lies primarily in detecting high values that may have negative health consequences. In that case the rejection region would be one-tailed, and the null and alternative hypotheses would be:

$$H_0 : \mu_{TC} \leq 200 \quad \text{and} \quad H_1 : \mu_{TC} > 200$$

In this situation, if the population of scores is normally and independently distributed, and we know the population variance, we again can use the z test. Accordingly, we turn to Appendix Table C.2. Again, the rejection region consists of those extreme values of z that are consistent with the alternative hypothesis. However, because our alternative hypothesis is directional, we concentrate our rejection region in just one tail of the sampling distribution; specifically, the region consists only of the largest 5% of the z distribution. Therefore, again assuming that $\alpha = .05$, we will reject H_0 if the z calculated from our data is greater than 1.645. In the one-tailed case, the p-value is determined only by the part of the distribution beyond the value of the test statistic in the direction consistent with the alternative hypothesis.

The choice between one- and two-tailed tests should be made before the data are collected; it is determined by the hypothesis of interest, not the observed data. To understand why, consider the following scenario. Suppose we originally hypothesized that the average *TC* score should be above 200; we have a one-tailed hypothesis. However, upon examining the data, we find that the sample mean is less than 200, and we now restate our hypotheses, testing at the .05 level for a significant difference in the direction opposite to that originally hypothesized. This procedure capitalizes on a chance outcome and will result in rejecting the null hypothesis if the results fall in the upper .05 of the normal distribution ($z \geq 1.645$) *or* if the results fall in the lower .05 of the distribution ($z \leq -1.645$). Thus, if we operate in such a fashion, the actual probability of a Type 1 error is the probability of $z > 1.645$ or $z < -1.645$, or .10. Switching the hypothesis from one- to two-tailed after looking at the data is also a bad idea even if α is held constant. That decision process also capitalizes on chance because it is determined by the data (which vary randomly from sample to sample) rather than the theoretical hypothesis of interest.

Why not always carry out the two-tailed test? Doing so would allow us to test for departures from the null hypothesis in both directions. There are two issues to consider. If the theoretical claim underlying the experimental question is strongly directional, then the most appropriate test is a one-tailed test. If the theoretical claim is simply one of a difference between conditions or from a specific value (e.g., $TC = 200$), then clearly a two-tailed test should be used. The second issue is statistical power. With $\alpha = .05$, a two-tailed test requires a cutoff of 1.96 in the right-hand tail of the normal distribution, whereas the one-tailed test requires a cutoff of 1.645. In other words, if the alternative hypothesis is that the population mean is greater than 200, the one-tailed test has a more lenient criterion for rejection. Therefore, as we illustrated with the binomial distribution of Chapter 4 (see Figure 4.5), when the null hypothesis is true the one-tailed test has more power against that alternative. When the null hypothesis is false, the one-tailed test also has more power than a two-tailed test, *if and only if* we choose the direction of H_1 accurately; otherwise, it has dramatically less power.

5.5.5 Hypothesis Tests and Confidence Intervals

To understand the relationship between confidence intervals and hypothesis tests, let's consider the usual decision rule for a two-tailed test: Assuming $\alpha = .05$, reject H_0 if

$$\frac{\bar{Y} - \mu_{hyp}}{\sigma_{\bar{Y}}} > 1.96 \quad \text{or} \quad \frac{\bar{Y} - \mu_{hyp}}{\sigma_{\bar{Y}}} < -1.96$$

Some algebra will show that this is equivalent to the rule, reject if

$$\mu_{hyp} < \bar{Y} - 1.96\sigma_{\bar{Y}} \quad \text{or} \quad \mu_{hyp} > \bar{Y} + 1.96\sigma_{\bar{Y}}$$

However, $\bar{Y} \pm 1.96\sigma_{\bar{Y}}$ are the lower and upper limits of a 95% confidence interval on μ. Therefore, the null hypothesis will be rejected at the .05 level (two-tailed) whenever the hypothesized value of the population mean is less than the lower bound, or more than the upper bound, of a 95% confidence interval. In the example of the seasonal change in depression scores, the value zero was below the lower limit of the 95% confidence interval, allowing us to reject the hypothesis of no mean change in the sampled population at the .05 level of significance. Note that the confidence interval permits evaluation of any null hypothesis; any hypothesized parameter value that falls outside the limits will lead to rejection by a significance test with α equal to 1 minus the confidence level, whereas null hypotheses that assert values of the parameter within the confidence interval will not be rejected.

A confidence interval provides several advantages over a hypothesis test:

1. *It provides a bounded estimate of the population parameter*, thus focusing attention on a range of plausible parameter values rather than on asking whether that parameter equals one specific value.
2. *The confidence interval permits tests of all possible null hypotheses simultaneously*, thus providing considerably more information than does the hypothesis test.
3. *The interval width provides information about the precision of the research*. A significant result, coupled with a vnlry narrow interval, may suggest that power was so great as to enable us to reject even a trivial effect. On the other hand, a nonsignificant result, together with a wide interval, suggests that our experiment lacked precision, pointing to the need for either a less variable measure, more careful application of experimental procedures, or a larger sample.

If you recall the discussion from Chapter 4 of factors that influence power, you will note that the width of the interval is influenced by the same variables that influence power. The narrower the interval, the more powerful a test of any null hypothesis will be. As n increases and as s decreases, the confidence interval narrows, and power increases. Furthermore, increasing α and decreasing confidence have parallel effects. An increase in α increases power at the cost of increasing the Type 1 error rate. There is a similar tradeoff between confidence level and the interval width; decreasing confidence yields a narrower interval, providing a more precise estimate but with less confidence in that estimate.

5.5.6 Using Software for Hypothesis Testing and Confidence Intervals

Inferences about the population, whether in the form of a confidence interval or hypothesis test, involve finding critical values of the normal distribution. We can use Appendix Table C.2 or the distributional calculators available in statistical software like R and SPSS. In Section 5.4.3, we used the *pnorm* function in R and the CDF.NORMAL function in SPSS to compute the probability of a score less than the observed z score. There are corresponding functions to do the opposite task, namely to identify the critical value of z that has a particular probability of scores falling below it.

In R, the *qnorm* function in the {stats} package performs that task. For example, the command *qnorm*(.95)[4] returns the critical value of z, 1.645, that has 95% of scores below it and 5% above it. Like the *pnorm* function described in Section 5.4.3, *qnorm* can also take different means and standard deviations as input: *qnorm*(.95, mean = 500, sd = 15) returns 524.7, the value of the variable at the 95th percentile of an $N(500, 15)$ population distribution.

In SPSS, we begin by creating a variable, x, with value 0.95. This value is the probability for which we seek a critical value. Next, use the *Transform* and *Compute Variable* pull-down menus and then select the *Inverse DF* function group. Within this group, choose the *Idf.Normal* function, clicking on the "up" arrow to move the function to the *Numeric Expression* box. The Idf.Normal function has three input variables, the probability (x) and the mean and standard deviation of the normal distribution. Entering Idf.Normal(0.95, 0, 1) and clicking OK will store the result, 1.645, in the variable named in *Target Variable*. This result is the critical value of z that has 95% of scores below it and 5% above it. We can also identify the x value with a certain proportion of data below it in the population. For example, Idf.Normal(0.95, 500, 15) will return 524.7, the value of x at the 95th percentile of an $N(500, 15)$ distribution.

5.6 The Power of the z Test

In planning an experiment, the researcher might wish to know what sample size would be needed to attain a certain degree of power in the null hypothesis test for a given effect size. Or another investigator might wish to know what the power would be if a different sample size were used in a replication of a previously published study. In the latter case, the researcher seeks to determine power for known values of n, σ (or an estimate), and a specific effect size (such as a difference between two means). Building on the developments in Chapter 4, we will provide an example of how power is calculated when n, σ, and a specific effect size are given. Further examples of the determination of power, and of n, will be presented in Chapter 6.

5.6.1 Determining the Power of the Normal Probability (z) Test

A situation very similar to the example of the analysis of cholesterol scores is one in which we have two or more scores for each participant, each in different conditions. For example, in the *Seasons* study, the researchers were interested in seasonal change. Accordingly, we might obtain a single score for each subject by subtracting a score in one season from that in another season, thus obtaining a single change score. We did this for 215 participants who identified as female, subtracting their spring Beck Depression scores from their winter depression scores; see the *D_Change Data* file; the link is on the *Seasons* page of the book's website (select sex = 1 in the data file). The mean change score was .557 and the *SE* was .266. Using the methods of our analysis of total cholesterol scores, and assuming H_0: $\mu_{change} = 0$ and H_1: $\mu_{change} > 0$, the value of the z statistic, 2.094, is significant with a p of approximately .018. Therefore, we reject the null hypothesis.

Now suppose that a research group in another part of the country wants to know whether the observed effects of seasonal change on depression scores can be replicated in their area, an area in which seasonal climates differ from those in Massachusetts. Further, suppose that their sample of female-identifying participants is limited to an n of 100. This is a smaller sample than the 215 tested by the University of Massachusetts researchers. Would this second group of researchers have reasonable power to reject the null hypothesis if it is false? To answer this question, we follow the same steps we outlined for the binomial test in Chapter 4.

The basic principles in computing power are the same that dictated the power calculations of Section 4.5.2 and Box 4.3. Power is an important concept to understand, so let's briefly review the logic of the calculations. We begin by determining the theoretical sampling distribution of Y, assuming H_0 is true. This distribution tells us how likely each outcome is, if H_0 really describes the population. Next, we use H_1 to identify the rejection region – the set of possible outcomes that are so unlikely, given the null hypothesis, that we would conclude that the null hypothesis could be rejected in favor of an alternative. Then, we assume a specific version of that alternative hypothesis, H_A, is actually true and we determine its theoretical sampling distribution. Any observed outcome that falls in the rejection region will lead us to reject the null hypothesis, and the sampling distribution of H_A allows us to calculate how probable those outcomes are, if H_A is true. The proportion of the sampling distribution for H_A that falls within the rejection region is the power of the test: The probability of observing an outcome that allows rejection of the null hypothesis when H_A is true.

To carry out the calculations for our proposed replication experiment, we need the standard error of the mean. To find the numerical value of the *SE*, we assume that $\sigma_{change} = 3.897$, the value of s_{change} obtained in the *Seasons* study, so that $SE = 3.897/\sqrt{100}$, or .390. We can now calculate the *a priori* power of that proposed study.

1. *Find the critical value of z that defines the rejection region.* Assume that $\alpha = .05$ and that the null hypothesis is H_0: $\mu_{change} = 0$ and the alternative hypothesis is H_1: $\mu_{change} > 0$. Then the decision rule is: reject H_0 if $z \geq 1.645$. We can use Appendix Table C.2, or the techniques described in Section 5.5.6, to find this critical value of z.
2. *Establish* H_A, *the specific alternative hypothesis.* Our best guess of the population effect is the mean seasonal change observed in the Massachusetts study, 0.557. Because that sample size is fairly large ($n = 215$), the effect is probably reasonably well estimated. It bears repeating that significant results obtained in small samples tend to overestimate the population effects, so for small samples we might compensate by using a value smaller than the observed effect. Because our rejection region under the null hypothesis was described in terms of standard deviation units – z scores – we convert .557 to a z score:

$$\begin{aligned} z &= \left(\mu_A - \mu_{hyp}\right) / \sigma_{\bar{Y}} \\ &= (.557 - 0) / .390 \\ &= 1.429 \end{aligned}$$

   Thus, the mean of the H_A distribution is 1.429; it falls 1.429 standard deviations above the mean of H_0.
3. *Compute power.* Power is the area under the H_A distribution that falls above the critical value of z specified by H_0, 1.645. We can use G*Power 3.1 to calculate power for a planned replication with a known sample size, $n = 100$, and an estimate of the expected effect, 1.429. In G*Power, we choose the Test Family of z tests, and the Generic z test in the Statistical Test list, as is shown in Figure 5.4. Next, we select "post hoc power" only because we have a fixed n in mind and an expected effect size, which G*Power calls the noncentrality parameter. (We will learn about noncentrality parameters in Chapter 6.) Entering these values, and clicking calculate, we see that power = .41. All of this simply means that the power of the planned replication with $n = 100$ is only .41; if we were

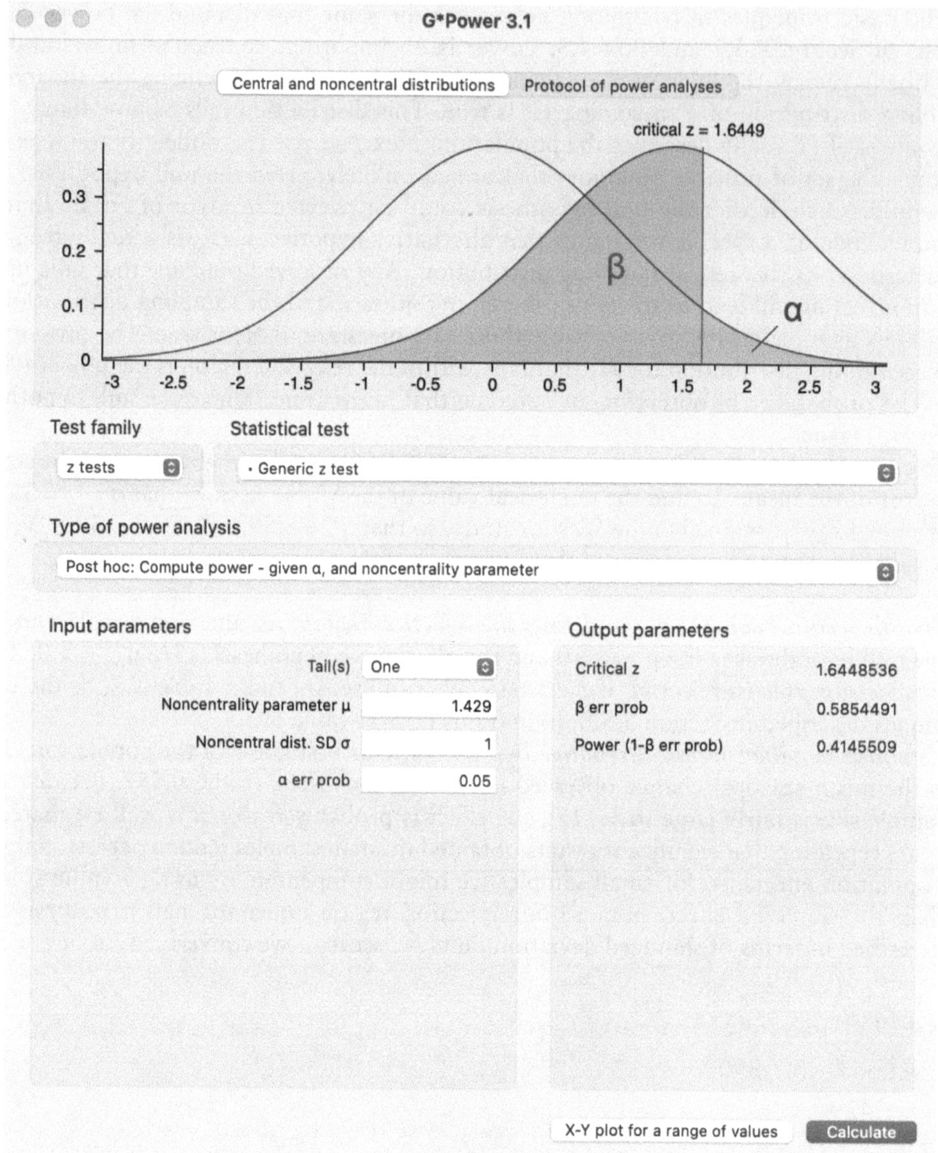

Figure 5.4 Null and alternative normal distributions in G*Power.

to repeat that replication many times, we would expect only 41% of them to provide a significant result.

The distributions at the top of the Figure 5.4 are helpful for understanding the power calculations. The left-hand distribution is the sampling distribution assuming the null hypothesis is true. It is centered on 0 because H_0: $\mu_{change} = 0$, and the standard deviation is 1 because we are working with z scores, not raw data. The distribution on the right is the sampling distribution assuming that the specific alternative, $H_A = 1.429$, is true. It has a mean of 1.429 and a standard deviation of 1. The rejection region is also shown. It is the shaded area labeled α that is under the null distribution and above the critical value of z, 1.645, which is marked with a vertical line. Finally, power is the unshaded area under the alternative distribution above $z = 1.645$. By eye, we can estimate that power is a bit less than 0.5 because the critical value that defines the rejection region falls above the mean of the alternative distribution. This visual check is consistent with the reported power of 0.41.

We can also compute power using the distributional calculators described in Section 5.4.3. For example, using the *pnorm* function in R, *pnorm*(1.645, mean = 1.429, sd = 1, lower.tail = FALSE) returns 0.414, the same as the power obtained in G*Power and visualized in Figure 5.4. This command calculates the area above the critical z score of 1.645 in a normal distribution with a mean of 1.429 and a standard deviation of 1, as shown in the right-hand distribution in Figure 5.4. The analogous calculation in SPSS is 1 – Cdf.Normal(1.645, 1.429, 1).

Whatever calculational method we choose, the result is the same: If the mean change score in the population is really 0.557, the planned experiment with $n = 100$ has only a .41 probability of resulting in a rejection of the null hypothesis that there is no change in depression over the two seasons. The low power serves to remind us of the critical effect that variability has on our inferences. Despite what many laboratory scientists would consider to be a large sample, power is low against what appears to be a reasonable alternative hypothesis (based on an actual study) and, accordingly, the Type 2 error rate is very high. The situation is considerably better – though hardly great – if we plan to replicate the study using the same sample size, $n = 215$; power against the specific alternative ($\mu = .557$) with an n of 215 instead of 100 is approximately .67. Clearly, the variability in Beck Depression scores makes it difficult to achieve high power to test hypotheses, even with relatively large samples.

The preceding developments detail the calculation of power using the z statistic and the normal distribution. However, it is important to understand those aspects of our procedure that are general to calculating the power of all statistical tests. The point is not that we need to know how to calculate power; computer software will usually do this for us. The point is that we should understand exactly what is meant by the power of an experiment, and the factors that affect it.

As we have just illustrated, although the relevant test statistic and sampling distribution vary across different situations, the steps involved in power calculations remain the same. Thus, the steps that we followed in computing power for a test based on the normal distribution were the same steps that we followed in computing power for a test based on the binomial distribution (see Box 4.3). Further, the factors affecting power are the same regardless of the test.

5.6.2 *Factors Affecting Power*

In general, the power of a test depends upon several factors, all of which have effects qualitatively like those noted in the discussion of power in Chapter 4.

1. *Power increases as α increases* because the increase in α requires an increase in the range of values included in the rejection region. If $\alpha = 0.10$, the critical value in Figure 5.4 shifts from 1.645 to 1.28, increasing the rejection region, and consequently the area above it.

2. *Power is affected by the nature of* H_0 *and* H_1. If the statement of H_1 is two-tailed, with $\alpha = 0.05$, the decision is to reject H_0 if $z < -1.96$ or $z > 1.96$. Then there would be two critical values in Figure 5.4, −1.96 and 1.96, and power would correspond to the sum of the areas to the right of 1.96 and to the left of −1.96 under the H_A distribution. As we can see in the H_A distribution of Figure 5.4, the probability of $z < -1.96$ if H_A is true is essentially zero, and the probability that $z > 1.96$ is less than the probability that $z > 1.645$. Therefore, the one-tailed test is more powerful against the specific alternative, $\mu_A = .557$. On the other hand, this one-tailed test has virtually no power against specific alternatives of the form $\mu < \mu_{hyp}$ whereas the two-tailed test has the same probability of rejecting H_0 against these alternatives as against those of the form $\mu > \mu_{hyp}$.
3. *Power is affected by the population variance and the sample size.* Decreased σ and increased n result in smaller standard errors. As the *SE* decreases, the sample mean is more likely to be close to the true parameter value. As we noted when considering confidence intervals, a smaller *SE* increases the probability of getting values of $\bar{Y}$ close to the true population mean. Accordingly, as σ decreases or n increases, we are more likely to reject the mean assumed under the null hypothesis, if that hypothesis is wrong.

Most null hypotheses are false. We should always consider whether n is so large that the detected effect is of little practical or theoretical importance. This is one reason why confidence intervals are an important part of our analyses. Very large sample sizes may sometimes result in rejection of a null hypothesis even if the effect is trivial, but the confidence interval, by providing a bounded estimate of the effect, enables us to assess its importance. In subsequent chapters, we will consider other measures of the effect size.

We can influence variability by our choice of measures and experimental design, as well as by controlling extraneous factors that might contribute to chance variability. How large an n we need will depend on the other factors we noted and the power we want, as well as the smallest size effect we want to be able to reject with that power. A sample size of as little as 40 would have provided more than the .41 power we calculated if the variance of the depression scores had been less, or if μ_A had been larger than .557. Many sources, including books (e.g., Cohen, 1988; Kraemer & Thiemann, 1987), software, and websites, enable researchers to calculate the sample size needed to have a certain level of power against a specified alternative. We will further demonstrate the use of power calculations for research planning in Chapter 6.

5.7 Validity of Assumptions

The validity of the inferences that are based on confidence intervals and hypothesis tests rests upon three key assumptions. Scores are assumed to be independently and normally distributed and to have a known standard deviation, σ. Let us consider each of these assumptions in turn.

5.7.1 The Independence Assumption

Two scores, Y_i and Y_j, are independent of each other if the probability of any value of one score is unrelated to the value of the other. In the notation of Chapter 3, we have two independent scores if $p(Y_i \mid Y_j) = p(Y_i)$. In other words, two scores are independent if knowing one score provides no information about the value of the other score. If scores are not independently distributed, the confidence interval for μ may be invalid and Type 1 error rates and power associated with tests of hypotheses about μ may be seriously affected. In the *Seasons* data, spring and winter scores are likely to be correlated, and therefore not independent; individuals who are more depressed than others in the winter will also tend to be so in the spring. For this reason we cannot treat the winter and spring samples as independent of each other.

Suppose that we had ignored the independence assumption for the spring and winter scores. For that data set, the confidence interval for the difference would have been overly wide and power would have been low relative to that in the correct analysis. The reason for this is that the standard error of the difference between two independent means is usually larger than that for two dependent means; Appendix 5.1 shows why this is so. By treating the means as independent when they are not, we use too large an estimate of the variability in this research design. In other research designs, the result of a failure to take nonindependence into account in the data analysis may result in an inflation of Type 1 error rate.

Of course, in our treatment of the depression scores, we did use the results for both seasons; however, we created a single change score for each participant. The mean and standard deviation we reported were based on these change scores. By computing the difference between the winter and spring scores for each subject, we incorporated information about the relation between the scores into the resulting change score. Further, because we have just one such difference score for each participant, we have satisfied the assumption of independence. Assuming that our subjects were randomly sampled, the change scores can be analyzed to provide inferences about μ_{change}, the mean of the population of change scores.

5.7.2 The Normality Assumption

Skewness and kurtosis statistics, together with various data plots described in Chapter 2, indicated that the distribution of depression change scores was symmetric, but not normal. However, the issue for any assumption is not whether it is correct but whether it is sufficiently close to being correct that our inferences are valid. In the example of the depression change scores, the departure from normality is not likely to be a problem. Our inferences assume that the sampling distribution of the mean change score is normal. Even if the population of scores is not normal, the central limit theorem implies that the sampling distribution of the mean is approximately normal because we have a large number (215) of change scores.

5.7.3 The Assumption of a Known Value of the Standard Deviation

Although we can be certain that the value of s calculated from our data is not exactly the same as σ, s is an unbiased estimate of σ, and a consistent one. Consistency implies that as the sample grows larger, the probability increases that s is close to σ. Because our sample size is large, using the sample value of the standard deviation in our calculations in place of the true (unknown) population value should not present a problem. Some evidence that violations of the normality and known-σ assumptions are not critical when n is large derives from a computer study we conducted. We drew 2,000 samples of 215 scores each from a population distribution with characteristics – mean, standard deviation, skewness, and kurtosis – like those of the sample of 215 scores. We then calculated a confidence interval for each sample. The proportion of samples yielding limits containing the mean of the simulated population was .945, quite close to the theoretical value of 95%. In terms of a two-tailed test of the null hypothesis, this implies a rejection rate of .055.

The close approximation of confidence and significance values to the theoretical values indicates that even if the population is not normally distributed, the normal probability function and an estimate of the population standard deviation can provide adequate inferences when the sample is large. This raises the question of how large is large enough. There is no simple answer to this. Using the population we simulated and drawing samples of size 30 instead of 215, .940 of the 2,000 confidence intervals contained the true value of the population mean, a reasonable approximation to the theoretical value of 95%. However, the results may not generally be this satisfactory with small samples, particularly if the population distribution deviates more markedly from normality.

5.8 Summary

Chapter 5 has continued the development of inferential concepts and principles begun in Chapter 4. Central to both chapters has been the emphasis on the distributions of random variables, but in Chapter 5 we have used a continuous distribution to illustrate several important points concerning estimation and hypothesis testing:

- Every statistic that can be calculated has a sampling distribution whose properties are important in assessing criteria for estimators, and affecting the width of confidence intervals, and influencing the power of hypothesis tests.
- The quality of population parameter estimates depends on several criteria. A good estimator is unbiased; the mean of its sampling distribution equals the parameter. A good estimator is consistent; as sample size increases, the estimator is more likely to be close to the estimated parameter value. And a good estimator is efficient relative to other estimators of the same parameter; its sampling distribution has a smaller variance about the parameter.
- The central limit theorem states that the sampling distribution of the mean will tend toward normality for large samples even when the underlying population distribution is not normal. However, the speed with which the sampling distribution approaches normality as a function of sample size will depend on how closely the underlying distribution resembles the normal.
- Confidence intervals about population means will be narrower, and hypothesis tests involving means will be more powerful, as the variability of the sampling distribution decreases; that is, as the standard error of the mean (the *SEM*) decreases. The *SEM* will decrease as sample size increases and as the sample variance decreases.

We illustrated the calculation of confidence intervals and hypothesis tests using probabilities obtained from the table of the normal distribution and from the normal distribution functions in R and in SPSS. Although in most instances, such calculations would be carried out with the t distribution, the concepts and principles are the same and extend to all the procedures in this book, and many others as well. The interpretation of p-values, confidence interval bounds, Type 1 errors, and power (or its complement, the probability of a Type 2 error) have been a focus of both Chapters 4 and 5. They are critical to interpreting the results of data analysis.

Appendix 5.1

The Variance of Linear Combinations

A *linear combination* of *n* observations, *L*, has the general form.

$$L = w_1 Y_1 + w_2 Y_2 + \ldots + w_n Y_n$$

where the weights, the w_i, are any real numbers. A special case is one in which $n = 2$ and the weights are either both 1, in which case *L* is a sum, or 1 and −1, in which *L* is a difference. We consider this case first.

Variances of Sums and Differences. The variance of a sum or difference of two random variables depends both on the variances of the variables and on their covariance. The covariance is a function of the variances and the correlation coefficient; as we stated in Chapter 2, the sample correlation coefficient of two variables, *X* and *Y*, is defined as

$$r_{XY} = \frac{\Sigma(X - \bar{X})(Y - \bar{Y}) / (n-1)}{s_X s_Y} \tag{5.13}$$

and the covariance is defined as the numerator, $\Sigma(X - \bar{X})(Y - \bar{Y})/(n - 1)$. We will denote the covariance as s_{xy}. First consider the variance of $X + Y$; by definition of the sample variance,

$$s_{X+Y}^2 = \frac{1}{n-1} \Sigma\left[(X+Y) - \left(\overline{X+Y}\right)\right]^2$$

But

$$\overline{X+Y} = (1/n)\sum_i (X_i + Y_i) = \bar{X} + \bar{Y}$$

Therefore,

$$\begin{aligned} S_{X+Y}^2 &= \frac{1}{n-1} \Sigma\left[(X - Y) + (\bar{X} - \bar{Y})\right]^2 \\ &= \frac{1}{n-1} \sum_i \left[(X_i - \bar{X}) + (Y_i - \bar{Y})\right]^2 \\ &= \frac{1}{n-1} \sum_i \left[(X_i - \bar{X})^2 + (Y_i - \bar{Y})^2 + 2(X_i - \bar{X})(Y_i - \bar{Y})\right] \\ &= s_X^2 + s_Y^2 + 2s_{XY} \end{aligned}$$

where $s_{xy} = r_{xy} s_x s_y$, the covariance of *X* and *Y*. Therefore,

$$s_{X+Y}^2 = s_X^2 + s_Y^2 + 2r_{XY} s_X s_Y \tag{5.14}$$

The variance of the difference scores has a similar form:

$$s_{X-Y}^2 = s_X^2 + s_Y^2 - 2r_{XY} s_X s_Y \tag{5.15}$$

The only difference in the two expressions is in the sign of the covariance term. Note that if the two variables are uncorrelated (i.e., $r = 0$), both the variance of $X + Y$ and $X - Y$ are $s_x^2 + s_y^2$, the sum of the variances of X and Y.

Analogous expressions hold when we consider population parameters. The population variances for $X + Y$ and $X - Y$ are

$$\sigma_{X+Y}^2 = \sigma_X^2 + \sigma_Y^2 + 2\rho_{XY}\sigma_X\sigma_Y \tag{5.16}$$

and

$$\sigma_{X-Y}^2 = \sigma_X^2 + \sigma_Y^2 - 2\rho_{XY}\sigma_X\sigma_Y \tag{5.17}$$

where the Greek letter rho (ρ) is the population correlation coefficient. If X and Y are uncorrelated, the variances of both $X + Y$ and $X - Y$ are $\sigma_x^2 + \sigma_y^2$.

The General Case. We previously defined L as $L = w_1Y_1 + w_2Y_2 + \ldots + w_nY_n$. The variance of L can be shown to be an extension of the formulas for two observations:

$$s_L^2 = \sum_j w_j^2 s_j^2 + \sum_{j \neq j'}\sum_{j'} w_j w_{j'} r_{jj'} S_j S_{j'} \tag{5.18}$$

For example, given two test scores, $T1$ and $T2$, and a final exam score, F, and supposing the grade is calculated by weighting the first two exams by 1/4 each and the final by 1/2, the variance of the grade is

$$\begin{aligned} s_{grade}^2 &= (1/16)\left[s_{T1}^2 + s_{T2}^2\right] + (1/4)s_F^2 + (2)(1/4)(1/4)\left(s_{T1}s_{T2}r_{T1,T2}\right) \\ &\quad + (2)(1/4)(1/2)\left(s_{T1}s_F r_{T1,F}\right) + (2)(1/4)(1/2)\left(s_{T2}s_F r_{T2,F}\right) \end{aligned}$$

The Variance of the Mean. An important consequence of these developments is that we can readily derive an expression for the variance of the mean of n independently distributed scores. The mean is $(1/n)(Y_1 + Y_2 + \ldots + Y_n)$. We view the variance of a score as the variance of scores in that position in the sample over many independent samples. Then $s_1^2 = s_2^2 = \ldots s_n^2 = s^2$ and each value of w is $1/n$. Because the scores are independently distributed, all covariances equal zero, and substituting into Equation 5.18,

$$\text{var}(\bar{Y}) = n \times s^2 / n^2 = s^2 / n$$

and the standard error of the mean is (as defined in Chapter 2).

$$SEM = \sqrt{s^2 / n}$$

Exercises

The answers to several exercises involve calculating z scores and then using Appendix Table C.2 or distributional calculators in R or SPSS. In some cases, it may be helpful to draw a normal curve, shading the area asked for. Also, in several exercises, we ask for the variance (or standard deviation) of means, or differences among means. We do this for cases in which the means are dependent and for cases in which they are independent. A review of Appendix 5.1 will be helpful in finding the answers to those exercises.

5.1 [*z* scores] A standard IQ test yields scores that are normally distributed with $\mu = 100$ and $\sigma = 15$. Y is a randomly selected score on the test.

a) What is the probability that a randomly selected student will score higher than 115?
b) (i) What is $p(Y > 130)$? (ii) $p(85 < Y < 145)$? (iii) $p(Y > 70)$? (iv) $p(70 < Y < 80)$?
c) What scores define the middle 80% of the distribution?
d) What is the 75th percentile (score such that it exceeds 75% of the scores)?
e) What is the probability that the mean IQ of a group of 10 randomly selected students will be greater than 105?

5.2 [*z* scores] On a new test of logical reasoning, the mean and standard deviations for a population of students from School A are $\mu = 170$ and $\sigma = 50$; for students from School B, $\sigma = 200$ and $\sigma = 60$.

a) What is the probability that a randomly School B student will have a score greater than 170?
b) What is the probability that the mean of a group of nine randomly sampled School B students will be greater than 170?
c) Assume that many pairs of School A and School B scores are drawn from their respective populations and a difference score, $d = B - A$, is calculated for each pair. What is the mean and standard deviation of the population of such difference scores (see Appendix 5.1)?
d) What is the probability that a randomly selected School B student will have a higher score than a randomly selected School A student? [Note: $p(B > A) = p(B - A > 0)$.].

5.3 [Linear combinations of variables] Assume that X and Y are independently and normally distributed variables. For X, $\mu = 30$ and $\sigma = 20$; for Y, $\mu = 20$ and $\sigma = 16$.

a) What is the probability of sampling an X score (i) < 25? (ii) > 60? (iii) between 15 and 40?
b) What is $p(X > \mu_Y)$?
c) Let $W = X + Y$. (i) What is the mean of W? (ii) What is the variance of W? (iii) What is $p(W > 35)$?
d) An individual's X score is at the 85th percentile (i.e., it exceeds .85 of the population of X scores); her Y score is at the 30th percentile of the Y distribution. What percent of the population of W scores does her W score exceed?

5.4 [Normal approximation to binomial] In this problem, we will use the normal probability distribution to test a hypothesis about a proportion.
A population of individuals has a disease that has been treated and symptoms are no longer present. However, 40% of this population suffers a recurrence of the symptoms within one year. A new drug developed to prevent recurrence of the disease is tried on a sample of 48 patients. We wish to determine whether the probability of failure (i.e., recurrence of symptoms) is less than 0.4.

a) Let π = the probability of failure in the population sampled. State H_0 and H_1.

b) Let p = probability of failure in the sample. If the null hypothesis is true, the mean of the sampling distribution of p is π (.4) and its variance is $\pi(1 - \pi)/n$. In the study, only 12 of the 48 participants suffered a recurrence of symptoms after one year. Use the normal probability distribution to test the null hypothesis. State your conclusions.
c) We might want a better sense of the true probability of recurrence of symptoms. Calculate a 95% confidence interval and interpret your result.
d) In parts (b) and (c), we used the normal probability (z) table to draw inferences about a population probability. (i) What assumption about the sampling probability of p is implied by our methods? (ii) What justifies this assumption? (iii) Would the assumption be as justifiable if the sample had only 10 people in it? Explain.

5.5 [Sampling from nonnormal distribution] A population of scores is uniformly distributed between 0 and 1. This means that all values between 0 and 1 are equally likely and that $F(y)$ (the probability of sampling a score less than y) equals y. For example, $p(Y < .8) = .8$. The mean and standard deviation of this uniformly distributed population are .5 and $1/\sqrt{12}$.

a) (i) What is $p(Y < .6)$? (ii) What is the probability that in a sample of two scores both are less than .6? Express your answer as a probability raised to a power. (iii) What is the probability that in a sample of 20 scores all are less than .6?
b) Assume that we draw many samples of 20 scores and calculate the mean of each sample. Describe the shape of the sampling distribution. What are its mean and standard deviation?
c) Based on your answer to part (b), what is the probability that the mean of a sample of 20 scores is less than .6?
d) Briefly state your justification for your approach to part (c). Would the same approach be appropriate in answering part (a) (iii)? Explain.

5.6 [Expected values and variances] We have a population in which $\pi = p(X = 1) = .2$ and $1 - \pi = p(X = 0) = .8$.

a) (i) Calculate the population mean, μ_x, and variance, σ_x^2 [$E(X)$ and $var(X)$]. Note that $E(X) = (\pi)(1) + (1 - \pi)(0)$ and $\text{var}(X) = E(X^2) - [E(X)]^2$. (ii) Assume that we draw samples of size 3 from this population. What would be the variance of the sampling distribution of the mean?
b) Assume we draw samples of size 3. If we define the outcome of the experiment as a value of Y where $Y = \Sigma X$, there are four possible outcomes. Complete the following table (S^2 is the variance with n in the denominator whereas s^2 has $n - 1$ in the denominator):

Y	$p(Y)$	$\bar{X}$	S_X^2	$S_{\bar{X}}^2$	$s_{\bar{X}}^2$
0	$.8^3 = .512$				
1	$(3)(.8^2)(.2) = .384$				
2	$(3)(.8)(.2^2) = .096$				
3	$.2^3 = .008$				

c) Using the entries in the above table, find (i) $E(Y)$, (ii) $E(\bar{X})$, (iii) $E\left(S_{\bar{X}}^2\right)$, and (iv) $E\left(s_{\bar{X}}^2\right)$.
d) How do $E(Y)$ and $E(\bar{X})$ compare with the value of $E(X)$ obtained in part (a)?
e) How do $E\left(S_{\bar{X}}^2\right)$ and $E\left(s_{\bar{X}}^2\right)$ compare with the value of $\text{var}(\bar{X})$ obtained in part (a)?

f) What do your answers to parts (d) and (e) say about which sample statistics are biased or unbiased estimators?

5.7 [Confidence interval, hypothesis testing, and power] A national survey of many college students in 1983 yielded a mean "authoritarianism" score of 52.8 and a standard deviation of 10.5. For all practical purposes, we may view these as population parameters.

a) Suppose we wish to examine whether authoritarian attitudes have increased in the decades since the survey by examining a random sample of 50 students. State H_0, H_1, and the rejection region assuming $\alpha = .05$.
b) Assume that the mean of the sample of 50 scores is 56.0. Carry out the significance test and state your conclusion.
c) Suppose the true population mean is now 57.00. What is the power of your significance test?
d) Based on your sample (and assuming the population variance has stayed the same), what is the 95% confidence interval for the current population mean of authoritarianism scores?
e) In part (d) you found the 95% confidence interval. What exactly *is* a 95% confidence interval? What exactly is supposed to happen 95% of the time?

5.8 [Properties of estimators] Two random samples are available from a population with unknown mean. Sample 1 has n_1 scores and has a mean of $\bar{Y}_1$, sample 2 has n_2 scores and a mean of $\bar{Y}_2$. Consider two possible estimates of the population mean, μ_y, that are based on both samples: One estimate is the unweighted mean of the sample means, UM, where

$$\text{UM} = \frac{\bar{Y}_1 + \bar{Y}_2}{2}$$

and the other is the weighted mean of the sample means, WM, where

$$\text{WM} = \frac{n_1\bar{Y}_1 + n_2\bar{Y}_2}{n_1 + n_2}$$

In answering the following questions, remember that $E(\bar{Y}_1) = E(\bar{Y}_2) = \mu_y$

a) Is UM an unbiased estimator of the population mean? Show why or why not.
b) Is WM an unbiased estimator of the population mean? Show why or why not.
c) Assume that $n_1 = 20$ and $n_2 = 80$, and the population variance, σ^2, equals 4. Calculate the variance of UM.
d) Calculate the variance of WM. Is UM or WM a better estimate of the population mean? Why?
e) Is either UM or WM a better estimator of the population mean than $\bar{Y}_1$ or $\bar{Y}_2$? Why?

5.9 [Linear combinations and effect size] Assume that we have a treatment (T) and a control (C) population for which μ_T is larger than μ_C. Assume that both populations are normally distributed and have the same variance, σ^2.

a) Let T and C be randomly sampled scores from their respective populations. Assume that μ_t is $.5\sigma$ larger than μ_c. Express the mean and variance of the sampling distribution of $T - C$ as a function of σ (see Appendix 5.1).

b) What is the probability that a randomly chosen score from the treatment population will be higher than a randomly chosen score from the control population? [i.e., what is $p(T - C > 0)$?] You don't need numerical values for μ_t, μ_c, and σ.
c) (i) what is the probability that the mean of nine randomly chosen scores from the treatment population will be larger than the mean of nine randomly chosen scores from the control population? (ii) What is the probability if μ_t is $.2\sigma$ larger than μ_c?

5.10 [Effect of stratified sampling on variance] A population of voters consists of equal numbers of conservatives and liberals. Furthermore, .9 of the liberals prefer the Democratic candidate in the upcoming election, whereas only .3 of the conservatives prefer the Democratic candidate.

a) What is P_d, the probability of sampling an individual from the entire population who prefers the Democratic candidate?
b) The variance of a proportion p is $p(1 - p)/n$ (see Appendix B for the proof). With this in mind, what is the variance of the sampling distribution of p_d, the proportion of Democratic voters in a sample of 50 individuals who are randomly selected from the population of voters?
c) Suppose you are a pollster who knows that the population is equally divided between liberals and conservatives, but you don't know what the proportion of Democratic voters is. You sample 50 individuals with the constraint that 25 are liberals and 25 are conservatives; this is referred to as stratified sampling.

i. From the information presented at the start of this problem, what is the variance of the sampling distribution of $p_{d|l}$, the proportion of Democratic voters in a sample of 25 liberals?
ii. What is the variance of the sampling distribution of $p_{d|c}$, the proportion of Democratic voters in a sample of 25 conservatives?
iii. The proportion of Democrats in the stratified sample is $p_d = (1/2)(p_{d|l} + p_{d|c})$. What is the variance of the sampling distribution of D when stratification is employed?
iv. In view of your answers to (b) and (c) (iii), discuss the effect of stratification.

5.11 [Confidence intervals and hypothesis testing] Instruction in problem-solving methods raised the scores of a sample of 225 students by an average of two points. A difference between the *before* and *after* scores was calculated for each student and the standard deviation of the differences was 13.6.

a) Find the standard error of the difference in the means.
b) Using the result in part (a), find the .95 confidence interval for the difference in the means.
c) Carry out the z test of the null hypothesis of no instructional effect; $\alpha = .05$.

5.12 [Comparing between- and within-participant designs] Assume that we estimate the population variances from a previous study to be $s^2_1 = 160$ and $s^2_2 = 240$.

a) Assume that there is a new experiment with n participants in each of two groups, one from each population. Express the variance of the difference between the means as a function of the variance estimates and n. (Appendix 5.1 should be helpful here.)

b) Using your results from part (a), calculate the n needed in each group to have a .95 confidence interval that is at most five points wide.

c) Suppose that instead of two groups of participants, we tested one group of n participants under both of the conditions in part (a). Still assuming the same variances, and assuming that the correlation of scores is 0.5, how many participants should we run now to have a .95 confidence interval of five points? (Note: The variance of difference scores is $s_d^2 = s_1^2 + s_2^2 - 2s_1s_2r_{12}$.)

d) Comparing your answers to (b) and (c), if you had a limited supply of participants, which design would you choose? How does the size of the correlation affect the difference between the two designs?

5.13 [Confidence interval and hypothesis testing] Following are summary statistics for the total cholesterol scores for the winter (*TC*1) and spring (*TC*2) seasons for males; the data are at the website.

$\bar{Y}_1 = 224.059 \quad \bar{Y}_2 = 218.818$

$s_1 = 40.794 \quad s_2 = 40.113$

$r = .855 \quad n = 220$

a) Find the standard error of the difference in the means.

b) Using the result in part (a), find the .95 confidence interval for the difference in the two seasonal means.

c) Carry out the z test of the null hypothesis of no seasonal effect; $\alpha = .05$.

5.14 [Sample size and power]

a) Given the statistics in Exercise 5.13, how large should n be to have a .95 confidence interval of four points?

b) Cohen (1988) has defined a standardized effect (d_z) as $\bar{d} / s_d$, where $\bar{d}$ is the difference between the means and s_d is the standard deviation of the difference scores. Calculate the standardized effect size for the data of Exercise 5.13.

c) Assume that we wish to replicate our study of cholesterol differences in a new sample of males. If we have only 100 subjects available, what is the power to detect the standardized effect calculated in part (a)? Assume a one-tailed test with $\alpha = .05$.

5.15 [Confidence interval with real data] In the *TC Data* file (see the *Seasons* page on the book's website), we created an educational level (*el*) variable. If *schoolyr* = 1, 2, or 3, *el* = 1; if *schoolyr* = 4, 5, or 6, *el* = 2; and if *schoolyr* = 7 or 8, *el* = 3. *el* = 1 corresponds to individuals with a high school education or less, *el* = 2 corresponds to those with education beyond high school but not including the bachelor's degree, and *el* = 3 corresponds to those with a college or graduate school education.

a) Calculate the standard error of the sampling distribution of the difference between the means of the *el* = 1 and the *el* = 2 groups.

b) Calculate a 95% confidence interval for the difference in the *TC* population means (*tc_mean*) between the *el* = 1 and the *el* = 2 groups. What can you conclude about this difference based on the confidence interval?

5.16 [Real data with outliers] In this exercise, we use the *mean_d* variable in the *Beck_D* file linked to the *Seasons* page on the book's website. This is an average of the four seasonal depression scores for those individuals who were tested in all four seasons. Note that there are missing values of the *mean_d* measure because not all individuals were tested in all four seasons.

a) Create a new variable, *el*, as described in problem 5.15. In SPSS, use the "*Recode in Different Variables*" function of the *Transform* menu, then choose the variable *schoolyr* to transform to *el*, and select ranges of *schoolyr* to assign to specified values of *el*. Each separate assignment must *added* to the list of *Old New Values* before clicking *Continue*. In R, one option is to use the *mutate* and *case_when* functions in the {dplyr} package.
b) Tabulate descriptive statistics separately for *el* = 1 and *el* = 2, and compare these. Then graph the two data sets in any way you choose, relating characteristics of the plots to the statistics. Comment on location, spread, and shape.
c) Using the statistics you obtained, construct a .95 confidence interval for $\mu_1 - \mu_2$ (*el* = 1 – *el* = 2 *Beck_D* population means) and decide whether the means differ significantly at the .05 level as a function of education level. Do you think the assumption of normality is valid?
d) Outlying scores frequently influence our conclusions. Considering the education levels *el* = 1 and 2 separately, what values would be outliers?
e) Redo parts (b) and (c) with the outliers excluded. How does this affect your earlier conclusions?

5.17 [Sample size planning] We plan a study of cholesterol levels in a population of patients. Based on the *TC* (total cholesterol) data in the present *Seasons* study, we assume that the population standard deviation is 30. We would like power = .80 to detect effects of small size (.2σ or 6 points) above a level of 200.

a) What are the null and alternative hypotheses for the proposed study?
b) What is the specific alternative hypothesis?
c) How many subjects should we recruit for our study? Assume α = .05.

Notes

1 Technically, $\hat{\theta}$ is a consistent estimate of θ if the probability approaches 1 as *n* increases that the absolute distance between $\hat{\theta}$ and θ is less than any arbitrarily chosen small value.
2 As noted earlier, because *n* is large, the results based on the normal distribution are quite similar to those based on the *t* distribution; the confidence limits using the *t* distribution are 218.952 and 230.415. Generally, we base inferences about population means on the *t* distribution because we usually do not know the value of σ.
3 Notice the form of this test statistic: it's the difference between the observed and hypothesized values, divided by a measure of variability for the observed statistic. We will see this general form again.
4 Don't confuse the confidence level with the input to the function: *qnorm*(.95) returns the *z* score that falls above 95% of the scores. This is the upper critical value, +1.645. Using *qnorm*(.05) or *qnorm*(.95, lower.tail = FALSE) provides the lower critical value, –1.645, which is also evident from the symmetry of the normal distribution.

Chapter 6

The *t* Distribution and Its Applications

6.1 Overview

In this chapter we will consider two possible ways to design an experiment with two conditions: The *independent-groups design* and the *correlated-scores design*. We introduce the t distribution and present assumptions and calculations related to its application to data from these two experimental designs. The major goals of the chapter are as follows:

- *To compare the independent-groups and correlated-scores designs*, focusing on their relative merits in those instances in which both designs are applicable.
- *To provide a description of the* t *distribution*. This will involve the definition of the t statistic, a description of the distribution, both when the null hypothesis is true (the *central* t *distribution*) and when it is false (the *noncentral* t *distribution*).
- *To present confidence intervals, hypothesis tests, and power analyses based on the* t *distribution*, including the assumptions and calculations for the two designs under consideration.
- *To present a standardized effect size measure and its confidence interval*. The difference between means is the *unstandardized*, or *raw*, *effect*. It is a function of the variability in the data, and therefore varies across experiments. Standardized effect sizes take that variability into account, providing a measure of effect size that is independent of the data scale.
- *To describe how power calculations are useful in planning the research*, including justifying a sample size and choosing a design.

The organization of this chapter is as follows. We first compare the independent-groups and correlated-scores designs. Following that, we define the t statistic and describe its distribution. We then present data sets and the related data analyses, including effect size estimation and power analyses, first for the independent-groups design, and then for the correlated-scores design.

Although later chapters will discuss analyses for designs that involve more than one factor and more than two conditions, the material in the present chapter will be of central importance for two reasons. First, we continue to develop inferential concepts that will be equally important in the subsequent chapters. Second, even when there are more conditions in an experiment, the researcher's hypotheses and questions often translate into a comparison of two means, or a comparison of a single mean against some hypothesized value. Some of the comparisons will be superficially more complex than those of this chapter but they

DOI: 10.4324/9781003453550-7

basically involve the same assumptions and calculations. Therefore, the material we will discuss here deserves close study.

6.2 Design Considerations: Independent Groups or Correlated Scores?

Consider a hypothetical experiment in which we wish to compare the effectiveness of a prescribed diet to reduce total cholesterol (*TC*) level with a control condition in which no diet has been prescribed. The first issue is how to design the experiment; that is, whether each participant will be tested in only one condition (i.e., control or treatment) or in both conditions. Before analyzing our hypothetical data set, we consider the relative merits of these two approaches.

In the *independent-groups*, or *between-subjects*, *design*, $2n$ participants are randomly divided into two groups, usually with the restriction that there are the same number of participants in each condition. In this case, all $2n$ scores are independent of each other. In the *correlated-scores*, or *paired-samples*, *design*, each of n participants is tested in each of two conditions, or two groups of n participants each are selected so that one member of each group is matched on one or more relevant dimensions (e.g., age, weight, gender identity, initial *TC* level) with one member of the other group. The design in which each participant is tested twice is often referred to as a *repeated-measures* or *within-participants design*, and the design involving matching of participants is often referred to as a *matched-pair design*. In these designs, a difference is calculated between the scores of each participant or between the scores of members of each matched pair, and the n difference scores are analyzed.

When both types of design are feasible, as they are in the study of effects of diet on *TC* level, the independent-groups design has two advantages. First, it involves more independent observations. Because each group provides an independent estimate of the population variance, each estimate is based on n independent scores, and altogether there are $2n$ independent scores. The correlated-scores design involves only n independent observations because only the n difference scores are independent. As we will see in Section 6.3, the power of the t test increases and the confidence interval is narrower as the number of independent observations increases. Therefore – *if all other things are equal* – the independent-groups design has a power advantage. However, all other things usually are not equal when comparing the independent-groups and correlated-scores designs.

A second potential advantage of the independent-groups design is practical. Participants are not required to return for a second session as is sometimes the case in the repeated-measures variation of the correlated-scores design, nor is there a need to obtain any additional measures to use to establish pairs, as in the matched-pairs variation.

Why, then, should we ever use a correlated-scores design? In the case of the repeated-measures design, a practical advantage is that the researcher requires half the number of participants to make the same number of observations as the corresponding independent-groups design. However, the more important motivation for using a correlated-scores design lies in a comparison of the denominators of the t statistics for the two designs. As we shall see, in the independent-groups design, the standard error of the difference between the means (*SE*) is based on an average of the two group variances, whereas in the correlated-scores design, the *SE* is based on the variance of the n

difference scores. The latter *SE* will generally be smaller. To see why, consider a data set in which the experimental and control scores are perfectly correlated. For example, the scores for 10 participants tested under both an experimental (*E*) and control (*C*) condition might be

Condition	*Subject*									
	1	*2*	*3*	*4*	*5*	*6*	*7*	*8*	*9*	*10*
E	5	10	3	7	12	13	9	4	8	15
C	7	12	5	9	14	15	11	6	10	17

Note that the difference between scores, *C* – *E*, is the same for each participant in this example. In the correlated-scores design, the standard deviation of the difference, divided by the square root of *n*, is the denominator of the *t* statistic that is used in significance tests. In this example, this standard deviation of the difference would be zero, and *any* difference between the *E* and *C* means would be significant. This would be true no matter how variable the scores within a condition were. Although we never have correlations this high, the correlations achieved by testing each participant under both conditions, or by matching participants in pairs, will usually reduce the *SE* of the mean difference considerably compared to the variability of the individual scores. Therefore, the *t* statistic will usually be larger in the correlated-scores design. Appendix 5.1 details how the correlation between scores affects the variability of their difference.

To sum up, both statistical and practical factors should influence the choice of design. With respect to statistical factors, although there are more independent observations in the independent-groups design, the denominator of the *t* statistic (the *SE* of the mean difference) will be smaller in the correlated-scores design. Therefore, the *t* test usually will be more powerful and the confidence interval narrower when the two groups of scores are correlated than when they are independent. With respect to practical considerations, in some situations the correlated-scores design is difficult or impossible to implement. It would make no sense to use both methods of arithmetic instruction on the same students, although participants could be matched on a pretest score. Furthermore, some variables such as socioeconomic status (SES) cannot be manipulated within participants. The researcher also should be aware of the possibility of "contrast effects"; some experimental treatments (e.g., one type or quantity of reward) have a very different effect when the same participant has been exposed to other treatments (e.g., other reward values) than when the participant has experienced only that treatment.

In this chapter we will run a hypothetical experiment on the effects of diet on *TC* level both ways. We first consider the analysis of data when an independent-groups design is used. In this design, we ask whether total cholesterol (*TC*) level is lower for a group on a prescribed diet than for a control group. Following that, we present data for a similar experiment, but using a correlated-scores design in which the *TC* scores are obtained for the same individuals before and after the diet. However, before we can analyze the data for these two designs, we first need to know something more about the *t* statistic and its distribution.

6.3 The t Distribution

6.3.1 The t Statistic

In Chapter 5, we considered statistics that had the general form $z = [(V - E(V)]/\sigma$, where V represents some statistic such as the sample mean, $\bar{X}$, or the difference between sample means, $\bar{X}_1 - \bar{X}_2$. $E(V)$ is the expected value of V, also referred to as the population parameter. For example, if $\bar{X}$ is the sample mean, $E(V)$ is the population mean, μ. The denominator of z is the standard error of the sampling distribution of V. In this chapter we consider a similar ratio, the t statistic; it has the general definition

$$t = \frac{V - E(V)}{s_V} \tag{6.1}$$

where V and $E(V)$ are defined as before. The statistic s_v is the sample statistic that estimates σ_v, the *SE* of the sampling distribution of V. Note that the t statistic is identical to the z statistic except that the denominator of Equation 6.1 is an *estimate* of the standard error of V rather than its actual value. An implication of this difference is that the denominator of the z statistic is a constant over replications of the experiment whereas the denominator of the t statistic will vary over samples. As in the case of z, it is assumed that scores are independently and normally distributed.

The t distribution is tabled in Appendix C.3. When we turn to that table, we find that the entries in the first column correspond to something labeled *df*, which stands for *degrees of freedom*. The concept of degrees of freedom is closely related to sample size but is not quite the same thing. Because degrees of freedom are an important concept, not only for the t distribution but also for other distributions that play a role in data analysis, they deserve further discussion.

6.3.2 Degrees of Freedom (df)

The degrees of freedom associated with any quantity are the number of *independent*, free to vary, observations upon which that quantity is based. The meaning of "independent observations" is best illustrated by an example. Suppose that we had 10 markers of different colors to pass out to a group of children. The first nine children would have some choice in colors, but the last child would not (i.e., no *df* remaining). Translating to a numerical example, suppose that we are asked to choose 10 numbers that sum to 50. We can freely choose any nine values, but the tenth number must be 50 minus the sum of the first nine. In this case, there are 10 scores but because only nine can vary independently, there are only 9 *df*. The constraint on the values that they must sum to 50 costs us a "degree of freedom."

The same type of constraint occurs when we calculate a sample standard deviation. Doing so requires us to subtract the mean from each score. Recall from Chapter 2 that the sum of the deviations of all scores about their mean is always zero; that is, $\Sigma(Y - \bar{Y}) = 0$. Rewriting, we see that $\Sigma Y = n\bar{Y}$. If the sample mean is 5 and n is 10, then the sum of the 10 scores must equal 50. Therefore, the sample standard deviation is based on $n - 1$, or 9, *df*.

At this point, it looks as if the *df* are always just $n - 1$. That is true if the statistic of interest involves only one restriction or constraint. But suppose we draw two samples from some population; one sample is of size n_1 and the other of size n_2. We want to estimate the population variance but we have two estimates, one from each sample. The estimates can be averaged, as we'll see; however, the point now is that there are two restrictions if two sample variances are computed: the sum of the n_1 scores in the first sample must equal $n_1\bar{Y}_1$ and the sum of the n_2 scores in the second sample must equal $n_2\bar{Y}_2$. There are $n_1 - 1$ *df* associated with the variance for the first sample and $n_2 - 1$ *df* associated with that for the second sample. The *df* associated with a statistic involving some combination of these two variances will be $df = (n_1 - 1) + (n_2 - 1) = n_1 + n_2 - 2$. The message is that the *df* are not necessarily the number of scores minus 1. Rather,

df = *number of independent observations*
= *total number of observations minus the number of restrictions on those observations.*

In the two-sample example, there are $n_1 + n_2$ observations and two restrictions caused by taking deviations about each of the sample means.

6.3.3 Critical Values and Areas Under the t Distribution

Appendix Table C.3 contains critical values of the *t* statistic corresponding to various levels of significance and degrees of freedom. At the head of each column are two numbers corresponding to levels of significance for one- and two-tailed tests; each row of the table corresponds to a different number of *df*. As an example of the use of the table, find the column corresponding to a one-tailed proportion of .025 (and a two-tailed proportion of .05) and the row for $df = 9$. The critical value in the cell is 2.262. We interpret this to mean that when there are 9 *df*, a *t* of 2.262 is exceeded by .025 of the sampling distribution of *t*; .05 of the distribution is greater than 2.262 or less than −2.262. Now look down the same column to the row labeled infinity. The critical value in that cell is 1.96. This means that the probability is .025 of exceeding 1.96 and the probability of $t > 1.96$ or $t < -1.96$ is .05. This is exactly the critical value in Table C.2, the normal distribution probability table. In general, the critical value of *t* decreases as the *df* increase, rapidly approaching the critical value in Table C.2 for the normal distribution. Table C.3 usually provides more accurate inferences than Table C.2, although there is little difference when sample sizes are large. This is because s_V is a consistent estimator of σ_V: as sample size increases, the sample standard error approaches the population standard error.

Assume that we want a .05 significance level for a test based on 35 *df* against a two-tailed alternative. We select the column labeled .05 for the two-tailed test and we want a row corresponding to 35 *df*. Table C.3 has values for 30 and 40 *df*, but not for 35, so we interpolate, taking a value halfway between 2.042 and 2.021, or 2.0315. The true value, 2.030, can be obtained from most statistical software packages. For example, in R we can use the *t* distribution function, *qt*(0.975, df = 35), to find the *t* statistic at the .975 quantile (i.e., with 2.5% of the area above it) when there are 35 *df* in the data. As expected, it returns 2.030. The approximation and the actual critical value yield very similar results. Alternatively, we can use the *pt* function in R to find the proportion of a specific *t* distribution below an observed value of *t*; input variables are the observed *t* and the *df*. For example, *pt*(2.030, df = 35) returns 0.97499. See Box 6.1 for more details.

With some knowledge of the *t* distribution, we are now ready to use it to draw inferences from data sets. We first consider the analysis of data from an independent-groups design.

Box 6.1

We have now encountered three different distributions, the binomial, normal, and *t* distributions. Appendix Tables C.1–C.3 provide critical values for specific tail areas. The R functions for these distributions accomplish the same purpose, with additional flexibility. They share a similar structure, varying primarily in terms of the input parameters that are appropriate for each distribution. In this table, replace the "*dist*" part of each function name with the relevant distribution abbreviation (e.g., *binom*, *norm*, or *t*) and enter specific parameter values as needed. Note: *ncp* is short for the noncentrality parameter, which we introduce in Section 6.8.1 and call δ. Also note that, in SPSS, the distributional functions for *t*, *CDF.T*, and *IDF.T* work similarly to those for R, as detailed in Chapter 5 for the normal distribution.

R function	*Input value*	*Distribution-specific parameters*	*Default parameter values*	*Returns*
pdist	*q* = score	*binom*: size = size of each sample, prob = π	lower.tail = TRUE	$p = P(y \le q)$
		norm: mean, sd	mean = 0, sd = 1, lower.tail = TRUE	
		t: df, ncp	ncp = 0, lower.tail = TRUE	
qdist	*p* = probability	*binom*: size = size of each sample, prob = π	lower.tail = TRUE	*q* = Critical value with area *p* below it
		norm: mean, sd	mean = 0, sd = 1, lower.tail = TRUE	
		t: df, ncp	ncp = 0, lower.tail = TRUE	

6.4 Data Analyses in the Independent-Groups Design

The data file, *Table 6_1TC_IG* (found by a link from the *Tables* page on the website for this book) contains data for a design in which there are two independently sampled groups of 36 scores each. The first column in the file, labeled "Treatment," designates two types of treatment: control and diet. The second column, labeled *TC_level*, contains the total cholesterol scores for each of the two groups whose names appear in the first column. The group means and variances are presented in Table 6.1.

Table 6.1 Total cholesterol (*TC*) statistics from an independent-groups design

Group	*Means*	*Variances*
Control	224.50	2246.51
Diet	207.71	1581.75

Ordinarily, we would begin our analyses by examining data plots and additional numerical summaries, but we will leave this data exploration stage as an exercise for the student. However, before proceeding to the inferential analysis of the *TC* data, we need to develop a formula for the standard error of the difference between the drug and control group means.

6.4.1 *The Standard Error (SE) of $\bar{Y}_1 - \bar{Y}_2$*

We earlier defined the *t* statistic as $t = [(V-E(V)]/s_v$. In the current example, the random variable, *V*, is the difference in *TC* level between the two group means: $\bar{Y}_c-\bar{Y}_d$ (control - diet means) or, in general, $\bar{Y}_1 - \bar{Y}_2$. S_v, the *SE* of the difference between the sample means, can be understood by considering a sampling experiment. Suppose we drew many independent pairs of random samples of sizes n_1 and n_2 from two independently and normally distributed populations of scores. If we carried out this sampling procedure, we could compute a difference between the means for each pair of samples drawn. The standard deviation of the sampling distribution of this difference, the *SE*, is the quantity to be estimated by the denominator of the *t* in the two-group study. The issue is how to calculate that estimate.

If the $n_1 + n_2$ *scores are independently distributed*, the variance of the sampling distribution of the difference between the two group means is the sum of the variances of the means (see Appendix B for a proof). Then the *SE* of the difference is

$$\sigma_{\bar{Y}_1-\bar{Y}_2} = \sqrt{\frac{\sigma_1^2}{n_1} + \frac{\sigma_2^2}{n_2}} \tag{6.2}$$

In addition to the assumption that the scores are drawn from independently normally distributed populations, using the *t* distributions of Appendix Table C.3 requires one additional assumption, that the two population variances are equal; that is, $\sigma^2_1 = \sigma^2_2 = \sigma^2$. This is usually referred to as the assumption of *homogeneity of variance*. Under this assumption, Equation 6.2 becomes

$$\sigma_{\bar{Y}_1-\bar{Y}_2} = \sqrt{\frac{\sigma^2}{n_1} + \frac{\sigma^2}{n_2}} = \sigma\sqrt{\frac{1}{n_1} + \frac{1}{n_2}} \tag{6.3}$$

We have a single population variance (σ^2) and two possible estimates of it, the variances of the two groups of scores sampled in an experiment. To obtain the best single estimate of the σ, we need to average the two group variances and take the square root of the result. Because variance estimates are consistent statistics, the estimate based on the larger group is more likely to be close to the true variance. Therefore, the best estimate of σ^2 is a weighted average of the two group variances. This is the *pooled-variance estimate*, or, s^2_{pooled}, and is calculated as

$$s^2_{pooled} = \left[\frac{n_1 - 1}{n_1 + n_2 - 2}\right]s_1^2 + \left[\frac{n_2 - 1}{n_1 + n_2 - 2}\right]s_2^2 \tag{6.4}$$

Note that the weight on each group variance in Equation 6.4 is obtained by dividing the *df* for that group by the sum of the *df* for the two groups, $[(n_1 - 1) + (n_2 - 1)]$. The *df* rather than the n are used in these weights because this yields an unbiased estimate of $\sigma^2_{\bar{Y}_1 - \bar{Y}_2}$. The pooled-variance estimate can also be written as

$$s^2_{pooled} = \frac{SS_1 + SS_2}{n_1 + n_2 - 2} \tag{6.5}$$

where the SS for a group is a sum of squared deviations of the scores about the mean of the group; that is,

$$SS_j = \sum_i \left(Y_{ij} - \bar{Y}_{ij}\right)^2 \tag{6.6}$$

This quantity is usually referred to as the *sum of squares*, and it will play an important role in the remaining chapters.

We can now state the expression for the estimate of the SE of the sampling distribution of the difference of two independent means:

$$s_{\bar{Y}_1 - \bar{Y}_2} = s_{pooled}\sqrt{\frac{1}{n_1} + \frac{1}{n_2}} \tag{6.7}$$

When $n_1 = n_2 = n$, as is frequently the case in experimental research, Equation 6.7 leads us to

$$s_{pooled} = \sqrt{\left(s_1^2 + s_2^2\right)/2} \tag{6.8}$$

and Equation 6.8 becomes

$$s_{\bar{Y}_1 - \bar{Y}_2} = s_{pooled}\sqrt{2/n} \tag{6.9}$$

For the TC data, from Equation 6.8,

$$s_{pooled} = \sqrt{(2246.51 + 1581.75)/2} = 43.75$$

and from Equation 6.9,

$$s_{\bar{Y}_1 - \bar{Y}_2} = 43.75\sqrt{2/36} = 10.31$$

With this estimate of the standard error, we can now proceed to draw inferences about the difference between the means of the control and diet populations.

6.4.2 Confidence Intervals for the Independent-Groups Design

The general form of the confidence interval, with confidence level $1 - \alpha$, is

$$p\left[V - \left(t_{df,\alpha/2}s_V\right) \leq E(V) \leq V + \left(t_{df,\alpha/2}s_V\right)\right] = 1 - \alpha \tag{6.10}$$

and the confidence limits are

$$\text{CI} = V \pm \left(t_{df,\alpha/2} s_V\right) \tag{6.11}$$

In other words, the CI is the observed statistic of interest +/− a critical value multiplied by a measure of variability for the statistic. For our *TC* example, $df = (2)(36 - 1)$, or 70, and, assuming we want 95% confidence, $\alpha/2 = .025$. Then Equation 6.11 becomes

$$\text{CI} = \left(\bar{Y}_1 - \bar{Y}_2\right) \pm \left(t_{70,.025} s_{\bar{Y}_1 - \bar{Y}_2}\right) \tag{6.12}$$

The difference between the means is 224.50 – 207.71 = 16.79 and its standard error (see Section 6.4.1) is 10.31. Turning to Appendix Table C.3, or using *qt*(.975, df = 70) in R or *Idf.T*(.975, 70) in SPSS (see Box 6.1), we find that the critical *t* value is 1.99. Putting our values together into Equation 6.12, we obtain the bounds for the 95% confidence interval:

$$CI = 16.79 \pm (1.99)(10.31) = -3.77, 37.35$$

Because these bounds encompass a value of zero we know that a two-tailed test at the .05 alpha-level will not reject the null hypothesis of no difference between the diet and control means. Also, the fact that the confidence interval is quite wide suggests that our test of that hypothesis may be low in power, a point we will explore further in this chapter.

6.4.3 *Interpreting Confidence Intervals*

Section 5.5 described the interpretation of confidence intervals in detail. Those points are briefly reviewed here because confidence intervals are often misinterpreted. Specifically, note the following points:

1. Ninety-five percent confidence *does not mean* that there is a .95 probability that the computed bounds enclose the true value of $\mu_{control} - \mu_{diet}$. The bounds either enclose the population mean difference or they do not.
2. Ninety-five percent confidence *does not mean* that if the experiment were repeated many times, 95% of the observed differences between means would lie between −3.771 and 37.353, the bounds calculated from the *TC* data.
3. Instead, ninety-five percent confidence *means* that if the experiment is repeated many times, and a new set of bounds is calculated each time, 95% of those bounds would contain the true value of $\mu_{control} - \mu_{diet}$.

In summary, calculating a confidence interval accomplishes several things:

1. The CI bounds provide a range of plausible values for the parameter being estimated. Higher confidence levels result in wider confidence intervals.
2. The CI width provides an index of precision of our estimate that is sensitive to the level of confidence, the variability, and the sample size.
3. The CI provides a test of an infinite set of null hypotheses. Those hypothesized values falling outside the interval can be rejected at $\alpha = 1$ – confidence level, whereas those inside the interval cannot be rejected.

6.4.4 The Student's t Test in the Independent-Groups Design

A direct test of H_0: $\mu_1 - \mu_2 = 0$ against the hypothesized alternative H_1: $\mu_1 - \mu_2 \neq 0$ follows from the preceding developments. The test statistic follows from Equation 6.1 and is

$$t = \frac{\left(\bar{Y}_1 - \bar{Y}_2\right) - \left(\mu_1 - \mu_2\right)_{hyp}}{s_{\bar{Y}_1 - \bar{Y}_2}} \tag{6.13}$$

Substituting numerical values, we obtain $t = 16.79 / 10.31 = 1.63$. This value of t is less than the critical value, 1.99, confirming that we cannot reject the null hypothesis. The exact p-value is .108 (see Section 6.4.5). Note that this implies that a one-tailed test at the .05 level would also fail to reject H_0 because p would equal .054 for that test, assuming we were lucky enough to select the appropriate directional test.

The validity of the confidence interval and the t statistic we calculated rest on certain assumptions. In Section 6.6, we restate these assumptions and describe procedures to respond to their violations.

6.4.5 Using Software for Student's t Test in the Independent-Groups Design

Independent-groups t tests are straightforward in both SPSS and R. To use either, begin by importing the data in the *Table 6_1 TC_IG.xlsx* Excel file. There are two columns, one identifying the treatment condition and the other for TC_Level.

In SPSS, use the *Analyze* and *Compare Means and Proportions* pull-down menus to select the *Independent Samples t Test.* Move TC_level to the *Test Variable* box and Treatment to the *Grouping Variable*; you will need to *Define Groups* by entering Control and Diet as Group 1 and Group 2, then click Continue and OK. The output will include descriptive statistics for each group, t test results for both equal- and unequal-variance assumptions (see Section 6.6.2), as well as several effect size measures and confidence intervals (see Section 6.7).

In R, assume the data are in a data frame called *dat*. Then the *t.test* function in the {stats} package will provide the t test results as well as a confidence interval and some descriptive statistics. The default analysis is the unequal-variance t (see Section 6.6.2); for the equal-variance assumption, simply change the *var.equal* option to be TRUE, as shown here: *t.test*(data = dat, TC_Level ~ Treatment, var.equal = TRUE).

6.5 Data Analyses in the Correlated-Scores Design

For reasons discussed earlier in this chapter, many comparisons of two conditions are made in a correlated-scores design in which each participant is tested in two conditions, or participants are matched on some dimension(s) related to the dependent variable. We illustrate the analyses of correlated-scores data with an artificial set of 36 pairs of total cholesterol (*TC*) scores.

The data file *Table 6_2TC_CS.xlsx* (from the *Tables* page on the book's website) has three columns. The first, labeled *Before*, contains the *TC* scores for the 36 subjects before receiving any treatment. The second column, labeled *After*, contains scores for the same individuals after one year on a diet prescribed by the researcher. The third column of scores, labeled *Change*, contains the difference in *TC* level, obtained by subtracting the *After*

Table 6.2 Descriptive statistics for data from a correlated-scores experiment

	Descriptive statistics			
	N	*Mean*	*SE of the mean*	*Std. deviation*
Before	36	224.50	7.90	47.40
After	36	208.17	10.69	64.12
Change	36	16.33	5.42	32.52

scores from the *Before* scores for each participant. A positive score represents an improvement in *TC* level, and a negative score means that *TC* level was worse (higher) after being on the diet. In a real study, participants might keep a diary describing the amounts and types of food eaten, and the researchers would probably test the participants at regular intervals – perhaps once a month – during the study. However, we will focus on a single set of change scores.

In the correlated-scores experiment, our primary interest is in the change scores: What effect did the prescribed diet have on changes in *TC* level? Is there much variability in the effects among individuals? What can we conclude about the potential effects of the treatment for a population like the participants in our study? A good place to begin is with the descriptive statistics and graphs described in Chapter 2. We leave this exploration of the data as an exercise but for convenience present means and standard deviations in Table 6.2.

Following exploration of the data, we wish to use the sample of data to draw inferences about the population. First, what is the mean of the population of change scores, and what are reasonable bounds on the mean change? Second, is that change significantly greater than zero? That is, can we conclude that, on the average, a population represented by our sample would show a decrease in *TC* level because of the prescribed diet? We next turn to these questions.

6.5.1 *Confidence Intervals in the Correlated-Scores Design*

Our estimate of the mean of the population of change scores is $\bar{Y}_{change}$, the average of the 36 change scores. From Table 6.2, this is 16.333. In order to calculate confidence interval bounds on the mean of the population of change scores (μ_{change}), we need this value, but also the standard deviation of the 36 change scores (s_{change}), the number of change scores (n), and the critical value of t. Calculations of confidence intervals and significance tests follow the procedures developed in Chapter 5, except that the critical value is obtained from the t rather than the normal probability distribution. The general form of the confidence interval in the one-sample case is:

$$p\left[\bar{Y} - t_{n-1,\alpha/2}\frac{s}{\sqrt{n}} \le \mu \le \bar{Y} + t_{n-1,\alpha/2}\frac{s}{\sqrt{n}}\right] = 1 - \alpha \tag{6.14}$$

where $\bar{Y}$ is the sample mean, $s/\sqrt{n}$ is the *SE* of the mean, and $t_{n-1,\alpha/2}$ is the value of t such that $\alpha/2$ of the distribution on $n - 1$ *df* lies to the right of it. In the example of the 36 *TC* change scores, $df = 35$, and therefore the critical t value for a 95% confidence interval is 2.03. From

Table 6.2, we substitute the other necessary values into Equation 6.14 to obtain the lower (LB) and upper (UB) 95% bounds on μ_{change}:

$$LB = 16.333 - (2.03)(5.421) = 5.328 \text{ and } UB = 16.333 + (2.03)(5.421) = 27.338$$

We have 95% confidence that the difference between the population means is between 5.33 and 27.34. Note that 0 is not in the interval, indicating that we can reject 0 as a plausible value for the mean TC change in the population; there is evidence that prescribed diet lowered TC compared to the control condition.

6.5.2 *The t Test in the Correlated-Scores Design*

The direct test of the null hypothesis parallels that developed in Chapter 5; the only difference is that we replace σ by an estimate, s, and use the t distribution as the basis for our test. We calculate

$$t_{n-1} = \frac{\bar{Y} - \mu_{hyp}}{s_Y / \sqrt{n}} \tag{6.15}$$

and, if the test is two-tailed, reject H_0 if $|t| > t_{n-1,\alpha/2}$. In the present example, $\mu_{hyp} = 0$, and the denominator is 5.421, the SE in Table 6.2. Therefore,

$$t = (16.333 - 0) / 5.421 = 3.013$$

a much larger value than the critical value of 2.03. Accordingly, we reject the null hypothesis that the treatment had no effect and conclude that the average change in the population is a reduction in total cholesterol level.

The logic of hypothesis testing does not require that the null hypothesis is one of zero change. We could choose a specific different value instead. For example, suppose we knew that various treatments previously investigated in other laboratories had yielded an average TC reduction of 6. If we wished to ask whether the mean change was significantly different from 6, we would substitute 6 for μ_{hyp} in Equation 6.15, and solve for t. Alternatively, we note that 6 falls within the 95% confidence interval bounds previously computed. Accordingly, that null hypothesis cannot be rejected by a two-tailed test at the.05 level. Therefore, we cannot conclude that our diet differs in its average effect from previously investigated treatments.[1]

6.5.3 *Using Software for Student's t Test in the Correlated-Scores Design*

Software makes this analysis easy. In SPSS, select the *Paired Samples t* test in the *Compare Means and Proportions* pull-down menu. Enter the paired variables (Before, After) in the desired order for the difference of interest, either Before then After or the reverse. The output provides some basic descriptive statistics, the significance test and confidence interval for the difference, and some estimates of effect size. In R, again assuming the data are in a data frame called dat, we can use the *t.test* function with the paired = TRUE option: *t.test*(dat$Before, dat$After, paired = TRUE). The output includes a test of significance and confidence bounds. While the default value for the difference is 0, it can be changed with the *mu* option. For example, *t.test*(dat$Before, dat$After, paired = TRUE, mu = 6) reports the *t* from Equation 6.15 assuming μ_{hyp} is 6.

6.6 Assumptions Underlying the Application of the *t* Distribution

Inferences based on data from the independent-groups design rest on the following assumptions:

1. The scores in each treatment population are independently distributed.
2. The scores in each treatment population are normally distributed.
3. The variances of the two populations of scores are the same; this is the *homogeneity of variance* assumption.

The independence assumption is usually valid except in situations in which there is some interaction among participants; this might happen when one participant responds in the presence of another. For example, suppose the scores were weights, rather than total cholesterol levels, and there were weekly weigh-ins observed by all participants. The success or failure of some individuals might encourage or discourage others from adhering to a diet. In that case, the scores would not be independent.

Although the independence assumption is usually not a concern in these designs, the normality assumption and the homogeneity of variance assumption are often violated, and their violations have implications for Type 1 and Type 2 error rates. In the next sections, we discuss those consequences and suggest some possible remedies.

6.6.1 The Normality Assumption

Consequences of Violating the Normality Assumption

When data are not normally distributed, particularly if the distributions are skewed, the Type 1 error rate may be distorted. For example, when the nominal Type 1 error rate is .05, the true probability might be .08 or .02, depending on the direction of skew. Although our data are frequently skewed, the distortion of Type 1 error rates is usually slight with moderately large samples thanks to the central limit theorem (see Chapter 5). "Moderately large" may be as small as 20 if $n_1 = n_2$ and if the two populations have symmetric distributions. More conservatively, except in cases of very extreme skew, n of 40 or more in both correlated-scores and independent-group designs will suffice to provide valid Type 1 error rates. Large data sets not only yield increased power and narrower confidence intervals, they also tend to provide greater inoculation against Type 1 error rate distortions from violations of the normality assumption.

Although the Type 1 error rate may approximate its nominal value, another important consideration is that there may be a loss of power when populations are skewed or have outliers. In such cases, the sampling distribution of $\bar{Y}_1 - \bar{Y}_2$ will tend to be long-tailed (i.e., a greater proportion of extreme scores than would be found with the normal distribution); therefore, estimates of the difference between population means will be less precise, and variability may be greater than when the normality assumption holds. We next consider possible ways of addressing the potential loss of power due to departures from normality.

Dealing With Violations of the Normality Assumption

There are several possible modifications of, and alternatives to, the standard inference procedures based on the t distribution. These will result in more valid Type 1 error rates when samples are small and skewed and often will yield greater power when there are departures from normality even in large samples. Three possible approaches are as follows:

1. *Transformations* of the data will often result in data distributions that more closely approximate the normal.
2. *Trimming outliers* will generally yield more powerful tests of the null hypothesis and narrower confidence intervals.
3. *Tests based on ranks* reduce the effect of extreme scores and often provide greater power.

First, we consider why, when, and how to transform data. In transforming data, the same operation is applied to every score; for example, the square root of each score might be calculated and the analyses would then be based on the scores on this square-root scale. Transformations have been used to transform skewed distributions into more nearly normal distributions, to equate group variances more closely, and to reduce the different effects treatments may have on different subjects' scores when data are obtained in several conditions from each subject. A transformation that achieves one purpose may not be best for other purposes; however, transformations that reduce heterogeneity of variance often also result in more normally distributed data.

Three very common transformations are (1) the *square-root* transformation; (2) the *logarithmic* transformation; and (3) the *reciprocal* – sometimes called the *inverse* – transformation. The square-root transformation is the weakest, causing the least change in the distribution, and the reciprocal is the most powerful of these transformations. In seeking to transform a skewed data distribution to normality, we first should view the results of the weakest possible transformation and, if that is not satisfactory, then investigate the effectiveness of more powerful transformations.

In Chapter 2, we viewed *Q–Q* plots (Figure 2.5) of multiplication response times and multiplication speeds. As we noted then, the distribution of response speed, a reciprocal transformation of response time, more closely approximated the normal distribution. In general, we strongly recommend the approach taken in Chapter 2: Before attempting transformations or other remedies, explore the statistics of the data, including indices of skew and kurtosis, as well as distributional plots. If the data appear skewed and a transformation is tried, examine the results again. Whether inferential methods are applied to transformed data, and if so, which transformation is used, should depend on looking at the data, not on the results of subsequent significance tests. A transformation should *not* be selected because it provides the largest value of the t statistic for testing the difference between means; rather, the criterion for choosing a transformation should be that it achieves a distribution that is more consistent with assumptions underlying the planned analyses.

Transformations are not always advisable. They work by changing relative distances among scores, compressing one tail of the distribution more than the other. The values on the new scale may not be as readily interpretable and the nature of relationships between the dependent and independent variables may be changed. For example, when there are more than two levels of the independent variable, a previously linear relationship between mean scores and the independent variable may now be curvilinear.

A further limitation is that confidence intervals on the transformed scale may be difficult to interpret. If a single sample of scores is transformed, the limits may be transformed back to the original scale. For example, consider a random sample of 10 scores with $\bar{Y} = 100.1$ and 95% CI for $\mu = [98.95, 101.28]$. If we transform *Y* by taking the square root, the CI is [9.9473, 10.0633]. These bounds can be squared, thus returning to the original scale except for rounding error: $9.9473^2 = 98.949$. However, if we obtain a CI on the difference between two means of data transformed by taking the square root of each score, retransforming those limits makes no sense. If the lower limit is negative, squaring the limits will yield two positive limits which will not contain zero, even if the difference being tested is not significant.

An alternative approach that can be useful in dealing with violations of the normality assumption is a *trimmed t test*. In Chapter 5, we saw that a trimmed mean can have a much smaller standard error than $\bar{Y}$ when the population distribution has a longer tail than the normal distribution (see Table 5.1). Arbitrarily trimming outliers and then calculating the conventional *t* statistic for the remaining scores is not appropriate because the test statistic will not necessarily be distributed as *t*. However, Tukey and McLaughlin (1963) proposed a statistic that has an approximate *t* distribution and takes advantage of the reduced standard error that results from trimming scores. Chapter 7 presents an example of this trimmed *t* test and compares the results with that of the test performed on the untrimmed data. Details of the calculations are also presented there. In general, with either skewed or symmetric long-tailed distributions, the trimmed *t* test will have more power and the related confidence intervals will be narrower than when calculated in the usual way. Furthermore, when scores are normally distributed, Type 1 and 2 error rates are little affected by trimming.

Finally, a third approach to dealing with violations of the normality assumption is the use of *tests based on ranks*. This class of tests is usually referred to as *nonparametric* or *distribution-free*. When the data are obtained from two independent groups, the *Wilcoxon rank-sum* or the equivalent *Mann–Whitney U* test can be applied. These essentially perform a *t* test on the mean ranks obtained by converting the data into ranks, which reduces the influence of outliers. If the null hypothesis is true, the mean rank in each group will be the same; the difference in mean rank across groups is larger when the larger raw scores occur more often in one group than the other. A detailed description of the Mann–Whitney *U* is presented in Section 7.7.1. Other nonparametric tests, including formulas, can be found in a number of sources (e.g., Gibbons, 1993; Hollander & Wolfe, 1999; Siegel & Castellan, 1988).

One caution is in order with respect to any nonparametric test for independent-groups data. These methods are applicable and provide a more powerful test of the hypothesis of identical distributions when the data distributions are not normal (Boneau, 1962; Zimmerman & Zumbo, 1993). However, contrary to some beliefs, there are important assumptions underlying these tests. In addition to the usual assumption of independence of scores, it is assumed that the population distributions have the same shape. If they do not, significant test results may reflect differences in variance, or in shape parameters, even though the averages are the same.

Which of these approaches – transformations, trimming, or tests of ranked data – will be appropriate in any situation depends on the data distribution as well as on the nature of the dependent variable. Transforming data makes sense if the data distribution is skewed and the original scale is arbitrary, as in some personality measures, or if the transformed scale makes as much sense as the original scale, as when response time is transformed into speed.

The Mann–Whitney *U* test rests on the assumption that the two treatment distributions have the same shape and has the added limitation that confidence intervals are difficult to construct. However, if the assumptions are met and all that is required is a significance test, this will provide a powerful alternative to the *t* test when data are not normally distributed and the population distributions are assumed to be the same shape. When the data distribution is symmetric but long-tailed, the trimmed *t* test will often be most powerful and has the added advantage that confidence intervals can easily be constructed.

6.6.2 *The Assumption of Homogeneity of Variance*

It is common to encounter violations of the assumption that variances are the same in both experimental conditions. The consequences of such violations depend on a couple of considerations.

Consequences of Violating the Assumption of Homogeneity of Variance

The denominator of the equation for the two-sample Student's *t* test is based on the pool of two variance estimates (see Equation 6.4); the underlying assumption is that the two group variances estimate the same population variance. If this is not true – if the population variances are heterogeneous – then the sampling distribution of the *t* statistic may not have a true *t* distribution. Table 6.3 gives a sense of what may happen in this case. We drew 100,000 pairs of samples of various sizes from two normal populations with identical means but different variances. Proportions of rejections for alphas equal to .01 and .05 are presented.

Table 6.3 Type 1 error rates for the Student's and Welch's *t* tests as a function of population variances and sample sizes

			Student's t		*Welch's t'*	
n_1	n_2	σ^2_1/σ^2_2	$\alpha = .05$	$\alpha = .01$	$\alpha = .05$	$\alpha = .01$
5	5	4	.059	.014	.050	.010
5	5	16	.073	.021*	.054	.013
5	5	100	.081*	.027*	.053	.012
15	15	4	.053	.011	.050	.010
15	15	16	.058	.014	.051	.011
15	15	100	.058	.015*	.049	.010
5	10	4	.117*	.038*	.056	.015*
5	10	.25	.021*	.003*	.046	.008
10	15	4	.086*	.024*	.050	.010
10	15	.25	.031	.005	.051	.010
10	20	4	.113*	.037*	.051	.011
10	20	.25	.018*	.002*	.049	.010
20	30	4	.083*	.023*	.049	.010
20	30	.25	.028	.004*	.050	.009

Note: Asterisks denote Type 1 error rates that are at least 50% larger or smaller than the nominal α level.

Two points about the results should be noted.

1. If the two sample sizes are equal, the difference between the empirical and theoretical Type 1 alpha rates tends to be relatively small, except when n is very small (i.e., $n = 5$) and the variance ratio is very large (i.e., $\sigma^2_1/\sigma^2_2 = 100$).
2. If the ns are unequal, whether the Type 1 error rate is inflated or deflated depends upon the relationship between sample size and population variance. In either case, the difference between the empirical and theoretical rejection rates can be substantial.

The relation between sample size and variance is that the denominator of the t is based on a weighted average of two variance estimates; the weights are proportions of degrees of freedom. Therefore, when the larger group is drawn from the population with the larger variance, the larger variance estimate receives more weight than the smaller estimate. The denominator of the t test tends to be large and the observed t small; the rejection rate is less than the nominal α-level. Conversely, when sample size and population variance are negatively related, the smaller variance estimate gets the larger weight; the denominator of the t statistic tends to be small and the t large; the rejection rate is inflated above the nominal α.

Unequal sample sizes should be avoided whenever possible. However, we recognize that there will be many cases in which sample sizes will differ, often markedly. For example, the two populations of interest may be vastly different in size, as when comparing patients with a rare disease to control participants. Likewise, the response rate to questionnaires may be quite different for two populations such as different socioeconomic status or for workers employed full versus part time. We do not advocate discarding data from the larger sample; this would increase sampling variability for statistics computed from that sample. Instead, we recommend an alternative to the standard t test.

Dealing With Unequal Variances: Welch's t Test

One alternative to the standard t test is a t that does not use the pooled estimate of the population variance. The denominator of this statistic is that of the z test for two independent groups with variance estimates instead of known population variances. We define

$$t' = \frac{\left(\bar{Y}_1 - \bar{Y}_2\right) - \left(\mu_1 - \mu_2\right)}{\sqrt{s_1^2 / n_1 + s_2^2 / n_2}} \tag{6.16}$$

This statistic is usually referred to as Welch's t (1938). If the scores have been drawn from normally distributed populations, t' is distributed approximately as t but not with the usual degrees of freedom. Instead, the degrees of freedom are

$$df' = \frac{\left(s_1^2 / n_1 + s_2^2 / n_2\right)^2}{\dfrac{s_1^4}{n_1^2\left(n_1 - 1\right)} + \dfrac{s_2^4}{n_2^2\left(n_2 - 1\right)}} \tag{6.17}$$

When the variances and sample sizes are the same in each group, Student's t will equal Welch's t', and their df and p-values will also be identical.

As we described in Section 6.4.5, the *t.test* function in the {stats} package of R calculates Welch's t' by default (*var.equal* = FALSE). For the data summarized in Table 6.1, $t' = 1.63$, with $df = 67.95$ and $p = .108$. Notice that this result, using Welch's t', is almost identical to the results of the Student's t test described in Section 6.4.4. This is because the group sizes are equal and the standard deviations are similar.

Many statistical packages, including SPSS, compute values of both t' and t when an independent-groups t test is performed. SPSS also reports a test of homogeneity of variance, Levene's (1960) test. In Levene's test, an average absolute deviation of scores about each group's mean is calculated and the ratio of these two averages is assessed using an F distribution. (We will describe the family of F distributions in Chapter 8.) The Brown–Forsythe (1974a) test is similar, except the group means are replaced by their medians; some statistical packages call both tests "Levene's" tests.

It may appear wise to test the homogeneity of variance assumption prior to choosing a t test. However, there are two important reasons to proceed directly to Welch's t'. First, Levene's test often has very low power (Zimmerman, 2004; Nordstokke & Zumbo, 2007; Delacre, Lakens, & Leys, 2017), and Table 6.3 demonstrates that the Student's t test is not robust when the groups have unequal variance. Second, selecting the t test based on Levene's test does not fully correct the distortion of the Type 1 error rate. Skipping the test for homogeneity of variance and proceeding directly to Welch's t' results in a more robust test with power that is nearly as high as Student's t. Thus, we advocate using Welch's t' for all independent samples t tests.

6.7 Measuring the Standardized Effect Size: Cohen's *d*

In the experiment on the effects of diet on total cholesterol level, the difference between the two group means provides a measure of the size of the effect. However, that measure is on the original data scale. Although this has the advantage of being meaningful to the researcher, it has accompanying disadvantages: It is difficult to evaluate the importance of differences between means, or to compare such differences with effects of the same independent variable in other groups or laboratories, or to combine results of several experiments.

The p-value associated with a hypothesis test is sometimes incorrectly interpreted as a measure of the importance of an effect. A p-value simply indicates the probability that data as extreme as those observed in our study, or more so, would be sampled from the population if the null hypothesis is correct. Very small effects, perhaps unimportant in any practical or theoretical sense, may be statistically significant because the sample sizes are large. Conversely, what appear to be large effects may occur just by chance in small samples, even if the null hypothesis is true. Finally, important real effects, even large differences, may not be statistically significant because the sample was too small, or variability too great, for the test to have had much power.

In view of these concerns, several expressions for effect size have been proposed (e.g., Cohen, 1988; Kraemer, 2005; McGraw & Wong, 1992; Rosenthal & Rubin, 1994, 2003; Shieh, 2013), and this remains an active area of research. In all cases, the estimated effect size (ES) is the difference between the sample means measured in standardized units: ES = $\frac{\bar{Y}_1 - \bar{Y}_2}{standardizer}$. Different assumptions about the most appropriate standardizer lead to variations in the effect size measure. Unfortunately, the same effect size measure has sometimes been given different names, and different measures have been labeled identically. In all cases, our guiding principles for choosing an effect size measure are that the measure should be appropriate for the design (independent vs correlated samples) and the equality, or inequality, of the population variances. As was true of the criteria for other estimators of population statistics, we also seek *unbiased and consistent* estimators of the population effect sizes (see Section 5.2). To these, we add one additional point: the standardizer should be the best available estimator of the population parameter of interest (Cumming, 2013).

In this section, we focus on variants of *Cohen's d*, which evaluates the effect of a treatment relative to its standard deviation. Estimates of d have several uses:

1. They provide indices of the importance of the effect, measured in standardized units.
2. As we shall see in Section 6.8, they provide information required for power calculations.
3. They are used in meta-analyses, analyses which combine results from several studies.

There is strong agreement that measures of effect size are important and should be reported along with the results of statistical tests. Most journals explicitly require that measures of effect size be reported, and the seventh edition of the *Publication Manual* of the American Psychological Association (2020) recommends their inclusion to aid readers' understanding of the importance of the effect. Moreover, because effect sizes are themselves point estimates of the population value, the Task Force for Statistical Inference (Wilkinson & the Task Force on Statistical Inference, 1999) recommends reporting a confidence interval for every effect size. As we shall see, the confidence intervals for these effect sizes are calculated from a *noncentral t* distribution, which is also important for power calculations.

6.7.1 *Estimating Cohen's* d[2]

Cohen's d for the independent-groups design.

Cohen's d for the *independent-groups design* is defined as

$$d = \left[(\mu_1 - \mu_2) - \Delta\right] / \sigma \tag{6.18}$$

where μ_1 and μ_2 are the actual population means, and Δ (Greek upper-case delta) is the population difference assuming H_0 is true; Δ is usually, but not always, 0. The choice of standardizer in the denominator of Equation 6.18 depends on whether the population variances are equal and whether there is a true control condition.

If we can assume homogeneity of variance, then there is good agreement that the standardizer in the denominator of Equation 6.18 is s_{pooled}, where s_{pooled} was defined by Equation 6.4. This equation for Cohen's d is also known as Hedges' g (Hedges & Olkin, 1985). In the TC example,

$$\hat{d} = (16.791 - 0) / 43.751 = .384$$

Like the other effect sizes for the independent samples design, Cohen's d is a positively biased estimator, particularly for small samples. If $N < 32$, d should be corrected by $\hat{d}_{corrected} = \hat{d} \times [1 - 3/(4N - 9)]$ (Goulet-Pelletier & Cousineau, 2018). An advantage of Cohen's d is that it is closely connected to power calculations, as we will see in Section 6.8.1.

When the variances of the two groups differ, we face another choice of standardizer that depends on the experimental design. If one of the conditions is truly a control condition, then one option is to use the standard deviation of that condition, say s_1, as the standardizer. This measure was suggested by Glass (1976) and endorsed by Cumming (2013) and others. If neither condition can readily be called the control condition, Cohen (1988) suggested replacing s_{pooled} with the root mean square of the two variances: $\sqrt{\frac{s_1^2 + s_2^2}{2}}$. Using the *TC* example data, this effect size estimate is $\widehat{d^*} = \frac{224.50 - 207.71}{\sqrt{\frac{2246.51+1581.75}{2}}} = 0.384$, a value equal to $\hat{d}$ in this case because $n_1 = n_2$. Cohen's $\widehat{d^*}$ is a positively biased estimator of the population effect size, especially for smaller sample sizes (Hedges & Olkin, 1985).

In contrasting two independent means, Cohen (1988) suggests .2, .5, and .8 as guidelines for small, medium, and large standardized effect sizes. Following this recommendation, the standardized effect size of the experimental diet on *TC* level falls somewhere between small and medium, regardless of which measure we choose.

Cohen's d *for the Correlated-Scores Design*

Cohen has described two possible measures of effect size in the *correlated-scores design*. One, d_z, is the ratio of the difference between the means to the standard deviation of the difference scores. Using our *TC* change score data to illustrate (Table 6.2), our estimate is $\hat{d}_z = 16.333/33.524 = .502$. Cohen's d_z is a positively biased estimator of the population effect size, and should be corrected by $\hat{d}_{corrected} = \hat{d} \times [1 - 3/(4N - 9)]$ if $N < 16$. An important feature of d_z is that it is closely linked to statistical power; we will elaborate on this point in Section 6.8.2. A limitation of d_z is that, for two independent means, it is always larger than $\hat{d}$: Lakens (2013) showed that $\hat{d} = \frac{d_z}{\sqrt{2(1-r)}}$ where r is the correlation of scores between conditions. Although that sounds like an advantage for d_z, the design of an experiment should not influence our estimate of the effect size in the population of interest.

The second measure of effect size in the *correlated-scores design* is $d = (\mu_1 - \mu_2)/\sigma$; this is estimated just as for the independent-groups design with heterogeneity of variance: The population means are estimated by the corresponding sample means and σ is estimated by averaging the two within-condition variances and taking the square root of the result. Using the *TC* change score data to illustrate, first converting the condition standard deviations to variances by squaring them, we compute our estimate of the pooled standard deviation by $\sqrt{(2{,}246.521 + 4{,}111.725)/2} = 56.384$. Then, $\hat{d} = (224.5019 - 208.1694)/56.384 = .29$, a relatively small effect.

Cohen preferred d to d_z because values of the former are comparable to those obtained from experiments involving the same conditions but in an independent-groups design. However, *dz* is more directly linked to power calculations, as we will discuss in Section 6.8.2. In any event, the two statistics are closely related; if s_{diff} is the standard deviation of the difference scores and s_{pool} is the square root of the average within-condition variance, then $\hat{d} = \left(s_{diff} / s_{pool}\right)\hat{d}_z$.

6.7.2 Confidence Intervals on Cohen's d

In considering the difference between means, we saw that the confidence interval provided a sense of the plausible range within which the population parameter may fall. The interval bounds are an important adjunct to any parameter estimate; narrower intervals indicate a more restricted range of plausible values of the parameter being estimated. In constructing intervals for the raw difference between means, we used the formula in Equation 6.12. That equation provides symmetric confidence bounds around a parameter estimate: The upper and lower bounds of the CI are equally distant from the estimate. The symmetry occurs because the confidence interval is based on the sampling distribution of the null hypothesis, which has a *central t* distribution. Central *t* distributions have a mean of 0 and are symmetric; they differ from one another only because of their degrees of freedom. In a central *t* distribution, the critical values that define the upper and lower 2.5% of the distribution have the same magnitude, differing only in their sign (e.g., +/−2.571 for 5 *df*, see Table C.3); they are shown with light gray lines in Figure 6.1.

In contrast, confidence intervals for Cohen's *d* are not based on the sampling distribution for the null hypothesis. Instead, they are based on a noncentral *t* distribution that does not have a mean of 0 and is not symmetric. We will have more to say about noncentral *t* distributions and their connection to Cohen's *d* in Section 6.8.1. For now, we only need to understand something about their shape. An example noncentral *t* distribution is shown in Figure 6.1, along with the central *t* with the same degrees of freedom (df = 5). The

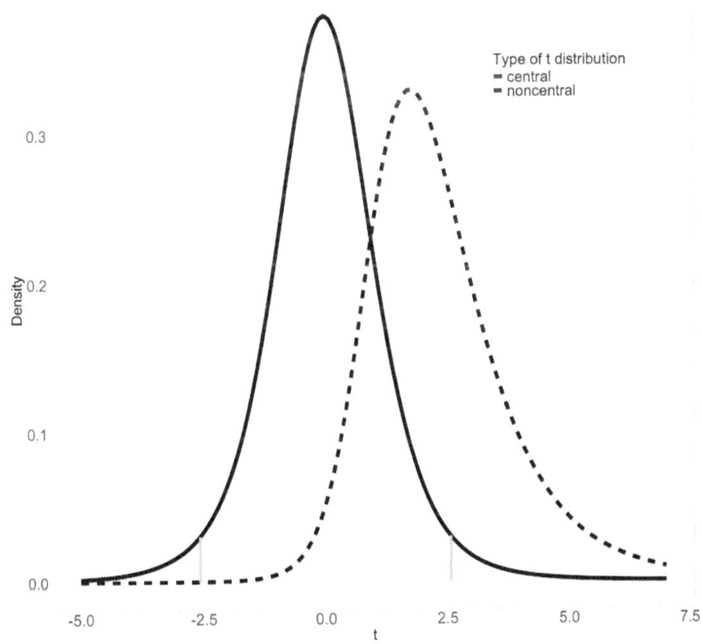

Figure 6.1 Plot of the central and noncentral *t* distributions with df = 5 and noncentrality parameter = 2.

asymmetry in the noncentral t is apparent: larger values of t are more likely than smaller values and the critical values of the noncentral t that define the upper and lower 2.5% of the distribution do not have the same magnitude. For the example distribution in Figure 6.1, they are 0.042 and 6.292 (these values can be found using a distributional calculator, see Box 6.1). This means that a confidence interval based on a noncentral t will also be asymmetric; the distance from the lower bound to d will be less than the distance from the upper bound to d. It also means that there's no simple equation like Equation 6.12 that can be used to compute the CI for Cohen's d. Instead, we must use software. Discussions of the rationale underlying the method may be found in several sources (e.g., Cumming & Finch, 2001; Hedges & Olkin, 1985; Steiger & Fouladi, 1997); Goulet-Pelletier and Cousineau (2018) provide a particularly clear explanation.

In SPSS, the independent samples t test provides three effect sizes estimates – Cohen's d with and without correction for bias, and Glass's delta that assumes a real control condition – and their confidence intervals. For the data in Table 6.1, SPSS reports bias-corrected $d = 0.38$ with a 95% CI of [–.08, .84]. Similar values are obtained in R using the *gethedgesg* function described by Goulet-Pelletier and Cousineau (2018).[3] This CI is wide: The population value of d may be anywhere from virtually zero to a large positive treatment effect. Thus, the confidence interval limits not only set bounds on plausible values but remind us that the point estimate by itself can be misleading.

For the correlated data example from Table 6.2, the paired sample t test in SPSS estimates d_z by default. For these data, $d_z = .50$ with a 95% CI of [.15, .85]. To estimate d instead, select the option to standardize using the average of variances. As expected, the estimate of d is smaller than the estimate of d_z. Here, the result is $d = .29$ with a 95% CI of [–.05, .62]. In R, the function provided by Fitts (2020) calculates the confidence interval for Cohen's d_z and other effect sizes. It finds $d_z = .50$ with CI limits of .16 and .88 for the example data in Table 6.2.

It should be noted that these estimates rest on the assumption of homogeneity of variance. In addition, because calculating the CI for d requires finding areas under the noncentral t distribution, it is also assumed that the population distribution is normal. In response to possible violations of these assumptions, Algina, Keselman, and Penfield (2005a, 2006) have proposed modifications of d that make use of the trimmed means and winsorized variances that are defined in Chapter 7.

6.7.3 *Two Cautions on Effect Size*

Although effect size estimates provide important information to the researcher, we offer some cautions on their use. First, Cohen's d can be mis-estimated. All versions of Cohen's d are positively biased, especially for small samples. With fewer than 32 observations (i.e., 16 participants/group for independent samples), $\hat{d}$ should be debiased as described in Section 6.7.1. In addition, estimates of Cohen's d usually assume that the group variances are homogeneous. When the variances for the two groups differ, $\hat{d}$ may be a poor estimate of d, and d^* may not fully compensate. This and other issues in using $\hat{d}$ as an index of effect size are discussed in several articles, together with possible solutions (Grissom & Kim, 2001; Kelley, 2005; Olejnik & Algina, 2000).

The second caution concerns Cohen's widely adopted benchmarks for effect sizes; namely, that d values of .2, .5, and .8 correspond to small, medium, and large effects, respectively. Although guidelines can be useful, it is important to emphasize that they are just that – guides, not mandates. Previous results, theoretical considerations, and practical goals are better bases for evaluating the importance of an effect size. We should ask: Is this effect large relative to those found in related studies in the research literature or in our laboratory? Smaller? About the same as in other studies of this type? Is our research goal – whether theoretical or applied – one for which even a small effect is important to recognize? In short, Cohen's *d* is just one perspective on the "importance" of an effect. By expressing the magnitude of an effect relative to variability, *d* provides useful information; however, it is entirely possible for an effect with a small value of *d* to have great practical or theoretical importance. There is a difference between statistical and practical significance.

6.8 Deciding on Sample Size

Too many experiments waste time, effort, and resources because the researcher has failed to collect enough data to have sufficient power to detect differences that may be present in the population. Power estimates should be used when deciding on the sample size. Analyses that do this have been referred to as *a priori* or *prospective power analyses*.

6.8.1 A Priori *Power Analyses in the Independent-Groups Design*

How might we decide on the sample size of a planned experiment with two independent groups? We begin by deciding on the smallest effect size that would be of interest. For example, if a current drug is known to reduce total cholesterol by, say, 20 mg/dL of blood, then researchers developing a new drug might assume that 20 is the minimum raw effect of interest: Smaller reductions would not be marketable. We also need to decide on an acceptable minimum level of power to detect that smallest effect size of interest. Understanding the role of these two requirements in calculating power involves understanding how *t* is distributed when the null hypothesis is false. The relevant distribution is called the noncentral *t* distribution and we discuss it next. Following that discussion, we will consider a specific example of its application to the decision about sample size.

The Noncentral t *Distribution*

As with the binomial (Chapter 4) and normal probability (Chapter 5) tests, we can conceive of two distributions, one when the null hypothesis is true and one when it is false by a specific amount. For example, in the *TC* experiment we might have one distribution corresponding to H_0: $\mu_{diet} - \mu_{control} = 0$ and an alternative distribution corresponding to H_A: $\mu_{diet} - \mu_{control} = 20$. As we noted in Section 6.7.2, the distribution of *t* assuming the null hypothesis is the central *t* distribution, a symmetric distribution with a mean of 0 (see Figure 6.1). As with other statistical tests, we first determine the rejection region in this distribution – the critical values of *t* that lead to rejection of the null hypothesis. That region can be obtained from Appendix Table C.3, the *qt* function in R, or the IDF.T function in SPSS (see Box 6.1). Once we know the rejection region, we can calculate power by finding the probabilities of the *t* values in that region, assuming the alternative distribution; that is, assuming the noncentral *t* distribution under a specific alternative hypothesis.

Figure 6.1 displays central (left) and noncentral (right) *t* distributions on 5 *df* (from $N = 6$). In the example in the figure, $\alpha = .05$, the alternative hypothesis was two-tailed, and the actual (standardized) effect was 0.82 (arbitrarily chosen for the sake of illustration).

Because the relation between $\hat{d}$ and t is simply $t = \hat{d}\sqrt{n}$, the distance between the means of the central and noncentral t distributions in Figure 6.1 is $(0.82)(\sqrt{6})$ or 2.0 units. The rejection region is the area to the left of the vertical line at −2.57 and to the right of the vertical line at 2.57. Power corresponds to the area under the right (noncentral t) above the positive critical value of t and below the negative critical t. In this example, power is only 0.37. Several points should be noted.

1. The noncentral t distribution has a different shape than the central t distribution; it is not just displaced. The location, variance, and shape of the noncentral t distribution are determined by the df and a *noncentrality parameter*, δ (the Greek letter delta). This is a measure of distance between the two distributions and incorporates information about variability (σ), sample size (n), and the raw effect size under the specific alternative hypothesis. In general, the noncentral t distribution tends to be skewed to the right (especially for very large effect sizes), although both distributions look more like the normal distribution as df increase.
2. Virtually none of the area under the noncentral t distribution is to the left of the left critical value. If the alternative hypothesis is that the effect is positive, there is usually very little power to reject alternatives in the other direction.
3. If we assume a larger specific effect in the alternative hypothesis, the distance between the two distributions will increase, as will the right skew in the noncentral t distribution. As a result of these two changes, more of the area of the noncentral t distribution will lie to the right of the rejection region, displaying increased power.

Deciding on Sample Size: An Example

We found that the value of t for a comparison of diet and control group total cholesterol (TC) means was 1.63. Testing against the one-tailed alternative, H_0: $\mu_{control} - \mu_{diet} > 0$, the corresponding p-value is slightly higher than .05. Suppose we decide to replicate the experiment. This requires us to state an effect size, a value of d under the specific alternative hypothesis. We also must state the power with which we wish to be able to reject H_0 given this effect, and the total sample size ($N = n_1 + n_2$) needed to achieve the desired power. Finally, we need to set the desired significance level, α, and decide whether the test will be one- or two-tailed. Let's consider how we might decide on the targeted effect size, d, and then how we might decide on the sample size.

What effect size is important? We may begin by asking what difference between the means would be of theoretical or applied importance. A small effect that distinguishes between two theories can be important. On the other hand, in testing a new treatment for a clinical disorder, we may wish to detect only an effect so large that it will clearly be worth the cost of further development or will clearly be better than existing treatments. In the example of the effect of a diet on TC level, after reviewing the relevant research literature, we might decide that a reduction of 20 points in the total cholesterol level would be important to detect. Therefore, we decide that our alternative hypothesis should be H_A: $\mu_{control} - \mu_{diet} > 20$, reflecting the minimum effect size of interest.

We now need an estimate of σ. In planning a replication of our TC experiment, we might base our estimate on our present data set; then $s_{pooled} = 43.751$. We could also examine estimates from related studies in the literature to see whether this is a reasonable estimate of the variance of change in total cholesterol level, and perhaps to improve our estimate. After some consideration, we decide that $s = 44$ is reasonable; dividing 20 by 44 yields $d = .47$. (Our original sample size was $N = 72$, so we do not need to correct d for bias.)

Other options exist for setting the effect size. We might use the estimate from our pilot study of .38. However, pilot studies tend to be modest in scale, and small samples can dramatically overestimate the true effect size (see Figure 4.2). For this reason, we prefer to think first in terms of the raw effect size that is of interest. Another possibility in our example is based on noting that .47 is close to .5, Cohen's (1988) medium effect size. Therefore, for the purposes of easy communication to a future audience, we might set $d = .5$. Further discussion of the considerations involved in choosing an effect size in an *a priori* power analysis may be found in Lenth (2001).

What power do we want? Ideally, we want a sample size that will provide maximum power to detect the effect of interest. Unfortunately, there are practical considerations. The required sample size increases either as the target effect size decreases or as the targeted power increases. The N needed, particularly for high power and small effect sizes, is much larger than most researchers realize. Suppose we decide on power = .8, a level of power that has been frequently suggested as reasonable. In that case, we would need N – 114, with $n_1 = n_2 = 57$. Increasing the desired power brings a cost; to have .90 power against the alternative, $d = .47$, we need 79 participants in each group. Our sample size will often be a compromise among considerations of the minimum effect size of interest, the power we would like to have to detect that effect, and the practical limitations of subject availability and the time required for data collection.

Calculating the Required Sample Size

We noted earlier that power calculations are based on the noncentral t distribution, whose location, variance, and shape are determined, in part, by a *noncentrality parameter*, δ. If our study involves two independent groups, δ is defined as

$$\delta = \frac{\mu_1 - \mu_2}{\sigma\sqrt{(1/n_1)+(1/n_2)}} \tag{6.19}$$

From Equation 6.18, we can express δ as a function of d and the group sizes

$$\delta = d / \sqrt{(1/n_1)+(1/n_2)} \tag{6.20}$$

G*Power 3.1 takes values of d and the ratio of the ns as input and uses them to calculate δ in doing power computations. As an example, Figure 6.2 presents our input (in the white boxes) to G*Power and the resulting output (in the shaded boxes). We requested the *a priori* t test option for independent groups with power = .9, $\alpha = .05$, the effect size = .47 (our specific alternative), and tested against a one-tailed alternative. The total N of 158 is more than twice the N of 72 in our original experiment.

Suppose we had decided that a practical limit on the total sample size was 150 participants. Generally, we will have to accept that power will suffer if there are constraints on the size of our sample. In this example scenario, the power penalty would not be severe because our maximum sample size is not much smaller than that needed for power of .9. On the other hand, suppose that we were dealing with a much smaller minimum effect size of interest. Even if we lowered our desired power to .8, if we set d to .25 – small by Cohen's (1988) standards – we would require 398 subjects and proceeding with only 150

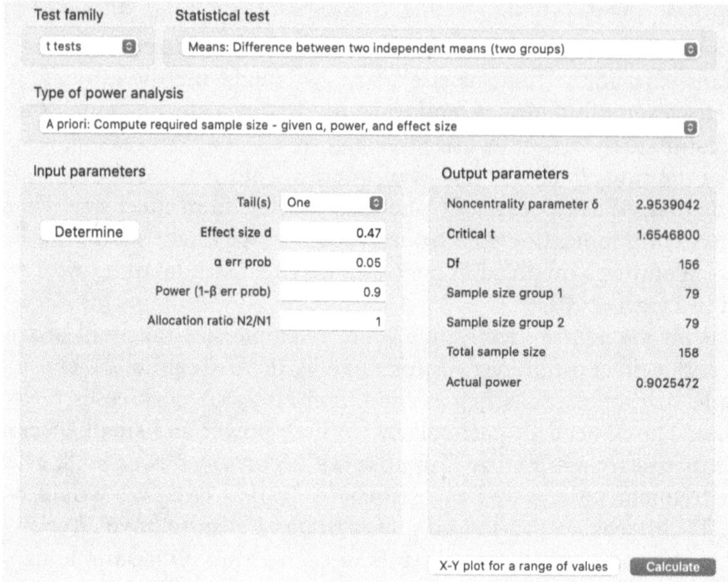

Figure 6.2 Determining sample size with G*Power 3.1.

participants would be foolhardy. In that case, a better strategy would be to redesign our experiment or choose a dependent measure with lower variability and thus an increased effect size (see Equation 6.18). Continuing with our current example, a small increase in effect size from .47 to .5 enables us to achieve *a priori* power of .9 with 140 subjects.

In summary, choosing a specific alternative hypothesis requires a statement of effect size that should be well justified and reasonable to others in your research community. Even moderate power to reject the null hypothesis when d is small requires quite large sample sizes. In many studies, power to detect an effect is too low because the sample sizes are too small. Despite articles that have made this point (e.g., Cohen, 1962; Greenwald, 1993), researchers continue to underestimate the sample size needed for adequate power. The negative consequences for the research literature are substantial: We may fail to detect real effects of interest, or we may grossly overestimate the true effect size (see Figure 4.2), potentially resulting in subsequent studies that are also underpowered.

6.8.2 A Priori *Power Analyses in the Correlated-Scores Design*

In the case of the correlated-scores design, the definition of δ refers to a population of difference scores, rather than two independent populations of scores. Assuming a sample of n difference scores, as in the example of the *TC* change scores, δ is defined as

$$\begin{aligned}\delta &= \frac{\mu_{difference}}{\sigma_{difference} / \sqrt{n}} \\ &= \frac{\mu_{difference}\sqrt{n}}{\sigma_{difference}}\end{aligned} \qquad (6.21)$$

where $\mu_{difference}$ is the mean of the sampled population of difference, or change, scores and $\sigma_{difference}$ is the standard deviation of that population. Just as δ for the independent-groups design is closely related to *d* (see Equation 6.20), δ for the correlated-groups design is closely related to d_z, where $d_z = \mu_{difference}/\sigma_{difference}$. Therefore, Equation 6.21 can be rewritten as

$$\delta = d_z \sqrt{n} \tag{6.22}$$

To do power calculations for a correlated-score design, G*Power requires d_z as an input. Choose the *difference between two dependent means* option, and where the input requests *effect size* d_z, enter the value of d_z as the specific alternative to the null hypothesized value. Again, a value of d_z may originate from any of several sources. A researcher might have pilot data that may be used to estimate a value of d_z. Knowledge of typical effect sizes in the relevant literature may provide the basis for choosing a value of d_z. Or, a researcher may select a value of d_z based on Cohen's guidelines for a small, medium, or large effect, or on the minimum effect size of interest. Let us consider these possibilities.

Using Pilot Data

If pilot data are used to estimate a value of d_z, d_z may be estimated as

$$\hat{d}_z = \left(\bar{Y}_1 - \bar{Y}_2\right) / s_{Y_1 - Y_2} \tag{6.23}$$

Or there is a simple relation between $\hat{d}_z$ and *t* in the correlated-scores case that is useful in estimating d_z:

$$\hat{d}_z = t / \sqrt{n} \tag{6.24}$$

Once an estimate is obtained, it may be inserted into G*Power 3.1 using the *t* test /dependent means module. A more conservative approach would be to use Fitts's (2020) function in R (or another method from Section 6.7.2) to obtain a confidence interval on d_z, and then enter into G*Power 3.1 a value of d_z from the lower range in the CI. This approach involves more steps, but it will produce a value of *n* that is more likely to guarantee the desired level of power because it recognizes that estimates of d_z are typically quite variable and, for small samples, overestimate the population effect size.

Using Knowledge of the Literature

In many circumstances, a researcher plans to do an experiment that has an established context in a research literature. A review of related studies using the same paradigm will provide the researcher with a range of effect sizes in the relevant literature. A prudent strategy would be to select a relatively small value of d_z from those observed and use it as the basis for power calculations. If the true population standardized effect is larger than the selected estimate of d_z, the calculated value of *n* and the corresponding power will be greater than what is targeted, but that is better than having less power than desired.

Using Cohen's Guidelines

If there is no empirical basis for selecting a value of d_z on which to base power calculations, the researcher may rely on Cohen's guidelines for small, medium, and large effect sizes. Although Cohen's guidelines refer to d rather than d_z, there is a relation between d_z and d that can be exploited to obtain an estimate of d_z from d. Comparing Equations 6.19 and 6.21, we see that d and d_z differ only in their denominators. The relation between the standard deviation of the treatment populations, σ, and the standard deviation of the population of difference scores is

$$\sigma_{Y_1-Y_2} = \sqrt{\sigma_1^2 + \sigma_2^2 - 2\rho\sigma_1\sigma_2} \tag{6.25}$$

where ρ (rho) is the population correlation between scores in the two conditions. Assuming homogeneous variances, this can be rewritten as

$$\sigma_{Y_1-Y_2} = \sigma\sqrt{2(1-\rho)} \tag{6.26}$$

Substituting Equation 6.26 into Equation 6.21 and solving for d_z,

$$\begin{aligned} d_z &= \frac{\mu_1 - \mu_2}{\sigma\sqrt{2(1-\rho)}} \\ &= d \,/\, \sqrt{2(1-\rho)} \end{aligned} \tag{6.27}$$

As an example of working from a value of d, suppose that we conducted an experiment with an independent-groups design. Despite a reasonable effect size of .5, the experiment failed to produce a significant effect. We decide to try again, but we choose a correlated-scores design for the follow-up. We wish to compute the n for .8 power assuming the same effect size found in the first experiment. To compute n for the correlated-scores design, we must translate our d-value of .5 to a value of d_z. We need to assume a value of ρ to do the conversion. In the absence of any information, we will assume a relatively low correlation of .3. Substituting in Equation 6.25, we compute $d_z = .42$. Using G*Power 3.1 for $d_z = .42$, $\alpha = .05$, power = .8, and a one-tailed test, we calculate that n should be 37.

Cohen's guidelines for small, medium, and large values of d have been widely adopted, so it may be useful to translate those values into corresponding values of d_z. To do so, we will present values of d_z for values of ρ equal to .2, .5, and .8. Entries in Table 6.4 are values of d_z calculated using Equation 6.27 for the nine combinations of values of d and ρ. The values in the rows correspond to Cohen's (1988) guidelines for small, medium, and large effects.

Table 6.4 Values of d_z corresponding to nine combinations of d and ρ

Values of d	*Values of ρ*		
	.2	.5	.8
.2	.16	.20	.32
.5	.40	.50	.79
.8	.63	.80	1.26

6.9 Post Hoc Power

Having failed to reject the null hypothesis, researchers sometimes suggest that power was too low to detect a population effect. As support, they may calculate power based on the observed effect, perhaps using the *post hoc* option in G*Power, or SPSS's *Observed Power* option. For example, suppose we had two groups of 10 subjects each and the observed t = 1.9, slightly short of the critical value of t = 2.1 required for a two-tailed test. Suppose that the researcher calculates d = .85. Noting that the effect is large by Cohen's guidelines, the researcher calculates post hoc power, which is only .44. With this information – a large effect and low power – the researcher claims that only the lack of power prevented a clear demonstration of statistical significance.

There are several problems with such *post hoc*, or retrospective, power calculations, as we described in Section 4.5.5. Suppose that an investigator fails to reject the null hypothesis. A post hoc power calculation based on the observed effect size suggests low power, so the investigator concludes that the lack of significance was due to low power. However, this argument fails to recognize that sampling error makes it unlikely that the observed effect size is the same as the population effect size. Therefore, the estimate of power may be in error; it is possible that power was higher than indicated by the estimate and that there really is no effect to be detected. In short, null results are ambiguous.

There are other reasons to avoid post hoc power calculations. One is that post hoc power provides no more information than the p-value (Hoenig & Heisey, 2001; Lenth, 2001). Observed power and the p-value are inversely related because they are both functions of the observed effect size and the sample size. Stated differently, if the result of the t test was not significant, post hoc power was either low or the null hypothesis is correct.

In sum, we strongly recommend *against* the practice of post hoc power computations. Instead, confidence intervals on both the raw and standardized effect sizes are much more useful. Confidence intervals on raw effect sizes are likely to be on a more meaningful scale; confidence intervals on standardized effect sizes provide comparability with other results. Post hoc power adds nothing to our understanding and the strategy of computing post hoc power misrepresents the essential ambiguity of a failure to reject the null hypothesis.

6.10 Summary

This chapter focused on inferential statistics based on the t distribution, using two research designs to illustrate the use of those statistics. The following points are important to keep in mind:

- There are both practical and statistical considerations in selecting a research design. When several designs are practical, testing each subject under several conditions, or testing pairs of subjects matched on some relevant measure, will usually result in less error variance than the independent-groups design although there will be fewer degrees of freedom.
- Confidence intervals provide information that is not directly available from the t statistic alone. The interval provides a measure of the precision of our estimate of the treatment effect, and at the same time provides a test of the universe of possible null hypotheses. The width of the interval – and therefore the precision of the estimate of the treatment effect – was shown to decrease with (1) increases in sample size, (2) decreases in variance,

and (3) increased alpha. Because it affects variability, the design of the experiment – for example, correlated scores or independent groups – is also a major factor.

- Effect size measures and their confidence intervals are an important addition to parameter estimates and significance tests: (1) they speak to the practical or theoretical significance of the effect, as distinct from statistical significance; (2) they permit scale-free comparisons of effects from different conditions or experiments; (3) they are used in calculating the power of the statistical test; and (4) they provide a basis for meta-analyses in which results from several experiments are combined to provide a "big picture" of the effectiveness of an independent variable.
- We have emphasized statistical power, arguing that it should be a major consideration in the design of the experiment. Power is influenced by the same factors that affect the width of confidence intervals: sample size, variance, alpha-level, and the experimental design. The use of freely available software, G*Power 3.1, was illustrated in decisions about sample size (*a priori*, or prospective, power).
- We presented arguments against calculating *post hoc* power based on the statistics of a completed experiment. The results of such analyses are misleading because they rest on the assumption that the sample statistics have the same value as the population parameters, and they are largely uninformative because post hoc power is a function of the p-value associated with the observed t statistic.
- The validity of statistical methods rests on the validity of assumptions. We considered the assumptions underlying the methods presented, emphasizing the consequences of their violations as a function of such factors as sample size, shape of the population distribution, and the ratio of treatment population variances. We then suggested alternative analyses in response to these violations of assumptions.

Exercises

6.1 [Computing confidence intervals and effect sizes] A sample of nine 30-day-old, protein-deficient infants are given a motor skills test. The mean for a normal population is 60. The data are

$$Y_1 = 40, 69, 75, 42, 38, 47, 37, 52, 31$$

a) Find a 90% confidence interval for the mean of the protein-deficient population.

b) Is the mean score of the protein-deficient children significantly below that of a normal population?

c) After 3 months of a normal diet, the scores of the nine children are

$$Y_2 = 48, 68, 77, 46, 47, 46, 41, 51, 34$$

Estimate the mean of the population after 3 months of a normal diet. Calculate the 90% confidence interval and test whether this sample mean is below the population value of 60. Assume $\alpha = .05$ for the significance test.

d) Calculate difference scores and test whether there has been an improvement from the first test to the second. Assume $\alpha = .05$.

e) Estimate Cohen's d for the comparison of the Y_1 and Y_2 means and use software to find the 95% CI for that estimate.

6.2 [Comparing correlated-scores and independent-groups designs] An investigator wants to determine whether the difficulty of material to be learned influences the anxiety of college students. A random sample of 10 students are given both hard and easy material to learn (order of presentation is counterbalanced). After completing part of each task, anxiety level is measured by a questionnaire. The anxiety scores are as follows:

Student	*1*	*2*	*3*	*4*	*5*	*6*	*7*	*8*	*9*	*10*	*11*	*12*
Hard task	48	71	65	47	53	55	68	71	59	31	80	77
Easy task	40	59	58	51	49	55	70	61	57	32	70	69

a) Find the 95% confidence interval for the difference in the population means corresponding to the two conditions.
b) Test whether anxiety is significantly different in the two difficulty conditions, using a correlated-scores *t* test. State H_0 and H_1 and indicate the rejection region for $\alpha = .05$.
c) Redo parts (a) and (b), assuming that the experiment had been done with two independent groups of 12 subjects each. What are the strengths and weaknesses of each design? Note any differences in results of the analyses and the reasons for them in your answer, as well as any other considerations that you feel are important.

6.3 [Comparison of power with *t* and standard normal distributions] For a correlated-scores design, we wish to test H_0: $\mu_d = 0$ against H_1: $\mu_d > 0$ at $\alpha = .05$. Using a sample of 16 subjects, we find $\bar{D} = 2.0$ and $s_D = 5.6$.

a) Carry out the *t* test.
b) Calculate the standardized effect size, d_z.
c) What *n* would be required to have power equal to or greater than .80?
d) Calculate power using the *t* distribution and also the standardized normal distribution (see Chapter 5 for a review of the method, or use G*Power) for $n = 16$ and 36, and for the *n* in your answer to part (c). How good an approximation are these results to the results you obtained using the *t* distribution? Is the approximation better or worse as *n* increases? Why might this be?

6.4 [Effect sizes, power, and confidence intervals; independent groups] In an independent-groups design, we have

Group 1	*Group 2*
$n_1 = 18$	$n_2 = 14$
$s^2_1 = 16$	$s^2_2 = 20$
$\bar{Y}_1 = 30.1$	$\bar{Y}_2 = 27.7$

a) Find the 95% confidence interval for $\mu_1 - \mu_2$, assuming equal variances. If we test H_0: $\mu_1 = \mu_2$ against a two-tailed alternative, what can we conclude?
b) Calculate the standardized effect size, *d*. Assuming this effect size, what power did the experiment have to reject the null hypothesis?
c) Suppose we wished to redo the study with equal *n* and want .8 power to reject H_0, assuming the effect size calculated in part (b). What size *n* would we need?
d) Using the *n* from part (c), and assuming the variances given in part (a), what would the width of the new confidence interval be?

6.5 [Comparing Student's *t* and Welch's *t*] In an independent-groups design we have

Group 1	*Group 2*
$n_1 = 21$	$n_2 = 11$
$s_1^2 = 8$	$s_2^2 = 30$
$\bar{Y}_1 = 30.2$	$\bar{Y}_2 = 27.0$

a) Test the null hypothesis at $\alpha = .05$ against a two-tailed alternative using the pooled-variance *t* test.
b) Test the null hypothesis at $\alpha = .05$ against a two-tailed alternative using the unequal variance (Welch's) *t* test.
c) Explain any differences in your conclusions in parts (a) and (b).

6.6 [Testing differences of differences] An arithmetic skills test is given to 8- and 10-year-old children taught with two different teaching methods. There are 10 children in each of the four cells of this research design. The means and standard deviations are given as follows:

		8 years	*10 years*
Method A	$\bar{Y}$	58	72
	s	2.7	2.1
Method B	$\bar{Y}$	53	60
	s	2.9	2.2

a) (i) Calculate a 90% confidence interval for the difference in population means for 8- and 10-year-olds taught with Method B ($\mu_{10,B} - \mu_{8,B}$), assuming equal variances. (ii) Assume you wish to test the null hypothesis against H_1: $\mu_{10,B} > \mu_{8,B}$. What can you conclude from the confidence interval?
b) We wish to test whether the difference between teaching methods is significantly greater at age 10 than at age 8. (i) State H_0 and H_1 in terms of the four population means. (ii) Calculate the numerator of the *t* statistic that reflects whether the teaching method difference is greater at age 10 than at age 8. (iii) Calculate the standard error of the quantity calculated in part (ii); see Equation 5.18 in Appendix 5.1. (iv) Carry out a *t* test of your null hypothesis, briefly reporting the conclusion.

6.7 [Independent groups and correlated scores *t* tests] Several researchers have compared reading under lab conditions where participants knew they would be tested for recall with natural reading without knowing they would be tested. In one such study, two groups of eight subjects each (lab, natural groups) were tested twice on the same materials, once on each of two different days. Free-recall percentages (correct responses) were the following:

Lab	Day 1	45	60	42	57	63	38	36	51
	Day 2	43	38	28	40	47	23	16	32
Natural	Day 1	64	51	44	48	49	55	32	31
	Day 2	21	38	19	16	24	27	22	35

a) For each group, find the 95% confidence interval for the population mean of the change in recall over the two days. For each group, is the change significant at the .05 (two-tailed) level?
b) We wish to compare the two groups on day 2. Assuming a two-tailed test, can we reject H_0 at the .05 level?
c) From part (a), we have a change score for each subject. We wish to test whether the amount of change is the same for the two populations of readers. State the null and alternative hypotheses. Do the test at the .05 level.

6.8 [Practical and statistical significance of effects in real data] The data for this problem are in the *TC_Data* file at the website.

a) Calculate the standardized effect size (Cohen's d) for the winter-spring difference in *TC* scores ($tc1 - tc2$) for the $Sayhlth = 2$ (very good) and for the $Sayhlth = 4$ (fair) group. How would you characterize the effects in terms of Cohen's guidelines?
b) Calculate the winter-spring confidence intervals for the two *Sayhlth* groups of part (a). In which is the confidence interval narrower? Also calculate the t statistic for each. Which has the larger t? The lower p-value?
c) Considering the various statistics, discuss the effects of seasons (winter versus spring) on total cholesterol level.

6.9 [Effect sizes and sample size planning, real data] In Exercise 6.8, we calculated an estimate of Cohen's d, a measure of the standardized effect size. When scores in two conditions (such as winter and spring) are correlated, an alternative measure is dz, the ratio of the difference between the means to the standard deviation of the difference.

a) Calculate d_z for each of the two *Sayhlth* groups we have been considering.
b) Your result for d_z should differ from the result obtained when calculating d. In general, what will influence the size of the difference between these two measures of effect size?
c) If we were to replicate the study for the two *Sayhlth* groups, assuming the effect sizes estimated from the present study, how many subjects would be needed in each condition to ensure power of at least .8 to reject the null hypothesis of no seasonal difference? Assume $\alpha = .05$ and a two-tailed test.

6.10 [Confidence intervals and hypothesis testing] A group of 18 subjects, the experimental group (E), reads a passage designed to increase support for lowering the legal drinking age to 18 years. A second group of 18 subjects, the control group (C), reads a passage unrelated to this topic. Statistics for the two groups are as follows, with higher means reflecting greater support:

	E	C
Mean	4.0	3.2
Variance	2.25	1.44

a) Calculate the 95% confidence interval. Based on the limits you calculated, is the difference between the means significant at the .05 level, two-tailed?
b) Given the purpose of the experimental passage, reconsider whether the difference is significant.

6.11 [Sample size planning for different designs]

a) Estimate Cohen's d for the statistics given in Exercise 6.10.
b) Assume your estimate of d is the actual population value. If we replicate the experiment, how large must each group be to have power = .8 against a one-tailed alternative?
c) Suppose we assume the same value of d but decide to replicate the experiment using a design with a single group of subjects who are tested before and after reading the passage. A reasonable estimate of the correlation between *before* and *after* scores is .5. What should n be in this experiment to have power = .8?

6.12 [Effect sizes and t tests] In studies testing whether a single sample mean differs significantly from zero, we have the following t values and sample sizes (n):

t	2.0	4.5	2.0	4.5	2.0	4.5
n	16	81	36	36	4	225

a) Calculate the value of Cohen's d statistic for each study.
b) Is there a relationship between t and d? Explain your answer.

6.13 [Significance and measures of importance] Consider the following for four two-group experiments. In each experiment, the difference between the means is 3.3, and the variance is 22 for group 1 and 27 for group 2. The groups are of equal size within each experiment and the group sizes are as follows:

Experiment =	1	2	3	4
n	4	9	16	25

a) Calculate t and d and find the p-value (two-tailed) and the post hoc power of the test. What are the relations among d, p, and post hoc power?
b) Which is the best index of the importance of the effect? Why?

The next two problems are open-ended but represent the task faced by the investigator with a large data set.

6.14 [Analysis of real data] The *Royer_acc* file at the website contains subtraction, addition, multiplication, and mean percent correct for third- to eighth-graders who had accuracy scores for all three arithmetic operations. Choose and test a reasonable hypothesis of interest that compares two conditions, being sure to state your hypotheses clearly. Support your conclusions with graphs, and any statistics – including significance test results, confidence intervals, and effect sizes – that you find relevant.

6.15 [Analysis of real data] Using the *Royer_rt* file, evaluate the same hypothesis as in Exercise 6.14, again supporting your answer with graphs and statistics. Does any aspect of your analysis differ from your approach in 6.14? If so, why?

6.16 [Challenge Problem: Violating assumptions]. Using R, simulate data from two conditions, Y_1 is normally distributed with a mean of 100 and standard deviation of 10; Y_2 is distributed uniformly between 0 and 200, so the mean is also 100. The *rnorm* and *runif* functions will accomplish this sampling. Simulate at least 1,000 experiments in each of the sample size combinations in this table:

n_1	n_2	*Proportion of significant differences using Student's t*	*Proportion of significant differences using Welch's t*
10	10		
100	10		
10	100		
100	100		

Using α =.05 and the *t.test* function, run Student's and Welch's *t* tests for each sample and save the *p*-values. Report the proportion of significant test results in each case, then explain your findings and their implications.

Notes

1 Because an increase in *TC* level is of no interest, a one-tailed test might be more appropriate. With 35 *df*, the critical *t* value for the one-tailed test at the .05 level is 1.690. Because $t = [16.333 - 6]/5.421 = 1.906$, we may conclude that the mean reduction in the population of *TC* scores under our treatment was greater than that observed in previous experiments. However, it is important to understand that decisions about the directionality of the alternative hypothesis should be made before carrying out the test (see the discussion in Chapters 4 and 5).

2 Our estimate of *d* for the independent-groups design is the statistic denoted by *g* by Hedges and Olkin (1985).

3 Fitts (2020) noted an error in the *df* used for the correlated scores calculations (using the design = "within" option), which should not be used. He provided an R script that solves the problem; it appears in his article and is included in the R script provided for this chapter on the book's website.

Chapter 7

Integrated Analysis I

7.1 Overview

This is the first of several integrated analysis chapters designed with two purposes in mind.

- First, to help integrate concepts and procedures presented in the preceding chapters.
- Second, to help bridge the gap from theory to application.

Using new examples, these chapters review and extend the analyses presented in the previous chapters. The presentation of the material in these chapters follows the form of a research report, including the method used to collect the data, the results of both exploratory and inferential analyses, and a discussion of the results.

In this chapter, we consider a hypothetical independent-groups experiment on the effect of a drug on memory. We first consider the rationale for the experimental design, and then begin the data analysis by examining key descriptive statistics and data plots. We then estimate the raw and standardized differences between the means of the two treatment groups, and construct confidence intervals and test hypotheses based on the t distribution. In response to evidence from the exploratory phase of possible violations of assumptions underlying these procedures, we calculate alternatives to the usual t test.

Following the analyses, we interpret the results, referring to the descriptive statistics and data plots obtained in the exploratory stage of the data analysis, as well as the results of hypothesis tests. In trying to understand our results, we not only have to account for causes of statistically significant effects but also consider whether nonsignificant trends represent chance variability or a lack of power due to factors such as small sample size, outliers, nonnormality of the distribution, and other violations of underlying assumptions.

7.2 Introduction to the Research

Assume that a pharmaceutical company has developed a new pill for slowing the typical memory decline that accompanies aging. Because the pill is expensive to develop, a small pilot study was designed to test its effects. If the results are promising, a larger, full-scale study will be conducted, using more participants and more measures of memory. The researchers decided to test two groups, each having 25 volunteers over 65 years of age. The experimental group took the memory pill once a day for 1 month prior to presentation of a list of 30 words and an immediate recall test. The control group was tested on the same list after taking a placebo for 1 month. The scores are in the Y column of the *IA1 _memory.xlsx* file.

DOI: 10.4324/9781003453550-8

7.3 Method

7.3.1 *Participants*

The pharmaceutical researcher posted a request for volunteers on the bulletin board of a local senior center. Volunteers were assessed in individual sessions to ensure that they were not suffering from any form of dementia. The assessment included a brief test of memory, an interview, and questions about facts that were judged to be common knowledge (e.g., the name of the vice president of the United States). From the pool of potential volunteers, 50 were chosen, ranging in age from 65 to 75.

An important point is that participants who volunteer for an experiment may differ in meaningful ways from people who chose not to volunteer. For example, self-selected volunteers may be younger, healthier, wealthier, or simply have more time available in their schedules. They may be better educated or more concerned about their memory. We must remember these potential biases in our sample because they limit the population to which our results will generalize.

7.3.2 *Experimental Design*

As we discussed in Chapter 1, there are several design options to consider. We could employ a within-participants (correlated scores) design, testing the participants prior to their taking the memory pill and then testing again later, sometime after taking the pill. The researcher decided against this, feeling that prior exposure to the testing situation might raise scores in the subsequent test, thus reducing the effect of the drug. Furthermore, a test-retest design would necessitate constructing a second memory test that was equated for difficulty with the first one. Still another possible problem was that age-related memory decline might accompany the period between test and retest, potentially masking any benefit of the pill. Finally, just knowing they were given a pill designed to aid memory might have motivated participants, producing an improvement unrelated to the pill itself. Modifications of the test-retest design could address these concerns; however, the researcher decided that the simplest procedure was to use a between-participants design, randomly assigning the volunteers to either an experimental pill condition or a control (i.e., placebo) condition. A single research assistant collected the data but did not know the condition to which each participant was assigned. As noted in Chapter 1, this is known as a double-blind procedure because neither the individual collecting the data nor the participants knew who was in which condition.

Further modifications of this independent-groups design are possible. Restricting the assignment of participants to ensure that the groups had equal numbers of men, women, and nonbinary individuals would equate any effects due to gender and facilitate investigating gender effects. A correlated-scores design in which participants were assigned so that the groups were equated for age, or for scores on the preliminary test given to all volunteers, would reduce error variance. However, these variables could also be used in regression analyses after the primary data were collected (e.g., analysis of covariance; see Chapter 24), or in a subsequent, larger experiment if this pilot study yielded promising results. Another possible design would involve a combination of the correlated-scores and independent-groups design; this would involve repeated testing of both experimental and control participants, controlling for any changes due to prior exposure to the test, or age-related memory decline. Although several of these designs, as well as others, might result in reduced error variance or greater control of nuisance variables that could bias the results, the experimenters opted for a relatively simple design for their pilot study.

7.3.3 *Procedure*

Each participant was given a bottle containing 28 pills. Half of the bottles contained the placebo and half the experimental pills. Each participant was assigned a different numbered bottle of pills and only the researcher knew the relationship between the bottle numbers and the experimental conditions. Participants were instructed to take one pill each day and to return to the laboratory on day 29. They were told that no side effects were expected but that, as a precaution, they should note any departures from their usual physical or emotional state.

On day 29, participants were seated, and the research assistant presented a list of words on a projection screen, one at a time. These 30 test words were common English nouns of five to seven letters in length. Immediately after they were read, the participants independently wrote down as many of the words as they could recall, in any order.

7.4 Exploring the Data

The score for each participant was the number of words recalled correctly out of 30. Minor spelling errors were ignored when the participant's intended word was clear. For the present, we consider only the column of scores labeled *Y* in the *IA1_memory* file. We begin our analysis by obtaining the descriptive statistics and data plots that were illustrated in Chapter 2. Table 7.1 presents several descriptive statistics obtained from the *summarise* function in the {dplyr} package of R, details of which are available at the book's website.

Although we have used R for this purpose, most statistical software packages will provide similar output.[1] The medians have similar values to the means in each condition, suggesting that the distributions are symmetric. The drug did result in numerically better average recall scores than the placebo taken by participants in the control group. However, the mean difference of about 1.5 words recalled is small and may not be of either practical or statistical significance. We will shortly calculate a standardized effect measure, Cohen's *d*, to obtain a better sense of the importance of the effect of the memory pill. Confidence intervals and significance tests will address the issue of statistical significance. First, however, we consider several of the other statistics in Table 7.1, and some data plots as well.

Before conducting any inferential tests, we should evaluate whether violations of test assumptions are a concern. We begin with an assessment of the normality of the data, including skewness and kurtosis. Recall that the standard errors of both skewness and kurtosis are a function of sample size and can be calculated easily (see Chapter 2). The SE_{skew} is 0.46 in each group, which is roughly the magnitude of skewness in the drug condition. In the control condition, the ratio of the skew statistic to its standard error is just over 2, which suggests some skewness in those data. There is stronger evidence of kurtosis in the data of both groups: The ratios of the transformed statistics to their standard errors (0.902) are greater than 2.5. This suggests that tails may be longer than in a normal distribution, a condition that might contribute to reduced power. We can also use the Shapiro–Wilk (1965)

Table 7.1 Statistics for the memory data (from the *summarise* function in R's {dplyr} package)

Method	*N*	*AvgScore*	*StdDev*	*SE*	*Median*	*Skewness*	*Kurtosis*	*Min*	*Max*	*Range*	*IQR*
Control	25	16.8	3.75	0.750	17	−1.010	5.52	5	24	19	4
Drug	25	18.4	3.32	0.663	19	−0.436	5.45	9	27	18	3

Table 7.2 Shapiro–Wilk tests of normality for the memory data (from R)

Condition	*Statistic*	*df*	*Sig*
Control	.905	25	.023
Drug	.892	25	.012

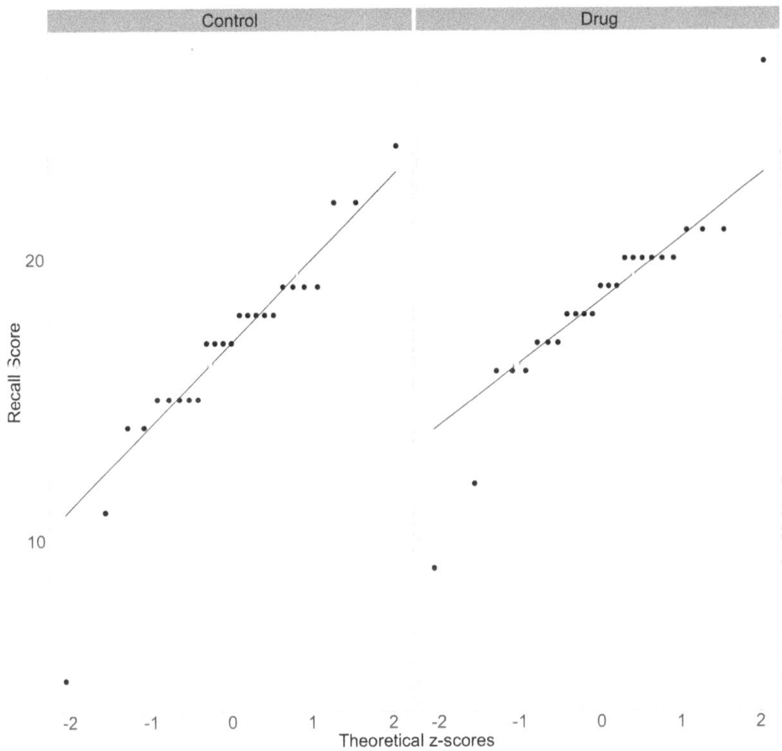

Figure 7.1 Q–Q plots of the memory data.

test of normality, which, loosely speaking, compares the variance of the data with the variance expected from rank-ordered data sampled from a normal distribution. In R, *shapiro.test* in the {stats} package reveals that both conditions show significant departures from normality, with *p*-values less than .03 (see Table 7.2).

We can learn more about the distributions by examining plots of the data. The *Q–Q* plots in Figure 7.1 confirm that the data are not normally distributed and indicate that this is largely attributable to several extreme scores, especially in the drug condition. This conclusion is subjective; there are no hard-and-fast rules for interpreting *Q–Q* plots. In this case, the extreme scores are evident in both panels because there are individual points that fall far above or below the line. To determine which of these scores are outliers in the technical sense (see Chapter 2), we turn to the box plots in Figure 7.2, which show the outliers as black dots (the gray dots are individual scores). These boxplots confirm that there are several outliers in the data set. The presence of outliers increases variability and reduces the power

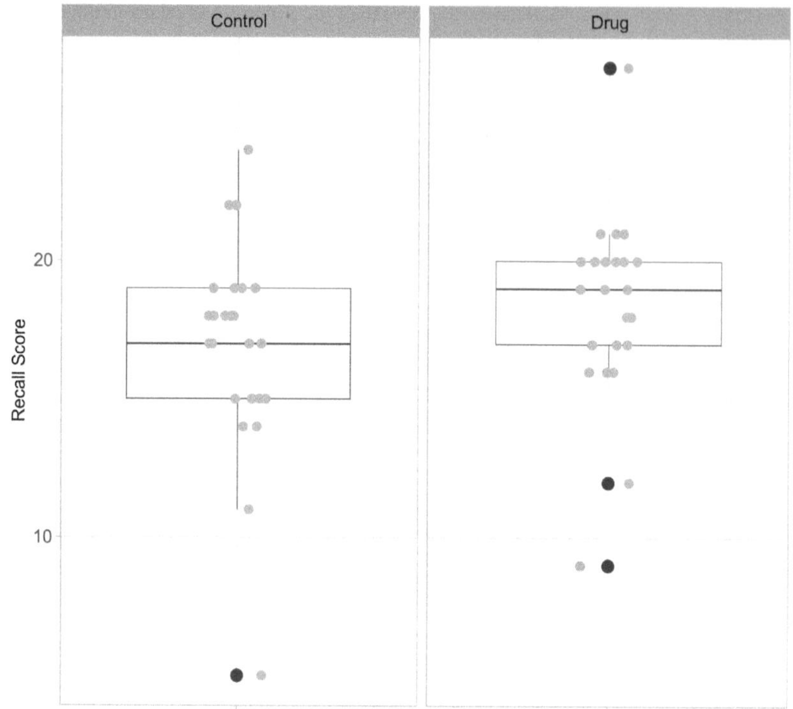

Figure 7.2 Box plots of the memory data.

of t tests. We should not just delete the outliers and then do the usual statistical tests because the variances based on trimmed data are biased estimates of the true population variances. However, after analyzing the entire data set, we will perform a rank-based analysis as well as a trimmed t test, using corrected variance estimates. For demonstration purposes, we will conduct multiple analyses of the same data, although it is important to emphasize that our purpose is not to "shop around" for the analysis that gives the result we prefer. In practice, only one of the procedures illustrated here should be selected, with the decision resting on an evaluation of the normality and homogeneity of variance assumptions.

In sum, our exploration of the data has shown that the sample means are consistent with our hypothesis that the drug improves memory and has revealed that outliers are present that may reduce the power of the standard t test. With this information in hand, we proceed beyond the descriptive statistics to inferences about the population. We will first calculate confidence intervals for, and tests of, the difference between the group means and then, with the outliers in mind, we will also run a rank-based analysis and an analysis employing the trimmed mean and winsorized variance.

7.5 Confidence Intervals and Hypothesis Tests

In R, we use the *t.test* function in the {stats} package, which calculates Welch's t' by default. The data are in a data frame called 'dat', which contains variables called 'Y' (our dependent measure) and 'Method' (the grouping variable). Using the command *t.test*(data = dat,

```
> t.test(data = dat, Y ~ Method) #two-sample Welch's t'

        Welch Two Sample t-test

data:  Y by Method
t = -1.5186, df = 47.291, p-value = 0.1355
alternative hypothesis: true difference in means between group Control and group Drug
is not equal to 0
95 percent confidence interval:
 -3.5332902  0.4932902
sample estimates:
mean in group Control    mean in group Drug
                16.84                 18.36
```

Figure 7.3 Results of Welch's *t* test on the means in Table 7.1.

Y ~ Method), which you can read as an instruction to compare Y as a function of Method using an unequal-variance independent groups *t* test, we obtain the result in Figure 7.3. A comparable analysis in SPSS is the *independent samples t test* in the *Compare Means and Proportions* section of the *Analyze* pull-down menu (see Section 6.4.5).

Three points are apparent from the figure:

1. The observed t' is negative because *t.test* compares the conditions in alphabetical order: Control – Drug has a mean difference of −1.52. It doesn't matter which condition is called Y_1 and which is Y_2 in Equation 6.16; the statistical conclusions are the same.
2. The confidence interval for the difference between the means is wide, ranging from −3.5 to 0.49, meaning a benefit of the drug that averages 3.5 items recalled, or a disadvantage of a half an item, on average. The true population difference either is or is not contained within the confidence interval. If we repeated this experiment many times, 95% of the resulting confidence intervals would include the population difference.
3. Presumably, we are interested only if the drug has an advantage. Therefore, our alternative to the null hypothesis is one-tailed and the reported *p*-value of .136 should be halved (or we could run a one-tailed test using *t.test*(data = dat, Y ~ Method, alternative = "less"); either way, the one-tailed test yields $p = .068$. Thus, by conventional criteria, we cannot reject the null hypothesis of no difference between the two population means.

7.6 The Standardized Effect Size (Cohen's *d*)

The difference between the group means is one measure of the effect of the drug. The standardized effect size provides an alternative measure that takes the variability into account. From Equation 6.19, $\hat{d} = \left(\bar{Y}_{Drug} - \bar{Y}_{Control}\right) / s_{pooled}$. When the *n*s are equal, the pooled variance is just the average of the two variances. From the values in Table 17.1, $s^2_{pooled} = (1/2)(3.115^2 + 3.749^2)$, or 12.522. The square root of the pooled variance is 3.539; therefore, $\hat{d} = (18.36 - 16.84) / 3.539 = .43$. By Cohen's (1988) guidelines, this is close to a medium-size effect for the comparison of two independent means. However, using the *gethedgesg* function in R (see Section 6.7.2), we found the 95% confidence interval bounds on the effect size to be −.13 and 1.02, indicating that the population effect was somewhere between a very small advantage of the placebo and a very large advantage of the memory pill. Once again, we lack strong evidence that the pill aids memory.

7.7 Reanalysis: Alternative Approaches

The results of analyses conducted to this point are ambiguous: On the one hand, they are suggestive of a positive effect of the new pill; on the other hand, the effect of the pill is not significant. We have found that the data are quite variable; in particular, there are several outliers in the data. As we noted previously, it's important not to keep running different tests in the hope of obtaining a smaller p-value; doing so inflates the α-level. Nonetheless, in the section we demonstrate a range of possible analyses; the researcher should choose an analysis strategy after exploring the data.

There are a couple of options for the researcher at this point. One is to conduct a Mann–Whitney U test, or the closely related Wilcoxon W test, on the data. These procedures cope better with outliers than the t test because they analyze the rank-ordered observations. Further, the data meet the conditions under which the U test is most appropriate; namely, the scores in the two conditions of the experiment are distributed similarly, with similar variances, kurtosis values, and skewness. The boxplots in Figure 7.2 are also similar except for location. (As for the Q–Q plots, there are no hard rules on the interpretation of boxplots. They simply provide a visual summary of the data.) Another option is to run a trimmed t test on the data, as described in Chapter 6. The trimmed t test is often more powerful than the usual t test when the tails of the distribution are longer than would be expected in a normal distribution, as occurs when there are outliers. We will discuss each test in turn.

7.7.1 *The Mann–Whitney U and Wilcoxon W Rank-Based Tests*

The Mann–Whitney U test assesses whether two conditions differ in their location by computing the number of times the rank of a score from one condition exceeds the rank of a score in the other condition. This should happen more frequently when the location difference is larger. The U test is closely related to the Wilcoxon W rank-sum test, which assesses whether the sum of the ranks in one condition exceeds the sum of ranks for the other condition. Again, the difference in rank-sums is greater when the location difference is larger.

For either test, we begin by ranking all the scores across both groups, assigning a rank of 1 to the smallest score and a rank of N to the largest score; ties are assigned the average of the ranks for that set (e.g., if two scores would receive ranks 9 and 10 if they weren't tied, then assign rank 9.5 to each). We can use the *rank* function in {base} R to accomplish this easily. The U test compares all possible pairs of ranks in the two groups, adding up the number of times the rank for the group 1 score exceeds that of the group 2 score. (Half-credit is given for each tied pair.) This total is called U. For our example experiment, we can use R's *wilcox.test* function in the {stats} package to compute U (which is reported as W); $U = 217$ and has an associated p-value of 0.063, which is not significant. Dividing the sum, U, by the number of comparisons made ($n_1 * n_2$), we get f, the proportion of pairs that support the conclusion that the scores in group 1 exceed the scores in group 2. Here, $f = 217 / (25*25) = 0.347$, a value that provides evidence suggestive of a difference between conditions: About 35% of the between-group comparisons favored group 1, and about 65% favored group 2.

If we run the Mann–Whitney U test in SPSS, by choosing the *Non-parametric Tests* option in the *Analyze* menu, and then selecting *Independent Samples*, the results appear different. They are shown in Figure 7.4, and one thing to notice is the reported $U = 408$, not the 217 that R produced. This is not an error; it simply reflects a different way of

Independent-Samples Mann-Whitney U Test

Y across Method

Independent-Samples Mann-Whitney U Test Summary

Total N	50
Mann-Whitney U	408.000
Wilcoxon W	733.000
Test Statistic	408.000
Standard Error	51.170
Standardized Test Statistic	1.866
Asymptotic Sig.(2-sided test)	.062

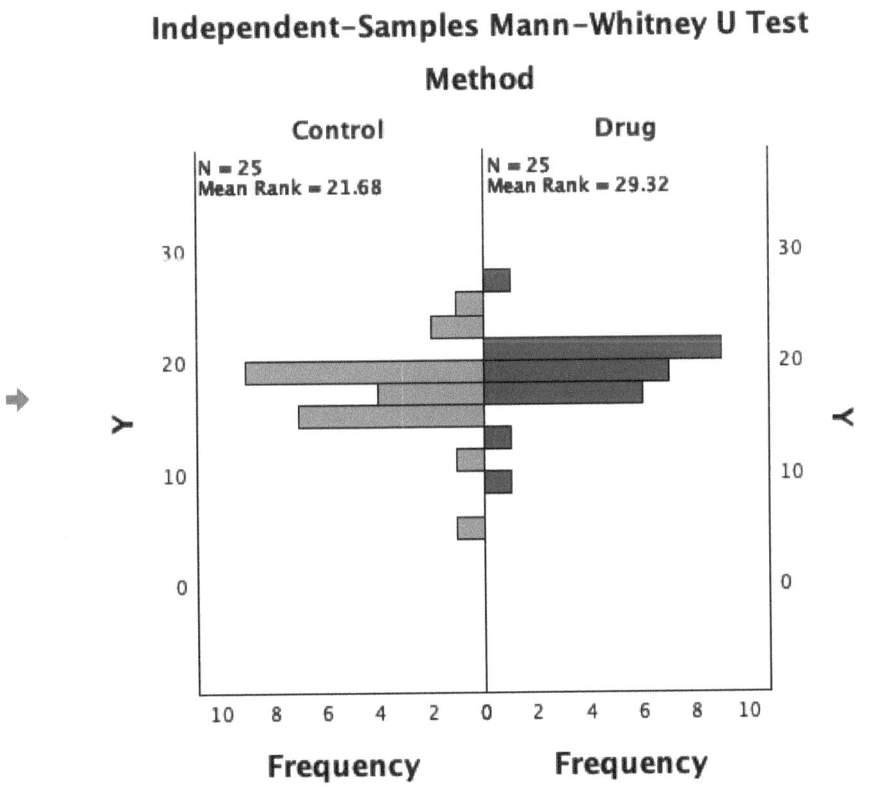

Figure 7.4 Mann–Whitney *U* test results for the memory data in SPSS.

summarizing the analysis. Logically, if we sum all the ranks in each condition, then if the groups differ in location, a larger rank-sum will be found in one condition than the other. The rank-sum for the drug condition, which we can find by multiplying the mean rank shown in the histogram in Figure 7.4 by the sample size (29.32*25), is 733, the value SPSS reports for Wilcoxon *W*. The Mann–Whitney *U* and Wilcoxon *W* are linearly related:

$$U = W - \frac{n_2(n_2+1)}{2} \tag{7.1}$$

where *W* is the sum of ranks in one condition, and n_2 is the sample size for the other condition. Using *W* from the drug condition, we find $U = 733 - (25*26)/2 = 408$, the value reported by SPSS. Of course, it is equally appropriate to compute *U* from the rank-sum of the control condition, which is (21.68*25) = 542. In that case, $U = 542 - (25*26)/2 = 217$, the value reported by R. Using either SPSS or R, we reach the same conclusion: The conditions do not differ significantly.

7.7.2 The Trimmed t Test

The other option for dealing with the outliers in the memory data set is the trimmed *t* test (Tukey & McLaughlin, 1963; Yuen, 1974) cited in Chapter 6. When the tails of the distribution are longer than would be expected in a normal distribution, this test often will prove more powerful than the usual *t* test. Further, the trimmed *t* test has the advantages of permitting confidence interval calculations and power computations. It also allows effect size calculations, although the result is difficult to interpret (Cumming, 2013). Of course, the trimmed *t* suffers from the criticism that a substantial fraction of the data is discarded, with a corresponding decrease in degrees of freedom.

The trimmed *t* test for two independent samples involves two separate treatments of the scores in each condition: (1) trim *k* scores from both the upper and lower ends of the sorted data values and calculate the means of the remaining values; (2) *winsorize* the scores by replacing each of the *k* lowest values with the next lowest value and replace the *k* highest scores with the next highest. The difference in the trimmed means is used to calculate the numerator of the trimmed *t*, and the winsorized data are used to calculate the standard error for the denominator. Note that winsorizing reduces the degrees of freedom in the data, so that standard error is based on $2(n - 1 - 2k)$ degrees of freedom.[2] Wilcox (1997) has recommended that *k* be 20% of the scores,[3] and an example of the resulting data is shown in the *IA1_memory* file. In that file, the *Method* column specifies the condition, *Y* contains the original scores, *T* shows the 20% trimmed scores, and *W* contains the winsorized values.

In R, we can use a simple function to calculate the trimmed means: *mean*(Y, trim = .2), where *Y* contains the data for a single condition, or we can use the *YuenTTest* function in the {DescTools} package to compute the trimmed *t* as well as the confidence interval for the difference in trimmed means (see Figure 7.5). Doing so for the memory data reveals that the 95% confidence interval for the differences in trimmed means is [−3.11, −0.09] and the trimmed $t = -2.17$ (because R evaluates conditions in alphabetical order, as previously noted). Thus, this analysis allows us to reject the null hypothesis that the drug has no effect. We have achieved this outcome by eliminating the influence of outliers and long tails in the data; Yuen (1974) showed that the trimmed *t* has greater power than Welch's *t* under those conditions.

```
Console  Terminal
R 4.3.1 · ~/
> YuenTTest(data=dat,Y~Method,trim=.2)

        Yuen Two Sample t-test

data:  Y by Method
t = -2.1685, df = 26.856, trim = 0.200, p-value = 0.03915
alternative hypothesis: true difference in trimmed means is not equal to 0
95 percent confidence interval:
 -3.11426642 -0.08573358
sample estimates:
trimmed mean in group Control    trimmed mean in group Drug
                     17.06667                      18.66667
```

Figure 7.5 Results of trimmed *t* analysis of memory data.

Frequently, there will be reason to suspect that both the normality and homogeneity of variance assumptions are violated. Yuen (1974) proposed that the means in the numerator of Welch's *t* (described in Section 6.6.2) be replaced by trimmed means, and that the variances in that statistic, as well as in the formula for the degrees of freedom, be replaced by winsorized variances. This combination of Welch's *t* and the trimmed *t* guards against distortion of Type 1 error rates due to heterogeneity of variance and provides increased power when the distribution of data is long-tailed.

7.8 Discussion of the Results

After our original analysis of the data, we estimated that the memory drug has a medium size advantage over a placebo in the population of older adults. However, the difference was not significant, and confidence intervals on both the raw and standardized effects were wide. At that point, we could have decided (1) that further development and testing of our memory pill is not worthwhile or (2) that there is sufficient indication of an effect to merit further testing and development. The second choice rests on two assumptions: (1) that an effect of the size estimated is worth the cost involved in further testing and possible marketing; and (2) that the failure to reject the null hypothesis reflects a lack of statistical power due to the presence of outliers in our small sample.

Further support for the assumption that the pill *may* provide a memory advantage was found by reanalyzing the data. The rank-based Mann–Whitney *U* test found that about 2/3 of the scores in the drug condition were higher than those in the control condition, suggestive of a drug-based memory improvement, although the *p*-value didn't allow rejection of the null hypothesis. The trimmed *t* test reduced the variance caused by the long tails and, particularly, by the outlying scores. Although that reanalysis enabled us to reject the null hypothesis, it is essential that we not keep running tests until we obtain the result we like. Doing so risks inflating the Type 1 error rate and drawing erroneous conclusions. In our example memory data, we should be cautious about overinterpreting these various results, and we would feel more confident if we could replicate the result. We should design a large-scale study if we wish to further investigate the value of the pill. This would mean deciding on a larger sample size, and possibly additional measures of memory. If we believe that even a medium size effect warrants manufacture of the pill, and we want one-tailed power ≥ .90 to reject the null hypothesis of no advantage if Cohen's $d \geq .5$, we can use G*Power with $\alpha = .05$ to find that a total sample size of 140 (70 in each group) is required for further experimentation with these independent and dependent variables. Section 6.8 discussed the role of effect size and power in deciding on sample size, and Figure 6.2 provided an example of the application of G*Power.

7.9 Summary

All too often, researchers are concerned only with whether they have a significant effect. The current example illustrates the importance of considering both descriptive and inferential statistics. The exploratory stage of our analysis provided several pieces of important information:

- The memory pill had a numerical advantage in the sample. Using the sample means and variances, we estimated that the population effect was medium in magnitude by Cohen's (1988) guidelines.
- Comparison of group variances, and significance tests comparing those variances, indicated that the assumption of homogeneity of variance underlying subsequent inferential analyses was not violated.
- A careful examination of the exploratory output indicated that the data departed significantly from the assumption of a normal distribution underlying the test and confidence interval based on the *t* distribution. Box plots of the data revealed the presence of outliers that could contribute to an increase in error variance and a resulting loss of power.

A *t* test on the full data set failed to reject the null hypothesis of no difference, and the confidence interval for the difference between the population means was wide. However, the results of the exploratory stage suggested that tests and confidence intervals based on ranked or trimmed data might provide stronger support for the memory pill. While the rank-based analyses were only marginally significant, a reanalysis using the trimmed *t* statistic resulted in a highly significant effect and a narrow confidence interval.

These are just some examples of the potential benefits of examining the data before carrying out statistical tests. We urge exploration of any data set, whether intended as a pilot study or as a more extensive study, and whether experimental or observational. As we stated both in Chapter 2 and in the overview to this chapter, such exploration may reveal patterns in the data that support our hypotheses or suggest others we have not considered; they may also reveal departures from assumptions underlying our planned inferential analyses; and they may reveal outliers, scores that depart markedly from the distribution of our data. Although outliers should never be routinely discarded without examining their possible causes, their presence may suggest the need for other analyses.

Our analysis also went further in other respects than is often the case. We calculated confidence intervals on both the raw and standardized effect sizes. These emphasized how variable our data were and how wide the possible range of effects might be. The standardized effect size also helped us to decide on the effect size to input into an *a priori* power analysis to determine the sample size for a larger, more decisive study of the effects of the drug we were studying.

In summary, after the data are collected, researchers should explore the data, examining carefully both statistics and various data plots relevant to the assumptions of subsequent analyses. Confidence intervals and test statistics then should be calculated. Following careful exploratory and inferential analyses, the researcher is prepared to interpret the results and make decisions based upon them, including plans for replications or follow-up studies.

Exercises

For this chapter, which integrates content from all of Part I, we do not provide a purpose for each problem. Instead, your understanding of the previous chapters will guide you to the solutions.

7.1 a) Scores of −1, 0, and 1 are assigned to underachievers, high achievers, and achievers, respectively, based on a test of motivation. If the proportions of the population in each category are 1/4, 1/2, and 1/4, what are the population mean and variance?

b) Assume that a sample of two students is drawn from the population. What are the possible values of the mean of their two scores? What is the probability of each value? Construct a table of the possible means and their probabilities. Note that this is the sampling distribution of the mean for samples of size 2.

c) What are the mean and variance of the sampling distribution constructed in part (b)?

7.2 Assume that the selection criteria for personnel for a mission to Mars are based on tests of motivation (M), physical fitness (P), and intelligence (I). The following population parameters have been established:

	M	P	I
μ	62	66	60
σ^2	28	20	13

M, P, and I are independently and normally distributed. The selection criterion (C) is a weighted average of each individual's score: $C = .25M + .25P + .5I$.

a) What are μ_C, the average criterion score in the population, and σ_C, the standard deviation of the population of criterion scores?

b) Those selected for training for the mission must be in the top 5% of the population with respect to C. What criterion score will serve as the cutoff?

c) The five-person crew of a previous mission had a mean criterion score of 64.5. Find the 90% confidence interval. Do the limits indicate that this crew's mean score was significantly below the current criterion score computed in part (b)?

7.3 In a study phase, each of 100 participants is shown a series of six objects, each for five seconds. In the following test phase, each participant is tested on six trials. On each trial, four objects are presented, one of which was present in the test phase. The participant's task is to choose the previously studied object on each test trial.

a) If a participant is guessing, what is the probability of a correct response on a single test trial?

b) Assuming guessing, state the probabilities of the possible outcomes (i.e., 0 correct, 1 correct, 6 correct) for a single participant.
c) Assuming guessing, what is the expected number of participants exhibiting each possible outcome?
d) Assuming guessing, what is the expected mean number correct for each participant?
e) What is the probability that by chance a participant will exceed the expected number correct?

7.4 In the experiment described in Exercise 7.3, the actual distribution of outcomes was

Number Correct	0	1	2	3	4	5	6
Number of Participants	12	30	32	19	6	1	0

a) Assign a plus to all participants who had more than the expected value in your answer to Exercise 7.3(d) and a minus to all others. Then, test whether participants have more correct responses than would be expected by chance. You can do this by using a normal approximation to the binomial distribution: calculate $z = \frac{(y - pn - 0.5)}{\sqrt{p(1-p)n}}$.

In part (a), we used the normal distribution to carry out a sign test of the null hypothesis of guessing. However, because we know the scores of the 100 participants, we could calculate the observed mean and variance, and perform a t test of the null hypothesis.

b) What is the observed mean of the 100 scores?
c) What is the variance of the 100 scores?
d) Calculate the t statistic and test the null hypothesis of chance performance.

7.5 Cobb and Hops (1973) compared reading scores of 6 control participants with scores for 12 experimental volunteers who were trained by reinforcement and shaping techniques. Pretest and posttest scores are in the file, *EX7_5.xlsx*. Group 1 is the experimental group.

a) Calculate the 95% confidence intervals for the change scores for each group. State whether the change for each group is significant.
b) Test the difference between groups in the average change score, assuming $\alpha = .05$ and the test is one-tailed.

7.6 The test in Exercise 7.5, part (b), is likely to have had low power because the group sizes were small. Suppose we replicate the study.

a) Estimate Cohen's d for the difference between groups in mean change scores for the *EX7_5.xlsx* data. Use this result to determine the group size needed to have power = .9, assuming $n_1 = n_2$ in a replication. Assume a one-tailed test.
b) Assuming the group size you obtained in part (a), what power would you have for a one-tailed test that the control group improved from pre- to posttest? Estimate the effect size based on the data at the website.

7.7 In this chapter, we analyzed the memory data from an independent-groups design in which some participants received a drug designed to improve memory and others received a placebo. We want to rerun the experiment, this time testing a group of participants before and after taking the drug. We have estimated Cohen's d to be 0.43 for the data from the previous experiment and wish to use those data to decide on the sample size for the new experiment.

a) Assuming the population variances are the same in both the experimental and control groups, we estimate that variance to be 12.5 from the data analyzed earlier. A reasonable estimate of the population correlation (ρ) is .6 for participants tested in both conditions. Given that the population variance of difference scores is $\sigma^2_1 + \sigma^2_2 - 2\sigma_1\sigma_2\rho_{12}$, and if $\sigma^2_1 = \sigma^2_2 = 12.5$ and $\rho = .6$, estimate the standard deviation of the difference scores.
b) From the independent-groups experiment, we estimate the difference between the two population means to be 1.52. Use this information and your answer to part (a) to estimate Cohen's d_z.
c) How many participants should we run in this test – retest design to have .8 power to detect a memory improvement corresponding to the value of d_z calculated in part (b)?

7.8 A new drug was administered to several patients to reduce anxiety. One question is whether this drug has a different effect on cognitive processes than a previously prescribed drug. A clinical researcher compared response times (*RT*s) on a decision-making task for 30 participants who had been taking the new drug (Group 1) with *RT*s for 20 participants (Group 2) who had been taking a drug that had long been prescribed for anxiety. The data are in the *EX7_8.xlsx* file.

a) Explore the data and describe any aspects that might affect the error rates of the t test comparing the group means.
b) Select a transformation of the data and explore the transformed data. Did the transformation address your concerns?
c) Test the hypothesis that the drug slows response times in cognitive tasks.

7.9 Consider alternative approaches to analyzing the data of Exercise 7.8.

a) Perform a trimmed t test using the *RT* data.
b) Compare the 95% confidence interval with that for the untrimmed data.
c) Perform a Mann–Whitney U test on the same *RT* data.

Notes

1 Recall from Chapter 2 that skewness and kurtosis can be calculated and reported in slightly different ways, resulting in slightly different values. The differences are more obvious for small samples. The kurtosis reported here is untransformed; to equal the $g2$ reported in SPSS, subtract 3 from the values in Table 7.1.

2 If you use a statistical package to analyze the W scores, it may report a different standard error than the one here. The reason is that it is dividing the sums of squared deviations by degrees of freedom based on the original data set, 48 in this example, rather than on the degrees of freedom adjusted for the fact that the data are winsorized.

3 Rosenberger and Gasko (1983, p. 311) have provided a definition of the trimmed mean that is general enough to apply even when k is not an integer. However, we suggest the simpler approach of rounding k to the nearest integer when necessary.

Part 2

Between-Participants Designs

Chapter 8

Between-Participants Designs

One Factor

8.1 Overview

In Chapter 8, we consider a basic research design in which there is a single independent variable with several levels that make up the conditions of the study, no participant is tested in more than one condition, and each individual contributes one score to the data set. Like any experimental design, the one-factor, between-participant design has advantages and disadvantages.

The primary advantage of the design is that it is simple. Data collection is simple: only one observation is taken from each participant. No additional measures are required for the purpose of matching participants in different conditions. Nor is there a need to be concerned about the order of presentation of treatments, or the interval between tests, as in designs in which participants are tested in several conditions. There are also fewer assumptions underlying the data analysis than in most other research designs. More complex designs involve additional assumptions that, if violated, increase the likelihood of drawing incorrect conclusions from our data. Finally, there are fewer calculations than in other designs, and decisions about how to draw inferences based on those calculations are less complicated.

One disadvantage of the between-participants design is that it requires more participants than designs in which individuals are tested in several conditions. A second disadvantage is that the design provides less control of nuisance variables, and therefore the error variance is larger than in other designs. Because participants in different conditions vary in characteristics such as ability and motivation, it is more difficult to assess the effects of experimental treatments or conditions than in designs in which such individual differences are better controlled.

In between-participants designs, participants may be either sampled from existing populations or assigned randomly to one of several experimental conditions or treatment levels. An example of the former is the *Seasons* study[1] in which individuals were sampled from populations differing with respect to various factors, including gender, educational level, and occupation. Strictly speaking, that study is an *observational study*. True experiments involve random assignment of participants to levels of an independent variable; the independent variable is said to be manipulated and the design is often referred to as *completely randomized*. Whether the levels of the independent variable are observed or manipulated, the data analysis has much the same form and the underlying assumptions are the same. Despite the similarities in statistical analysis, observational studies do not allow the researcher to infer a causal link between the factors (e.g., occupation) and the dependent measure (e.g., total cholesterol level). Only experiments involving random assignment to conditions and at least one manipulated variable permit causal conclusions.

DOI: 10.4324/9781003453550-10

We view each group of scores as a random sample from a *treatment population*. The first question of interest is whether the means of these treatment populations vary. To address this question, we introduce the *analysis of variance*, or *ANOVA*, in which the total variability in the data set is partitioned into two components, one reflecting the variance of the treatment population means, and a second that reflects only the effects of nuisance variables.

In addition to testing whether the treatment population means are equal, we want some way of evaluating the practical or theoretical importance of the independent variable. Therefore, following the development of the ANOVA, we focus on several measures of importance. We also consider the role of statistical power in the research design and relate power to measures of importance.

Throughout this chapter, we make certain assumptions to justify the calculations presented. When those assumptions are violated, Type 1 and Type 2 error rates may increase. Therefore, the chapter also discusses such consequences and alternative procedures that may improve the situation.

In summary, the main concerns of this chapter are as follows:

- *Testing the null hypothesis that the treatment population means are equal.* This involves the ANOVA for the one-factor between-participants design.
- *Measures of the importance of the independent variable.* These are derived from the ANOVA table.
- *The power of the test of the null hypothesis* and the relationship between power and the decision about sample size.
- *The assumptions underlying the ANOVA, measures of importance, and power of the significance test*, including the consequences of violations of assumptions and alternative methods that can be used in the face of violations.

8.2 An Example of the Design

An example of an experiment, together with a data set, will make subsequent developments more concrete. Table 8.1 presents data from a hypothetical memory study in which 40 participants were randomly divided into four groups of 10 each. Each participant studied a list of 20 words and was tested for recall a day later. Ten participants were taught and instructed to use a memory strategy called the method of loci, in which each object on the list was associated with a location on campus; 10 participants were told to form an image of each object on the list; 10 others were told to form a rhyme with each word; and 10 others – the control group – were just told to study the words.[2]

Figure 8.1 presents the group means and 95% confidence intervals for those means. The three groups that were instructed to use a memory strategy had higher average recall scores than the control group, although the widths of the confidence intervals indicate that the data were quite variable. There is also some indication that the method of loci may be superior to the other two experimental methods. However, differences among the four means may just reflect differences in the effects of nuisance variables. By chance, the average ability or motivational level in one group of students may be higher than in the others; or other differences between individuals or between the conditions in which they were tested (e.g., the time of day, the temperature in the room) may account for the apparent differences among experimental conditions. A major goal of the data analysis is to separate out the effects of the instructional method from the effects of nuisance variables.

Table 8.1 Recall scores from a hypothetical memory study

	Control	*Loci*	*Image*	*Rhyme*	
	11	10	13	16	
	4	18	16	9	
	8	6	3	7	
	3	20	6	10	
	11	15	13	9	
	8	9	10	14	
	2	8	13	16	
	5	11	9	3	
	8	12	5	9	
	5	12	19	12	
$\bar{Y}_{.j}$ =	6.5	12.1	10.7	10.5	$\bar{Y}_{..}$ = 9.95
s_j^2 =	10.056	19.433	25.567	16.722	

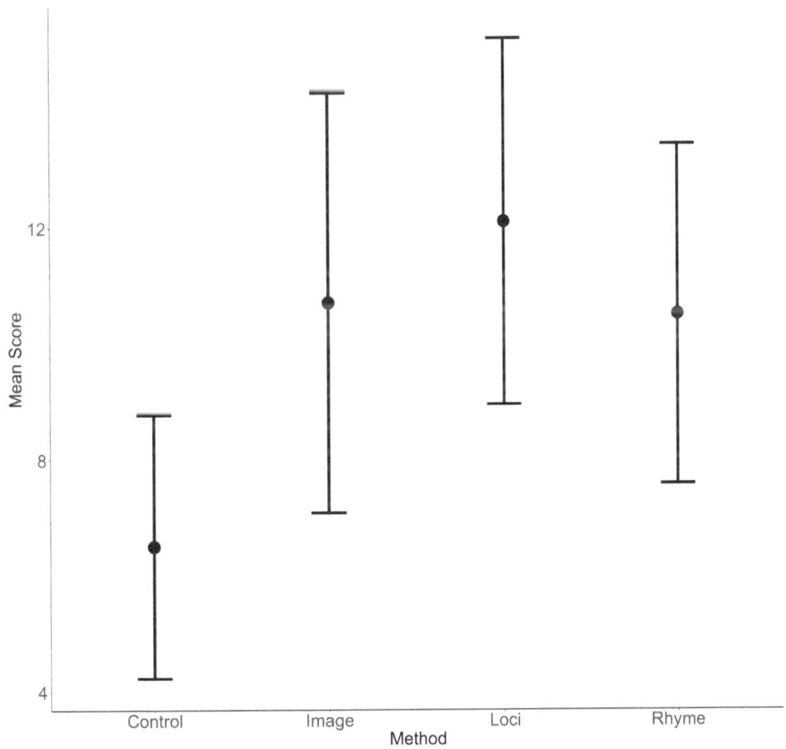

Figure 8.1 Means and confidence interval bars for the data of Table 8.1.

At this point, it would be wise to explore the data further, calculating additional statistics and plotting other graphs as described in Chapter 2. However, we will leave that as an exercise for the reader and proceed to address the question of whether the differences in Figure 8.1 reflect true differences in the effects of the four study methods, or merely error variance. We begin by developing a framework for the analysis of variance.

8.3 The Structural Model

We view the various groups of scores in a study as random samples from populations selected for investigation. Then the question of whether the four study methods of Table 8.1 differ in their effectiveness can be rephrased as: Do the means of the four treatment populations differ? To answer this question, we need a way of linking the observed data to the hypothetical populations, of relating sample statistics to population parameters. We begin by constructing a *structural model*, a model that relates the scores to population parameters.

We start by assuming that the participants in the experiment are identical in ability, motivation, and any other characteristics that would affect their scores. We further assume that they are identically treated – e.g., tested at the same moment in time, and under the exact same conditions. Under these very unrealistic assumptions, everyone in an instructional population, and therefore everyone in an instructional group, would have the same score; that score would be the treatment population mean. We can represent this state of affairs with the following notation:

$$\begin{aligned} Y_{11} &= Y_{21} = Y_{31} = \ldots = Y_{i1} = \ldots = Y_{n1} = \mu_1 \\ Y_{12} &= Y_{22} = Y_{32} = \ldots = Y_{i2} = \ldots = Y_{n2} = \mu_2 \\ Y_{13} &= Y_{23} = Y_{33} = \ldots = Y_{i3} = \ldots = Y_{n3} = \mu_3 \end{aligned}$$

where there are n participants in a group, Y_{ij} represents the ith score in the jth group, and μ_j is the mean of the jth population of scores. For example, Y_{52} would refer to the score of the fifth participant in the second group.

Of course, this is not realistic; the scores of individuals in an instructional group will vary, and therefore will differ from the instructional population mean, because of nuisance variables such as motivation level, prior relevant experience, interest in the topic, and conditions at the time of testing. We can represent this complication by saying that the score of the ith participant in the jth group will differ from the treatment population mean, μ_j, by some amount – an *error component*, ε_{ij}. This means that an individual's score equals the mean of the treatment population plus an error component. That is,

$$\begin{aligned} Y_{ij} &= \mu_j + \left(Y_{ij} - \mu_j\right) \\ &= \mu_j + \varepsilon_{ij} \end{aligned} \tag{8.1}$$

Note that ε_{ij} can be positive or negative; that is, nuisance variables can raise the score above the population mean or lower it below that mean.

We can rewrite Equation 8.1 in a way that more directly expresses the relation between a score and the effect of the condition under which that score was obtained. First, we define one more population parameter, μ, the mean of all the treatment populations; i.e., $\mu = \Sigma\mu_j/a$, where a is the number of levels of the independent variable. Equation 8.1 is unchanged if we add and subtract μ from the right side:

$$Y_{ij} = \mu + \left(\mu_j - \mu\right) + \varepsilon_{ij} \tag{8.2}$$

Let $\alpha_j = (\mu_j - \mu)$; because α_j (Greek alpha) is the difference between the mean of the *j*th treatment population and the grand mean of all the populations, it represents the effect of the *j*th treatment on the scores in the *j*th population. Therefore, we can rewrite Equation 8.2:

$$Y_{ij} = \mu + \alpha_j + \varepsilon_{ij} \tag{8.3}$$

Equation 8.3 is a *structural equation*: It defines the structure of a score obtained in a one-factor between-participants experiment. In words, the structure of a score is

score = *grand mean* + *treatment effect* + *error component*

The parameters in Equation 8.3 are rather abstract and not very useful unless we tie them to statistics that we can calculate from our data. To do this, we need to estimate the population means, the treatment effects, and the error terms. We have the following parameter estimates:[3]

Parameter	μ	μ_j	α_j	ε_{ij}
Estimate	$\bar{Y}_{..}$	$\bar{Y}_{.j}$	$\bar{Y}_{.j} - \bar{Y}_{..}$	$Y_{ij} - \bar{Y}_{.j}$

where Y_{ij} is the score of the *i*th person in the *j*th group, $\bar{Y}_{.j}$ is the mean of all the scores in the *j*th group, and $\bar{Y}_{..}$ is the mean of all the scores in the data set. For example, in Table 8.1, Y_{23} is 16, the score of the second person in the image condition; $\bar{Y}_{.4}$ is 10.5, the mean of the rhyme condition; and $\bar{Y}_{..}$ is 9.95, the grand mean.

With the structural equation as a basis, we now can begin to calculate the necessary terms to draw inferences from our data.

8.4 The Analysis of Variance (ANOVA)

The ANOVA involves partitioning the variability of all the scores into two components, called *sums of squares*. These in turn are divided by their degrees of freedom to form *mean squares*, which are estimates of population variances. The ratio of mean squares, the *F ratio*, provides a test of the hypothesis that all the treatments have the same effect. In what follows, we consider each of these aspects of the ANOVA.

8.4.1 *Sums of Squares*

As Equation 8.3 implies, scores can vary because of the effects of the independent variable (these contribute systematically to the α_j) and because of the effects of uncontrolled nuisance variables (these contribute randomly to the ε_{ij}). If we can separate those two sources of variance in our data, we will have the basis for deciding how much, if any, of the variance is due to the independent variable.

The structural equation suggests an approach to partitioning variability. If we rewrite Equation 8.3 by subtracting μ from both sides, we can express the deviation of a score from the grand mean of the population as consisting of a treatment effect and error; that is,

$$Y_{ij} - \mu = \alpha_j + \varepsilon_{ij}$$

Replacing the parameters in the preceding equation by the estimates we presented earlier, we have

$$Y_{ij} - \bar{Y}_{..} = \left(\bar{Y}_{.j} - \bar{Y}_{..}\right) + \left(Y_{ij} - \bar{Y}_{.j}\right) \quad (8.4)$$

The next step in partitioning the total variability is to calculate the terms in Equation 8.4, and then square and sum them to get the sums of squares. For the data set of Table 8.1, the left side of Equation 8.4 leads to the *total sum of squares*:

$$SS_{total} = \sum_{j=1}^{4}\sum_{i=1}^{10}\left(Y_{ij} - \bar{Y}_{..}\right)^2 = (11 - 9.95)^2 + \ldots + (12 - 9.95)^2 = 819.9$$

The first term to the right of the equal sign in Equation 8.4 is also squared and summed for each participant, yielding the *treatment sum of squares*:

$$SS_{method} = 10\sum_{j=1}^{4}\left(\bar{Y}_{.j} - \bar{Y}_{..}\right)^2 = 10\left[(6.5 - 9.995)^2 + \ldots + (10.5 - 9.95)^2\right] = 173.9$$

and finally we obtain the *residual sum of squares* which can be calculated either directly as

$$SS_{residual} = \sum_{j=1}^{4}\sum_{i=1}^{10}\left(Y_{ij} - \bar{Y}_{.j}\right)^2 = (11 - 6.5)^2 + \ldots + (12 - 10.5)^2 = 646.0$$

or as the difference between the total and treatment sums of squares:

$$SS_{residual} = SS_{total} - SS_{method} = 819.9 - 173.9 = 646.0$$

The preceding results are based on the example of Table 8.1. In general, we designate the independent variable by the letter *A*, and we assume the experiment includes *a* levels of *A* with *n* scores at each level. Then,

$$\begin{array}{ccccc} \sum_{j=1}^{a}\sum_{i=1}^{n}\left(Y_{ij} - \bar{Y}_{..}\right)^2 & = n\sum_{j=1}^{a}\left(\bar{Y}_{.j} - \bar{Y}_{..}\right)^2 & + \sum_{j=1}^{a}\sum_{i=1}^{n}\left(Y_{ij} - \bar{Y}_{.j}\right)^2 \\ SS_{total} & = SS_A & + SS_{S/A} \end{array} \quad (8.5)$$

where *S/A* represents "participants within levels of *A*" to remind us that the residual term reflects the variability of the scores within each level of *A*.[4] A general proof that $SS_{total} = SS_A + SS_{S/A}$ is presented in Appendix 8.1.

8.4.2 Degrees of Freedom (df)

The three terms in Equation 8.5, the sums of squares, are numerators of variances (see Equation 2.4) and, as such, must be divided by their corresponding degrees of freedom (*df*) to be converted into variances, called *mean squares*. The *df* associated with a particular *SS* term is the number of independent observations contributing to that estimate of variability. For our three *SS* terms, we have the following *df*:

1. *The total degrees of freedom,* df_{total}. The *total sum of squares,* SS_{total}, is the numerator of the variance of all *an* scores. Therefore, $df_{total} = an - 1$.
2. *The between-groups degrees of freedom,* df_A. The *between-groups sum of squares,* SS_A, is *n* times the numerator of the variance of the *a* group means about the grand mean and is therefore distributed on $a - 1$ *df*.
3. *The within-groups degrees of freedom,* $df_{S/A}$. The *within-groups sum of squares,* $SS_{S/A}$, is the sum, or "pool" of the numerators of each of the group variances. Because each of the *a* group variances is distributed on $n - 1$ *df*, $SS_{S/A}$ is distributed on $a(n - 1)$ *df*. Note that the degrees of freedom partition neatly, in the same way as the sums of squares:

$$\begin{aligned} an \quad 1 - a(n-1) + (a-1) \\ df_{tot} = df_{S/A} + df_A \end{aligned} \tag{8.6}$$

Equation 8.6 demonstrates that the degrees of freedom are partitioned into two parts that correspond to the sums of squares. This partitioning of the degrees of freedom provides a partial check on the partitioning of the total variability. Although the partitioning in a one-factor design is simple, keeping track of the number of distinguishable sources of variance becomes more difficult in more complex designs. There are also designs in which it is a challenge to analyze the relations among the factors in the design. Therefore, when designs have many factors, it is wise to find the degrees of freedom associated with each source of variability and to check whether the *df* sum to the total number of scores minus one.

8.4.3 Mean Squares, Expected Mean Squares, and the F Ratio

The ratio of a sum of squares to its degrees of freedom is called a *mean square*. In the one-factor design, the relevant mean squares are the *A* mean square, where

$$MS_A = SS_A \,/\, df_A$$

and the *S/A* mean square,

$$MS_{S/A} = SS_{S/A} \,/\, df_{S/A}$$

Under the assumptions summarized in Box 8.1, the ratio $MS_A/MS_{S/A}$ has a sampling distribution called the *F* distribution *if the null hypothesis is true*. It provides a test of the null hypothesis that the treatment population means are all equal; that is, the *F* statistic tests the *null hypothesis*:

Box 8.1 Parameter Definitions and Assumptions

1. *The parent population mean*, μ. This is the grand mean of the treatment populations selected for this study and is a constant component of all scores in the a populations. It is the average of the treatment population means:

$$\mu = \sum_{j=1}^{a} \mu_j / a$$

2. The effect of treatment A_j, α_j. This equals $\mu_j - \mu$ and is a constant component of all scores obtained under A_j but may vary over treatments (levels of j).

   2.1 Because the deviation of all scores about their mean is zero, $\Sigma_j \alpha_j = 0$.
   2.2 If the null hypothesis is true, all $\alpha_j = 0$.
   2.3 The population variance of the treatment effects is $\sigma_A^2 = \sum_{j=1}^{a} \alpha_j^2 / a$.

3. *The error*, ε_{ij}. This is the deviation of the ith score in group j from μ_j and reflects uncontrolled, or chance, variability. It is the only source of variation within the jth group, and if the null hypothesis is true, the only source of variation within the entire data set. We assume that

   3.1 The ε_{ij} are independently distributed; i.e., the probability of sampling some value of ε_{ij} does not depend on other values of ε_{ij} in the sample.
   3.2 The ε_{ij} are normally distributed in each of the a treatment populations. Also, because $\varepsilon_{ij} = Y_{ij} - \mu_j$, the mean of each population of errors is zero; i.e., $E(\varepsilon_{ij}) = 0$.
   3.3 The distribution of the ε_{ij} has variance σ^2_e (error variance) in each of the a treatment populations; i.e., $\sigma^2_1 = \ldots = \sigma^2_j = \ldots = \sigma^2_a$. This is the assumption of *homogeneity of variance*. The error variance is the average squared error; $\sigma^2_e = E(\varepsilon^2_{ij})$.

$$H_0 : \mu_1 = \mu_2 = \ldots = \mu_j = \ldots = \mu_\alpha = \mu$$

or, equivalently,

$$H_0 : \alpha_1 = \alpha_2 = \ldots = \alpha_j = \ldots \alpha_a = 0$$

To understand the logic of the F test, we need to consider the relationship of the mean squares to the population variances. This requires us to determine the expected values of our two mean square calculations.

Suppose we draw a samples of n scores from their respective treatment populations, and calculate MS_A and $MS_{S/A}$. These MS values are sample statistics that would vary across replications of this experiment; this implies that they each have a sampling distribution. The means of these two sampling distributions, one for MS_A and one for $MS_{S/A}$, are the expected values of the mean squares, or the *expected mean squares* (EMS). Given the structural model of Equation 8.3, and assuming that the ε_{ij} are independently distributed with variance, σ^2_e,

Table 8.2 Sources of variance (*SV*) and expected mean squares (*EMS*) for the one-factor between-participants design

SV	*EMS*
A	$\sigma^2_e + n\theta^2_A$
S/A	σ^2_e

Note: $\theta^2_A = \sum_j (\mu_j - \mu)^2 / (a - 1)$. We use the θ^2 (theta squared) notation rather than σ^2 to remind us that the treatment component of the *EMS* involves division by degrees of freedom; the variance of the treatment population means would be $\sigma^2_A = \sum_j (\mu_j - \mu)^2 / a$.

the *EMS* of Table 8.2 can be derived (Kirk, 1995; Myers & Well, 1995). Let's consider each expected mean square in turn to understand the information provided by MS_A and $MS_{S/A}$.

$E(MS_A)$ states that the *between-groups mean square*, MS_A, estimates *error variance*, σ^2_e, plus n times the variance in the treatment population means, θ^2_A (if there is any effect of the treatment). This result should make intuitive sense when you examine the formula for MS_A:

$$MS_A = \frac{n\sum_j^a (\bar{Y}_{.j} - \bar{Y}_{..})^2}{a - 1} \tag{8.7}$$

Equation 8.7 states that MS_A is the variance of the condition means times the sample size, n. Even if there were no differences among the treatment population means, the sample means would differ just by chance because there are different individuals with different characteristics in each group. The error variance, σ^2_e, reflects this. If there is also an effect of the treatment, the μ_j will differ and their variability will also be reflected in the value of MS_A.

$E(MS_{S/A})$ states that the *within-groups mean square*, $MS_{S/A}$, is an estimate of error variance. Again, this result may be understood intuitively by examining how $MS_{S/A}$ is calculated:

$$\begin{aligned} MS_{S/A} &= \frac{SS_{S/A}}{a(n-1)} \\ &= \frac{\sum_j \sum_i (Y_{ij} - \bar{Y}_{.j})^2}{a(n-1)} \end{aligned} \tag{8.8}$$

Equation 8.8 may be rewritten as

$$MS_{S/A} = \left(\frac{1}{a}\right)\sum_j \left[\frac{\sum_i (Y_{ij} - \bar{Y}_{.j})^2}{n - 1}\right]$$

The expression in the square brackets on the right side is the variance of the *j*th group of scores, and the entire right side is an average of the *a* group variances. Because participants within a condition are treated identically, they should differ only due to error (see

Equation 8.1). If we assume that error variance is equal in each treatment population, $MS_{S/A}$ is an average of a estimates of the population variance, σ^2_e.

Given our understanding of what MS_A and $MS_{S/A}$ estimate, we are in a position to understand the logic of the F test, where $F = MS_A/MS_{S/A}$. First, *assume that the null hypothesis is true* and, also, that there is *homogeneity of variance*; that is

$$\mu_1 = \mu_2 = \ldots = \mu_j = \ldots = \mu_a \text{ and } \sigma_1 = \sigma_2 = \ldots = \sigma_j = \ldots = \sigma_a$$

Under these assumptions, MS_A is an estimate of the error variance common to the a treatment populations. In terms of $E(MS_A)$, $\theta^2_A = 0$ so $E(MS_A)$ is an estimate of error variance, σ^2_e. Thus, *if the null hypothesis is true*, MS_A and $MS_{S/A}$ both estimate the same population error variance and their ratio should be about 1. Of course, it would be surprising if two independent estimates of the same population variance were identical; that is, the ratio of MS_A to $MS_{S/A}$ has a distribution of values. More precisely, *if H_0 is true*, the ratio, $MS_A/MS_{S/A}$, is distributed as F on $a - 1$ and $a(n - 1)$ *df*.

But what if the null hypothesis is, in fact, false? For example, suppose that the method of study does affect recall in the example of Table 8.1. Then the means of the groups of scores in Table 8.1 will differ not only because the scores in the different groups differ by chance, but also because the groups were studied by different methods. In other words, if H_0 is false, $\theta^2_A > 0$ so $E(MS_A) = \sigma^2_e + n\theta^2_A$. The within-group variance does not change: $MS_{S/A}$ should not be affected by the independent variable because all participants in a group receive the same treatment. Therefore, when H_0 is false, the ratio $MS_A/MS_{S/A}$ is expected to be greater than 1.

In summary, under the assumptions of the null hypothesis, homogeneity of variance and independently distributed scores MS_A and $MS_{S/A}$ are two independent estimates of the population error variance, σ^2_e. If we also assume that the population of scores is normally distributed, the ratio of two independent estimates of the same population variance has an F distribution. Therefore, under the assumptions summarized in Box 8.1, the ratio $MS_A/MS_{S/A}$ has an F distribution. Because the numerator is based on an estimate of the variance of a population means, it has $a - 1$ *df*. The denominator has $a(n - 1)$ *df* because the variance estimate for each group is based on $n - 1$ *df* and $MS_{S/A}$ is an average of a variance estimates.

Appendix Table C.5 presents critical values of the F distribution. As an example of its use, suppose we have three groups of 11 participants each. Then the numerator $df = df_1 = a - 1 = 2$, and the denominator $df = df_2 = a(n - 1) = 30$. Turning to the column headed by 2 and the block labeled 30, if $\alpha = .05$, we would reject the null hypothesis of no difference among the three treatments if the F we calculate is greater than 3.32. Interpolation may be needed for degrees of freedom not listed in the table. Alternatively, we can easily use R to find the critical value for any combination of degrees of freedom. Recall from Box 6.1 in Chapter 6 that R includes several distributional calculators. The F distributional abbreviation is *f*, and so, for example, the command *qf*(.95, df1 = 2, df2 = 30) will return the critical value of $F(2, 30)$ with 95% of the area below it and 5% above, which is 3.3158. Of course, most analyses performed with software will report the *p*-value associated with the empirical value of F.

8.4.4 The ANOVA Table

Panel *a* of Table 8.3 summarizes the developments so far, presenting the formulas for sums of squares, degrees of freedom, mean squares, and the F ratio for the one-factor between-participants design. For any data set, most statistical software packages present

Table 8.3 The analysis of variance for the one-factor between-participants design (*a*) General form of the ANOVA

Source	*df*	*SS*	*MS*	*F*
Total	*an* − 1	$\sum_{j=1}^{a}\sum_{i=1}^{n}\left(Y_{ij}-\bar{Y}_{..}\right)^2$		
A	*a* − 1	$n\sum_{j=1}^{a}\left(\bar{Y}_{.j}-\bar{Y}_{..}\right)^2$	SS_A/df_A	$MS_A/MS_{S/A}$
S/A	*a*(*n* − 1)	$\sum_{j=1}^{a}\sum_{i=1}^{n}\left(Y_{ij}-\bar{Y}_{.j}\right)^2$	$SS_{S/A}/df_{S/A}$	

(*b*) ANOVA of the data of *Table 8.1*

Source	*Sum of squares*	*df*	*Mean square*	*F*	*p-value*
Method	173.90	3	57.967	3.230	.034
Error	646.00	36	17.944		
Total	819.90	39			

this table with numerical results in some form. Panel *b* presents the output for the data of Table 8.1. The results are significant at the .05 level, indicating that there are differences among the means of the populations defined by the four different study methods. Figure 8.1 hints that this is due to the poorer performance of the control condition, although reaching that conclusion would require additional testing. There are a range of interesting questions that the omnibus *F* test leaves unanswered. For example, are all three experimental methods significantly superior to the control condition? Do the means of the three experimental methods differ from each other? We will consider such comparisons of means within subsets of conditions, called *contrasts*, in Chapter 10.

We might also ask whether the differences among the four population means are large enough to be of practical significance. As we noted when discussing effect size measures in Chapter 6, statistical significance is not the same as practical or theoretical significance.

8.4.5 *Using Software for the One-Factor Between-Participants ANOVA*

We can use R or SPSS to obtain the ANOVA table for the data of Table 8.1. In either package, begin by importing the data file, Table 8_1 Memory Data.xlsx, which has two columns: one showing the *Method* for each participant and the other showing the resulting *Score*. In R, the imported data can be assigned to a data frame, which we call MemDat for this example. Then, we use the *aov* function in the {stats} package to run the calculations and the *summary* function in the {base} package to organize the output into an ANOVA table: *summary*(*aov*(data = MemDat, Score ~ Method)). The "Score ~ Method" tells *aov* to run the ANOVA on the data in Score as a function of Method.

In SPSS, select the *Compare Means and Proportions* submenu of the *Analyze* pull-down list, and then choose *One-Way ANOVA*. Move the Score variable to the box labeled "Dependent List" and Method to the box labeled "Factor," then click OK. If Method fails to appear in the list of available variables, a few more steps are required. SPSS only allows numeric variables to serve as the factor in an ANOVA. Because the Method variable identifies the treatment conditions by their names (i.e., strings of characters), we must first create a new variable that replaces those names with numeric values. Using the *Recode into Different Variables* option in the *Transform* menu, move Method to the *Input-Variable* position in the middle box, and type a new variable name in the Output Variable box on the right. Let's call it *FactorA*. Then, click "change" and "Old and New Values." In the window that pops up, enter each *Method* level name (e.g., Control) and a numeric *New Value* (e.g., 1), then press "Add." Once all levels of *Method* have been assigned new values, click "Continue" to see *FactorA* in the Data View. Now the ANOVA can be run as described, using *FactorA* as the Factor variable.

8.5 Measures of Importance

The p-value in Table 8.3 informs us that the effects of the method of memorization are statistically significant, if $\alpha = .05$. In addition, however, we need some indication of the practical or theoretical importance of our result. Generally, we seek a measure that assesses the magnitude of the effect of our treatment, A, relative to error variance. We will find that the *EMS* analyses that guided the logic of the F test are also very useful in thinking about appropriate ways in which to assess the importance of an effect. We will consider several possible measures in this section. Several sources also present discussions of these and other measures (e.g., Kirk, 1996; Maxwell, Camp, & Arvey, 1981; Olejnik & Algina, 2000, 2003).

8.5.1 Measuring Strength of Association in the Sample: η^2 (Eta-Squared)

Greek-letter designations are usually reserved for population parameters, but η^2 is a sample statistic that is often used as a measure of association between the dependent and independent variables (Cohen, Cohen, West, & Aiken, 2003). It describes the proportion of variability in the sample that is primarily attributable to the treatment factor as

$$\eta^2 = \frac{SS_A}{SS_{total}} \tag{8.9}$$

Referring to Table 8.3, we have

$$\eta^2_{method} = 173.9/819.9 = .212$$

Eta-squared has the advantage of being easily calculated and easily understood as a proportion of sample variability. However, the value of η^2 is influenced not only by the relative magnitudes of the treatment effect and error variance, but also by design choices and sample size details that we don't associate with a population effect: n, df_A, and $df_{S/A}$. That is, η^2 is influenced by the number of levels of an experimental factor (df_A) and even by the spacing of the levels (Okada & Hoshino, 2017). In addition, σ^2_e contributes to the numerator of η^2, meaning that η^2 is a positively biased estimator. For these reasons, other statistics that measure importance are often preferred (e.g., Kroes & Finley, 2023). We turn to such estimates now, bearing in mind that our results rest on the assumptions underlying the derivation of the *EMS*; i.e., independence of the ε_{ij} and homogeneity of variance.

8.5.2 *Estimating Strength of Association in the Population: ω^2 (Omega-Squared)*

Whereas η^2 describes the sample- and design-specific strength of association between the dependent and independent variables by forming a ratio of sample sums of squares, ω^2 is a measure of the strength of association in the population; unlike η^2, it is a ratio of population variances:

$$\omega^2 = \frac{\sigma_A^2}{\sigma_e^2 + \sigma_A^2} \tag{8.10}$$

The numerator of the ratio is the variance of the treatment population means (the μ_j) or, equivalently, the variance of the treatment effects (the α_j):

$$\begin{aligned} \sigma_A^2 &= \frac{\sum_j^a \left(\mu_j - \mu\right)^2}{a} \\ &= \frac{\sum_j^a \alpha_j^2}{a} \end{aligned} \tag{8.11}$$

The denominator of ω^2 is the total population variance; that is, the treatment population error variance, σ^2_e, plus the variance of the treatment population means, σ^2_A. Thus, ω^2 assesses the magnitude of the treatment effect relative to the total variance in the design. We cannot know the ratio described by Equation 8.10 but we can derive estimates of σ^2_A and σ^2_e and therefore of ω^2. We begin with the *EMS* equations of Table 8.2:

$$E\left(MS_A\right) = \sigma_e^2 + n\theta_A^2 \tag{8.12}$$

and

$$E\left(MS_{S/A}\right) = \sigma_e^2 \tag{8.13}$$

To obtain an estimate of σ^2_A we first subtract Equation 8.13 from Equation 8.12, and divide by n; then we have

$$\frac{MS_A - MS_{S/A}}{n} = \hat{\theta}_A^2$$

where the "hat" above θ^2_A means "is an estimate of." Because the numerator of ω^2 as defined by Equation 8.10 involves σ^2_A, not θ^2_A, and noting that $\sigma^2_A = [(a - 1)/a] \times \theta^2_A$, our estimate of σ^2_A is

$$\hat{\sigma}_A^2 = \left(\frac{a-1}{a}\right)\left(\frac{MS_A - MS_{S/A}}{n}\right) \tag{8.14}$$

We now have estimates of the numerator and denominator of ω^2, therefore, substituting into Equation 8.10, we have an estimate of ω^2 for the one-factor between-participants design:

$$\hat{\omega}^2 = \frac{[(a-1)/a](1/n)(MS_A - MS_{S/A})}{[(a-1)/a](1/n)(MS_A - MS_{S/A}) + MS_{S/A}} \tag{8.15}$$

Okada (2013) has shown that $\hat{\omega}^2$ very slightly underestimates ω^2, with good efficiency and consistency.

We may write Equation 8.15 in a different form, one which allows us to calculate $\hat{\omega}^2$ from knowledge of the F ratio, a, and n. The advantages are that the expression is somewhat simpler and, perhaps more importantly, because most research reports contain this information, we can estimate the strength of association for data collected by other investigators. We begin by defining $F_A = MS_A/MS_{S/A}$. Then, multiplying the numerator and denominator of Equation 8.15 by an, and dividing by $MS_{S/A}$, we have

$$\hat{\omega}^2 = \frac{(a-1)(F_A - 1)}{(a-1)(F_A - 1) + na} \tag{8.16}$$

Let's review what Equations 8.15 and 8.16 represent. If we replicate the experiment many times, the average value of the right-hand term will approximately equal ω^2, the proportion of the total variance in the a treatment populations that is attributable to the variance of their means. We say "approximately equal" because the expected value of a ratio is not the same as the ratio of expected values. The approximation is reasonably accurate and the expression is much simpler than that for the exact expression.

One other aspect of Equation 8.16 should be noted. Because the numerator and denominator of the F reflect two independent estimates of the population error variance, when the null hypothesis is true or the effects of A are very small, the F may be less than 1. Then, $\hat{\omega}^2$ would be less than 0. Because a variance cannot be negative, we conclude that $\omega^2 = 0$; that is, none of the total population variance is attributable to the independent variable.

We can apply Equation 8.16 to the memory data in Table 8.1. In that experiment, $a = 4$, $n = 10$, and (from Table 8.3) $F = 3.230$. Then, inserting these values into Equation 8.16,

$$\hat{\omega}^2 = \frac{(3)(2.23)}{(3)(2.23) + 40} = .143$$

This is very close to the value of.146 noted earlier for adjusted R^2. That the values of R^2_{adj} and ω^2 are so close is not unusual; Maxwell et al. (1981) reported that the two rarely differ by more than .02. With respect to assessing the importance of either measure, Cohen (1988) suggested that values of .01, .06, and .14 may be viewed as small, medium, and large, respectively. According to those guidelines, the proportion of variability accounted for by the study method may be judged to be large. Again, however, we caution that the importance attached to any value must be assessed in the context of the research problem and the investigator's knowledge of the research literature.

8.5.3 Cohen's f

In Chapter 6, we presented Cohen's d, a measure of the standardized effect for designs in which two means are compared. Cohen's f (1988) is a similar measure for situations in which the variance of more than two means is of interest. The parameter f is defined as

$$f = \sigma_A / \sigma_e \tag{8.17}$$

We can estimate f by substituting the estimate in Equation 8.14 in the numerator and $MS_{S/A}$ in the denominator. Then we have

$$\hat{f} = \sqrt{\frac{(a-1)(MS_A - MS_{S/A})}{anMS_{S/A}}} \tag{8.18}$$

which can also be written as

$$\hat{f} = \sqrt{(a-1)(F_A - 1)/an} \tag{8.19}$$

For the data of Table 8.1, substituting the F value from Table 8.3 into Equation 8.19, we have

$$\hat{f} = \sqrt{(3)(2.23)/40} = .409$$

Cohen has suggested that values of f of .1, .25, and .4 be viewed as small, medium, and large, respectively. Therefore, as with ω^2, the guidelines for f suggest that the standardized variance of the four reading method estimates is large. That ω^2 and f lead to the same conclusion about the size of the variance of effects follows from the relationship between them; given an estimate of f, we can also calculate an estimate of ω^2, and vice versa. The relations are

$$f^2 = \frac{\omega^2}{1-\omega^2} \quad \text{and} \quad \omega^2 = \frac{f^2}{1+f^2}$$

A useful property of f is that it is closely related to the noncentrality parameter of the F distribution; specifically,

$$\lambda = Nf^2 \tag{8.20}$$

The parameter, λ (lambda), determines areas under the noncentral F distribution, and therefore the power of the F test. The relation between f, λ, and power will be developed further in Section 8.7.

8.5.4 Measures of Importance: Limitations

In an introductory chapter to an edited collection aptly titled, "What if there were no significance tests?" Harlow (1997, pp. 5–6) reported that 11 of the book's other 13 chapters "were very much in favor" of reporting measures such as R^2, ω^2, and f, and the remaining two contributors "at least mildly endorsed such use." Similar support for measures such as these can be found in the American Psychological Association's guidelines for statistical usage (Wilkinson & Task Force, 1999), which urge researchers to report effect size statistics. Nevertheless, there are potential pitfalls. Values of these statistics may depend on the experimental design, the choice and number of levels of the independent variable, the dependent variable, and the population sampled. Estimates of ω^2 and f imply homogeneous variances and independence of observations (cf. Grissom & Kim, 2001, who discuss the variance assumption and suggest alternative approaches for the two-group case). Another concern is that squared coefficients tend to be small, and it is sometimes easy to dismiss an effect as trivial because of a small value of ω^2.

These arguments suggest that we must be careful when interpreting these measures, generalizing the results of any one study, or making comparisons across studies that differ with respect to the factors just cited. In addition, we should treat guidelines such as those set forth by Cohen (1988) as mere labels, not as definitive boundaries between important and unimportant effects. Even a very small advantage of one therapy over another may be important. In theoretical work, a small effect predicted by a theory may be important support for that theory. Funder and Ozer (2019) provide many psychological examples that can help ground our understanding of effect sizes. In summary, if care is taken in interpreting measures of strength, statistics such as $\hat{f}$ and $\hat{\omega}^2$ are useful additions to the test statistics usually computed.

8.5.5 *Using Software for Measures of Effect Size*

SPSS reports several measures of effect size, including η^2, as part of the ANOVA described in Section 8.4.5 if the box marked "estimate effect size for overall tests" is checked. Two versions of ω^2 are shown. The fixed-effects version equals 0.143, the same value as we obtained using Equation 8.16. The difference between fixed- and random-effects variables will be described in Chapter 13.

Using *R*, η^2 can be obtained with the *eta_squared* function in the {effectsize} package: *eta_squared*(*aov*(data = MemDat, Score ~ Method)).[5] The same package includes the *omega_squared* function, which calculates ω^2. Like *eta_squared*, the *omega_squared* function takes the output of *aov* as input: *omega_squared*(*aov*(data = MemDat, Score ~ Method)) returns a value of 0.14 for these data. Note that there is also a *cohens_f* function in the {effectsize} package but it returns a sample-specific version of f, $f = \sqrt{\frac{\eta^2}{1-\eta^2}}$ rather than f based on *EMS* as developed in Section 8.5.3.

8.6 When Group Sizes Are Not Equal

In the developments so far, our formulas have assumed that there are the same number of scores in each group. In this section, we present an example with unequal *n*s, and present formulas for sums of squares, expected mean squares, and measures of importance for this case.

The *n*s in conditions in a study may vary for one of several reasons. The populations may be equal in size but data may be lost from some conditions, perhaps because of a malfunction of equipment or a participant's failure to complete the data-collection session. Often, individuals can be replaced, but sometimes that is impossible. In other instances, the treatments may affect the availability of scores; for example, animals in one drug condition may be less likely to survive the experiment than animals in another condition. In still other instances, usually when we collect data from existing populations, conditions may differ naturally in the availability of individuals for participation. For example, in clinical or medical settings, there may be different numbers of individuals in the diagnostic categories of interest.

Unequal n complicates calculations in the one-factor design, which might tempt some researchers to discard scores from some conditions to equate n. This is not a good idea for a few reasons. Discarding data to equalize the group sizes will reduce error degrees of freedom and, consequently, power. Discarding participants also may misrepresent the relative size of the populations sampled. If so, the effects of some conditions may be weighted too heavily or too lightly in the data analysis. Finally, computational ease should not be a consideration when software programs are available to handle the calculations.

8.6.1 *The ANOVA With Unequal* n

The ANOVA for unequal group sizes is a straightforward modification of the equal n case, at least in the one-factor between-participants design. (Complications arise when more than one factor is involved; these will be treated in Chapters 9 and 23.) Table 8.4 presents the ANOVA formulas and expected mean squares for the unequal n case; the squared deviations in the SS_A formula and the n_j are weighted by the group size. Note that if the n_j are equal, these formulas reduce to the formulas in Table 8.3.

Table 8.5 presents statistics based on Beck Depression scores for four groups of male participants from the University of Massachusetts Medical School research on seasonal effects; the statistics are based on scores averaged over the four seasons. For the purposes of this example, we excluded some participants (those having no or only some high school education, and those with vocational training or an associate's degree). The remaining groups are HS (high school diploma only), C (some college), B (bachelor's degree), and GS (graduate school).[6]

The statistics of Table 8.5 and the box plots of Figure 8.2 indicate that the groups differ in their average depression score. Both means and medians are noticeably higher for those participants who had only a high school education; participants with a graduate school

Table 8.4 The analysis of variance for the one-factor between-participants design with unequal group sizes

Source	*df*	*SS*	*MS*	*F*	*EMS*
A	a − 1	$\sum_{j=1}^{n} n_j(\bar{Y}_{.j}-\bar{Y}_{..})^2$	SS_A/df_A	$MS_A/MS_{S/A}$	$\sigma_e^2+\frac{1}{a-1}\sum_j n_j\alpha_j$
S/A	N − a	$\sum_{j=1}^{a}\sum_{i=1}^{n_j}(Y_{ij}-\bar{Y}_{.j})^2$	$SS_{S/A}/df_{S/A}$		σ^2_e
Total	N − 1	$\sum_{j=1}^{a}\sum_{i=1}^{n_j}(Y_{ij}-\bar{Y}_{..})^2$			

Note: n_j is the number of scores in the jth group and $N=\sum_{j=1}^{a} n_{j.}$.

Table 8.5 Summary statistics for Beck Depression scores in four educational levels (the data are in the Table 8_5 *Male_educ.xlsx* file; go to the *Seasons* page on the book's website)

	Level of education			
	High School	*Some College*	*Bachelor's*	*Grad School*
No. of cases	19	33	37	39
Median	6.27	2.88	2.26	3.03
Mean	6.90	3.67	3.33	4.85
Variance	34.5	5.97	9.86	26.2
Skewness	.757	.351	1.96	1.22
Kurtosis	2.83	2.04	7.92	3.50

Note: HighSchool = high school diploma only; Bachelor's = bachelor's degree; GradSchool = graduate school.

Table 8.6 ANOVA of the depression means in Table 8.5

Source	*SS*	*df*	*MS*	*F*	*p*
Education	186.501	3	62.167	3.562	.016
Error	2,164.061	124	17.452		
Total	2,350.562	127			

education have lower scores but they are higher than those in the remaining two categories. Variances are also highest in the *HS* and *GS* groups; the differences among the variances as well as among the *H*-spreads in the figure warn us that heterogeneity of variance may be an issue. We also note that both the skew statistics and the long tails at the high end of depression scores in the figure indicate that the populations are unlikely to be normally distributed. A *Q–Q* plot (see Chapter 2) would confirm this.

Applying the formulas in Table 8.4, or using software, we obtain the ANOVA results in Table 8.6; these reveal that the means of the four groups differ significantly. However, the characteristics of the data revealed by our preliminary exploration (Table 8.5, Figure 8.2) indicate that the assumptions of the analysis of variance are violated. In Section 8.8, we discuss those assumptions, consider alternative analyses that respond to violations of the assumptions, and apply one such analysis to the depression scores.

8.6.2 *Measures of Importance With Unequal* n

As in the equal *n* design, $\eta^2 = SS_A/(SS_A + SS_{S/A})$. For the Beck Depression data, substituting values from Table 8.6, $\eta^2 = 186.501/2{,}350.562$, or .079. Remember that this estimate is positively biased, and that it can be obtained using the *eta_squared* function in the {effectsize} package in R or using SPSS (see Section 8.5.5). The formulas for ω^2 and Cohen's f undergo a very slight modification. We replace the n in Equation 8.14 by the average n, $\bar{n}$, where $\bar{n} = \sum_j n_j / a = N / a$. We can simplify things further by replacing $a\bar{n}$ by N, the total sample size. Then the equations estimating, σ^2_A, ω^2, and Cohen's f apply with no further changes.

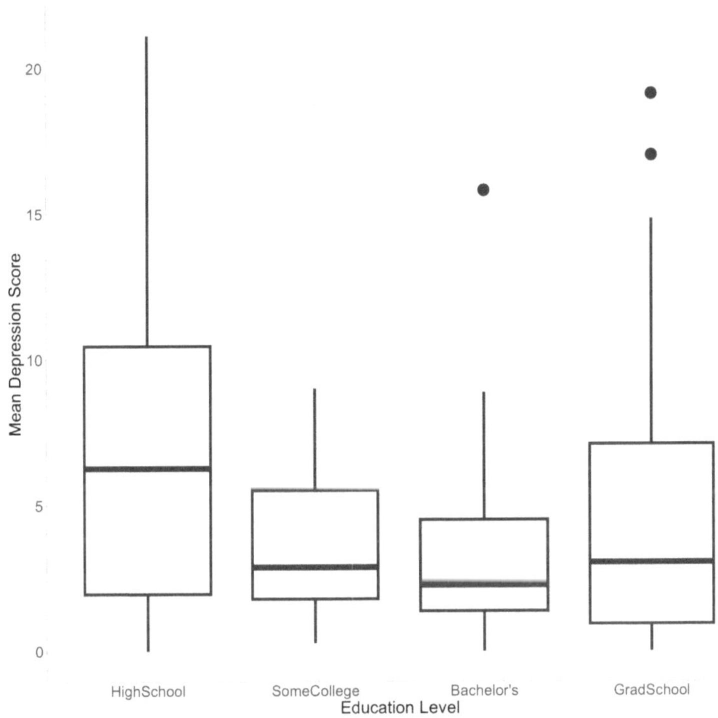

Figure 8.2 Box plot of Beck Depression scores as a function of educational level.

To estimate the population variance of the Beck Depression means as a function of educational level, substitute the mean squares from Table 8.6 into Equation 8.14, and with $\sum_j n_j = 128$,

$$\hat{\sigma}_A^2 = (3)(62.167 - 17.452)/128 = 1.048$$

We now can estimate ω^2 using Equation 8.16 with N replacing an. Then

$$\begin{aligned}\hat{\omega}^2 &= \frac{(a-1)(F_A-1)}{(a-1)(F_A-1)+N} \\ &= \frac{(3)(2.562)}{(3)(2.562)+128} = .057\end{aligned}$$

The same estimate can be obtained from the *omega_squared* function in R's {effectsize} package, or using SPSS as described in Section 8.5.5. We use the same variance estimate to estimate Cohen's f:

$$\begin{aligned}\hat{f} &= \hat{\sigma}_A / \hat{\sigma}_e \\ &= \sqrt{1.048/17.245} = 0.245\end{aligned}$$

Whether we view ω^2 or f, Cohen's guidelines suggest that the effect is of medium size.

In leaving the analysis of the effects of educational level on Beck Depression scores, we again caution that our exploration of the data suggested that both the normality and homogeneity of variance assumptions were violated, making suspect the results of significance tests and estimates of effect size. We will return to the general issue of violations of assumptions, and possible remedies, in Section 8.8.

8.7 Deciding on Sample Size: Power Analysis in the Between-Participants Design

Together with the practical constraints imposed by the available access to participants and time, statistical power should be a primary consideration in deciding on sample size. In order to incorporate this into our decision about sample size, we need to decide on a value of power and we need a value of the minimum effect size that would be of interest. As we saw in our treatment of the power of t tests in Section 6.8, there are several ways that we might proceed.

One possibility is that we use Cohen's guidelines to establish an effect size. For example, suppose that in designing the memory experiment described in Section 8.2, we had decided that we want power equal to at least .90 to reject an effect that was large by Cohen's guidelines; then $f = .4$. How many participants should we have included in the study? An easy way to answer this question is to use G*Power 3.1. Figure 8.3 shows the screen involved in the calculations. We selected *F tests* from the *Test Family* menu, the ANOVA for the one-way design from the *Statistical test* menu, and the *a priori* option from the *Type of power analysis* menu. We set the *Input Parameters* as shown in the left-hand column. In the *Output Parameters* column, we find that the required total N is 96, or 24 in each group. The other output results should be self-explanatory except for the noncentrality parameter λ. Recall that the noncentral t distribution has a noncentrality parameter, δ (see Section 6.8.1). The analog to δ for the F distribution is a noncentrality parameter is called λ; $\lambda = N f^2$, or $96 \times .16$. This parameter is an index of the noncentral F distribution's distance from the central F distribution; when $\lambda = 0$, power equals the Type 1 error rate and as λ increases, so does the power of the test.

An alternative to using Cohen's guidelines is to base our assumed effect size on results from a pilot study, keeping in mind that pilot studies tend to have small samples and often overestimate the population effect size. In this case, we would recommend using Equation 8.18 or 8.19 to obtain an estimate of f. We would then insert that f into G*Power. For the data of Table 8.1, $f = .409$. If we require power = .9, set $\alpha = .05$; then with four groups the required N is 92, slightly less than when we entered $f = .4$.

Finally, in some cases, we might have no single data set on which to base an estimate of f. However, practical or theoretical considerations, or a review of published results from several related studies, might suggest reasonable values of the treatment population means and of the population standard deviation. For example, in planning an experiment involving three groups, we might decide that the most likely values of the population means are 50, 60, and 70, and that the population standard deviation is about 20. Using G*Power 3.1, we enter $\alpha = .05$, power = .9, and the number of groups = 3, and then select the *determine* button. This brings up a panel in which we enter the hypothesized population means and the population standard deviation. Select "Calculate and transfer to main window." G*Power calculates

$$\sigma_A = \sqrt{\left[(50-60)^2 + (60-60)^2 + (70-60)^2\right]/3} = 8.165$$

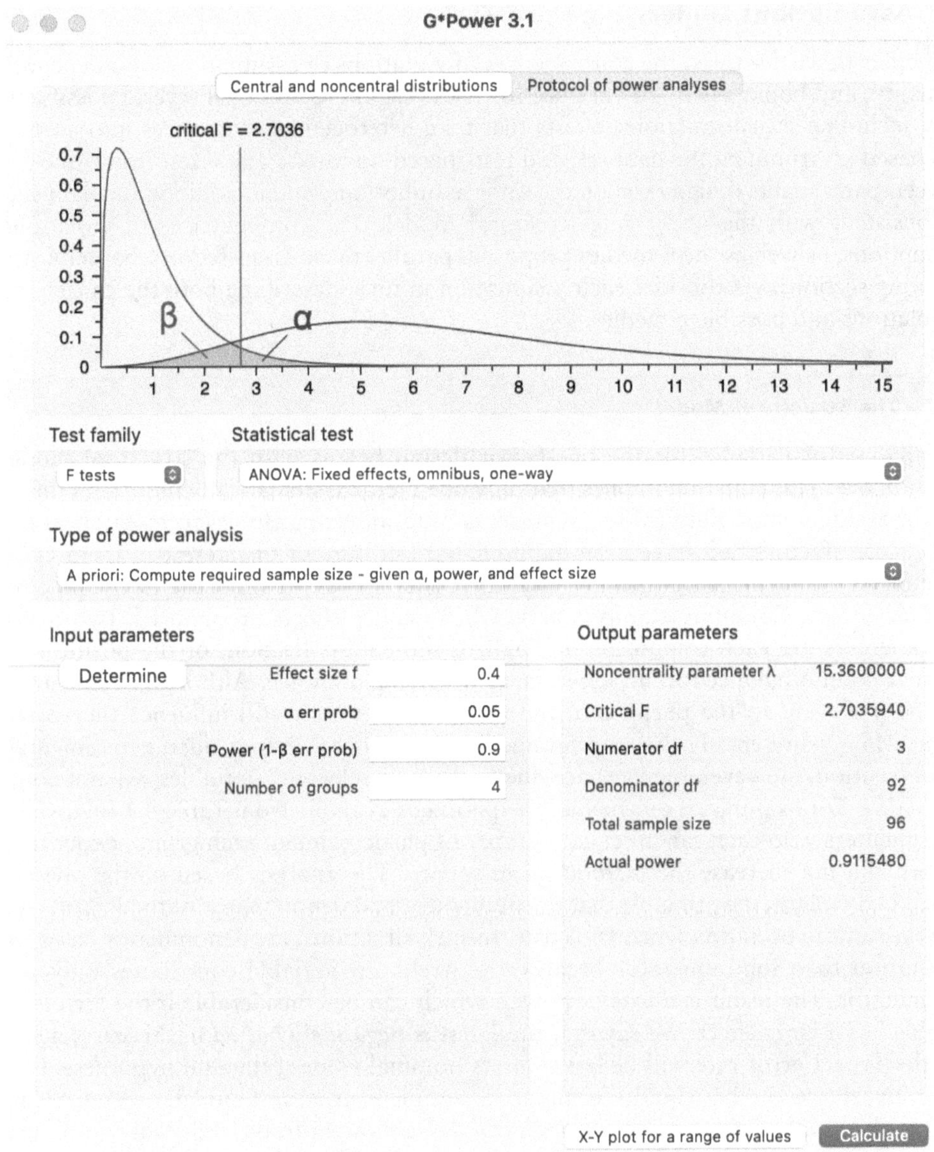

Figure 8.3 G*Power 3.1 calculation of the *N* needed to have power = .90 when the effect size, *f*, equals .4.

and divides by the standard deviation of 20 to yield $f = .408$. Transferring this value to the main window, the required total *N* is 81, or 27 in each of the three groups.

Post hoc, or retrospective power, is also available in G*Power. However, as we discussed in Chapter 6, we have reservations about reporting post hoc power based on the observed set of statistics. Confidence intervals on the raw and standardized effects will prove more informative and be less misleading.

8.8 Assumptions Underlying the *F* Test

In Chapter 6, we discussed the consequences of violations of assumptions of independence, normality, and homogeneity of variance for the *t* test and considered several possible remedies, including transformations, *t* tests that take heterogeneous variances into account, *t* tests based on trimming the data set, and tests based on ranks. The *F* test in the one-factor between-participants design rests on the same assumptions, and in addition the data should be consistent with the underlying structural model. The consequences of violations of assumptions, as well as the remedies proposed, parallel those discussed in Chapter 6. In the following sections, we consider each assumption in turn, describing both the consequences of violations and possible remedies.

8.8.1 *The Structural Model*

The analysis of variance for the one-factor design begins with the structural model of Equation 8.3. This equation implies that only one factor systematically influences the data, and the residual variability ($MS_{S/A}$) represents random error. However, researchers sometimes ignore factors that have been manipulated but are not of interest in themselves. If those factors contribute significant variability, the one-factor model is not valid for the research design. Common examples arise when gender effects are ignored, two different experimenters are each responsible for half of the data collection, or the position of an object is counterbalanced in an experiment involving a choice. Although these variables may be irrelevant to the purpose of the research, they may still influence the scores. If so, the $MS_{S/A}$ represents both error variance and variance due to gender, experimenter, or object position. However, the variance due to these "irrelevant" variables will not contribute to MS_A. For example, if our memory experiment example from Table 8.1 involved two experimenters who each ran an equal number of participants in each group, experimenter identity will not increase the *method* mean square. The analysis based on the one-factor model then violates the principle that the numerator and denominator of the *F* ratio should have the same expectation when H_0 is true. In such situations, the denominator has a larger expectation than the numerator because the irrelevant variable contributes only to the denominator. The result is a loss of power, which can be considerable if the "irrelevant" variable has a large effect. We say that the *F* test is *negatively biased* in this case, meaning that the Type 1 error rate will be less than its nominal value if the null hypothesis is true. Generally, the researcher should formulate a complete structural model, one which incorporates all systematically varied factors, even those thought to be irrelevant or uninteresting. In the examples cited, this would mean viewing the study as involving two factors, the independent variable of interest and experimenter (or gender, or object position), and carrying out the analysis described in Chapter 9.

8.8.2 *The Independence Assumption*

When only one observation is obtained from each participant, and participants are randomly assigned to treatments or randomly sampled from distinct populations, the assumption that the scores are independently distributed is likely to be met. However, there are exceptions that sometimes are not recognized by researchers. For example, suppose we want to compare attitudes on some topic for younger and older college students. Further suppose that before being tested, participants engage in three-person discussions of the relevant topic. The scores of individuals who were part of the same discussion group will tend to be positively correlated. Another example comes from educational research, where a disruptive child can affect learning for other students in the classroom (Glass & Stanley, 1970). If these failures of the independence assumption are ignored (and it has been in some studies; see Anderson & Ager, 1978, for a review), there will be a *positive bias* – an inflation of Type 1 error rate – in an F test of the relevant effect (Myers, DiCecco, & Lorch, 1981; Myers & Well, 1995). A class of analyses referred to as multilevel, nested, or hierarchical (e.g., Raudenbush & Bryk, 2002) provides a general approach to this type of data analysis issue.

8.8.3 The Normality Assumption

Violations of the normality assumption are relatively common and merit attention because they can reduce the power of the F test.

Consequences of Violating the Normality Assumption

The Type 1 error probability associated with the F test is often described to be little affected by sampling from nonnormal populations unless the samples are quite small and the departure from normality extremely marked (e.g., Donaldson, 1968; Lindquist, 1953, pp. 78–90; Scheffé, 1959). This is true even when the independent variable is discretely distributed, as it is whenever rating data or response frequencies are analyzed. In all but the most skewed distributions, simulation studies indicate that Type 1 error rates are relatively unaffected when such measures are submitted to an analysis of variance if the distributions are highly similar across groups (Bevan, Denton, & Myers, 1974; Hsu & Feldt, 1969; Lunney, 1970).

Although in many instances the Type 1 error rate is relatively unaffected by departures from normality, loss of power and reduction of effect size are concerns when distributions are long- or heavy-tailed (kurtotic), skewed, or include outliers. In each of these situations, variability is high relative to the normal distribution. Thus, procedures that address the increased variability often have more power than the conventional F test. Several potential remedies, familiar from Chapter 6, are considered next.

Dealing With Violations of the Normality Assumption: Tests Based on Trimmed Data

As we explained in Chapter 6, merely deleting scores is not a valid procedure. However, a ratio of mean squares distributed approximately as F can be constructed by an approach like the trimmed t test illustrated in Chapter 7 (see Section 7.7). In essence, we estimate the variance of the condition means from the trimmed data, and we use the winsorized data to estimate the error variance. An example should help us understand how this is done. Table 8.7 presents three groups of 11 scores each; the Y scores are the original values, sorted in ascending order within each group. Recognizing the presence of some outlying scores in each group's tail, we trimmed the lowest and highest two scores in each group, yielding the T data. We then replaced the deleted scores with the closest remaining scores to create the winsorized, W, set. We chose to delete two from each group's tail because this is roughly 20% of 11, the number of Y scores. Two ANOVAs, one on the T scores and one on the W

Table 8.7 An example of original (*Y*), trimmed (*T*), and winsorized data (*W*)

	Y	5	6	8	8	10	10	10	10	10	11	14
Group 1	T			8	8	10	10	10	10	10		
	W	8	8	8	8	10	10	10	10	10	10	10
	Y	5	9	11	11	11	11	12	13	13	13	16
Group 2	T			11	11	11	11	12	13	13		
	W	11	11	11	11	11	11	12	13	13	13	13
	Y	9	10	10	10	10	11	11	11	12	12	15
Group 3	T			10	10	10	11	11	11	12		
	W	10	10	10	10	10	11	11	11	12	12	12

scores, provide the values needed for the trimmed *F* test. The test is described and illustrated in Box 8.2.

Box 8.2 The Trimmed *F* Test Applied to the Data of Table 8.7

After performing the ANOVAs on the *T* and *W* scores, do the following:

1. From the ANOVA of the *T* data, get the between-groups mean square, $MS_{A,tk}$ (*tk* refers to trimming *k* scores from each tail); $MS_{A,tk} = 9.19$.
2. From the ANOVA of the *W* data, get the within-groups sum of squares, $SS_{S/A,wk}$; $SS_{S/A,wk} = 27.455$. This is divided by the degrees of freedom for the trimmed data set, $a(n - 1 - 2k) = (3)(10 - 4) = 18$. Therefore, the winsorized error mean square, $MS_{S/A,wk}$, is $27.455/18 = 1.523$.
3. The trimmed *F* statistic is $F = MS_{A,tk}/MS_{S/A,wk} = 9.19/1.523 = 6.03$, and is distributed on 2 and 18 *df*. Then $p = .01$.

For comparison purposes, the *F* statistic on 2 and 30 *df* for the *Y* data was 2.531; the corresponding *p*-value was .096. Clearly, in this example, the trimmed *t* test led to a considerably lower *p*-value.[7] However, a word of caution is in order. In our example, the variances were roughly homogeneous and the distributions were roughly symmetric. Under other conditions, particularly if the group sizes vary, Type 1 errors may be inflated (Lix & Keselman, 1998). We will consider such cases in Section 8.8.4.

Dealing With Violations of the Normality Assumption: Tests Based on Ranked Data

In the *rank-transform F test* (Conover & Iman, 1981) and the *Kruskal–Wallis* H *test* (1952), all scores are ordered with a rank of 1 assigned to the lowest score in the data set and a rank of *N* assigned to the largest, where *N* is the total number of scores. In case of ties, the median rank is assigned; for example, if the five lowest scores are 1, 4, 7, 9, and 9, they would receive ranks of 1, 2, 3, 4.5, and 4.5 respectively. The usual one-way ANOVA is performed on the ranks to execute a rank-transform *F* test and the test statistic, F_r, is evaluated on $a - 1$ and $N - a$ *df*. For the data in Table 8.7, we first create a new variable containing the ranks of the *Y* values using the *rank*(Y) function in {base} R, or by selecting the *Rank Cases* function within the *Transform* pull-down menu in SPSS. Then run the ANOVA to find F_r (2,30) = 3.8, with $p = 0.034$.

The Kruskal–Wallis *H* test is also run on the ranks. In R, the *H* test is available with the *kruskal.test* function in the {stats} package. In SPSS, select *Analyze*, then *Nonparametric Tests*, then *Legacy Dialogs*, and finally *K Independent Samples*. Applying the Kruskal–Wallis *H* test to the data of Table 8.7, we find that $H = 6.47$, $p = 0.039$.[8] As with the trimmed *F* test, the result is a lower *p*-value than that associated with the usual *F* test of the *Y* data. *H* and F_r will generally provide similar *p*-values. This is not surprising given that they are related by the following equation:

$$F_R = \frac{(N-a)H}{(a-1)(N-1-H)}$$

Both tests are more powerful alternatives than the usual *F* test when the populations are not normally distributed but have the same values of variance, skewness, and kurtosis; that is, when the populations have the same, nonnormal distribution. Furthermore, they are only slightly less powerful than the *F* test when the distributions are normal. However, if the treatment populations do not have identical distributions, then *H* and F_r tests may reject the null hypothesis because of differences in the shapes or variances of the distributions. Therefore, the tests are not appropriate as tests of location when heterogeneity of variance is suspected (Oshima & Algina, 1992; Vargha & Delaney, 1998).

8.8.4 *The Homogeneity of Variance Assumption*

Variances may differ across conditions for one of several reasons. One possible cause of heterogeneity of variance is an interaction of an experimental treatment with individual characteristics. For example, a drug tested for its effects on depression may result in a higher variance than, but the same mean score as, a placebo. This would suggest that some individuals had improved but others had been adversely affected by the drug. A second possible reason for unequal variances is that some populations are more variable than others on a particular task. For example, although older adults tend to have longer reaction times on many tasks, they also tend to have a higher variance in those *RT*s. Still another possible factor in findings of heterogeneity of variance are floor, or ceiling, effects. That is, variability may be reduced in one condition relative to another because of a lower, or upper, limit on performance due to the measuring instrument. Finally, variances tend to be correlated with means, usually positively; the normal distribution is the sole exception in which the means and variances are independently distributed. For all these reasons, variances are often unequal, or *heterogeneous*, in the populations sampled in our research. In what follows, we summarize some consequences of the failure of this assumption and we then consider alternatives to the standard *F* test. Again, the problem and possible solutions are like those described for *t* tests in Chapter 6.

Consequences of Heterogeneity of Variance

When there are the same number of scores in all conditions, heterogeneous variances usually will cause Type 1 error rates to be slightly inflated. The inflation is usually less than .02 at the .05 level, and less than .005 at the .01 level, provided the ratio of the largest to smallest variance is no more than 4 to 1, and n is at least 5. Even larger ratios may not be a problem, but this will depend upon sample size, the number of groups, and the shape of the population distributions. The results of computer simulations employing these factors are discussed in several articles, including those by Clinch and Keselman (1982), Tomarken and Serlin (1986), and Cribbie, Fiksenbaum, Keselman, and Wilcox (2012).

When there are different numbers of scores in each condition, simulation studies clearly demonstrate that heterogeneous variances are a problem. Sampling from heavy-tailed and skewed distributions and using variance ratios of largest to smallest as high as 16:1, Lix and Keselman (1998) found that error rates were as high as .50 in some conditions. Sampling from sets of either three or four normally distributed populations, Tomarken and Serlin found that at a nominal .05 level, the actual Type 1 error rate was as low as .022 when the group size was positively correlated with the variance (i.e., larger groups associated with greater variances) and as high as .167 when the correlation was negative. This is because the average within-group variance (i.e., the error term, $MS_{S/A}$) is increased when the largest groups have the largest variances and, conversely, is decreased when the largest groups have the smallest variances. The same consequences of unequal variances were present in t tests, as we described in Chapter 6. When the variances are heterogeneous, a positive relation between group size and variance will negatively bias the F test whereas a negative relation will positively bias the test.

There is evidence in the research literature that extreme variance ratios do occur (Wilcox, 1987), and simulation studies make clear that heterogeneity of variance can inflate Type 1 error rates or deflate power, depending upon various factors such as sample sizes and the type of distribution sampled. That leaves us with two questions. First, for a given data set, how do we decide whether to abandon the standard ANOVA for some remedial procedure? Second, if we do decide that unequal variances are a threat to the validity of the standard F test, what alternative should we use? As we will see, the answers to these questions parallel those for t tests described in Chapter 6.

Detecting Heterogeneity of Variance

As always, we urge that researchers begin the data analysis by examining summary statistics and plots of the data. Box plots for the Beck Depression data as a function of educational level were presented in Figure 8.2 and, as we noted there, differences among the groups in shape and spread are quite evident. The range of variances in Table 8.5 exceeds the 4:1 rule of thumb and suggests that the alpha-level reported in Table 8.6 may not be the actual probability of a Type 1 error. For confirmation of this, we may wish to test whether the variances are homogeneous. Several tests of homogeneity of variance have been proposed. Some are overly sensitive to violations of the normality assumption (Bartlett, 1937; Cochran, 1941; Hartley, 1950; Levene, 1960) or lack power relative to other procedures (Box, 1953; Games, Keselman, & Clinch, 1979; Scheffé, 1959).

We recommend the Brown–Forsythe test (Brown & Forsythe, 1974a) based on deviations from the median. Sampling studies indicate it has only a slightly inflated Type 1 error rate and good power relative to various competitors even when ns are not equal and distributions depart markedly from the normal (Games et al., 1979). In this test, the absolute deviation of each score from its group median, $|Y_{ij} - \tilde{Y}_{.j}|$, is computed, and these values are then submitted to the analysis of variance. Although these deviations do not directly represent the variance, their variance is an index of the spread of scores. For the depression scores summarized in Table 8.5, the *leveneTest* function in the {car} package of R, when used with the *center = median* option, provides the Brown–Forsythe test.[9] In SPSS, the test is available by clicking the "options" button in the standard ANOVA window and checking the box for "homogeneity of variances test." With either software package, the test statistic is 4.511 which, on 3 and 124 *df*, is significant ($p = .005$). This indicates that the mean absolute deviation varies significantly as a function of education level, confirming our sense that the spread of scores was indeed a function of the educational level.

Once we conclude that the population variances are not equal, the next question is: What shall we do about it? One possible response is to seek a transformation of the data that yields homogeneity of variance on the scale resulting from the transformation. As we discussed in Chapter 6, data transformations can make it difficult to interpret the results; in practice, they are used infrequently. A second possibility is to compute an alternative to the usual F test. We consider this approach next.

Dealing With Heterogeneity of Variance: Welch's F Test

Several alternatives to the standard F test of the equality of the a population means have been proposed (Alexander & Govern, 1994; Brown & Forsythe, 1974b; James, 1951, 1954; Welch, 1951), but no one test is best under all conditions. When the data are normally distributed and *ns* are equal, most of the procedures are reasonably robust with respect to Type 1 error rate; however, the standard F is slightly more powerful if the population variances are equal. When the variances are not equal, the choice of test depends upon the degree of skew and kurtosis, whether outliers are present, the degree of heterogeneity of variance, the relation between group sizes and group variances, and the total N (Clinch & Keselman, 1982; Coombs, Algina, & Oltman, 1996; Cribbie et al., 2012; Grissom, 2000; Lix, Keselman, & Keselman, 1996; Tomarken & Serlin, 1986).

Although there is rarely a clear-cut choice for any given data set, simulation studies have concluded that the *Welch test*, F_w, provides a good alternative when both Type 1 error rates and power are considered (Delacre, Leys, Mora, & Lakens, 2019; Lix et al., 1996). F_w performs well relative to various competitors except when the data are highly skewed (skew > 2.0) or group sizes are less than 10 (Lix et al., 1996; Tomarken & Serlin, 1986). Furthermore, the test is readily available in statistical packages. For example, in R we can use the *welch.test* function in the {onewaytests} package; in SPSS, simply click the "options" button in the standard ANOVA, check the box for "Welch test," and continue as normal. Alternatively, the equations for Welch's F test are in Box 8.3.

For the depression data summarized in Table 8.5, we find $F_w\,(3,55) = 2.53$ with $p = 0.066$. The resulting p-value is considerably higher than the .016 we obtained using the standard F calculations. The discrepancy can be accounted for by noting that the correlation between n_j and s^2_j is negative, −.59. There are only 19 participants in the group having only a high school education (HS), whereas the other groups all have at least 33 participants. Because the larger groups have smaller variances, they have more weight in the denominator of the F test; that small denominator contributes to a larger F statistic with a resulting inflated probability of a Type 1 error. The Welch test has compensated for this by taking the inequalities in group sizes and variances into account.

Box 8.3 Formulas for the Welch (F_w) Test

$$F_w = \frac{A}{B}$$

where $A = \frac{1}{a-1}\sum w_j\left(\bar{Y}_j - \bar{Y}_{..}\right)^2$

$$B = 1 + \left[\frac{2(a-2)}{a^2-1}\right]\sum\frac{\left[1-\left(w_j / u\right)\right]^2}{n_j - 1}$$

and $w_j = n_j / s_j^2; u = \sum w_j \bar{Y}_{..} = \sum w_j \bar{Y}_{.j} / u$

$$df_1 = a - 1$$

$$\frac{1}{df_2} = \left[\frac{3}{a^2-1}\right]\sum\frac{\left[1-\left(w_j / u\right)\right]^2}{n_j - 1}$$

A Robust F Test Based on Trimmed Means

The normality and homogeneity of variance assumptions often are violated in the same data set. This is particularly a problem when *ns* are unequal, as in the Beck Depression data we analyzed, or when the experimental conditions have differently shaped distributions. A promising approach is to apply the Welch's *F* test to means based on data from which the highest and lowest 20% have been trimmed (Cribbie et al., 2012; Keselman, Wilcox, Othman, & Fradette, 2002; Lix & Keselman, 1998). The test is described in Box 8.4. In R, the *welch.test* function in {onewaytests} computes F_w using trimmed means and winsorized variances when the *rate* = 0.2 option is employed to trim *rate*% of the data from each tail (here, 20%). For the depression data, this robust *F* test statistic equals 1.58 with 3 and 30 degrees of freedom. The reduction in the degrees of freedom resulted in a larger *p*-value of 0.24.

Box 8.4 Welch's (1951) *F* Test with Trimmed Means and Winsorized Variances

Replace the n_j in Box 8.2 by $h_j = n_j - 2k_j$, where k_j is the number of scores trimmed from each tail of the *j*th group. For example, in the HS group of the depression analysis, $n_j = 19$. If we trim approximately 20% from each tail, $k = 4$ and $h = 19 - (2)(4) = 11$.

The $\bar{Y}_{.j}$ are replaced by the trimmed means.

The s^2_j are replaced by the winsorized group variances.

With these substitutions, the formulas for the Welch's *F* test in Box 8.3 apply directly.

8.9 Summary

This chapter introduced the analysis of variance in the simplest possible context, the one-factor between-participants design. The developments in this chapter will be relevant in the analyses of data from other designs. These developments included:

- *The analysis of variance.* We illustrated the idea of a structural model that underlies the data and directs the partitioning of variability in the data. The structural model is the basis for determining what the variance calculations estimate in terms of population variance parameters. The expected mean squares (*EMS*), in turn, justify the error terms for tests of the null hypothesis and are involved in estimating measures of the magnitude of effects.
- *Measures of importance.* We defined and applied to data several statistics that indicate the importance of the independent variable. Confidence intervals also were presented that provide a range of plausible values of the parameter being estimated.
- *A priori power* and sample size. We illustrated how sample size for a multi-group study can be determined, once the values of α, the desired power, and the effect size of interest are selected.
- *Assumptions underlying the significance test and the estimates of measures of importance.* We reviewed these assumptions, discussed the consequences of their violation, and described several solutions to violations.

Appendix 8.1

Partitioning the Total Variability in the One-Factor Design

The following developments involve two indices of summation: i indexes a value from 1 to n within each group, where n is the number of individuals in a group; j indexes a value from 1 to a, where a is the number of groups. Appendix A provides an explanation of the use of this notation, using several examples.

Squaring both sides of Equation 8.1 yields

$$\left(Y_{ij}-\bar{Y}_{..}\right)^2=\left(Y_{ij}-\bar{Y}_{.j}\right)^2+\left(\bar{Y}_{.j}-\bar{Y}_{..}\right)^2+2\left(Y_{ij}-\bar{Y}_{.j}\right)\left(\bar{Y}_{.j}-\bar{Y}_{..}\right)$$

Summing over i and j, and applying the rules of Appendix A, we have

$$\sum_j^a\sum_i^n\left(Y_{ij}-\bar{Y}_{..}\right)^2=\sum_j^a\sum_i^n\left(Y_{ij}-\bar{Y}_{.j}\right)^2+n\sum_j^a\left(\bar{Y}_{.j}-\bar{Y}_{..}\right)^2+2\sum_j^a\sum_i^n\left(Y_{ij}-\bar{Y}_{.j}\right)\left(\bar{Y}_{.j}-\bar{Y}_{..}\right)$$

Rearranging terms, we can show that the right-most (cross-product) term equals 0:

$$\begin{aligned}2\sum_j^a\sum_i^n\left(Y_{ij}-\bar{Y}_{.j}\right)\left(\bar{Y}_{.j}-\bar{Y}_{..}\right)&=2\sum_j^a\left(\bar{Y}_{.j}-\bar{Y}_{..}\right)\sum_i^n\left(Y_{ij}-\bar{Y}_{.j}\right)\\&=2\sum_j^a\left(\bar{Y}_{.j}-\bar{Y}_{..}\right)(0)=0\end{aligned}$$

The last result follows because the sum of deviations of scores about their mean is zero.

Exercises

8.1 [Properties of variance] A data set has three groups of five scores each. Because the scores involve decimal values, each score is multiplied by 100.

a) How will the mean squares and F ratio be affected (relative to an analysis of the original data set)?
b) In general, what happens to a variance when every score is multiplied by a constant?
c) Suppose we just add a constant, say, 10, to all 15 scores. How would that affect the mean squares and F ratio?
d) Suppose we add 5 to all scores in the first group, 10 to all scores in group 2, and 15 to all scores in group 3? Should MS_A change? $MS_{S/A}$? Explain.

8.2 [Sample size influence on ANOVA] Following are summary statistics from a three-group experiment. Present the ANOVA table when (a) $n_1 = n_2 = n_3 = 10$ and (b) $n_1 = 6$, $n_2 = 8$, and $n_3 = 10$; the totals, or sums of scores, for the groups, the $T_{.j}$, and the variances are as follows:

	A_1	A_2	A_3
Totals	30	48	70
Variances	3.2	4.1	5.7

8.3 [Relationship between Student's t and F] The data are

A_1 : 27 18 16 33 24 A_2 : 23 33 26 19 38

a) Perform the ANOVA.
b) Next, do a t test. How are the results of parts (a) and (b) related?

8.4 [Using the F distribution] The F ratio is basically a test of the equality of two population variance estimates, one in the numerator and another in the denominator. Therefore, it is applicable to the following problem. We have samples of reading scores from 5 students taught with Method A and 11 taught with Method B. We form a ratio of the variances of the two samples, s^2_A/s^2_B.

a) If many samples of sizes 5 and 11 are drawn, (i) what is the proportion of F values greater than 2.61 that we should expect? (ii) less than 4.47?
b) What assumptions are implied in your approach to answering part (a)?

8.5 [Consequences of violating ANOVA assumptions] The file *EX8_5.xlsx* at the website contains three groups of 15 scores.

a) Explore the data; examine statistics and graphs relevant to assessing the normality and homogeneity of variance assumptions. What are the implications for a significance test?
b) Calculate the F and Kruskal–Wallis H tests for these data and comment on the outcome, relating your discussion to your answer to part (a).

8.6 [Exploring the effect of data transformations] Continue using the data in *EX8_5.xlsx* to answer this problem.

a) A nonparametric test is only one way to reduce the effect of the straggling right tail of the data. Explore the data after transformation by taking (1) the square root of each score and (2) the natural log of each score. Does either one better conform to the assumptions underlying the F test? Explain.

b) Carry out the ANOVA with the transformation you selected in part (a). How do the results compare with those for the original F test in Exercise 8.5?

c) Find the confidence intervals for the three means, using the Y data. Then do the same with the group means for the transformed scores. Transform the means of the transformed scores back to the original scale. For example, if you had selected the square-root transformation, you would square the transformed means; if you had selected the log transformation, you would raise e to the power of the mean on the log scale (for example, if the mean on the log scale = 3, on the original scale we would have $e^3 = 20.09$). Do the same for the 95% confidence limits for each of the three means. Compare the results for the original and transformed data.

8.7 [Effects of sample size on measures of importance] The following are the results of two experiments, each with three levels of the independent variable.

Table 1

Source	*df*	*MS*
A	2	80
S/A	27	5

Table 2

Source	*df*	*MS*
A	2	42.5
S/A	12	5

a) For each of the two tables, calculate the Fs and estimates of ω^2_A.

b) What does a comparison of the two sets of results suggest about the effect of the change in n upon these two quantities?

c) Calculate η^2_A for each table. How does the change in n affect the value of η^2_A?

d) Suppose $F = 1$. (i) What must the value of ω^2_A be? (ii) What must the value of η^2_A be (as a function of a and n)?

e) Comment on the relative merits of the various statistics calculated as indices of the importance of A.

8.8 [Calculating effect size and power from published results] The result of an ANOVA of a data set based on three groups of 10 scores each is:

Source	*df*	*SS*	*MS*	*F*
A	2	192	96	3.2
S/A	27	810	30	

a) Is there a significant A effect if $\alpha = .05$?

b) Estimate Cohen's f for these results.

c) Assuming this is a good estimate of the true effect of A, what power did the experiment have? What new information does this post hoc power estimate provide?

d) How many participants would be required to have power – .8 to detect a medium-sized effect? Use Cohen's guidelines.

8.9 [Using theory to predict effect size] According to a mathematical model for the experiment in Exercise 8.8, the predicted means are 10 in condition 1, 14 in condition 2, and 18 in condition 3. If the theory is correct, what sample size would be needed to achieve .8 power to reject the null hypothesis of no difference among the means?

Assume that the error mean square is a good estimate of the population variance, and $\alpha = .05$.

8.10 [Calculating effect sizes and power from raw data] In a study of the relative effectiveness of three methods of teaching elementary probability, students read one of three texts: the Standard (*S*), the Low Explanatory (*LE*), and the High Explanatory (*HE*). The data – scores on a test after a single reading – are in the file *EX8_10.xlsx*.

a) Explore the data. Are there any indications of departures from the underlying assumptions?
b) Test the null hypothesis that the texts do not differ in their effects.
c) Estimate ω^2 and Cohen's f. Verify that $\omega^2 = f^2/(1 + f^2)$ and $f^2 = \omega^2/(1 - \omega^2)$.
d) Based on these results, if you were to replicate the study, how many participants would you run to have power = .8?

8.11 [Effects of data transformations] The *Sayhlth* file linked to the *Seasons* page on the website categorizes individuals by employment category; 1 = employed full time; 2 = employed part time; 3 = not employed. These categories will be the independent variable in this exercise and the *Beck_D* score will be the dependent variable in the following analyses. The *Beck_D* score is an average of the four seasonal Beck Depression scores and is available only for those participants whose scores were recorded in all four seasons. The distribution of *Beck_D* scores tends to be skewed and, as in most nonnormal distributions, heterogeneity of variance is often a problem.

a) Explore the *Beck_D* data in each *employed* category, looking at relevant graphs and statistics. Comment on the validity of the ANOVA assumptions.
b) Run an ANOVA on the *Beck_D* scores as a function of employment status. What do you conclude about the effects of employment?
c) Does the Welch's F test confirm or contradict your conclusion?
d) Calculate Cohen's f. How would you characterize the effect sizes? In general, what can you say about the effect of employment status on depression scores?

8.12 [Using R with real data] The {palmerpenguins} package in R contains a data set called *penguins* (Horst, Hill, & Gorman, 2020). Install the package and use ?penguins to see the variables included.

a) These data were collected on three islands in the Palmer Archipelago in Antarctica during 2007–2009. Does the weight of the penguins differ across the islands?
b) Select the Adelie species of penguins and determine if their weight varies across islands. Explain how these results fit with those of part (a).

Notes

1 See the *Seasons* data set on the website for this book.
2 The data are also in a file labeled *Table 8_1 Memory Data*; a link is on the *Tables* page on the website for this book.
3 These are *least-squares estimators*; that is, we can show that if these statistics are calculated for many samples, their variance about the parameter being estimated is less than that for any other estimator.
4 The SS_A and the $SS_{S/A}$ are frequently called the $SS_{between}$ and SS_{within}, reflecting variability between and within cells. We prefer these more specific labels because they are essential to understanding more the complex experimental designs and analyses that are presented in later chapters.

5 For the one-factor design, η^2 is the same as R^2, the squared correlation between the observed scores (Y_{ij}) and their predicted values $(\bar{Y}_{.j})$. The adjusted-$R^2 = \frac{SS_A - (a-1)MS_{S/A}}{SS_{total}}$ corrects for the positive bias of η^2.
6 The data may be found in the *Table 8_5 Male_educ.xlsx* file; go to the *Tables* page of the book's website.
7 Keselman et al. (2004) and Wilcox, Peterson, and McNitt-Gray (2018) argued for further refinement of the trimmed *t* test and associated confidence intervals using bootstrapping methods that go beyond the scope of this book. See Field and Wilcox (2017) for detailed R-based examples of this approach for a variety of experimental designs.
8 H is distributed as a χ^2 with degrees of freedom equal to $a - 1$. In R, the χ^2 distributional calculator abbreviation for χ^2 is *chisq*, and so, by analogy to the examples in Box 6.1, we calculate 1 – *pchisq*(6.4687, df = 2) to confirm that $p = .039$.
9 Be careful not to use the *bf.test* function in the {onewaytests} package for this purpose. That function tests for the equality of means, not variances.

Chapter 9

Multi-Factor Between-Participants Designs

9.1 Overview

In this chapter, we consider between-participants designs that include two or more factors. In multi-factor designs, the experimental conditions are formed by creating every combination of levels of the independent variables. There are two advantages of such designs. (1) The obvious advantage is economy; the effects of each of several factors can be studied in the same experiment. (2) The second advantage is that the combined, or *interaction*, effects between the variables can be studied. A specific example may help illustrate these two points. Wiley and Voss (1999) were interested in how students learn from written material. In their experiment, students read about the Irish potato famine of the first half of the nineteenth century. One factor in the design was the presentation format: The material was either presented in a single, textbook-like chapter ("text format") or divided among eight hyperlinked sources like one might see online ("web format"). The second factor was the instruction participants received: They were told to write either a narrative (N), a summary (S), an explanation (E), or an argument (A) about what produced changes in Ireland's population between 1800 and 1850. Thus, there were eight conditions in the experiment corresponding to the two formats combined with the four types of instruction. Following reading, students were tested on the material.

Wiley and Voss had several hypotheses that could be tested in their design. One hypothesis was that the more difficult web format would lead to a deeper understanding by forcing readers to integrate material obtained from several sources. Therefore, one question was whether the formats differed in their average effects on learning. A second hypothesis was that the argument instruction would promote "more conceptual understanding." Therefore, a second question was whether the instructions would differ in their average effect. Finally, a third question was whether the size of any difference in performance between the argument instruction and the other instructions would depend upon the format. This is a question of whether there is an *interaction* between format and instructions.

These questions point to the two main goals of this chapter:

- To extend the models, assumptions, and analyses of Chapter 8 to deal with multi-factor between-participants designs.
- To introduce the concept of interaction, providing illustrations of its analysis and interpretation.

DOI: 10.4324/9781003453550-11

9.2 The Two-Factor Design: The Structural Model

9.2.1 *The Model Equation*

In these designs, there are two independent variables, A and B, with a levels of A, b levels of B, and n scores in each of the ab cells. A score will be represented as Y_{ijk}, the ith score at the jth level of A and the kth level of B. We want to test hypotheses about population means and therefore need a model that relates the observed scores to the means of the populations formed by the combinations of the variables A and B and to the error component of each score. These population parameters and the sample statistics that estimate them are presented in Table 9.1.

Consider a specific combination of levels of A and B, say, A_j and B_k. In terms of the *Wiley–Voss* experiment described in Section 9.1, this would be a cell formed by the combination of one of the two formats and one of the four instructions. Because the same combination of treatments is applied to everyone in that cell, the scores in the population corresponding to A_j and B_k should vary only because of error variance due to individual differences in factors such as ability, motivation, and physical state, or differences in other factors that can affect performance. Stating this more formally,

$$Y_{ijk} = \mu_{jk} + \varepsilon_{ijk} \tag{9.1}$$

where μ_{jk} is defined in Table 9.1 as the mean of the population formed by A_j and B_k, and ε_{ijk} is the error component of the ith score in the cell formed by A_j and B_k in the experiment.

Now consider the possibility that scores in different combinations of treatments may differ systematically. It is useful to express the deviation of a score from the grand mean of all the populations by subtracting μ from both sides of Equation 9.1:

$$Y_{ijk} - \mu = \left(\mu_{jk} - \mu\right) + \varepsilon_{ijk} \tag{9.2}$$

Table 9.1 Population parameters and estimates for a two-factor design

Population parameters	*Estimates*
μ_{jk} = mean of the population of scores at A_j and B_k	$\bar{Y}_{.jk} = \sum_i Y_{ijk}/n$
$\mu_{j.} = \sum_k \mu_{jk}/b$ = mean of the populations in condition A_j	$\bar{Y}_{.j.} = \sum_i \sum_k Y_{ijk}/nb$
$\mu_{.k} = \sum_j \mu_{jk}/a$ = mean of the populations in condition B_k	$\bar{Y}_{..k} = \sum_i \sum_j Y_{ijk}/na$
$\mu = \sum_j \sum_k \mu_{jk}/ab$ = mean of all ab populations	$\bar{Y}_{...} = \sum_i \sum_j \sum_k Y_{ijk}/nab$
$\varepsilon_{ijk} = Y_{ijk} - \mu_{jk}$ = error component of Y_{ijk}	$e_{ijk} = Y_{ijk} - \bar{Y}_{.jk}$

The μ_{jk} may vary for any of three reasons; namely, they correspond to different levels of *A* and *B* and to different combinations of those levels. We can represent this idea by the following identity:

$$(\mu_{jk} - \mu) = (\mu_j - \mu) + (\mu_k - \mu) + [(\mu_{jk} - \mu) - (\mu_j - \mu) - (\mu_k - \mu)] \tag{9.3}$$

A simpler notation is

$$\mu_{jk} - \mu = a_j + \beta_k + (\alpha\beta)_{jk} \tag{9.4}$$

Equation 9.4 states that $\mu_{jk} - \mu$ is a sum of three effects. The first of these, α_j, is the *main effect* of treatment *A*; its value represents the extent to which the average score in the population defined by the treatment A_j differs from the mean of all the scores in the *ab* populations. Similarly, β_k is the main effect of treatment *B* and reflects the extent to which the average score in population *B* differs from the average of all the scores in the *ab* populations. Finally, $(\alpha\beta)_{jk}$ is the *interaction effect* of *A* and *B*. The interaction is the difference between μ_{jk} and μ that remains after removal of the main effects of *A* and *B*; that is,

$$(\alpha\beta)_{jk} = (\mu_{jk} - \mu) - (\mu_j - \mu) - (\mu_k - \mu) \tag{9.5a}$$

This interaction effect is more often represented by simplifying the right-hand term in Equation 9.5a:

$$(\alpha\beta)_{jk} = \mu_{jk} - \mu_j - \mu_k + \mu \tag{9.5b}$$

From Equation 9.4, we can substitute for $\mu_{jk} - \mu$ in Equation 9.2. Recognizing that nuisance variables will also contribute to the observed data, we have the structural model underlying tests of hypotheses for the two-factor design:

$$Y_{ijk} - \mu = \alpha_j + \beta_k + (\alpha\beta)_{jk} + \varepsilon_{ijk} \tag{9.6}$$

In words, the variability among scores in the populations has four possible sources: the effects of manipulating *A*; the effects of manipulating *B*; the joint, or interaction, effects of *A* and *B*; and the error component. A more detailed summary of the model components and related assumptions is presented in Box 9.1.

Box 9.1 Components of the Structural Model

1. *The main effect of treatment* A, $\alpha_j = \mu_j - \mu$. Factor A is assumed to have fixed effects; that is, the *a* levels have been arbitrarily selected and are viewed as representing the population of levels.[1] Then $\sum_j \alpha_j = 0$. The *F* test of the *A* main effect tests the null hypothesis that

$$H_0 : \mu_{1.} = \mu_{2.} = \ldots = \mu_{j.} = \ldots = \mu_{a.}$$

or, equivalently,

$$H_0 : \alpha_1 = \alpha_2 = \ldots = \alpha_i = \ldots = \alpha_a = 0$$

2. *The main effect of treatment B*, $\beta_k = \mu_k - \mu$. Factor B is also a fixed-effect variable and so $\sum_k \beta_k = 0$. The *F* test of the B main effect tests the null hypothesis that

$$H_0 : \mu_{.1} = \mu_{.2} = \ldots = \mu_{.k} = \ldots = \mu_{.b}$$

or, equivalently,

$$H_0 : \beta_1 = \beta_2 = \ldots = \beta_k = \ldots = \beta_b = 0$$

3. The *interaction effect of* A_j *and* B_k, $(\alpha\beta)_{jk} = \mu_{jk} - \mu_j - \mu_k + \mu$. Because both *A* and *B* have fixed effects, $\sum_j (\alpha\beta)_{jk} = \sum_k (\alpha\beta)_{jk} = 0$. The relevant null hypothesis is

$$H_0 : (\alpha\beta)_{11} = (\alpha\beta)_{12} = \ldots = (\alpha\beta)_{jk} = \ldots = (\alpha\beta)_{ab} = 0$$

4. *The error component*, $\varepsilon_{ijk} = Y_{ijk} - \mu_{jk}$. The errors are independently and normally distributed, with mean zero and variance σ^2_e, within each treatment population defined by a combination of levels of *A* and *B*.

Equation 9.6 provides the basis for the analysis of variance for the two-factor design. We will develop this relation between the analysis of variance and the structural model in Section 9.3. However, before doing so, we will try to provide a better understanding of what the main and interaction effects represent.

9.2.2 Understanding Main and Interaction Effects

Assume that we have a 2 × 4 design (e.g., factor *A* has 2 levels and factor *B* has 4 levels, like the *Wiley–Voss* example). Further assume that the population means are those in Panel *a* of Table 9.2. Note that the main effects (α_j and β_k) are calculated by subtracting the *grand mean* (obtained by averaging over all scores) from the *marginal means* (those obtained by averaging all the means in a row or column). For designs with equal *n*, the main effects are independent or *orthogonal*. That is, knowing how the means of one factor vary across levels tells us nothing about how the means vary on the other factor (see Chapter 3).

In Panel *b*, we have the population means after the main effects have been subtracted. For example, in the A_1B_2 population, the cell mean after removal of the main effects that have contributed to it is 50 – 5 – (–1) = 46. Note that although the marginal means are now identical, the adjusted cell means still vary. The reason for this variation is the presence of interaction effects. If we subtract the grand mean, μ, from each of the values in Panel *b*, we obtain the results in Panel *c* of Table 9.2; these are the interaction effects associated with each cell. In summary, one definition of interaction effects is that they are the difference

Table 9.2 Treatment population means and effects

(*a*) Original population *means*

	B_1	B_2	B_3	B_4	μ_j	$\alpha_j = \mu_j - \mu$
A_1	65	50	47	58	55	5
A_2	43	48	51	38	45	−5
μ_k	54	49	49	48	$\mu = 50$	
$\beta_k = \mu_k - \mu$	4	−1	−1	−2		

(*b*) Population means after removing the *A* (α_j) and *B* (β_k) main effects

	B_1	B_2	B_3	B_4	*Mean*
A_1	56	46	43	55	50
A_2	44	54	57	45	50
Mean	50	50	50	50	$\mu = 50$

(*c*) Interaction effects: $(\alpha\beta)_{jk} = (\mu_{jk} - \mu) - \alpha_j - \beta_k$

	B_1	B_2	B_3	B_4	*Mean*
A_1	6	−4	−7	5	0
A_2	−6	4	7	−5	0
Mean	0	0	0	0	

between the cell mean and the grand mean, after removing the main effects of the independent variables.

It is useful to compare the pattern of means in Panel *a* of Table 9.2 with the pattern in Panel *a* of Table 9.3. The means in Table 9.3 show the same main effects of *A* and *B* as the means in Table 9.2, as seen by comparing the corresponding values of α_j and β_k of the two tables. However, Table 9.3 does not present an interaction; in that case, we say that *A* and *B* have *additive effects* because the mean of each cell is determined by *adding* the main effect of *A* and the main effect of *B* to the grand mean. The absence of interaction effects is shown by subtracting the row and column effects from each mean. For example, in the A_1B_1 cell, $59 - 5 - 4 = 50$; doing the same for all cells, the values are all also 50, as shown in Panel *b*. An important point that is implicit in this comparison of Tables 9.2 and 9.3 is that the magnitudes of the main effects and interaction are unrelated, or *orthogonal* to one another, in a design where *n* is equal for all conditions. Thus, the presence of one or both main effects does not tell us anything about the magnitude of the interaction, and vice versa.

Another way of comparing the two tables is particularly useful for understanding the meaning of an interaction. In Panel *a* of Table 9.4, the effect of *A* is computed at each level of *B* for the data of Table 9.2; this is done by subtracting the A_2 mean from the A_1 mean. These *simple effects of A* are also computed in Panel *b* for the data of Table 9.3. The key observation is that the values of the simple effects of *A* differ over levels of *B* in

Table 9.3 Treatment population means with no interaction effects

	B_1	B_2	B_3	B_4	μ_j	$\alpha_j = \mu_j - \mu$
A_1	59	54	54	53	55	5
A_2	49	44	44	43	45	−5
μ_k	54	49	49	48	$\mu = 50$	
$\beta_k = \mu_k - \mu$	4	−1	−1	−2		

(*b*) Population means after removing the *A* (α_j) and *B* (β_k) main effects

	B_1	B_2	B_3	B_4	*Mean*
A_1	50	50	50	50	50
A_2	50	50	50	50	50
Mean	50	50	50	50	$\mu = 50$

Table 9.4 Simple effects of A at each level of *B* ($\mu_{1k} - \mu_{2k}$) for the data of Tables 9.2 and 9.3

(*a*) Population means from Table 9.2 with interaction effects

	B_1	B_2	B_3	B_4	
A_1	65	50	47	58	55
A_2	43	48	51	38	45
$\mu_{1k} - \mu_{2k}$	22	2	−4	20	10

(*b*) Population means from Table 9.3 with no interaction effects

	B_1	B_2	B_3	B_4	μ_j
A_1	59	54	53	53	55
A_2	49	44	43	43	45
$\mu_{1k} - \mu_{2k}$	10	10	10	10	10

Panel *a*, indicating that *A* and *B* interact. In contrast, the simple effects of *A* are constant over levels of *B* in Panel *b*, indicating no interaction. Thus, *an interaction means that the size of the effect of factor* A *differs over levels of factor* B or, equivalently, that the effect of factor *B* depends on the level of factor *A*. The same observation may be made graphically. The means for Panel *a* of Table 9.4 are graphed in the left panel of Figure 9.1; the means for Panel *b* of Table 9.4 are graphed in the right panel of Figure 9.1. The obvious difference in the two graphs is that the curves on the left are not parallel, whereas the curves on the right are. Thus, with respect to the population means, an *interaction is a departure from parallelism.* We say that the interaction is a significant source of variance when the size of the effects of one variable differs significantly across levels of the other variable.[2]

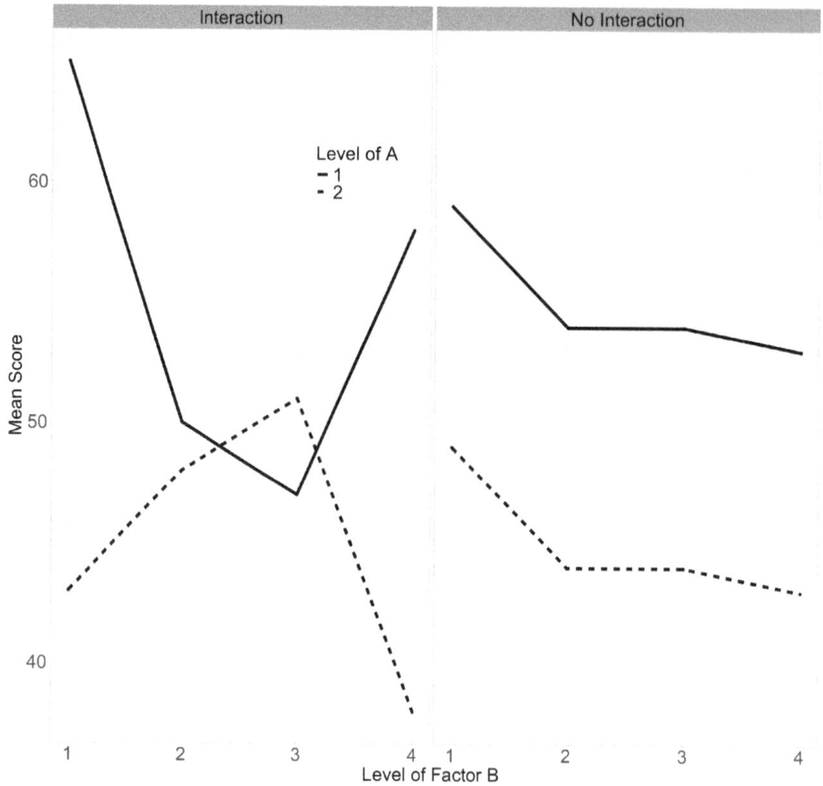

Figure 9.1 Plots of the population means in Tables 9.2 and 9.3.

9.3 Two-Factor Designs: The Analysis of Variance

The analysis of the variability in the data for the two-factor design follows the same logic developed for the one-factor design presented in Chapter 8. The structural model (Equation 9.6) suggests a way to partition the deviation of a score from the grand mean, $Y_{ijk} - \bar{Y}_{...}$, into several components. We derive these components in Section 9.3.1, and then develop formulas for the sums of squares based on them in Section 9.3.2. We then construct mean squares and, subsequently, F statistics to test null hypotheses about main and interaction effects.

9.3.1 *Components of the Scores*

In Section 9.2.1, we assumed the structural model

$$Y_{ijk} = \mu + \alpha_j + \beta_k + (\alpha\beta)_{jk} + \varepsilon_{ijk}$$

After subtracting μ from both sides and substituting treatment population means, we have

$$Y_{ijk} - \mu = (\mu_j - \mu) + (\mu_k - \mu) + (\mu_{jk} - \mu_j - \mu_k + \mu) + \varepsilon_{ijk} \tag{9.7}$$

Substituting the estimates of these parameters from Table 9.1, we have the basis for the ANOVA:

$$\left(Y_{ijk}-\bar{Y}_{...}\right)=\left(\bar{Y}_{.j.}-\bar{Y}_{...}\right)+\left(\bar{Y}_{..k}-\bar{Y}_{...}\right)+\left(\bar{Y}_{.jk}-\bar{Y}_{.j.}-\bar{Y}_{..k}+\bar{Y}_{...}\right)+\left(Y_{ijk}-\bar{Y}_{.jk}\right) \quad (9.8)$$

In other words, an individual score may differ from the grand mean for four distinct reasons: (1) the effect of the level of factor A to which they are exposed; (2) the effect of the level of factor B to which they are exposed; (3) the effect of the combination of the levels of factors A and B; and (4) measurement error and other random noise. Put more simply,

$$\begin{aligned}score - grand\ mean = main\ effect\ of\ A + main\ effect\ of\ B + interaction\ effect \\ + residual\ error\end{aligned}$$

9.3.2 Sums of Squares

Equation 9.8 forms the basis for the sums of squares; these are important components in tests of null hypotheses. Squaring both sides of Equation 9.8 and summing yields the sums of squares (*SS*) formulas in Panel *a* of Table 9.5. As a first step, we partitioned the total sum of squares into two components: a between-cells sum of squares and a within-cell (*S/AB*) sum of squares. The between-cell variability itself is usually not of interest because it has several possible sources. For example, in the *Wiley–Voss* data set, the eight cell means may differ because they represent different formats, different instructions, or different combinations of formats and instructions. Although software packages do not include the between-participants variability, SS_{cells}, we include it because it is involved in our conceptualization and calculation of the SS_{AB} and the $SS_{S/AB}$. The components of the between-cells sum of squares in Table 9.5 correspond to the main and interaction sources of variance that we wish to test.

The formulas presented in Table 9.5 define the sums of squares. Although statistical software usually will be available to perform the calculations, the formulas are presented to remind us that the sums of squares are indices of variability. For example, the SS_{total} is $abn - 1$ times the variance of all the scores, the SS_A is $bn(a - 1)$ times the variance of the *A* marginal means, and the SS_{cells} is $n(ab - 1)$ times the variance of the *ab* cell means. The tests of null hypotheses corresponding to the *A*, *B*, and *AB* sources of variance (*SV*) test whether those variances are greater than chance.

9.3.3 Degrees of Freedom

A formula for degrees of freedom is associated with each of the sources of variance. The df_{total} are $abn - 1$ because this *SV* represents the variability of all *abn* scores about the grand mean. The SS_{cells} is distributed on $ab - 1$ *df* because it involves the variability of the *ab* cell means about the grand mean. The *df* for the main effects have the same form as in Chapter 8; these *SV* reflect the variance of the *a* (or *b*) means about the grand mean and therefore one *df* is lost. The interaction degrees of freedom are

$$df_{AB}=(ab-1)-(a-1)-(b-1)=(a-1)(b-1)$$

Table 9.5 The analysis of variance (ANOVA) table for the two-factor between-participants design (*a*) and expected mean squares (*b*)

(*a*) ANOVA

Source	*df*	*SS*	*MS*	*F*
Total	$abn-1$	$\sum_j^a\sum_k^b\sum_i^n\left(Y_{ijk}-\bar{Y}_{...}\right)^2$		
Between cells	$ab-1$	$n\sum_j^a\sum_k^b\left(\bar{Y}_{.jk}-\bar{Y}_{...}\right)^2$		
A	$a-1$	$nb\sum_j^a\left(\bar{Y}_{.j.}-\bar{Y}_{...}\right)^2$	SS_A/df_A	$MS_A/MS_{S/AB}$
B	$b-1$	$na\sum_k^b\left(\bar{Y}_{..k}-\bar{Y}_{...}\right)^2$	SS_B/df_B	$MS_B/MS_{S/AB}$
AB	$(a-1)(b-1)$	$SS_{cells}-SS_A-SS_B$	SS_{AB}/df_{AB}	$MS_{AB}/MS_{S/AB}$
S/AB	$ab(n-1)$	$SS_{total}-SS_{cells}$	$SS_{S/AB}/df_{S/AB}$	

(*b*) Expected mean squares

Source	*EMS*
A	$\sigma_e^2+nb\sum_j\frac{\alpha_j^2}{a-1}=\sigma_e^2+nb\theta_A^2$
B	$\sigma_e^2+na\sum_k\frac{\beta_k^2}{b-1}=\sigma_e^2+na\theta_B^2$
AB	$\sigma_e^2+n\sum_j\sum_k\frac{(\alpha\beta)_{jk}^2}{(a-1)(b-1)}=\sigma_e^2+n\theta_{AB}^2$
S/AB	σ_e^2

Note: $\alpha_{j\cdot}=\mu_{j\cdot}-\mu$, $\beta_k=\mu_{\cdot k}-\mu$, and $(\alpha\beta)_{jk}=(\mu_{jk}-\mu)-\alpha_j-\beta_k=(\mu_{jk}-\mu_j-\mu_k+\mu)$. The θ^2 notation serves as a reminder that $\Sigma\alpha_j^2/(a-1)$ is not a population variance; the variance of the treatment population means has *a* as the denominator.

reflecting the adjustment of cell variability for the variability due to *A* and *B*. In practice, we can generate the degrees of freedom for an interaction just by multiplying the degrees of freedom for the interacting variables.

The $df_{S/AB}$ may be thought of as the difference between df_{total} and df_{cells};

$$df_{S/AB}=(abn-1)-(ab-1)$$

These degrees of freedom may also be viewed as the result of summing the degrees of freedom for variability within each cell; there are *ab* cells, each with $n-1$ *df*, yielding $ab(n-1)$

df. These two ways of thinking about degrees of freedom, as a difference between the total *df* and the cell *df*, or as a sum over cells, are equivalent: $(abn - 1) - (ab - 1) = ab(n - 1)$.

9.3.4 Mean Squares (MS), Expected Mean Squares (EMS), and F Ratios

As in the one-factor design, the *MS* of Table 9.5 are ratios of *SS* to *df*. Conceptually, however, the mean squares for the main effects are simple functions of variances. For example, MS_A is the variance of the *a* marginal means in the *A* conditions, multiplied by *nb*, the number of scores upon which each mean is based. Similarly, MS_B is the variance of the *b* marginal means in the *B* conditions, multiplied by *na*, the number of scores upon which each mean is based. The error mean square, $MS_{S/AB}$, is an average of the within-cell variances; it can be calculated as

$$MS_{S/AB} = (1/ab)\sum_j \sum_k s_{jk}^2$$

where s^2_{jk} is the variance of the *n* scores in the cell defined by A_j and B_k.[3]

All three *F* ratios are formed by dividing by $MS_{S/AB}$. This is justified by the expected mean squares (*EMS*; see Panel *b* of Table 9.5). As we stated in Chapter 8, forming a ratio of two mean squares follows the rule that the numerator and denominator *MS* must have the same expectation when the null hypothesis corresponding to the numerator is true.

The *EMS* are derived by assuming the structural model of Equation 9.6, independence of the scores, and homogeneity of the population variances. In addition, if the treatment populations are normally distributed and the null hypothesis is true, the ratio of mean squares is distributed as *F*.[4]

We now have both a conceptual framework and formulas on which to base tests of hypotheses about main and interaction effects. We next apply this framework to the analysis of the inference verification test (*IVT*) data in Table 9.6. The complete data set, reported by Wiley and Voss (1999), includes other measures and may be found in the *Wiley* file among the *Data Sets* pages on the website for this book.

9.3.5 The Wiley–Voss Example

Before considering the ANOVA, we should get some sense of the effects of our variables. Looking at the two marginal format means in the right-most column of Table 9.6, we find that performance for the web format ($\bar{Y}_{Web}$) was better than that for the text format ($\bar{Y}_{Text}$). This difference between the web and text formats is largely due to the argument instructional condition; although the web format has a higher mean than the text format in all instructional conditions, the differences between web and text cell means are small except in the *Argument* column. Turning next to the marginal instructional means ($\bar{Y}_N$, $\bar{Y}_S$, $\bar{Y}_E$, and $\bar{Y}_A$), we find the *IVT* mean to be higher in the argument condition than in any of the others. Again, however, we must qualify this: The advantage of the argument condition is quite pronounced for the web format, but rather small in the text format. Whether we view the data as showing that the difference between format means depends on instructions, or as showing that the differences among instructional means depend on format, our focus should be on the interaction of format and instructions. This is clearer in the bar graph of Figure 9.2. Although web learning has an

Table 9.6 Inference scores (percent correct) from Wiley and Voss (1999) with summary statistics

Format	*Instructions*				
	Narrative	*Summary*	*Explanation*	*Argument*	
	70	50	70	70	
	80	90	80	70	
	80	60	70	60	
Text	70	80	60	60	
	60	70	60	70	
	50	80	80	90	
	80	80	70	90	
	80	70	60	80	
$\bar{Y}_{Text,k}$ =	71.25	72.5	68.75	73.75	$\bar{Y}_{Text}$ = 71.56
$s^2_{Text,k}$ =	126.79	164.29	69.64	141.07	
	100	70	60	100	
	80	70	60	90	
	60	80	80	100	
Web	60	50	80	80	
	60	90	80	90	
	70	60	60	100	
	90	100	80	70	
	90	70	80	90	
$\bar{Y}_{Web,k}$ =	76.25	73.75	72.5	90	$\bar{Y}_{Web}$ = 78.13
$s^2_{Web,k}$ =	255.36	255.36	107.14	114.29	
$\bar{Y}_{j,Instruct}$ =	73.75	73.13	70.63	81.88	$\bar{Y}_{..}$ = 74.84

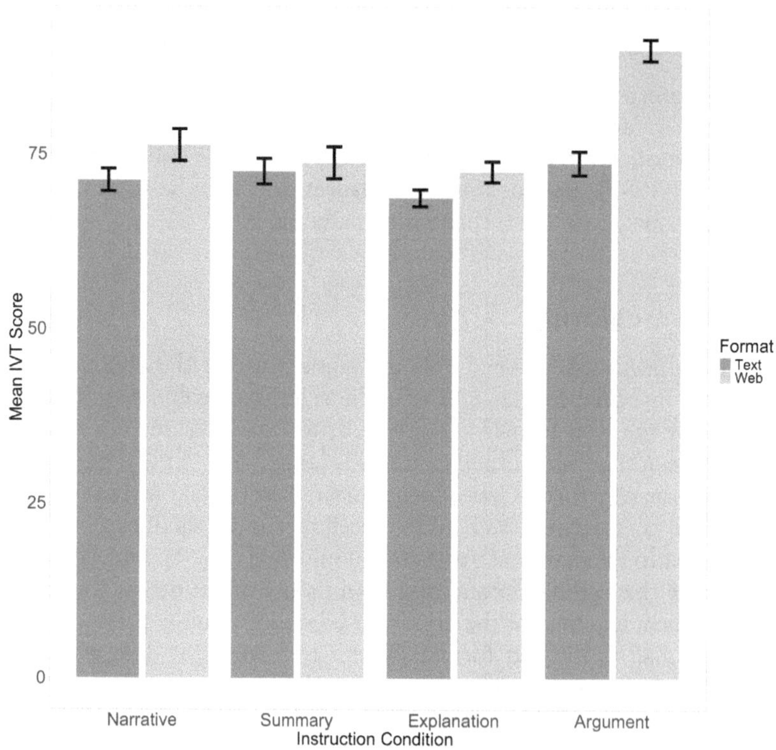

Figure 9.2 Bar graph of the Wiley and Voss (1999) *IVT* data (error bars are +/− *SEM*).

advantage in all four instructional conditions, that advantage is clearly larger in the argument condition than in any of the other three.

In addition to examining the means, we also checked for any departures from the assumptions underlying the hypothesis tests we wished to perform. Box plots and the Levene tests failed to reveal any violations of the assumption of homogeneity of variance severe enough to undermine conclusions based on the tests of main and interaction effects. Nor were any outliers present in the box plots. Finally, plots of residuals (deviations of scores from the cell means) and significance tests revealed no departure from normality. In summary, the ANOVA appears to provide appropriate tests for the *IVT* data.

The results of the analysis of variance for the data are presented in Table 9.7. As the experimenters hypothesized, a significantly higher proportion of inferences were correctly verified by participants in the web than in the text format condition. This means that the average of the four populations of *IVT* scores obtained under the web format differs from the average of the four populations of *IVT* scores obtained under the text format. In terms of Table 9.6, it is the *marginal means*, $\bar{Y}_{Web}$ (78.13) and $\bar{Y}_{Text}$ (71.56), that differ significantly. We can conclude that, *averaging over levels of instruction*, the effects of the two formats differ significantly. However, this does not provide information about the effects of the formats at any particular level of instructions.

The experimenters also were interested in whether instructions would affect performance. They reported that the effect was "marginally significant" because the p-value was .07, short of the .05 level usually required for statistical significance. We will consider further the question of the effect of instructions later in this chapter when we calculate various measures of effect size for the *Wiley–Voss* data.

The F test of the *Format* × *Instructions* interaction tests the null hypothesis that the effects of one variable are the same under all levels of the other variable. One statement of the null hypothesis of no interaction is that the difference between the text and web population means is the same under all types of instructions. This may be represented as

$$H_0 : \left(\mu_{Text,\ N} - \mu_{Web,\ N}\right) = \left(\mu_{Text,\ S} - \mu_{Web,\ S}\right) = \left(\mu_{Text,\ E} - \mu_{Web,\ E}\right) = \left(\mu_{Text,\ A} - \mu_{Web,\ A}\right)$$

where, for example, $\mu_{Text,N}$ is the mean of the population of scores obtained under the text format and the narrative instructions.

An equivalent statement of the null hypothesis of no interaction is that the effects of instructions are the same at the two format levels. We could state this null hypothesis as

$$H_0 : \left(\mu_{Text,N} - \mu_{Text,S}\right) = \left(\mu_{Web,N} - \mu_{Web,S}\right)$$

Table 9.7 The analysis of variance (ANOVA) table for the *Wiley–Voss* data

SV	*df*	SS	*MS*	F	P
Total	63	10,998.44			
Between cells	7	2,360.94			
Format (*F*)	1	689.06	689.06	4.47	.039
Instructions (*I*)	3	1,142.19	380.73	2.47	.071
FI	3	529.69	176.56	1.14	.337
S/FI	56	8,367.50	154.24		

and

$$\left(\mu_{Text,\,S}-\mu_{Text,\,E}\right)=\left(\mu_{Web,\,S}-\mu_{Web,\,E}\right)$$

and

$$\left(\mu_{Text,E}-\mu_{Text,A}\right)=\left(\mu_{Web,E}-\mu_{Web,A}\right)$$

Both forms of H_0 are ways of stating that the effect of one factor is the same at each level of the other factor (i.e., parallel functions, as in the right panel of Figure 9.1).

The difference between the observed text and web means under argument instructions appears considerably larger than the other differences, as shown in the means of Table 9.6 and the bar graph of Figure 9.2. Nevertheless, the *F* test of the interaction fell well short of significance. This raises several questions. Given that the *Instructions* and the *Format* × *Instructions* sources of variance were not significant, is it proper to test more specific hypotheses related to those sources such as whether the narrative and argument means differ? If so, should we have different criteria for significance than for the usual *t* test? And what should our error term be? We will consider those questions, and attempt to further clarify the concept of interaction, in Chapter 10.

In sum, the analyses of the *Wiley–Voss* data are somewhat ambiguous. The effect of instructions is not significant. Graphically, there appears to be an interaction, but the *F* test does not confirm its reliability. The only significant result is the effect of format, showing better performance with the web presentation. However, there is some doubt about how to interpret this seemingly straightforward result because of the question about whether the effect of format is due to the results from the argument condition.

At this point, we complete our presentation of the basic ANOVA by extending it to designs with more than two factors.

9.3.6 *Using Software for Two-Factor Between-Participants ANOVA With Equal* n

Two-factor between-participants ANOVAs are straightforward when the sample sizes are equal in all cells. In R, the *aov* function in the {stats} package works the same way as for one-factor designs; the only modification required is to include both factor names. For example, *summary*(*aov*(data = DFname, DV ~ A*B)) requests an ANOVA on the data in the data frame called *DFname*: The *aov* requests the ANOVA and *summary* requests the output in the form of an ANOVA table. Here, the dependent variable, *DV*, is analyzed as a function of factors *A*, *B*, and their interaction.

In SPSS, we select the *General Linear Model* from the *Analyze* pull-down menu, and then *Univariate*. Move the dependent variable from the list on the left to its box on the right, and move the factor variables to the box labeled "Fixed Effects" before clicking okay. By default, the ANOVA will include all main effects and interactions.

9.4 Three-Factor Between-Participants Designs

9.4.1 *The General Case*

Extending the two-factor design to the three-factor design is straightforward; the only new concept is the three-factor interaction. Therefore, we will present the general case of the three-factor design concisely so that we may reinforce the basic concepts already developed for the simpler one- and two-factor designs.

The general case of the three-factor between-participants design involves a levels of A, b levels of B, c levels of C, and n scores in each of the abc cells. The relevant indices are

$$i = 1,2,\ldots,n; j = 1,2,\ldots,a; k = 1,2,\ldots,b; \text{ and } m = 1,2,\ldots,c$$

The structural model looks much like that for the two-factor experiment except that there are now three two-factor interactions (AB, AC, and BC) and there is the added three-factor interaction:

$$Y_{ijkm} = \mu + \alpha_j + \beta_k + \gamma_m + (\alpha\beta)_{jk} + (\alpha\gamma)_{jm} + (\beta\gamma)_{km} + (\alpha\beta\gamma)_{jkm} + \varepsilon_{ijkm} \tag{9.9}$$

Definitions in terms of population means, together with estimates of those means, are presented in Table 9.8. The only new definition is that of the interaction effect for the cell $A_jB_kC_m$; this is the difference between the cell mean and the grand mean, adjusted for all main and first-order interaction effects that contribute to the cell.

The sums of squares follow directly from the parameter estimates by squaring each term in the Estimate column of Table 9.8 and summing over the indices. The results of this process are presented in Table 9.9 together with the degrees of freedom. The only new df term is $(a - 1)(b - 1)(c - 1)$, the df_{abc}. This follows by subtracting the main and two-factor df from $abc - 1$, the between-cells df.

The mean squares are obtained, as usual, by dividing sums of squares by degrees of freedom. $MS_{S/ABC}$, the average within-cell variance, is the error term against which all main and interaction terms are tested. As in other designs, the expected mean squares provide the rationale for this choice of error terms; these are presented in Table 9.10. The *EMS* have been derived from the structural model (Equation 9.9) under the usual assumptions that the scores in the abc treatment populations are independently distributed and that the population variances all equal σ^2_e. In addition, all three factors are assumed to have *fixed effects*; that is, the levels have been arbitrarily selected and not randomly sampled from a universe of treatment levels.

Table 9.8 Parameters of the structural model for a three-factor design

Source	*Population parameter*	*Estimate*
A	$\alpha_j = \mu_{j..} - \mu_{...}$	$\bar{Y}_{.j..} - \bar{Y}_{....}$
B	$\beta_k = \mu_{.k.} - \mu_{...}$	$\bar{Y}_{..k.} - \bar{Y}_{....}$
C	$\gamma_m = \mu_{..m} - \mu_{...}$	$\bar{Y}_{...m} - \bar{Y}_{....}$
AB	$(\alpha\beta)_{jk} = (\mu_{jk.} - \mu_{...}) - \alpha_j - \beta_k$	$\bar{Y}_{.jk.} - \bar{Y}_{.j..} - \bar{Y}_{..k.} + \bar{Y}_{....}$
AC	$(\alpha\gamma)_{jm} = (\mu_{j.m} - \mu_{...}) - \alpha_j - \gamma_m$	$\bar{Y}_{.j.m} - \bar{Y}_{.j..} - \bar{Y}_{...m} + \bar{Y}_{....}$
BC	$(\alpha\gamma)_{.km} = (\mu_{.km} - \mu_{...}) - \beta_k - \gamma_m$	$\bar{Y}_{..km} - \bar{Y}_{..k.} - \bar{Y}_{...m} + \bar{Y}_{....}$
ABC	$(\alpha\beta\gamma)_{jkm} = (\mu_{jkm} - \mu_{...}) - \alpha_j - \beta_k$	$\bar{Y}_{.jkm} - \bar{Y}_{.j..} - \bar{Y}_{..k.} + \bar{Y}_{...m}$
	$-\gamma_m - (\alpha\beta)_{jk} - (\alpha\gamma)_{jm} - (\beta\gamma)_{km}$	$-\bar{Y}_{.jk.} - \bar{Y}_{.j.m} - \bar{Y}_{..km} - \bar{Y}_{....}$
Error	ε_{ijkm}	$Y_{ijkm} - \bar{Y}_{.jkm}$

Table 9.9 Degrees of freedom and sums of squares in a three-factor design

Source	*df*	*SS*
Total	*abcn* − 1	$\sum_i\sum_j\sum_k\sum_m\left(Y_{ijkm}-\bar{Y}_{....}\right)^2$
Between cells	*abc* − 1	$n\sum_j\sum_k\sum_m\left(\bar{Y}_{.jkm}-\bar{Y}_{....}\right)^2$
A	*a* − 1	$nbc\sum_j\left(\bar{Y}_{.j..}-\bar{Y}_{....}\right)^2$
B	*b* − 1	$nac\sum_k\left(\bar{Y}_{..k.}-\bar{Y}_{....}\right)^2$
C	*c* − 1	$nab\sum_m\left(\bar{Y}_{...m}-\bar{Y}_{....}\right)^2$
AB	(*a* − 1)(*b* − 1)	$nc\sum_j\sum_k\left(\bar{Y}_{.jk.}-\bar{Y}_{.j..}-\bar{Y}_{..k.}+\bar{Y}_{....}\right)^2$
AC	(*a* − 1)(*c* − 1)	$nb\sum_j\sum_m\left(\bar{Y}_{.j.m}-\bar{Y}_{.j..}-\bar{Y}_{...m}+\bar{Y}_{....}\right)^2$
BC	(*b* − 1)(*c* − 1)	$na\sum_k\sum_m\left(\bar{Y}_{..km}-\bar{Y}_{..k.}-\bar{Y}_{...m}+\bar{Y}_{....}\right)^2$
ABC	(*a* − 1)(*b* − 1)(*c* − 1)	$n\sum_j\sum_k\sum_m\begin{pmatrix}\bar{Y}_{.jkm}+\bar{Y}_{.j..}+\bar{Y}_{..k.}+\bar{Y}_{...m}\\-\bar{Y}_{.jk.}-\bar{Y}_{.j.m}-\bar{Y}_{..km}-\bar{Y}_{....}\end{pmatrix}^2$
S/ABC (within cells)	*abc*(*n* − 1)	$SS_{tot}-SS_{betw.cells}$

9.4.2 *Extending the* Wiley–Voss *Example*

To illustrate the concepts and analyses, we add a hypothetical third factor to the design of the *Wiley–Voss* experiment. Assume that there are only two levels of instruction (*I*), Summary and Argument, and two formats (*F*), Text and Web. Further assume that half of the participants are older adults and half are young adults; call this factor Age (*A*). Assume that there are 10 scores in each of the eight cells; i.e., $n = 10$ and $N = 80$. The means for this hypothetical experiment are presented in Table 9.11. The $MS_{S/cells}$ is presented as having a value of 154, but the relevant condition variances on which that value is based have been omitted. Note that we are dealing with the simplest case of a three-factor design, one in which there are only two levels of each of the three variables. The simplicity of the design enables us to concentrate on basic concepts. The design has the added advantage that it is a very common research design.

Main Effects

In the ANOVA, there will be three sources of main effects: instructions, format, and age. We can view these main effects by calculating the marginal means separately for each factor. For example, the test of the format source of variance involves a comparison of the text

Table 9.10 Expected mean squares (*EMS*) for the three-factor design

SV	*EMS*
A	$\sigma_e^2 + nbc\sum_j \alpha_j^2 / (a-1)$
B	$\sigma_e^2 + nac\sum_k \beta_k^2 / (b-1)$
C	$\sigma_e^2 + nab\sum_m \gamma_m^2 / (c-1)$
AB	$\sigma_e^2 + nc\sum_j\sum_k (\alpha\beta)_{jk}^2 / (a-1)(b-1)$
AC	$\sigma_e^2 + nb\sum_j\sum_m (\alpha\gamma)_{jm}^2 / (a-1)(c-1)$
BC	$\sigma_e^2 + na\sum_k\sum_m (\beta\gamma)_{km}^2 / (b-1)(c-1)$
ABC	$\sigma_e^2 + n\sum_j\sum_k\sum_m (\alpha\beta\gamma)_{jkm}^2 / (a-1)(b-1)(c-1)$
S/ABC	σ_e^2

Note: The parameters of the structural model are defined in Table 9.8.

Table 9.11 Means for a hypothetical extension of the *Wiley–Voss* experiment

		Summary	*Argument*	*Mean*
Older Adult	Text	71.25	73.75	72.50
	Web	76.50	90.00	83.25
	Mean	73.875	81.875	77.875
Young Adult	Text	70.50	74.00	72.25
	Web	88.25	89.75	89.00
	Mean	79.375	81.875	80.625
Averaging over Age	Text	70.875	73.875	72.375
	Web	82.375	89.875	86.125
	Mean	76.625	81.875	79.25

Note: Each cell contains 10 scores; i.e., $n = 10$, $N = 80$. The error mean square, $MS_{S/cells} = 154$.

and web marginal means. These means are obtained by averaging over the four combinations of instructions and age. As can be seen in the right-hand column in the bottom panel of Table 9.11, the text and web means are 72.375 and 86.125. The significance test is a test of the null hypothesis that, *averaging over the four populations corresponding to the combinations of age and instructions*, there is no difference between the population text and web means.

To test this null hypothesis, we can calculate MS_{Format} by computing the variance of the two marginal means, 94.53125, and then multiplying by the number of scores each mean is based on, 40, to obtain MS_{Format} = 3,781.25. To construct the F test, divide MS_{Format} by $MS_{S/cells}$: F = 3,781.25/154 = 24.55, which is significant on 1 and 72 degrees of freedom. In R, 1 – *pf*(24.55, 1, 72) reveals that the p-value is 0 to five decimal places.

An astute reader might wonder whether the effect of format could be tested by a simple t test for independent means. In fact, we can calculate a t statistic:

$$t = \frac{\bar{Y}_{web} - \bar{Y}_{text}}{\sqrt{MS_{S/cells}\left(\frac{1}{n_{web}} + \frac{1}{n_{text}}\right)}} = \frac{86.125 - 72.375}{\sqrt{154\left(\frac{1}{40} + \frac{1}{40}\right)}} = 4.955$$

Alternatively, noting that when the numerator has one degree of freedom, $F(1, \text{df}_2) = t^2(\text{df}_2)$; we could have calculated

$$F = \frac{\left(\bar{Y}_{Web} - \bar{Y}_{Text}\right)^2 / \left(\frac{1}{n_{Web}} + \frac{1}{n_{Text}}\right)}{MS_{S/cells}} \tag{9.10}$$

The instruction and age sources of variance can be tested in a similar manner.[5] We encourage the reader to compute the mean squares for instruction and age and confirm that they match the values in Table 9.12.

First-Order (Two-Factor) Interactions

There are three possible two-factor interactions in a three-factor design. In the *Wiley–Voss* example, they are *Format* × *Instructions* (*FI*), *Format* × *Age* (*FA*), and *Instructions* × *Age* (*IA*). The *FA* interaction is of particular interest because age should not have an effect in the text condition, but it might in the web condition. The interpretation of this interaction is essentially the same as if F and A were the only factors in the experiment, except that in this case, the relevant means are obtained by averaging over the levels of the third

Table 9.12 The analysis of variance (ANOVA) table for the hypothetical data in Table 9.11

SV	*df*	*SS*	*MS*	*F*	*P*
Total	79				
Between cells	7	5,127.50	732.5		
Format (*F*)	1	3,781.25	3,781.25	24.55	.000
Instructions (*I*)	1	551.25	551.25	3.58	.05
Age (*A*)	1	151.25	151.25	<1	
FI	1	101.25	101.25	<1	
FA	1	180.00	180.00	1.17	
IA	1	151.25	151.25	<1	
FIA	1	211.25	211.25	1.37	
S/FIA	72		154.00		

variable, instructions. These means are in the right-most column in the upper two panels of Table 9.11. A better way to see the possible interaction is to redisplay the means:

		Age Group	
		Younger	*Older*
Format	Web	89.00	83.25
	Text	72.25	72.50

The pattern of means indicates that the advantage of the young adults over the older adults is, as hypothesized, greater when information is presented in the web format. The significance test of this interaction is a test of the null hypothesis that, *averaging over the two levels of instructions*, there is no difference in the magnitude of the effect of format as a function of age group.

Calculating the interaction sum of squares is generally too tedious to do by hand. However, in the case of interactions based on 1 *df*, we can represent the interaction as a difference between differences. This makes hand calculations more manageable; much more importantly, expressing a 2 × 2 interaction as a difference between simple effects is a better way to understand what is being tested. Let I indicate the interaction effect; then,

$$I_{FA} = (89.00 - 72.25) - (83.25 - 72.50) = 6.00$$

In words, the effect of format is 6 points greater for young adults than for older adults. Rearranging terms, this *FA* interaction can be rewritten as a difference between the sums of diagonal cell means:

$$I_{FA} = (89.00 + 72.50) - (83.25 + 72.25)$$

If we calculate the averages of the two diagonal sums, we have the basis for a test of the difference between two means; these are (89.00 + 72.50)/2, or 80.75, and (83.25 + 72.25)/2, or 77.75. As with our example in testing the main effect of format, we can test the *FA* interaction by comparing these two means via a t test:

$$t = \frac{80.75 - 77.75}{\sqrt{154\left(\frac{1}{40} + \frac{1}{40}\right)}} = 1.081$$

Alternatively, applying Equation 9.10, we can construct an F test:

$$F = \frac{(80.75 - 77.75)^2 / \left(\frac{1}{40} + \frac{1}{40}\right)}{154} = 1.169 = 1.081^2$$

The degrees of freedom for the error term are $abc(n - 1)$, or 72 in this design, assuming 10 participants in each of the eight cells. The F of 1.17 is not significant on 1 and 72 *df* and therefore we lack sufficient evidence to reject the hypothesis of no interaction. In other words, we cannot conclude that the advantage of the web format over the text format is significantly greater for young adults than for older adults.

The tests of the *FI* and *IA* interactions follow from the example of the test of the *FA* interaction.

The Second-Order (Three-Factor) Interaction

The eight cell means in Table 9.11 are plotted in Figure 9.3. We assigned instructions to the x-axis, with the older adults' means in one panel and the young adults' in the other, as well as different lines for the two formats. However, this assignment of variables in the plot is arbitrary. We could have had the two formats, or the two types of instructions, in different panels. We will soon discuss some factors that may influence the decision when plotting means from a three-factor experiment. For now, let's focus on the interpretation of the second-order interaction.

In the left-hand panel of Figure 9.3, we have plotted the interaction of format and instructions for the older adults; we designate this interaction by *FI/O*. In the $2 \times 2 \times 2$ design, it is helpful to think of the three-factor interaction as a contrast of two-factor interactions. For example, in the older adult panel on the left, the advantage of the web over the text format is larger in the argument than in the summary condition. However, the opposite is true in

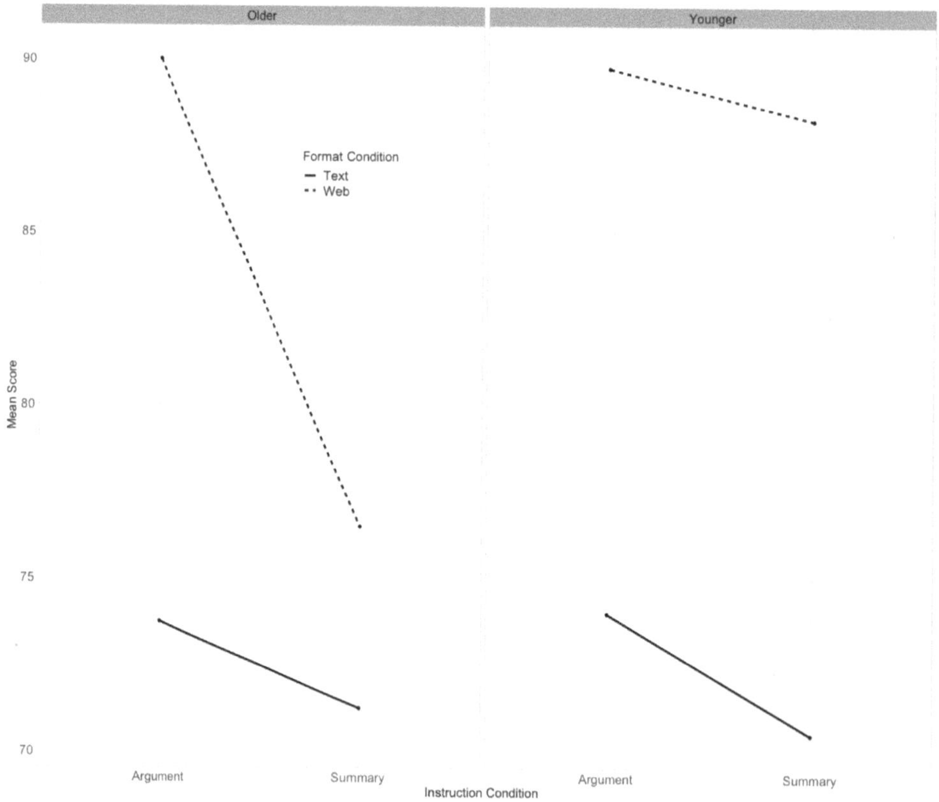

Figure 9.3 A plot of the means in Table 9.11.

the young adult panel on the right; there, the advantage of the web format over the text format is slightly larger under summary than under argument instructions. Looking at the actual means, the *FI* interaction in the older adult condition is

$$I_{FI/O} = (90.00 - 73.75) - (76.50 - 71.25)$$

The corresponding *FI* interaction in the young adult group is

$$I_{FI/Y} = (89.75 - 74.00) - (88.25 - 70.50)$$

The *FIA* interaction is the difference between the two interactions; that is,

$$I_{FIA} = [(90.00 - 73.75) - (76.50 - 71.25)] - [(89.75 - 74.00) - (88.25 - 70.50)]$$

We can rewrite this as

$$I_{FIA} = (90.00 + 71.25 + 74.00 + 88.25) - (73.75 + 76.50 + 89.75 + 70.50)$$

The average of the four terms to the left of the minus sign is 80.875 and the average of the four terms to the right of the minus sign is 77.625. We again have a basis for converting an interaction into a comparison of two means and therefore can test the three-way interaction with either a t test or an F test.

$$t = \frac{80.875 - 77.675}{\sqrt{154\left(\frac{1}{40} + \frac{1}{40}\right)}} = 1.171$$

or

$$F = \frac{(80.875 - 77.675)^2 / \left(\frac{1}{40} + \frac{1}{40}\right)}{154} = 1.372 = t^2$$

These significance tests are both relevant to the null hypothesis that the magnitude of the *FA* interaction does not differ for older and younger adults. Based on 1 and 72 *df*, the F test does not approach significance. We conclude that the three-factor (or "three-way") interaction is not significant. We cannot conclude that the *FI* population interaction differs as a function of the age of the participant. Nor does the *FA* interaction differ as a function of instructions, nor the *IA* as a function of the format. No matter which simple two-factor interactions are contrasted, the numerator of the F ratio always is equivalent to a contrast of the same two sets of four means.

9.4.3 More on 2^3 Interactions

A significant three-way interaction means that the simple interaction effects of any two variables vary as a function of the level of the third variable. Researchers often understand this to mean that whenever the plot of the *AB* combinations looks different at different levels of *C*, the three-factor interaction is likely to be significant. However, plots like the one in Figure 9.3 can be misleading with respect to the three-factor interaction. The following set of means should help convey this point.

	C_1		C_2	
	B_1	B_2	B_1	B_2
A_1	22	11	34	23
A_2	20	14	23	17

Panel (a) of Figure 9.4 presents a plot of the eight cell means under consideration. If these were population means, would you think that there is a second-order interaction? The pattern of means looks different at C_1 than at C_2: The lines cross in the C_1 panel, but not in the C_2 panel. As a result, students usually believe that an *ABC* interaction is present. In fact, if we calculate the interaction contrast, we find it is exactly zero, so there cannot be an *ABC* interaction. For example, calculating the *AB* interaction contrast at each level of *C*, and subtracting,

$$I_{ABC} = I_{AB/C_1} - I_{AB/C_2} = [(22-11)-(20-14)]-[(34-23)-(25-19)] = 0.$$

Sometimes plotting the data in different ways is helpful. In Panel (b) of Figure 9.4, the data from Panel (a) have been replotted. Several points are now clearer than in Panel (a). It should be evident that there is no *BC* interaction, something that was not at all clear in Panel (a). It also appears that there is an *AB* interaction because the difference between the B_1 and B_2 lines is greater in the A_1 panel than in the A_2 panel. Finally, it appears that there is no second-order interaction because the *BC* interaction contrast is zero in both panels. Of course, these are idealized data points, lacking the variability present in real data. However, the point still stands that it is often helpful to plot data in several ways. Different patterns may become evident, making clearer why certain effects in the ANOVA were significant whereas others were not.

The example of Panel (a) of Figure 9.4 demonstrates that the pattern of means can be deceptive. However, some patterns will clearly signal the possibility of a three-factor interaction. If the lines in an *AB* plot are approximately parallel (i.e., there is no *AB* interaction) at one or more levels of *C*, but there is an interaction at least at one other level of *C*, an *ABC* interaction is indicated. Also, if the lines in one panel converge whereas those in other panels diverge, an *ABC* interaction is indicated. If the two *AB* plots are the same (or displaced by a constant amount), except for one point, there is reason to expect a three-factor interaction. Plotting the data several ways could reveal these patterns.

9.4.4 *Using Software for Higher-Order Between-Participants ANOVA With Equal n*

For three-factor between-participants ANOVA, SPSS works exactly as for two-factor ANOVAs (see Section 9.3.6). The only difference is that more factors are included in the "fixed effects" box. In R, the analysis command is also modified only to reflect the additional factor: Rather than the DV ~ A*B for the two-factor design, the three-factor design is DF ~ A*B*C.

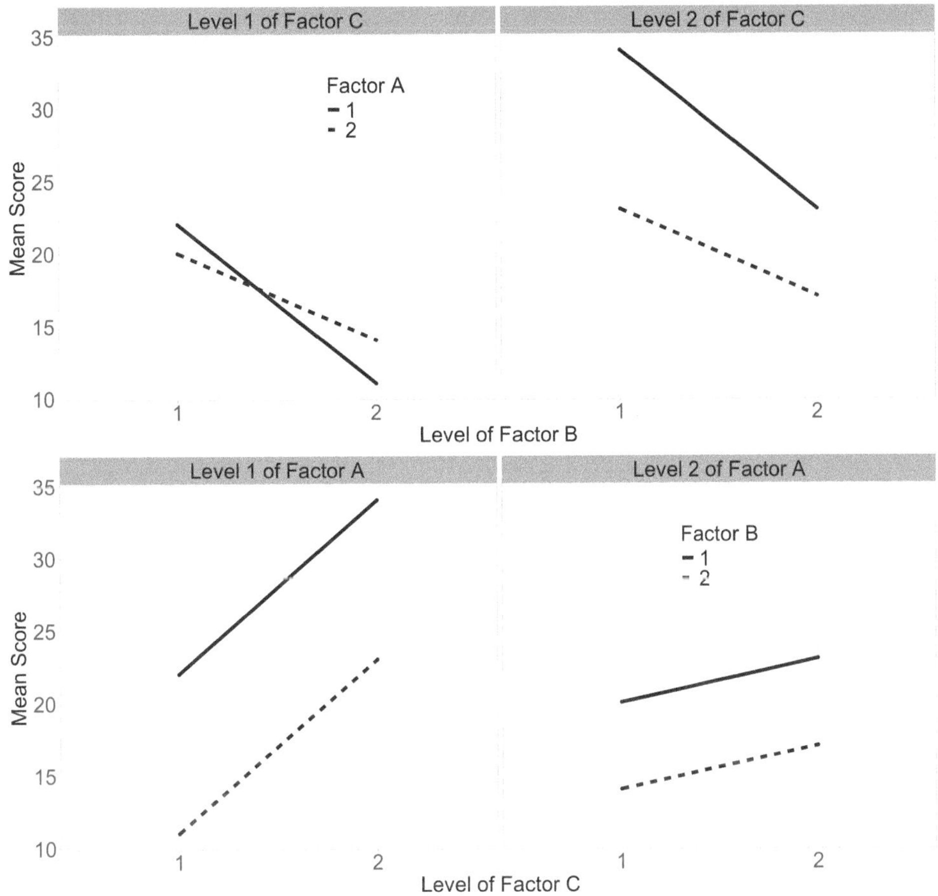

Figure 9.4 Two ways of plotting a three-factor interaction.

9.5 More Than Three Independent Variables

The analyses of data from between-subject designs involving more than three factors are in all respects straightforward generalizations of the material presented for two- and three-factor designs. However, designs with four or more factors become unwieldy in a couple of respects. Each variable and each possible combination of variables is a potential contributor to the total variability, and so is the variability among scores within each cell of the design. As the number of factors increases, the number of possible effects to be tested increase rapidly; with f factors there are $2^f - 1$ possible effects to test. For example, with five factors there are 31 possible effects: 5 main effects, 10 two-way interactions, 10 three-way interactions, 5 four-way interactions, and 1 five-way interaction.

The calculations of higher-order interactions follow directly from what we learned for two- and three-factor interactions. As might be guessed, the degrees of freedom for any higher-order interaction are a product of the degrees of freedom for the variables entering into the interaction. For example, an $ABCD$ interaction would have $(a - 1)(b - 1)(c - 1)(d - 1)$ df.

However, although the calculations are simple, the interpretation of such higher-order interactions is usually difficult. We can say that a significant four-way interaction indicates that the interaction of any three variables is a function of the level of the fourth variable, but that is not very enlightening. Unless we have prior grounds for expecting such interactions to be significant or can attribute the interaction to some subset of cell means, care should be taken before making too much of the result.

9.6 Measures of Effect Size

In Chapter 8, we introduced η^2 and ω^2 as measures of effect size. Both measures assessed the magnitude of an effect against the total variability in the experiment. There are problems with this general approach to measuring effect size when we consider multi-factor designs. Namely, the introduction of more factors into a design will generally result in an increase in total variability. As a result, the assessment of the magnitude of the effect of some factor, A, will decrease as the number of factors in the design increases. In short, the use of η^2 or ω^2 as measures of importance does not allow comparisons across designs with different numbers of factors. This outcome is undesirable because a population effect size should not be influenced by the design of the experiment used to assess it.

One solution that has been offered for this problem is to use either *partial* η^2 or *partial* ω^2 to assess effect size in a design. Partial η^2 is defined as $SS_{Effect}/(SS_{Effect} + SS_{S/cells})$. Partial ω^2 is defined similarly as $\sigma^2_{Effect}/(\sigma^2_{Effect} + \sigma^2_e)$. These measures do not depend on the number of factors in a design because they explicitly exclude other factors from consideration. However, partial η^2 and partial ω^2 do not solve the problem of comparability of measures across designs.

To create effect size measures that are comparable across designs, we must account for the nature of the factors in the design. We will follow the lead of Olejnik and Algina (2003), who proposed a distinction between extrinsic and intrinsic factors in a design (also, see Cohen, 1973). An *extrinsic factor* is a variable that is manipulable and therefore independent of participant characteristics and other factors in a design. For example, the dosage level of a drug or the number of presentations of an item in a memory experiment are examples of extrinsic factors. An *intrinsic factor* is a variable that cannot be manipulated, although it might be measured or controlled (e.g., by blocking on the factor). For example, in a study of close relationships, intrinsic factors might include gender identity, marital status, or family size. The total variance in an experiment varies as a function of the number of extrinsic factors in the design. In contrast, total variance is not influenced by the number of intrinsic factors in a design because those factors contribute variability regardless of whether they are measured or controlled.

The distinction between extrinsic and intrinsic factors implies that the effect size statistic for a factor A should meet two criteria:

1. The statistic should not be affected by the contribution of extrinsic factors.
2. The baseline for assessing an effect size should include both random variance and variability associated with intrinsic factors.

Let us see how these criteria can be applied to the two measures of effect size we have been considering.

9.6.1 *Eta-Squared (η^2) for the Multi-Factor Between-Subjects Design*

The first criterion is readily met. For example, in a three-factor experiment in which all the factors have been manipulated, the statistic would be

$$\eta_p^2(Effect) = \frac{SS_{Effect}}{SS_{Effect} + SS_{S/cells}} \tag{9.11}$$

As we have seen, this statistic is referred to as *partial* η^2. We use the subscript "p" in the notation to distinguish partial η^2 from the classical η^2 defined in Chapter 8. η^2_p is an appropriate measure of effect size in a design that contains only extrinsic factors. It is not appropriate, however, if the design contains intrinsic factors, as in the next example.

Assume that one of the three factors in the experiment is the participant's experience with the task. Factors such as experience, age, and level of ability are intrinsic to the subject so they should be retained in the denominator of our calculation of η^2. Olejnik and Algina (2003) propose a statistic they call *general eta-squared*, which we will notate with a "g" in the subscript to distinguish it from classical and partial eta-squared. In a design with two extrinsic factors (A, B) and one intrinsic, C, η^2_g for A now would be

$$\eta_g^2(A) = \frac{SS_A}{SS_A + SS_C + SS_{AC} + SS_{BC} + SS_{ABC} + SS_{S/cells}} \tag{9.12a}$$

which is equivalent to

$$\eta_g^2(A) = \frac{SS_A}{SS_{total} - SS_B - SS_{AB}} \tag{9.12b}$$

The reasoning behind this equation is that if we were comparing this statistic with one based on a one-factor design, the error variance in the one-factor design would include variability due to the intrinsic factor, C, and its interaction with other factors. Therefore, to have comparable values of η^2 for the two designs, we include the intrinsic sources of variability in the denominator of the statistic for the two-factor design.

Now suppose we wanted a measure of the effect size for the intrinsic factor, C, in a three-factor design. Then,

$$\eta_g^2(C) = \frac{SS_C}{SS_C + SS_{AC} + SS_{BC} + SS_{ABC} + SS_{S/cell}} \tag{9.13a}$$

which can be rewritten as

$$\eta_g^2(C) - \frac{SS_C}{SS_{total} - SS_A - SS_B - SS_{AB}} \tag{9.13b}$$

We may summarize developments as follows:

The denominator of η^2_g includes the sums of squares for the effect of interest, for the within cell error term, and for all intrinsic effects and their interactions.

This rule holds whether the effect of interest is a main or interaction effect. As one further example, consider the design of Table 9.11 in which *Format* and *Instructions* are extrinsic factors and *Age* is an intrinsic factor. If we wanted eta-squared for the *Format* × *Instruction* interaction,

$$\eta_g^2(FI) = \frac{SS_{FI}}{SS_{FI} + SS_A + SS_{FA} + SS_{IA} + SS_{FIA} + SS_{S/FIA}} = \frac{SS_{FI}}{SS_{Total} - SS_I - SS_F}$$

Using the results in Table 9.12, $\eta^2_g(FI)$ = 101.25/(16,215.5 – 551.25 – 151.25) = .007. Clearly, the interaction makes only a small contribution. Suppose that instead of age, our third factor had been some manipulated variable – perhaps the time allowed for studying the material. In that case, the general eta-squared formula reduces to partial eta-squared:

$$\eta_g^2(FI) = SS_{FI} / (SS_{FI} + SS_{S/cells}) = \eta_p^2(FI)$$

and the proportion of variance when all factors are extrinsic is now 101.25/(101.25 + 11088), or .009, slightly larger than before, but still small.

9.6.2 *Omega-Squared (ω^2) for the Multi-Factor Between-Subjects Design*

Although η^2_g is easily calculated and interpreted, it is an overestimate of the population variance attributable to a factor. As we discussed in Chapter 8, $\hat{\omega}^2$ is more satisfactory in that respect. Most commonly, *partial* ω^2 has been reported. In a multi-factor design in which we wish to measure the effect of factor *A*, this parameter would be defined as

$$\omega_p^2(A) = \sigma_A^2 / (\sigma_A^2 + \sigma_e^2)$$

However, we prefer to again follow the approach suggested by Olejnik and Algina (2003). Therefore, we will focus our discussion on *general* ω^2. Because our estimate of ω^2_g requires estimates of population variances, we first define a general formula for the estimate of the variance of any main or interaction effects, assuming that all levels of all variables have been arbitrarily selected; that is, that these are *fixed-effect variables*, regardless of whether they are extrinsic or intrinsic. For such between-participants designs, the general formula is

$$\hat{\sigma}^2_{Effect} = df_{Effect} \times (MS_{Effect} - MS_{S/cells})/N \tag{9.14}$$

where *N* is the total number of scores. For example, for a three-factor between-participants design, the estimate of the variance of the *AB* population interaction effects is

$$\hat{\sigma}^2_{AB} = (a-1)(b-1)(MS_{AB} - MS_{S/ABC})/abcn$$

Using the example of Table 9.11, the estimates of the population variances for the sources listed in Table 9.12 are

$\hat{\sigma}^2_F$	$\hat{\sigma}^2_I$	$\hat{\sigma}^2_A$	$\hat{\sigma}^2_{FI}$	$\hat{\sigma}^2_{FA}$	$\hat{\sigma}^2_{IA}$	$\hat{\sigma}^2_{FIA}$
45.341	4.966	0	0	.325	0	.716

Estimates have been set to zero when calculations based on Equation 9.14 had negative results, and σ^2_e is estimated by $MS_{S/cells}$.

Once the estimates of population variances have been calculated, we can calculate $\hat{\omega}^2$. We will follow the notational conventions introduced for η^2 to distinguish three variants of ω^2; namely, ω^2 without a subscript denotes the classical ω^2 introduced in Chapter 8, ω^2_p denotes partial ω^2, and ω^2_g denotes general ω^2. We need only a slight revision of the rule we formulated for calculating an estimate of general ω^2, $\hat{\omega}^2_g$.

The denominator of $\hat{\omega}^2_g$ includes the estimates of variances for the effect of interest, for the within cell error term, and for all intrinsic effects and their interactions.

Some examples should clarify this rule. Assume that we have three extrinsic factors; say, format, instructions, and time exposed to the material. Then, the estimate *of general* $\hat{\omega}^2$ for format is

$$\hat{\omega}^2_g(F) = \frac{\hat{\sigma}^2_F}{\hat{\sigma}^2_F + \hat{\sigma}^2_e} = \frac{(f-1)(MS_F - MS_{S/cells})/N}{(f-1)(MS_F - MS_{S/cells})/N + MS_{S/cells}}$$

Substituting either the previously calculated estimates of the variances or values of mean squares from Table 9.12, $\hat{\omega}^2_g(F) = .227$.

Now assume the design of Table 9.11 in which one factor, Age, is intrinsic. In that case,

$$\hat{\omega}^2_g(F) = \frac{\hat{\sigma}^2_F}{\hat{\sigma}^2_F + \hat{\sigma}^2_A + \hat{\sigma}^2_{FA} + \hat{\sigma}^2_{IA} + \hat{\sigma}^2_{FIA} + \hat{\sigma}^2_{error}}$$

Now the estimated variance due to experience and all its interactions contributes to the denominator. Substituting numerical estimates,

$$\hat{\omega}^2_g(F) = 45.341/(45.341 + 0 + .325 + 0 + .716 + 154) = .227$$

Because Age and its interactions contribute little variance in this particular experiment, the estimate is unchanged from when we assumed that all factors were extrinsic. In either case, Cohen's suggested guidelines (see Chapter 8) indicate that format has a large effect.

To summarize developments thus far, we have presented an approach to computing measures of effect size that has as a goal comparability across designs. We endorse the approach advocated by Olejnik and Algina (2003) based on distinguishing extrinsic and intrinsic factors or, equivalently, manipulated and measured variables. Although this approach has not been widely adopted, general η^2 and general ω^2 do a better job of achieving the goal of comparability across experimental designs than do the more conventional statistics of partial η^2 and partial ω^2.

9.6.3 Cohen's f for the Multi-Factor Between-Subjects Design

As described in Chapter 8, Cohen (1988) defined f as

$$f = \sigma_{effect}/\sigma_{error} \tag{9.15}$$

for example, based on our previously calculated variance estimates,

$$\hat{f}_{Format} = \sqrt{45.341/154} = .54$$

As with $\hat{\omega}^2$, this is a large effect according to Cohen's guidelines. However, Equation 9.15 does not take into consideration whether the remaining factors are intrinsic or extrinsic. To increase compatibility across different designs, we might define f in a way consistent with the Olejnik and Algina (2003) approach to η^2 and ω^2. As with general ω^2, general f would involve a denominator that incorporates intrinsic effects, in addition to random error.[6] As with η^2_g and ω^2_g, such a statistic would make sense in comparing effect sizes in two experiments, one of which involved a design in which intrinsic factors were controlled and one of which involved a design in which this was not the case.

Despite the advantage of f_g when comparing values of f based on data from different designs, we will limit our use to the classical formula for f in Equation 9.15. The reason for this is that f was viewed by Cohen as a parameter dictating the power of the F test and it is used in this way by G*Power 3.1. Because the denominator of F is based solely on the within-cell error variance, a value of f that includes intrinsic variation in its denominator would underestimate the power of the F test.

9.6.4 *Using Software for Effect Sizes*

In SPSS, it is easy to obtain estimates of η^2_p for each effect in the experiment. When setting up the analysis, simply click on the "Options" button and then select "Estimates of effect size." The output will include η^2_p in the ANOVA table. As we have argued, partial eta-squared values generally are not comparable across different experimental designs, whereas general eta-squared values are comparable. Although it is tempting to report the partial η^2 values readily available in a computer output, investigators should calculate and report values of general η^2 (which must be done by hand). Hand calculations are also required for ω^2_g or f.

In R, the *eta_squared* function in the {effectsize} package reports η^2_p for each effect by default, and it will calculate general η^2 if the intrinsic (measured) variable is identified using the optional input *generalized = "VarName"* (where VarName would be Age in our example). The *omega_squared* function in {effectsize} reports ω^2_p. Note that the *cohens_f* function in {effectsize} computes a sample-specific f that is different from Equation 9.15. See Section 8.5.5 for details.

9.7 *A Priori* Power Calculations

Assume that we wish to replicate the *Wiley–Voss* study. Further assume that our primary interest is whether the instructional effects vary. Cohen's $f = .262$ for instructions in Table 9.7. Suppose we want power = .8 to reject H_0 if the standardized effect is medium, as this value suggests. Figure 9.5 shows the input to G*Power 3.1 and the resulting output. The total N of 179 translates to at least 22 participants per condition, which is more than twice the number in the *Wiley–Voss* experiment.

9.8 Unequal Cell Frequencies

When cell frequencies are not equal, the ANOVA presented so far in this chapter has a problem. In this section, we describe the nature of the problem and briefly introduce the different analyses that are available. Chapter 24 provides more detailed coverage within a regression framework.

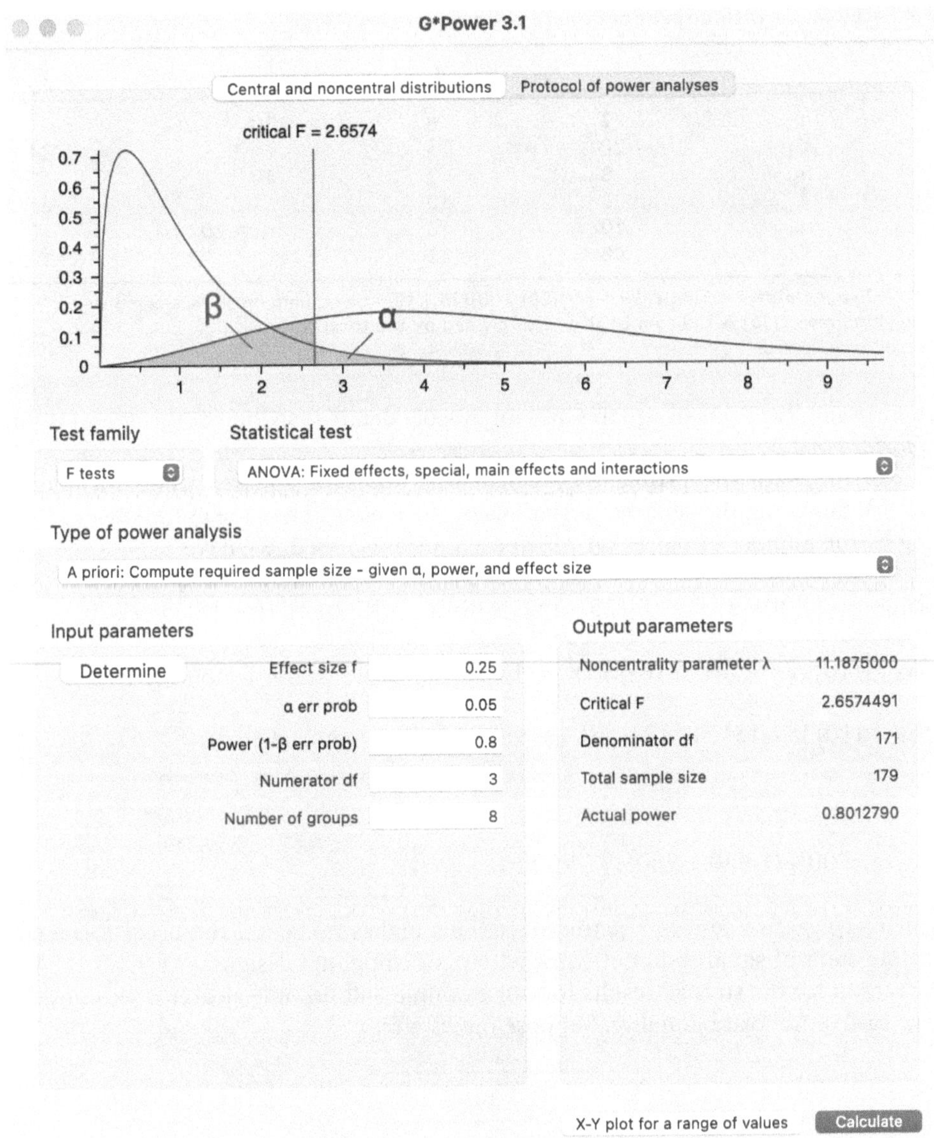

Figure 9.5 G*Power 3.1 screen for *a priori* power in the multi-factor between-participants design.

9.8.1 The Problem

In the developments thus far in this chapter, we considered only cases in which the number of scores is the same in each of the *ab* cells. As we stated in Chapter 6, when *n*s are not equal, heterogeneity of variance influences both Type 1 and 2 error rates. Furthermore, the sums of squares for main and interaction sources usually will not add to the SS_{cells} when each is calculated by ignoring the other effects. This is because when cell sizes are unequal and

Table 9.13 Example with disproportionate cell frequencies

		B_1	B_2	$n_{j.}$	$\bar{Y}_{.j.}$
A_1	n_{1k}	2	8	10	
	$\bar{Y}_{.1k}$	20	25		24
A_2	n_{2k}	8	2	10	
	$\bar{Y}_{.2k}$	5	10		6
	$n_{.k}$	10	10	$n_{..} = 20$	
	$\bar{Y}_{..k}$	8	22		$\bar{Y}_{...} = 15$

Note: $\bar{Y}_{.j.} = \Sigma_k n_{jk}\bar{Y}_{.jk}/n_{j.}$; for example, 24 = [(2)(20) + (8)(25)]/10. The column means are computed in a similar way. The grand mean (15) is the sum of all scores divided by the total *N*.

disproportional, the sums of squares are not independently distributed; the design is said to be *nonorthogonal.*

We can illustrate the problem by calculating sums of squares using the means in Table 9.13. Summing the squared deviations of the means about the grand mean and multiplying by the number of scores on which each mean is based, we have

$$SS_{cells} = (2)(20-15)^2 + (8)(25-15)^2 + (8)(5-15)^2 + (2)(10-15)^2 = 1700$$

$$SS_A = (10)\left[(24-15)^2 + (6-15)^2\right] = 1{,}620$$

$$SS_B = (10)\left[(8-15)^2 + (22-15)^2\right] = 980$$

And

$$SS_{AB} = 1700 - (1{,}620 + 980) = -900$$

Of course, a negative sum of squared deviations makes no sense; the usual formulas for calculating sums of squares do not work with nonorthogonal designs.

The reason for the strange results for our example will become clearer if we consider an extreme case of nonorthogonality. Suppose the *n*s were

	B_1	B_2
A_1	0	8
A_2	8	0

Now SS_A and SS_B are identical; both are based solely on the difference between the A_1B_2 and A_2B_1 cell means, and therefore the *A* and *B* main effects are perfectly correlated. In Table 9.13, the correlation is not perfect, but it is still high. The magnitude of both SS_A and SS_B will still depend primarily on the difference between the A_1B_2 and A_2B_1 means.

Figure 9.6a contains a graphic representation of the situation when cell frequencies are equal and the variability in the experiment partitions neatly into independent (i.e., orthogonal) components. The square represents the SS_{total}, which is partitioned into non-overlapping circles representing SS_A, SS_B, and SS_{AB}, with the remaining area representing SS_{error}. In contrast, Figure 9.6b shows overlapping circles representing the covariance of effects that

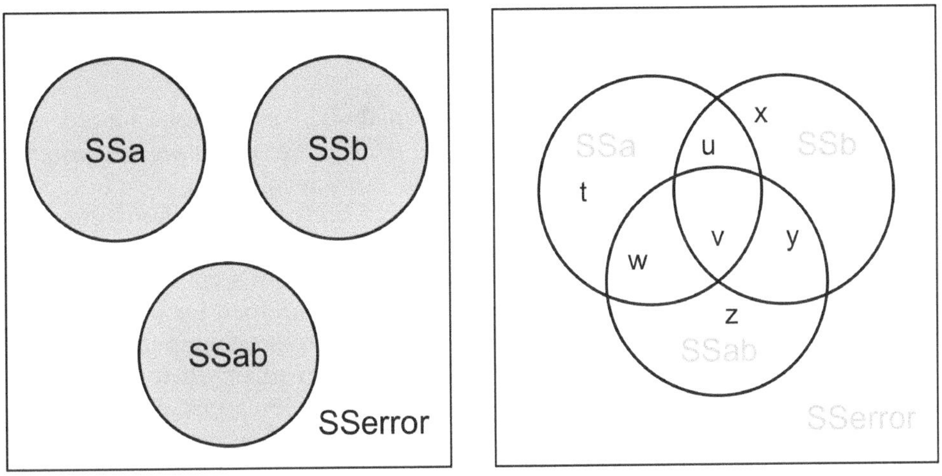

Figure 9.6 The partitioning of variability in a two-factor experiment with (a) orthogonal or (b) nonorthogonal design.

results from nonorthogonality (unequal cell frequencies). In a nonorthogonal design, the variability in the experiment is not partitioned into independent components. Covariances can be positive or negative so that subtracting the covariance from a sum of squares might result in a smaller quantity (if the covariance is positive) or a larger one (if the covariance is negative). The presence of correlations among effects poses choices in data analysis and interpretation.

9.8.2 *Three Types of Sums of Squares*

There are three approaches available in most software packages for analyzing data when effects covary because cell frequencies are unequal. We will briefly describe each.

Type III sums of squares involve adjusting each main and interaction effect for the contributions of the others. For example, the adjusted SS_A would consist only of the area labeled t in Figure 9.6b. The Type III analysis is the default in SPSS. It weights all the cell means equally and therefore is appropriate if it is assumed that the sampled populations are of equal size and the unequal ns reflect chance variation. Type III sums of squares should be calculated when the data come from true experiments in which the independent variables have been manipulated, and the loss of data is due to factors such as the random failure of participants in various conditions to appear for the experiment.

Type II sums of squares involve adjusting an effect of interest for effects at the same or lower order, and for higher-order effects that do not include the effect of interest. In the two-factor design, this requires adjusting the sum of squares for each main effect for variability due to the other main effect, but not for the interaction. In Figure 9.6b, the Type II SS_A would be represented by the areas labeled t and w. This analysis implies that there are no interaction effects. Therefore, the interaction mean square can be averaged ("pooled"; see the next section) with the within-cell error term, yielding more error degrees of freedom and potentially more power. However, such pooling runs the risk of failing to detect interaction effects that may exist and of inflating the error term and thus increasing the rate of Type 2 errors when testing main effects. As a rule, Type II sums of squares should not be calculated unless there is strong *a priori* reason to assume no interaction effects, and a clearly nonsignificant interaction sum of squares.

Type I sums of squares involve a hierarchical analysis. For example, suppose we are interested in the effects of educational level upon income. We might wish to control for other factors such as the industry sector in which the participant is employed (e.g., finance, hospitality, etc.). In that case, we would first remove the sum of squares due to industry sector, and then calculate the sum of squares due to education. In terms of Figure 9.6b, if we represent education by A and sector by B, the sum of squares for sector would correspond to the full circle for SS_B and the education sum of squares, adjusted for sector, would correspond to the areas t and w. Type I sums of squares rest on the assumption that cell sizes represent population sizes, an assumption that is likely to be met when the independent variable is observed rather than manipulated (e.g., income level, occupation, or clinical diagnosis) or when some treatments are more aversive than others (e.g., participants may find some diets more difficult to maintain in a study of cholesterol level).

Although the Type III sum of squares is often the default, all three types of analyses are available in most software packages. In the next section, we dropped scores randomly from the eight cells of the *Wiley–Voss* data set and used software to analyze the data to note the similarities and differences among the three types of sums of squares. Appendix 9.1 details how to obtain different *SS* in both R and SPSS.

9.8.3 A Numerical Example

The data set is linked to the *Wiley–Voss* page on the book's website and has the file name *Wiley_Unequal_N.xlsx*. The means and cell frequencies are in Table 9.14. Table 9.15 presents the results of several analyses, each after choosing a different model.

Table 9.14 Cell means (and frequencies) for an experiment with unequal *n*

Format	*Instructions (A)*			
	A_1 *(Narrative)*	A_2 *(Summary)*	A_3 *(Explanation)*	A_4 *(Argument)*
B_1 (text)	68.000 (5)	75.714 (7)	70.000 (4)	73.750 (8)
B_2 (web)	76.250 (8)	75.000 (2)	76.667 (8)	90.000 (8)

Table 9.15 Sums of squares (*SS*) for data with unequal cell frequencies

Source	*df*	*Type III SS*	*Type II SS*	*Type Ia SS*	*Type Ib SS*
Instructions (*A*)	3	801.57 (*t*)	854.97 (*t* + *w*)	683.77 (*t* + *w* + *v* + *u*)	854.97 (*t* + *w*)
Format (*B*)	1	567.35 (*x*)	1004.54 (*x* + *y*)	1004.54 (*x* + *y*)	833.33 (*x* + *y* + *v* + *u*)
AB	3	368.60 (*z*)	368.60 (*z*)	368.60 (*z*)	368.60 (*z*)
S/AB	40	5409.76	5409.76	5409.76	5409.76

Note: The Type Ia *SS* is obtained when *A* is the first factor listed in the software package; the Type Ib *SS* is the result of entering *B* first. The letters in parentheses refer to the corresponding areas in Figure 9.6b. See the text and Appendix 9.1 for further explanation.

The following points should be noted about the results in Table 9.15:

1. The within-cell error term is the same in all analyses in the table. It is obtained by calculating the sums of squared deviations of scores about their cell means, and then summing across cells; the degrees of freedom are $\sum_j(n_j - 1) = N - a$.
2. The interaction term is also invariant across analyses. In all four cases, it is the sum of squares adjusted for the contributions of all other effects. This is $SS_{Between\ Groups} - (SS_A + SS_B)$ where $SS_A + SS_B$ is obtained from either Type I analysis.
3. For the Type Ia *SS*, the *A* sum of squares was calculated while ignoring all other sources, and the *B* sum of squares was calculated after adjusting for (removing variability due to) *A* effects. For the Type Ib *SS*, the *B* sum of squares was calculated while ignoring all other sources, and the *A* sum of squares was calculated after adjusting for (removing variability due to) *B* effects.
4. The Type Ia sum of squares for *B* equals the Type II sum of squares for *B*. This is because both analyses adjust the *B* sum of squares for variability due to *A*. Similarly, the Type Ib sum of squares for *A* equals the Type II sum of squares for *A* because both adjust the *A* variability for the contribution of *B*.

The set of possible analyses was performed to illustrate the differences in results. However, when analyzing data from an experiment, only one of these analyses should be performed. Which one will depend upon whether variability in cell frequencies is assumed to be due to chance, in which case Type III is appropriate; whether there is strong reason to believe that the interaction effects are negligible, in which case Type II is appropriate; or whether it is assumed that there is a causal relationship among the factors in the study, in which case a Type I analysis is appropriate.

9.9 Pooling in Factorial Designs

Pooling is the process by which two or more mean squares are averaged. This is often done when the investigator believes that some source of variance contributes only error variance and therefore pools that mean square with the error mean square. It is also done unintentionally when the investigator fails to consider a factor in the design. Because pooling affects tests of sources of interest, it merits a closer look.

9.9.1 *What Is Pooling?*

When two or more sources of variance are pooled, the sums of squares are added together and divided by the sum of the degrees of freedom. For example, in a two-factor design, the pool of the *AB* term and the *S/AB* term is

$$MS_{pool} = \frac{SS_{AB} + SS_{S/AB}}{df_{AB} + df_{S/AB}} \tag{9.16}$$

Equation 9.16 can be rewritten as the weighted average of the two mean squares:

$$MS_{pool} = \left(\frac{df_{AB}}{df_{AB} + df_{S/AB}}\right) MS_{AB} + \left(\frac{df_{S/AB}}{df_{AB} + df_{S/AB}}\right) MS_{S/AB} \tag{9.17}$$

This form of the equation raises the question: When is it proper to average two mean squares? The answer is that if the two mean squares estimate the same population variance, or variances, pooling is proper. In the example of Equation 9.17, the assumption is that both MS_{AB} and $MS_{S/AB}$ estimate σ^2_e (i.e., $\sigma^2_{AB} = 0$). The advantage of pooling two or more estimates of the population error variance is that MS_{pool} is distributed on more *df* than $MS_{S/AB}$; therefore, *F* tests based on the pooled error term may be more powerful than tests based on $MS_{S/AB}$. Of course, we never know for certain that $\sigma^2_{AB} = 0$. If it is not, we may lose power in using the pooled error term to test a false null hypothesis about a main effect. To see why, look again at Equation 9.17. If, contrary to the assumption upon which pooling is based, $E(MS_{AB}) = \sigma^2_e + n\theta^2_{AB}$, then the weighted average of this expectation and of σ^2_e will be larger than σ^2_e. As a result, the *F* test of a main effect will be *negatively biased*; the expectation of the error term involves more than just σ^2_e and there will be too many Type 2 errors. Perhaps surprisingly, when the null hypothesis about the main effect is true, there may be an *increase* in Type 1 errors. The reason for this is that if MS_{AB} and $MS_{S/AB}$ do not both estimate σ^2_e, the ratio of MS_A (or MS_B) to MS_{pool} will not be distributed as *F*, and the tail area may be larger than the nominal α. In view of these considerations, it is not clear when, if ever, to pool. We consider this issue next.

9.9.2 When (If Ever) to Pool

One possible approach is to apply a sometimes-pool rule; interaction terms are tested against the within-cell error term, and the two mean squares are pooled if *p* is greater than some criterion value because a large value of *p* is assumed to reflect a clear lack of difference, justifying pooling. With respect to the designs of the current chapter, a study by Mead, Bancroft, and Han (1975) is relevant. For a design with two fixed-effect factors and equal cell frequencies, even with the criterion α for the preliminary test of the *AB* source set at .50, the sometimes-pool rule often resulted in a loss of power when the null hypothesis about the main effect was false, and an increase in Type 1 error rate when it was true. Therefore, in designs in which all factors have fixed effects, we recommend never pooling. We will consider whether the sometimes-pool rule is advisable in other designs when we discuss those designs.

9.9.3 Unintended Pooling

Researchers often pool terms without realizing they have done so. Typically, in such cases there are one or more treatment variables and then one "nuisance" variable that the researcher regards as irrelevant. For example, the position of a reward may appear equally often in all experimental conditions of a discrimination study, or each of several experimenters may run an equal number of participants in each condition. Ignoring such factors essentially pools them with the within-cell error term. As we noted in discussing Mead et al.'s (1975) results, such pooling runs the risk of a loss of power if the null hypothesis is false, or an increased Type 1 error rate if it is true. The message is simple: Include all factors in the analysis, whether they are of interest or not.

9.10 Advantages and Disadvantages of Between-Participants Designs

We will consider further tests of hypotheses in between-participants designs in the next chapter. However, this is a good point at which to review the major advantages and disadvantages of between-participants factorial designs:

1. Assuming equal cell frequencies, the analysis of the data from the between-participants design is much simpler than for most other designs.
2. For any given number of scores, the error degrees of freedom in the analysis of the between-participants design will be larger than for any comparable design.
3. The requirements of the underlying model are more easily met by between-participants designs than by other designs, and violations of assumptions are less likely to affect the distribution of the F ratio.
4. However, there is one major disadvantage in using between-participants designs. Because the error variance is largely due to individual differences, designs that permit the adjustment of the error variance for the differences among individuals often will enable more powerful tests and more precise estimates of population parameters, and often will do so with fewer observations.

Because of the large error variance associated with between-participants designs, these designs are most useful whenever participants are relatively similar with respect to the dependent variable, or whenever a large N is available to compensate somewhat for the variability among individuals. Also, there are many experiments in which the nature of the independent variable (e.g., the method of instruction or the educational level of participants) will constrain the design. However, there are many situations in which several different designs are feasible and desirable. In Chapter 12, we will introduce the idea of design efficiency and consider the efficiency of alternative designs, and in Chapters 13–16, we will consider those designs and their analyses more closely.

9.11 Summary

In this chapter, we considered experiments in which each subject contributed one score to a cell in a design in which all possible combinations of two or more factors were represented. Within this context, we covered the following topics:

- *The extension of the structural model to multi-factor designs.* Building on that model, we partitioned the total variability and degrees of freedom into components representing main effects, interactions, and error sources of variance.
- *The interpretation of interaction.* An interaction occurs when the magnitude of an effect varies over levels of another variable. One of the advantages of a multi-factor design is the ability to examine interactions among variables.
- *Measures of effect size.* We extended measures of importance to multi-factor designs, introducing modifications of the eta-squared and omega-squared measures presented in Chapter 8. In addition, we again considered Cohen's f.

- *Analyses when cell frequencies are unequal.* We explained why this may cause problems for the standard ANOVA, and briefly described alternative ways of calculating sums of squares, depending on what is assumed about population sizes and the presence of interaction, as well as the purpose of the study.

Thus far, we have focused on the omnibus *F* test. However, in most instances, other comparisons of means are of more interest. We turn next to that topic.

Appendix 9.1

Obtaining Type I, II, and III Sums of Squares Using Software

Using R. Several different functions are available for calculating ANOVAs. The *aov* function in the {stats} package calculates Type I sums of squares. For two-factor designs, *aov*(data = DFname, DV ~ A*B) will provide Type Ia SS on the data in the data frame named "DFname" where the dependent variable is called "DV" and the factors are "A" and "B." Using A*B tells *aov* to include both main effects and their interaction. Changing the model to DV ~ B*A will provide Type Ib SS.

Type II sums of squares are the default in the *Anova* function in the {car} package. For two-factor designs, the command *Anova*(*lm*(data = DFname, DV ~ A*B)) returns an ANOVA table of results; using DV ~ B*A will return the same result.

Type III sums of squares are an option in the *Anova* function; either type = 3 or type = "III" can be used. Importantly, we must also specify the effects of interest by using a *contrasts* option, contrasts = list(A = "contr.sum", B = "contr.sum")), which states that we want to compare the cell means of A averaged over levels of B and the cell means of B averaged over levels of A. Contrasts will be described in depth in Chapter 10. The full command, *Anova*(*lm*(data = DFname, DV ~ A*B, contrasts = list(A = "contr.sum", B = "contr.sum")), type = "III"), will return an ANOVA table.

Using SPSS. Choose the *General Linear Model* from the *Analyze* pull-down menu, and then select *Univariate.* Move the dependent variable from the variable list to the box on the right, and move the factor variables to the "Fixed effects" box. Next, click on the "Model" button. At the bottom of the window that pops up is a selector for the Sum of Squares option, which defaults to Type III. Change this option to select Type I or Type II SS, and then click "Continue" and "OK" to run the ANOVA. Note that Type Ia and Type Ib SS depend only on the order in which the factors are listed in the "Fixed effects" box: if Factor A is listed first, then Type Ia SS are computed; if Factor B is listed first, then Type Ib SS are computed.

Exercises

9.1 [Comparing ANOVA and sums of squares calculations]

a) Perform an analysis of variance on the following data set (also available from the Exercises page on the book's website in the file *EX9_1*).

B_1				B_2			
A_1	A_2	A_3	A_4	A_1	A_2	A_3	A_4
24	22	31	28	52	26	20	51
22	14	33	31	25	18	25	40
36	24	43	29	37	17	15	54

b) Estimate the population main effects (the α_j and β_k) and the interaction effects [the $(\alpha\beta)_{jk}$] for the data in part (a).
c) Using the estimated effects, calculate the main and interaction sums of squares. How do the results from part (a) compare?

9.2 [*EMS* and power]

a) State the *EMS* for the data in Exercise 9.1. Coefficients for variance components should be numbers, not letters.
b) Based on the *EMS*, estimate Cohen's *f* for *B*.
c) The number of observations in this study was quite small. Assuming the estimate in part (b), what total sample size (*N*) would be needed to have .8 power to test *B*?

9.3 [Understanding means squares and simple effects] A researcher publishes the following cell means and variances; $n = 10$. We wish to verify the results of the analysis and perform further tests.

Means				*Variances*		
	B_1	B_2	B_3	B_1	B_2	B_3
A_1	2.6	4.3	6.5	2.75	5.00	5.50
A_2	4.3	3.6	3.4	1.75	2.25	3.75

a) Carry out the ANOVA and present the tabled results.
b) Test the simple effects of *A* at B_3.
c) Test the simple effects of *B* at A_2.
d) Briefly justify your choices of error terms for parts (b) and (c).

9.4 [Understanding means squares and simple effects] Suppose an article contains the following table of cell means from a between-participants design with two factors and six scores in each cell:

	B_1	B_2	B_3
A_1	12	16	14
A_2	18	14	12

In addition, the article reports that only the *AB* interaction is significant, $F = 8.0, p = .002$.

a) Reconstruct the entire ANOVA summary table.
b) Test the simple effect of *B* at A_2. What assumption is needed to justify your test?

9.5 [Calculating general eta-squared]

a) Carry out the ANOVA for the data set in the file *EX9_5* on the book's website.
b) Assuming *A* and *B* are both manipulated variables, calculate general η^2 for *B*.
c) Repeat part (*b*), now assuming that *A* represents three different personality types.

9.6 [Calculating omega-squared]

a) For the data set in the file *EX9_5*, calculate general ω^2 for *A*, *B*, and *AB*, assuming that *A* and *B* are both manipulated variables.
b) Repeat part (a), now assuming that *A* represents three different personality types.

9.7 [ANOVA on real data] In the Wiley and Voss (1999) data set (in the *Wiley* file on the website), the variable *causal* represents the number of causal connections participants introduced into essays based on the information obtained. [Note: *format* = 1 and 2 corresponds to *Text* and *Web*, respectively; *instruct* 1 through 4 correspond to *Narrative* (N), *Summary* (S), *Explanation* (E), and *Argument* (A), respectively.]

a) Are the assumptions underlying the analysis of variance met by the *causal* scores? Justify your answer from any relevant statistics and plots.
b) Carry out an ANOVA with *format* and *instruct* as the independent variables.
c) Transform the data by *log*(*causal* + 1). How has this affected the results in parts (a) and (b)?

9.8 [Implications of hypotheses for observed data patterns] Consider each of the following sets of hypotheses. Which sources of variance should be significant? Plot means consistent with the theory.

a) In a bar press experiment, we hypothesize that $Y = K \times D \times P$ where Y = bar pressing rate (the dependent variable), K is a constant, D = hours of deprivation, and P = number of practice trials.
b) In impression formation studies, we give participants some information on the attractiveness (A) and intelligence (I) of an individual and then ask them to rate the individual. We believe that the rating, R, equals $(A + I)/2$.
c) Patients in a mental hospital are divided into experimental groups based on their socioeconomic status (*SES*, three levels) and the kind of treatment they receive (*T*, two levels, psychotherapy and behavior therapy). The investigator predicts that (1) psychotherapy will be less effective than behavior therapy, and (2) psychotherapy will be more effective the higher the *SE* level of the patient, but that this will not be true for behavior therapy. In fact, no effect of *SE* level is predicted for the behavior therapy patients.

9.9 [The relationship of t and F tests for interactions] Assume that 40 participants are divided into good and poor readers based on a pretest. They then read either intact or scrambled text and are tested for their recall. The means and variances (in parentheses) are as follows:

	Text	
Reading ability	*Scrambled*	*Intact*
Good	63 (145)	67 (130)
Poor	37 (155)	54 (170)

a) Construct the numerator of a t test of the *Ability* × *Text* interaction.
b) What is the denominator of that t test?
c) Calculate the t and report the level of significance. What is the value of the corresponding F statistic?

9.10 [Calculating effect sizes from published data]

a) Calculate general η^2 for the *Text* effects in Exercise 9.9.
b) Estimate general ω^2 for the *Text* effects.

9.11 [Consequences of ignoring a systematic variable] The file *EX9_11* at the website contains six groups of five scores in a two-factor experiment. There are three levels of *A*, amount of reward, and two experimenters each ran half of the participants at each level of *A*.

a) Present the ANOVA table, including the *EMS*.
b) Estimate general ω^2 for *A*, treating *A* and *E* as extrinsic variables.
c) Possible effects due to using more than one experimenter are often ignored. Write out the ANOVA table including *EMS*, assuming that *E* has no effect in the population.
d) Estimate ω^2 for *A* for this analysis.
e) Which is the more appropriate model? Why?

9.12 [Interpreting data in three-factor design] Assume that the following numbers represent population means in a three-factor design. Which sources contribute to the variance among the population of means? Explain your answer.

	C_1				C_2		
	A_1	A_2	A_3		A_1	A_2	A_3
B_1	21	27	30	B_1	9	5	10
B_2	17	23	26	B_2	23	19	24
B_3	14	20	23	B_3	22	18	23
B_4	16	22	25	B_4	22	18	23

9.13 [Intrinsic and extrinsic variables; effect sizes] Eighty participants listened to a lecture (*L*) on either the general impact of global warming in Alaska or the impact on wildlife in Alaska. This was then followed by a movie (*M*) on the impact either of global warming in Alaska or on wildlife in Alaska. They then rated their agreement on a series of questions about the environment and global warming. There were equal numbers of participants who identified as male and female. The results of the analysis of variance are as follows:

Source	*df*	*MS*	*F*	*P*
Movie (*M*)	1	.445	7.18	.009
Lecture (*L*)	1	.273	4.40	.039
Gender (*G*)	1	.245	3.77	.056
ML	1	.341	5.50	.022
LG	1	.076	1.17	.283
MG	1	.089	1.37	.246
MLG	1	.142	2.29	.135
S/MLG	72	.065		

a) Calculate general η^2 for the lecture and gender terms.
b) Calculate general ω^2 for the movie term.

9.14 [Relationships among effect sizes, *p*-values, and power] The *Wiley.xlsx* file on the website contains several dependent variables. Here we analyze the *svt* (sentence-verification task) measure.

a) Plot the cell means as a function of instruct and format, including standard error bars.
b) Perform the ANOVA.

c) Calculate the standardized effects (Cohen's f) for the main and interaction effects.
d) Compare the relative sizes of the f values with the p-values for the ANOVA. If there is a difference in ordering, discuss why this has happened.
e) What N would you need to have .8 power to reject H_0 if the *Format* f value was .25 (medium, by Cohen's guidelines)? What N would you need for the *Instruction* effect? Is there a difference? If so, why?

9.15 [Calculating effect sizes and power] Assume that in a multi-factor experiment, $a = 4$, $b = 3$, and $c = 2$, and $N = 96$. Given that $MS_A = 56.8$ and the $F = 2.84$:

a) What is n? $MS_{S/ABC}$?
b) Estimate σ^2 and f^2 for the A effect.
c) Assuming the effect size is medium (i.e., $f = .25$), what is the power to test the A effect (at $\alpha = .01$) if the study is redone with $n = 8$?

9.16 [Generating intuition about effects from data patterns] The file *EX9_16* on the website contains data modeled after that collected in a study by Bless, Bohner, Schwarz, and Strack (1990). These investigators manipulated the *mood* (1 = happy, 2 = sad) of their participants and presented them with a *message* (1 = strong, 2 = weak) designed to influence their attitude about student service fees. Participants' *focus* (1 = content, 2 = language) was also varied. The dependent measure in our data set is the recommended fee (in dollars) after reading the message.

a) Describe the pattern of the eight cell means. What main and interaction effects are suggested by these means?
b) Perform an ANOVA on the data in the file. Discuss the results with reference to the plot of the means in part (a).

9.17 [Consequences of unbalanced design] Following is a data set with unequal ns. The cell totals / ns ($T_{\cdot jk} / n_{jk}$) are

	B_1	B_2
A_1	20/2	40/4
A_2	16/8	4/2

a) Calculate SS_{cells}. The equation is $SS_{cells} = \sum_j \sum_k n_{jk}(\bar{Y}_{\cdot jk} - \bar{Y}_{...})^2$.
b) In the same way you did part (a), calculate SS_A and SS_B. For example, $SS_A = \sum_j n_{j\cdot}(\bar{Y}_j - \bar{Y})^2$. Then subtract these from SS_{cells} to get SS_{AB}. Do you see any problem with this procedure? Explain.
c) Calculate $\hat{\alpha}_1$ and $\hat{\alpha}_2$. This requires finding the marginal (row) means for A_1 and A_2 and subtracting the grand mean. If you have done this correctly, $\sum_j n_{j\cdot}\hat{\alpha}_j = 0$ ($n_{\cdot j}$ is the total n for row j).
d) Subtract $\hat{\alpha}_j$ from each cell mean in row j. Look at the table of means that results. Using these adjusted means, what are SS_B and SS_{AB}? How does this result compare with your answer in part (b)?

9.18 [Consequences of disproportional cell frequencies] Consider the following table of cell means. We will adjust these means for the effects of A under different assumptions about cell frequencies and, by observing what happens to the column means, infer the consequences of equal, proportional, and disproportional cell frequencies for ANOVA. The cell means are as follows:

	B_1	B_2	B_3
A_1	12	8	22
A_2	8	6	13
A_3	1	3	16

a) Assume that the cell frequencies are all the same. (i) Calculate the row and column means. (ii) Calculate estimates of the row effects ($\hat{\alpha}_j$). (iii) Subtract each value of $\hat{\alpha}_j$ from the three cell means in the corresponding row. How do the column means of this adjusted (for *A* effects) matrix compare with those in part (i)?

b) Assume the original cell means presented earlier. This time, however, assume the following cell frequencies (n_{jk}):

	B_1	B_2	B_3
A_1	4	8	12
A_2	3	6	9
A_3	1	2	3

Redo (i)–(iii) of part (a). Be careful in calculating the row, column, and grand means; the values being averaged must be weighted by their corresponding frequency. For example, the mean of the first row is [(4)(12) + (8)(8) + (12)(22)] / 24. Similarly, the grand mean is the weighted average of the 12 cell means (or of the three row or column means).

c) Again assume the original set of means. This time, the n_{jk} are as follows:

	B_1	B_2	B_3
A_1	5	10	4
A_2	5	5	10
A_3	10	3	5

Again do (i)–(iii).

d) Review your answers to this problem and draw a conclusion about the effects of equal, proportional, and disproportional cell frequencies about the partitioning of variability in ANOVA.

9.19 [The utility of *EMS*] Eighty students taking a class in statistics are divided into 20 study groups, each consisting of two women and two men. Ten of the study groups are instructed in class by one method, and the remaining 10 groups are taught by a different method. Much of the analysis is summarized next:

Source	*df*	*MS*	*EMS*
Method (*M*)	1	3627	$\sigma^2_e + 4\sigma^2_{g/m} + 40\theta^2_m$
Sex (*X*)	1	128	$\sigma^2_e + 2\sigma^2_{gx/m} + 40\theta^2_x$
MX	1	123	$\sigma^2_e + 2\sigma^2_{gx/m} + 20\theta^2_{mx}$
Groups/method (*G/M*)	18	420	$\sigma^2_e + 4\sigma^2_{g/m}$
GX/M	18	32	$\sigma^2_e + 2\sigma^2_{gx/m}$
Residual	40	30	σ^2_e

a) Calculate the appropriate F statistics for M, X, and MX. Report whether each is significant.

The investigator reanalyzes the data, pooling the GX/M with the residual error term. The resulting sources of variance are as follows:

Source	*df*	*MS*	*EMS*
Method (M)	1	3,627	
Sex (X)	1	128	
MX	1	123	
G/M	18	32	
Pooled residual			

b) What assumption would justify this reanalysis?
c) In view of your answer to part (a), fill in the *EMS* column. Also provide the *df* and *MS* for the pooled residual.
d) Recalculate the F ratios and report whether each is significant. What are the consequences of the reanalysis?
e) What is the justification for the reanalysis?

9.20 [Using R with real data] The {palmerpenguins} package in R contains a data set called *penguins* (Horst et al., 2020). Install the package and use ?penguins to see the variables included. Three species of penguins were studied over a period of 3 years.

a) Does bill length differ across species? Be sure to consider other possible sources of variance in your response.
b) The sample sizes vary across species and year. Redo your analysis from part (a) assuming different types of sums of squares. Which analysis is most justified?

Notes

1 This assumption means that the selected levels of A are the levels of interest; we cannot generalize our conclusions to other possible levels of A from our statistical analysis. We will consider random effects factors beginning in Chapter 13.

2 These statements are true from a statistical perspective, and conclusions about the presence or absence of an interaction of the factors on the dependent variable (DV) may be made with confidence. However, in psychological research, as well as many other disciplines, the dependent measure provides an indirect way to assess the underlying, *latent*, variable of interest. For example, we may use reaction times (RT) as a measure of memory strength. The function mapping the latent variable to the DV is unknown, and a statistically significant interaction in the DV does not necessarily entail a significant interaction in the latent variable. Caution is needed when drawing conclusions about interactions of the experimental factors on the latent variable. See, e.g., Loftus (1978) and Wagenmakers, Krypotos, Criss, and Iverson (2012) for detailed discussion.

3 Note that the mean square for the interaction is related to the variance of the cell means, but not as simply as is the case for the main effects. The variability associated with both main effects must be removed from the variability in the cell means; this happens at the level of the sums of squares calculations.

4 Strictly speaking, if the null hypothesis is true, we have the central F distribution whose cutoffs are presented in Appendix Table C.5. However, if the null hypothesis is not true, we have members of the noncentral F distribution family. The noncentral F distribution can be used to perform power calculations.

5 Note that the t test computed in this example appropriately uses an error term based on the within-cell variances of the three-factor design. This is not the error term that would be calculated by a software program like SPSS or R if the user were to simply request a t test comparing the web and text formats. Rather, a simple t test that ignores the other two factors in the design would compute an error term that would be inflated by effects involving the two ignored factors. The degrees of freedom provide a good clue to this difference: the df in the t-test will be larger than the denominator df in the F test because in the full ANOVA some df are used by other experimental factors rather than being left for the error term.
6 Let γ be the sum of all variance estimates involving an intrinsic factor; then, general f is $\hat{f}_g(Effect) = \sqrt{\hat{\sigma}^2_{Effect}/(\hat{\sigma}^2_e + \gamma)}$.

Chapter 10

Contrasting Means in Between-Subjects Designs

10.1 Overview

In Chapter 8, we learned how to test the omnibus null hypothesis that the means of all populations sampled in an experiment are equal. Although a significant F provides evidence that the population means are not all equal, it does not reveal the source of the differences among the means. For example, in the study of the effects of educational level on mean depression scores (see Table 8.5), we are left with questions such as the following: Does the mean for the population having only a high school education differ from that for the population having some college education? Does the mean for the population with only a high school education differ from the mean for a population with more than a high school education; that is, from the mean of the other three populations? Within the context of a multi-factor designs we might ask these same questions, and whether such differences depend on the gender identity of the participant.

Answering such questions requires calculating a test statistic and evaluating its significance. The calculations involve minor modifications of the t statistics presented in Chapter 6. Evaluating significance, however, is an issue when several tests are performed on a set of group means. Just as the probability of at least one head increases as the number of coin tosses increases, so does the probability of at least one Type 1 error increase as more significance tests are performed. The set of contrasts tested is referred to as a *family*, and the probability that the family contains at least one Type 1 error is referred to as the *familywise error rate*, or *FWE*. There are many factors that influence the *FWE* and many methods for controlling it.

The primary goals of this chapter are as follows:

- To extend the t test for independent groups to testing contrasts of any form within a between-participants design.
- To introduce a distinction between controlling the probability of a Type 1 error for an individual comparison, *EC* (for "error rate per comparison"), and controlling the probability of a Type 1 error within a family of comparisons, *FWE*.
- To distinguish among several different kinds of families of contrasts and present appropriate methods for controlling the *FWE* for each kind of family.
- To explain the relationship between ANOVA and test of contrasts.

We begin by developing these issues and applications in the context of a single-factor design. Once the procedures for testing contrasts are established, we will extend them to multi-factor designs with emphasis on analyzing interactions.

DOI: 10.4324/9781003453550-12

10.2 Definitions and Examples of Contrasts

In Chapter 6, we reviewed procedures based on the *t* distribution for comparing a pair of means in the context of either testing hypotheses or constructing confidence intervals to estimate the difference between two means. In fact, the procedures introduced in Chapter 6 may be extended to more complex comparisons. In this chapter, we develop procedures for evaluating any type of contrast among means. We begin by defining a contrast.

A *contrast* of population means is denoted by the Greek letter *psi* (Ψ) and is defined as a *linear combination* of the means; that is,

$$\psi = \sum_j w_j \mu_j \tag{10.1}$$

where μ_j denotes a population mean, w_j refers to a numerical weight, at least one w_j is not zero, and $\Sigma_j w_j = 0$.[1] To illustrate, consider the data set of Table 8.1. In this hypothetical experiment, one group was taught to memorize words by the method of loci, a second group was told to form an image of each word, a third group was told to form a rhyme, and the fourth, a control group, was given no special instructions. We might wish to compare the average of the three strategies with the control condition. The comparison of these population means might be represented by

$$\psi_1 = \frac{\mu_{loci} + \mu_{image} + \mu_{rhyme}}{3} - \mu_{control} \tag{10.2a}$$

Rewriting Equation 10.2a to be explicit about the weights on the means and ordering the means according to their sequence in the data file gives

$$\psi_1 = (-1)\mu_{control} + \left(\frac{1}{3}\right)\mu_{loci} + \left(\frac{1}{3}\right)\mu_{image} + \left(\frac{1}{3}\right)\mu_{rhyme}$$

Arguing that the image and loci strategies both involve imaging, the experimenter might ask whether their mean differs from the mean of the rhyme condition; then the contrast is

$$\psi_2 = \left(\frac{1}{2}\right)(\mu_{loci} + \mu_{image}) - \mu_{rhyme} \tag{10.2b}$$

Again, we may rewrite the contrast as

$$\psi_2 = (0)\mu_{control} + \left(\frac{1}{2}\right)\mu_{loci} + \left(\frac{1}{2}\right)\mu_{image} + (-1)\mu_{rhyme}$$

Lastly, the researcher may be interested in potentially more subtle differences between the two imaging conditions:

$$\psi_3 = \mu_{loci} - \mu_{image} \tag{10.2c}$$

Being explicit about the weights on the means, Equation 10.2c may be written

$$\psi_3 = (0)\mu_{control} + (1)\mu_{loci} + (-1)\mu_{image} + (0)\mu_{rhyme}$$

To construct a point estimate of a particular contrast on population means, $\hat{\psi}$, we simply substitute sample means for the corresponding population means:

$$\hat{\psi} = \sum_j w_j \bar{Y}_{.j} \tag{10.3}$$

For the contrasts in Equation 10.2, the point estimates are

$$\hat{\psi}_1 = (-1)\bar{Y}_{control} + \left(\frac{1}{3}\right)\bar{Y}_{loci} + \left(\frac{1}{3}\right)\bar{Y}_{image} + \left(\frac{1}{3}\right)\bar{Y}_{rhyme} \tag{10.4a}$$

$$\hat{\psi}_2 = (0)\bar{Y}_{control} + \left(\frac{1}{2}\right)\bar{Y}_{loci} + \left(\frac{1}{2}\right)\bar{Y}_{image} + (-1)\bar{Y}_{rhyme} \tag{10.4b}$$

$$\hat{\psi}_3 = (0)\bar{Y}_{control} + (1)\bar{Y}_{loci} + (-1)\bar{Y}_{image} + (0)\bar{Y}_{rhyme} \tag{10.4c}$$

Each of these equations estimates a different contrast of the four population means and therefore provides the basis for testing a different null hypothesis. Note that the zero weights in the second and third contrasts are not necessary for purposes of expressing the relevant contrast. However, it will be necessary to provide explicit weights for the means of all conditions when software is used to execute contrast computations, so it is good to get into the habit of providing weights for all conditions.

10.3 Calculations for Hypothesis Tests and Confidence Intervals on Contrasts

In Chapter 6, we presented calculations for the t statistic for two independent groups; that is, for *pairwise comparisons*. In this section we extend those calculations to contrasts involving more than two groups, such as those in Equations 10.4a and 10.4b. We also discuss the selection of weights when *n*s are unequal and extend Welch's t' test of means when variances are heterogeneous to contrasts involving more than two groups.

10.3.1 Calculations When ns Are Equal and Variances Are Equal

A straightforward extension of the t test of Chapter 6 provides a test of contrasts in a one-factor design. We illustrate the test of contrasts using the data from the hypothetical memory experiment; those data were presented in Table 8.1 and are summarized in Table 10.1. The weights in the fourth row of the table are used for a test of the null hypothesis corresponding to Equation 10.2b:

$$H_0 : (1/2)(\mu_{loci} + \mu_{image}) - (1)\mu_{rhyme} = 0$$

Table 10.1 Means and variances of the data in Table 8.1 with contrast weights

	Method			
	Control	*Loci*	*Image*	*Rhyme*
Mean	6.5	12.1	10.7	10.5
Variance	10.056	19.433	25.567	16.722
n	10	10	10	10
Weight	0	.5	.5	−1
Weight × 2	0	1	1	−2

Given the information in Table 10.1, we can now proceed to calculate the *t* statistic. The formula is

$$t = \hat{\psi} / s_{\hat{\psi}} = \frac{\sum_j w_j \bar{Y}_j}{\sqrt{MS_{S/A} \sum_j \frac{w_j^2}{n_j}}} \tag{10.5}$$

Note that the denominator involves $MS_{S/A}$, an average of all four within-group variances. Even though the control group mean is not included in the contrast, including the variance of the control group in the calculation of the error term is justified if we can assume homogeneous variances. The advantage of using $MS_{S/A}$ is that the estimate of error variance will be more precise and our test will be more powerful because the error degrees of freedom are based on four groups rather than three.

For the example of Table 10.1, we will assume homogeneity of variance. Therefore, given Equation 10.5 and the statistics in Table 10.1, we calculate

$$\hat{\psi} = (1/2)(12.1 + 10.7) - (1)(10.5) = 0.9$$

$$MS_{S/A} = (10.056 + 19.433 + 25.567 + 16.722) / 4 = 17.944$$

and so the standard error of the contrast is

$$s_{\hat{\psi}} = \sqrt{17.944\left(\frac{0^2}{10} + \frac{.5^2}{10} + \frac{.5^2}{10} + \frac{(-1)^2}{10}\right)} = \sqrt{2.6916} = 1.6406$$

Finally,

$$t = \frac{\hat{\psi}}{s_{\hat{\psi}}} = \frac{0.9}{1.6406} = 0.548$$

The *df* associated with the *t* test are the *df* on which the estimate of the error variance is based; these are the *df* associated with $MS_{S/A}$, or 36. The choice of an appropriate critical value of *t* will be deferred until we discuss the concept of familywise error rate. However, the computed *t* value of .548 is clearly not significant by any reasonable criterion.

The computations illustrated for the hypothesis test might have been used to compute a confidence interval to estimate the contrast, $(1/2)(\mu_{loci} + \mu_{image}) - (1)\mu_{rhyme}$. In Chapter 6, we learned how to construct an interval estimate on the difference between two population means. The generalization of that procedure to any contrast is

$$\hat{\psi} \pm (t_{critical})(s_{\hat{\psi}}) \tag{10.6}$$

That is, the confidence interval is a point estimate for a statistic plus or minus the product of a measure of variability for that estimate and a critical value. Again, discussion of the selection of a critical value of *t* will be deferred for now. The contrast and standard error are computed the same way as for the hypothesis test, and the standard error is, again, based on 36 *df* in our example.

We have been considering a contrast where some of the weights on the means are fractions (i.e., 1/2 and 1/2). When using software to do such computations, it is sometimes useful to express fractional weights in decimal form (e.g., .5). To avoid a repeating decimal place (e.g., 1/3 converts to .33333. . .), the weights in a contrast can be multiplied by a constant to convert the weights to integers. For example, multiplying the weights in our example by 2 converts the weights to 1, 1, and –2 (see row 4 of Table 10.1). Multiplying the weights of a contrast by a constant has no effect on the value of a *t* test because both the numerator and denominator of the *t* ratio are multiplied by the same constant. However, a change in the weights of a contrast will affect the point estimate for the contrast, $\hat{\psi}$, as well as its standard error, $SE_{\hat{\psi}}$, both of which are multipled by the constant. Therefore, confidence intervals for the contrast are also affected: The point estimate and the width of the interval will be multiplied by the constant. The upper and lower bounds of a confidence interval should be returned to its original scale by dividing the two boundary values by the constant. The reader should compute a *t* test and confidence interval for our example contrast, using the weights 1, 1, and –2 to verify that the value of the *t* ratio is unaffected by the changed weights, whereas the confidence interval boundaries are multiplied by 2.

Most statistical software packages will perform the calculations illustrated in this section and will do so even if group sizes or variances are not equal. For example, SPSS's *One-Way ANOVA* module (in the *Analyze* menu) has a *Contrasts* option that enables the user to select group weights. The process in R is detailed in Box 10.1. To take advantage of available software, the researcher must become comfortable with specifying the weights on each condition mean and with expressing the weights (as integers, in SPSS). Finally, if the interest is in constructing confidence intervals and the contrast weights are all integers, it is important to remember to convert the bounds on each interval back to the original scale.

10.3.2 Weighting Means When ns Are Unequal

When the group sizes differ, a key question is whether the sizes of the populations also differ in the same proportions or whether the populations are equal in size. If we can assume the populations are equal in size and the variability in the *n*s is random, then we compute contrasts and CIs exactly as if the *n*s were equal. However, if the unequal group sizes in the

Box 10.1 Computing Contrasts in R

1. Begin with the data in a data frame. Here, we'll call it *dat*.
2. Create a vector of weights for each contrast of interest. For example, psi1 <– *c*(1, –1/3, –1/3, –1/3) compares the first group against the average of the other three. When there are *a* levels of factor *A*, *a* – 1 contrasts may be tested. For example, if *a* = 4, three independent contrasts are possible.
3. Combine the contrast vectors into a temporary matrix by rows, adding a row in which every weight is 1/*a*: temp.mat <– *rbind*(constant = 1/*a*, psi1, psi2, psi3). The constant row represents the equally weighted grand mean of the dependent variable (dv).
4. Invert the temporary matrix using the *solve* function and assign it a new name: mat <– *solve*(temp.mat).
5. Delete the first column of the matrix: mat <– mat[, –1].
6. Run the analysis using the *lm* or *aov* function, saving the results: model.out <– *lm*(data = dat, dv ~ iv, contrasts = list(iv = mat)). This command applies the contrasts in mat to the independent variable named in iv.
7. Review the results using the *summary* function: *summary*(model.out). The output will be a table of estimated contrast values, $\hat{\psi}$ ("estimate"), their standard errors, and corresponding, *df*, *t*, and *p*-values. The Intercept estimate will equal the grand mean.

sample reflect populations that are different in size, then the weights on the means in the contrast must take into account the variation in population size.

Equal population sizes. Suppose we wish to compare the speeds of solving arithmetic problems in four different grades (e.g., Royer et al., 1999; see the *Royer rt_speed* data file on the book's website). Table 10.2 presents mean variances and class sizes (*n*). Although there are different numbers of students in the different grades, this is likely to be due to chance; that is, there is no reason to view the four sampled populations as unequal in size. Because it is safe to assume the populations are equal in size, any means that are averaged should receive equal weight in a contrast. For example, when testing the average speed of fifth-graders against the combined average of the sixth-, seventh-, and eighth-graders, the weights would be –1, 1/3, 1/3, 1/3 or, equivalently, –3, 1, 1, and 1. Equal weighting of means that are averaged on one side of a contrast usually will also be appropriate in the analysis of data from any true experiment in which the independent variable is manipulated. Using the integer weights, we can use Equation 10.5 to calculate the statistic for the contrast of fifth-graders' mean speed against that of the combined mean of the sixth-, seventh-, and eighth-graders. Using integer weights, you should verify that $t = \hat{\psi} / s_{\hat{\psi}} = .680 / .130 = 5.233$. Because the estimate of variability in the calculation of the standard error is $MS_{S/A}$, the *df* associated with the test are (90 – 4) or 86. Again, we defer selection of a critical value of *t* until our discussion of the *FWE*. Of course, we could also construct a confidence interval on the contrast. If we used the integer weights and resulting values of .680 for the contrast and .130 for the standard error, the upper and lower boundary values would be divided by 3 to return the

Table 10.2 Summary information for the *Royer* response time data

	Grade				
	5	6	7	8	
Mean	0.350	0.560	0.586	0.583	
Variance	0.033	0.028	0.031	0.038	
n	23	26	21	20	N = 90

interval to our original scale. Using fractional weights in R, the resulting estimate of $\hat{\psi}$ is 0.227 and $SE_{\hat{\psi}}$ is .043 (see Box 10.1).

Unequal population sizes. In contrast to most experimental designs, in many observational studies, differences in group sizes reflect differences in population sizes. A case in point is the *Seasons* study, which we cited previously. In Chapter 8, we tested the omnibus null hypothesis that mean depression scores were equal for four populations defined by their education level. The four groups were males with only a high school education (*HS*), some college experience (*C*), a bachelor's degree (*B*), or graduate school experience (*GS*). Panel *a* of Table 10.3 presents group sizes, means, and variances.

Assume that one question of interest was whether the mean depression scores differed between males with a high school education and all other males. Because the relative group sizes suggest that the *HS* population is considerably smaller than the others, we assume that the four populations vary in size. Then the mean of the last three populations, which we will denote as $\mu_{>HS}$ ("greater than high school"), would be a weighted average; that is,

$$\mu_{>HS} = \frac{w_C\mu_C + w_B\mu_B + w_{GS}\mu_{GS}}{w_C + w_B + w_{GS}}$$

and the null hypothesis of interest is

$$H_0 : \mu_{HS} - \frac{w_C\mu_C + w_B\mu_B + w_{GS}\mu_{GS}}{w_C + w_B + w_{GS}} = 0$$

Because the *t* test of a contrast is not affected when all weights are multiplied by a constant, we can simplify things by multiplying the expression by $w_C + w_B + w_{GS}$, yielding the contrast

$$\psi = (w_C + w_B + w_{GS})\mu_{HS} - (w_C\mu_C + w_B\mu_B + w_{GS}\mu_{GS}) \tag{10.7}$$

and we can test the null hypothesis that this contrast equals zero.

Unless we know the actual sizes of the populations, we now need values of the *w*s. In many situations, the simplest and most reasonable will be the group sizes. Panel *c* of Table 10.3 presents output for Contrasts 1 and 2; the weights for the two contrasts are presented in Panel *b*. The weights for Contrast 1 would be appropriate if we assumed that the populations are of equal size; that is not the case in this example, but Contrast 1 provides a comparison with Contrast 2, which uses weights based on the assumption that the group

Table 10.3 Summary statistics (*a*), contrast coefficients (*b*), and test results (*c*), (*d*) for depression scores as a function of educational level

(*a*) Statistics

Educational level	*n*	*Mean*	*Variance*
HS	19	6.903	34.541
C	33	3.674	5.97
B	37	3.331	9.861
GS	39	4.847	26.218

(*b*) Contrast coefficients

Contrast	*Educational level*			
	HS	*C*	B	*GS*
Equal weights	1	−1/3	−1/3	−1/3
Sample-size weights	109	−33	−37	−39

(*c*) Contrast test results (*from R*)

Contrast	*Value of contrast*	*Std. error*	*t*	*df*	*Sig. (2-tailed)*
Equal weights	2.9525	1.0389	2.842	124	0.005
Sample-size weights	318.922	113.204	2.817	124	0.006

(*d*) Welch's t test results

Contrast	*Value of contrast*	*Std. error*	*t*	*df*	*Sig. (2-tailed)*
Equal weights	2.9525	1.394	2.118	20.712	0.047
Sample-size weights	318.922	152.261	2.095	20.712	0.049

sizes reflect the population sizes. In Contrast 2, the weights on the means were calculated with Equation 10.7: The weight on the *HS* mean is the sum of the *n*s for the other three conditions; the weights on *C*, *B*, and *GS* are computed as −1 times the group size. Results are reported when equal variances are assumed and when they are not; we will discuss the calculations for the latter case in Section 10.3.3. For now, note that the assumption about variances can greatly influence the values of t and p.

Although the results for the two contrasts are very similar in this case, equal weighting and weighting by frequency can yield very different results; the distribution of group sizes is the critical factor. The reason results were similar in this example is that if we divide 109, −33, −37, and −39 by 36, we get 3.028, −.917, −1.028, and −1.083 – not very different from 3, −1, −1, and −1, which is equivalent to 1, −1/3, −1/3, and −1/3.

We should point out that the issue of weights arises only with contrasts in which one or both subsets of conditions are based on at least two means. When testing pairwise comparisons (by far the most common situation), the weights will always be 1 and –1 for the two means involved in the comparison, and 0 for all other means.

10.3.3 Testing Contrasts When Variances Are Not Equal

To this point, we have assumed that our treatment populations have the same variance and have therefore used $MS_{S/A}$ in computing the standard error of any contrast. If the assumption of homogeneity of variance is incorrect, we are not justified in using $MS_{S/A}$ because doing so might bias the test. In fact, relatively small differences in the variances of the contrasted groups can inflate the Type 1 error rate.

There are two situations in which $MS_{S/A}$ is an inappropriate error term. In one case, the variances corresponding to the conditions involved in the contrast are very similar but different from the variances of those conditions not included in the contrast (that is, those having zero weight). The standard t is appropriate here, but the denominator should be based only on the variances corresponding to the included conditions. Degrees of freedom are lost, due to the omitted group, but the standard error of the contrast is a valid denominator for the t test. For example, if there are three groups with variances 20, 21, and 5, and the means of the first two groups are to be compared, the variance of 5 should not be included in the denominator because the estimate of the standard error of the contrast will be too small, and the Type 1 error rate will be inflated.

The second situation is one in which there is heterogeneity of variance within the set of means that are to be contrasted. In this case, an extension of Welch's t test (t'; Welch, 1947; see Section 6.6.2) should be calculated. Recall that when variances are assumed to be homogeneous, the condition variances are pooled; however, when variances are assumed to be heterogeneous, the variances are not pooled and should be weighted differently according to the different weights on the corresponding means in the contrast. Also, the df are adjusted downward when unequal variances are assumed. The difference between the standard t results and those for t' are illustrated in Panels c and d of Table 10.3 where two sets of results are presented for each of the two contrasts of the HS depression mean with the mean of the other three groups. One result is obtained assuming equal variances, and the t is calculated as in Equation 10.5. The second result is obtained when equal variances are not assumed.

To obtain the result when equal variances are not assumed, calculate

$$t' = \hat{\psi} / s_{\hat{\psi}} \tag{10.8}$$

where

$$s_{\hat{\psi}} = \sqrt{\sum_{j=1}^{a} \frac{w_j^2 s_j^2}{n_j}} \tag{10.9}$$

and the degrees of freedom are

$$df' = \frac{s_{\hat{\psi}}^4}{\sum_j \frac{w_j^4 s_j^4}{n_j^2 (n_j - 1)}} \tag{10.10}$$

(Notice that these are generalized versions of Equations 6.16 and 6.17.) In summary, as can be seen in Table 10.3, the value of the test statistic depends both on whether the population variances are assumed to be equal and whether the populations are assumed to be equal in size.

10.4 Extending Cohen's *d* to Contrasts

In Section 6.8, we introduced Cohen's d, a measure of the effect when there are two levels of the independent variable. This measure can be extended to contrasts in which several group means are involved. Assuming homogeneous variances, the general form of the standardized effect size for a contrast is

$$d = \hat{\psi} / \sqrt{MS_{S/A}} \tag{10.11}$$

Using the memory data summarized in Table 10.1, and the contrast illustrated there,

$$d = {}^{0.9}\!/\!\sqrt{17.944} = 0.21$$

Thus, the standardized contrast is of small size according to Cohen's (1988) guidelines. Without consideration of confidence bounds, it is difficult to evaluate this statistic, but it does leave open the possibility that power may have been low in the original test.

10.5 The Proper Unit for the Control of Type 1 Error

10.5.1 Defining a Family of Tests

As we stated in the Introduction, the probability of a Type 1 error increases with the number of significance tests. Therefore, if the probability of each significance test is set without regard to how many tests might be conducted, the error rate for the entire collection of tests may rise to an unacceptable level. Statisticians and researchers generally are agreed that the proper unit for control of the Type 1 error rate is not the individual test but a set of contrasts called a *family*. Before we address the question of how to limit the Type 1 error rate for the family, we should clarify the idea of a family of contrasts.

It is useful to distinguish between the *error rate per contrast* (EC) – the probability that a single contrast results in a Type 1 error – and the *familywise error rate* (FWE) – the probability that a set, or family, of contrasts will contain at least one Type 1 error.[2] For a family of K independent tests,

$$\begin{aligned} FWE &= p(\text{at least one Type 1 error in the family}) \\ &= 1 - p(\text{no Type 1 errors in the family}) \\ &= 1 - p(\text{no Type 1 error on a single test})^K \end{aligned}$$

The probability of a Type 1 error on a single test is the EC. Therefore,

$$FWE = 1 - (1 - EC)^K \tag{10.12}$$

If a family consists of six independent tests each conducted at *EC* = .05, substitution in Equation 10.12 results in $FWE = 1 - (1 - .05)^6 = .265$; that is, even if the population means are all equal, the probability is .265 that one or more of the six tests will be significant. If the six tests are not independent, the exact value of *FWE* is difficult to calculate, but it is still greater than .05. In general, the larger the family, the more the *FWE* exceeds the *EC*. This suggests that to control *FWE*, we will need to adjust *EC* downward by an amount that will depend upon the size of the family of comparisons. This line of reasoning requires that we decide how to specify a family of contrasts. We will consider three alternatives.

An investigator working in a research area over a period of years might perform hundreds of experiments and test thousands of hypotheses. We might consider these thousands of tests to form a single family and set *FWE* equal to .05. This is not a reasonable specification of a family of tests, in part because it would result in an *EC* that would be infinitesimally small. In that case, the Type 1 error rate would be well controlled, but the Type 2 error rate would soar to unacceptable levels. That experimenter could be confident that significant results revealed real effects but would miss finding many real effects. Because lowering the *EC* results in a reduction of power, the definition of family must be based on a compromise between concerns about Type 1 and Type 2 errors.

Another, more reasonable, choice for a family of comparisons would be the set of tests conducted to analyze the results of an experiment. This approach to specifying families of comparisons can be criticized on the grounds that more complex experimental designs have more possible tests that could be performed. Simple experiments with few factors entail smaller families of tests than for more complex, multi-factor experiments with many conditions. Defining a family of comparisons in this way is also contrary to the strong arguments in favor of controlling for systematic, but uninteresting, sources of variance that would otherwise end up in the error variance: if those factors count toward the family of comparisons, then the power of the test of interest is limited by the reduction in *EC*.

This brings us to a third and more reasonable approach to specifying families of comparisons; namely, to identify families of tests with sources of variance in an experimental design. For example, in an experiment with three factors, a set of tests to understand the *AB* interaction would constitute one family of comparisons, and a set of tests to understand the main effect of C would constitute another. *FWE* would be controlled independently for the two families.

Identifying a family with the set of comparisons conducted to analyze a single source of variance seems a reasonable approach to defining families of contrasts in a few senses. First, it strikes a balance between control of Type 1 errors and power considerations by keeping the size of the family manageable. Related to this, compared to the two previously considered definitions of a family, this approach results in more consistency in the size of a family. Finally, there is a clear, substantive basis for the definition of a family when a family is identified with a source of variance.

Although the identification of a family of comparisons with a source of variance does a good job of addressing some important issues associated with controlling *FWE*, a researcher generally has many decisions to make regarding which contrasts to perform when analyzing a source of variance. For example, suppose we are comparing the effects of four different drugs on depression. There are six possible pairwise comparisons. In addition, we could compare each one of the drugs with the average of the other three, and with the average of two of the other three. That leads to a total of 22 possible significance tests. Many of the tests may be of interest, but we would suffer a substantial loss of power for each test if we conducted such a large set of tests while controlling *FWE* at a reasonable level. In short, the specification of a family of contrasts should involve a compromise between seeking as much information as possible and keeping the number of contrasts low enough to control *FWE* while still having reasonable power. In view of this, we should think hard about which hypotheses are of interest before we collect the data. We need to focus both the research design and, critically, the power of our significance tests on those questions that are of most interest to us. Whenever possible, use significance tests to ask specific questions of interest rather than general (omnibus) questions.

10.5.2 Different Types of Families of Tests

In the sections that follow, we will describe several different methods for controlling *FWE*. The reason for the different methods has to do with the goals of procedures for controlling *FWE*. Any procedure for controlling familywise error rate attempts to (1) control *FWE* at a specified level (e.g., .05 or .10), while (2) maintaining good power to test individual contrasts. To meet these goals, a method must consider both the size of the set of contrasts and the relations among those contrasts (e.g., independent or correlated). It is therefore useful to distinguish four categories of families of contrasts that differ in both these respects; these distinctions will help to organize the different procedures for controlling *FWE*.

1. *Planned contrasts.* A sensible research strategy is to include only those conditions in an experiment that are of interest, and to plan the contrasts to be tested before collecting the data. Focusing both the design and planned tests on only those contrasts that are important conserves time and effort in data collection and, by limiting the size of the family of contrasts, ensures more powerful tests of the contrasts when the familywise error rate is controlled.
2. *All pairwise comparisons.* Not all research can be planned to the point of specifying a set of contrasts *a priori*. A researcher must therefore be able to conduct tests based on an examination of the patterns of observed means. A simple, principled approach to analyzing a source of variance is to conduct the $(1/2)(a)(a - 1)$ tests of differences between all pairs of group means to try to summarize patterns of effects.
3. *All comparisons with a control condition.* A relatively common experimental design includes a single control condition that serves as a baseline for evaluation of several experimental conditions. Assuming a conditions including the control, there will be $a - 1$ significance tests of interest.
4. *Post hoc contrasts.* Despite thoughtful planning of contrasts, unanticipated differences may appear between means that may require more complex tests than simple pairwise comparisons. For example, it may appear that the mean across three experimental conditions is greater than the mean of a control condition. A researcher would, of course, want to explore those differences. However, the *FWE* is considerably larger in this situation than in the three preceding situations because the size of the family of tests is considerably larger, as we shall see.

To recap, the categories of families differ in size and in the relations among the contrasts that comprise them. Thus, different methods for controlling *FWE* are appropriate for these different types of families. It will be useful to keep in mind that the same t test and confidence interval calculations may be applied in every case that we will consider. All that changes from one situation to another is the method of determining a critical value of t.

10.6 Controlling the *FWE* for Families of *K* Planned Contrasts Using Methods Based on the Bonferroni Inequality

We begin with some very general methods that are applicable whenever several tests are planned. These methods apply not only to tests of differences among means, but also to tests of hypotheses about other parameters, such as proportions or correlations. *Note that it is not necessary that the omnibus* F *be significant prior to testing planned contrasts with the methods described in this section.* In fact, power is lost by requiring a significant F before carrying out planned tests with these methods. What is critical is that the contrasts are decided on before the data are collected and a method for evaluating significance of the tests is used that maintains the familywise error rate at or below a reasonable level, presumably .05 or .10.

Equation 10.12 describes the relation between FWE and EC when the K tests are independent. Because this condition rarely holds, a more general statement of the relation is

$$FWE \leq 1-(1-EC)^K \tag{10.13}$$

In other words, the FWE is equal to *or less than* the term on the right with the inequality holding when the tests are not independent. Furthermore, if K tests are conducted with error rates $EC_1, EC_2, \ldots, EC_K$,

$$FWE \leq \sum_K EC_K \tag{10.14}$$

where EC_K is the probability of a Type 1 error for the Kth contrast. The relationship expressed in Equation 10.14 is known as the *Bonferroni inequality*, and it is the basis for several procedures for testing planned contrasts. From the inequality, it follows that if each of the K contrasts that make up the family is tested at $EC = FWE/K$, the probability of a Type 1 error for the family cannot exceed the FWE. If, for example, the family contains five planned contrasts, FWE will not be larger than .05 if each contrast in the family is tested at the .01 level.

To illustrate methods based on the Bonferroni inequality, we reconsider the memory experiment results summarized in Table 10.1. Table 10.4 contains an analysis of variance of those data, including results of tests of the three contrasts we considered earlier (Equations 10.2a–10.2c). The significance values for the three contrasts in Table 10.4(c) are the ECs. According to this criterion, and assuming $\alpha = .05$, the average of the three experimental methods yields significantly better recall than the control condition (Contrast 1), but neither of the other comparisons are significant. However, the reported p-values do not take into consideration the fact that three tests were performed on the means. Therefore, we will consider how each of two methods controls the FWE for this set of three tests.

10.6.1 The Dunn–Bonferroni Method (Dunn, 1961)

The Dunn–Bonferroni method for controlling *FWE* follows from Equation 10.14. If there are *K* contrasts, the *FWE* will not exceed a nominal value if the *EC* is set at that value divided by *K*. For example, assuming an *FWE* of .05, we test the three contrasts in Table 10.1 at the .0167 (.05/3) alpha-level. The *t* statistics are those reported in Table 10.4; Equation 10.5 provides the formula for the *t* when variances are assumed equal, and Equations 10.8–10.10 provide the formulas for the *t* and *df* when variances are not assumed equal. If the exact *p*-value is available, we can control the *FWE* by comparing *p* with *FWE*/*K* and rejecting H_0 only for those contrasts where $p < FWE/K$. Looking at Table 10.4 for our example, we would evaluate each of our three contrasts at the .0167 level.[3] Equivalently, the *p.adjust* function in R's {base} package, with the method = "bonferroni" option, multiplies an input vector of *p*-values by *K*, where *K* = the number of tests (i.e., the number of *p*-values); the resulting adjusted *p*-values may be compared to any desired α. For example, *p.adjust*(c(.005, .587, .465), method = "bonferroni") returns three adjusted *p*-values (0.015, 1.000, and 1.000), for which only the hypothesis corresponding to the first *p*-value is significant when α = .05.

Table 10.4 Output for tests of three contrasts of means in a memory experiment

(a) ANOVA

	Sum of squares	*df*	*Mean square*	*F*	*Sig*
Method	173.900	3	57.967	3.230	.034
Residuals (Error)	646.000	36	17.944		

(b) Contrast coeffic*ients*

Contrast	*Method*			
	Control	*Loci*	*Image*	*Rhyme*
1	3	−1	−1	−1
2	0	1	1	−2
3	0	1	−1	0

(c) Contrast *tests*

Contrast	*Estimate*	*Std. error*	*t*	*df*	*Sig. (2-tailed)*
Assumes equal variances					
1	−13.80	4.640	−2.974	36	.005
2	1.80	3.281	0.549	36	.587
3	1.40	1.894	0.739	36	.465
Does not assume equal variances					
1	−13.80	3.902	−3.537	21.949	.002
2	1.80	3.345	0.538	20.466	.596
3	1.40	2.121	0.660	17.672	.518

Confidence intervals based on the *FWE* have a form like those we encountered in Chapter 6 (see Equation 6.11); in general,

$$CI = \hat{\psi} \pm t_{FWE/K} s_{\hat{\psi}} \tag{10.15}$$

Consider Contrast 1 in Table 10.4. Inserting the means from Table 10.1, we have

$$\hat{\psi}_1 = (3)(6.5) + (-1)(12.1) + (-1)(10.7) + (-1)(10.5) = -13.80$$

Assuming equal variances, the standard error of this contrast is given in Table 10.4 as 4.640; it is calculated as in the denominator of Equation 10.5:

$$s_{\hat{\psi}} = \sqrt{MS_{S/A} \sum_j w_j^2 / n}$$

The critical value of t, t_{FWE}, can be obtained using the *qt* function in R assuming $df = 36$ and two-tailed $p = 0.0167$ (see Box 6.1): *qt*(.0167 / 2, df = 36, lower.tail = FALSE) returns a t value of 2.51. Alternatively, the critical t can be interpolated from Appendix Table C.6.

Substituting values into Equation 10.15, the confidence bounds for ψ_1 are

$$CI = -13.80 \pm (2.51)(4.640) = [2.154, 25.446]$$

We are not quite done. Recall that our original coefficients were 1, −1/3, −1/3, and −1/3 and we multiplied by 3 to use integer weights. We return to the original scale by dividing the interval bounds by 3, finding .717 and 8.48. We would follow the same procedure for constructing confidence intervals for the other two contrasts in our example.

When confidence intervals are based on the *FWE*, they are interpreted somewhat differently than when they are based on the *EC*. It may help to understand the distinction if we assume many random replications of the memory experiment. In each replication, a set of three confidence intervals, one for each of the planned contrasts in Table 10.1, is calculated. We expect that in 95% of the replications, all three intervals will contain the true (population) value of the contrast. Thus, we are 95% confident that all the confidence intervals in the family are accurate statements; that is, we are 95% confident that all three intervals contain the value of the population parameter being estimated. These intervals based on the *FWE* are referred to as *simultaneous confidence intervals*.

10.6.2 Hochberg's Sequential Method (1988)

Sequential methods (also referred to as stepwise, or multistage) test contrasts in several stages. Some of these methods such as Duncan's (1955) and Newman–Keuls (Keuls, 1952; Newman, 1939) fail to adequately control the Type 1 error rate and therefore will not be considered. A limitation of all sequential methods is that confidence limits cannot be calculated. However, sequential methods confer a power advantage over the Dunn–Bonferroni procedure if the researcher is solely interested in a set of hypothesis tests.

There are several sequential methods for controlling *FWE* for a family of planned hypothesis tests. The simplest of these are Holm's sequentially rejective method (1979), and

Hochberg's step-up method (1988). The Hochberg method is the more powerful of the two; it is described in Box 10.2. The power of this procedure can be increased, but at the cost of greater complexity (Hommel, 1988; Rom, 1990).

Box 10.2 Hochberg's (1988) Sequential Testing Method

1 Rank order the K contrasts according to their p-values with p_1 being the smallest and p_K being the largest. In the example of Table 10.4, the contrasts would be ordered: ψ_2, ψ_1, and ψ_3.
2 If $p_K \leq FWE$, all K null hypotheses are rejected. If not, consider the next largest p-value, p_{K-1}. If $p_{K-1} \leq FWE / 2$, reject the null hypothesis corresponding to the $K - 1$ contrast and all remaining contrasts. If this test is not significant, test whether $p_{K-2} \leq FWE / 3$, and so on.
3 Using the example of Table 10.4 and assuming $FWE = .05$, we first compare .587 with .05. This fails and we compare the next largest p-value, .033, against .05 / 2 or .025. This fails and we compare the last p-value, .005, against .05 / 3, or .0167. This is the only significant contrast, so we can reject only H_0: $\psi_2 = 0$.
4 Equivalently, in R the *p.adjust* function in {base}, with the method = "hochberg" option, multiplies an input vector of p-values by K, where $K = 1$ for the largest p-value, $K = 2$ for the next largest p-value, and so on. These adjusted p-values may be compared to any preferred ψ. For the contrasts in Table 10.4, *p.adjust*(.033, .005, .587, method = "hochberg") returns 0.066, .015, and .587. With $\alpha = .05$, we can reject only H_0: $\psi_2 = 0$.

To summarize, the choice between the Dunn–Bonferroni and the Hochberg procedures depends on whether the researcher wants confidence intervals. The Hochberg method is more powerful when conducting hypothesis tests but does not yield confidence intervals. Both the Šidák and Hochberg methods are available in SPSS, but the results are based on comparisons of members of all pairs of group means.

10.7 Testing All Pairwise Contrasts

10.7.1 The Studentized Range Statistic

We present several methods for controlling *FWE* across a family of pairwise comparisons; all the methods are based on q, *the Studentized range statistic.* This statistic is the range of a set of observations from a normally distributed population, divided by the estimated standard deviation of the population. If the observations are group means,

$$q = \frac{\overline{Y}_{\max} - \overline{Y}_{\min}}{s_{\overline{Y}}} \tag{10.16}$$

where $\bar{Y}_{max}$ and $\bar{Y}_{min}$ are the largest and smallest means in a set of a ordered means and $s_{\bar{Y}}$, the standard error of the mean, is

$$s_{\bar{Y}} = \sqrt{MS_e / n} \tag{10.17}$$

assuming homogeneity of variance and equal *n*s. Critical values of q can be found in Appendix Table C.8 as a function of the *FWE*, a (the number of means), and the *df* associated with the error mean square. Alternatively, R has two distribution functions in the {stats} package for the Studentized range. Analogous to the functions described in Box 6.1, *qtukey* (0.95, nmeans = 4, df = 36) returns the critical value with area .05 above it (3.809 in this example), and *ptukey*(q, nmeans = a, df = $df_{S/A}$) returns the area under the distribution above q.[4]

The Studentized range statistic is closely related to t; the two statistics differ only in their denominators, so it is a simple matter to derive the relationship between t and q. If we were to carry out a t test of the difference between the largest and smallest means, assuming equal variances and group sizes in a one-factor design, the statistic would be.

$$\begin{aligned} t &= \frac{\bar{Y}_{\text{max}} - \bar{Y}_{\text{min}}}{\sqrt{MS_{S/A}(2/n)}} \\ &= q / \sqrt{2} \end{aligned} \tag{10.18}$$

Making explicit the relation between t and q makes it clear that the difference between the two classes of procedures is in the criterion for evaluating contrasts, rather than in the procedures for computing contrasts. Further, the relationship can be useful in comparing the results of procedures based on t (such as the Dunn–Bonferroni) with those based on q and has also been the basis for dealing with unequal *n*s and unequal variances.

10.7.2 Tukey's (1953) HSD Test

Tukey's *HSD* (honestly significant difference) test controls the *FWE* for the set of all possible pairwise comparisons. It is a simultaneous method for testing hypotheses or constructing confidence intervals, meaning that a single critical value is used to evaluate all contrasts in a set. If the procedure is carried out without the help of software, it is helpful to compare differences against a critical difference. Specifically, a critical value of q is selected from Appendix Table C.8 (or using *qtukey* in R), and that value is multiplied by the standard error of the mean to find the critical difference between the two means that must be exceeded for significance. Once the critical difference is computed, it is a simple matter to evaluate any difference among group means. If the differences are ordered from largest to smallest, testing may stop once a nonsignificant difference is found because the remaining comparisons must also be nonsignificant.

A complete example of the procedure is presented in Box 10.3 using the summary statistics for the memory study that were presented in Table 10.1. Box 10.3 presents the necessary steps in testing all pairwise contrasts and for constructing simultaneous confidence intervals. The test is available in several statistical packages. For example, in R the *TukeyHSD* function in the {stat} package takes as input the results of an ANOVA from the *aov* function and returns as output a table of pairwise comparisons, confidence bounds, and exact *p*-values. For the data in Table 10.1, *TukeyHSD*(*aov*(data = dat, Score ~ Method), which = "Method") returns the same results as the method described in Box 10.3, and the *conf.level* option allows selection of different confidence levels.[5]

Box 10.3 Applying Tukey's *HSD* Test to the Memory Data of Table 10.1

Hypothesis tests

1. Order the means in Table 10.1 from smallest to largest:

Method	*Control*	*Rhyme*	*Image*	*Loci*
Mean	6.5	10.5	10.7	12.1

2. From Appendix Table C.8, find the value of q required for significance when $FWE = .05$, $a = 4$, $n = 10$, and $df = a(n - 1) = 36$. That value is approximately 3.81. Alternatively, use the *qtukey* function in the {stats} package in R: *qtukey* (.95, nmeans = 4, df = 36) returns 3.81.
3. Calculate the standard error of the mean using the values of the variance in Table 10.1. Averaging the variance (assuming equal n), $MS_{S/A} = 17.944$, and

$$s_{\overline{Y}} = \sqrt{MS_{S/A} / n} = 1.340$$

4. We can now calculate a critical difference between means as $d_{crit} = SEM \times q_{crit} = 1.34 \times 3.81 = 5.1054$. All pairwise differences greater than this difference will be judged significant. For example, the largest difference (between the control and loci means) is 5.6 and is therefore significant. No other difference is larger than 5.1054, so this is the only significant difference when the FWE is controlled.

Confidence intervals

1. To construct confidence intervals, find the critical value of q and the standard error of the mean as earlier.
2. For any particular comparison of conditions, compute the difference between the means and compute an interval by: $\hat{\psi} \pm q_{(1-confidence),a,df} s_{\overline{Y}}$. For example, the 95% confidence limits on the difference in mean recall of the control and loci populations are

$$5.6 \pm (3.81)(1.34) = .495, 10.705$$

In many studies, only a few of the possible pairwise comparisons will be of interest. In such cases, is the Tukey test or the Dunn–Bonferroni procedure the better choice for controlling *FWE*? Dunn (1961) demonstrated that when all possible pairwise comparisons are tested, the Tukey procedure has the narrower confidence interval and is the more powerful test. However, the advantage of the Tukey procedure declines as the *FWE* decreases, the *df* increases, or *K* decreases. Furthermore, if only a subset of all possible pairwise comparisons are planned and tested, a point is reached at which the Dunn–Bonferroni procedure is more powerful. For example, if there are four groups but only four or fewer of the possible six comparisons are tested, the Dunn–Bonferroni method requires a smaller value of *t* for significance than does Tukey's method.

If the researcher plans to test a subset of all possible pairwise comparisons, the relative power of the Dunn–Bonferroni and Tukey methods is easily assessed by comparing the critical values of the two procedures when both are expressed as *t* statistics: Calculate the ratio of critical values of the Dunn–Bonferroni to the Tukey method:

$$D-B \text{ to } Tukey\ ratio = \frac{t_{\text{FWE},K}}{q_{\text{FWE},K} / \sqrt{2}} \tag{10.19}$$

where *K* is the number of comparisons, and the *t* and *q* statistics are the values required for significance when *K* comparisons are made. When the ratio is less than 1, the Dunn–Bonferroni requires a smaller critical value and will therefore be the preferred method.

In summary, if the researcher carefully plans only those pairwise comparisons that are truly of interest, power may be gained by using methods that focus only on the planned comparisons. Equation 10.19 provides a basis for deciding between the methods.

10.7.3 The Fisher–Hayter Test

The tests considered so far do not require a preliminary test of the omnibus null hypothesis; they control the *FWE* at or below its nominal level. In fact, requiring a significant omnibus *F* prior to these tests is likely to result in a loss of power. However, there are procedures for controlling the *FWE* that include an omnibus *F* test as a first stage; pairwise comparisons are conducted only if the initial *F* test is significant. In Fisher's (1935) *LSD* (least significant difference) procedure, the pairwise comparisons are tested by the usual *t* test at the .05 level in a second stage. However, because the *LSD* test has been shown to have an inflated *FWE* for $a > 3$, Hayter (1986) modified the test. The resulting Fisher–Hayter test maintains the *FWE* at or below its nominal level. In this test, a significant *F* test in the first stage is followed by tests of all pairwise comparisons using a standard *t* test, but each test is evaluated against the criterion $q_{FWE,a-1,df_e} / \sqrt{2}$. Note that in entering Appendix Table C.8, the column corresponding to $a - 1$ means provides the critical value, and if using *qtukey* in R (see Section 10.7.1), the *nmeans* parameter should also be $a - 1$.

Note that the Fisher–Hayter test is a sequential procedure involving two steps. As with other sequential testing methods, power is gained relative to simultaneous tests but at the loss of the ability to construct simultaneous confidence intervals. Therefore, the choice between the Fisher–Hayter test and the Tukey (or Tukey–Kramer) test depends on whether such confidence intervals are desired. Another consideration is whether variances are assumed to be homogeneous. The Fisher–Hayter test was derived under that assumption and therefore a test such as the Games–Howell or the Dunnett T3 should be used if homogeneity of variance is in doubt.

10.7.4 When ns Are Unequal: The Tukey–Kramer Test

Tukey's *HSD* procedure assumes an equal number of participants in each condition. However, it is common for n to vary across conditions. For example, in the *Royer* study, the number of students in the fifth–eighth grades varied; see Table 10.2 for the group means, variances, and *n*s. A modification of Tukey's *HSD* test suggested by Kramer (1956) applies in such situations. The standard t statistic is calculated and compared with $q_{FWE,a,df_e} / \sqrt{2}$. Box 10.4 illustrates the test as applied to a comparison of the fifth- and sixth-grade mean speeds.

Box 10.4 An Example of the Tukey–Kramer Test When *ns* Are Not Equal

1. Find the critical value of q. In the example of the *Royer* speed data, $a = 4$ and the error df are 86. If $FWE = .05$, we can use the *qtukey*(.95, nmeans = 4, df = 86) to find the critical q value is 3.705. [a]
2. Calculate the critical value of $t : t = q_{FWE,a,df_e} / \sqrt{2} = 3.710 / 1.414 = 2.623$.
3. To test the difference between the fifth- and sixth-grade mean speeds, calculate the usual t statistic presented in Chapter 6 (with s_{pooled} replaced by $MS_{S/A}$). The Tukey–Kramer t equals the mean difference $\left(\hat{\psi}\right)$ value divided by the std. error $\left(s_{\hat{\psi}}\right)$;

$$t = \frac{\bar{Y}_{Grade6} - \bar{Y}_{Grade5}}{\sqrt{MS_{S/Grade}\left(\frac{1}{n_6} + \frac{1}{n_5}\right)}} = \frac{.560 - .350}{\sqrt{(.032)\left(\frac{1}{23} + \frac{1}{26}\right)}}$$

$$= .21 / .051 = 4.101$$

which clearly exceeds 2.623.

[a]Harter, Clemm, and Guthrie (1959) provide a method of nonlinear interpolation when the exact df are not in the table and R is not available. For example, when the $df = 86$, find $q_{.05,4,60} = 3.74$ and $q_{.05,4,120} = 3.69$ from the table, and the reciprocals of 86 (.0116), 60 (.0167), and 120 (.0083). The critical value for $df = 86$ is then given by

$$q_{.05,4,86} = 3.74 - \left(\frac{.0167 - .0116}{.0167 - .0083}\right)(3.74 - 3.69) = 3.710$$

10.7.5 When Variances Are Unequal

Both the Tukey *HSD* test and the Tukey–Kramer test assume equal variances across treatment conditions so an alternative test is needed when the homogeneity of variance assumption is violated. In Chapter 8, in our analysis of the effects of educational level upon mean depression scores of males in the *Seasons* study, we found that the variances were quite heterogeneous. Several methods have been proposed to deal with this problem. Most use Welch's t' and df' (see Section 10.3.3) but differ in the criterion against which t' is evaluated. In the Games and Howell test (1976), t' is compared with $q_{FWE,a,df'}/\sqrt{2}$. The procedure is illustrated in Box 10.5 using the *Seasons* depression data.

Box 10.5 The Games–Howell Procedure for Testing All Pairwise Comparisons When Variances Are Not Equal

1. Compute Welch's t' and df', using Equations 6.15 and 6.16. For the *HS* and C statistics of Table 10.3, Panel *a*, we have:

$$t' = \frac{\bar{Y}_{HS} - \bar{Y}_C}{\sqrt{\frac{s_{HS}^2}{n_{HS}} + \frac{s_C^2}{n_C}}} = \frac{6.903 - 3.674}{\sqrt{\frac{34.541}{19} + \frac{5.970}{33}}} = 2.284$$

and

$$df' = \frac{\left(\frac{s_{HS}^2}{n_{HS}} + \frac{s_C^2}{n_C}\right)^2}{\frac{s_{HS}^4}{n_{HS}^2(n_{HS}-1)} + \frac{s_C^4}{n_C^2(n_C-1)}} = \frac{\left(\frac{34.541}{19} + \frac{5.970}{33}\right)^2}{\frac{34.541^2}{(19^2)(18)} + \frac{5.970^2}{(33^2)(32)}} \approx 22$$

2. Obtain the critical value of t from Appendix Table C.8 or R's *qtukey* function. For our example with 22 *df*, interpolate in Appendix Table C.8 between $df = 20$ and $df = 24$, with $a = 4$, $FWE = .05$. The critical q value is approximately 3.93. Then

$$t_{.05,4,22} = 3.93/\sqrt{2} = 2.779$$

3. Because $2.284 < 2.779$, we cannot reject H_0: $\mu_5 = \mu_6$. In similar fashion, values of t' and df' can be calculated for each of the remaining five pairwise comparisons. Note that the critical value of t must be recalculated for each test because the df' are likely to change for each comparison.
4. Alternatively, use the *games_howell_test* function in the {rstatix} package in R to see all pairwise comparisons simultaneously. For example, when the data from Table 10.3 are in a data frame called *dat*, the command *games_howell_test*(dat, mean_d ~ educ.level, detailed = TRUE) will return all 6 pairwise comparisons, including each t', df', adjusted p-value, and confidence interval, The *conf.level* option allows selection of difference confidence levels, and thus *FWEs*.

The Games–Howell method generally does a good job of controlling *FWE* with reasonable power. However, there are some circumstances under which *FWE* may be a bit inflated with Games–Howell. If the variances are fairly homogeneous and the group sizes are less than 50, the *FWE* for the Games–Howell method may sometimes be as high as .07 when the nominal probability is .05 (Dunnett, 1980; Games, Keselman, & Rogan, 1981). Even when the variances are not homogeneous, the *FWE* may be inflated with *n*s less than 6. However, the *FWE* is close to the nominal level under most other conditions. Furthermore, the test is more powerful than any of the several competitors that have been proposed and has narrower confidence intervals for each comparison. If the researcher is concerned about the possible inflation of the *FWE*, Dunnett's T3 test (Dunnett, 1980) appears to be the most powerful of several alternatives that maintain the *FWE* at less than or equal to the nominal value. The test requires tables of the Studentized maximum modulus distribution. Miller has described the procedure (1981, pp. 70–75) and has provided tables of the distribution. The test is also available in several statistical packages; in R, the *dunnettT3Test* function in the {PMCMRplus} package will provide the adjusted *p*-values for all pairwise comparisons. For the data summarized in Table 10.3, none of the comparisons are significant by this test.

10.7.6 Pairwise Comparisons: Summing Up

The rather detailed set of recommendations for controlling *FWE* over families of pairwise comparisons results from an attempt to satisfy two criteria. The first is that we want a method that adequately controls the Type 1 error rate over the family. We have seen in previous chapters that two factors that often affect Type 1 error rates are heterogeneous variances and unequal *n* across conditions. We again find that the same two factors require adjustments of our procedures for controlling familywise error rates.

Given that we can identify alternative procedures that satisfactorily control Type 1 error rates, our second consideration is to choose the procedure that has the most power. Seaman, Levin, and Serlin (1991) simulated tests of all pairwise comparisons under conditions of equal *n*s and equal variances. Over the conditions examined in their study, Seaman et al. found that the Fisher–Hayter method had a power advantage over the Tukey *HSD* method that varied between 2% and 9%; Tukey, in turn, had a power advantage of 2%–3% over the Dunn–Bonferroni method. The other distinction among procedures that is relevant to power concerns is the distinction between simultaneous and sequential methods. Sequential tests have more power than simultaneous tests, although confidence intervals are not available when sequential methods are used.

Box 10.6 summarizes our recommendations with respect to the control of *FWE* over families of pairwise comparisons.

Box 10.6 Recommendations for Controlling *FWE* on Families of Pairwise Comparisons

1. If the researcher wants to construct confidence intervals to estimate the differences between group means:

   a) If the variances are homogeneous and *n*s are equal, use Tukey *HSD*.

   b) If the variances are unequal, use Games–Howell (or Dunnett T3 if the *n*s are less than 6).
   c) If the *n*s are unequal, use Tukey–Kramer.
   d) If only a subset of all pairwise comparisons are planned, use Equation 10.19 to determine whether Tukey *HSD* or Dunn–Bonferroni will have more power.

2. If the researcher only wants to conduct hypothesis tests:

   a) If the variances are homogeneous, use Fisher–Hayter.
   b) If the variances are unequal, use Games–Howell (or Dunnett T3 if the *n*s are less than 6).

We have excluded from our discussion of pairwise comparison procedures that are sometimes used; namely, Fisher's *LSD* test (1935), the Student–Newman–Keuls test (Keuls, 1952; Newman, 1939), and Duncan's multiple range test (1955). We advise against the use of these procedures because they yield *FWE*s that often are considerably greater than the nominal value. We have also excluded several procedures that maintain the *FWE* at or below its nominal level and have slightly more power than the Fisher–Hayter method under some combinations of number of groups and group size (e.g., Peritz, 1970; Ramsey, 1978, 1981; Shaffer, 1979, 1986; Welsch, 1977). Based on various sampling studies, the very slight power advantage of these methods (usually 1% or 2%) does not warrant the added complexity they usually entail. Descriptions of these methods, together with results of sampling experiments, may be found in the article by Seaman et al. (1991); multiple comparison procedures are also reviewed by Zwick (1993), Shaffer (1995), and Toothaker (1993).

Finally, we note that most of the pairwise comparison procedures we have considered, as well as the Dunn–Bonferroni and Dunn–Šidák tests, are available in various statistical software packages. Although it is tempting to run several of these tests, we urge researchers to select one procedure in advance and base conclusions on the results of that test.

10.8 Comparing *a* − 1 Treatment Means With a Control: Dunnett's Test

Dunnett (1955, 1964) proposed a test for studies in which the researcher plans to contrast each of several treatments with a control. If these are the only comparisons of interest, methods that control the *FWE* for a family consisting of *all* pairwise comparisons will be overly conservative; power will be lost and simultaneous confidence intervals will be wider than necessary. The Dunn–Bonferroni procedure with $K = a - 1$ will be an improvement but will still offer less power and wider intervals than the Dunnett test.

Assuming that the group sizes are equal and that variances are homogeneous, the test is quite simple. Box 10.7 illustrates the procedure using the data from the memory study. However, if the group sizes are not equal, replace $2/n$ in the equation for t by $1/n_j + 1/n_C$, and use the Dunn–Bonferroni procedure with $K = a - 1$. If any of the a group variances differ, use Welch's t' and again use the Dunn–Bonferroni procedure.

Box 10.7 Dunnett's Test Comparing the Experimental Means With the Control Mean in the Memory Study

1. Compute the usual t statistic comparing the control with each experimental group; e.g., to compare the control and loci means

$$t = \frac{\bar{Y}_{Loci} - \bar{Y}_C}{\sqrt{MS_{S/A}\left(\frac{2}{n}\right)}} = \frac{12.1 - 6.5}{\sqrt{17.944\left(\frac{2}{10}\right)}} = 2.96$$

and for the comparison of the control group mean with the rhyme and image means, $t = 2.11$ and 2.22, respectively.

2. Evaluate the three t statistics against the critical value of $d_{FWE,a,df}$ in Appendix Table C.8, where a is the number of means including the control and df is the number of degrees of freedom associated with the ANOVA error term. In the present example, FWE (two-tailed) =.05, $a = 4$, and the error $df = 36$; the critical value is 2.48. Only the control and loci means differ significantly. At this time, there is no working distribution calculator function in R for the critical values of Dunnett's d.
3. The confidence intervals have the same form as in the two-independent-group examples of Chapter 6; the bounds are

$$\left(\bar{Y}_j - \bar{Y}_C\right) \pm s_{\hat{\psi}} d_{FWE,a,df}$$

For example, for comparison of the loci mean with the control mean, the bounds are

$$(12.1 - 6.5) \pm \sqrt{17.944\left(\frac{2}{10}\right)} \times 2.48 = .90, 10.30$$

4. Alternatively, the *DunnettTest* function in the {DescTools} package in R will provide the differences, confidence bounds, and *FWE*-adjusted *p*-values. For example, if the memory data are in a data frame called *dat*, then *DunnettTest*(data = dat, Score ~ Method, control = "Control") returns a confidence interval of [0.95, 10.25] for the difference in the loci and control means. The small differences from the CI computed by hand are likely due to rounding and approximation errors.

10.9 Controlling the Familywise Error Rate for Post Hoc Contrasts

Sometimes observed patterns in the data suggest the presence of effects that had not been anticipated and that are not adequately captured by the set of all possible pairwise comparisons. When the corresponding null hypotheses are tested to determine whether these effects are significant, we should be quite conservative in evaluating the result. In testing contrasts "after the fact" we are capitalizing on chance and, in effect, investigating the family of *all* possible outcomes. Therefore, the methods we present are quite conservative because they control for the probability of at least one Type 1 error in a very large set of possible contrasts.

10.9.1 Scheffé's Method

Assuming that the populations are normally distributed and have equal variances, Scheffé's (1959) method maintains the *FWE* at its nominal level when the family consists of all possible contrasts associated with a source of variance. Using the fifth- to eighth-grade multiplication speeds in the study by Royer et al. (1999) as an example, assume that we had not anticipated the pattern of means in Table 10.2. After viewing the data, we observe that the sixth–eighth grades had very similar means, each higher than the fifth-grade mean. We might wish to test whether the mean of the fifth-grade response times differs significantly from that of the three combined sixth- to eighth-grade times. Box 10.8 describes the Scheffé procedure and illustrates its application to the contrast of the fifth-grade mean with the average of the other three means.

Box 10.8 Scheffé's Method to Test H_0: $(1/3)(\mu_6 + \mu_7 + \mu_8) - \mu_5 = 0$ (Royer Speed Data)

1. Calculate the *t* statistic to test the contrast of interest (see Equation 10.5).
2. Compare the computed value of *t* with $S = \pm\sqrt{df_1 \cdot F_{FWE,df_1,df_2}}$ where df_1 and df_2 are the numerator and denominator degrees of freedom.
3. For the arithmetic experiment, $df_1 = 3$ and $df_2 = 36$; if $FWE = .05$, the critical *F* is approximately (from Appendix Table C.5) $F_{FWE,df_1,df_2} = 2.88$. Therefore,

$$S = \pm\sqrt{(3)(2.88)} = 2.94$$

4. Reject the null hypothesis if $t > S$ or $t < -S$. To test the null hypothesis, $t = 5.23$. Because $5.21 > 2.94$, we reject H_0.
5. The formula for the confidence interval bounds is

$$\hat{\psi} \pm s_{\hat{\psi}}\sqrt{df_1 \cdot F_{FWE,df_1,df_2}}$$

where $\hat{\psi} = .680$ and $s_{\hat{\psi}} = \sqrt{MS_{S/A}\left(\Sigma w_j^2 / n_j\right)} = .130$. Therefore, the bounds on $(\mu_6 + \mu_7 + \mu_8) - 3\mu_5$ are $.680 \pm (.130)(2.94) = [.298, 1.062]$.

6. To return to the original scale, these bounds must be divided by 3; the bounds on $(1/3)(\mu_6 + \mu_7 + \mu_8) - \mu_5$ are .099 and .354.
7. Alternatively, use the *ScheffeTest* function in the {DescTools} package in R. It takes an *aov* model as input, as well as a vector or matrix of the contrast or contrasts of interest, and returns $\hat{\psi}$ and confidence interval bounds. For example, *ScheffeTest*(*aov*(data = dat, M_Speed ~ grade), contrasts = list(grade = *c*(−1, 1/3, 1/3, 1/3))) provides the same values as in step 6.

It is instructive to compare the confidence interval presented in Box 10.8 with the results we would have obtained if our contrast had been planned. Assume that the contrast was one of three planned for the experiment. In that case, we could have used the Dunn–Bonferroni method to compute the confidence interval. In contrast to the interval limits in Box 10.8, the Dunn–Bonferroni limits are

$$\hat{\psi} \pm t_{FWE/K} s_{\hat{\psi}}$$

The contrast is .680 and its standard error is .130 if integer weights are used (see Box 10.8), and the t required for significance at the .05/3 = .0167 level (two-tailed) is 2.51. Substituting these values (and dividing the resulting limits by 3 to return to the original scale), we find the Dunn–Bonferroni limits to be .118 and .335. The Dunn–Bonferroni interval is narrower than the Scheffé interval in Box 10.8, revealing the price we pay in precision of estimation and power when contrasts are not planned. Whenever possible, it is a good strategy to plan all contrasts that might conceivably be of interest, and then use the Dunn–Bonferroni or Fisher–Hayter method. Although the power of these methods decreases as the number of planned contrasts increases, a rather large number of comparisons must be planned before the Scheffé criterion requires a smaller value of t for significance (see Perlmutter & Myers, 1973, for a more detailed comparison of the Dunn–Bonferroni and the Scheffé methods).

Experimenters who have used both the standard ANOVA tests and the Scheffé procedure have sometimes been surprised to find that the omnibus null hypothesis is rejected by the ANOVA test but that no contrasts are significant by the Scheffé criterion. The source of this apparent contradiction is that the overall F test has the same power as the *maximum possible contrast* tested by the Scheffé procedure. That contrast may be of little interest, so it may not have been tested. It could be something like $(11/37)\mu_1 + (26/37)\mu_2 - (17/45)\mu_3 - (28/45)\mu_4$. In summary, although rejection of the omnibus null hypothesis indicates that at least one contrast is significant by the Scheffé criterion, there is no guarantee that any obvious or interesting contrast will be significant.

As with all the tests we have so far considered (except the Fisher–Hayter), there is no logical necessity that the Scheffé tests of contrasts be preceded by a significant omnibus F. On the other hand, if the omnibus F test is not significant, no contrast will be significant. Thus, there is little point in expending energy on a series of post hoc Scheffé tests unless first determining that the F test is significant.

10.9.2 The Brown–Forsythe Method When Variances Are Not Equal

Brown and Forsythe (1974b) proposed that Welch's t' and df' (Box 10.4) be used with a criterion similar to Scheffé's S when the test is post hoc and the assumption of homogeneity of variance is questionable. The only difference is that the critical value of S against which t' is evaluated is based on df' (see Equation 10.10).

10.10 Controlling the Familywise Error Rate in Multi-Factor Designs

We have been considering the control of FWE in the context of examples taken from one-factor designs. Although the calculations and methods for control of error rates are the same in multi-factor designs, there are several additional issues. Consider a two-factor design with four levels of B (e.g., type of drug) and two levels of A (e.g., age group). We may wish to compare the B marginal means to determine the relative efficacy of the different drugs. We might use the Tukey *HSD* procedure to control *FWE*. We may also wish to compare the means of the B conditions within each level of A (that is, the simple effects of drug at each age). Is each level of A a family with the *FWE* set at .05? Or should *FWE* be set at .025 at each level of A so that the *FWE* is .05 for the complete set of tests? And should the comparisons of marginal means and of simple effects be considered separate families or one family? Suppose we are also concerned with testing interaction effects. Is this still another family of tests? Or should all the tests be considered a single family, thus controlling Type 1 error rates simultaneously for all hypotheses tested in the experiment, but sacrificing power? There are no generally agreed-upon answers to such questions. We will make some recommendations; however, depending on their designs and the questions they wish to address, investigators may decide on different approaches than the one we take in this section. Whatever the approach to controlling *FWE*, it is important that any research report is clear about exactly how the *FWE* was controlled so that readers may perform their own evaluation of the significance of results.

In the remainder of this section, we illustrate some common tests of contrasts, and our recommendations for controlling the *FWE*, using a simple 2 × 4 set of means, each based on six scores. Table 10.5 contains the cell means and the A and B marginal means, together with the ANOVA summary.

10.10.1 Testing Hypotheses About Marginal Means

The results of the analysis of variance in Table 10.5 reveal that the drug type (B) has significant effects, and the significant interaction suggests that the sizes of these effects are different in the two age groups (A). Let us consider the effects of B first. Most likely, we would wish to compare pairs of drug means. Tukey's *HSD* method provides a way to control the

Table 10.5 Cell means (*a*) and ANOVA (*b*) (*a*) Cell means

Age group	*Drug type*				
	B_1	B_2	B_3	B_4	*Mean*
A_1	12	6	5	15	9.5
A_2	16	2	9	3	7.5
Mean	14	4	7	9	

(*b*) ANOVA (*n* = 6)

Source	*df*	*SS*	*MS*	*F*	*p*
Age group (*A*)	1	48	48	1.280	.265
Drug (*B*)	3	638	212.67	5.653	.003
AB	3	528	176	4.693	.007
S/*AB*	40	1,500	37.5		

familywise error rate for the six pairwise tests we will perform. Setting the *FWE* at .05 and turning to Appendix Table C.8 or using *qtukey*(.95, nmeans = 4, df = 40), we find that the critical value of *q*, the Studentized range statistic, is 3.79 when $a = 4$ and the error $df = 40$. The standard error of the mean is $SEM = \sqrt{MS_{S/AB} / n} = 1.768$. Note that $n = 12$, not 6, because both age groups contribute to each mean at each level of *B*. As in Box 10.3, we multiply the critical value of *q* by the *SEM*, yielding a critical difference of 6.700. The difference between the marginal B_1 and B_2 means, and between the marginal B_1 and B_3 means, exceed this value; therefore, controlling the *FWE* at .05, only these two pairwise differences are significant.

Sometimes the researcher may have an *a priori* hypothesis that specified that only certain comparisons, whether pairwise or more complex, would be significant. In that case, the Dunn–Bonferroni method should be followed. The per comparison error rate would be the desired family-wise error rate divided by *K*, the number of tests; this would be the critical *p*-value for our significance tests.

Another possible scenario is that the researcher observes the means in Table 10.5 and decides at that point that the difference between the B_1 and B_2 means is large enough to warrant further investigation. Or, after viewing the means, the researcher decides that the B_1 mean is different enough from the other means that it should be tested against the average of the other three means. Such post hoc contrasts should be tested by the Scheffé procedure described in Box 10.8.

Two points should be noted about the developments so far. First, we have identified a family as a set of comparisons related to a single source of variance. If we had several levels of *A*, we also would have controlled the *FWE* at .05 for all comparisons of those marginal means. The second point is that we have assumed homogeneity of variance. As we discussed earlier in this chapter, modifications of the usual tests are indicated when this assumption is not met. In particular, the Tukey *HSD* method is not appropriate because the various means have different standard errors. In such a case, even if all pairwise comparisons are of interest, the Bonferroni procedure should be applied with the standard error of the difference between means based on the average within-cell variance of the cells involved in each comparison.

In sum, contrasts to analyze main effects within multi-factor designs follow the same recommendations and procedures presented earlier in the context of the one-factor design.

10.10.2 Testing Hypotheses About Simple Effects

Having found a significant difference between the marginal B_1 and B_2 means, we may ask whether this difference is significant in either or both age groups. Also, it is possible that there are other significant differences in one of the two age groups. Therefore, a reasonable next step is to compare the drugs at each level of *A*. The issue is what familywise error rate will be acceptable. We view the set of all contrasts of simple effects as a single family and therefore recommend that the *FWE* at each level of *A* be .05 divided by the number of levels of *A*. In the example of Table 10.5, pairwise comparisons within each of the two age groups would be tested with the critical value of *q* at the .025 level. Although Appendix Table C.8 contains only .01, .05, and .10 values for the Studentized range, both the *qtukey* and *TukeyHSD* functions in R allows selection of an alpha-level. The required value of *q* is 4.196 when $a = 4$ and the error $df = 40$.

It may seem that a first step would be to perform an omnibus *F* test at each level of *A*. However, the Studentized range statistic when applied to the largest difference between means is a test of the omnibus null hypothesis that all means are equal.[6] Therefore, we proceed directly to pairwise comparisons at each level of A. Assuming that the eight population variances are homogeneous, the critical difference is the standard error (across all eight cells) times the critical value of q; i.e., $d_{crit} = 2.5 \times 4.196$, or 10.49. At A_1 the difference between the B_4 and B_2 means, and that between the B_4 and B_3 means, are the two largest, but neither exceeds the critical value. Accordingly, we cannot reject the hypothesis that the B population means at A_1 are equal. At A_2, however, there are significant differences; the B_1 mean differs significantly from the B_2 and the B_4 means. Our analysis of the simple effects of the drugs has provided important information. Although we cannot be sure that other differences among the drug effects do not exist (because we cannot accept the null hypothesis), we have found that in the second age group drug B_1 is clearly superior to drugs B_2 and B_4.

We have assumed that all pairwise comparisons within each level of a second factor are to be performed. There may be circumstances in which only some of the possible comparisons within each level are of interest. For example, assume that we have *a priori* hypotheses about three contrasts at A_1 and two more contrasts at A_2. In that case, the Dunn–Bonferroni procedure with $K = 5$ is appropriate; assuming the contrasts are equally important, each test would be performed with $\alpha = .01$. If one of the contrasts was – *a priori* – more important than the others, the *FWE* could be distributed unequally (e.g., with $\alpha = .04$ for that contrast and .0025 for each of the four others). On the other hand, assume that the same contrasts appear to be of interest only after viewing the cell means. In that case, the Scheffé procedure, described in Box 10.8, is appropriate with $df_1 = (a - 1)(b - 1)$.

10.10.3 The Relation Between Tests of Interactions and Tests of Simple Effects

It is a common misconception that tests of simple effects are performed only to help us understand the causes of a significant interaction. Simple effect tests are sometimes useful for understanding interactions, particularly when differences among means are significant at one level of a factor but not at others, as in our analysis of the means in Table 10.5. However, the results of interaction tests and corresponding tests of simple effects are not always consistent. We may have a significant interaction but fail to find any significant tests of simple effects. Or the interaction may fail to be significant even if there is a significant simple effect. These apparent discrepancies are due to differences in the power of tests of interactions and of simple effects due to differences in the criteria for significance and differences in the cell frequencies and variances involved.

A common result that sometimes puzzles researchers is illustrated by the following cell means:

	B_1	B_2
A_1	15	5
A_2	8	6

Assume nine scores in a cell and an average cell variance of 100. Then, the t test of interaction is

$$t_{interaction} = [(15-5)-(8-6)]/(10\sqrt{4/9}) = 1.20$$

The interaction is not significant. However, if we compare the simple effects of B at A_1 we have

$$t_{B/A_1} = (15-5)/(10\sqrt{2/9}) = 2.12$$

which is significant at the .021 level. The test at A_2 does not yield a significant result (t = 0.42). Even controlling the *FWE* at .025 for each test comparing simple effects, we conclude that there is a difference between the B means at A_1. We now have an apparent contradiction: The test of the interaction does not provide evidence of an interaction in the population. However, the significant result of the test at A_1, coupled with a nonsignificant result of the test at A_2, suggests that there is an interaction with B having effects at A_1 but not at A_2. The reason for this inequality is that the standard error of the interaction involves the variance of differences among four means, whereas the standard error for the simple effect involves the variance of only two means. Therefore, it is possible to have a pattern of test results for simple effects that is not consistent with the result of the test of interaction because the power of the test of the simple effects is greater than the power of the test of the interaction.

10.10.4 Testing Hypotheses About Interaction

Various proposals have been made about the proper follow-up analyses to perform to understand a significant interaction (Games, 1973; Marascuilo & Levin, 1972, 1973; Tukey, 1991). Remember that an interaction occurs when the simple effect of one variable differs across levels of another variable. So, a 2 × 2 interaction compares one simple effect to another. Thus, the most direct approach to understanding an interaction effect is to test the embedded 2 × 2 interactions. By assessing all the 2 × 2 interactions, we can identify the simple effects that differ from one another.

For example, in Table 10.5, we might test the interaction involving the two age groups and the B_1 and B_4 drugs. This interaction effect size is (12 + 3) – (15+ 16) = –16. Dividing this by its standard error, $\sqrt{MS_{S/AB}(4/n)}$, we have the t statistic or, squaring, we have the F. If this test had been planned, the alpha-level would be set at .05. However, if such analyses are performed, they are typically post hoc. In that case, we have two options for controlling the familywise error rate. Assuming homogeneous variances, we can apply Scheffé's method; we calculate $S = \sqrt{(a-1)(b-1)F_{.05,(a-1)(b-1),ab(n-1)}}$. In our example, $(a - 1)(b - 1) = 3$ and $ab(n - 1) = 40$. The F required for significance at the .05 level with 3 and 40 *df* is 2.84. Therefore, S = 2.92. An alternative is to view the set of six possible 2 × 2 interaction contrasts as a family and use the Bonferroni criterion. With the *FWE* set at .05, the alpha-level for each t test is .05/6, or .0083. Assuming a two-tailed test, the critical value of t = 2.78, slightly smaller than the critical value of S. As in this example, the Bonferroni criterion will often be slightly more powerful, though its advantage will be lost as the number of interaction contrasts increases. When variances are homogeneous, as they are here, the choice between methods rests on a comparison of the critical values (see Perlmutter & Myers, 1973). If variances are heterogeneous, the Bonferroni method should be used, with the error term and error *df* based on the contrast tested.

10.10.5 Cautions When Using Software to Perform Further Analyses

In most software packages, simple effects can readily be tested by splitting the file. For example, in R we can use the *filter* function in the {dplyr} package to pull out the data for a particular level of a factor. The analysis of interest can then be performed on the reduced data set. Likewise, in SPSS, we can select the *Data* menu, and then indicate that the file is to be split by levels of A. There are two caveats to keep in mind when using this approach to conduct contrasts on subsets of observations in a data file. First, the error terms for the tests will be based just on the subset of observations selected for analysis. If variances are homogeneous in the experiment and the researcher wishes to take advantage of the increased number of df that result from pooling the variances across all conditions (i.e., using $MS_{S/AB}$ in the error term of each contrast), the researcher will need to recompute the denominator of the contrast. The second caveat is that the reported p-values are unlikely to reflect control of the familywise error rate. In general, the researcher must ensure that the appropriate criterion for significance has been applied, either through options in the software or, when these are lacking, by comparing the test statistic against the appropriate criterion.

10.11 The Sum of Squares Associated With a Contrast

The t statistic used throughout this chapter provides one approach to testing hypotheses about contrasts. An alternative, equivalent, test is based on the idea that any contrast of the group means corresponds to a component of SS_A, the sums of squares for the variable, A. That component sum of squares, $SS_{\hat{\psi}}$, will be distributed on 1 df and the contrast can be tested by dividing it by $MS_{S/A}$. We develop this approach now to emphasize the continuity between the analysis of variance and tests of contrasts.

Consider the square of the t statistic that was presented in Equation 10.5 for testing contrasts:

$$t^2 = \frac{\hat{\psi}^2}{s_{\hat{\psi}}^2} = \frac{\left(\sum w_j \bar{Y}_{.j}\right)^2}{MS_{S/A}\sum \dfrac{w_j^2}{n_j}} = \frac{\left(\sum w_j \bar{Y}_{.j}\right)^2 \Big/ \left(\sum \dfrac{w_j^2}{n_j}\right)}{MS_{S/A}}$$

The numerator, $\left(\sum w_j \bar{Y}_{.j}\right)^2 / \left(\sum \dfrac{w_j^2}{n_j}\right)$, is distributed as a sum of squares on one degree of freedom. We will denote this sum of squares by $SS_{\hat{\psi}}$; that is,

$$SS_{\hat{\psi}} = \frac{\left(\sum w_j \bar{Y}_{.j}\right)^2}{\sum w_j^2 / n_j} \tag{10.20}$$

Then

$$t^2 = \frac{SS_{\hat{\psi}}}{MS_{S/A}} \tag{10.21}$$

but a sum of squares on 1 *df* is a mean square, so that

$$t^2_{a(n-1)} = \frac{SS_{\hat{\psi}}}{MS_{S/A}} = F_{1,a(n-1)} \tag{10.22}$$

Let's reconsider the contrasts in our memory experiment. The means of the four conditions are shown in Table 10.1, and the ANOVA and several contrasts are displayed in Table 10.4. Applying Equation 10.20, we obtain the sum of squares for each of these three contrasts. For example, for the first contrast,

$$SS_{\hat{\psi}_1} = \frac{\left[(3)(6.5)+(-1)(12.1)+(-1)(10.7)+(-1)(10.5)\right]^2}{\frac{3^2}{10}+\frac{(-1)^2}{10}+\frac{(-1)^2}{10}+\frac{(-1)^2}{10}} = 158.7$$

If we assume equal condition variances, we may construct an F test, using Equation 10.22: $F(1, 36) = 158.7 / 17.944 = 8.8442$. The square root, $\sqrt{8.8442} = 2.974$, equals the t for this contrast shown in Table 10.4(c). Analogous calculations for the other contrasts in Table 10.4(b) yield $SS_{\hat{\psi}_2} = 5.4$ and $SS_{\hat{\psi}_3} = 9.8$. An important point is that the SS_{ψ} sum to 173.9, the SS_A shown in Table 10.6. That is,

$$\begin{aligned} &158.7 + 5.4 + 9.8 = 173.9 \\ &SS_{\hat{\psi}_1} + SS_{\hat{\psi}_2} + SS_{\hat{\psi}_2} = SS_A \end{aligned}$$

Think of the treatment sum of squares as a pie and each contrast as a piece of the pie. In the example of Table 10.4, the pieces do not overlap, and together they account for the whole pie. Here, the contrast of the control group with the average of the three treatment conditions (ψ_1) is the biggest piece, accounting for over 90% (158.7/173.9) of the variability among the condition means.

Not every set of contrasts result in sums of squares that add to SS_A. In the memory experiment example, the contrasts making up the set had a particular property that resulted in their accounting for different, nonoverlapping, portions of the variability, for independent pieces of the pie. When this is the case, the contrasts are said to be *orthogonal*. The maximum number of orthogonal contrasts in each set is equal to $a - 1$, the degrees of freedom of the treatment sum of squares. The sums of squares of $a - 1$ orthogonal contrasts will always add up to SS_A.

Table 10.6 ANOVA on memory data from Table 10.1 with contrasts from Table 10.4(*b*)

Source	*df*	*SS*	*MS*	*F*
A	3	173.9	57.967	3.23
ψ_1	1	158.7	158.7	8.844
ψ_2	1	5.4	5.4	0.301
ψ_3	1	9.8	9.8	0.546
S/A	36	646.0	17.844	

Before defining orthogonality more precisely, let us consider an example in which it does not occur. Instead of testing the two imaging conditions against one another (as in ψ_3), the experimenter decides to compare the image and control conditions. In other words, she tests

$$H_{01}: \mu_{control} - \mu_{image} = 0$$

as well as ψ_1 and ψ_2 as described previously. If the mean of the control condition differs from that of the image condition, there is a good chance that the control condition will also differ from the average of the treatment conditions because that average includes the image condition. In other words, there's a positive relation between ψ_1 and our new hypothesis. This lack of independence between the two contrasts is called *nonorthogonality*. It becomes evident in our example when we calculate the sums of squares corresponding H_{01}, which equals 88.2. Adding the sums of squares for the tests of ψ_1, ψ_2, and H_{01} totals 252.3. The sum is clearly greater than SS_A, 173.9. These pieces of the pie overlap.

We do not have to add the sums of squares to determine whether two contrasts are, or are not, orthogonal. Consider two contrasts, ψ_p and ψ_q, such that

$$\psi_p = \sum_j w_{jp}\mu_j \quad \text{and} \quad \psi_q = \sum_j w_{jq}\mu_j$$

If there are n scores at all levels of A, the criterion for orthogonality is

$$\sum_j w_{jp}w_{jq} = 0 \tag{10.23}$$

For example, we know that the first two contrasts in Table 10.4(b) are orthogonal because

$$(-3)(0) + (1)(1) + (1)(1) + (1)(-2) = 0$$

If the ns vary across treatment conditions, the criterion for orthogonality becomes

$$\sum_j \frac{w_{jp}w_{jq}}{n_j} = 0 \tag{10.24}$$

Several points about orthogonality deserve emphasis. First, a set of $a - 1$ orthogonal contrasts can be thought of as asking $a - 1$ logically independent questions that collectively "use up" all the degrees of freedom and variability associated with the independent variable. Note that just as you can divide up a pie in different ways, it is possible to define different sets of orthogonal contrasts on the same source of variance. Also, whether two contrasts are orthogonal depends on the contrast weights, not on the values of the means being contrasted. One way of thinking about this is that orthogonality depends on what questions are presented by the contrasts, not on what the answers turn out to be.

The second point is that we choose to test contrasts because they are of substantive interest, regardless of whether they are orthogonal. For example, researchers commonly test sets of pairwise comparisons. These are often of interest and, if they are, they should be tested even though they are not orthogonal. That said, it is rarely the case that all pairwise comparisons are equally interesting or important. The number of contrasts associated with a factor in the experimental design should be constrained by the *df* associated with that source of variance: Testing more contrasts than there are *df* always means that there is some redundancy in the questions, in the sense that that the contrasts must share some variance (their pieces of the pie overlap). Also, as the number of contrasts increases, the power of each one decreases because the *FWE* must be controlled.

Summarizing, we have developed the idea that a sum of squares based on multiple degrees of freedom can be broken into component parts, each of which may be tested as a separate contrast. One way to conduct such analyses is to define $a - 1$ orthogonal contrasts, testing them either with a t or an F ratio. No matter how they are tested, orthogonal contrasts partition the *SS* for a source of variance into independent components, and if there are $a - 1$ orthogonal contrasts then together they account for all the variance associated with the source.

10.12 Summary

This chapter developed the following points:

- Contrasts are specific comparisons on a set of means that allow researchers to pose detailed questions of a data set. They capture components of the source of variance they are intended to analyze. If a set of contrasts is orthogonal, they partition the variance associated with the source of variance, asking logically independent questions.
- Procedures for conducting hypothesis tests and constructing confidence intervals on contrasts are straightforward extensions of the t test procedures covered in Chapter 6.
- In an experimental design of any complexity, there are many possible contrasts that might be of interest to a researcher. The probability that at least one Type 1 error will occur in a set, or family, of contrasts increases as a function of the size of the family. It is therefore important to control the probability of a Type 1 error across a family (i.e., *FWE*).
- Just as there are many types of human families, there are many kinds of families of contrasts of means. These include the family of comparisons planned prior to data collection; the family of all possible comparisons of members of pairs of group means; the family of all comparisons of experimental group means with a control mean; and the family of post hoc contrasts determined after viewing the data.
- Different methods have been developed for controlling the *FWE*, depending on the kind of family and on whether confidence intervals are desired. Most of these methods involve the usual t statistic, or its close relative, the Studentized range statistic, or – when heterogeneity of variance is suspected – Welch's t'.

Table 10.7 provides a summary of many comparisons presented in this chapter. This summary integrates several considerations affecting the choice of procedures for controlling Type 1 error. One reason for the many different procedures in Table 10.7 is that the control of Type 1 errors is influenced by considerations such as whether the assumption of homogeneity of variance has been met and whether *n*s are equal across conditions. It is paramount that a given procedure controls Type 1 errors at close to the nominal level. Given that this criterion is met, power considerations are the second major reason for the

Table 10.7 Recommended procedures for controlling *FWE*

Family type	*Simultaneous methods*[a]		*Sequential methods*	
	Equal variances	*Unequal variances*	*Equal variances*	*Unequal variances*
Planned	Dunn–Bonferroni	Dunn–Bonferroni using Welch's t'	Hochberg	Hochberg using Welch's t'
All pairwise	Tukey *HSD* (equal n) or Tukey–Kramer (unequal n)[b]	Games–Howell or Dunnett T3	Fisher–Hayter	
Exptl vs control	Dunnett (equal n) or Dunn–Bonferroni (unequal n)	Dunn–Bonferroni using Welch's t'		
Post hoc	Scheffé	Scheffé using Welch's t'		

[a] Only the simultaneous methods allow the construction of simultaneous confidence intervals.
[b] Assuming K pairwise tests, the Bonferroni method will be more powerful than the Tukey under some conditions; Equation 10.19 provides the basis for the choice.

many procedures presented in Table 10.6. Given two procedures that adequately control Type 1 error rates, we prefer the method that results in more powerful tests. Sequential methods provide more powerful tests, so they are preferred over simultaneous methods when hypothesis tests are conducted. However, sequential methods are not applicable to the construction of confidence intervals because there is no ordering within a set of confidence intervals. Finally, we emphasize again that contrasts should be planned whenever possible because a set of planned contrasts is almost always smaller than a set of unplanned contrasts. Thus, tests of planned contrasts generally have more power than tests conducted on other kinds of families. But perhaps the more important benefit of planning contrasts is that the careful thought that is required to specify the key research questions will probably lead to research designs that are more closely focused on those questions.

Exercises

10.1 [Testing contrasts] There are five treatment conditions in a problem-solving study, each with $n = 20$. Two groups, F_1 and F_2, are given instructions designed to facilitate problem solving. The third group is a control group given neutral instructions. The fourth and fifth groups, I_1 and I_2, are given instructions designed to interfere with problem solving. The data are

	F_1	F_2	C	I_1	I_2
$\bar{Y}_j$	14.6	14.9	13.8	11.8	11.7
s^2_j	3	4	5	4	4

Test each of the following hypotheses with $\alpha = .05$. State H_0 and H_1.

a) The average of the facilitation group population means is greater than the mean of the control population.

b) The average of the interference population means is different from the mean of the control population.
c) The average of the facilitation means is not the same as the average of the interference means.

10.2 [Planned vs post hoc contrasts]

a) Assume that all three tests in Exercise 10.1 had been planned prior to data collection. Using the Dunn–Bonferroni method, construct 90% simultaneous confidence intervals for the three contrasts. Re-evaluate whether the null hypotheses in parts (b) and (c) should be rejected, using the Dunn–Bonferroni criterion with *FWE* = .10.
b) Assume the contrasts were decided on after viewing the means. Use the Scheffé method to construct simultaneous confidence intervals. Re-evaluate whether the null hypotheses in parts (b) and (c) of Exercise 10.1 should be rejected with *FWE* = .10.

10.3 [Implications of controlling *FWE* for power] The following is suggested by a study conducted by Fenz and Epstein (1967). In a study of conflict in parachutists, galvanic skin response (GSR) measures of stress were obtained for five different groups of five participants who differed with respect to when the measures were taken: 2 weeks before the jump (*BJ*-2), 1 week before (*BJ*-1), on the day of the jump prior to jumping (*DJ*-*P*), and on the day of the jump after jumping (*DJ*-*A*). There was also a control group who did not jump. The MS_{error} for the ANOVA = 4.0, and it is reasonable to assume homogeneity of variance. The means were

	BJ-2	*BJ-1*	*DJ-A*	*DJ-P*	*C*
Mean =	5	5	7	9	2

a) Assume that the investigator had planned to compare each of the four experimental groups with the control (C). With α =.05 (two-tailed), test the difference between the *DJ*-*P* and C means, using (i) the Dunn–Bonferroni procedure and (ii) the Dunnett procedure.
b) Assume that the experimenter tested all possible pairwise comparisons. Redo the test in part (a), using the appropriate procedure for controlling the *FWE* at .05.
c) Comment on the relative power of these three procedures, justifying your conclusion by citing relevant information in your preceding answers. Explain why these situations give rise to the differences in power that you indicate.
d) Calculate the confidence intervals obtained with each of the three procedures and relate the results to your answer to part (c).

10.4 [Benefits of planned contrasts] A sample of students learning statistics is divided into three groups of 10. One group receives training on relevant concepts *before* reading the text, a second receives the training *after* reading the text, and a third is a no-training *control*. Summary statistics on a test are as follows:

	Before	*After*	*Control*
$\bar{Y}_j$	20	14	13
s^2_j	72	62	76

a) We want to test whether the mean of the *before* population is higher than the average of the other two populations combined. In answering the following parts, assume that $FWE = .05$. (i) State the null and alternative hypotheses. (ii) What is the estimate of the variance of the sampling distribution of $\hat{\psi}$ (assume homogeneity of variance)? (iii) Calculate the t statistic appropriate for testing H_0.

b) Evaluate the test statistic you just calculated, assuming (i) the test was the sole contrast tested and had been planned before viewing the data; (ii) the test was a result of viewing the data.

10.5 [Comparing methods of controlling *FWE*] We have five group means, each based on 10 scores, with $MS_{S/A} = 4.0$. The means are

A_1	A_2	A_3	A_4	A_5
8.6	9.5	9.2	8.0	10.4

a) We plan five contrasts with $FWE = .05$. Test the contrast of A_5 against the average of the other four groups. State the criterion required for significance, and whether H_0 can be rejected.

b) What is the result of the significance test if we decided on the contrast in part (a) after inspecting the data?

c) Find the confidence intervals corresponding to the tests in parts (a) and (b). Explain the difference in widths.

d) Suppose we did all possible pairwise tests. Actually calculate the test for A_1 against A_2. What is the criterion statistic? What conclusion do you reach about H_0?

e) Suppose the only contrast we planned pitted the average of A_1 and A_2 against the average of the remaining three groups. Do the calculations and report the results, showing the criterion statistic.

10.6 [Planned vs post hoc contrasts] In an attitude-change study, four groups of participants are presented with persuasive messages about a topic. Two groups read the messages; a positive message for one group and a negative message for the other. Two other groups receive the messages by viewing a videotape. A fifth, control, group receives no message. Each group has its attitude assessed by a questionnaire in which larger scores mean a more positive attitude. There are seven participants in each group and $MS_{S/A} = 20$. The group means are as follows:

A_1	A_2	A_3	A_4	A_5
Video/positive	Video/negative	Read/positive	Read/negative	Control
71	42	63	47	52

a) Determine which experimental conditions differ significantly from the control, using the Dunnett test with $FWE = .05$.

b) Test the hypothesis that the difference between the positive and negative messages is the same whether they are read or are presented by videotape. Assume this is the only planned comparison.

c) By how much would two groups have to differ before they would be considered significantly different by the Tukey test with $FWE = .05$?

10.7 [Comparing ways of controlling *FWE* in real data] The *male_educ* file on the *Seasons* page contains mean (over seasons) for four of the *schoolyr* categories (3 = only high school, 5 = some post high school, 7 = bachelor's degree, 8 = graduate school). In what follows, assume that all pairwise differences are tested.

a) Test the difference between the *schoolyr* = 3 and *schoolyr* = 5 *beck_d* means, using (i) the Tukey–Kramer method, and (ii) the Dunn–Bonferroni method, assuming all pairwise comparisons and $FWE = .05$. Assume homogeneous variances. (iii) Compare the 95% confidence intervals.

b) (i) Perform the Games–Howell test of the difference in part (a) and compare the results with those in part (a). Which of these procedures should be used with these data? (ii) Calculate the Games–Howell confidence interval.

10.8 [Comparing CIs for different methods of controlling *FWE* in real data] The *Sayhlth* file on the *Seasons* page of the website contains Beck Depression scores as a function of several factors.

a) Test whether employment status significantly affects *Beck_D*, the mean (over seasons) of the Beck Depression scores.

b) Calculate all simultaneous confidence intervals ($FWE = .05$) by the Tukey–Kramer method. Assume homogeneous variances.

c) Redo part (b), using the Dunn–Bonferroni method. Assume homogeneous variances.

10.9 [Testing assumptions]

a) Is the assumption of homogeneity of variance reasonable for the data in Exercise 10.8? Support your conclusion with statistical evidence.

b) Assume that we wish to know whether the mean depression score for fully employed individuals (category 1) differs from that of those who are not fully employed (categories 2 and 3 in the *Sayhlth* file). Test whether the difference is significant, assuming this is the sole comparison tested and was planned prior to the collection of data.

10.10 [Contrasts of contrasts] Ninety children, varying in age ($A_1 = 5$, $A_2 = 7$, and $A_3 = 9$), are taught by one of three mnemonic methods (methods for memorizing; B_1, B_2, and B_3). All participants are then shown a series of objects and their recall is scored. Thus we have nine groups of 10 participants each. The cell means and variances are as follows:

	Means			*Variances*		
	A_1	A_2	A_3	A_1	A_2	A_3
B_1	44	58	78	75	79	84
B_2	56	66	83	61	82	85
B_3	52	70	79	90	71	77

a) Perform an ANOVA, using these statistics.
b) B_2 and B_3 both involve the use of imagery whereas B_1 involves repeating the object names. Therefore, a contrast of the B_1 mean against the average of the B_2 and B_3 means is of interest. Calculate the 95% confidence interval for this contrast. Does the contrast differ significantly from zero?
c) Test whether the contrast in part (b) is different at A_1 from at A_3.

10.11 [A typical scenario, start to finish]

a) The file *EX10_11* on the *Exercises* page of the website contains a 3 × 3 data set. Table and plot the marginal and cell means. Describe the pattern of means.
b) Carry out an ANOVA on the data and present the results in a table.
c) With FWE =.05, calculate confidence intervals for pairwise comparisons of the A means; state which – if any – comparisons are significant.

10.12 [Adding a control condition] Bless et al. (see Exercise 9.16 and the data in *EX9_16*) also collected data from a control group (n = 10) that received no message but were asked to assess a fee. Assume that the mean for the control group is 48 and the standard deviation is 4.5. Test the difference between the control group and each of the eight experimental groups. Which of the experimental groups differed significantly (α = .10) from the control group? Be explicit about the method for controlling the FWE and the selection of the error term(s).

10.13 [Working with real data] The R {datasets} package includes a file called ToothGrowth, which reports the lengths of cells responsible for tooth growth in guinea pig as a function of vitamin C dose and delivery method. Use ?ToothGrowth to learn more.

a) Explore the data graphically and numerically and determine whether the experimental factors affected cell length.
b) How many pairwise comparisons are possible in this experiment? If a Bonferroni correction is used, what α_{EC} results when α_{FWE} = .05?
c) Choose the most appropriate method for controlling FWE and compare all cells means.
d) Assume that the researchers expected the delivery method to matter more for the two lower doses than at the highest dose. (i) Write that hypothesis in the form of a contrast. (ii) Choose the appropriate method and test the contrast.
e) Does your answer in part (d) require any specific statistical outcome from part (a), such as a significant main effect?

10.14 [Sums of squares and orthogonal contrasts] Each cell in the following table contains a mean based on 10 scores:

	A_1	A_2	A_3
B_1	20	10	6
B_2	6	10	8

a) Find the sums of squares accounted for by each of the following contrasts of the A marginal means: $\psi_1 = \mu_1 - (1/2)(\mu_{2.} + \mu_{3.})$; $\psi_2 = \mu_{2.} - \mu_{3.}$.

b) Are the two contrasts orthogonal? Consider only the coefficients in your answer.
c) Find SS_A and compare it with the sum of the two sums of squares found in part (a).
d) Do either of the prior contrasts vary as a function of B? Find the SS terms associated with each of the relevant significance tests. Add these terms and compare them with SS_{AB}.

10.15 [Sums of squares and orthogonal contrasts] The following group means are each based on 10 scores:

A_1	A_2	A_3
24	16	14

a) Calculate SS_A.
b) Calculate the sum of squares for each of the following contrasts: (i) $\hat{\psi}_1 = \overline{Y}_{.1} - \overline{Y}_{.2}$; (ii) $\hat{\psi}_2 = (1/2)(\overline{Y}_{.1} + \overline{Y}_{.2}) - \overline{Y}_{.3}$; (iii) $\hat{\psi}_3 - \overline{Y}_{.1} - \overline{Y}_{.3}$. What should be true of the relation between SS_A and the sums of squares for $\hat{\psi}_1$ and $\hat{\psi}_2$? Why?
c) We can remove the effect associated with $\hat{\psi}_1$ from the data by setting the means at A_1 and A_2 equal to their average. The adjusted means are

A_1	A_2	A_3
20	20	14

Redo part (b), (ii) and (iii). Are either of the sums of squares different from those calculated for the original (unadjusted) means? Explain, emphasizing the relation of the results to the concept of orthogonality.

10.16 [Sum of squares in unbalanced designs]

a) Suppose the group sizes in Exercise 10.15 were not equal; the n_j are 8, 10, and 12, respectively. Returning to the original means, calculate SS_A. Then calculate the sums of squares for $\hat{\psi}_1$ and $\hat{\psi}_2$. Now, what is the relation between the sums of squares for $\hat{\psi}_1$ and $\hat{\psi}_2$ and SS_A?
b) Redefine the ψ_2 contrast so that it is orthogonal to that for ψ_1 for the samples sizes stated in this exercise. Calculate $SS_{\hat{\psi}_2}$. Does $SS_{\hat{\psi}_1} + SS_{\hat{\psi}_2} = SS_A$?

10.17 [Standardized contrasts] Consider the group means in Exercise 10.3. Assume that $MS_{S/A} = 900$.

a) Assuming $n = 10$ in all three groups, calculate the standardized contrast, $\hat{\psi}_S$, for part (b, ii) of Exercise 10.3.
b) Repeat part (a) but assume the unequal ns of Exercise 10.3 and define the contrast as in part (b). Assume that the error mean square still equals 900.

Notes

1 The requirement that the weights sum to zero ensures that the contrasts deal with differences among means.

2 A third type of error rate, the *false discovery rate* (*FDR*), is the proportion of null hypothesis rejection decisions that are Type I errors (Benjamini & Hochberg, 1995). Controlling the *FDR* at, say, .05, requires a compromise between limiting the *EC* to .05 and limiting the *FWE* to .05. Unless all null hypotheses are true, controlling *FDR* entails an *FWE* that exceeds .05.

3 Šidák (1967) proposed that H_{ok}: $\psi_k = 0$ be rejected if $p_k \leq 1 - (1 - FWE)^{1/K}$. Because $1 - (1 - FWE)^{1/K} > FWE / K$, this method has more power and a narrower confidence interval than the original Dunn–Bonferroni procedure. However, the difference is very small.

4 Users of SPSS can select the *Transform* option from the main menu, then *Compute Variable*; double-click on the *cdf.srange* option. Then, insert values for *q* (perhaps several trial values), *a*, and the error *df*. For example, in the left-hand panel, you may have a variable labeled *p* and in the right-hand panel, 2*(1 – CDF.SRANGE(5.05, 4, 19)). The *p* column of your data form should now show the value .02.

5 *TukeyHSD* works exactly for balanced (equal *n*) designs and provides sensible intervals for mildly unbalanced designs.

6 See Myers (1979), for a discussion of the relative power of the *F* and Studentized range tests of the omnibus null hypothesis.

Chapter 11

Integrated Analysis II

11.1 Overview

In this chapter, we review the material presented in the preceding chapters dealing with the analysis of data from between-participants designs. We introduce a hypothetical two-factor experiment, and then analyze the data, including exploratory statistics, the ANOVA, and follow-up tests. Finally, we discuss the results.

11.2 Introduction to the Experiment

The effects of the organization of factual material upon memory for the material have been studied by several researchers. In one classic study (Myers, Pezdek, & Coulson, 1973), participants read a series of 25 sentences, each of which related the name of one of five fictional countries with an attribute; there were five categories of attributes (e.g., climate, principal industry, language). The sentences were presented to participants in five paragraphs, with each paragraph on a separate page. For the Name organization group, a paragraph consisted of five sentences describing five different attributes of a single country; each attribute was from a different category. For the Attribute organization group, each paragraph presented attributes for a single category for each country; for example, a page might state the different climates of the five countries, or the different languages. For the Random organization, the 25 facts were randomly placed, five to a paragraph. Subjects read each paragraph for 40 seconds and after reading all five paragraphs, wrote down as many of the 25 facts as they could recall. The Attribute organization resulted in significantly better recall than either of the other two types of organization; there was little difference between the Name and Random results.

In a hypothetical follow-up study, we imagine using the same materials but allowing participants as much study time as needed to produce perfect recall. Each of the three organizational groups is then divided into three delay groups to be retested after a delay of either 1, 2, or 3 days. With respect to the main effect of organization of the 25 facts, there are several possibilities: (1) the order of the Name, Attribute, and Random organization means may be the same as in the original study; (2) when perfectly memorized, the differences among the three organizations may be eliminated in the delayed tests; (3) the groups that performed poorest in the original study (the Name and Random organizations) may do better than the Attribute groups because participants are likely to spend more time

DOI: 10.4324/9781003453550-13

reaching the criterion of perfect recall and thus have the material better stored in memory. To examine these possibilities, we will conduct pairwise tests of the differences among the three organizations.

Although recall scores should be worse the longer the retest is delayed, the interaction of organization and delay is of interest. Will the effect of delay be smaller in the Attribute condition than the others? Alternatively, if learning in the Name and Random organizations required more effort than in the Attribute condition, then perhaps they will show smaller effects of delay. There questions are appropriately addressed with a planned contrast, which we evaluate in the Results section.

11.3 Method

11.3.1 Participants

The experimenters decided that the interaction of organization and delay of test was of primary interest, so they planned the experiment to have good power to detect an interaction effect of medium size (Cohen's $f = .25$). Using the *a priori* option in G*Power 3.1 (see Figure 11.1), they specified 4 *df* for the numerator of the *F* test of the interaction and 9 for the number of conditions in the experiment. They found that 196 participants would yield power of .8, which corresponds to approximately 22 participants per condition. The 196 participants were recruited from students in undergraduate psychology courses.

11.3.2 Experimental Design

The design was a 3 (organization of the texts) × 3 (delay of the test) with participants randomly assigned to the nine cells of the design with the constraint that there were 22 in each cell.

11.3.3 Procedure

The experiment had three phases. In the first phase, participants studied on their own time a five-page booklet with five facts on each page, organized according to the organizational condition to which they had been assigned. They were told to study the materials until they had memorized all 25 facts and then return on an assigned day. In the second phase, participants were tested in groups of nine; they remained in the lab until they could correctly reproduce all 25 facts they had previously studied. They were told that they could leave when they had produced a written sheet with all 25 facts correct, and that the purpose of the experiment was to see how much time was required to recall material they had previously studied. They were then assigned a day (1, 2, or 3 days later) on which they were to return to participate in "another experiment." In the third phase, participants were tested for recall of the 25 facts. After completion of this phase, during debriefing they were asked whether they had studied the list between the second and third phase; none reported having done so.

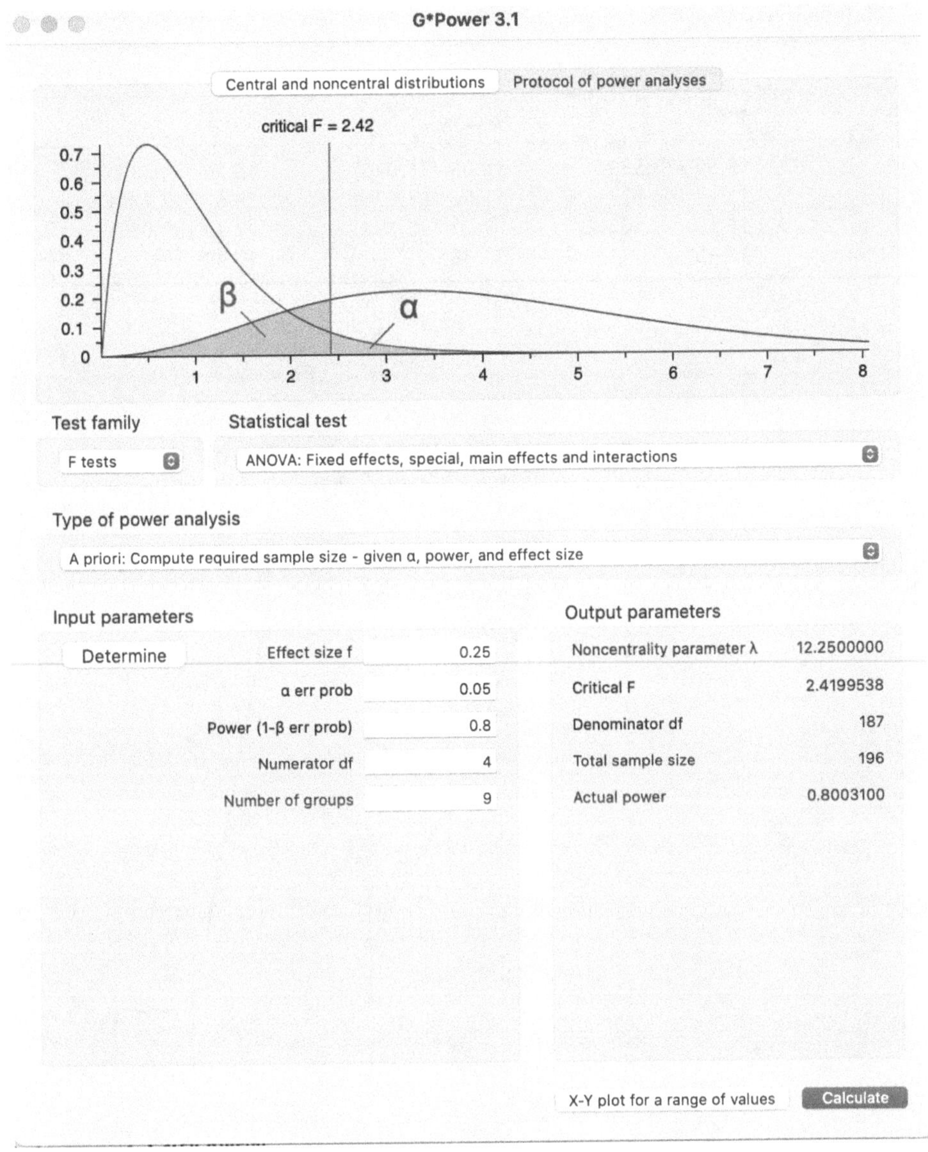

Figure 11.1 Using G*Power 3.1 to compute *n* for .8 power to test the interaction.

11.4 Results and Discussion

The score for each participant was the number of facts recalled correctly in the third experimental phase. Spelling errors were ignored if the category, attribute, and country name were correctly linked. The data are in a file on the *Tables* page of the website labeled *Table 11_1IA2 Data*. The means and variances are presented in Table 11.1, and the means and standard errors are plotted in Figure 11.2.

Table 11.1 Means and variances (in parentheses) for a hypothetical text recall experiment

Organization	*Delay*			
	1 day	*2 days*	*3 days*	*Mean*
Attribute	19.59 (33.587)	18.05 (26.045)	15.36 (54.338)	17.67
Name	16.27 (43.827)	11.86 (67.076)	8.68 (37.275)	12.27
Random	16.50 (35.500)	7.23 (33.708)	7.23 (30.089)	10.32
Mean	17.45	12.38	10.42	13.42

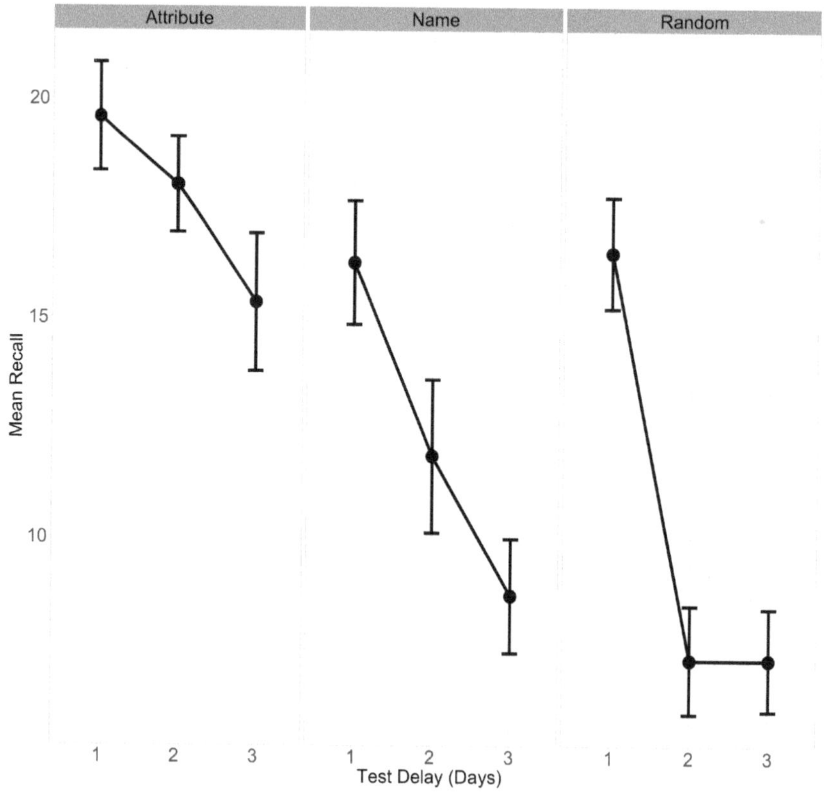

Figure 11.2 Means and standard error bars for the text memory experiment.

11.4.1 Exploring the Data

Several points should be noted about the means. With respect to organizational effects, the Attribute organization yields better recall than the other two in all the delay conditions. In addition, the Name groups perform similarly to the Random groups, except after a 2-day delay when the Name condition has an advantage of over four words. With respect to delay effects, recall deteriorates as a function of delay of test, as expected; however, the drop is only about four items in the Attribute condition from days 1 to 3, whereas it is more than seven items in the Name and Random conditions for the same period. These different rates of memory decay suggest an interaction.

We used the Shapiro–Wilk test to examine possible violations of the normality and homogeneity of variance assumptions. (In R, use {dplyr} to first *group_by*(Organization, Delay) and then *summarise*(SW = shapiro.test(Recall)$p) to see just the *p*-values of the Shapiro–Wilk test for each condition.) In eight of the nine groups, the data departed significantly from normality. However, the groups are large enough that we assume that the ratios of mean squares will be distributed approximately as *F*.[1]

To examine the possible violation of the homogeneity-of-variance assumption, we used the Brown–Forsythe test of heterogeneity of variance based on absolute deviations about the group medians (see Section 8.8.4), finding the *p*-value to be .075. Considering that the ratio of the largest to the smallest of the nine group variances is only about 2 to 1 (see Table 11.1), and that with 8 and 189 *df* the tests are powerful enough to detect even trivial differences among the population variances, we conclude that variance heterogeneity is not a problem.

11.4.2 The ANOVA

Table 11.2 contains the results of the analysis of variance of the recall scores. The effects of organization and delay are both significant, whereas the interaction *p*-value falls short of the .05 level. The η^2 values are ratios of the effect sum of squares to the sum of the effect and error sums of squares and the *f* estimates were calculated using the formulas in Table 9.13. These statistics reinforce the sense that the main effects contribute most of the variance among the nine cell means. By Cohen's guidelines, according to which $f = .4$ indicates a large effect, both organization and delay make large contributions to the population variance whereas their interaction contribution falls about halfway between small ($f = .1$) and medium ($f = .25$).

Let us analyze these effects further for a better understanding of the causes of the differences between organizations. Although the interaction of delay and organization failed to reach significance, we will also carry out our planned comparison of the delay effects in the Attribute condition against the average of the Name and Random organizations. This test is justified because it was planned in advance of data collection (see Section 10.6); had we looked at the data before developing the hypothesis, then the resulting post hoc test would capitalize on chance and risk inflating the Type 1 error rate.

Table 11.2 Analysis of variance of the recall scores summarized in Table 11.1

Source	*Sum of squares*	*df*	*Mean square*	*F*	*p*	*Partial* η^2	$\hat{f}$
Organization (*O*)	1,912	2	956.1	23.806	.000	.201	.480
Delay (*D*)	1,738	2	869.1	21.641	.000	.186	.457
OD	364	4	90.9	2.263	.064	.046	.160
Error	7,590	189	40.2				
Total	11,604	197					

R squared = .346 (adjusted *R* squared = .318).

11.4.3 *A Planned Contrast*

The experimenter predicted that delay would have a different effect for the Attribute condition compared to the other conditions, although the direction of the effect was not clearly specified. A planned contrast to test the prediction is

$$\hat{\psi} = \left(\overline{day3} - \overline{day1}\right)_{Attribute} - \frac{\left(\overline{day3} - \overline{day1}\right)_{Name} + \left(\overline{day3} - \overline{day1}\right)_{Random}}{2}$$

Substituting the means from Table 11.1, we find $\hat{\psi} = 4.2$, and applying Equation 10.5, $t(189) = 1.79$. The p-value is found using a distributional calculator. Because the test is two-tailed, we calculate twice the probability of obtaining a t at least at large as 1.79; in R, the command 2*(1 – *pt*(1.79, df = 189)) reveals that $p = .074$. We conclude that the experiment provides no clear evidence in favor of the hypothesis that the effect of test delay is different for the Attribute organization than the average of the Name and Random organizations.

11.4.4 *Post Hoc Tests and Confidence Intervals*

Turning to the effects of organization, the results of the Tukey *HSD* test in Table 11.3 reveal that the significant variance among the three Organization means is largely due to the advantage of the Attribute condition over the Name and Random conditions; the difference between the latter two means is not significant.[2] The confidence intervals can be obtained as stated earlier in Box 10.3. The formula for the bounds and its application to the Attribute-Name contrast is

$$\begin{aligned} CI &= \hat{\psi} \pm \left(q_{\alpha,a,ab(n-1)} / \sqrt{2}\right)\left(MS_{S/AB}\sqrt{2/ab}\right) \\ &= 5.39 \pm (3.35/\sqrt{2})(1.103) \end{aligned}$$

Completing the calculation, the lower bound of the confidence interval is 2.79 and the upper bound is 8.00.

Because statistical significance does not always mean that an effect is of practical or theoretical importance (particularly when the large number of error degrees of freedom provide power to detect even very small – perhaps uninteresting – effects), we can compute estimates of Cohen's d from the output in Table 11.3. Each value is the mean difference divided by the square root of 40.16, the error mean square. For the three contrasts in Table 11.3, they are 0.85, 1.16, and 0.31. The estimates of d are quite large for comparisons of the Attribute mean with the other two organizational means, and relatively small for the comparison of the Name and Random means, providing further confirmation that the significant variance of the organizational means is largely due to the differences between the Attribute and the other two means.

Other contrasts are possible. Noting that the Name and Random organization means are similar to each other but markedly lower than the Attribute mean, we might wish to conduct a post hoc test of the difference between their average and that for the Attribute participants.

Most statistical packages will enable such tests. For example, in SPSS the *Helmert* option calculates the confidence interval for the contrast between each level of the factor

Table 11.3 Tukey's *HSD* test of pairwise comparisons of the organizational means

```
> m<-pairs(emmeans(memout,~Org)) #Tukey's HSD for pairwise comparisons of Organization
NOTE: Results may be misleading due to involvement in interactions
> m
 contrast estimate  SE  df t.ratio p.value
 A - N        5.39 1.1 189   4.889  <.0001
 A - R        7.35 1.1 189   6.661  <.0001
 N - R        1.95 1.1 189   1.772  0.1819

Results are averaged over the levels of: Delay
P value adjustment: tukey method for comparing a family of 3 estimates
> confint(m) #confidence intervals for those differences
 contrast estimate  SE  df lower.CL upper.CL
 A - N        5.39 1.1 189    2.788     8.00
 A - R        7.35 1.1 189    4.742     9.95
 N - R        1.95 1.1 189   -0.651     4.56

Results are averaged over the levels of: Delay
Confidence level used: 0.95
Conf-level adjustment: tukey method for comparing a family of 3 estimates
```

Note. The standard error for these tests = $\sqrt{MS_{error}(2/66)}$ the use of the error mean square from the ANOVA rests on the assumption of homogeneity of variance, and 66 is the number of scores at each organizational level.

and the mean of the subsequent levels. Similarly, in R, the command *contrasts*(dat$Org) <- *contr.helmert*(3) assigns contrasts to the Org factor of the data frame called dat, which has three levels. In this case, however, the comparisons are between each level of the factor and the mean of the *previous* levels, thus R's "contr.helmert" contrasts are sometimes called reverse Helmert or difference contrasts. Using named contrasts like this requires careful attention to the order of the levels within the factor of interest. Additionally, because the meaning of "helmert" (and other named contrasts) can vary across software, it is wise to confirm that the contrasts being calculated are those you intended.

The results of the Helmert contrasts are summarized in Table 11.4. The significance level reported there is not adjusted for the fact that the test was post hoc. Therefore, to evaluate significance, use the Scheffé criterion as described in Box 10.8:

1. From Table 11.4, calculate t by dividing the contrast by the standard error (6.37/.955) = 6.67.
2. Compare the computed value t of 6.67 with $S = \sqrt{df_1 \cdot F_{.05,df_1,df_2}} = \sqrt{2F_{.05.2,189}} = 2.47$. The result is significant.
3. The confidence interval bounds also fail to take into consideration that the test is post hoc. From Chapter 10, the bounds are

$$\begin{aligned} CI\ bounds &= \hat{\psi} \pm s_{\hat{\psi}} S \\ &= 6.37 \pm (.955)(2.47) \\ &= 4.01, 8.73 \end{aligned}$$

Table 11.4 Post hoc contrasts of organizational conditions

Contrast	*Estimate*	*SE*	*df*	*Lower.CL*	*Upper.CL*
A_v_NR	6.371	0.955	189	4.487	8.26
N_v_R	1.95	1.103	189	-0.222	4.13

Note: *A = Attribute, NR = average of Name and Random, N = Name, R = Random*

Because the contrast was selected after viewing the results, the bounds are based on the family of all possible contrasts and are therefore somewhat wider than those reported by software (Table 11.4). We should also note that the weights for the first contrast were 1, –.5, and –.5. Had we used other weights, the bounds would have required adjustment. For example, if we had used 2, –1, and –1, the values of the contrast and its standard error would have been twice those in Table 11.4. Accordingly, the bounds would have been divided by two to return to the original data scale.

There is one more point to notice about these Helmert contrasts: They form an orthogonal set. With three levels of a factor, there are two Helmert contrasts, level 1 versus the mean of levels 2 and 3, and level 2 versus level 3; the contrast weights are 1, –.5, –.5 and 0, 1, –1. Using Equation 10.23, we can confirm their orthogonality: $(1)(0) + (-.5)(1) + (-5.)(-1) = 0$. Because they are orthogonal contrasts, their *SS* neatly partition the *SS* for the Organization factor in the design:

$$SS_{A\,v\,NR} = \frac{(6.371)^2}{\dfrac{1^2 + (-.5)^2 + (-.5)^2}{66}} = 1785.944$$

$$SS_{N\,v\,R} = \frac{(1.95)^2}{\dfrac{(1)^2 + (-1)^2}{66}} = 125.4825$$

Summing, we find 1785.944 + 125.4825 = 1911.427, which is within rounding error of $SS_O = 1912$ shown in Table 11.2.

To sum up, the statistical analyses demonstrate that the type of organization influences recall; recall is better in the Attribute condition than in the Name or Random conditions, which do not differ.

11.5 Summary

In this chapter, we used data from a hypothetical two-factor experiment to review developments in Chapters 8–10.

- As always, we first explored the data, noting various trends in the means and examining the spreads of scores and the shapes of distributions in the different conditions of the experiment.

- Following the ANOVA, we first tested a planned contrast that compared the effect of delay across different types of organization during the study phase.
- We tested pairwise comparisons of the means for the qualitative variable, Organization, using Tukey's *HSD* method to control the familywise error (*FWE*) rate.
- Finally, we computed confidence intervals for two orthogonal contrasts using Scheffé's method to control the *FWE* rate and showed their sums of squares fully accounted for the sums of squares for the factor.

Exercises

11.1–11.3 The website contains the files *EX11_1–EX11_3* on the *Exercises* page of the book's website. Each file contains a data set with three groups of scores. In each case, explore the data and based on your exploration, decide on the appropriate test of the omnibus null hypothesis. The test might be the usual *F* test, an *F* test based on trimmed data, the Welch's *F* test, a test based on a transformation, or a nonparametric test. Justify your choice and then carry out the test you have chosen.

11.4 An experiment on the effects of persuasion on attitude change yielded the following data:

	Movie	*Lecture*	*Movie + lecture*	*Control*	*Neutral movie*	*Neutral lecture*
Mean	5.4	3.0	6.4	−1.4	.8	−.6
Variance	13.1	11.8	115.1	15.1	16.3	16.6

There were five participants in each cell. In what follows, *FWE* = .05, two-tailed.

a) Test the mean of the left-most two conditions (movie, lecture) against the mean of the right-most two conditions (neutral movie, neutral lecture) assuming that it is one of four planned contrasts.
b) Assume we wish to test each of the conditions against the control. Perform the test of the movie against the control mean.

11.5 Now assume the following means and variances for the experiment of Exercise 11.4:

	Movie	*Lecture*	*Movie + lecture*	*Control*	*Neutral movie*	*Neutral lecture*
Mean	5.4	3.0	6.4	−1.4	.8	−.6
Variance	7.2	8.0	8.8	24.1	15.3	20.6

a) Redo part (a) of Exercise 11.4.
b) Redo part (b) of Exercise 11.4.

11.6 The *Exercises* page on the website contains a data set, *EX11_6*. *Employment* = 1 (fully), 2 (part time), and 3 (unemployed). The scores are participants' ratings of their happiness.

a) Explore the data. Is there any evidence that the assumptions underlying *F* and *t* tests are violated?

b) Perform the usual ANOVA and also calculate Welch's F to test the null hypothesis that the population mean happiness scores do not differ.
c) Assuming that all pairwise differences between the means are considered, calculate the confidence interval for the difference between the means of groups 1 and 2.

11.7 a) The *Format* × *Age* interaction in Table 9.12 may be viewed as a contrast; that is, as a sum of weights times means. State the weights and calculate the sum of squares for this contrast. The cell means are in Table 9.11.
b) Do the same for the three-way interaction *of Format* × *Age* × *Instructions*.

11.8 A researcher sampled the political attitudes of college students and older retirees. From each sample, participants were chosen so that there were two age groups of 50, each further divided into five equal-size groups based on their political attitudes from very liberal (*LL*) to very conservative (*CC*). Each of the 100 participants was then given a description of a hypothetical political candidate and asked to rate their support for such a candidate. The group means and variances (in parentheses) are

	Very liberal (LL)	*Liberal (L)*	*Moderate (M)*	*Conservative (C)*	*Very conservative (CC)*
Retirees	16.1 (107)	14.7 (99)	13.5 (124)	13.8 (113)	12.4 (118)
Students	18.5 (102)	16.9 (107)	8.5 (96)	3.2 (115)	1.4 (119)

Let A = political attitude and B = age group.

a) Present the ANOVA table.
b) Present the *EMS*.
c) Calculate general ω^2 for A; i.e., the estimated proportion of variance due to political attitude.

11.9 a) The researcher in Exercise 11.8 computes simultaneous confidence intervals for all pairwise differences. Assuming $FWE = .05$, what is the CI for the difference between the L and M marginal means? Is there a significant difference between the means?
b) The researcher tests the difference between the mean of the two combined liberal groups and the mean of the combined two conservative groups. State the null hypothesis and calculate the test statistic.
c) Consider each of the following situations for part (b). In each of the following cases, state the critical value for the test statistic and your conclusion ($\alpha = .05$): (i) The test was decided on before the data were collected and it is the only follow-up test. (ii) The researcher also decides *a priori* to test two additional hypotheses. (iii) After viewing the means, the researcher decides on the contrast in part (b).
d) Before collecting the data, the researcher wants to know whether the difference between moderates and the combined L and LL participants is different for old and young participants. (i) State the null hypothesis. (ii) Calculate the confidence interval and state your conclusion about the null hypothesis.

Notes

1 Given the violation of the normality assumption, we considered two alternatives to the usual ANOVA. One is the rank transform method described in Chapter 8. However, this test is valid only if there are no interaction effects (Akritas, 1990; Hora & Conover, 1984). We did perform an

ANOVA based on trimmed means and winsorized variances (see Chapter 8) and arrived at the same conclusions as in the analysis of the original data, although the trimmed data distribution more closely approximated the normal.

2 As noted in Chapter 10, this and several alternative post hoc tests are available in many statistical packages. The R commands are shown in Table 11.3, where *memout* is an *aov* object. In SPSS, select *Analyze*, then *General Linear Model*, and then *Univariate*. Next select the independent (e.g., organization, delay) and dependent variables (e.g., recall). To test pairwise differences among the organizational levels, select the *post hoc* option, and then select the test from the array presented. In this example, we chose *Tukey*.

Part 3

Repeated-Measures Designs

Chapter 12

Comparing Experimental Designs and Analyses

12.1 Overview

The preceding chapters were primarily concerned with analyses of data from a single experimental design, the completely randomized design. In a completely randomized design, each participant is randomly assigned to one of several experimental conditions. This is the simplest experimental design; as such, it provides a good foundation for developing the basic concepts of statistical inference and data analysis associated with ANOVA. However, random assignment of participants to experimental conditions is only one of several approaches to designing experiments. Other experimental designs and statistical analyses offer important advantages compared to the completely randomized design; most notably, they often have a smaller error variance and, consequently, greater power. In the current chapter, we will introduce these alternatives and compare them to the completely randomized design and to one another. Our goal in this chapter is to consider general issues in choosing among alternative designs. Therefore, we will present only the simplest versions of the designs and we will assume simple, highly constrained, structural models. Development of the details of the new designs will be left to subsequent chapters.

The organization of the chapter is as follows:

- *Factors influencing the choice of a design* will be discussed first. We identify the general considerations involved in selecting an approach to a research question.
- *Blocking designs* are a straightforward extension of the completely randomized design. Participants are divided into blocks based on a *concomitant variable* that is related to the dependent measure, resulting in smaller within-cell variance.
- *Analysis of covariance* (ANCOVA) also makes use of concomitant measures. The difference is that ANCOVA uses a concomitant variable as a *covariate* to provide a statistical adjustment to remove some of the error component of each of the scores in the data set. To the degree that the covariate and the dependent variable are correlated, the error variance is reduced.
- The *repeated-measures design*, sometimes referred to as a *within-participants design*, will be presented next. In the simplest version, each participant is tested in every condition. Because the variability due to individual differences can be removed from the error variance, this design provides still another way of dealing with the effects of nuisance variables.
- We end the chapter with *Latin square designs*. Participants are each tested in several counterbalanced conditions, with an equal number of participants receiving each possible order of the treatments. This strategy removes variability due to individual differences as well as removing the variability contributed by different treatment orders.

DOI: 10.4324/9781003453550-15

A central theme in comparing these alternative designs and statistical analyses is that they partition the same total variance in different ways, all with a shared goal of reducing error variance. A second, closely related theme of the chapter is that designs may be compared with respect to how effectively they reduce error variance. If design *A* results in less error variance than design *B*, we say that design *A* is *relatively efficient* compared to design *B*. Finally, a third theme of the chapter is that there are trade-offs in choosing among designs. Generally, more efficient designs are more complex designs, and more complex designs have more things that can go wrong with them.

12.2 Factors Influencing the Choice Among Designs

An important purpose of this chapter is to provide an understanding of the factors involved in deciding among experimental designs when a choice is available. One classification of such design factors has been proposed by Keren (1993; also see Chapter 1). He suggested three factors that should influence the choice of design; these are (1) theoretical, (2) methodological, and (3) statistical.

12.2.1 *Theoretical Issues*

Theoretical issues take priority in determining design choices in the sense that the theoretical question motivating an experiment often places clear constraints on what types of designs are, and are not, appropriate. Theoretical considerations will typically dictate whether the researcher should take one observation per participant or multiple observations. Keren (1993) provides an example. Suppose that a theory predicts the overall proportion of choices of one option versus another; in this instance, a between-participants design is adequate. However, if the theory predicts sequences of choices – for example, the probability of option *A* being selected, given that on the preceding trial option *B* had been selected – the researcher clearly needs to test participants in a series of trials in which both options are available. More generally, any study in which our theory predicts the effects of a series of responses upon subsequent responses requires a within-participants design.

As Keren (1993) pointed out, theoretical issues often reduce to consideration of context effects. For example, experiments with rats (Crespi, 1944; Zeaman, 1949), monkeys (Schrier, 1958), and college students (Schnorr, Lipkin, & Myers, 1966) have shown that the effects of amount of incentive depend on whether each participant is tested at several different levels of incentive (i.e., a within-participants design) or at only a single incentive level (i.e., a between-participants design). In this example, the use of a within-participants design is appropriate to test hypotheses about contrast effects, the effect of one incentive level as a function of whether the participant has previously been tested with a smaller or larger amount. The between-participants design is appropriate if interest centers on the absolute, rather than the relative, effect of incentive.

12.2.2 *Methodological Issues*

The choice of a research design is often affected by practical concerns. For example, if a between-participants design is considered, we must be able to recruit enough participants to ensure sufficient statistical power. However, if the task is boring or unpleasant, it may be difficult to recruit and retain participants. As another example, if we wish to test the same participants under several conditions, then completing the procedure may require either that participants be available for long time periods or that they be willing to participate in multiple sessions.

Developmental studies provide one example of some of the practical concerns in choosing between designs. Suppose we want to study the development of math skills taught by an experimental method to children in all grades in an elementary school. Should we test children in different grades simultaneously (*cross-sectional design*) or repeatedly test the same children as they proceed through the grades (*longitudinal design*)? There are trade-offs in making this choice. Cross-sectional studies avoid the possible loss of data as participants move from the area or change schools. They also avoid the possibility that the study will be disrupted, or learning affected, by changes over time in factors other than the variable of interest (e.g., an event producing a marked social or economic change such as a natural disaster, a pandemic, or military conflict). On the other hand, the longitudinal study eliminates variability due to differences among individuals in different age groups. For example, some children in the higher grades may have entered from another school and not been initially taught by the method of interest. At the least, this complicates the analysis of cross-sectional data because their data must be removed from analyses.

Even in laboratory studies in which multiple measures are obtained in a single session, there are potential issues. Possible practice or fatigue effects may be confounded with the effects of the independent variable when participants are tested under several conditions, or if different participants are tested at different times. In addition, repeated-measures designs may result in *carry-over effects* such that a participant's performance in each condition may be affected by their experience in earlier conditions. As we noted in our discussion of theoretical issues, such context effects may be of interest; however, if they are not, the researcher must carefully weigh the possibility of carry-over effects.

12.2.3 *Statistical Issues*

Designs differ with respect to the degree to which they control nuisance variables; this translates into differences in error variance. Designs with less error variance translate into more powerful statistical tests and narrower confidence intervals given the same number of observations. Thus, an important basis for comparing designs is with respect to their error variances; the ratio of the error variances of two designs is a measure of their *relative efficiency*. In this chapter, we will consider the relative efficiency of various pairs of designs and explain why some designs are more efficient than others.

In sum, the choice of a research design must balance theoretical, methodological, and statistical considerations. Theoretical considerations and practical constraints sometimes dictate a single, appropriate design. However, the researcher often has choices, even after theoretical and methodological issues are considered. In that case, the relative efficiency of alternative designs should be a major consideration in selecting the research design.

In the remainder of this chapter, we will discuss the methodological and statistical pros and cons of the four experimental designs cited earlier. We will omit further consideration of theoretical issues in selecting a design because these are related to specific research questions.

12.3 The Treatments × Blocks Design

12.3.1 The Design

Blocking usually refers to a design in which the participants are matched with respect to a measure that is correlated with the dependent variable. For example, in an experiment comparing three methods of teaching fractions to elementary school students, we might take steps to match the distributions of arithmetic skills of the children assigned to the three methods. One way to do this would be to administer a pretest designed to measure basic arithmetic skills of 36 children, and then divide the children into four blocks based on their pretest scores. The nine children scoring highest would then be randomly assigned to one of the three instructional methods, so that within this block there would be three students in a cell. The same would be done with the next highest scoring nine students, and so on until we had four blocks based on pretest scores.

Although this is a common type of blocking design, blocking could be done based on other variables that we assume will be related to performance on the dependent variable. If two experimenters are involved in testing participants, and if there is reason to believe that responses may be influenced by the person administering the test, half of each experimental group could be tested by each experimenter. In this case, experimenter is the blocking variable. Similarly, if there is reason to believe that the time of day at which testing takes place can affect performance, we can either test all participants at the same time of day or, if that is impractical, we can establish time of day as a blocking variable. In that case, if we had four testing times, 1/4 of each treatment group would be tested at each time.

Regardless of how it is accomplished, the primary purpose of blocking is to reduce error variance, thus increasing the power of the hypothesis test and decreasing the width of confidence intervals. We next consider how this occurs.

12.3.2 Blocking vs Completely Random Assignment

Suppose two investigators independently run the experiment we described in which 36 students were divided into three treatment groups, each group taught by one of three instructional methods. Experimenter *CR* randomly assigns each of the 36 participants to one of the three groups; the only restriction is that there are 12 students taught by each method. Experimenter *TB* follows the blocking procedure described in the preceding section, using pretest scores to divide the students into four blocks with equal *n* and then randomly assigning each of the nine students within a block to one of the three instructional methods, with the constraint that there are three students in a block taught by each method. The first design is a one-factor completely randomized design, with the data analyzed as described in Chapter 8. The data from the blocking design are analyzed as a two-factor between-participants design, as in Chapter 9.

The blocking design should result in a smaller error variance than the one-factor *CR* design based on the same total number of participants. This statistical benefit occurs because the three participants within each treatment × block ($A \times B$) cell will vary less in arithmetic knowledge than will the 12 participants in a level of *A* in the *CR* design, so we can expect that their scores on the posttest will also be less variable. For example, consider the following set of 12 posttest scores, grouped into three levels of ability as measured on a pretest:

Low				*Medium*				*High*			
7	12	8	5	8	12	13	14	14	9	16	15

The variances of the posttest scores for low, medium, and high ability groups are 8.67, 6.92, and 9.67, respectively; the average within-group variance is 8.42 (i.e., $MS_{S/AB}$). However, if the 12 scores were not grouped by pretest level, their variance would be 12.63 (i.e., $MS_{S/A}$) Thus, the error variance will be smaller in a treatments × blocks design than in a completely randomized design, *provided that the pretest scores are correlated with the posttest scores.*

To further understand the effect of blocking, we consider the average result of many independent replications of two experiments differing in their design. Table 12.1 presents the average sums of squares and mean squares for 5,000 independent replications of the two experimental designs under consideration. The results in the table were obtained by using a simulation to draw 5,000 samples of 36 pretest and posttest scores from a population in which the correlation between the two measures was 0.4. In the *CR* design, the 36 posttest scores then were divided randomly among the three treatment levels; in the blocking design, the posttest scores were divided among four blocks in order of their pretest scores, and then randomly divided among the treatments within each block. To simulate treatment effects, a constant was added to all posttest scores at a given treatment level in both designs, with a different constant for each level. An ANOVA was then performed for the data in each design and the proportion of the 5,000 samples that resulted in a rejection at the.05 level was recorded as power. The numerical results presented in Table 12.1 are averages over the 5,000 replications of the simulated experiment. There are several points to note about the results:

1. The average SS_{total} is virtually identical for the two designs, the very small difference being due to chance variability involved in performing a finite number of simulations.
2. The treatment variability, SS_A, is smaller in the blocking than in the *CR* design. To see why, consider the expected mean square for treatments, $E(MS_A)$. As we saw in Chapter 8, with 12 scores at each level $E(MS_A) = \sigma_e^2 + 12\theta_A^2$. But as the preceding numerical example demonstrated, we expect the error variance, σ^2_e, to be smaller in the $T \times B$ design than in the completely randomized design. Therefore, we expect the MS_A to be smaller in the blocking design.
3. The $T \times B$ design partitions both the *df* and *SS* for error in the CR design ($df_{S/A}$, $SS_{S/A}$) into variance components for the *B*, *AB*, and *S/AB* terms. The reduction in *SS* for error is proportionally larger than the reduction in *df*, resulting in a smaller *MS* for error in the $T \times B$ design ($MS_{S/AB} < MS_{S/A}$).
4. Power, the proportion of significant results, is greater in the $T \times B$ ANOVA, despite the fact that the MS_A and the error *df* – factors that affect power – are smaller in that design.

Power is greater in the $T \times B$ design than in the *CR* design because $MS_{S/AB}$ is smaller than $MS_{S/A}$. To understand why the two error terms differ in magnitude, consider the sources

Table 12.1 Average results of analyses of variance for completely randomized and blocking designs, based on 5,000 replications

Completely randomized design					*Treatments (A) × blocks (B) design*				
SV	*df*	*SS*	*MS*	*Power*	SV	*df*	*SS*	*MS*	*Power*
A	2	1,163.43	581.71	.550	*A*	2	1,123.40	561.70	.595
S/A	33	4,736.78	143.54		*B*	3	1,058.74	352.91	
Total	35	5,900.21			*AB*	6	746.19	124.37	
					S/AB	24	2,974.16	123.92	
					Total	35	5,902.49		

of random variation in the two designs. The error term of the *CR* design is influenced by measurement error and individual differences, including differences in arithmetic skills. We can expect that differences in arithmetic skills will be responsible for much variation in performance in the experiment, so by measuring those differences and incorporating them into the experimental design via blocking, we remove that variability from the error term of the $T \times B$ design. This conceptual difference in the two designs is realized in differences in how they partition the same total variance: whereas the *CR* design distinguishes just two sources of variability, the $T \times B$ design distinguishes four. The two additional terms of the $T \times B$ design – the block (*B*) source of variance and, to a lesser extent, the interaction (*AB*) – capture much of the variance contributing to $MS_{S/A}$ of the *CR* design. In short, by incorporating the pretest scores as a systematic source of variance in the design, we are left with a smaller residual error variance.

12.3.3 Relative Efficiency

We have presented an argument that the error term will be smaller and, consequently, power will be higher in the $T \times B$ design than in the *CR* design. However, the argument was based on a single example, so it is fair to question the generality of that conclusion. Will the $T \times B$ design always have more power than the corresponding *CR* design? If not, what factors determine the relative power of the two designs? When the $T \times B$ design does have a power advantage, is the added time, effort, and expense involved in executing the design worth the trouble? Key to answering these questions is the concept of *relative efficiency* (*RE*).

Two designs may be compared to determine their relative efficiency by computing the ratio of their MS_{error} terms. In our current example of the $T \times B$ and *CR* designs, the relative efficiency of the blocking design to the randomized design is

$$RE_{T\times B\,to\,CR} = \frac{MS_{S/A}}{MS_{S/AB}} \tag{12.1}$$

Treating the MS_{error} terms in Table 12.1 as if they came from a single experiment, and substituting them into Equation 12.1, we find the relative efficiency of our hypothetical experiment to be 1.158. A value greater than 1 indicates that the $T \times B$ design is more efficient than the *CR* design. The *RE* of these designs is greater than 1 if the pooled mean square due to blocking (the *B* and *AB* sources) is greater than the error mean square; i.e., $RE > 1$ if $\left[(SS_B + SS_{AB})/a(b-1)\right] > MS_{S/AB}$. This condition will hold if there is a correlation between the blocking variable and the dependent variable. Typically, it is a simple matter for a researcher to identify a blocking variable that is related to the dependent variable, so the efficiency of the $T \times B$ design will generally be greater than that of the *CR* design.

12.3.4 How Many Blocks?

If a blocking design is chosen, the researcher must decide how many blocks to use. As the number of blocks increases, the scores in the cells representing the combinations of treatments and blocks should be less variable, yielding a still smaller error term. However, the increase in blocks involves a loss of error degrees of freedom; if we had six instead of four blocks, n would equal 2, and the error $df = ab(n - 1) = 3 \times 2 \times (2 - 1) = 6$. There is a trade-off; the reduction in error variance with more blocks leads to more power, but the loss of degrees of freedom with more blocks leads to less power. Feldt (1958) published a table of the optimal number of blocks; that is, the number such that power is maximized. This optimal number of blocks depends on several factors:

1. *The correlation between X, the concomitant variable, and Y, the dependent variable.* If there is no correlation, blocking costs error degrees of freedom with no corresponding reduction in error variance because *Y* will not be affected by factors affecting *X*. As the population correlation coefficient increases, so does the optimal number of blocks.
2. *The total sample size, N.* Although increasing the number of blocks decreases the error degrees of freedom, this loss has less effect on power when the sample size is large; that is, when there are many error degrees of freedom to begin with.
3. *The number of levels of the independent variable, A.* As the number of levels of *A* decreases, the number of blocks that can be profitably used increases because the loss in error degrees of freedom is smaller when *a* is lower.

In sum, the efficiency of the blocking design depends on *N*, *a*, and ρ, the correlation between concomitant variable and the dependent variable. This last factor is particularly important. Blocking requires more time, effort, and often expense than random assignment. That cost will result in more powerful tests and more precise estimates of effects only to the extent that the concomitant variable is a good predictor of performance.

We have spent considerable space on the blocking design to illustrate certain concepts structured in a format that should be familiar from the preceding chapter. However, there are alternatives available to the researcher. One such alternative is the analysis of covariance; this analysis makes use of concomitant data in a different way from that of the blocking analysis. We introduce this approach to increased efficiency next.

12.4 The Analysis of Covariance

In both the blocking design and the analysis of covariance (ANCOVA), a concomitant variable is used to account for variance that would otherwise be treated as random (i.e., error). As we have seen, the blocking design reduces error by incorporating the concomitant variable, *B*, into the design and thus extracting variance associated with *B* from the error term. In ANCOVA, the concomitant variable, *X*, is not incorporated into the experimental design; rather, participants are assigned randomly to conditions without regard to their scores on *X*. However, the correlation of *X* with the dependent variable, *Y*, is used to remove from SS_{total} the variance that is predictable from *X*. The resulting, *adjusted* SS_{total} is then partitioned into variability attributable to *A* and *S/A*. Thus, ANCOVA is like the blocking design in that it also uses the concomitant variable to reduce the error variance; however, it accomplishes the reduction statistically instead of incorporating *X* into the design of the experiment.

ANCOVA is best developed within a regression framework and therefore we postpone most details until Chapter 24. For now, we will present ANCOVA conceptually, using a numerical example.

12.4.1 ANCOVA: A Numerical Example

Table 12.2 presents a data set. The independent variable, *A*, represents three instructional methods. As in the completely randomized design, 12 students were randomly assigned

Table 12.2 Data to illustrate analysis of covariance

	Y	53	59	69	57	55	63	54	81	59	63	59	71
A_1	X	66	84	60	78	59	72	69	80	81	55	65	81
	Y	79	74	80	69	66	65	72	71	62	73	69	79
A_2	X	100	60	80	68	63	66	85	72	47	89	78	85
	Y	60	70	80	57	50	83	97	91	86	78	52	81
A_3	X	54	85	72	41	55	74	74	85	70	69	55	71

Note: Y is the dependent variable and X is the covariate.

to each method. As in the blocking ($T \times B$) design, a pretest of basic arithmetic skills was administered because the pretest scores should be highly correlated with performance on the posttest. In Table 12.2, the X values are the pretest scores and the Y values are the posttest scores. Our interest is in the effects of the instructional methods on the posttest score.

Table 12.3 presents the results of two analyses: An analysis of variance in which the pretest scores are not relevant, and an analysis of covariance in which the pretest scores are used to account for some of the variance in the original data set. Note that the adjusted total sum of squares and error sum of squares in the ANCOVA are smaller than the unadjusted terms in the ANOVA. Those reductions can be explained by reference to correlations between the X and Y data. The adjusted total sum of squares is the portion of the total sum of squares that *cannot* be predicted from X. Although we do not usually calculate the terms in Table 12.3 this way, we could obtain the adjusted sum of squares (3,614 in our example) by first calculating r, the correlation between all the X and Y scores. The square of r is the proportion of the variance of the Y scores that is predictable from our knowledge of X (see Chapter 24). Therefore, the proportion of the total sum of squares that is *not* predictable from X is $1 - r^2$, and the amount of the total variability not predicted is $SS_{total(adj)} = (1 - r^2)SS_{total(y)}$. In our example, $r^2 = .2415$ and $SS_{total(adj)} = (1 - .2415) \times 4764.75$, or 3613.905. This equals the adjusted total sum of squares in the ANCOVA results of Table 12.3. Similarly, the ANCOVA error variance is adjusted depending on the within-condition correlation of the covariate and dependent variable; as a result, the error term is reduced relative to the ANOVA.

Chapter 24 presents ANCOVA as a special case of multiple regression. For now, the important point to understand is that the increased efficiency of the covariance analysis is a function of the variability in Y predictable from X that would otherwise show up in the error term of an ANOVA. This reduction in error is achieved with the loss of only one error degree of freedom, but at the cost of more complexity and more assumptions.

12.4.2 *Comparing ANCOVA to the CR and T × B Designs*

We have indicated that when its assumptions are met, ANCOVA can provide more power than the *CR* analysis. But how does ANCOVA compare with analyses based on blocking? One simple answer is that ANCOVA "costs" only one error degree of freedom, whereas a blocking design is more costly as the number of blocks increases. As stated earlier, Feldt (1958) derived the optimal number of blocks to be used in the blocking design. Using those numbers of blocks, he concluded that the one-way ANCOVA was almost always more efficient than a treatments × blocks design. However, the differences were often small, especially for larger sample sizes ($N \leq 50$) and, as we will see in Chapter 24, ANCOVA comes with some stringent assumptions that the blocking design avoids.

Table 12.3 Analyses of the data of Table 12.2

Source	*Sums of squares*	*df*	*Mean squares*	*F*	*p*
Analysis of variance					
A	953	2	476.3	4.123	.025
S/A	3,812	33	115.5		
Total	4,765	35			
Analysis of covariance					
A	1,122	2	560.8	7.201	.003
S/A	2,492	32	77. 9		
Total	3,614	34			

Maxwell, Delaney, and Dill (1984) offered two alternative design approaches that promise greater power than simply choosing between a blocking design and ANCOVA. One analysis is a hybrid approach in which the assignment of participants to treatments is based on their concomitant scores together with an analysis of covariance on the dependent measure (Maxwell et al., 1984). The other approach is an alternative ranks method that uses a method for assigning participants to blocks that results in a closer matching of blocks on the concomitant variable than does the *TB* design (Dalton & Overall, 1977).

When used properly, ANCOVA and the variants just described can be an excellent tool for reducing error variance and thereby increasing power. In practice, ANCOVA is often used not only in designs involving random assignment of participants, but also in designs that involve pre-established groups. In such circumstances, the covariate may differ across groups, leading to possible difficulties of interpretation. We will consider interpretative issues and provide further details on the analysis and its underlying assumptions in Chapter 24.

12.5 Repeated-Measures (RM) Designs

12.5.1 The Design

The designs discussed in the preceding sections are between-participants designs in which N participants are distributed among the a conditions, and each participant contributes exactly one score to the data set. In many studies, however, participants contribute scores in several conditions. In these repeated-measures designs, instead of an participants, there are an observations comprised of n participants each tested in all a conditions.

12.5.2 The ANOVA

Suppose a researcher wishes to compare the tastes of four light beers. Having different participants each rate only a single beer requires many participants for only a few seconds each. A more efficient use of participants is to have a smaller number each rate all four. The beers are presented for a single taste in a different random order to each of eight participants with a short interval in between, during which participants rinse their mouths. Table 12.4 presents ratings (Panel *a*) and ANOVA results (Panel *b*).

Table 12.4 Data and analysis for a repeated-measures design
(*a*) the data

Subject	*Beer 1*	*Beer 2*	*Beer 3*	*Beer 4*	*Mean*
1	2	5	3	3	3.25
2	4	6	5	4	4.75
3	5	7	4	5	5.25
4	3	4	3	4	3.50
5	6	7	6	5	6.00
6	2	5	4	3	3.50
7	4	5	6	4	4.75
8	3	6	4	6	4.75
Mean	3.63	5.63	4.38	4.25	4.47

(*b*) the ANOVA

SV	*df*	*Sum of squares*	*Mean square*	*F*
Participants (*S*)	7	26.219	3.746	6.091
Beer (*A*)	3	16.844	5.615	9.136
Residual	21	12.906	.615	

In general, this participants × treatments design involves *a* levels of the treatment variable and *n* participants; here $a = 4$ and $n = 8$. Mean squares are calculated as in the two-way completely randomized or treatments × blocks designs. The term labeled "residual" in Table 12.4 is computed as an interaction of $S \times A$;[1] the only unusual aspect of the calculations is that there is only one score per cell instead of a mean of several scores, as in a between-participants design. The residual term serves as the error term to test the effect of *A*, as we will justify in Section 12.5.3.

The design summarized in Table 12.4 is usually referred to as a one-way, repeated-measures design. In the case of the one-way, completely randomized design, recall that we were able to partition the total variance into just two sources of variance – the main effect of *A* and an *S/A* term that served as the error term. In the case of the repeated-measures design, however, the fact that each participant participates in all conditions of the experiment makes it possible to distinguish the main effect of participants from the interaction of participants with treatments (labeled "residual" in Table 12.4b). This finer partitioning of the variance in the repeated-measures design usually results in a substantial reduction of the error term (i.e., "residual") relative to the completely randomized design. For example, if the data in Table 12.4 had been collected in a CR between-participants design, the $MS_{S/A}$ would be 1.397, more than double the repeated-measures error term.

Earlier (see Section 12.3.2), we attributed the greater efficiency of the $T \times B$ design to the fact that blocking removed some individual difference variability from the error term, thus reducing the error term relative to the completely randomized design. The use of a concomitant variable in ANCOVA was similarly presented as statistically removing some individual difference variability from the error term, relative to the completely randomized design. In the case of the repeated-measures design, we are essentially blocking by participants, with the result that differences in the performances of participants are entirely removed from the error term of that design. Thus, the participants × treatments design will typically result in a smaller error term than any of the designs we have considered to this point.

The *Participants* source of variance (*S*) provides an indication of the potential increase in power if an *RM* design is used instead of a *CR* design. In many types of research, we would expect to find considerable individual differences in performance. This variability would be a component of the error term of the ANOVA for a *CR* design. Thus, the repeated-measures design will be more efficient than the completely randomized design (and other between-participants designs) when there is substantial individual variability. However, if there is little variability among participants' mean scores, it would suggest that we had not greatly reduced error variance relative to the *CR* design. In such a situation, the repeated-measures design could actually be less efficient than the completely randomized design because the *RM* design has fewer error degrees of freedom. In the example of Table 12.4, there are only 21 error degrees of freedom compared to 28 in a completely randomized design with the same number of scores. To better understand the issue of the efficiency of the *RM* design compared to the *CR* design, we next consider the expected mean squares.

12.5.3 *Expected Mean Squares and Relative Efficiency*

Panel *a* of Table 12.5 presents the results of 5,000 independent simulations of two experimental designs. In each simulated replication of the completely randomized experiment, there were three groups of 10 scores; each of the 30 scores consisted of an error component and an individual difference component. The three groups of scores each had a different treatment effect added. For the replications of the repeated-measures experiment, the three scores for a simulated participant contained different errors of measurement, but all three scores had the same value of the individual difference error. The two types of errors were

Table 12.5 Results of analyses of variance for completely randomized and repeated-measures designs, based on 5,000 replications (*a*) and expected mean squares (*b*)

(*a*) Simulation results

Completely randomized design					*Repeated-measures design*				
SV	*df*	*SS*	*MS*	*Power*	*SV*	*df*	*SS*	*MS*	*Power*
A	2	589.38	294.69	.235	A	2	391.53	195.77	.684
S/A	27	3,666.55	135.90		Subjects	9	3,024.17	336.02	
Total	29	4,255.93			Residual	18	650.68	36.15	
					Total	29	4,066.38		

(*b*) Expected mean squares

Completely randomized design			*Repeated-measures design*		
SV	*df*	*EMS*	*SV*	*df*	*EMS*
A	$a - 1$	$\sigma_e^2 + \sigma_S^2 + n\theta_A^2$	A	$a - 1$	$\sigma_e^2 + n\theta_A^2$
S/A	$a(n - 1)$	$\sigma_e^2 + \sigma_S^2$	Subjects	$n - 1$	$\sigma_e^2 + a\sigma_S^2$
			Residual	$(a - 1)(n - 1)$	σ_e^2

drawn from the same population as in the completely randomized simulations. Three treatment effects were added, as in the completely randomized experiment.

There are several points to note about the results in Panel *a* of Table 12.5.

1. Summing the average values of $SS_{Subjects}$ and $SS_{Residual}$ the result is 3674.85, within sampling error of the average $SS_{S/A}$ (3,666.55). This is because the average error sum of squares for the *CR* design contains variability due to measurement errors and individual differences. However, in the *RM* design, the variability due to differences in performance across participants winds up in the participants source of variance with the consequence that the $SS_{Residual}$ term in the *RM* analysis is usually much smaller than the $SS_{S/A}$ term of the *CR* analysis.
2. The SS_A is smaller in the *RM* design. This is because individual differences no longer contribute to the differences among the observed means in the *RM* design.
3. Despite having a smaller numerator mean square and fewer error degrees of freedom, the *RM* design resulted in considerably more power; that is, the proportion of the 5,000 replications resulting in a rejection of the false null hypothesis was .684 as opposed to only .235 for the *CR* design. This power advantage reflects the fact that much of the error variance in the *CR* design was due to individual differences, and these did not contribute to the *RM* error term.

In summary, when both designs are practical, the repeated-measures design will usually be more efficient, both in the use of participants and with respect to error variance. The relative efficiency promises greater power and narrower confidence intervals. However, one caution is in order. We have assumed a very simple structural model in presenting our results. There are alternative, more complicated, and often more realistic models for data obtained with the repeated-measures design. We will consider this topic and other aspects of the design and analysis of related data in the next few chapters.

12.5.4 Advantages and Disadvantages of the RM Design

We have established that the repeated-measures design is usually a relatively efficient design compared to the between-participants designs we considered earlier in the chapter. An advantage of this efficiency is that fewer participants are required to achieve the same degree of statistical power in the *RM* design. This is important if the population is limited in size, as with many clinical populations; or when participants are difficult to recruit, as when the task is very boring or dangerous; or when the experiment involves expensive animals such as monkeys. Even without these constraints on participant availability, the *RM* design may prove more practical than a between-participants design. For example, if it takes very little time to obtain a score from a participant, it may be more efficient to run one participant in several conditions than to run several participants each in a different condition.

Repeated-measures designs make efficient use of participants, both in the sense of requiring fewer participants than between-participants designs, and in the sense of having less error variance. However, not all independent variables lend themselves to such designs. For example, participant variables such as gender identity or clinical category must be treated as between-participants factors. Also, many experimental manipulations such as surgical procedures or instructional methods do not lend themselves to within-participant manipulation. For example, in an experiment designed to compare the effectiveness of different methods of teaching mathematics, knowledge achieved by being exposed to one of the methods cannot be miraculously expunged so that it can be relearned by a second method.

Although the between-participants designs we considered in previous chapters may involve more error variance, they are relatively simple. Scores in different groups can be assumed to be independent and the within-cell variance can be used as the error term for testing any effect. We pay for the *RM* design's smaller error variance with some additional complexity:

- *Carry-over effects*: The influence of a treatment at one point in time upon a score in a later condition can complicate the interpretation of treatment effects.
- *Scores will be correlated* because each participant contributes several scores; as we will see in the next few chapters, this will have implications for the validity of the *F* test of treatments.

In short, like any other design, the *RM* design is not appropriate for all situations. However, when it is a suitable choice, its high efficiency makes it a very attractive option.

12.6 The Latin Square Design

12.6.1 An Example of the Design

Suppose we wish to measure accuracy on five tasks of varying structure. We intend to present each task for a block of trials, recording percent correct for each block. Because a trial block requires little time to complete, we decide to present the tasks in a series of five blocks such that there will be a score for each participant in each trial block. We have at least two options with respect to the design. If we have *n* participants, we may construct *n* random sequences of the five trial blocks, assigning each participant to a sequence. For five participants, the sequences in this repeated-measures design might look like the left side of Table 12.6. Note that some treatments appear more often than others in some sequential positions. For example, the fifth condition (A_5) is presented first to two participants and second to another two. If there is an improvement with time in the situation, performance on this task would be at a disadvantage because its average position is early in the sequence of trial blocks. Although randomization ensures that the average sequential positions of the tasks will be equal over replications of the experiment, the practice effect does contribute to error variance, rendering the design less efficient than it would be if we could somehow remove these practice effects from the error variance.

The right side of Table 12.6 presents a counterbalanced design in which each treatment appears once in each row (participant) and column (like the numbers in a game of Sudoku). Columns may represent blocks of trials, as in the preceding example, or sets of materials, as in many studies of language processing. The column variable is an independent variable in this *Latin square design* and we can now separate its effects from the error variance. Therefore, this design has the potential to be more efficient than the repeated-measures design in which treatments are randomly assigned to trial blocks, or material sets. Although the design on the right is potentially a very efficient one, therefore promising considerable power and precision of estimates, there are some potential drawbacks, as in the following:

Table 12.6 Design layouts for five participants with random sequences and Latin squared sequences (*A* refers to a treatment condition, *C* to a trial block, and *S* to a participant)

Random sequences						*Latin squared sequences*					
Trial block						*Trial block*					
Participant	C_1	C_2	C_3	C_4	C_5	*Participant*	C_1	C_2	C_3	C_4	C_5
S_1	A_5	A_3	A_2	A_1	A_4	S_1	A_1	A_3	A_4	A_2	A_5
S_2	A_2	A_1	A_4	A_3	A_5	S_2	A_2	A_4	A_5	A_3	A_1
S_3	A_4	A_5	A_3	A_1	A_2	S_3	A_5	A_2	A_3	A_1	A_4
S_4	A_3	A_5	A_4	A_2	A_1	S_4	A_4	A_1	A_2	A_5	A_3
S_5	A_5	A_1	A_4	A_3	A_2	S_5	A_3	A_5	A_1	A_4	A_2

1. As we will see when we consider the ANOVA, the $df_{error} = (a-1)(a-2)$. If $a = 4$, the error $df = 6$, and power may be quite low. For this reason, it is often recommended that there be at least five levels of *A*. In most psychological research, there are several participants in each row of the square, providing an additional source of variance, a potential error term with more degrees of freedom than in the design in which there is only one participant in each row.
2. It is not possible to analyze interactions among the three factors. Furthermore, if such interactions are present in the population, they reduce the efficiency of the design and may bias the test.
3. There may be effects of one treatment level on a subsequent one. Such carry-over effects differ from trial block effects in that the latter involve a practice, acclimation, or fatigue effect associated with the position in time, whereas carry-over effects represent the effect of exposure to one treatment upon responses to a subsequent one. There are modifications of the basic Latin square design that are balanced so that the residual effect can be estimated (e.g., Williams, 1949), but the analysis becomes complicated.

Despite these limitations and potential drawbacks, the basic Latin square is a component of designs frequently used in research. Often, each row of the square represents a group of participants, rather than a single individual. We will consider that variation of the design in Chapter 15; here we focus on the basic Latin square design.

12.6.2 Analyzing the Data

Suppose we are interested in studying the effects of the structure of a display upon the probability of detecting a target on a screen within some fixed interval of time. The display might be an open field (A_1) or be segmented into one to four vertical areas (A_2, A_3, A_4, and A_5). Participants are tested in each of five counterbalanced blocks, and the proportion of correct detections is recorded. Panel *a* of Table 12.7 presents the design and the data. The five columns on the left represent the five trial blocks, with C_1 being the first and C_5 being the last. For example, the first participant (S_1) was first tested in a trial block with the A_1 (open field) display, then in a block with A_3 (two segments), and so on. On the right side of the panel, the scores have been rearranged so that all the scores from the same treatment level are in the same column.

There are three independent variables in this design: *S*, *A*, and *C*. The sums of squares for these main effects can be calculated by the usual formulas. For example, the SS_A is *c* times

Table 12.7 Data and ANOVA for a Latin square design

(a) Data

	Trial Blocks					*Levels of Factor A*				
	C_1	C_2	C_3	C_4	C_5	A_1	A_2	A_3	A_4	A_5
S_1	(A_1) 58	(A_3) 58	(A_4) 73	(A_2) 63	(A_5) 71	(C_1) 58	(C_4) 63	(C_2) 58	(C_3) 73	(C_5) 71
S_2	(A_2) 49	(A_4) 54	(A_5) 53	(A_3) 60	(A_1) 57	(C_5) 57	(C_1) 49	(C_4) 60	(C_2) 54	(C_3) 53
S_3	(A_5) 85	(A_2) 79	(A_3) 83	(A_1) 85	(A_4) 84	(C_4) 85	(C_2) 79	(C_3) 83	(C_5) 84	(C_1) 85
S_4	(A_4) 77	(A_1) 73	(A_2) 74	(A_5) 86	(A_3) 82	(C_2) 73	(C_3) 74	(C_5) 82	(C_1) 77	(C_4) 86
S_5	(A_3) 56	(A_5) 65	(A_1) 50	(A_4) 64	(A_2) 59	(C_3) 50	(C_5) 59	(C_1) 56	(C_4) 64	(C_2) 65
Mean	65	65.8	66.6	71.6	70.6	64.6	64.8	67.8	70.4	72.0

(b) ANOVA

Source	*df*	*SS*	*MS*	*F*
A	4	217.84	54.46	3.80
C	4	177.44	44.36	3.10
S	4	3,074.64	768.66	53.64
Residual	12	171.92	14.33	
Total	24	3,641.84		

the sum of the squared deviations of the five treatment (A) means around the grand mean, and $df_A = a - 1$. Similarly, SS_C *is a* times the sum of the squared deviations of the trial block (C) means around the grand mean, and $df_C = c - 1$. In this Latin square design, $a = c = n$ (see Table 12.6), so we can write the degrees of freedom as $(a - 1)$ for each of the independent variables. The *residual* sum of squares is obtained by subtraction of the three sums of squares for treatments, rows, and columns from the total sum of squares. The residual *df* are also calculated as a difference: The total *df*, $a^2 - 1$, minus the sum of the *df* for the main effects, $3(a - 1)$, equals $(a - 1)(a - 2)$. Note that we have not mentioned the calculation of *SS* terms for any interaction, because the analysis assumes there are no interactions among the variables. If that assumption is wrong, any interaction variability will contribute to the residual variance with the consequence that tests of the three main effects will be negatively biased. We will have more to say about calculations in an extended discussion of this design and the related analysis in Chapter 15.

12.6.4 *Efficiency of the Latin Square Relative to the Repeated-Measures Design*

The error degrees of freedom for the repeated-measures design with a participants and a treatment levels is $(a - 1)(a - 1)$, whereas that for the Latin square design is $(a - 1)(a - 2)$. However, despite fewer error degrees of freedom, the Latin square design will be more efficient than the repeated-measures design if the column variability makes a large contribution to the total variability. We can see this in the following estimate of the mean square error for the repeated-measures design (MSE_{RM}) derived from the Latin square ANOVA results:

$$estMSE_{RM} = \frac{(a-1)MSE_{LS} + MS_C}{a} \tag{12.2}$$

where the terms on the right are the residual and column mean squares from the Latin square ANOVA. Notice that if $MS_C = 0$, $estMSE_{RM} = ((a - 1)/a)MSE_{LC}$ and the repeated-measures design is more efficient. However, as MS_C increases, the point at which $estMSE_{RM} > MSE_{LC}$ is quickly reached. Clearly, despite the loss of four error degrees of freedom, the Latin square design will usually provide greater power and more precise estimates of population parameters than will the repeated-measures design.

12.7 Summary

Chapter 12 has introduced four new approaches to designing experiments and analyzing the data. Each approach seeks to reduce error variance with corresponding benefits for power and parameter estimates.

- The *treatments × blocks design* attempts to remove some sources of individual differences relevant to performance by establishing blocks of participants using a concomitant variable.
- *ANCOVA* assigns participants randomly to conditions but uses a concomitant variable to statistically remove some sources of individual differences.
- The *repeated-measures design* involves testing participants in all the conditions of the experiment with the result that individual difference variance may be entirely removed from the error term of the design.
- The *Latin square design* is a design in which the sequence of conditions is counterbalanced across participants. This procedure permits removing not only the contribution of individual differences from the error mean square, but also the contribution of temporal (sequential) effects due to fatigue or practice.

When executed successfully, both the $T \times B$ design and ANCOVA reduce error relative to the completely randomized design; the *RM* design reduces error further; and the Latin square design reduces error further still.

Although we have emphasized the increased efficiency of these four designs and analyses, relative efficiency is not the only consideration in selecting a research design. We have noted that designs differ in how much effort and expense they require to implement. In addition, more complex designs typically involve more stringent assumptions and more complex statistical models. We will consider these issues thoroughly when we study each of these alternative design.

Exercises

12.1 [Comparing *CR* and $T \times B$ designs] A teacher compared three types of instructional methods (A) on mathematics final exam test scores (Y). In the research design, students were first divided into four blocks (B) of 15 each based on a pretest score (X), then assigned to the three conditions. In summary, the design had three levels of A

roughly matched for ability and four ability blocks, with five students in each $A \times B$ cell. The data are in the file *EX12_1* at the website.

a) In analyzing the data, the teacher ignored the X data and also did not consider B as a factor in the design. Assuming that she had accomplished her purpose by equating the three instructional conditions, she analyzed the data as if from a one-factor between-participants design with 20 scores at each level. Do this analysis and explain the potential problem with it.
b) Perform an analysis taking B into consideration and state your conclusions.

12.2 [Relative efficiency of $T \times B$ and CR designs]

The teacher in Exercise 12.1 could have randomly assigned 60 students to the three conditions without blocking. We can estimate the error variance for this design from the results for the blocking design, using this equation:

$$est\left(MS_{S/A}\right) = \left[1 - \frac{a(b-1)}{abn-1}\right] MS_{S/AB} + \frac{SS_B + SS_{AB}}{abn-1}$$

a) Calculate the estimated error variance for a completely randomized design.
b) Use Equation 12.1 to estimate the relative efficiency of the $T \times B$ and CR designs.

12.3 [Effect sizes and power]

a) From the ANOVA of the treatments × blocks design (Exercise 12.1, part (b)), estimate Cohen's f for the A source.
b) Using the estimate of σ^2_A from part (a) and the error variance estimated in Exercise 12.2, estimate Cohen's f for the A source for the CR design.
c) Using the two estimates of f, compare the power of the two designs for this study.
d) You should have found that the $T \times B$ design provides a more powerful test of A. How many students would be needed in the CR design to match the power obtained with the $T \times B$?
e) Find the N needed to achieve .8 power for the $T \times B$ design. Then, using this value and the relative efficiency estimated in Exercise 12.2, find the N needed if you had used a completely randomized design.

12.4 [Deciding between designs] Blocking is not always preferable to completely random assignment of participants to treatments. What factors should influence the decision to divide participants into blocks?

12.5 [Comparing ANCOVA and CR designs] Analysis of covariance (ANCOVA) provides another approach to analyzing the data of Exercise 12.1. Consider the summary tables from the one-factor ANOVA and from the ANCOVA (ignoring B in both analyses).

ANOVA			*ANCOVA*		
Source	*df*	*SS*	*Source*	*df*	*SS*
A	2	861.233	A	2	1,006.147
S/A	57	10,096.700	S/A	56	9,873.610
Total	59	10,957.933	Total	58	10,879.757

a) In this chapter, we stated that $SS_{total(adj)} = (1 \times r^2)SS_{total(y)}$, or $r^2 = 1 \times SS_{total(adj)}/SS_{total(y)}$. Calculate the correlation for the 60 XY pairs and show that this relation holds for the two totals in the preceding tables.

b) A similar relationship holds between the two error sums of squares: $SS_{S/A(adj)} = (1 \times r^2_{S/A})SS_{S/A(y)}$. The r in this case is an average of three correlations, one from each level of A. However, the best estimate is not obtained by adding the three correlation coefficients and dividing by three. Instead, $\left[r^2_{S/A} = \left(\Sigma_j SP_j\right)^2\right]\left[\left(SS_{S/A(X)}\right)\left(SS_{Y(S/A)}\right)\right]$ where SP_j is the sum of cross-products; $SP_j = \Sigma_i\left(X_{ij} - \bar{X}_{.j}\right)\left(Y_{ij} - \bar{Y}_{.j}\right)$, the numerator of the covariance of X and Y in group A_j. The sums of squares terms are as usually defined; they are the sums of squared deviations of scores about their group means, summed over groups. Although most software has an option for calculating these terms, for convenience, we provide them here. Calculate $r^2_{S/A}$ and use it to verify the relation between it and the two error sums of squares. Here are the necessary sums of cross-products and sum of squares.

	A_1	A_2	A_3
SS_X	1,736.000	3,591.750	1,874.200
SS_Y	2,624.550	2,418.950	5,053.200
SP_{XY}	-749.000	2,010.750	5.800

12.6 [*RM* analysis] The following data set consists of three scores for each of four participants:

	A_1	A_2	A_3
S_1	10	14	16
S_2	8	8	17
S_3	9	9	13
S_4	7	6	9

Calculate the SS_{total}, SS_S, SS_A, and SS_{SA}; $SS_{SA} = SS_{total} \times (SS_S + SS_A)$. Divide SS_S, SS_A, and SS_{SA} by their degrees of freedom to obtain the mean squares. Then, assuming the expected mean squares of Table 12.5, Panel *b*, calculate the *F* test for *A*.

12.7 [Effect sizes and power]

a) Cohen defined his effect size, *f*, for the repeated-measures design as σ_A / σ_W; σ^2_W is the variance within treatment populations; $\sigma^2_W = \sigma^2_S + \sigma^2_e$. From the *EMS* of Table 12.5, estimate the variances based on the data in Exercise 12.6; then estimate *f*.

b) In G*Power 3.1, select the following options: *F test*, *ANOVA: Repeated-measures, within factors*, and *Post hoc*. Then enter $\alpha = .05$, total sample size = 4 (the number of participants), number of groups = 1, number of measurements = 3 (number of within-participant conditions), correlation among repeated measures = .75, and epsilon = 1. Insert your estimate of *f* from part (a). Calculate post hoc power, assuming these values.

c) Now use G*Power's option for *ANOVA Fixed effects Omnibus oneway*. Assuming the value of f calculated in part (a), estimate post hoc power for the completely randomized design.
d) Assuming we wished to redo the experiment to detect a medium sized effect ($f = .25$) with power = .9 and $\alpha = .05$, how many observations would be required for each design?
e) How does the relation between the required sample sizes change as r decreases?

12.8 [Comparing *LS* and *RM* designs] Five experts taste each of five wines and rate each on a 100-point scale. The order of presentation is counterbalanced so that over the five participants, each wine appears in each of the five possible positions in the sequence of tastings. Let *A* be the wines and C be the position in time. The data, also present on the website in the file *EX12_8*, are organized by wines (*A*) and position in time (*C*):

A_1	A_2	A_3	A_4	A_5	C_1	C_2	C_3	C_4	C_5
82	83	80	84	81	82	83	80	84	81
89	84	91	84	79	84	79	89	84	91
88	86	87	88	84	86	87	88	84	88
88	86	89	95	77	77	88	86	89	95
86	86	78	78	79	78	78	79	86	86

a) Often researchers ignore the blocking factor, C, and carry out the following ANOVA:

Source	*df*	*SS*	*MS*	*F*	*p*
Subjects (*S*)	4				
Wines (*A*)	4				
SA	16				

Complete the table.

b) A similar analysis can be performed, ignoring *A*, and having participants *C* and *SC* as the sources. Do this analysis.
c) Subtract SS_c from part (b) from SS_{SA} in part (a). Call the result $SS_{Residual}$. Also subtract SS_A from part (a) from SS_{SC} in part (b). How does this compare with the residual sum of squares?
d) As we did for Table 12.7, we can carry out a complete Latin square ANOVA. The sources and *df*s are as follows:

Source	*df*	*SS*	*MS*	*F*	*P*
Subjects (*S*)	4				
Wines (*A*)	4				
Time (*C*)	4				
Residual					

Complete the table, testing *A* and *C* against the residual mean square. What are the differences between the results of this analysis and those in parts (a) and (c)?

12.9 [Relative efficiency of *LS* and *RM* designs] Applying Equation 12.2 to the data in Exercise 12.8, estimate the error term if each sequence of ratings had been randomized independently for each participant.

Note

1 As for between-participants designs, the *S* denotes "Subjects," a traditional term for "participants." We retain this notation in the ANOVA tables and as subscripts to sources of variance to be consistent with most software.

Chapter 13

One-Factor Repeated-Measures Designs

13.1 Overview

Our introduction to the repeated-measures design in Chapter 12 emphasized its efficiency, both with respect to the need for fewer participants and its smaller error variance relative to the completely randomized design. In developing our case for the small error variance of the repeated-measures design, we assumed that participants and treatments (or trials) did not interact. The structural model corresponding to this assumption is called an *additive model*. It implies that if we exclude measurement error and plot the performance of the population of participants as a function of treatment level (or trial), the functions will be parallel; in other words, the treatment effects will be the same for all participants in the population.

The alternative to the additive model is a *nonadditive model* that includes a participants × treatments interaction term. That is, it is assumwwwed that treatment effects are different for different participants. This is probably the more realistic model for most situations. However, because the additive model enables us to introduce certain ideas in a simpler context, we will begin developments with that model.

To summarize, the goals of this chapter are as follows:

- To present two alternative models of the structure of the data: The additive model in which participants × treatments ($S \times A$) interaction effects are not present in the population, and a nonadditive model in which they are assumed to be present. We consider possible problems arising from nonadditivity and possible solutions to those problems.
- To explain fixed and random-effects variables, and their implications.
- To introduce an additional assumption of the repeated-measures ANOVA, *sphericity*, and describe the consequences of nonsphericity.
- To define several measures of the effect of the independent variable.
- To discuss the factors that influence decisions about sample size when planning experiments and the use of G*Power 3.1 to estimate the required sample size.
- To present tests of contrasts for the repeated-measures design.
- To describe the problem of different types of missing observations in repeated-measures designs.
- To present nonparametric procedures, based on ranks, that are designed to test hypotheses when the data do not conform to the assumptions underlying the analysis of variance.

DOI: 10.4324/9781003453550-16

13.2 The Additive Model in the One-Factor Repeated-Measures Design

13.2.1 The Structural Model

We assume that the *n* participants are a random sample from an infinitely large population of participants. We view Y_{ij}, the score of the *i*th participant tested in the *j*th condition, or on the *j*th trial, as being composed of a *true score*, μ_{ij}, and measurement error, ε_{ij}; that is,

$$Y_{ij} = \mu_{ij} + \varepsilon_{ij} \tag{13.1}$$

If the different treatment levels all have the same effect on the scores and if all participants in the population have the same scores, all the true scores will equal the grand mean of the population, μ. However, different treatments may have different effects and different participants will score differently because of differences in ability, experience, or motivation. Therefore, the parameter μ_{ij} in Equation 13.1 can be viewed as being equal to the population mean plus the effects due to the *j*th treatment and the *i*th participant. Acknowledging this, Equation 13.1 is rewritten as

$$\begin{aligned} Y_{ij} &= \mu + (\mu_i - \mu) + (\mu_j - \mu) + \varepsilon_{ij} \\ &= \mu + \eta_i + \alpha_j + \varepsilon_{ij} \end{aligned} \tag{13.2}$$

The parameters of this structural equation are defined in the upper panel of Box 13.1, and related assumptions are in the lower panel. The model of the data represented by Equation 13.2 is referred to as *additive* because scores are viewed as reflecting the sum of participant (η_i; eta-sub-*i*) and treatment (α_j) effects; no interaction is assumed to contribute to the data.

Box 13.1 Definitions and Assumptions for Parameters of the Additive Model

Parameter	*Definition*
$\mu_{i\cdot}$	$\Sigma_j \mu_{ij} / a$
$\mu_{\cdot j}$	$E(\mu_{ij})$
μ	$\Sigma_j \mu_{\cdot j} / a$
η_i	$\mu_{i\cdot} - \mu$
α_j	$\mu_{\cdot j} - \mu$

The following conditions hold for α_j, η_i, and ε_{ij}:

1 The α_j, η_i, and ε_{ij} are distributed independently of each other.
2 If *A* is a fixed-effect variable,

2.1 $\Sigma_j \alpha_j = \Sigma_j (\mu_{\cdot j} - \mu) = 0$.
2.2 The variance of the treatment effects is $\sigma_A^2 = \sum_j \alpha_j^2 / a$.
2.3 The null hypothesis about the effects of treatments is H_0: $\alpha_1 = \alpha_2 = \ldots = \alpha_j = \ldots = \alpha_a = 0$ or, equivalently, $\sigma^2_A = 0$.

3 We assume that the participants in the experiment are a random sample from an infinite population. We further assume that the population of η_i values is distributed independently and normally with mean zero and variance $\sigma^2_s = E(\eta^2_i)$. Because the n participants are viewed as a random sample from an infinite population,

3.1 The sum of the n values of η_i sampled in the experiment is unlikely to sum to zero; that is, $\Sigma_i \eta_i \neq 0$.

3.2 However, the average value of all such effects for the population of participants will be zero; that is, $E(\eta_i) = 0$.

4 The error component, ε_{ij}, is assumed to be distributed independently and normally with mean $E(\varepsilon_{ij}) = 0$ and variance $\sigma^2_e = E(\varepsilon^2)$.

13.2.2 Advantages of Additivity

Although we will consider both additive and nonadditive models and the analyses related to them, analyses based on data consistent with the additive model have these advantages:

1. *Simplicity of interpretation*. The interpretation of treatment effects is simple under the additive model because such effects are the same for all participants. When effects of treatments vary over participants, interpretation of the effects of the independent variable is unclear. It may be necessary to distinguish among subpopulations that differ in the effect of treatments.
2. *Design efficiency*. The presence of interactions increases variability and therefore reduces the power of statistical tests and the precision of estimates.
3. *Estimation of population variances*. Some measures of the importance of independent variables require estimates of components of variance; for example, σ^2_s, the variance of the population of participant means. As will be seen when expected mean squares are considered, unbiased estimates are available assuming the additive model, but not when a nonadditive model is assumed.
4. F *tests*. Under the nonadditive model, only approximate F tests of some effects are possible.

In summary, tests of the effects of the independent variable and estimates of several measures of importance rest on the underlying structural model. Understanding the relation between the model, expected mean squares, and various statistics requires that we understand the distinction between two types of variables, fixed effects and random effects. We consider this distinction next.

13.3 Fixed and Random Effects

Generally, the treatment levels in a study are based on theoretical considerations and the researcher's understanding of the domain. For example, in a study of the effects of dosage levels on treatment efficacy, a researcher might select four equally spaced dosages ranging across levels thought likely to differ in their effects. In a study of memory, a researcher might compare three strategies that are based on different theoretical principles. When the levels of an independent variable have been systematically chosen, as in these examples, we view the population of treatment levels as consisting of *only* those levels included in the study; we say that the independent variable has *fixed effects*. In contrast, the participants in a study are not systematically selected; rather, they are usually viewed as having been randomly sampled from a wider population of potential participants. For this reason, the participant variable is said to have *random effects*. Items in a study, such as pictures that are rated for some characteristic, or words whose recognition times are recorded, also are usually viewed as having random effects.

The distinction between fixed- and random-effects variables has implications for how we define the population parameters estimated by our data, and for the generality of conclusions based on the data analysis. With respect to parameter definitions, we see in Box 13.1 that the variance of the population treatment means, σ^2_A, is a variance for only those treatment levels included in the study, whereas the participant variance, σ^2_s, is defined for the population of participants. With respect to the generality of conclusions, although we are always free to draw conclusions about a drug dosage or a strategy that was not in the study, such conclusions are extra-statistical generalizations. They may be based on extrapolation from the function relating depression scores to drug dosage, or the similarity we perceive between some strategy not included in the study and those that were. Conclusions about treatments that were not included in the study are not necessarily less correct than those about the included treatments, but they are not grounded in the data analysis like conclusions about the treatments included in the study. In contrast, the variance of the mean scores of participants in the experiment estimates the variance of the sampled population and permits generalizations beyond the immediate sample. Similarly, if the items used as stimuli in a study were randomly sampled from a stimulus population of interest, then our statistical analysis will permit generalization to other items from that population.

The distinction between fixed- and random-effects variables has implications not only for the generality of conclusions but also for the expected mean squares. In turn, this influences our estimates of population parameters and the calculation of the *F* tests in the more complex designs of Chapters 14 and 15.

13.4 The Additive Model

13.4.1 *The ANOVA and Expected Mean Squares for the Additive Model*

An example of the use of a repeated-measures design is the *Seasons* study carried out by researchers at the University of Massachusetts Medical School. Table 13.1 presents Beck Depression scores for each season for each of 14 men under the age of 35 who served in the *Seasons* study, and for whom scores were available in all four seasons.[1] Each row of the data set represents a different participant and each column represents a season. Figure 13.1 plots the means together with the 95% confidence interval (CI) bounds obtained as

$$CI = \bar{Y}_{.j} \pm SEM \times t_{.025,39} \qquad (13.3)$$

where *SEM* is the standard error of the mean and equals $\sqrt{MS_{error} / n}$, where n is the number of observations on which each condition mean, $\bar{Y}_{.j}$, is based. The t is the two-tailed critical value for $p = .05$, where the *df* are those on which the ANOVA error term is based.

Table 13.1 Seasonal depression scores for males under 35 years of age

Participant	*Winter*	*Spring*	*Summer*	*Fall*	*Mean*
1	7.500	11.554	1.000	1.208	5.316
2	7.000	9.000	5.000	15.000	9.000
3	1.000	1.000	0.000	1.000	0.500
4	0.000	0.000	0.000	0.000	0.000
5	1.059	0.000	1.097	4.000	1.539
6	1.000	2.500	0.000	2.000	1.375
7	2.500	0.000	0.000	2.000	1.125
8	4.500	1.060	2.000	2.000	2.390
9	5.000	2.000	3.000	5.000	3.750
10	2.000	3.000	4.208	3.000	3.052
11	7.000	7.354	5.877	9.000	7.308
12	2.500	2.000	0.009	2.000	1.627
13	11.000	16.000	13.000	13.000	13.250
14	8.000	10.500	1.000	11.000	7.625
Mean	4.290	4.712	2.585	4.943	4.133

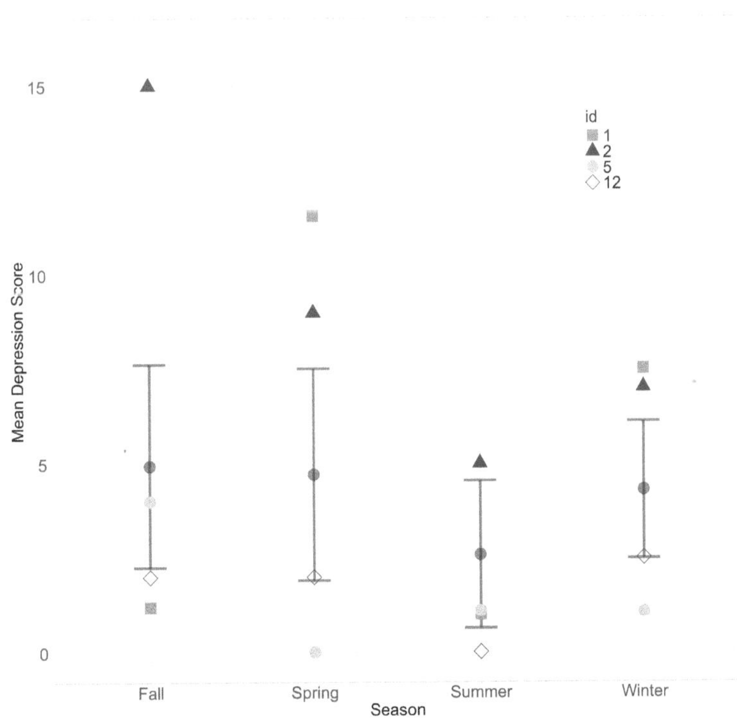

Figure 13.1 Means of depression scores of Table 13.1 with confidence intervals, as well as four individual participants' data.

Table 13.2 ANOVA of the data in Table 13.1 and expected mean squares

SV	*df*	*SS*	*MS*	*F*	*P*	*EMS*
Subjects (*S*)	$n - 1 = 13$	$a\sum_i \left(\bar{Y}_{i.} - \bar{Y}_{..}\right)^2 = 779.100$	$SS_S/df_S =$ 59.931	$MS_S/MS_{SA} =$ 11.295	0.000	$\sigma_e^2 + a\sigma_S^2$
Seasons (*A*)	$a - 1 = 3$	$n\sum_j \left(\bar{Y}_{.j} - \bar{Y}_{..}\right)^2 = 47.778$	$SS_A/df_A =$ 15.926	$MS_A/MS_{SA} =$ 3.001	0.042	$\sigma_e^2 + n\theta_A^2$
Residual (*SA*)	$(n - 1)(a - 1) = 39$	$SS_{total} - SS_S - SS_A = 206.960$	$SS_{SA}/df_{SA} =$ 5.307			σ_e^2
Total	$an - 1 = 55$	$\sum_i \sum_j \left(Y_{ij} - \bar{Y}_{..}\right)^2 = 1033.838$				

Note: $\theta_A^2 = \sum_j \left(\mu_j - \mu\right) / (a - 1)$.

From Table 13.2, $SEM = \sqrt{5.307/14} = .616$; $t_{.025,39} = 2.023$ and $SEM \times t_{.025,39} = 1.246$. It is apparent from Figure 13.1 that the mean depression scores are quite similar except for the mean in the summer season, which is noticeably lower. However, as the confidence intervals indicate, there is considerable variability within each condition. There are also individual differences across the seasons, as can be seen in the four randomly chosen participants' data that are plotted in Figure 13.1; this will be important in Section 13.5.

Before proceeding to the ANOVA of the data, we explored the data using the methods described in Chapter 2. We will not take the space to reproduce the results, but we note that plots revealed that the data were extremely skewed to the right, and statistical tests showed a noteworthy departure from normality. These results indicate that we may want to consider alternatives to the conventional ANOVA, and we will do so later in the chapter. For now, we ignore this violation of assumptions and perform the tests of the *Seasons* effects.

Table 13.2 presents the results of an analysis of variance of the data, together with the expected mean squares (*EMS*). Equation 13.2 and the assumptions about its component parameters provide the basis for the *F* tests. Consistent with the three terms in the equation, we have three sources of variance representing participants, treatment, and residual error variability. The mean squares for the participants and treatment terms are calculated as in any two-factor design, and the residual sums of squares and *df* are obtained by subtraction from the totals. Assuming the additive model, the *F* tests are justified by the expected mean squares displayed in the table. The *EMS* indicate that, under the additive model, $MS_{residual}$ is the appropriate error term for testing null hypotheses about both *S* and *A*. In either test, if the null hypothesis is true, the numerator and denominator of the *F* ratio have the same expected value.

13.4.2 *Using Software for the Repeated-Measures ANOVA*

The mechanics of running the repeated-measures ANOVA are not complicated. However, the structure of the data file is important and differs across software. For SPSS, the data file should be structured so that each participant's responses appear on a single row, with columns for each measure; this is called a "wider" format. In R, the data file should be structured in a "longer" format, so that each observation has its own row and the columns indicate participant identity and the condition associated with the observation. Transforming between the two formats is not difficult but can take some practice. As a general rule, the data files on the website are in wider format.

In SPSS, select the *General Linear Model* option within the *Analyze* menu, and then choose *Repeated Measures*. Within the window that pops up, enter Season as a name for the within-subject factor, and type 4 in the box for *Number of Levels*, then click Add and then Define. In the window that appears, move the variables Winter, Spring, Summer, and Fall to the *Within-Subjects Variables* (*Season*) box in that order. You should see Winter(1), Spring(2), Summer(3), and Fall(4) to indicate the order of the levels. Click on Continue to see the analysis results. SPSS splits the ANOVA table into an analysis of the between-participants effects, where the row labeled "Error" shows the Subjects effects, and an analysis of the within-participants effects, where the results from the additive model are provided in the rows labeled "Sphericity Assumed."

In R, the ANOVA is straightforward once the data are in the necessary format, with one observation per row (so-called tidy data). Converting the data in the *Male_D under 35.xlsx* file from wider to longer format (or the reverse) is easy with the *pivot_longer* and *pivot_wider* functions in the {tidyr} package. (Use *?pivot_longer* or *?pivot_wider* for help in R.) We begin by adding a participant identifier, id, using the *mutate* function in {dplyr} on the WideFormat data: *mutate*(id = row_number()). Then select the relevant data columns for our analysis: *select*(id, Winter, Spring, Summer, Fall). Finally, we use the *pivot_longer* function, renaming the specified columns to the overarching label "Season" and saving the result in a new data frame: LongForm <- WideForm %>% *pivot_longer*(cols =!id, names_to = "Season"). The result has three columns: id, Season, and value (which contains the observed score). Of course, this transformation is reversible. We can return to the wider format using *pivot_wider*(LongForm, names_from = Season) or simply use our original WideForm data frame. Using the longer format, the ANOVA is mostly familiar: *summary*(*aov*(data = LongForm, value ~ Season + Error(id/Season)). The main change from the between-participants analyses of earlier chapters is that we must specify the error term as the interaction of participant and Season.

13.5 The Nonadditive Model for the $S \times A$ Design

In many studies, the effects of treatments may vary over participants with the result that the additive model of Equation 13.2 fails to provide a valid description of the structure of the data. For example, the effects of rate of presentation of text material upon comprehension may depend upon such individual factors as reading ability, familiarity with the topic of the text, current state of alertness, and motivation to perform well in the experiment. In such cases, although analyses can be carried out, F tests may be biased and estimates of population parameters may be lacking or imprecise. In what follows, we develop the nonadditive model and its consequences. Then we consider ways of correcting possible bias in the F tests of treatments.

13.5.1 The Structural Equation

For the nonadditive model, we add an interaction component to the additive model of Equation 13.2. Therefore, the equation for the nonadditive case is

$$Y_{ij} = \mu + \alpha_j + \eta_i + (\alpha\eta)_{ij} + \varepsilon_{ij} \tag{13.4}$$

Assumptions about the distribution of the terms that were in Equation 13.2 are unchanged. The interaction effect associated with the *ij*th cell is defined as

$$\begin{aligned}(\eta\alpha)_{ij} &= (\mu_{ij} - \mu) - \eta_i - \alpha_j \\ &= (\mu_{ij} - \mu) - (\mu_{i.} - \mu) - (\mu_{.j} - \mu) \\ &= \mu_{ij} - \mu_{i.} - \mu_{.j} + \mu\end{aligned} \tag{13.5}$$

Because the independent variable, *A*, is a fixed-effect variable, the sum of the *a* interaction effects for any participant in an experiment equals 0; i.e., $\sum_{j=1}^{a}(\eta\alpha)_{ij} = 0$. However, because participants are a random-effects variable, and therefore only *n* of the infinitely possible population effects at any level of *A* are sampled, $\sum_{i=1}^{n}(\eta\alpha)_{ij}$ rarely will equal zero, although the expectation of all population values is zero; i.e., $E(\eta\alpha)_{ij} = 0$. These properties of the $S \times A$ interaction effects have implications for the *EMS*, and therefore for the *F* tests. We consider this next.

13.5.2 *Expected Mean Squares* (EMS)

Table 13.3 presents the sources of variance (*SV*), degrees of freedom (*df*), and *EMS* under the additive and nonadditive models. The numerical values of the sums of squares and the mean squares are the same for the two models. Furthermore, in both cases, the *A* source of variance would be tested against MS_{SA}. However, there is a difference between the two cases: The interaction variance, σ^2_{SA}, contributes to the *A* and *SA* terms in the nonadditive case.

Why does the *SA* variance contribute to MS_A, the variance among the treatment means, but not to MS_S, the variance among the participant means? The answer lies in the distinction between fixed- and random-effects variables: using *EMS*, we will see that the variances of fixed-effects variables can be neatly partitioned, whereas random-effects variables tend to spread out, interacting with other variables and putting their proverbial "fingers" in other parts of the variance pie. Table 13.4 may help clarify this. The table presents the true scores and parameters for a toy population of four participants with interaction effects calculated as in Equation 13.5. All ε_{ij} have been set equal to zero to simplify our discussion. Note that the treatment population (column) means are identical; therefore, the null hypothesis is true. Also note that the average interaction effect in each column and in each

Table 13.3 Expected mean squares and *F* ratios for the $S \times A$ design

SV	*df*	*Additive EMS*	*Additive F ratio*	*Nonadditive EMS*	*Nonadditive F ratio*
S	$n - 1$	$\sigma^2_e + a\sigma^2_S$	MS_S / MS_{SA}	$\sigma^2_e + a\sigma^2_S$	
A	$a - 1$	$\sigma^2_e + n\theta^2_A$	MS_A / MS_{SA}	$\sigma^2_e + \sigma^2_{SA} + n\theta^2_A$	MS_A / MS_{SA}
SA	$(n - 1)(a - 1)$	σ^2_e		$\sigma^2_e + \sigma^2_{SA}$	

Table 13.4 Data for a toy population of four participants with interaction effects present

	A_1		A_2		A_3			
	Y_{i1}	$(\eta\alpha)_{i1}$	Y_{i2}	$(\eta\alpha)_{i2}$	Y_{i3}	$(\eta\alpha)_{i3}$	μ_i	η_i
S_1	8	−1	10	1	9	0	9	−5
S_2	10	3	9	2	2	−5	7	2.5
S_3	11	−1	12	0	13	1	12	2.5
S_4	9	−1	7	−3	14	4	10	5
μ_j	9.5		9.5		9.5			

Note: S_1 and S_2 are sampled from the population, and constitute the "experiment."

row is zero; this is an algebraic result of the definition of an interaction effect. Now assume an "experiment" in which S_1 and S_2 are selected by random sampling. Taking the "data" from the two participants in the experiment, we find that the means at the levels of A, the $\bar{Y}_{.j}$, are 9, 9.5, and 5.5. Although the treatment population means, the $\mu_{.j}$, are identical, the sample means, the $\bar{Y}_{.j}$, are not. This is not because of error of measurement, because there is none in this artificial data set. It is entirely due to the fact that the *sampled* interaction effects are different at the three levels of A. By randomly sampling participants, we have also sampled their interactions with the levels of factor A; sometimes those sampled interactions will raise certain treatment means, and in other samples they could reduce the very same treatment means. This random sampling of only some of the population interaction effects at each treatment level is why σ^2_{SA} contributes to the variability among the $\bar{Y}_{.j}$, and therefore to MS_A. Which treatment means will be most raised or lowered relative to their population values will depend upon the pattern of interaction effects that have been sampled from each treatment population.

One other point follows from Table 13.4. Because A is a fixed-effects variable, each participant mean includes the average of all interaction components in the corresponding row. That average is zero and therefore the interaction variance does not contribute to differences among participant means and, accordingly, does not contribute to MS_s.

Returning to the seasonal depression data shown in Table 13.1 and Figure 13.1, we can see that these data are not consistent with the additive model. Four participants were randomly selected from the sample, and their individual depression scores are shown in Figure 13.1. These data reveal large differences in the pattern of participants' depression levels over the seasons. For example, Participant 2 has a much higher depression score in the fall than the summer, whereas Participant 1's depression scores are lowest in the fall and winter and highest in the spring. These differences are random Participant × Season (SA) interaction effects. The EMS for the nonadditive model in Table 13.3 show that these SA interaction effects influence both the MS_A and the MS_{SA}. Despite this unwanted influence, the F ratio for the effect of A is computed in the same way for the additive and nonadditive models. Although the computations in the ANOVA are the same, nonadditivity has several consequences, which we will consider next.

13.5.3 *Consequences of Nonadditivity*

To briefly summarize, there are several undesirable consequences of nonadditivity:

1. *No proper F test of Participants*. Given the *EMS* of Table 13.3, a test of the *Participants* source of variance against the *SA* term would be negatively biased. However, participants usually differ greatly in their average responses, and so this test is rarely performed.
2. *Unbiased estimates of some measures of effect size are not possible* because it is not possible to isolate estimates of each of the variance components contributing to the effect size measures. Looking at the *EMS* in Table 13.2, we cannot algebraically isolate σ^2_e, σ^2_s, or σ^2_{SA}.
3. *The sphericity assumption will usually be violated*. This assumption, also called *circularity*, is an assumption about the variances and covariances of the data in the different conditions. We will define this and more fully discuss the consequences of violations in Section 13.6. For now, we merely note that biased *F* tests may result when variances of the differences between levels of within-participant conditions are not equal.
4. *The efficiency of the design is reduced* because the interaction effects contribute added "noise" to the data set.

This last point can be seen by considering the ratio of *EMS* under the two models. If additivity holds, that ratio is

$$\frac{E(MS_A)}{E(MS_{SA})} = 1 + \frac{n\sum_j(\mu_j - \mu)^2 / (a-1)}{\sigma_e^2} = 1 + \frac{n\theta_A^2}{\sigma_e^2}$$

However, if the nonadditive model is valid, the ratio is smaller:

$$\frac{E(MS_A)}{E(MS_{SA})} = 1 + \frac{n\sum_j(\mu_j - \mu)^2 / (a-1)}{\sigma_e^2 + \sigma_{SA}^2} = 1 + \frac{n\theta_A^2}{\sigma_e^2 + \sigma_{SA}^2}$$

The *SA* interaction effects will reduce the precision of parameter estimates and the power of the significance test. However, such interaction variance will usually be less than the individual difference variance associated with completely randomized designs. Therefore, even when the data do not conform to the additive model, the repeated-measures design usually will yield more powerful tests of the null hypothesis than will the completely randomized design with the same number of observations.

13.6 The Sphericity Assumption

In order for MS_A/MS_{SA} to have an *F* distribution, the data must meet an assumption that we did not encounter in the analysis of between-participants designs. This is usually referred to as the *sphericity*, or *circularity*, assumption (Huynh & Feldt, 1976; Rouanet & Lepine, 1970). In the following pages, we define and provide examples of sphericity; in Sections 13.6.2 and 13.11, we consider alternatives to the standard *F* test when the assumption is violated.

13.6.1 Sphericity Defined

Consider the typical repeated-measures design with n participants, each of whom is tested under a conditions. For each participant, we can calculate $(1/2)(a)(a - 1)$ differences, based on the number of pairs of treatments. For example, when there are three treatments, we can construct three pairings and calculate the difference between the scores in each pair. We have done that in Table 13.5. The left-most three columns of the table contain the scores for five participants at three levels of A, and the right-most three columns contain all possible difference scores for each participant, with the value of the variances, s^2_d, at the bottom of the columns. These data fit the assumption of sphericity because the three values of the variance are the same; *sphericity exists when the population variances of all possible difference scores are equal.*[2] If this assumption is violated, Type 1 error rates will be higher than the nominal alpha-level.

Nonsphericity, or heterogeneity of variance of difference scores, is similar in both form and consequences to heterogeneity of variance in the between-participants designs of Chapters 8–11. In Chapter 8, the error term, $MS_{S/A}$, was the average of the group variances. If the null hypothesis is true and if there are n scores at each treatment level, then $E(MS_A) = E(MS_{S/A})$ even if the group variances are very different from each other. Even though this heterogeneity of variance does not affect the ratio of expected mean squares, it does affect the sampling distribution of the ratio of mean squares. More precisely, when the null hypothesis is true and group sizes are equal, heterogeneity of variance inflates the probability of sampling large F values; the F test is positively biased.

We have a similar situation in the repeated-measures design. The error term, MS_{SA}, is always one-half of the average of all values of s^2_d. This relation between the variances of the difference scores and MS_{SA} will hold for any number of levels of A and will be true regardless of the values of the s^2_d. Therefore, MS_{SA}, and consequently the ratio of expected mean squares, will not be affected by differences among the variances of difference scores. However, the distribution of the F ratio is affected and the Type 1 error rate is inflated if the variances are very different, analogous to the case when group variances differ in the between-participants design. A test of the sphericity assumption was derived by Mauchly (1940) and is available in many computer packages (for example, in SPSS's *Repeated Measures* analysis and as part of the *anova_test* function in R's {rstatix} package). However, if the population distributions are not normal, the Mauchly test tends to yield significant results even when the violation of the sphericity assumption is small (Keselman, Rogan,

Table 13.5 Data exhibiting sphericity

	A_1	A_2	A_3	$Y_{i3}-Y_{i2}$	$Y_{i2}-Y_{i1}$	$Y_{i3}-Y_{i1}$
S_1	21.050	7.214	26.812	19.598	−13.836	5.760
S_2	6.915	29.599	16.366	−13.233	22.684	9.451
S_3	3.890	21.000	41.053	20.053	17.110	37.163
S_4	11.975	12.401	18.896	6.495	.426	6.921
S_5	31.169	34.786	31.872	−2.914	3.617	.703
Mean	15.000	21.000	27.000	6.000	6.000	12.000
s^2	124.000	132.000	100.000	208.000	208.000	208.000

Note: In general, $MS_{SA} = (1/2)\Sigma s^2_d/a$.

Mendoza, & Breen, 1980; Rogan, Keselman, & Mendoza, 1979). Therefore, we do not recommend relying on the outcome of Mauchly's test.

As we will see in the next section, one straightforward solution to a violation of sphericity is an adjustment to the degrees of freedom used in the *F* test. The magnitude of the adjustment varies continuously with the degree of nonsphericity, which is another reason to avoid relying on the binary outcome of Mauchley's test.

13.6.2 Dealing With Nonsphericity

There are two data-analysis strategies that protect the researcher against inflation of Type 1 error rates due to nonsphericity. The first of these is the *multivariate analysis of variance*, or *MANOVA*, which is beyond the scope of this book. Harris (2001) and Morrison (2004) are two excellent sources on MANOVA. The second strategy for dealing with nonsphericity is a *univariate F test with epsilon-adjusted degrees of freedom*. We focus on this latter solution because it is effective and easy to apply.

When the assumption of sphericity is violated, the Type 1 error rate is inflated if the conventional *F* test is evaluated on the usual degrees of freedom. That is, the test is positively biased. One way to controlling the Type 1 error rate in this situation is to conduct the *F* test with degrees of freedom that are adjusted downward. When the sphericity assumption is violated, the ratio of mean squares is still distributed approximately as *F*, provided that the numerator and denominator degrees of freedom are multiplied by a factor, ε (epsilon); that is, $df_A = (a - 1)\varepsilon$ and $df_{SA} = (a - 1)(n - 1)\varepsilon$. The value of epsilon ranges between 1, when the sphericity assumption is met, and $1/(a - 1)$ when the assumption is severely violated. In general, as nonsphericity increases, epsilon and the degrees of freedom decrease, with the result that a larger value of *F* is required for significance. In this way, the epsilon adjustment compensates for the inflation of Type 1 error rate caused by the failure of the sphericity assumption.

One estimator of ε, $\hat{\varepsilon}$ ("epsilon-hat"), was derived by Box (1954), and later extended to the mixed designs of Chapter 14 by Greenhouse and Geisser (1959). The Greenhouse–Geisser $\hat{\varepsilon}$ tends to be conservative, providing a larger adjustment to the *df*. Subsequently, Huynh and Feldt (1976) developed a more liberal estimator, $\tilde{\varepsilon}$. Both estimators require calculations involving the elements of the variance-covariance matrix. Fortunately, common statistical programs do these calculations.

Figure 13.2 presents the results of an ANOVA on the data from Table 13.1 conducted in R with the *anova_test* function of the {rstatix} package. At the top of the figure, we see that depression levels vary significantly over the season, $F(3, 39) = 3.001$, $p = .042$. Mauchly's test of sphericity is included automatically and is nonsignificant in this example, $p = .408$. Nonetheless, we also see two estimates of epsilon in the lower section of Figure 13.2. The Greenhouse–Geisser (*GGe*, $\hat{\varepsilon}$) estimate of ε is less than 1. Contrary to the results of the Mauchly test, $\hat{\varepsilon}$ indicates that the sphericity assumption is violated, and this causes a reduction in the degrees of freedom required to evaluate the *F* statistic; the adjusted *df* are also reported with the "Sphericity Corrections" as DF[GG] = 2.49, 32.43. However, consistent with the results of the Mauchly test, the Huynh-Feldt (*HFe*, $\tilde{\varepsilon}$) estimate is actually slightly greater than 1 (it would be reported as 1; there are no "free" degrees of freedom). These differences in the estimates lead to slightly different conclusions about the effects of seasons, as can be seen in the bottom line of Figure 13.2. Using the *H–F* estimate of ε, the variation among the mean depression scores for the four seasons is significant at the .05 level;

```
> seasons.out<-anova_test(data=LongForm,value ~ Season + Error(id/Season))
> seasons.out
ANOVA Table (type III tests)

$ANOVA
  Effect DFn DFd     F     p p<.05   ges
1 Season   3  39 3.001 0.042     * 0.046

$`Mauchly's Test for Sphericity`
  Effect     W     p p<.05
1 Season 0.648 0.408

$`Sphericity Corrections`
  Effect   GGe       DF[GG] p[GG] p[GG]<.05   HFe       DF[HF] p[HF] p[HF]<.05
1 Season 0.832 2.49, 32.43 0.053            1.045 3.13, 40.75 0.042         *
```

Figure 13.2 R analysis of the data of Table 13.1: ANOVA results ($ANOVA), Mauchly's test of sphericity ($'*Mauchly's . . .*'), and sphericity corrections ($'*Sphericity . . .*').

however, using the *G–G* estimate results in a failure to reject the null hypothesis. In cases where these corrections indicate different statistical inferences, we recommend relying on the *G–G* estimate because it is slightly more conservative.

The epsilon adjustment does not apply if *A* only has two levels. In that case, there is only one set of *n* difference scores and, therefore, only one variance of difference scores. Homogeneity of variance of difference scores becomes an issue only when *A* has more than two levels. In that case, the inflation in Type 1 error rate can be large if the *df* are not adjusted. Therefore, the univariate *F* should not be evaluated without an adjustment.

13.7 Measures of Effect Size

Our analysis of the *Seasons* data, shown in Table 13.2, informed us that differences among the seasonal depression means are statistically significant. However, there are additional questions to be asked of the data. For one, we would like some indication of the practical or theoretical importance of our result. There are several statistics that might be computed to answer this question. In this section, we will consider two measures of effect size based on the proportion of variability accounted for by our treatment, *A*. The measures, η^2 and ω^2, also were discussed in the context of between-participants designs in Chapters 8 and 9, and developments here will be similar. We will also consider Cohen's *f* because of its relevance to power calculations.

13.7.1 η^2 (Eta-Squared), the Proportion of Sample Variability

In a one-factor between-participants design, η^2 is calculated as

$$\eta^2(A) = \frac{SS_A}{SS_A + SS_{S/cells}} \tag{13.6}$$

In the one-factor repeated-measures design, investigators often calculate *partial* η^2:

$$\eta_p^2(A) = \frac{SS_A}{SS_A + SS_{SA}} \quad (13.7)$$

From the sums of squares column of Table 13.2, $\eta^2_p(A) = 47.778 / (47.778 + 206.96) = .188$. We can request this value using the *effect.size* = *"pes"* option in the *anova_test* function in the {rstatix} package in R. It is also the computation performed in SPSS if an effect size measure is requested in the *General Linear Model/Repeated-Measures* module. However, a limitation of Equation 13.7 is that it is not comparable to Equation 13.6 because Equation 13.6 includes variability due to individual differences, whereas Equation 13.7 omits such variability. Therefore, following the recommendation of Olejnik and Algina (2003), we recommend calculating *general* η^2:

$$\eta_g^2(A) = \frac{SS_A}{SS_A + SS_{SA} + SS_S} \quad (13.8)$$

Again turning to the *SS* column in Table 13.2, $\eta^2_G(A) = 47.778 / (47.778 + 779.1 + 206.96) = .046$. This is the default effect size reported by the *anova_test* function (labeled "ges"; see Figure 13.2). In the one-factor repeated-measures design, the denominator of the general statistic is the total sum of squares, provided that the numerator represents a factor with fixed effects. Note that η^2 and *general* η^2 are identical in the one-factor repeated-measures design; however, the denominator of the general statistic usually is not the total sum of squares for designs with multiple factors, so the two statistics are different in multi-factor designs. We will consider such designs in the next chapter.

As we stated in Chapter 8, η^2 has the advantage of being easily calculated and easily understood as a proportion of *sample* variance, but it tends to overestimate the proportion of *population* variance due to the independent variable. The next statistic to be considered provides a somewhat better estimate of the effect of the independent variable in the population.

13.7.2 ω^2_A (Omega-Squared), the Proportion of Population Variance

Typically, researchers have reported estimates of *partial* ω^2_A, which is the variance among the population treatment means divided by the sum of that variance and the population error variance:

$$\omega_p^2(A) = \frac{\sigma_A^2}{\sigma_A^2 + \sigma_e^2} \quad (13.9)$$

As with η^2, we recommend calculating a statistic that takes individual differences into account so that values of ω^2 may be compared across designs. *General* ω^2 is defined as

$$\omega_g^2(A) = \frac{\sigma_A^2}{\sigma_A^2 + \sigma_e^2 + \sigma_S^2} \quad (13.10)$$

If the additive model applies to a data set, the expected mean squares in Table 13.2 suggest the following estimates of the variances:

$$\hat{\sigma}_A^2 = \left(\frac{MS_A - MS_{SA}}{n}\right)\left(\frac{a-1}{a}\right)$$

$$\hat{\sigma}_S^2 = \left(\frac{MS_S - MS_{SA}}{a}\right)$$

and

$$\hat{\sigma}_e^2 = MS_{SA} \tag{13.11}$$

Substituting these estimates into Equation 13.10 and multiplying the numerator and denominator by *an*, we have

$$\hat{\omega}_g^2(A) = \frac{(MS_A - MS_{SA})(a-1)}{(MS_A - MS_{SA})(a-1) + n(MS_S - MS_{SA}) + anMS_{SA}} \tag{13.12}$$

We can obtain this estimate from published research reports even when mean squares are not reported if we are provided the *F* ratios and degrees of freedom. Dividing the numerator and denominator of Equation 13.12 by MS_{SA}, and noting that $F_A = MS_A / MS_{SA}$ and $F_s = MS_s / MS_{SA}$ we have

$$\hat{\omega}_g^2(A) = \frac{(a-1)(F_A - 1)}{(a-1)F_A + nF_S + (a-1)(n-1)} \tag{13.13}$$

If F_A is less than one, the estimate is assumed to be zero.

For the data set of Table 13.1, we substitute the *F* ratios and values of *a* and *n* into Equation 13.13, and find that

$$\hat{\omega}_g^2(A) = \frac{(4-1)(3.001-1)}{(4-1)(3.001) + (14)(11.295) + 39} = .030$$

We estimate that the variation in seasons accounts for about 3% of the total population variance.

In many circumstances, the nonadditive model will be a more accurate description of the data. In that case, the formula for general ω^2 must include σ^2_{SA}:

$$\omega_g^2(A) = \frac{\sigma_A^2}{\sigma_A^2 + \sigma_e^2 + \sigma_S^2 + \sigma_{SA}^2} \tag{13.14}$$

We still use Equation 13.12 or 13.13 to estimate general ω^2 because it is not possible to obtain estimates of all four variance terms; the resulting estimate is positively biased. Under the nonadditive model, Equation 13.12 estimates

$$\omega_g^2(A) = \frac{\sigma_A^2}{\sigma_A^2 + \sigma_e^2 + \sigma_S^2 + [(a-1)/a]\sigma_{SA}^2} \tag{13.15}$$

Equation 13.12 underestimates the denominator of general ω^2 and consequently overestimates the proportion of variance due to A. However, the overestimate will be small except when the interaction variance is large and a is small.

In addition to the problem of obtaining an unbiased estimate, a possible criticism of general ω^2 is that the estimates are usually small, leading researchers to disregard effects that may in fact be of practical or theoretical import. However, values of any index of effect size should be judged within the context of the researcher's experience. Furthermore, although the same data set will provide larger values of η^2, that statistic is inflated by the presence of error variance. Even if the population treatment means are identical, the sample means almost surely will differ, yielding a value of the partial η^2 greater than zero.

13.7.3 Cohen's f

Cohen's f is defined as the ratio of the standard deviation of effects to the standard deviation of the populations of scores within each level of A:

$$\begin{aligned} f &= \sigma_A / \sigma_W \\ &= \sigma_A / \sqrt{\sigma_e^2 + \sigma_S^2 + \sigma_{SA}^2} \end{aligned} \tag{13.16}$$

where σ_w is the standard deviation of the scores within any treatment population. Its square, the within-treatment variance, combines participant and error variance, making it comparable to the Cohen's f that we first introduced in Chapter 8 as a measure of effect size in a between-participants design. For additive data (i.e., when $\sigma^2_{SA} = 0$), an estimate of Cohen's f is

$$\hat{f}(A) = \sqrt{\frac{(MS_A - MS_{SA})(a-1)}{n(MS_S - MS_{SA}) + anMS_{SA}}} \tag{13.17}$$

Equation 13.17 may also be used when the data are nonadditive, but it will overestimate Cohen's f for the same reason that general ω^2 is overestimated by Equation 13.12 for nonadditive data.

Our interest in Cohen's f is that it has become a common measure of effect size. Also, it is the preferred measure of effect size in many programs for computing power, including G*Power 3.1.

13.8 Deciding on Sample Size: Power Analysis in the Repeated-Measures Design

As with other designs, considerations of statistical power should be a major factor in deciding on sample size. We need several pieces of information to compute *a priori* power for the repeated-measures design. The steps for using G*Power 3.1 are presented in Box 13.2 and an output from the program is displayed in Figure 13.3.

To begin, we select the F test family, identify the test option as *ANOVA: Repeated measures, within factors*, and indicate that we want to do an *a priori power* calculation. We must decide what value of power we seek, the alpha-level we will use, and the design

Box 13.2 Using G*Power 3.1 to Determine Sample Size in the Repeated-Measures Design

Figure 13.3 contains the screen illustrating these selections and entries.

1. Select the *F* test family; the test option *ANOVA: Repeated-measures, Within factors* and the *a priori* option for the type of test.
2. Enter the Type 1 error rate (α) and the desired power. We will set these to .05 and .9, respectively.
3. Enter an epsilon value. We will assume sphericity; that is, we set $\varepsilon = 1.0$. A conservative approach would be to use the minimum value of $1 / (a - 1)$.
4. Enter the number of levels of the independent variable for *Number of measurements*; there are four in the example of the *Seasons* study. *Number of groups* is the number of levels of between-participants factors, and it is one in this case.
5. Enter an estimate of the correlation between the repeated-measures conditions.
6. Select the *calculate* button.

of the experiment we are planning. The power level and alpha-level are entered directly into G*Power; the design information requested is the number of between-participants conditions (i.e., "number of groups") and the number of observations per participant (i.e., "number of measurements"). For our example, assume that we have a one-factor repeated-measures design with four levels of factor *A*. A repeated-measures design without any between-participants factors has one group of participants; thus, "1" is entered for the number of groups in our example, and "4" is entered for the number of repetitions.

Recall from our discussion in Chapter 8 that the power of the *F* test is determined, in part, by the magnitude of the effect of *A*, as well as the error variance associated with the test of *A*. Cohen's *f* is the input requested by G*Power 3.1 that provides information about the magnitude of the effect of *A*. With respect to error variance, Cohen's *f* also includes some information about the error variance; recall that the denominator of *f* is the within-condition variance. However, the error variance in a repeated-measures design is a function of both the variance within conditions and the correlation of scores between conditions. Thus, G*Power requests a value for the correlation between the repeated-measures, ρ.

Finally, we also must provide enough information to calculate the degrees of freedom associated with the test of *A*. This information is needed because it establishes the critical value of *F* for our test of *A*. The values of the degrees of freedom are determined, in part, by the number of groups and the number of repetitions (scores per participant) in the design. Also, if the sphericity assumption is violated, the degrees of freedom must be adjusted downward by the correction factor, ε. Thus, G*Power 3.1 requests an input for ε. If it is assumed that the sphericity assumption will be met, the input should be 1; if the worst case violation of sphericity is assumed, the input should be $[1 / (a - 1)]$; an intermediate choice might be the average of these best- and worst-case scenarios. In the event that pilot data

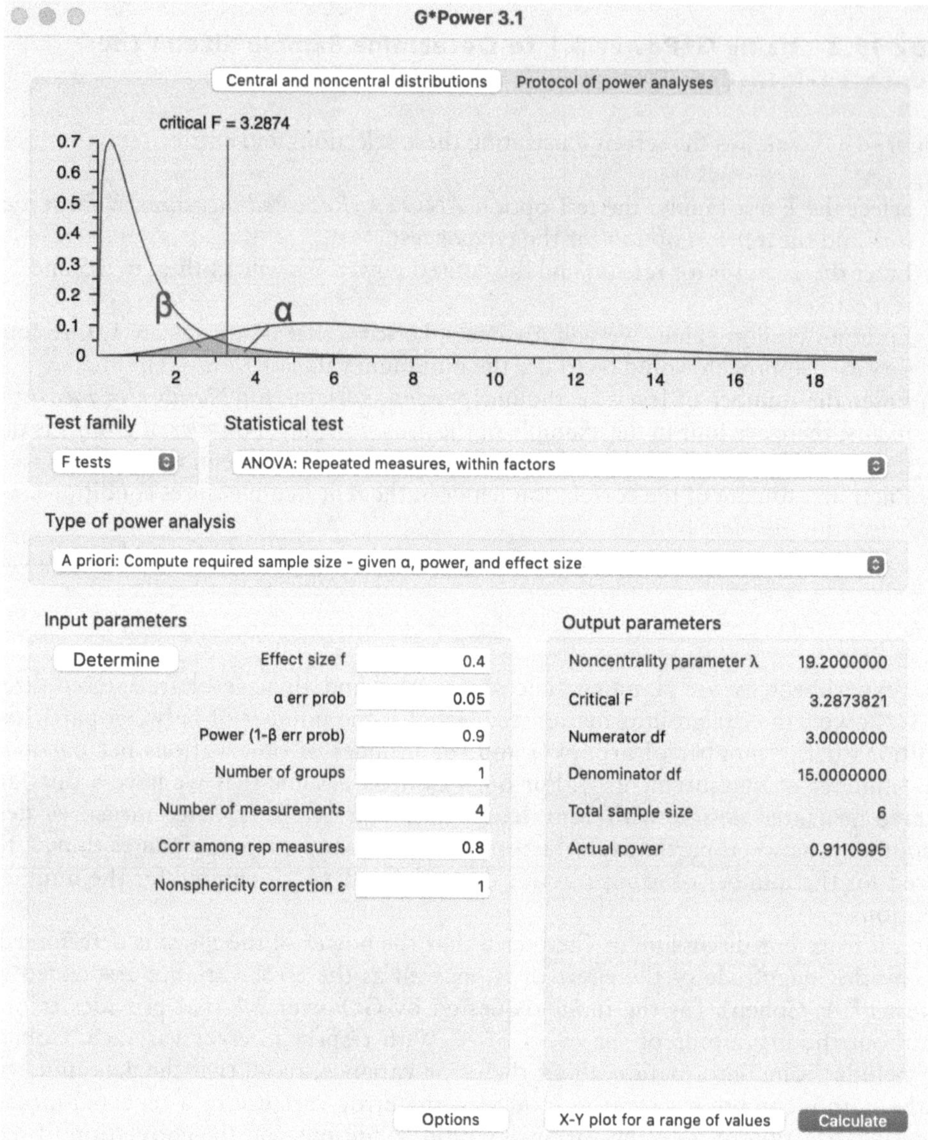

Figure 13.3 G*Power's screen for *a priori* power analysis for a repeated-measures design.

are available, the value of the Greenhouse–Geisser correction from an analysis of the data may be used.

How is a researcher to select values of Cohen's f and of ρ? One approach to providing values of f is to follow Cohen's guidelines and use .1, .25, and .4 for small, medium, and large effects, respectively. For repeated-measures designs, we suggest that a reasonable range of values of ρ is .2, .5, and .8 for small, medium, and large values, respectively. If a researcher wants .90 power to detect a large effect of A and assumes a strong correlation between conditions, the inputs to G*Power will be $f = .4$ and $\rho = .8$. For our example, the total sample size required is only 6. The required sample size will be larger if we assume a smaller effect of A, or a weaker correlation among conditions, or that the sphericity assumption is not met. For example, if we assume $f = .1$, $\rho = .2$, and $\varepsilon = .3334$ (worst case for a design with four levels of A), the required sample size is 633! As these examples make clear, researchers should do their best to make reasonable assumptions and, when in doubt, be conservative in their assumptions.

When pilot data are available, those data can be used to select values of f, ρ, and ε. Alternatively, a knowledge of the relevant literature and/or theoretical considerations may provide estimates of these parameters. G*Power 3.1 also provides help with calculating f. Select the *determine* button (see Figure 13.3) and, in the window that pops up, enter the estimates of the variance among the population means as the *Variance explained by special effect* and the within-population variance as *Variance within groups*. Then select the button labeled *Calculate and transfer to main window*. If the two variances were 4 and 25, a value of $f = .4$ (i.e., $\sqrt{4/25}$) would appear in the main window. Assuming that the other values – e.g., alpha, power, the correlation – have been inserted, selecting the *Calculate* button will yield the required sample size.

The output of G*Power 3.1 should be self-explanatory, except for the noncentrality parameter, λ. Lambda characterizes the noncentral F distribution and is defined as

$$\lambda = \varepsilon naf^2 / (1 - \rho) \tag{13.18}$$

Inserting the output value of n, and the input values of ε, a, f, and the correlation (from the left column of Figure 13.3), we have (1)(6)(4)(.16)/.2 = 19.2, the value of λ in Figure 13.3. This value can be used in some programs to calculate power. For example, in R, we can use an F distributional calculator, 1 – *pf*(q = 3.287, ncp = 19.2, df1 = 3, df2 = 15), to obtain the result 0.9111, the same power as reported in Figure 13.3.

13.9 Testing Single *df* Contrasts

13.9.1 Pairwise Comparisons

In most studies, pairwise comparisons or more complex contrasts are of more interest than the omnibus F test. Contrasts are distributed on one degree of freedom, so sphericity is not an issue because there is only a single contrast score for each participant. In the example of seasonal effects upon depression, we might wish to test all pairwise comparisons. The calculations are straightforward and, in any event, paired F or t tests can be carried out by statistical software. There are two issues, however:

1. *The error terms for the six possible pairwise comparisons.* Should each of the six comparisons in our example be tested against MS_{SA} or against the variance of the comparison under consideration?
2. *The control of familywise error (FWE) rate.* Should Tukey's *HSD* procedure be used, as described in Chapter 9, or is an alternative preferred?

With respect to the error term issue, MS_{SA} is distributed on more error degrees of freedom, $(a - 1)(n - 1)$, and therefore the F test with this denominator is potentially more powerful. Nevertheless, the default error term in statistical packages is usually the variance of the contrast being tested. For example, to test the difference between winter and summer mean depression scores, the mean ($\bar{d}$) and standard deviation (s_d) of the 14 differences are calculated, then

$$t = \bar{d} / \left(s_d / \sqrt{n}\right)$$
$$= 1.705 / (2.622 / \sqrt{14}) = 2.433$$

(Some software packages report the *F* statistic, which is just the square of the above *t*.) The reason for dividing by the standard error of the contrast instead of a denominator based on MS_{SA} is that if there is nonsphericity in the data set, the variances of difference scores may vary greatly across comparisons. Because MS_{SA} is a function of the average of these variances, it will be too small when testing some of the comparisons and too large when testing others. Boik (1981) showed that when the denominator of the *t* (or *F*) is based on MS_{SA}, even small departures from sphericity can create serious distortions in Type 1 and 2 error rates and in the widths of confidence intervals.

With respect to the question of the method for controlling *FWE*, Tukey's *HSD* procedure would seem to be the natural approach. However, the test performs badly if there is nonsphericity. Maxwell (1980) compared the Type 1 error rates and powers of various methods of controlling the *FWE* and concluded that the Bonferroni approach, with the error term based on the contrast being tested, provided the best solution. In the *Seasons* example, we would calculate a different standard error of the mean difference for each of the six comparisons, compute the six *t* tests, and compare each *t* value against the critical value for $FWE = .05$, $df = 13$, and $K = 6$, which is 3.107. Alternatively, if computer software is used to compute each *t* test, the *p*-value in the output may be compared against .05/6, or .008.

13.9.2 A Contrast Involving More Than Two Means

More complex contrasts are sometimes of interest. For example, we might have hypothesized prior to data collection that summer depression scores would be lower than the average of the other three seasons because summer provides sunshine and vacations, both of which may enhance participants' moods. Then from Table 13.1,

$$\hat{\psi} = (1/3)\left(\bar{Y}_{winter} + \bar{Y}_{spring} + \bar{Y}_{fall}\right) - \bar{Y}_{summer}$$
$$= (1/3)(4.290 + 4.712 + 4.943) - 2.585 = 2.063$$

The test statistic is

$$t = \frac{\hat{\psi}}{s_{\hat{\psi}} / \sqrt{n}} \quad (13.19)$$

where $S_{\hat{\psi}}$ is calculated by first obtaining for each participant

$$\hat{\psi}_1 = (1/3)\left(Y_{i(winter)} + Y_{i(spring)} + Y_{i(fall)}\right) - Y_{i(summer)}$$

and $S_{\hat{\psi}}$ is the standard deviation of the contrast scores (the $\hat{\psi}_1$) of the n participants. For example, the contrast score for the first participant is $\hat{\psi}_1 = (1/3)(7.5 + 11.554 + 1.208) - 1 = 5.754$.

Given the individual contrast scores, a one-sample t test can be performed. The same result can also be obtained without first obtaining the contrast scores by using the contrast option available in many statistical packages. The result frequently will be a value of F, the square of the t, or 7.872 in our example. Whether t or F is reported, the p-value for our example will be .015, indicating that the average of summer depression scores is significantly lower than the average for the other three seasons. Note that because the test involves a single group of difference scores, sphericity is not an issue.

There is the question of the control of *FWE* in this situation. Typically, multiple contrasts will be performed to understand the pattern of means associated with the effect of a source of variance. Again, the Bonferroni procedure should be used to control the *FWE*. This requires that the researcher plan the set of contrasts that will be conducted to analyze the effect of *A*. In the absence of such planning, the appropriate procedure for controlling *FWE* is the Roy–Bose method (1953), a multivariate extension of the Scheffé method.[3] Because use of this procedure will result in a substantial loss of power compared to the Bonferroni procedure, we strongly recommend that researchers make every effort to plan their comparisons.

13.10 The Problem of Missing Data in Repeated-Measures Designs

Missing data can be a problem in any kind of research. They are particularly troublesome in repeated-measures designs because when the *general linear model* (*GLM*) modules of the standard statistical packages are used to conduct ANOVAs, all cases that have missing scores in any condition are completely dropped from the analysis.

There are many possible causes of missing data. There may be some equipment failure, or a participant may not show up for a session or may fail to respond to a questionnaire item or stimulus presentation. What are we to do? If we drop participants who have scores in most but not in all conditions, or drop a condition because it contains too many missing observations, a good deal of effort and a lot of information may be lost. More importantly, cases with complete data may differ systematically from those with missing data. If missing data occur nonrandomly, conducting analyses with only complete cases will provide biased estimates of population parameters and standard errors and thereby invalidate our inferences.

We can distinguish among three general types of missing data: data *missing not at random* (*MNAR*, or nonignorable missing data), data *missing at random* (*MAR*), and data *missing completely at random* (*MCAR*). To illustrate the distinction, consider a data set like the one presented in Table 13.1 in which depression scores are collected at different seasons of the year, except assume there are 28 participants, 14 who identify as men and 14 who identify as women.

1. Missing data on Y are *MNAR* if the probability of missing scores on Y is related to the value of Y. As examples, some participants may have missing depression scores because they are too depressed to provide answers to some of the items on the depression questionnaire, or heavier participants may be less willing to report their weight. In this sense,

we can think of missing data as an extreme form of biased measurement error. In the examples we've given, if we only use data from participants without any missing scores, we are likely to underestimate depression and weight in the population. In general, if data are MNAR, analyzing only complete cases will result in biased parameter estimates and invalid inferences.

2. Missing data on a variable *Y* are *MAR* if the probability of missing data is not associated with the value on *Y after controlling for the other variables in the data set*. For example, missing depression scores would be MAR if men are less likely to respond to items on the depression scale than women, but the tendency to have missing depression data is not associated with level of depression for either gender identity. If we use only complete cases in our analyses, MAR missing data can also result in biased results, though the problem is not as severe as for MNAR data. Here, for example, if men tend to have lower depression scores than women and more men are dropped from analyses because of missing data, the overall mean depression score will be higher because of the missing data.
3. Missing data on a variable *Y* are said to be *MCAR* if the probability of missing data on *Y* is unrelated to the value of *Y* or to the values of any other variables in the data set. Perhaps data were not collected on a few trials because of a power outage. Or a participant did not show up for a session because of a traffic accident. In our illustration, missing depression scores would not be more likely to occur for more depressed or less depressed individuals, nor for one group or another. Participants with complete data can be thought of as randomly selected from the original sample. If we analyze only complete cases, the presence of MCAR missing data results in less power, but no bias.

If the missing data are MAR or MNAR, the pattern of missing observations may provide useful information about our research question. We can code the missing data by 1s and the non-missing data by 0s, and then look to see whether there are systematic tendencies for the missing data to be associated with particular values on the other variables. SPSS has a missing values analysis module as an optional add-on that is able to impute missing scores under a variety of assumptions. Similar functionality is available in R (e.g., in the {mice} package); see Gomila and Clark (2022) for clear examples and solutions to missing data. In addition, stand-alone, free software is also available for imputing missing data; for example, NORM (Schafer, 1999), which can be downloaded, along with some documentation, from https://scholarsphere.psu.edu/resources/156307cb-2a99-488b-b069-40b46fdfe633. Of course, the best solution to the problem of missing data is to design the test materials and data collection procedures in ways that produce as few nonresponses as possible.

In addition to the availability of procedures to estimate missing scores, a set of statistical procedures has been developed to deal with designs that produce correlated scores. These procedures are called, depending on the context, mixed-effects, random coefficient, multilevel, or hierarchical linear models. Good discussions may be found in several books and articles (e.g., Baayen, Davidson, & Bates, 2008; Luke, 2004; Quené & van den Bergh, 2004; Raudenbush & Bryk, 2002).

Readers interested in pursuing the topic of missing scores can find several excellent references. Cohen et al. (2003) has a readable chapter on the topic, and Allison (2002) offers an excellent, more comprehensive introduction. Two standard, more advanced texts on the problem of missing data are Little and Rubin (1987) and Schafer (1997). Also, a special section of the December 2001 issue of *Psychological Methods* was dedicated to new approaches to missing data in psychological research. The details of the estimation procedures are beyond the level of this book, but we will discuss the topic a bit more when we study correlation and regression.

13.11 Nonparametric Procedures for Repeated-Measures Designs

In previous chapters, we have discussed the implications of violations of the assumptions of statistical tests. We have found that such violations may result in distortions of Type 1 error rates or in loss of power. In such situations, we have found that nonparametric procedures often have advantages over the conventional *F* and *t* tests. For example, in Chapter 8, two tests based on ranks were presented, the Kruskal–Wallis *H* test and the rank-transformation test (Iman & Conover, 1981). These tests are often more powerful than the *F* test when the treatment population distributions have heavy tails or are skewed. Similar analyses of ranked scores can be applied to data from repeated-measures designs. We will consider these as well as a test for situations in which all scores are either zeros or ones.

13.11.1 *Friedman's χ^2 (Chi Square) Test*

In the first procedure we consider, the scores for each participant are assigned ranks from 1 (for the lowest) to *a* (for the highest); tied scores are assigned the median, or midrank, of their ranks. The test statistic is approximately distributed as chi square under the null hypothesis that, for each participant, all possible sequences of ranks are equally likely. If this null hypothesis is true, differences between conditions in the mean ranks are due to chance rather than to the effects of the independent variable.

Several programs calculate Friedman's (1937) χ^2; for example, SPSS's *nonparametric* module (select *Legacy Dialogs*, then *K Related Samples*) takes the original data as its input and yields a statistic distributed approximately as χ^2 when Friedman's χ^2 is selected from the available options. Similarly, in R the *friedman_test* function in the {rstatix} package provides the same calculations. For the *Seasons* depression data of Table 13.1, the result is $\chi^2 = 6.738$, $p = .081$.

13.11.2 *The Rank-Transformation F Test (F_r)*

An ANOVA based on ranks has been proposed by Iman, Hora, and Conover (1984; also see Hora & Iman, 1988). There are two steps:

1. Assign ranks to all *an* scores from smallest to largest, assigning midranks in case of ties. Note that, unlike Friedman's procedure, each participant's scores are not ranked separately.
2. Do the standard $S \times A$ ANOVA on the rank values. This means that once the Y_{ij} have been converted to R_{ij}, the transformed values can be submitted to any program that analyses data from a repeated-measures design.

When the R_{ij} transforms of the data of Table 13.1 are analyzed, $F = 3.44$ with $p = .026$ (see Exercise 13.11). The Greenhouse–Geisser and Huynh–Feldt epsilon estimates are .83 and 1, almost identical to those in the ANOVA of the original data, but the *p*-values are slightly lower, .035 and .026.

13.11.3 Which Test: F, Friedman's χ^2, or F_R?

When the distributions of scores in the treatment populations have the same shape and variance, the usual F test, χ^2_F, and F_R all test the null hypothesis of equal treatment population means. If those populations can also be assumed to be normal, or to have short tails (as when the data are ratings from a scale with only a few points), the F test will be most powerful. The relative power of the F and the rank-based tests changes when treatment populations are skewed or heavy tailed. In a study comparing F_R with F_F, an F test on data ranked by Friedman's method, Iman et al. (1984; also see Hora & Iman, 1988; Kepner & Robinson, 1988) found that with heavy-tailed or skewed distributions, both F_R and F_F usually have Type 1 error rates close to the nominal .05 level, and both are more powerful – often considerably so – than the F test on the original scores. As for the relative powers of F_R and χ^2_F (or of F_R and F_F), this depends upon several factors including the number of conditions, the shape of the treatment population distribution, the within-participant correlation, and the variability of participant means. The power advantage moves to the Friedman tests as the correlation or participant effects increase. Because of the influence of so many factors, there is no simple rule of thumb, but we recommend F_R unless a is more than 5 and the data are very skewed. In both tests, the epsilon adjustment should be used to reduce degrees of freedom when nonsphericity is present.

13.11.4 Paired Scores: The Wilcoxon Signed-Rank (WSR) Test

Neither F_R nor χ^2_F has good power when $a = 2$. When there are only two conditions, the Wilcoxon signed-rank, or *WSR*, test (1949) is an excellent alternative not only to these tests, but also to the standard t test. In essence, the *WSR* ranks the absolute value of the differences between the conditions, then multiplies each rank by the sign of the difference before summing the signed ranks. The resulting statistic is compared to its distribution under the null hypothesis (or a normal-approximation thereof) to obtain a p-value. The *WSR* is only slightly less powerful than the correlated-scores t test when the data are normally distributed and can be considerably more powerful when the difference scores are symmetrically (but not necessarily normally) distributed with heavy tails (Blair & Higgins, 1985). However, two cautions are in order. First, if the distribution of difference scores is skewed, Type 1 error rates may be inflated for the *WSR* test. Second, power is lost and sometimes paradoxical results are observed when difference scores are zero because these scores are discarded in the *WSR* test. The t test will have a clear power advantage in this situation because difference scores of zero are retained in the t test.

The *WSR* test is available in SPSS and in R. SPSS relies on a normal-approximation to the test distribution, which is quite good for samples as small as 20. Select *Analyse*, followed by *Nonparametric Tests*, *Legacy Dialogs*, and then *2 Related Samples*, and click on *Wilcoxon*. For smaller samples, formulas, examples, and tables of p-values are available in several books on nonparametric statistics; e.g., Bradley (1968) and Lehmann (1975). In R, the *wilcox.test* function in the {stats} package performs the same test.

13.11.5 Zero-One Data: Cochran's (1950) Q Test

A common research situation is one in which each participant responds on several trials or under several different conditions, and each response is classified in one of two ways. For example, suppose we record a success or failure for each participant on each of four mathematical problems that varied in their conceptual distance from a practice problem. The question is whether the probability of success depends upon the problem type. In general, $Y_{ij} = 1$ or 0, indicating a success or failure by participant i in condition j. If p_j is the probability of a success in the population of responses under A_j, the null hypothesis is

$$H_0 : p_1 = p_2 = \ldots = p_j \ldots = p_a$$

The Q statistic is defined as

$$Q = SS_A \,/\, MS_{A/S} \tag{13.20}$$

$MS_{A/S}$ is the average of n variances where each variance is based on the a scores for a participant. Q is distributed as chi square when n is large and the population correlation for any pair of conditions is the same as for any other pair. Therefore, the null hypothesis is rejected when Q exceeds the critical value of χ^2 on $a - 1$ *df*.

The χ^2 distribution rests on the assumption that the variable of interest is normally distributed. Under the central limit theorem, this assumption is essentially true for proportions when n is large. For the Q test, the effective n does not include any participants who have either all successes or all failures. Based on a review of several simulation studies, Myers, DiCecco, White, and Borden (1982) recommended that the effective number of participants be at least 16. When n is small, empirical rejection rates of true null hypotheses are less than the nominal alpha, and power is quite low.

Again, statistical packages often include the Q test. For example, SPSS provides a Q test in the *Nonparametric Tests / Legacy Dialogs* under *k related samples*. In R, one option is the *CochranQTest* in the {DescTools} package. The 0/1 data can also be submitted to a repeated-measures ANOVA. In fact, Myers et al. (1982) report that the F and Q tests have very similar Type 1 error rates for $n \geq 16$; for smaller n, the F test's Type 1 error rate may be inflated.

13.11.6 Nonparametric Tests and Assumptions

Although it is often assumed that nonparametric tests are assumption-free, this is not the case. The Friedman and rank-transformation tests require the usual assumptions of ANOVA for repeated-measures, but with respect to the ranks rather than to the original scores. The *WSR* test is a test of the null hypothesis that the median is zero only if the population of difference scores is symmetric. Lastly, Cochran's Q test rests on the assumption that the correlations for pairs of treatments are the same in the population and requires a sufficient sample size to warrant use of the χ^2 tables to evaluate significance. Thus, deciding whether to analyze the data of a repeated-measures design with ANOVA versus one of the nonparametric tests requires that the researcher pay close attention to the characteristics of the data and their correspondence (or lack thereof) with the assumptions of the statistical procedures under consideration.

13.12 Summary

In this chapter, we considered the analysis of data from a repeated-measures design in which participants were tested on several trials, or at several randomly sequenced levels of an independent variable. Within that context, we addressed the following topics:

- *The distinction between additive and nonadditive structural models* and the problems that may be encountered when data include participant × treatment interactions.

- *The assumption of sphericity*, which can be violated in nonadditive data, and a solution based on downward adjustments to the degrees of freedom.
- *Measures of effect size*, including eta-squared, omega-squared, and Cohen's *f*.
- *The role of power considerations in deciding on sample size* in the repeated-measures design.
- *Tests of contrasts* in the one-factor repeated-measures design.
- *Treatment of missing data*, including distinctions among types of missing data and their implications for estimates of population parameters.
- *Nonparametric tests* when the assumptions of the analysis of variance are violated.

The design considered in this chapter involved only one factor other than participants. In the next chapter, we build on this simple design, adding other factors either as between-participants or within-participants variables.

Exercises

13.1 [Additivity and *EMS*] The following data set consists of three scores for each of four participants:

	A_1	A_2	A_3
S_1	12	14	15
S_2	9	8	10
S_3	10	9	12
S_4	8	6	7

a) Carry out the ANOVA.
b) Assuming additivity, present the *EMS*.
c) Use your answer to part (b) to estimate general ω^2_A.
d) Assume these data are true scores, free of measurement error. Are they additive?

13.2 [Sphericity and MS_{SA}] Consider the data set in the *EX13_2.xlsx* file on the *Exercises* page of the book's website; follow the link from *Data Files*.

a) For each participant, calculate the three difference scores d_{12}, d_{13}, and d_{23}, where d_{ij} represents the difference, $Y_{ij} - Y_{ik}$. Find the variances of each set of difference scores.
b) Perform an ANOVA on the data and show that MS_{SA} = (1/2) × (average of the three variances calculated in the preceding two parts).

13.3 [Error variance for contrasts]

a) Analyze the data set of *EX13_3.xlsx* on the *Exercises* page of the website. Then use the error mean square (sphericity assumed) as the basis for a *t* test of the difference between the A_1 and A_2 means.
b) Repeat the *t* test using the variance of d_{12}.
c) Which analysis do you think should be preferred? Why?

13.4 [Sphericity] Huynh and Feldt (1970) present the following variance-covariance matrix: terms of the diagonal are variances; off-diagonal entries are covariances. Does it satisfy sphericity? Explain.

$$\begin{array}{c} \\ A_1 \\ A_2 \\ A_3 \end{array} \begin{array}{c} \begin{array}{ccc} A_1 & A_2 & A_3 \end{array} \\ \begin{bmatrix} 1.0 & .5 & 1.5 \\ & 3.0 & 2.5 \\ & & 5.0 \end{bmatrix} \end{array}$$

13.5 [Dealing with nonsphericity] Consider the following data set:

	A_1	A_2	A_3	A_4
S_1	1.8	2.2	3.2	2.4
S_2	2.4	1.5	1.9	2.7
S_3	1.9	1.7	2.5	3.5
S_4	2.7	2.6	2.4	3.1
S_5	4.7	4.0	1.1	1.8
S_6	3.6	3.1	4.2	5.4
S_7	4.4	4.2	4.1	4.9
S_8	5.8	6.1	6.4	6.6

a) Carry out the ANOVA on these data and find the lower and upper bounds on the *p*-value, assuming sphericity and nonsphericity, respectively. Assuming $\alpha = .05$, can you reach a conclusion with respect to the *A* source of variance?
b) Assume that we planned all pairwise comparisons for the preceding data set. Find the 95% confidence interval for $\bar{Y}_{.4} - \bar{Y}_{.2}$, controlling for the *FWE*.

13.6 [Using *EMS*] An educational psychologist wishes to develop a measure of articulation that can be used in examining the relation between reading comprehension and the ability to pronounce words. She has 40 third-graders read aloud each of 20 words, and measures the time required for the response. A *participants* × *words* ANOVA yields the following results:

Source	*df*	*MS*	F
Participants (S)	39	208,305.017	244.158
Words (W)	19	739.141	.866
SW	741	853.157	

One measure of the reliability of a measuring instrument is r_{11}, the proportion of the total variance attributable to differences among the participants.

a) Because the variability due to words is clearly negligible, obtain an error mean square by pooling the *W* and *SW* mean squares.
b) Estimate σ^2_s and σ^2_e.
c) Using the results from parts (a) and (b), calculate r_{11}.

13.7 [Repeated-measures contrasts] Each of five participants is tested at four equally spaced points in time on a visual detection task. The numbers of errors for each test are as follows:

	Time			
Participant	*1*	*2*	*3*	*4*
1	9	6	7	5
2	11	8	6	6
3	6	8	7	5
4	13	10	10	9
5	12	8	9	6

a) The experimenter plans to test whether the mean at Time 1 differs significantly from the combined mean for the other three times. State the null hypothesis and carry out the test with $\alpha = .05$.

b) Calculate a confidence interval for the contrast in part (a).

13.8 [Computing effect sizes]

a) Perform the ANOVA on the following data set, response times (in milliseconds) obtained under four different conditions for eight participants:

Participant	A_1	A_2	A_3	A_4
1	2036	2220	2211	2316
2	2034	2042	2094	2077
3	2198	2612	2272	2348
4	2593	2629	2652	2647
5	2347	2408	2416	2479
6	2308	2352	2463	2358
7	2454	2501	2475	2461
8	2462	2394	2491	2659

b) Assuming *A* is an extrinsic factor, estimate general ω^2_A and general ω^2_s.

c) Estimate Cohen's *f*, using Equation 13.17.

d) Estimate Cohen's *f*, using the *Determine* option in G*Power 3.1. The *Variance explained by special effect* is the variance of the means, 2,459 (i.e., MS_A/n); the *Variance within groups* is the average within-condition variance, 36,993 [i.e., $(SS_S + SS_{SA})/(df_s + df_{SA})$]. Enter these values to calculate *f*. (Note: There should be one group and four repetitions; values of the correlation and epsilon are irrelevant when the *Determine* option is used.)

e) Your two estimates should be different. Why is this?

13.9 [Power calculations] We wish to rerun the experiment of Exercise 13.8 with power = .9 to detect a medium-sized effect: $f = .25$.

a) Create a 3 × 3 table with cells containing the *n* needed for the following combinations of conditions: $\varepsilon = .6, .8$, or 1.0; correlation = .4, .6, .8.

b) Briefly explain the effects of varying these factors on the required *n* and the reason for the effects.

13.10 [Friedman's chi-squared test] For the data of Exercise 13.8, test the effects of A, using Friedman's χ^2.

13.11 [Cochran's Q test] Twenty people underwent a 1-week program aimed to help them quit cigarette smoking. The researchers running the program checked on the progress of the participants after 3, 6, and 9 months. The results follow with a 1 signifying that the individual has smoked at least once during the preceding 3-month period and a zero indicating that the individual has not smoked during that period.

	Participants																			
Period	*1*	*2*	*3*	*4*	*5*	*6*	*7*	*8*	*9*	*10*	*11*	*12*	*13*	*14*	*15*	*16*	*17*	*18*	*19*	*20*
1	1	0	0	0	1	0	0	0	0	0	0	0	1	1	0	0	1	1	1	0
2	1	1	1	1	1	0	1	0	0	1	1	0	1	0	0	1	0	0	1	1
3	1	1	1	1	1	0	0	1	1	1	1	1	0	1	0	1	1	1	1	0

The investigators want to know whether there has been a significant change in the percentage of smokers over the three periods in the follow-up study. Perform an analysis to answer this question and state your conclusion.

13.12 [Repeated-measures ANOVA on real data] The file *EX13_12.xlsx* on the *Exercises* page of the book's website contains hours/weekday of exposure to daylight for the oldest group of participants in the *Seasons* study. This was of interest to the researchers because it is believed that exposure to daylight affects mood.

a) For the participants who identify as female (i.e., *sex* = 1), perform an analysis of variance on the *DIRWDC* (direct exposure to daylight during weekdays) scores across seasons; 1 = winter, 2 = spring, 3 = summer, 4 = fall. Interpret the results.
b) Calculate general ω^2 for the effects of seasons.

Notes

1 The complete data set is on the *Seasons* page on the book's website; go to *Data Files*, then *Seasons*. The data in Table 13.1 are also in the file *Male_D under 35*.

2 For clarity, we have created toy data in which the sample variances of difference scores are identical. In a real data set, there would be some differences due to error variance; however, if the population variances were homogeneous, we would not expect marked differences among the sample variances.

3 The Roy–Bose method is as follows. Assume a contrast – e.g., $C = \bar{Y}_1 - (1/2)(\bar{Y}_2 + \bar{Y}_3)$. C is considered to be significantly different from zero if $C > (n - 1)(a - 1)F_{FWE(a-1),(n+a-1)}/(n + a - 1)$.

Chapter 14

Multi-Factor Repeated-Measures and Mixed Designs

14.1 Overview

In this chapter, we extend our discussion of repeated-measures designs to designs involving two or more factors. These factors may all be within-participants or a mix of within- and between-participants factors. We refer to the latter as *mixed designs*; they also are called *split-plot designs* in reference to their early appearance in agricultural experiments. In both pure repeated-measures designs and mixed designs, it is important to distinguish between variables that have fixed effects and those that have random effects. As we discussed in Chapter 13, this distinction has implications not only for the interpretation of results but also for expected mean squares and, consequently, for hypothesis tests and measures of effect size.

The chapter is organized as follows:

- We first consider pure repeated-measures designs in which all independent variables except participants have fixed effects.
- We then turn to mixed designs in which each participant is at only one level of A but is tested at all levels of B and both A and B are fixed-effect factors.
- We then extend each of these designs to include more than two factors.
- Following these developments, we introduce analyses involving a random-effects variable (in addition to participants) and consider the implications for the expected mean squares and the F test.
- Finally, procedures for comparing the treatment means within the designs are developed, and effect size measures and *a priori* power calculations are presented for both classes of design.

14.2 The $S \times A \times B$ Design With A and B Fixed

In a conventional, two-factor repeated-measures design, n randomly sampled participants are each tested at every combination of levels of the other factors. Therefore, if there are two within-participants factors, A and B with a and b levels, respectively, each participant is tested ab times. For example, consider a hypothetical experiment modeled on a study of factors affecting facial recognition (Murray, Yong, & Rhodes, 2000). These investigators presented participants with photos of several faces differing with respect to the type of *distortion* (three levels) and the *orientation* of the face (we describe three levels, a simplification of their experiment). Each of six participants saw examples of each of the

DOI: 10.4324/9781003453550-17

nine combinations of orientation and distortion, with the order of presentation of the nine combinations randomly determined for each participant. Participants rated each face for bizarreness, with normal being 1 and very bizarre being 7.

14.2.1 *The Structural Model and Expected Mean Squares* (EMS)

We assume that Y_{ijk}, the score from participant *i* in the cell formed by A_j and B_k, is composed of a true score and error of measurement; i.e., $Y_{ijk} = \mu_{ijk} + \varepsilon_{ijk}$. We may rewrite this in terms of the grand mean and the deviation of the true score from that mean:

$$Y_{ijk} = \mu + \left(\mu_{ijk} - \mu\right) + \varepsilon_{ijk}$$

Under a general nonadditive model, we assume that individual differences (η_i), the effects of *A* (α_j) and of *B* (β_k), and their interactions contribute to the difference between μ_{ijk} and μ. Substituting these effects for $\mu_{ijk} - \mu$, we have

$$Y_{ijk} = \mu + \eta_i + \alpha_j + \beta_k + (\eta\alpha)_{ij} + (\eta\beta)_{ik} + (\alpha\beta)_{jk} + (\eta\alpha\beta)_{ijk} + \varepsilon_{ijk} \quad (14.1)$$

The parameters and their variances are defined in Table 14.1 and the sums of squares (*SS*) formulas and *EMS* are in Table 14.2. The *EMS* are like those for the *S* × *A* design of

Table 14.1 Components of the structural model for the S × A × B design

Population parameter	*Definition*	*Variance*
η_i	$\mu_i - \mu$	$\sigma_S^2 = E\left(\eta_i^2\right)$
α_j	$\mu_j - \mu$	$\sigma_A^2 = \sum_{j=1}^{a} \alpha_j^2 / a$
β_k	$\mu_k - \mu$	$\sigma_B^2 = \sum_{k=1}^{b} \beta_k^2 / b$
$(\eta\alpha)_{ij}$	$\left(\mu_{ij} - \mu\right) - \eta_i - \alpha_j$	$\sigma_{SA}^2 = E\left[\sum_{j=1}^{a} (\eta\alpha)_{ij}^2 / a\right]$
$(\eta\beta)_{ik}$	$\left(\mu_{ik} - \mu\right) - \eta_i - \beta_k$	$\sigma_{SB}^2 = E\left[\sum_{k=1}^{b} (\eta\beta)_{ik}^2 / b\right]$
$(\alpha\beta)_{jk}$	$\left(\mu_{jk} - \mu\right) - \alpha_j - \beta_k$	$\sigma_{AB}^2 = \sum_{k=1}^{b} \sum_{j=1}^{a} (\alpha\beta)_{jk}^2 / ab$
$(\eta\alpha\beta)_{ijk}$	$\left(\mu_{ijk} - \mu\right) - \eta_i - \alpha_j - \beta_k - (\eta\alpha)_{ij} - (\eta\beta)_{ik} - (\alpha\beta)_{jk}$	$\sigma_{SAB}^2 = E\left[\sum_{k=1}^{b} \sum_{j=1}^{a} (\eta\alpha\beta)_{ijk}^2 / ab\right]$
ε_{ijk}	$Y_{ijk} - \mu_{ijk}$	$\sigma_e^2 = E\left(Y_{ijk} - \mu_{ijk}\right)^2$

Note: $\mu_{ijk} = E(Y_{ijk})$, $\mu_{ijk} = E(\mu_{ijk})$, $\mu_{ij} = \Sigma_k \mu_{ijk}/b$, $\mu_{ik} = \Sigma_j \mu_{ijk}/a$, $\mu_i = \Sigma_k \Sigma_j \mu_{ijk}/ab$, $\mu_j = E(\Sigma_k \mu_{ijk}/b)$, $\mu_k = E(\Sigma_j \mu_{ijk}/a)$, and $\mu = E(\mu_i) = \Sigma_j \Sigma_k \mu_{jk}/ab$. The random effects – η_i, $(\eta\alpha)_{ij}$, $(\eta\alpha\beta)_{ijk}$, and ε_{ijk} – are assumed to be normally distributed with mean zero and variance, as noted in the table.

Table 14.2 ANOVA for an $S \times A \times B$ design; A and B have fixed effects

Source	*df*	*SS*	*EMS*	*F*
S	$n-1$	$ab\sum_{i=1}^{n}\left(\bar{Y}_{i..}-\bar{Y}_{...}\right)^2$	$\sigma_e^2+ab\sigma_S^2$	
A	$a-1$	$nb\sum_{j=1}^{a}\left(\bar{Y}_{.j.}-\bar{Y}_{...}\right)^2$	$\sigma_e^2+b\sigma_{SA}^2+nb\theta_A^2$	MS_A / MS_{SA}
SA	$(n-1)(a-1)$	$b\sum_{j=1}^{a}\sum_{i=1}^{n}\left(\bar{Y}_{ij.}-\bar{Y}_{...}\right)^2$ $-SS_S-SS_A$	$\sigma_e^2+b\sigma_{SA}^2$	
B	$b-1$	$na\sum_{k=1}^{b}\left(\bar{Y}_{..k}-\bar{Y}_{...}\right)^2$	$\sigma_e^2+a\sigma_{SB}^2+na\theta_B^2$	MS_B / MS_{SB}
SB	$(n-1)(b-1)$	$a\sum_{k=1}^{b}\sum_{i=1}^{n}\left(\bar{Y}_{i.k}-\bar{Y}_{...}\right)^2$ $-SS_S-SS_B$	$\sigma_e^2+a\sigma_{SB}^2$	
AB	$(a-1)(b-1)$	$n\sum_{j=1}^{a}\sum_{k=1}^{b}\left(\bar{Y}_{.jk}-\bar{Y}_{...}\right)^2$ $-SS_A-SS_B$	$\sigma_e^2+\sigma_{SAB}^2+n\theta_{AB}^2$	MS_{AB} / MS_{SAB}
SAB	$(n-1)(a-1)(b-1)$	$SS_{total}-SS_S-SS_A-SS_{SA}$ $-SS_B-SS_{SB}-SS_{AB}$	$\sigma_e^2+\sigma_{SAB}^2$	

Note: $\theta_A^2=\Sigma_j\left(\mu_j-\mu\right)^2/(a-1)$, $\theta_B^2=\Sigma_k\left(\mu_k-\mu\right)^2/(b-1)$, and $\theta_{AB}^2=\Sigma_j\Sigma_k\left[\left(\mu_{jk}-\mu\right)-\alpha_j-\beta_k\right]^2/(a-1)(b-1)$.

Chapter 13. As was the case there, A is tested against the SA mean square. Similarly, B and AB are tested against SB and SAB respectively.

14.2.2 An Example of the S × A × B ANOVA

Table 14.3 (the data are also on the website in the file *Table 14.3 SAB Data* on the *Tables* page) presents the data for our hypothetical experiment on face perception. The scores in Table 14.3 are hypothetical averages of ratings of several photos in each condition, but the pattern of the means plotted in Figure 14.1 is similar to the pattern in the Murray et al. (2000) article.

Table 14.4 presents the results of the ANOVA of the $S \times A \times B$ data in Table 14.3. The degrees of freedom and sums of squares can be verified using the formulas in Table 14.2. The F tests are based on the nonadditive model, assuming A and B both have fixed effects and participants are a random sample from an infinitely large population of participants. As the *EMS* in Table 14.2 dictate, A is tested against SA, B is tested against SB, and AB is tested against SAB. With respect to the results, *Orientation* (A), *Distortion* (B), and

Table 14.3 Data for a two-factor repeated-measures experiment (*A* is orientation, *B* is distortion)

	B_1			B_2			B_3		
Participants	A_1	A_2	A_3	A_1	A_2	A_3	A_1	A_2	A_3
1	1.18	2.40	2.48	4.76	4.93	3.13	5.56	4.93	5.21
2	1.14	1.55	1.25	4.81	4.73	3.89	4.85	5.43	4.89
3	1.02	1.25	1.30	4.98	3.85	3.05	4.28	5.64	6.49
4	1.05	1.63	1.84	4.91	5.21	2.95	5.13	5.52	5.69
5	1.81	1.65	1.01	51.01	4.18	3.51	4.90	5.18	5.52
6	1.69	1.67	1.04	5.65	4.56	3.94	4.12	5.76	4.99

Cell and marginal means

	A_1	A_2	A_3	$\bar{Y}_{..k}$
B_1	1.315	1.692	1.487	1.498
B_2	5.020	4.577	3.412	4.336
B_3	4.807	5.410	5.465	5.227
$\bar{Y}_{.j.}$	3.714	3.89	3.455	$\bar{Y}_{...} = 3.687$

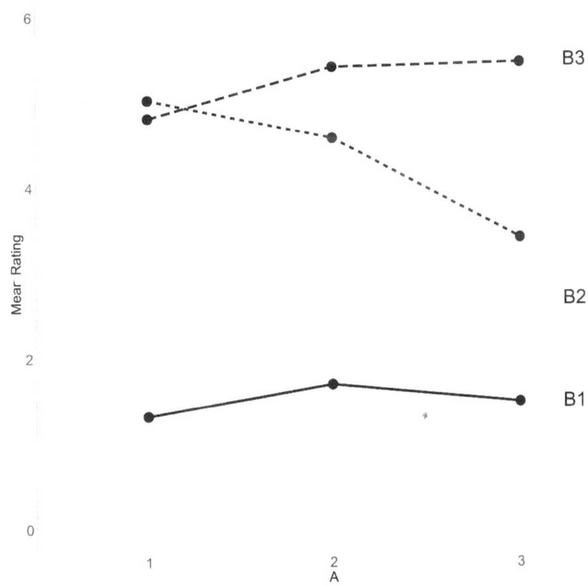

Figure 14.1 Plot of the means in Table 14.3. *A* is orientation in degrees and *B* is the type of distortion

their interaction are all significant. The large *F* for *Distortion* is primarily due to the difference between the average bizarreness rating of the unaltered photo (B_1) and the means for the two altered photos. Adjusting the degrees of freedom by the Greenhouse–Geisser (*G–G*) or Huynh–Feldt (*H–F*) estimates of epsilon did not change the *p*-values markedly.

Table 14.4 ANOVA of the data of Table 14.3

					Significance		
Source	*df*	*SS*	*MS*	*F*	*p*	*G–G*	*H–F*
S	5	.544	.109				
A (orientation)	2	1.749	.874	9.23	.005	.006	.005
SA	10	.947	.095				
B (distortion)	2	136.554	68.277	302.56	.000	.000	.000
SB	10	2.257	.226				
AB	4	8.560	2.140	7.75	.001	.011	.003
SAB	20	5.522	.276				
Total	53	156.133					

Note: *G–G* refers to the Greenhouse–Geisser correction for nonsphericity and *H–F* refers to the Huynh–Feldt correction.

14.2.3 *Repeated-Measures Designs With Three or More Factors*

It is straightforward to extend the $S \times A \times B$ design by including additional factors. For the current example, one possible extension might be to include a third within-participants factor of exposure duration (C, long or short) to the faces in the photos. This added factor will double the number of conditions in the experiment: photos will represent every combination of the two exposure durations with the three orientations and the three distortions. As before, every participant will see examples of all 18 conditions presented in random order.

Adding C increases the number of sources of variance from 7 to 15. There will be three main effects of interest (A, B, C), three interactions of interest (AB, AC, BC), and a three-way interaction (ABC); the F tests of these seven terms each involve a different error mean square (i.e., an interaction of participants with the term being tested). Conceptually and computationally, there is nothing new here; we discussed the interpretation of three-way interactions in Chapter 8. Furthermore, the underlying assumptions of the analysis are the same as in the $S \times A$ and $S \times A \times B$ designs. Sphericity is still assumed and software will adjust degrees of freedom for violations of this assumption.

14.3 Mixed Designs With *A* and *B* Fixed

Designs that incorporate a mixture of between- and within-participants factors are called *mixed designs*. Our development relies on the univariate analysis of variance, although we note that there are other approaches to such designs.[1] Mixed designs involve at least one between-participants variable and one within-participants variable. They are a compromise between a desire to employ within-participants factors to reduce error variance (and thus increase power and the precision of estimates) and the reality that certain variables simply cannot be treated as within-participants factors. Examples of variables that are inherently between-participants factors are those whose levels are observed rather than manipulated (e.g., individual differences variables such as gender identity, age, and clinical diagnostic category) and manipulated variables that entail carry-over effects (e.g., training method).

As an example of a mixed design, Table 14.5 (the data are also in the file *Table 14_5 Mixed Data* on the website) presents a data set for a hypothetical experiment in which one group of students was taught algebra by a standard instructional method (A_1), a second group was given additional problems (A_2), and a third group received additional problems from a computer that provided immediate feedback (A_3). All three groups were tested at the end of the instructional period, and then were tested once every two weeks until four different tests had been given. Assume that the tests were equated for difficulty so that any differences could be attributed to the passage of time. This design permits us to compare the instructional methods (A) and to see the time course (B) of performance following the end of instruction for each method. Notice that the design is a mixture of the between-participants design of Chapter 8 and the within-participants design of Chapter 13. If we average the four test scores for each participant, we have a between-participants design and can conduct the ANOVA exactly as in Chapter 8. If, on the other hand, we retain the four test scores but ignore the instructional factor, we have an $S \times B$ design in which 18 participants have scores at the four levels of B.

Table 14.5 Data for a design with one between-participants (A) and one within-participants (B) factor

Method of instruction		*Time of test*				$\bar{Y}_{ij.}$
		B_1	B_2	B_3	B_4	
A_1	S_{11}	82	48	41	53	56
	S_{21}	72	70	51	45	62
	S_{31}	43	35	30	12	30
	S_{41}	77	41	61	31	50
	S_{51}	43	43	21	29	34
	S_{61}	67	39	30	40	44
	$\bar{Y}_{.1k}$	64	46	39	35	$\bar{Y}_{.1.} = 46$
A_2	S_{12}	71	53	50	62	59
	S_{22}	89	67	76	68	75
	S_{32}	82	84	83	71	80
	S_{42}	56	56	55	45	53
	S_{52}	64	44	44	52	51
	S_{62}	76	74	64	74	72
	$\bar{Y}_{.2k}$	73	63	62	62	$\bar{Y}_{.2.} = 65$
A_3	S_{13}	84	80	75	77	79
	S_{23}	84	72	63	81	75
	S_{33}	76	54	57	61	62
	S_{43}	84	66	61	77	72
	S_{53}	67	69	55	69	65
	S_{63}	61	67	55	61	61
	$\bar{Y}_{.3k}$	76	68	61	71	$\bar{Y}_{.3.} = 69$
	$\bar{Y}_{..k}$	71	59	54	56	$\bar{Y}_{...} = 60$

14.3.1 *The Structural Model and EMS for the Mixed Design*

In designs such as the one in Table 14.5, *n* participants are tested at A_1, *n* other participants are tested at A_2, and so on. All *an* participants are tested at each of the *b* levels of the independent variable *B*. Therefore, $abn - 1$ *df* must be accounted for in the analysis of variance. With respect to notation, we refer to Y_{ijk} where *i* indexes the participant ($i = 1, 2, \ldots, n$), *j* indexes the level of the between-participants variable ($j = 1, 2, \ldots, a$), and *k* indexes the level of the within-participants variable ($k = 1, 2, \ldots b$). In Table 14.5, $n = 6$, $a = 3$, and $b = 4$.

As always, the structural model underlies the partitioning of the total variability in the ANOVA. The model equation for the data in Table 14.5 is

$$Y_{ijk} = \mu + \alpha_j + \eta_{i/j} + \beta_k + (\alpha\beta)_{jk} + (\eta\beta)_{ik/j} + \varepsilon_{ijk} \tag{14.2}$$

There are some important differences between this equation and Equation 14.1 for the $S \times A \times B$ design. These differences rest on the distinction between *nested variables* and *crossed variables*. Participants are nested within levels of *A* because there are different participants at each level of *A*; the same is true for *SB* interaction effects. We have indicated these nested relationships by inserting a slash before the *j* in the subscripts for participants and *SB* effects within levels of *A* [$\eta_{i/j}$, $(\eta\beta)_{ik/j}$]. Because a participant is tested at only one level of *A*, there are no *SA* interaction effects. Also, because the *SB* effects are nested in levels of *A*, there are no *SAB* effects in the model. In contrast, there are *SB* interaction effects because participants cross with *B*; that is, all participants are tested at all levels of *B*. The parameters and their variances are defined in Table 14.6.

Table 14.7 contains the sources of variance, *df* and *SS* formulas, and the *EMS*. The sources of variance follow from Equation 14.2. In Panel *a*, we have organized them into between- and within-participants terms to emphasize the relationship to designs encountered

Table 14.6 Components of the structural model for the mixed design

Population parameter	*Definition*	*Variance*
$\eta_{i/j}$	$\mu_{ij} - \mu_j$	$\sigma^2_{S/A} = E\left(\eta^2_{i/j}\right)$
α_j	$\mu_j - \mu$	$\sigma^2_A = \sum_{j=1}^{a} \alpha_j^2 / a$
β_k	$\mu_k - \mu$	$\sigma^2_B = \sum_{k=1}^{b} \beta_k^2 / b$
$(\eta\beta)_{ik/j}$	$(\mu_{ik} - \mu_j) - \eta_{i/j} - \beta_k$	$\sigma^2_{SB/A} = E\left[\sum_{k=1}^{b} (\eta\beta)^2_{ik/j} / b\right]$
$(\alpha\beta)_{jk}$	$(\mu_{jk} - \mu) - \alpha_j - \beta_k$	$\sigma^2_{AB} = \sum_{k=1}^{b}\sum_{j=1}^{a} (\alpha\beta)^2_{jk} / ab$
ϵ_{ijk}	$Y_{ijk} - \mu_{ijk}$	$\sigma^2_e = E\left(Y_{ijk} - \mu_{ijk}\right)^2$

Note: $\mu_{ijk} = E(Y_{ijk})$, $\mu_{jk} = E(\mu_{ijk})$, $\mu_{ij} = \sum_k \mu_{ijk} / b$, $\mu_{ik} = \sum_j \mu_{ijk} / a$, $\mu_i = \sum_k\sum_j \mu_{ijk} / ab$, $\mu_j = E\left[\sum_k \mu_{ijk} / b\right]$, $\mu_k = E\left[\sum_j \mu_{ijk} / a\right]$, and $\mu = E(\mu_i)$. The random effects – $\eta_{i/j}$, $(\eta\beta)_{ik/j}$, *and* ε_{ijk} – are assumed to be normally distributed with mean zero and variance as noted in the table.

Table 14.7 ANOVA formulas (*a*) and *EMS* (*b*) for the mixed design

(*a*) Partitioning degrees of freedom and sums of squares

Sources of variance	*df*	*Sums of squares*
Total	$abn-1$	$\Sigma_k\Sigma_j\Sigma_i\left(Y_{ijk}-\bar{Y}_{...}\right)^2$
Between-subjects (S)	$an-1$	$b\Sigma_j\Sigma_i\left(\bar{Y}_{ij.}-\bar{Y}_{...}\right)^2$
A (Instructions)	$a-1$	$bn\Sigma_j\left(\bar{Y}_{.j.}-\bar{Y}_{...}\right)^2$
S/A	$a(n-1)$	SS_S-SS_A
Within-subjects (W Ss)	$an(b-1)$	$SS_{total}-SS_S$
B (Time)	$b-1$	$an\Sigma_k\left(\bar{Y}_{..k}-\bar{Y}_{...}\right)^2$
AB	$(a-1)(b-1)$	$n\Sigma_j\Sigma_k\left(\bar{Y}_{.jk}-\bar{Y}_{...}\right)^2-SS_A-SS_B$
B × S/A	$a(n-1)(b-1)$	$SS_{WSs}-SS_B-SS_{AB}$

(*b*) Expected mean squares

Source	*EMS*
A	$\sigma_e^2+b\sigma_{S/A}^2+bn\theta_A^2$
S/A	$\sigma_e^2+b\sigma_{S/A}^2$
B	$\sigma_e^2+\sigma_{SB/A}^2+na\theta_B^2$
AB	$\sigma_e^2+\sigma_{SB/A}^2+n\theta_{AB}^2$
B × S/A	$\sigma_e^2+\sigma_{SB/A}^2$

previously. If we ignore the factor *B*, the design is a one-factor design with participants nested within levels of *A*. As in the one-factor between-participants design, *an* – 1 *df* are divided between the *A* and participants-within-*A* sources. If we ignore *A*, we have the *S* × *B* design of Chapter 13; there is a participants term (*S*), a *B* term, and an *SB* interaction. However, part of the *SB* variance is potentially due to the interaction of *A* and *B* and so the *SB* variability is partitioned into an *AB* and a *B* × *S/A* term, as shown in Table 14.7.

From the *EMS* of Panel *b* in Table 14.7, we see that the between-participants factor, *A*, is tested against the *S/A* source. This is the same as in a one-factor between-participants design. Differences among the α means reflect not only the effects of *A* and errors of measurement, but also individual differences because there are different participants at each level of *A*. It is important to recognize that the term σ^2_e has a different meaning in the current design than it had in Chapter 8. In the one-factor design of Chapter 8, σ^2_e encompassed both variability due to individual differences and measurement errors. However, in the current design, individual differences contribute only to between-participants factors; therefore, σ^2_e denotes only measurement errors and $\sigma^2_{S/A}$ denotes individual differences.

In the within-participants portion of the design, the *B* and *AB* terms are tested against the residual error, $MS_{B\times S/A}$. This is because the nested interaction effects of participants and *B* contribute to both the *B* and *AB* variance. This is analogous to the way we tested effects in the repeated-measures designs in Chapter 13 and Section 14.2 of the current chapter.

14.3.2 Assumptions in the Mixed Design

As in any repeated-measures design, the univariate *F* test requires the assumption of sphericity; it is assumed that within each level of *A* the population variances of difference scores for all pairs of *B* levels are the same. Violations of this assumption will result in inflation of Type 1 error rates. However, as in the pure within-participants design, the epsilon-adjusted *F* test provides an approximately correct *p*-value. Whether testing *B* or *AB*, the numerator and denominator *df* are multiplied by the Greenhouse–Geisser or Huynh–Feldt estimate of epsilon, reducing the *df* and thus requiring a larger *F* value for significance than would be required without the adjustment.

Consider the *b* populations of scores at a level of *A*. Our second assumption, which is less critical than the sphericity assumption, is that the variances and covariances of these populations are the same at all levels of *A*. Generally, if the group sizes are equal, the ratios of mean squares will be distributed as *F* although the degrees of freedom for within-participants terms may require correction if the sphericity assumption is violated.

14.3.3 ANOVA of Data From a Mixed Design

Figure 14.2 displays a plot of the means from our hypothetical study of the effects of instructional method and time of test; curves are plotted for each instructional group (*A*) as a function of time (*B*). The average curve declines over time, although there is an increase

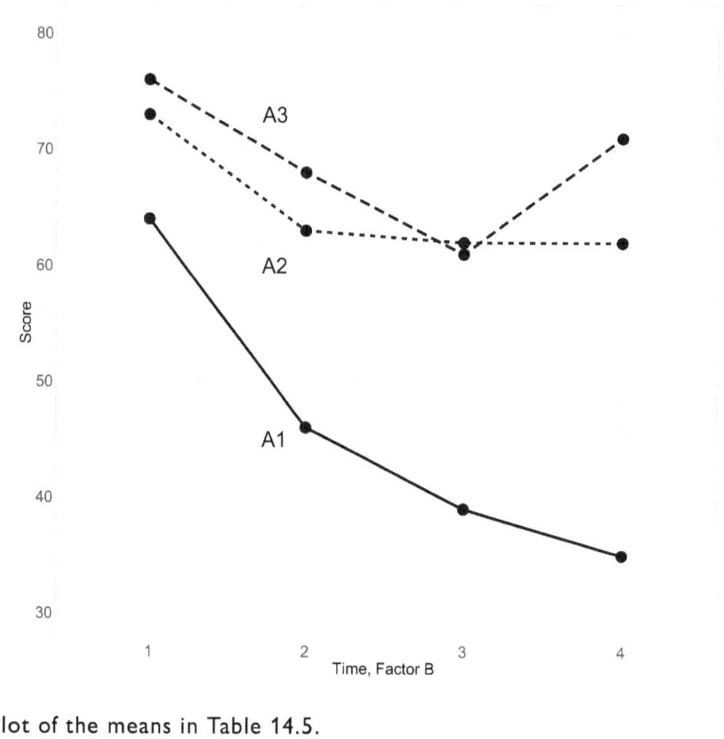

Figure 14.2 Plot of the means in Table 14.5.

Table 14.8 ANOVA of the data in Table 14.5

Source	*df*	*Sum of squares*	*Mean square*	*F*	*P*	*G–G*	*H–F*
A (Method)	2	7,248	3,624.00	7.76	0.005		
S/A	15	7,010	467.33				
B (Time)	3	3,132	1,044.00	18.72	0.000	0.000	0.000
AB	6	1,056	176.00	3.16	0.011	0.027	0.015
B × S/A	45	2,510	55.78				

for the A_3 method at the fourth measurement time. The decline seems steepest between the first and second test positions. There seems to be little difference in performance between groups A_2 and A_3, but both clearly have higher averages than the A_1 group.

Table 14.8 contains the results of an ANOVA of the data in Table 14.5. The denominators of the tests of the main and interaction effects follow the *EMS* of Table 14.7. Presumably because of the relatively poor performance of the A_1 group, the *A* main effect is significant. The significant *B* source of variance reflects the decrease in average scores over time. The *AB* interaction is also significant, reflecting the steeper decline in the A_1 curve and perhaps the upturn in the A_2 curve.

14.3.4 *Mixed Designs With Additional Fixed Factors*

Just as we were able to do with the $S \times A \times B$ design, we might also add factors to our mixed design. In our hypothetical experiment, we had three methods of instruction in algebra and we tested participants at four different times; method was a between-participants variable and time was a within-participants variable. At each of the four tests, we might include two types of problems. For example, we might compare problems presented as stories with problems expressed directly in terms of variables such as *X* and *Y*. Now we again have participants nested within levels of *A* (method), but participants have scores for each of eight combinations of *B* (time of test) and *C* (type of problem).

Table 14.9 presents sources of variance (*SV*), *df*, and *EMS* for our expanded design. There are four groups of sources. The first includes *A* and *S/A* as in the mixed design with only one within-participants factor. The only modification in the components of the *EMS* is in the multipliers of $\sigma^2_{S/A}$ and θ^2_A; the rule is that components of variance are multiplied by the number of scores on which each mean is based. There are *bc* scores for each participant and *bcn* scores at each level of *A*; therefore, these are the coefficients. The remaining groupings involve the within-participants factors (*B*, *C*, and *BC*). *A* interacts with each of these, as does *S/A*. Note that the result of crossing *S/A* with each within-participants factor is a nested interaction. Coefficients of the components of the *EMS* are determined as previously described. In short, the extension of the two-factor mixed design to a three-factor mixed design is straightforward.

14.3.5 *Pretest-Posttest Designs*

A common, special case of the mixed design is one in which there is a between-participants treatment (*A*), and a pretest and a posttest score are obtained from each participant. Participants are assigned randomly to the levels of *A* and the pretest scores are obtained before the

Table 14.9 Expected mean squares and error terms for a design with one between- and two within-participants factors

SV	*df*	*EMS*	*Error term*
A	$a - 1$	$\sigma^2_e + bc\sigma^2_{S/A} + nbc\theta^2_A$	*S/A*
S/A	$a(n - 1)$	$\sigma^2_e + bc\sigma^2_{S/A}$	
B	$b-1$	$\sigma^2_e + c\sigma^2_{SB/A} + nac\theta^2_B$	SB/A
AB	$(a-1)(b-1)$	$\sigma^2_e + c\sigma^2_{SB/A} + nc\theta^2_{AB}$	SB/A
SB/A	$a(n-1)(b-1)$	$\sigma^2_e + c\sigma^2_{SB/A}$	
C	$c - 1$	$\sigma^2_e + b\sigma^2_{SC/A} + nab\theta^2_C$	SC/A
AC	$(a - 1)(c - 1)$	$\sigma^2_e + b\sigma^2_{SC/A} + nb\theta^2_{AC}$	SC/A
SC/A	$a(n - 1)(c - 1)$	$\sigma^2_e + b\sigma^2_{SC/A}$	
BC	$(b - 1)(c - 1)$	$\sigma^2_e + \sigma^2_{SBC/A} + na\theta^2_{BC}$	SBC/A
ABC	$(a - 1)(b - 1)(c - 1)$	$\sigma^2_e + \sigma^2_{SBC/A} + n\theta^2_{ABC}$	SBC/A
SBC/A	$a(n - 1)(b - 1)(c - 1)$	$\sigma^2_e + \sigma^2_{SBC/A}$	

Note: *A*, *B*, and *C* are all assumed to have fixed effects.

treatment is applied, so that there are no systematic differences in the pretest scores across levels of *A*. The posttest scores reflect the effects of the treatment, if there are any.

Although pretest-posttest designs are common, they are often analyzed in a less than optimal way. One common analysis is a mixed factors ANOVA with time of test (pre versus post) as the within-participant factor. With this analysis, the *F* test for the *A* main effect will be very conservative because the pretest scores cannot be affected by the treatment. Another common strategy is to compute a gain score (posttest – pretest) for each participant and then conduct a one-factor between-participants ANOVA; this analysis is equivalent to testing the interaction effect in the mixed design. Both approaches lack power relative to conducting an ANCOVA in which the covariate is the pretest score and the dependent variable is the posttest score. For this reason, we recommend using ANCOVA to analyze the results from a pretest-posttest design (see Chapter 24).

14.4 Designs With More Than One Random-Effects Factor: The Fixed- vs Random-Effects Distinction Again

Recall from Section 13.3 that the distinction between random- and fixed-effects refers to how levels of a factor are sampled. In the case of a fixed-effects factor, the levels of the factor are either chosen in such a way that they exhaust all possible levels or they are chosen selectively from among all possible levels. An example of the former is taking observations from all four seasons in the *Seasons* data set. Examples of the latter are common, such as our hypothetical experiment in which face orientation and face distortion are manipulated. For both factors, the experimenter intentionally chose the specific orientations and particular distortions to be included in the experiment. In both cases, the levels of the variables were presumably chosen to be a representative range of variation on each dimension.

In the case of a random-effects factor, the levels of the factor do not exhaust all possible levels of the variable nor are they selected in a systematic way. Instead, levels are a randomly chosen subset of a much larger potential set of observations. The sole example of a random-effects factor that we have encountered to this point is participants.

However, there are many types of experiments in which stimuli are a factor and should be treated as a random-effects variable. As one example, suppose a software company has designed a new font which they believe will make reading easier than more commonly used fonts. The human factors group in the company performs an experiment in which participants are presented with the same set of words in each of the two fonts; the total number of presentations are randomly sequenced for each participant. The task is to say each word aloud as soon as it is presented; time to initiate pronunciation of each word is recorded to the nearest millisecond. The words are chosen from the population of words that are relatively low in English language frequency on the assumption that any differences in the time to recognize the words are more likely to show up when the task is relatively difficult. The software company is not interested in whether there is a difference in recognition time for only those words included in the experiment. Rather, the words used in the experiment are considered a random sample from a much larger population of low-frequency words. There is little point in marketing the new font unless the company is convinced that any observed advantage holds for the population of words sampled in the experiment.

Our example is not unusual. We often want to generalize our results to stimuli beyond those used in the study. Examples include sentences differing in syntactic construction in a psycho-linguistic study, different concrete and abstract words in a memory experiment, pictures of faces illustrating different emotions in a personality experiment, and photographs of different species of insect in a discrimination-learning experiment for pigeons. In these examples, it is important to distinguish the stimuli (i.e., sentences, words, pictures, and photographs) from the categories from which they were sampled (i.e., different types of syntactic construction, concrete and abstract concepts, types of emotions, and species of insect). The researcher's substantive interest is in the categorical variables, which are fixed-effects factors; however, the stimuli are distinct sources of variability in the experiment and they are most appropriately considered random-effects factors.

Treating a factor as *random* instead of as *fixed* does not change the sources of variance, sums of squares, or degrees of freedom when an ANOVA is performed. However, it does change the expected mean squares, so that now different F tests may be required. To see how and why the expected mean squares are affected by the designation of factors as random or fixed effects, we must first look at the rules for generating expected mean squares.

14.5 Rules for Generating Expected Mean Squares

To illustrate the rules for generating expected mean squares, we will use the design summarized in Table 14.9. Recall that A is a between-participants factor, whereas B and C are within-participants factors; A, B, and C are all assumed to have fixed effects, whereas S is a random-effects factor.

There are six components of variance that might potentially contribute to any of the expected mean squares in Table 14.9. These are the variances due to participants ($\sigma^2_{S/A}$), the interaction of participants and B ($\sigma^2_{SB/A}$), and error variance (σ^2_e), as well as components corresponding to the fixed effects in the structural equation (θ^2_A, θ^2_B, and θ^2_{AB}). Referring to Table 14.9, only a few of these terms contribute to any given expectation and therefore play a role in the selection of the error term or in the estimation of effect sizes. Box 14.1 presents "rules of thumb" for generating *EMS* that apply to many common designs.

Box 14.1 Rules of Thumb for Obtaining Expected Mean Squares

1. Decide for each variable, including participants, whether it has fixed or random effects. Assign a letter to represent the variable (e.g., *S*, *A*, *B*, *C*) and a letter to represent the number of levels (e.g., *n*, *a*, *b*, c).
2. Each *EMS* contains σ^2_e.
3. For each *EMS*, list for consideration all components whose subscripts include all the letters in the source under consideration. For example, for the *EMS* for *AB* in Table 14.10, we would list $nc\theta^2_{AB}$, $c\sigma^2_{SB/A}$, $n\theta^2_{ABC}$, and $\sigma^2_{SBC/A}$. Note that a component has three parts:

   a) A coefficient such as *bc* or *nac* representing the number of scores at each level of the variable.
   b) A σ^2 or θ^2 term depending on whether the subscripted variable is assumed to have random or fixed effects.
   c) Subscripts designating the variable under consideration.

4. Define *essential* subscripts. If there is a slash, as in *SB/A*, only those letters to the left of the slash are considered essential. For example, only *SB* is essential in the subscript *SB/A*. (In some designs, there can be several levels of nesting – e.g., students within classes within schools; only the letters to the left of the left-most slash are essential subscripts.)
5. Among the essential subscripts, if any letters that are not part of the source designation represent fixed effects, delete that component from the expected mean square. For example, under Rule 4 we would have listed $\sigma^2_{SBC/A}$ when considering the expectation for the *AB* source. Among the essential subscripts – *S*, *B*, and *C* – *C* is not part of the source designation (i.e., *AB*) and does have fixed effects. Therefore, $\sigma^2_{SBC/A}$ does not contribute to the *EMS* for *AB*.

The box illustrates the rules using the *EMS* for the *AB* source of variance in the design of Table 14.9. As a second example, consider how the rules of thumb might be used to find $E(MS_B)$. First, list σ^2_e because it contributes to every *EMS*. Then add every additional component of variance that has subscripts containing *B*, multiplying each component by its appropriate coefficient. The result is

$$\sigma_e^2 + nac\theta_B^2 + nc\theta_{AB}^2 + c\sigma_{SB/A}^2 + na\theta_{BC}^2 + n\theta_{ABC}^2 + \sigma_{SBC/A}^2$$

Rules 4 and 5 dictate deletion of the $nc\theta^2_{AB}$, $na\theta^2_{BC}$, and $n\theta^2_{ABC}$ components because once *B* is deleted from the subscripts, we are left with *A*, *C*, or both; these subscripts denote fixed-effects variables. We also delete $\sigma^2_{SBC/A}$ because when we ignore *B* we still have *C* among the essential subscripts. However, $nc\theta^2_{SB/A}$ is retained because when *B* is deleted from the essential subscripts of that component, we are left only with *S*, a subscript that denotes a random-effects variable. The result is given in the *B* line of Table 14.9.

The design in Table 14.9 involves three fixed-effects factors and one random-effects factor. Readers should verify that the expected mean squares presented for all the designs that we have considered to this point in the book are consistent with the rules of thumb. However, in all the cases considered up until now, the only random-effects factor was participants. The rules still apply when there is a second random-effects factor (e.g., stimuli); in fact, they are particularly important in such cases because without the rules to generate the *EMS* the selection of error terms to test effects of interest is unclear. We provide two examples in the following subsections.

14.5.1 S × A × B *Design With Two Random-Effects Factors*

In Section 14.4, we described a human factors experiment in which both participants and words (stimulus items) would be considered random-effects factors. Imagine that we conducted such an experiment involving n participants, b words, and a fonts. A summary of the appropriate analysis of this design is presented in Table 14.10. The sources of variance and their *df* are what you would expect, but the expected mean squares for S, A, and SA are unexpected from the designs encountered to this point. We will work through the expected mean square for A to illustrate the application of the rules for this term.

There are eight distinct variance components in this design: measurement error, σ^2_e; variability due to participants, σ^2_S; variability due to words, σ^2_B; variability due to the interactions of participants with A, σ^2_{SA}, with B, σ^2_{SB}, and with AB, σ^2_{SAB}; variability due to the interaction of words with A, σ^2_{AB}; and variability associated with the fixed-effects factor A, θ^2_A. The coefficients associated with each of these variance components is, as always, the product of the number of levels of all of the factors that are not represented in the subscript on the variance term (e.g., nb for A; a for SB). Consider how the rules apply to the generation of the expected mean square for the effect of interest, A.

First, list measurement error and every component that includes A in its subscript, along with the appropriate coefficients:

$$\sigma^2_e + \sigma^2_{SAB} + b\sigma^2_{SA} + n\sigma^2_{AB} + nb\theta^2_A \tag{14.3}$$

Next, apply Rules 4 and 5 in Box 14.1. Here these rules do not result in any deletions, so that Equation 14.3 is our final result for the expected mean squares for A. Compare this

Table 14.10 Expected mean squares for an $S \times A \times B$ design when B has random effects

Source	*df*	*EMS*[a]
S	$n-1$	$\sigma^2_e + \underline{a\sigma^2_{SB}} + ab\sigma^2_S$
A	$a-1$	$\sigma^2_e + \underline{\sigma^2_{SAB}} + b\sigma^2_{SA} + \underline{n\sigma^2_{AB}} + nb\theta^2_A$
SA	$(n-1)(a-1)$	$\sigma^2_e + \underline{\sigma^2_{SAB}} + b\sigma^2_{SA}$
B	$b-1$	$\sigma^2_e + a\sigma^2_{SB} + na\sigma^2_B$
SB	$(n-1)(b-1)$	$\sigma^2_e + a\sigma^2_{SB}$
AB	$(a-1)(b-1)$	$\sigma^2_e + \sigma^2_{SAB} + n\sigma^2_{AB}$
SAB	$(n-1)(a-1)(b-1)$	$\sigma^2_e + \sigma^2_{SAB}$

[a] Underlined components are absent if B is assumed to have fixed effects.

to what we found when B was a fixed-effects variable; in that case, the SAB and AB terms were dropped and the expected mean squares assumed its conventional form. In the current example, we find that the value of MS_A depends on not three, but five variance components; the sample means for the two font conditions may vary because of measurement error, because they are based on different combinations of participants, words, and fonts (SAB), different combinations of participants and fonts (SA), different combinations of fonts and words (AB), and different fonts (A). Looking at the *EMS* in Table 14.10, we can see that the expected mean squares for S and for SA are also affected by the inclusion of words as a random-effects factor in the design.

14.5.2 Mixed Design With Two Random-Effects Factors

The $S \times A \times B$ example can undergo further complications. We might have several groups of participants, perhaps differing in reading ability or grade level. In general, we conceive of an experiment with a between-participants variable, A, having fixed effects (e.g., grade level); a within-participants variable, B, also having fixed effects (e.g., font); and a second within-participants variable, C, having random effects (e.g., words). All combinations of the three factors appear in the design. To summarize the design under consideration: there are a groups of n participants each and all participants are tested on the same c items at each of b levels of a fixed-effects factor, B. We presented this design in Table 14.9 under the assumption that all three factors – A, B, and C – had fixed effects. Table 14.11 reproduces much of Table 14.9; the difference is that because C is now assumed to have random effects, the *EMS* now include additional terms. Those added terms are underlined in Table 14.11.

Table 14.11 Expected mean squares for a mixed design in which C is assumed to have random effects; A and B have fixed effects

SV	*df*	*EMS*[a]
A	$a - 1$	$\sigma_e^2 + \underline{b\sigma_{CxS/A}^2} + bc\sigma_{S/A}^2 + \underline{bn\sigma_{AC}^2} + nbc\theta_A^2$
S/A	$a(n - 1)$	$\sigma_e^2 + \underline{b\sigma_{CxS/A}^2} + bc\sigma_{S/A}^2$
B	$b - 1$	$\sigma_e^2 + c\sigma_{BxS/A}^2 + \underline{na\sigma_{BC}^2} + \underline{\sigma_{BCxS/A}^2} + nac\theta_B^2$
AB	$(a - 1)(b - 1)$	$\sigma_e^2 + c\sigma_{BxS/A}^2 + \underline{n\sigma_{ABC}^2} + \underline{\sigma_{BCxS/A}^2} + nc\theta_{AB}^2$
BxS/A	$a(n - 1)(b - 1)$	$\sigma_e^2 + \sigma_{BCxS/A}^2 + c\sigma_{BxS/A}^2$
C	$c - 1$	$\sigma_e^2 + b\sigma_{CxS/A}^2 + nab\sigma_C^2$
AC	$(a - 1)(c - 1)$	$\sigma_e^2 + b\sigma_{CxS/A}^2 + nb\sigma_{AC}^2$
CxS/A	$a(n - 1)(c - 1)$	$\sigma_e^2 + b\sigma_{CxS/A}^2$
BC	$(b - 1)(c - 1)$	$\sigma_e^2 + \sigma_{BCxS/A}^2 + na\sigma_{BC}^2$
ABC	$(a - 1)(b - 1)(c - 1)$	$\sigma_e^2 + \sigma_{BCxS/A}^2 + n\sigma_{ABC}^2$
BCxS/A	$a(n - 1)(b - 1)(c - 1)$	$\sigma_e^2 + \sigma_{BCxS/A}^2$

[a] Underlined terms are not present if C has fixed effects. If C has random effect, quasi-F statistics test A, B, and AB; those tests and the associated degrees of freedom follow from the developments in Section 14.6.3.

The underlined variance components are dictated by the rules of thumb introduced in Box 14.1. As an example of the application of the rules, consider the *A* source of variance. $\sigma^2_{C\times S/A}$ contributes to $E(MS_A)$ because *A* appears in the subscripts, and the essential subscripts, *S* and *C*, both represent variables having random effects. σ^2_{AC} also contributes to the expected mean square for *A* because, ignoring the *A* subscript, the remaining letter, *C*, represents a variable assumed to have random effects.

We have now seen two examples in which the designation of a factor as having random instead of fixed effects results in changes to the expected mean squares in the ANOVA, and therefore the appropriate *F* tests. *We emphasize that the identification of sources of variance, the computation of df, and the computations of* SS *and* MS *quantities are unaffected by the random- vs fixed-effects distinction. However, our conceptualization of what the* MS *calculations estimate in terms of population parameters is directly affected, as we have seen. These effects on the* EMS, *in turn, have direct implications for our F tests.*

14.6 Constructing Unbiased *F* Tests in Designs With Two Random Factors

Consider the $S \times A \times B$ design whose analysis is summarized in Table 14.10. The question of interest in this design is whether there is an effect of type of font, *A*, on participants' recognition of words, *B*. If we test MS_A against MS_{SA} we cannot know whether a significant result is because $\theta^2_A > 0$ (our null hypothesis is false) or because $\sigma^2_{AB} > 0$. To attribute a significant result to *A*, the error term needs to have an expectation of $\sigma^2_e + b\sigma^2_{SA} + n\sigma^2_{AB} + \sigma^2_{SAB}$. As can be seen in Table 14.10, there is no single source of variance that has this expectation.

The same problem is encountered for the mixed design summarized in Table 14.11. The effects of interest in this example are the main effects of *A* and *B* and the *AB* interaction. Examining the *EMS* terms for these three terms, it is apparent that no single *MS* in Table 14.11 provides an appropriate error term to test any of these effects. If we test MS_A against $MS_{S/A}$ for example, we again run the risk of inflating the Type 1 error rate.

We consider three possible responses to the problem of constructing an unbiased test of effects of interests in such designs. To anticipate, the approaches discussed in Sections 14.6.1 and 14.6.2 usually do not provide satisfactory solutions to the problem; however, we consider them because they help clarify the issue.

14.6.1 *Preliminary Tests of Interaction*

We may sometimes be fortunate to find that an effect of interest does not vary in magnitude across participants (i.e., the additive model). For example, if the effect of type of font is the same for all of the participants in our repeated-measures design, then the interaction, σ^2_{SA}, equals zero and may be dropped from all of the *EMS* terms in Table 14.10. In that event, the *EMS* for *A* simplifies to:

$$E(MS_A) = \sigma_e^2 + \sigma_{SAB}^2 + n\sigma_{AB}^2 + nb\theta_A^2$$

and MS_{AB} provides an appropriate error term to test *A* because

$$E(MS_{AB}) = \sigma_e^2 + \sigma_{SAB}^2 + n\sigma_{AB}^2$$

A similar situation results if the effect of A is constant over stimuli. In that case, σ^2_{AB} equals zero and MS_{SA} then provides an appropriate error term for the test of A. This suggests a preliminary test of the SA and AB terms against MS_{SAB}. For example, if the test of MS_{SA} against MS_{SAB} does not yield a significant result, we might conclude that σ^2_{SA} is zero and therefore MS_A could be tested against MS_{AB}.

We stated at the outset that this approach is not completely satisfactory. One limitation is that both the SA and AB interactions will often be significant; in those instances, the approach is not applicable. Less obviously, even when a nonsignificant interaction is observed, it is possible that the preliminary test will produce a Type 2 error; that is, we wrongly conclude that σ^2_{SA} equals zero. In that case, the Type 1 error rate in testing A against AB will be increased beyond the nominal level. In studies related to this issue, Type 1 error rates have ranged from .07 to .11 when the nominal α level is .05 (Janky, 2000). Therefore, preliminary tests to justify F tests of A involve risks we ordinarily would prefer to avoid.

14.6.2 *Separate Tests Against* MS_{SA} (F_1) *and* MS_{AB} (F_2)

A frequent approach to the problem of lacking a single mean square to serve as an error term is to form two F ratios: MS_A is tested against both MS_{SA} and MS_{AB} with significance being required on both tests to conclude that A has an effect. The presumed logic of this approach is that the test against MS_{SA} allows generalizability to the participant population and the test against MS_{AB} allows generalizability to the item population. It is true that our goal is to generalize our results to both the population of participants and the population of items from which we have sampled. However, even if both F tests are significant, we cannot claim that kind of generality. To see why, suppose our null hypothesis is true (i.e., $\theta^2_A = 0$) but A interacts with S and B. Then the ratio of expected mean squares when we test A against SA is

$$\frac{\sigma_e^2 + \sigma_{SAB}^2 + b\sigma_{SA}^2 + \sigma_{AB}^2}{\sigma_e^2 + \sigma_{SAB}^2 + b\sigma_{SA}^2}$$

which is greater than 1. Similarly, if we test A against AB the $S \times A$ variance will spuriously inflate the F ratio. In short, even if there is no effect of A in the population, the F_1 and F_2 tests could both be significant if A interacts with both participants and items. This would lead to the erroneous conclusion that there were A main effects. A more defensible approach is to calculate an F statistic such that the ratio of mean squares is 1 when the null hypothesis is true. We next consider such a statistic.

14.6.3 *Quasi-F (F′) Tests*

We have established that a test of A in the $S \times A \times B$ design summarized in Table 14.10 requires that we identify an error term whose expectation is $\sigma^2_e + b\sigma^2_{SA} + n\sigma^2_{SB} + \sigma^2_{SAB}$. Given the expected mean squares of Table 14.10, no single mean square has the appropriate expected value, but we can form a combination of mean squares with the appropriate expectation. The combination $(MS_{AB} + MS_{SA} - MS_{SAB})$ meets that criterion; thus, we can create a *quasi*-F *ratio* whose numerator and denominator are equal if the null hypothesis is true. That ratio is:

$$F_1' = \frac{MS_A}{MS_{AB} + MS_{SA} - MS_{SAB}} \tag{14.4}$$

A second possible *F* test may be constructed, using the same four *MS* terms:

$$F_2' = \frac{MS_A + MS_{SAB}}{MS_{AB} + MS_{SA}} \tag{14.5}$$

The rationale, again, is that the expectations of numerator and denominator differ only with respect to the null hypothesis component. In Equation 14.5 the ratio of *EMS* is

$$\frac{(2)\left(\sigma_e^2 + \sigma_{SAB}^2\right) + b\sigma_{SA}^2 + n\sigma_{AB}^2 + bn\theta_A^2}{(2)\left(\sigma_e^2 + \sigma_{SAB}^2\right) + b\sigma_{SA}^2 + n\sigma_{AB}^2}$$

which equals 1 if the null hypothesis ($\theta^2_A = 0$) is true. That the ratio of *EMS* terms equals 1 is a necessary but not a sufficient condition for the ratio of mean squares to be distributed as *F*. An important condition is that both the numerator and denominator must be distributed as χ^2 variables divided by their degrees of freedom. Satterthwaite (1946) has shown that under the usual assumptions of analysis of variance, a linear combination of mean squares has approximately this sampling distribution.

The denominator of F'_1 and the numerator and denominator of F'_2 are special cases of *combinations of mean squares* (*CMS*) that have the general form

$$CMS = w_1V_1 + w_2V_2 + \ldots + w_kV_k + \ldots + w_KV_K \tag{14.6}$$

where w_k is any real number, and V_k is either a mean square or a variance. Such terms have degrees of freedom that are more complicated than the usual ones. The general form of the degrees of freedom for any *CMS* is

$$df_{CMS} = (CMS)^2 / \sum_k \left(w_k^2 V_k^2 / df_k\right) \tag{14.7}$$

where df_k are the degrees of freedom associated with the *k*th mean square. For F'_1 in Equation 14.4, the numerator degrees of freedom are $a - 1$. The denominator of F'_1 is

$$CMS = (1)(MS_{AB}) + (1)(MS_{SA}) + (-1)(MS_{SAB})$$

and inserting this and the degrees of freedom for each of the three mean squares into Equation 14.7, the denominator *df* for F'_1 are

$$df_{denominator} = \frac{(MS_{SA} + MS_{AB} - MS_{SAB})^2}{MS_{SA}^2 / df_{SA} + MS_{AB}^2 / df_{AB} + MS_{SAB}^2 / df_{SAB}} \tag{14.8}$$

For F'_2 the appropriate *df* for the numerator are

$$df_{mum} = \frac{(MS_A + MS_{SAB})^2}{MS_A^2 / df_A + MS_{SAB}^2 / df_{SAB}} \tag{14.9}$$

and the denominator *df* are

$$df_{den} = \frac{(MS_{AB} + MS_{SA})^2}{MS_{AB}^2 / df_{AB} + MS_{SA}^2 / df_{SA}} \tag{14.10}$$

Simulation studies by Hudson and Krutchkoff (1968) and Davenport and Webster (1973) indicate that F'_2 has a slight power advantage over F'_1. However, F'_1 involves a simpler numerator and requires calculation of only denominator degrees of freedom. Software capable of calculating quasi-*F* statistics usually uses this form; see Section 14.6.5 for details.

14.6.4 A Numerical Example

To illustrate the computations described in the preceding subsection, we use a hypothetical data set based on the previously described experiment for testing the effect of type of font on word recognition. Our hypothetical experiment involves eight participants ($n = 8$), six words ($b = 6$), and two fonts ($a = 2$); time to report the word on a screen is measured in milliseconds. The data are in Table 14.12 and are also available on the website in the files labeled *Table 15_12 Font_S* (cases are participants) and *Table 15_12 Font_I* (cases are items). The ANOVA results are presented in Table 14.13. Referring to the formulas for *EMS* for the case in which participants and *B* are assumed to have random effects (Table 14.11), we find that the *B* and *AB* sources are tested against their respective interactions with participants; only the *A* source requires a quasi-*F* test. The note at the bottom of Table 14.13 defines the appropriate error term and corresponding degrees of freedom for this test. Although the *F*

Table 14.12 Response times (in msec) to items as a function of font

	Participant	*I1*	*I2*	*I3*	*I4*	*I5*	*I6*	$\bar{Y}_{i1.}$
	1	552	509	487	541	574	554	536
	2	628	735	816	842	614	652	715
	3	590	585	662	372	726	489	571
Font 1	4	496	594	549	547	587	590	561
	5	682	685	631	777	634	789	700
	9	746	775	774	693	817	777	764
	7	743	702	611	632	748	630	678
	8	569	581	588	514	595	623	578
$\bar{Y}_{.1k}$		626	646	640	615	662	638	$\bar{Y}_{.1.} = 638$
	1	713	588	335	624	513	767	590
	2	567	672	887	820	603	587	689
	3	734	598	774	217	1034	615	662
Font 2	4	508	672	644	521	593	598	589
	5	643	652	682	662	841	768	708
	9	890	900	871	717	888	844	852
	7	951	788	576	683	684	606	715
	8	406	672	579	551	725	623	593
$\bar{Y}_{.2k}$		677	693	669	599	735	676	$\bar{Y}_{.2.} = 676$

Note: Font and items are both within-participants factors.

Table 14.13 ANOVA of the data in Table 14.12

SV	*df*	*Sum of squares*	*Mean square*	*F*	*P*
Participants (*S*)	7	602,282	86,040		
Font (*A*)	1	34,846	34,846	11.451	.132
SA	7	32,356	4,622		
Items (*B*)	5	74,839	14,968	.721	.612
SB	35	726,103	20,746		
AB	5	19,601	3,920	.713	.618
SAB	35	192,451	5,499		

Note: To test *A*: $F'_1 = MS_A / (MS_{AB} + MS_{SA} - MS_{SAB})$ = 34,846/3,043 = 11.451. The error degrees of freedom are $(MS_{AB} + MS_{SA} - MS_{SAB})^2 / [(MS^2_{AB} / df_{AB}) + (MS^2_{SA} / df_{SA}) + (MS^2_{SAB} / df_{SAB})] = 1.33$. *B* and *AB* are tested against *SB* and *SAB* respectively.

value for the test of font is 11.45, it is based on only 1 and 1.33 *df* so the effect of font is not significant by this test.

14.6.5 *Using Software for Quasi-F Statistics*

The calculation of a quasi-*F* and the *df* for a combination of mean squares is cumbersome. Fortunately, some software packages will calculate both the quasi-*F* statistic and the degrees of freedom.

In SPSS, the data must be organized differently than for the more conventional repeated-measures design with all fixed-effects factors. Rather than each participants' data appearing on a single row, similar to Table 14.12, the quasi-*F* analysis in SPSS requires that each observation appear in a separate row of the data file. In this "long format," there is a column for each of the factors in the design (*Subject*, *Font* (*A*), and *Item* (*B*)), and a column for the dependent measure (e.g., *RT*). The file labeled FontUni contains the data of Table 14.12 in this format. To calculate F'_1, select the *General Linear Model* from the *Analyze* pull-down menu, then choose *Univariate*. Move *RT* to the *Dependent Variable* box, move *A* to the box for *Fixed Factors*, and move both *B* and *S* to the box for *Random Factors*, then click *OK*. The results will include F'_1 and a definition of the error term for each factor in terms of the *EMS*. Take care to confirm the appropriate error term is used for each effect of interest: For the example data in Table 14.12, SPSS computes F'_1 properly for *A* but computes an erroneous *F'* for *B*.

In R, the data must also be in the long format, with one observation per row. We begin by running the ANOVA. One option for doing so is the *aov_ev* function in the {afex} package, in which specification of the within- and between-participants factors is particularly straightforward. For example, the data in FontUni (and Table 14.12) can be analyzed with this command: *aov_ez*(data = dat, "Subject", "RT", *within* = c("Font", "Item")). Any between-participants factor could be included by adding the option *between* = *c*("FactorName"). The *aov_ez* function provides Greenhouse–Geisser corrected *df* and *p*-values by default; using the option *anova_table* = *list*(correction = "none") yields the uncorrected values. The function also reports standard *F* values by default, on the default assumption that all factors are fixed-effects variables. *F'* for the effect of font (factor *A*) must be calculated by extracting the *MS* terms from the output, and then applying Equation 14.4.

14.6.6 Some Other Issues in Using Quasi-F Tests

The construction of quasi-F ratios provides a general solution to the problem of testing effects in designs with multiple random-effects factors. However, researchers often encounter an obstacle to this approach. In many studies, observations are missing and therefore not all $S \times A \times B$ combinations are available. If so, it will be impossible to calculate MS_{SAB} and, therefore, the quasi-F statistic cannot be calculated. A conservative remedy in this situation is to calculate a minimum quasi-F (*minF'*; Clark, 1973; Forster & Dickinson, 1976):

$$minF' = MS_A / (MS_{AB} + MS_{SA}) \tag{14.11}$$

When observations are missing, *minF'* and the denominator *df* are calculated as a function of F_1 and F_2; the formulas are presented in Box 14.2. The test based on *minF'* is somewhat conservative because it lacks an estimate of MS_{SAB}.

Box 14.2 Calculating the Minimum Quasi-F Statistic (*minF'*)

1. Find the average of the scores in each $S \times A$ combination; presumably there will be a few scores in each combination even if some values are missing. Compute

$$F_1 = MS_A / MS_{SA}$$

2. Find the average of the scores in each $A \times B$ combination. Compute

$$F_2 = MS_A / MS_{AB}$$

3. Compute

$$minF' = \frac{F_1 F_2}{F_1 + F_2}$$

and

$$df_{error} = \frac{(F_1 + F_2)^2}{F_1^2 / df_{AB} + F_2^2 / df_{SA}}$$

The numerator *df* are $a - 1$.

What about the effects of violations of the sphericity assumption upon the distribution of F'? Somewhat surprisingly, Maxwell and Bray (1986) found that nonsphericity does not inflate the Type 1 error rate for F'. Their article presents an interesting discussion of the reasons for this.

Finally, a comment on the power of quasi-F tests. We know that the number of participants in an experiment is an important determinant of the power of statistical tests. This is because the more participants there are, the less error there is in sample means as estimates of the corresponding population means. When a researcher is planning an experiment that will necessitate quasi-F tests, the number of participants remains an important design consideration. However, the number of items sampled in the experiment also becomes a consideration. Interactions of A with both participants and items will contribute to the error variance against which treatment effects are evaluated. Therefore, there should be enough observations to have sufficient degrees of freedom associated with both participants and items to ensure powerful tests of A. An added benefit of including many participants and items in a design is that the distribution of F' more closely approximates F as a, b, and n increase.

14.7 Fixed or Random Effects?

It should be clear by now that designating effects as fixed or random has important implications for significance testing, as well as for the degree to which we can generalize our results. We will soon discover that the designation also has direct implications for parameter estimation (e.g., effect size). Therefore, we need to consider further the decision about classifying effects.

The decision to classify a variable as having fixed or random effects is not always a simple one. At one extreme, we have variables that clearly should be viewed as having fixed effects. The levels of the variable have been intentionally selected for inclusion in the experiment and, because of the way in which they have been selected, there is no basis for viewing them as a random sample of levels from a population of levels. This class includes most manipulated variables such as the type of distortion or the orientation of the photo in the Murray et al. (2000) experiment, drug dosing levels, and observed characteristics of individuals such as gender identity and clinical category. It also would include a variable such as seasons because the four seasons exhaust the population of possible seasons.

At the other extreme, we have random sampling from some well-defined population. This is rarely realized in practice, and it is therefore difficult to determine the population to which our results can be generalized. Can we reasonably view the participants in our experiment to be a random sample of adults, college students, college students interested in psychology, or college students who attend the particular university at which the study was run? The answer depends on the sampling process but also on the research topic. In studies of sensory processes like visual acuity, we might generalize to the population of adults having normal vision because we expect those processes to be "hard wired" within the human visual system. In studies of human learning, we might define the population more narrowly, reserving judgment about whether our conclusions will hold for populations having a markedly different average level of ability from that characterizing the institution in which our study was run. When in doubt, generalizations should be restricted to the more narrowly defined population.

Even though the population is rarely as well defined as we would like, it should be clear from the preceding comments that we do view participants as a random-effects variable. Our justification is that other individuals are provided an equal opportunity to participate and might well serve if replications of the experiment were run.

Classifying experimental stimuli such as words and pictures presents greater difficulty. For many experiments, we can argue on much the same grounds that we presented in discussing participants that the stimuli are a random sample from a (possibly ill-defined) population of potential items. That is, the specific stimuli selected were not chosen in such a way that they were the only items with an opportunity to be included in the experiment; rather, there were many other items that had an equal opportunity of being included under the sampling procedure used. In many other experiments, however, the choice of stimuli is so constrained that the items selected for the experiment virtually exhaust the stimuli that meet the sampling constraints. In studies involving responses to words, for example, restrictions are often placed on the grammatical class, length in both syllables and letters, familiarity, and number of associates of each word. The experimenter may find it difficult to identify items that meet those restrictions. Under such conditions, it is not clear that stimuli should be treated as having random effects. Two rough guidelines may be helpful. First, under the existing constraints, could independent investigators produce other samples of items? Assuming there is a reasonably large population of items, the second question is: Was there an equal likelihood that all members of the population could be included in the study? If the answer to this question is also positive, then it is reasonable to treat the stimuli as having random effects with all that this implies for our data analyses and the scope of our conclusions.

14.8 Understanding the Pattern of Means in Repeated-Measures Designs

Table 14.4 reveals that the orientation of pictures of faces (A), the type of distortion (B), and the AB interaction all have significant effects on bizarreness ratings. Ordinarily, to better understand the source of effects, we follow these overall F tests with tests and confidence intervals that more precisely target differences among the means. We might begin by testing comparisons among the marginal means. Comparisons of marginal means are of most interest when there clearly is no interaction, because then there is no evidence that the effects of one factor vary as a function of the level of the other factor. When an interaction is present, or even when there is some reason to anticipate an interaction prior to data collection, other tests should be of greater interest. Tests of simple effects, such as the difference between the B_2 and B_3 means at the A_2 orientation, might clarify the pattern of means. Other possibilities include tests of interaction contrasts. For example, we may decide to test whether the difference between the B_2 and B_3 means is significantly greater at A_2 than at A_1.

In this section, we illustrate approaches to analyzing marginal means to understand main effects, and cell means to understand interactions. We first consider examples from repeated-measures designs, and then present examples from mixed designs. We also discuss contrasts in designs with multiple random-effects factors. Finally, we consider the control of familywise error rate (*FWE*). Before dealing with the details of these analyses, we emphasize that there are no definite answers to the questions of which analyses to carry out or how to control the error rate. As we discussed in Chapter 10, we may have some *a priori* reason to focus interest on differences among specific cell means. Perhaps more commonly, we want to analyze a pattern of means that was not anticipated. In these circumstances, we will suggest some systematic approaches that might be used to understand the data, or researchers may design specific *post hoc* contrasts to summarize the pattern. Such considerations will influence the choice of contrasts and the alpha-level set for each test.

14.8.1 Comparing Marginal Means

In a multi-factor design, the main effect of a factor may be of interest if its interpretation is not qualified by the presence of an interaction. In that event, the researcher will want to analyze the pattern of marginal means. Although complex contrasts can be tested if theoretical considerations or the observed pattern of means warrant, researchers most often compute pairwise comparisons among the marginal means. As we discussed in Section 13.9.1, the safest approach to testing any contrast in a repeated-measures design is to use an error term based only on the data involved in the contrast (cf. Boik, 1981). Therefore, if we decided to carry out paired comparisons on the marginal means of *B* for the data in Table 14.3, our error term would *not* be the *SB* term from the overall ANOVA presented in Table 14.4; rather, the error term would be the *SB* interaction based only on the two levels of *B* involved in the comparison. A simple way in which to conduct each comparison is to select only the two levels of *B* you wish to include in the analysis, and then perform the usual ANOVA on that subset of data. The *B* source of variance in the resulting table will reflect the difference between the marginal means of the scores at the two levels of *B* selected for comparison, and the *SB* error term will be based on only those scores.

What should be our criterion for evaluating the contrasts we conduct? If *k* differences were selected *a priori* and were the only ones tested, we would set the *FWE* criterion at .05/*k*. If all pairwise comparisons are of interest, *k* would be the number of possible pairwise comparisons. In that case, the criterion *FWE* is .05/3 for our example.

One other question that arises is what our criterion should be if we test comparisons among both the *B* marginal means and among the *A* marginal means. As we discussed in Chapter 10, we consider comparisons among the levels of a variable to constitute a family; therefore, if we were to do the three pairwise comparisons of the *A* marginal means and the three comparisons of the *B* marginal means, each test would be carried out at the .05/3 level.

14.8.2 Comparing Cell Means

We are often interested in the pattern of cell means, particularly if we have predicted an interaction or observed one after the initial ANOVA. One common approach to analyzing cell means is to conduct all pairwise comparisons of levels of *A* within each level of *B* or vice versa. For the example of Table 14.4, we might compare the three levels of distortion within each level of orientation. There are nine possible such tests based on Table 14.3; that is, at each level of *A* we compare B_1 vs B_2, B_1 vs B_3, and B_2 vs B_3. If we wish to keep the *FWE* at or under .05 for the full set of tests, we require an alpha-level of .05/9, approximately .005, for significance of any one comparison (i.e., the Dunn–Bonferroni method). We selected *Compare Means* then *Paired-Samples t Test* from SPSS's *Analyze* menu; then we selected the nine pairs we wanted to test. Only two of these comparisons were not significant at the .005 level: $A_1B_2 - A_1B_3$ ($p = .546$) and $A_2B_2 - A_2B_3$ ($p = .025$). Note that these are the two smallest differences in Figure 14.1. This accounts for the interaction; it apparently is largely due to the lack of a difference in the distortion effects of B_2 and B_3 at A_1 and possibly at A_2. These results should focus the researcher on the question of why these effects are so different from the others that were tested.

Before turning to an alternative analysis of the cell means, it is useful to consider the nature of the *t* tests and provide an example of a confidence interval. Following are results for two of the nine contrasts in the SPSS output:

	Mean	Std. deviation	Std. error mean	95% Confidence interval of the difference			df	Sig. (2-tailed)
				Lower	Upper	t		
$A_2B_3 - A_2B_2$	0.833	0.642	0.262	0.159	1.508	3.177	5	.025
$A_3B_3 - A_3B_2$	2.053	0.950	0.388	1.056	3.050	5.294	5	.003

There are several points to note: First, the standard deviations of the differences are different for the two tests. This is because each difference is tested against a measure of the variability of that difference, not against an error term from the ANOVA (Table 14.4). Second, the test is conservative not only because of the criterion alpha (i.e., .05/9) but also because the degrees of freedom are $(n - 1)$ or 5 in this example. Third, the confidence interval does not take into consideration the fact that we have nine tests. The appropriate CI should be based on the t required for significance at the .005 level, not the .05 level. From Appendix Table C.3, or using *qt*(.005/2, *df* = 5, lower.tail = FALSE) in R, the critical t value at the .005 level (two-tailed) with 5 *df* is 4.773. Therefore, the CI bounds for the comparison between the A_2B_2 and A_2B_3 means $(\bar{Y}_{23} - \bar{Y}_{22})$ are $0.833 \pm (0.262)(4.773) = -0.418$ and 2.084. Using the appropriate critical value of t, 0 is within the CI and thus cannot be rejected as a possible value for the population difference.

14.8.3 *Tests of Interaction Contrasts*

A more direct way of understanding the reason for the significant *AB* interaction in Table 14.4 might be to test interaction contrasts. In Figure 14.1, we note that $\bar{Y}_{13} - \bar{Y}_{12}$ is smaller than either $\bar{Y}_{23} - \bar{Y}_{22}$ or $\bar{Y}_{33} - \bar{Y}_{32}$. This suggests that we focus on these interaction contrasts: $\psi_1 = [(4.81 - 5.02) - (5.41 - 4.58)]$ and $\psi_2 = [(4.81 - 5.02) - (5.47 - 3.41)]$. One way to test these two interaction contrasts is to use the *Transform* option in SPSS, or *mutate* in the {dplyr} package in R, to create two new "scores" for each participant: $(A_2B_3 - A_2B_2) - (A_1B_3 - A_1B_2)$ and $(A_3B_3 - A_3B_2) - (A_1B_3 - A_1B_2)$. We can then use one-sample t tests to assess these contrasts. For ψ_1, $t = 1.87$ and for ψ_2, $t = 4.917$.[2] If we had selected the two interaction contrasts before viewing the data, we would have used the Dunn–Bonferroni procedure to control the *FWE* at .05 by evaluating each contrast at alpha at .05/2 = .025. However, having determined that these were of interest after viewing Figure 14.1, and noting that there are nine possible 2 × 2 contrasts, we set alpha at .05/9 = .0056. With 5 *df* the critical value of t is 4.773 and therefore only the test of ψ_2 has a significant result. The *AB* interaction result in Table 14.4 seems to largely reflect this interaction embedded within the larger design.

14.8.4 *Mixed Designs*

Tests of contrasts in designs involving both between- and within-participants factors follow the approaches previously established. When analyzing effects of between-participants factors, the methods available are those described for completely randomized designs in Chapters 10 and 11. Software makes it easy to test contrasts to understand main effects of between-participants factors. For example, users of SPSS may request "options" in the

Repeated Measures module and have a choice of several multiple comparison procedures. The procedures each report all pairwise comparisons but use different methods to control *FWE*. The researcher should be warned, however, that some of the options are not wise choices because they lack power as methods for controlling error rate across the set of all pairwise comparisons; for example, Scheffé is not a good choice of procedures in this context.

If a researcher is interested in analyzing the main effects of a within-participants factor or an interaction involving a within-participants factor, the available options are the same as for any repeated-measures design. Software will make pairwise comparisons of means at different levels of within-participants factors. In addition, as described in the preceding subsections, interaction contrasts may be performed to analyze how a within-participants comparison (e.g., pairwise or more complex contrast) varies over levels of a between-participants factor. As we discussed in the context of repeated-measures designs, contrasts should be planned and the Dunn–Bonferroni procedure used to control *FWE*. If contrasts are unplanned, the control of the *FWE* with Dunn–Bonferroni is justified if the set of contrasts consists of all possible contrasts of the same form (i.e., all pairwise or all embedded 2 × 2 interactions).

14.8.5 Designs With Multiple Random-Effects Factors

In the examples of contrasts presented thus far, we have dealt only with designs in which all factors have fixed effects except participants. When a second random-effects factor is included in a design, we found that tests of effects of interest often necessitated use of quasi-*F* ratios. This is because when both participants and items are considered random effects, terms must be tested against both participant and item variability simultaneously. Similarly, when we conduct contrasts within designs in which both participants and items have random effects, the error terms to test the contrasts must also account for participant and item variability, so quasi-*F* tests are still in order.

Table 14.12 presented an example of a repeated-measures design with eight participants, six words, and two fonts where the task was to name each word as quickly as possible. Our interest in this experiment was whether the type of font influenced naming times. The test of font required a quasi-*F* ratio where the denominator was based on a combination of three mean squares: $(MS_{S \times Font} + MS_{Words \times Font} - MS_{S \times Words \times Font})$. Imagine an extension of this simple design where four types of fonts are tested, instead of just two. Also, suppose that we included three levels of illumination of the computer screen to manipulate the visual conditions under which the words must be perceived. Thus, the extended design includes two factors that have random effects (i.e., participants and the within-participants factor of words) as well as two within-participants factors that have fixed effects (i.e., font and illumination). Our substantive interest is in the effects of type of font and level of illumination on performance in the word-naming task. Tests of effects of font, illumination, and their interaction would still require quasi-*F* ratios, as in the simpler design. If significant *F* tests resulted, we would surely want to make detailed comparisons between specific fonts, levels of illumination, or combinations of fonts and illumination levels. How are we to carry out such comparisons?

As before, our general recommendation is that for any given comparison a researcher wishes to make, only the data from the conditions to be compared should be included in the analysis. For example, if the researcher wished to compare the results for Gothic font with Times Roman font, data on those two fonts would be included, but the data for the other two fonts would be excluded from the analysis. An ANOVA would be conducted specifying two levels of font; all levels of the other variables (i.e., participants, words, illumination) would be fully represented in the analysis. An *F* test of font would be conducted as in the overall ANOVA; the mean square for font would be tested against the combination ($MS_{S\times Font}$ + $MS_{Words\times Font}$ − $MS_{S\times Words\times Font}$). A significant result would allow the researcher to conclude that Gothic and Times Roman differed in their legibility.

It is important to understand that the combination of mean squares that is the error term is the same as in the ANOVA on the complete data set, but its value will differ because it is based on a subset of the data. If the sphericity assumption is met in the ANOVA on the complete data set, that would justify use of the error term from the overall analysis. In our experience, however, the sphericity assumption is frequently violated in the overall analysis.

14.9 Effect Size

14.9.1 The S × A × B Design

Common measures of effect size include partial eta-squared and partial omega-squared. However, as was noted in Chapter 13, both measures suffer from the limitation that their denominators change across designs. Because it is desirable to have measures of effect size that are comparable across designs, we have argued for the use of general eta-squared and general omega-squared as measures of effect size (Olejnik & Algina, 2003).

The denominator of general eta-squared is the sum of several sums of squares terms; namely, SS_{Effect} plus the sum of every *SS* term that includes participant variability. For example, in the $S \times A \times B$ design, general eta-squared for *A* is computed as

$$\eta_g^2 = \frac{SS_A}{SS_A + SS_S + SS_{SA} + SS_{SB} + SS_{SAB}} = \frac{SS_A}{SS_{Total} - SS_B - SS_{AB}} \tag{14.12}$$

However, as noted in Chapter 13, general eta-squared tends to overestimate the proportion of population variance due to a variable. For that reason, we prefer general omega-squared as a measure of effect size.

The definition of general omega-squared (ω^2_g) for the $S \times A \times B$ design is like that in Chapter 13 for the one-factor within-participants ($S \times A$) design. For example, if *A* and *B* are extrinsic factors, we define ω^2_g as

$$\omega_g^2(A) = \frac{\sigma_A^2}{\sigma_e^2 + \sigma_A^2 + \sigma_S^2 + \sigma_{SA}^2 + \sigma_{SB}^2 + \sigma_{SAB}^2} \tag{14.13}$$

Unfortunately, estimating $\omega^2_g(A)$ presents a problem because the interaction terms cannot be estimated from the expected mean squares in Table 14.2. An approximate solution is obtained if we assume that $\sigma^2_{SAB} = 0$. With this modification, the *EMS* of Table 14.2 provide estimates of the variances in Equation 14.13. Those estimates are presented in Table 14.14. Substituting into Equation 14.13, multiplying numerator and denominator by *abn* and collecting all the MS_{SAB} terms, we have

$$\hat{\omega}_g^2(A) = \frac{(a-1)(MS_A - MS_{SA})}{(a-1)(MS_A - MS_{SA}) + n\left[(MS_S + aMS_{SA} + bMS_{SB}) + (ab-1-a-b)MS_{SAB}\right]} \tag{14.14}$$

Table 14.14 Estimates of variance components based on the *EMS* of Table 14.2

Source	Estimator
A	$(a-1)(MS_A - MS_{SA}) / abn$
S	$(MS_S - MS_{SAB}) / ab$
SA	$(MS_{SA} - MS_{SAB}) / b$
SB	$(MS_{SB} - MS_{SAB}) / a$
SAB	MS_{SAB}

Note: Estimates rest on the assumption that $\sigma^2_{SAB} = 0$.

Table 14.15 Estimates of general omega-squared (ω^2_g) for a design with one between-participants (*A*) and one within-participants factor (calculations are based on the data of Table 14.5)

Source	Formula for the estimate	Estimate
A	$\dfrac{(a-1)(MS_A - MS_{SA})}{(a-1)(MS_A - MS_{SA}) + an(MS_{S/A} - MS_{B\times S/A}) + abnMS_{B\times S/A}}$	.356
B	$\dfrac{(b-1)(MS_B - MS_{B\times S/A})}{(b-1)(MS_B - MS_{B\times S/A}) + an(MS_{S/A} - MS_{B\times S/A}) + abnMS_{B\times S/A}}$	.206
AB	$\dfrac{(a-1)(b-1)(MS_{AB} - MS_{B\times S/A})}{(a-1)(b-1)(MS_{AB} - MS_{B\times S/A}) + an(MS_{S/A} - MS_{B\times S/A}) + abnMS_{B\times S/A}}$	.059

Note: Estimates are based on the assumption that $\sigma^2_{BS/A} = 0$.

To the extent that $\sigma^2_{SAB} > 0$, the expected value of the denominator of Equation 14.14 will underestimate the denominator of Equation 14.13, and therefore will overestimate $\omega^2_g(A)$; the denominator that is estimated is the same as in Equation 14.13 except that $(1-1/a - 1/b - 1/ab)\sigma^2_{SAB}$ replaces σ^2_{SAB}. Note that the error of estimation is smaller when *a* and *b* are larger.

What if *B* represents a random-effects factor? Such factors are usually extrinsic; for example, stimuli in a perception experiment or items in a psycholinguistic study. Therefore, Equation 14.13 still applies, but the estimates of population variances will differ from those on which Equation 14.14 is based. Still assuming that MS_{SAB} estimates only error variance (i.e., $\sigma^2_{sab} = 0$), estimates of the components of Equation 14.13 can be derived from Table 14.10. We leave the formula as an exercise.

14.9.2 Omega-Squared in the Mixed Design

Estimation of ω^2_g in the mixed design follows the approach taken in the $S \times A \times B$ design. For convenient reference, Table 14.15 presents the formulas for *A*, *B*, and *AB* for a design in which participants are nested within levels of *A* and are tested at all levels of *B* where *A* and *B* are both assumed to have fixed effects.

The right-most column of the table contains numerical estimates of ω^2_g for the *A*, *B*, and *AB* sources. The definition of general omega-squared is exemplified by the formula for the *A* source:

$$\omega_g^2(A) = \frac{\sigma_A^2}{\sigma_A^2 + \sigma_{S/A}^2 + \sigma_e^2 + \sigma_{B\times S/A}^2} \tag{14.15}$$

We estimated ω^2_A from the expected mean squares in Table 14.7. As can be seen in Table 14.15, our estimates of the remaining denominator components rest on the assumption that $\sigma^2_{BS/A} = 0$. If $\sigma^2_{BS/A} > 0$, the expectation of the denominator includes $[(b - 1) / b]$ $\omega^2_{BS/A}$ instead of $\sigma^2_{BS/A}$. If B is a random-effects factor, Equation 14.15 still applies, but the expected mean squares are slightly different from those in Table 14.7 and the estimates of the population variances consequently differ from those in Table 14.15. Estimates of population variances were presented in Table 14.11 for a design in which there are two within-participants factors, one of which is assumed to have random effects.

14.10 *A Priori* Power Calculations

14.10.1 The S × A × B Design

Suppose that we are planning a two-factor repeated-measures design and we want to determine how many participants to include in the experiment. The design will yield tests of the main effects of *A* and *B* and of their interaction. If one of these tests is of particular interest, we might focus on it in calculating the number of participants we need to achieve good power to test that effect. Or if one effect is expected to be smaller than the others, power computations based on that effect will ensure that the power of the other two tests will be at least as good.

Power calculations are readily performed with G*Power 3.1. If we wish to compute the required sample size to test the main effect of *B* we select *ANOVA: Repeated-measures, within factors* and specify that we want to do an *a priori* power analysis. Several items of information must be provided. Suppose that we assume a small effect of .10, set alpha at .05, we want .90 power to detect an effect, and our design includes three levels of *A* and four levels of *B*. There is only one "group" of participants because there are no between-participants factors; there are four "repetitions" or levels of *B*. We will assume that the average correlation of scores between conditions is .5, and that the sphericity assumption is violated and $\varepsilon = .8$. For this combination of values, the required sample size is 210 participants. This is quite large but if we are willing to sacrifice some power, perhaps by setting power to .8, the required *N* is reduced to 161. Alternatively, we can obtain power of .9 with only 35 participants if we have reason to assume a medium effect size of .25.

14.10.2 The Mixed Design

Power calculations can also be performed for mixed designs using G*Power 3.1. For example, suppose we are planning an experiment in which there will be three groups (A) of n participants, each of whom will be tested in three different conditions (B). We decide that large effects of *A* are important to detect and therefore set $f = .4$. We want power = .8 to detect the effects of *A* if the null hypothesis is false. From previous research, we have estimated the within-participants correlation coefficient to be.6. How many participants will we need, assuming $\alpha = .05$? We select the *a priori* type of power analysis and *repeated-measures, between factors* option. After we enter the parameters, G*Power returns a value of 48 for the total sample size, or 16 participants in each of the three groups. The actual power is .807.

We might ask how much power a sample size of 48 would give us to test the B factor, again assuming a large effect of.4. This time we select the *post hoc* power (because we have a planned N of 48), select the *repeated-measures, within factors* option, and enter our parameters. Assuming sphericity (i.e., set $\varepsilon = 1$), power is 1. If we had selected the *a priori* option with the same parameters and asked what n was needed for power to equal .8, we would have needed only 12 participants, 4 in each group.

The difference between the ns needed to achieve power in testing the between- and within-participants factors reflects the fact that variability is greater between participants than within participants. Furthermore, this depends on the size of the correlation of scores across repeated-measures. As ρ increases, the n required to achieve a particular level of power to test B decreases. In contrast, as ρ increases, the n required to achieve a particular level of power to test the between-participants factor, A increases. This pattern occurs because each participant's score is the average of b scores, and the variance of the population average is the sum of the variances plus a covariance term (see Appendix 5.1). Both effects can be observed by substituting values of $\hat{\rho}$ into the G*Power screens for the between and within factors.

14.11 Summary

Chapter 14 extended the repeated-measures designs of the previous chapter by inserting additional within-participants factors. It also introduced mixed designs involving one or more between-participants factors in addition to the within-participants factors. Within this context, we covered the following topics:

- *The structural model, EMS, and ANOVA for multi-factor repeated-measures designs* in which all factors, except participants, had fixed effects.
- *The structural model, EMS, and ANOVA for mixed designs* that include both within-participants and between-participants factors, again assuming all factors other than participants had fixed effects.
- *The consequences of additional random effects factors for the expected mean squares.* We presented general rules for determining the *EMS* and considerations for deciding whether a factor has random or fixed effects.
- *The need for quasi-*F *ratios in the context of random effects factors.*
- *Test of contrasts and control of* FWE.
- *Measures of effect size in multi-factor repeated-measures and mixed designs*, noting that unbiased estimates of effect size are often not possible.
- *Power computations in repeated-measures and mixed designs.*

The designs considered so far in this book undergo further modifications in many experiments. In the next chapter, we will consider some frequently encountered variations of these designs.

Exercises

14.1 [*EMS* and quasi-*F* tests] A company is selecting a new software package from four competitors. Ten randomly sampled workers (W) are observed on each of five randomly sampled occasions (O) with each of the four available programs (P). A score is obtained for each worker with each program on each occasion and an analysis of variance is carried out. The results are as follows:

Source	df	MS
W	9	2580
P	3	2610
O	4	690
WP	27	330
WO	36	370
PO	12	640
WPO	108	320

a) Write out the *EMS* for the above table, first specifying which factors have random effects and which have fixed effects.
b) Calculate a quasi-*F* test of the *P* source. Let $\alpha = .05$.
c) Calculate an alternative to the quasi-*F* test of *P*. What is assumed in doing this test?

14.2 [*EMS* for additive and nonadditive models] In research on personality, there has been much discussion of the relative importance of traits and situations. The basic research design involves n participants and t tasks representing a random sample of situations. Measures are obtained for each participant (S_i) on each task (T_j) on each of b randomly sampled occasions (O_k).

a) Assuming the completely additive model, $Y_{ijk} = \mu + \eta_i + \alpha_j + \beta_k + \varepsilon_{ijk}$, present expressions for the *SV*, *df*, and *EMS*.
b) Present expressions for estimates of the variance components for participants, tasks, and occasions.
c) Assume we have evidence from previous studies that participants and tasks interact. State the revised model, the revised ANOVA table (*SV*, *df*, and *EMS*), and revised estimates of the variance components.
d) Assuming the model of part (c), state the formula for estimating $\omega^2_g(S)$.

14.3 [Comparing ANOVAs under different assumptions] In the following data set, *B* represents statements that are rated before (A_1) and after (A_2) reading a persuasive communication.

	A_1			A_2		
	B_1	B_2	B_3	B_1	B_2	B_3
S_1	2	3	4	8	9	8
S_2	3	3	4	6	7	9
S_3	7	4	6	4	9	9
S_4	2	2	4	7	9	8
S_5	1	2	2	5	4	5

a) Assume *B* is fixed and do an $S \times A \times B$ ANOVA.
b) Find the mean at A_1 and at A_2 for each participant. Now do an ANOVA for this $S \times A$ design. How does this compare with the results in part (a)?
c) Now assume *B* is a random-effects variable. Present the *EMS*.

d) Test the A source of variance under the model of part (c).
e) What is the problem with the ANOVA of part (b) when B has random effects?

14.4 [Comparing ANOVAs for factor B fixed vs random] The design is an $S \times A \times B$ with $n = 10$, $a = 3$, and $b = 5$. The following table summarizes part of the results.

Source	*df*	*MS*
S	9	250
A	2	64
B	4	71
SA	18	24
SB	36	32
AB	8	44
SAB	72	10

Find the values of the F ratios and p-values for A, B, and AB in the following conditions:

a) Both A and B have fixed effects.
b) A has fixed effects but B represents randomly sampled items.

14.5 [Calculating general ω^2] Find the values of general ω^2 for A in the following conditions (assume the table in Exercise 14.4 and that SAB estimates only error variance):

a) Both A and B have fixed effects and are both extrinsic factors.
b) Both A and B are extrinsic factors but now B represents randomly sampled items.
c) A is an extrinsic factor, B is an intrinsic factor (e.g., personality type), and B has random effects.

14.6 [SS and orthogonality for contrasts] In Section 14.8.3, we described two tests of interactions embedded within the design of Table 14.3 where $n = 6$.

a) Are those two contrasts orthogonal? Explain.
b) The ψ_1 contrast may be represented by the following weights on the nine cell means:

A_1B_1	A_1B_2	A_1B_3	A_2B_1	A_2B_2	A_2B_3	A_3B_1	A_3B_2	A_3B_3
0	−1	1	0	1	−1	0	0	0

Ignore all cells having zero weights. Find the sum of squares on 3 df for the remaining four cells (A_1B_2, A_1B_3, A_2B_2, and A_2B_3).
c) Construct two contrasts that are orthogonal to ψ_1. From the means in Table 14.3, find the sum of squares associated with each, add the three terms, and compare your total with your answer in part (b).

14.7 [Comparing ANOVAs under different assumptions] Consider the following data set; A is a between-participants factor, and B is a within-participants factor.

		B_1	B_2	B_3
	S_{11}	23	16	12
A_1	S_{12}	27	17	14
	S_{13}	22	12	9
	S_{21}	32	27	22

		B_1	B_2	B_3
A_2	S_{22}	33	25	18
	S_{23}	34	32	23

a) Present the complete ANOVA table with all numerical results; assume that *A* and *B* both have fixed effects.
b) Find the mean score for each participant. You now have a one-factor, completely randomized design. Perform an ANOVA using the mean scores as the data. (i) How does the *F* test of *A* compare with that calculated in part (a)? (ii) How does MS_A in this analysis compare with that obtained in part (a)? Explain the relation.
c) $SS_{SB/A}$ is equivalent to the result of calculating the $S \times B$ sum of squares separately at each level of *A* and then summing the *a* terms. To demonstrate this, ignore the A_2 data and calculate the sum of squares for $S \times B$ at A_1 ($SS_{SB/A1}$). Do the same thing at A_2 and check the sum of the two terms against $SS_{SB/A}$ calculated in part (a). Confirm that $MS_{SB/A}$ is the average of the two $S \times B$ mean squares.

14.8 [Consequences of between-participants factor with random effects] A_1 and A_2 in Exercise 14.7 might have been two litters of three animals; in that case, we would assume that *A* has random effects. Assume that *B* represents trials and is a fixed-effect variable.

a) State the expected mean squares for the various sources of variance.
b) Recalculate any *F* ratios that are not the same as in Exercise 14.7.

14.9 [Consequences of within-participants factor with random effects] Now assume that A_1 and A_2 in Exercise 14.7 have fixed effects and the levels of *B* correspond to three problems sampled randomly from some very large population of problems.

a) Present the *EMS* under this model.
b) Recalculate *F* tests where necessary.

14.10 [*df* and error terms in four-way designs] An investigator interested in children's attention to violent acts on television runs an experiment with 120 participants: half who identify as male and half as female (gender, *X*) at each of three age levels (age, *A*). Each child views six scenes differing with respect to the level of violence (*V*, three levels) and the type of character; half the scenes involve animal cartoon characters and the other involve human characters (*C*, two levels). The dependent variable is a measure of attention during presentation of the scene.

a) State the *SV*, *df*, and error terms for this design.
b) In an alternative design, each of the 120 children might view only three scenes involving only one type of character; *C* would be a between-participants variable. Present the *SV*, *df*, and error terms for this design. What tests will be affected by this change in design? In what way?
c) Suppose the children are available for only short periods of time but the investigator has access to large numbers of participants. What are the advantages and disadvantages of a design in which each participant is tested only once in some combination of *V* and *C* versus the original design [part (a)]?

14.11 [*EMS* and simple effects] In a study of the development of the concepts of conservation of quantity and weight, two standardized tasks (*T*, two levels) are presented to each of 72 children. Mastering the first task requires that a child grasp the notion of conservation of quantity whereas the second task depends on conservation of weight. The score is the number of trials required for the mastery of the task. Both age (A, three levels: 5, 7, and 9 years) and school type (X, two levels: public school, home schooled) are included as major variables in the design.

a) Present the sources of variance associated with the design of this study as well as the *df* and *EMS* (using numbers where possible).
b) State the error term and its *df* for tests of the following simple effects: (i) The effect of age for the conservation-of-quantity task (T_1); (ii) the effect of age for all home-schooled participants on the T_1 task; (iii) the effects of task on the scores of all home-schooled children.

14.12 [Confidence intervals for contrasts] Twenty-four participants were divided into two groups. A group heard a tape of music by either Mozart or Albinoni (group, $g = 2$). All participants were tested on measures of spatial relations, arousal, mood, and enjoyment after listening to the music and after a period of silence (condition, $c = 2$). The means are

	Condition	
Group	*Music*	*Silence*
Mozart	14.8	11.0
Albinoni	9.8	11.4

a) Only F ratios were reported. We wish to be able to have some sense of the range of likely values of the effects reported. Find the 95% confidence interval on the difference between the group means if the reported F for *groups* is 6.20.
b) The F for *groups* × *conditions* was 16.89. Find the 95% confidence interval for the interaction.

14.13 [Interpreting ANOVA results and data patterns] In a pilot study of the effects of diet upon the ability to withstand physical stress, 12 volunteers were divided into three groups of four participants, with each group given a different diet. They then underwent a battery of physical tests on each of four successive days. A score was obtained on each day. The data are in the file *Ex14_13* at the book's website.

a) Plot the data to assess how the groups differed over time.
b) Perform the ANOVA. Discuss the results in terms of the plot in part (a).

14.14 [Confidence intervals for contrasts]

a) For the *EX14_13* data set, calculate the 95% simultaneous confidence intervals on the pairwise differences among diet means (averaging over the four days), using Tukey's *HSD* (1953) procedure. Report any significant differences.
b) After inspecting the data, the researcher notes that diet C yields better performance than A or B on day 4. Calculate a confidence interval for the difference between that mean and the average of the other two means, taking into consideration the fact that the test is post hoc ($FWE = .05$). Is the difference significant?

14.15 [Consequences of random effects factor] Two groups (A_1, A_2) of four participants each are tested six times; there are two levels of B and three levels of C crossing to yield six scores for each participant. The data are in the *EX14_15* file on the *Exercises* page of the book's website.

a) Report the results of tests of the A and AB terms, assuming all factors are fixed.
b) Suppose C represents randomly chosen items. Again test the A and AB sources. How does this change the tests and the results?

14.16 [ANOVA on real data] We wish to test whether the differences between the *SVT* (sentence verification) and *IVT* (inference verification) from the *Wiley–Voss* study are affected by the *format* and *instruct* manipulations. The two measures can be found in the *EX14_16* file on the *Exercises* page of the website.

a) Table the cell means and describe the resulting pattern. What effects are suggested?
b) Perform an ANOVA with *format* and *text* as between-participants variables and test (*SVT* vs *IVT*) as the within-participants variable. Discuss the results, relating your discussion to the table of means.

14.17 [Cohen's f, confidence intervals, and power for real data] In Exercise 14.16, there is a significant interaction between *format* and the type of test (*SVT* vs *IVT*). We may better understand this effect if we calculate some additional measures.

a) Estimate Cohen's f for the effect of *format* on the difference between test scores.
b) Calculate a 95% confidence interval for the interaction of *format* and *test*.
c) We wish to have a measure of the effect of instructions on the difference between *SVT* and *IVT* scores. Estimate Cohen's f statistic for that difference (i) in the *text* and (ii) in the *web* condition.
d) Using Cohen's f from part (c), determine how many participants would be required for power = .8 in the text condition.

Notes

1 The alternatives include MANOVA and what are referred to as *mixed models* (also called mixed-effects models, hierarchical models, and multi-level models). Although applicable to many designs including mixed-factors designs, a full treatment is beyond the scope of this book; however, a brief introduction is provided in Chapter 22.

2 The same results can be obtained by selecting only the cells involved in the 2 × 2 interaction contrast and performing an ANOVA. For example, the F test of the AB interaction for the ψ_1 contrast yields $F = 3.497$, the square of 1.870. However, using interaction contrast scores permitted us to test several interaction contrasts at once, and without the extraneous output that is part of the output of the repeated-measures ANOVA.

Chapter 15

Nested and Counterbalanced Variables in Repeated-Measures Designs

15.1 Overview

Repeated-measures designs can be very efficient because they allow the removal of variability due to individual differences; however, they are not without their problems. As we have previously seen, because each participant contributes data in at least two different treatment conditions, we must be concerned with effects of the ordering of conditions and with repetition effects that may occur if the same items are repeatedly presented in different conditions. In the present chapter, we discuss two modifications of the standard repeated-measures design that address these issues.

An example will be useful for contrasting the standard repeated-measures designs of previous chapters with the hierarchical and Latin square designs discussed here. Suppose we want to investigate the effect of font type on the time taken to name words. In the most basic repeated-measures design, the levels of the font (A) and item (B) factors are *crossed*; that is, we record naming times for the *same* b words presented in each of the a fonts. The ab trials would be sequenced randomly for each participant, perhaps with the restriction that the same word could not be presented on two successive trials. This approach would fail to account for the naming time advantage that would occur on test trials involving repetitions of the same items, over and above any effect of the a font conditions.

Chapter 15 will introduce and elaborate two alternative ways of designing such an experiment to eliminate the problem of repeating stimulus items in repeated-measures designs:

- Designs in which different stimuli are used within each treatment level. These stimuli are said to be *nested* within treatment level, and the designs are often referred to as *hierarchical* or *multi-level designs*.
- Designs in which factor levels are *counterbalanced* with respect to some potential source of error variance, such as position in time or sets of stimuli. These designs are referred to as *Latin square designs*.

15.2 Nesting Stimuli Within Factor Levels

One way to eliminate any potential effects of repeated testing of the same stimuli is to assign different stimuli to each treatment level. For example, in our example study of the effect of different fonts on word naming times, a hierarchical design would *nest* the stimulus words within level of font. Each participant would be tested with a random sequence of ab words, with b different words for each font. The researcher would probably make some attempt to

DOI: 10.4324/9781003453550-18

match the sets of words for their average length and familiarity or for other relevant dimensions. As another example, if the dependent variable is a measure of the effect of the mood depicted in pictures of faces, there might be different faces in the various mood conditions, perhaps matched for age, race, and gender.

The advantage of a hierarchical design is that it precludes the possibility that responding to stimuli in one treatment condition will influence responses to later presentations of the same items in other treatment conditions. The disadvantage is that differences in responses to the various sets of stimuli will be due in part to differences among the sets themselves, despite efforts to match the sets. For example, differences among the mean response times for the fonts will be partially attributable to differences among the sets of words presented in the different fonts. In other words, the hierarchical design eliminates effects due to stimulus repetition but at the cost of increased error variance. In the ANOVA for this hierarchical design, the $ab - 1$ *df* associated with trials are partitioned into components associated with the main effect of font ($a - 1$ *df*) and the effect of items nested within font ($a(b - 1)$ *df*). There is no font × item interaction because different items are used in each font condition.

In the font and mood examples, the independent variable was manipulated, and the use of a hierarchical design was the researcher's choice. In other experiments, the independent variable is observed. For example, the researcher may be interested in naming times for words that are or are not ambiguous, or words that are high or low in English-language frequency. In such cases, there is no choice; the items are naturally nested within levels of the independent variable.

15.2.1 An Example of the Design

Table 15.1 presents means for a data set from a hypothetical experiment in which eight participants were asked to categorize the mood depicted in each of 15 pictures. There were five pictures in each of three categories: sad, happy, and neutral. The time to push a button indicating a categorization response was recorded; the means of the eight response times for each of the 15 pictures are presented in Table 15.1 and the raw data are on the *Tables* page

Table 15.1 Mean response times (in ms) for items nested within moods

Response time (RT) means (averaged over participants)					
Sad faces		*Happy faces*		*Neutral faces*	
Item	*RT*	*Item*	*RT*	*Item*	*RT*
1	1,419.25	6	1,286.75	11	1,588.13
2	1,698.75	7	1,643.25	12	1,894.63
3	1,359.88	8	1,478.88	13	1,701.88
4	1,503.00	9	1,472.63	14	1,807.00
5	1,047.25	10	1,336.88	15	1,541.00
Mean	1,405.63	Mean	1,443.68	Mean	1,706.53

The grand mean = 1,518.61

Note: Each mean is an average of the scores of the eight participants. The complete data set is on the website in the file *Table 15_1Nested Data.*

of the website in the file *Table 15_1Nested Data*. In this example, items (B) are nested within moods (A) because the faces are different within each mood.

The study is a specific example of a general design in which there are n participants, each of whom is tested with b different stimuli at each of a levels of A; in the prior example, $n = 8$, $a = 3$ (moods), and $b = 5$ (faces within each mood). B is nested within levels of A (i.e., B/A) and is assumed to have random effects; A is assumed to have fixed effects. The distinction between fixed and random effects in this example stems from the fact that we have purposely chosen the three moods to be included in the experiment but have randomly sampled faces representing each mood. In what follows, we develop the analysis of data from this hierarchical design.

15.2.2 Partitioning the Sums of Squares and Degrees of Freedom

In the example represented by the data of Table 15.1, there are $a \times b$, or 3×5, pictures of faces so there are $ab - 1$, or 14, degrees of freedom for items. Part of the variability among the scores on the ab items is due to the a moods. This accounts for $a - 1$, or 2, of the 14 df for items. The remaining $a(b - 1)$, or 12, df account for the B/A term (faces within moods). In addition, each of n, or 8, participants is shown all of the pictures representing all three moods, so participants cross with A: SA, with $(n - 1)(a - 1) = 14$ df. Participants also cross with faces (SB), but because a different set of faces was used for each mood, the SB interactions are nested within levels of A and are written SB/A. This notation expresses the fact that the combination of participants and items is different for each of the levels of A. The SB interactions each have $(n - 1)(b - 1) = 28$ df, and there are three levels of A in which those interactions nest, so the SB/A $df = 3(28)$ or 84. Summing all these degrees of freedom, $2 + 12 + 14 + 84$, plus 7 more for the participants, equals 119, fully accounting for the total degrees of freedom in the experiment, $nab - 1$. Our final partitioning of the variance and df is summarized in Table 15.2.

15.2.3 Expected Mean Squares and Quasi-F Tests

The EMS shown in Table 15.2 follow the rules of thumb presented in Box 14.2. To illustrate, consider the expectation of the mean square for mood. As usual, error variance, σ^2_e, contributes to the expected mean square. In addition, because participants are assumed to be a random sample from the population of potential participants and no subscripts representing fixed effects other than A are present, σ^2_{SA} also contributes. Similarly, the only

Table 15.2 Sources of variance, *df*, and *EMS* for the hierarchical design of Table 15.1

SV	*df*	*EMS*
Participants (S)	$n - 1$	$\sigma^2_e + ab\sigma^2_S$
Mood (A)	$a - 1$	$\sigma^2_e + b\sigma^2_{SA} + n\sigma^2_{B/A} + nb\theta^2_A$
SA	$(n - 1)(a - 1)$	$\sigma^2_e + b\sigma^2_{SA}$
Faces within mood (B/A)	$a(b - 1)$	$\sigma^2_e + n\sigma^2_{B/A}$
SB/A	$a(n - 1)(b - 1)$	σ^2_e

Box 15.1 Quasi-*F* Test of *A* for the Data of Table 15.1 (*B* Nested in Levels of *A*)

To test the effects of mood (A),

$$F' = MS_A / (MS_{SA} + MS_{B/A} - MS_{SB/A})$$

Substituting the *MS* values from Table 15.3 into this equation, we have

$$F' = 1{,}073{,}858 / (98{,}662 + 260{,}647 - 41{,}107) = 3.375$$

The degrees of freedom for the numerator of this test are $a - 1 = 2$ and the denominator degrees of freedom are

$$df_{den} = \frac{(MS_{SA} + MS_{B/A} - MS_{S\times B/A})^2}{MS_{SA}^2 / df_{SA} + MS_{B/A}^2 / df_{B/A} + MS_{SAB}^2 / df_{S\times B/A}}$$

$$= (98{,}662 + 260{,}647 - 41{,}107)^2 / \begin{pmatrix} 98{,}662^2 / 14 + 260{,}647^2 / \\ 12 + 41{,}107^2 / 84 \end{pmatrix} = 15.88$$

Based on 3 and 16 *df*, the F' value of 3.375 is significant at α =.05; therefore, we can conclude that the moods displayed in the pictures of faces affects the time required to categorize the faces.

essential subscript in the B/A term, B, represents items which we assume have been randomly sampled from their population. Therefore, $\sigma^2_{B/A}$ is also present. Finally, the factor of interest, A, is represented by θ^2_A, with θ used instead of σ to remind us that we are dealing with a fixed effect. In other words, the *EMS* indicate that the means of the levels of A may differ for four distinct reasons: measurement error (σ^2_e); the possibility that the effects of A may differ across participants ($b\sigma^2_{SA}$); the possibility that the effect of items may differ ($n\sigma^2_{B/A}$); and the possibility that the means across the different levels of A may differ ($nb\theta^2_A$), that is, there is an effect of A.

The *EMS* show clearly that no single source of variance provides an appropriate error term to test A. Rather, we must form a combination of mean squares to create the error term, then compute a quasi-F' ratio. The relevant calculations are presented in Box 15.1, using the *MS* values from Table 15.3.

Researchers often fail to appreciate the importance of computing quasi-F' ratios. One approach frequently taken with data like those in Table 15.1 is to reduce the problem to the repeated-measures design of Chapter 12, ignoring the random effects of items. This can be done by averaging over items so that each participant has a single mean score at each treatment level, and then using the means as the input to the ANOVA. However, as we discussed in Chapter 14, this analysis is inappropriate because it ignores variability among items that potentially contributes to differences among the treatment means (factor A in Table 15.2). Considering the *EMS* in Table 15.2 again, the repeated-measures analysis would eliminate

Table 15.3 ANOVA of the data of Table 15.1

SV	df	SS	MS
Participants (S)	7	11,495,860	1,642,266
Mood (A)	2	2,147,746	1,073,858
SA	14	1,381,266	98,662
Items (B/A)	12	3,127,761	260,647
SB/A	84	3,452,994	41,107

all sources of variance involving factor *B* (items) from the variance partitioning without removing those effects from the data. In particular, the *EMS* for *A* would still include the random effects of *B*/*A*, meaning that using the MS_{SA} as the error term for MS_A would result in a positively biased test. Ignoring item variability does not make it disappear. Instead, it inflates the probability of a Type 1 error because the error term does not include the variance of the scores on the items whereas the numerator of the *F* does.

15.2.4 Using Software for the Hierarchical Design

Several statistical packages provide analyses for data from designs in which one or more variables are nested within levels of other variables. For example, in R, we can use the familiar *aov* function in the {stats} package. The function statement in *aov* that specifies the dependent and independent variables also specifies the structure of the design. For example, $Y \sim A + B$ specifies an additive model and $Y \sim A*B$ specifies a the nonadditive model (factors *A*, *B*, and their interaction are included). One simple way to specify nested variables is to use the %*in*% operator: $Y \sim S*A + B$ %in% *A* specifies a model that includes factors *S*, *A*, and their interaction, as well as factor *B*, which is nested in levels of *A*. Putting this all together, for data in a data frame called dat, the command *summary*(*aov*(data = dat, *Y* ~ *S***A* + *B* %*in*% *A*)) will provide the ANOVA table for the nested design in Table 15.1. Note that *aov* assumes all factors are fixed, and therefore additional calculations are required to test the main effect of *A* with *F′* (see Box 15.1).

In SPSS, we also begin with a familiar process: with data in long form (one observation per row, as in *Table 15_1 uni.xlsx* on the website), choose the *Univariate* option from the *General Linear Model* in the *Analyze* pull-down menu. Move variables to their appropriate boxes: *Y* is the dependent variable, *A* is a fixed-effect variable, and *S* and *B* are random-effects variables. Next, click on the *Model* button to build a structural model that is appropriate for this analysis. Select the radio button labeled "Build terms," and choose Main Effects from the "Build Term(s) Type" pull-down menu in the center of the window. Select *A* and *S* from the Factor list and move them to the Model box. Now select "Interaction," *A* and *S*, and move them to the Model box, which will now contain three components: *A*, *S*, and *A***S*. To add the nested terms, *B*/*A* and *SB*/*A*, to the model, select the "Build custom terms" radio button. Then, select *B* and move it down to the "Build Term" box, then click the "(Within)" button followed by selecting factor *A* and moving it down to the same box. You should now see *B*(*A*) in the "Build Term" box; click "Add" to include it in the model. Use a similar strategy to build the *S***B* interaction (you will need the "By *" button) and nest it within *A*, then add *S***B*(*A*) to the model. Finally, click Continue and OK to run the analysis. The results will display F' and adjusted df based on the error terms shown beneath the ANOVA table. Note that the reported effects for A, SA, and B/A are accurate but that the error term for S is not correct; this is not a concern because the main effect of participants is not usually of primary interest.

15.2.5 Estimating Omega-Squared in the Hierarchical Design

As in previous chapters, we wish to estimate the proportion of variance due to the factor of interest that is relative to variances due to participants and intrinsic factors, as well as error variance. Therefore, we define general omega-squared as

$$\omega_g^2(A) = \sigma_A^2 / \left(\sigma_e^2 + \sigma_A^2 + \sigma_S^2 + \sigma_{SA}^2\right) \tag{15.1}$$

Assuming n participants, a levels of A, and b levels of B, our estimates of the parameters in Equation 15.1 are

$$\hat{\sigma}_A^2 = \left(\frac{a-1}{a}\right)\left(\frac{MS_A - (MS_{SA} + MS_{B/A} - MS_{S\times B/A})}{bn}\right)$$

$$\hat{\sigma}_S^2 = (MS_S - MS_{S\times B/A})/ab$$

$$\hat{\sigma}_{SA}^2 = (MS_{SA} - MS_{S\times B/A})/b$$

and

$$\hat{\sigma}_e^2 = MS_{S\times B/A} \tag{15.2}$$

Substituting values from Table 15.3 into Equation 15.2, the variance estimate for A is 12,594.3, the variance estimate for S is 106,743.9, the variance estimate for SA is 11,511, and the variance estimate for error is 41,107. Putting these estimates together into Equation 15.1 gives an estimate for $\omega^2_g(A)$ of .073. This result indicates that the independent variable, mood, contributes moderate variance relative to the combined variance due to participant effects and error variance.

15.3 Adding a Between-Participants Variable to the Within-Participants Hierarchical Design

15.3.1 An Example of the Design

Nesting designs often involve a between-participants factor. For example, suppose that a researcher is interested in whether problem difficulty affects performance differently for experts and novices. Assume that there are five experts and five novices, and each participant is required to solve 12 problems, four of which are easy, four of intermediate difficulty, and four hard. In this example, problems are nested within difficulty levels and participants are nested within levels of experience. The data are presented in Table 15.4. They are also in the website file *Table 15_4nested mixed data*; follow links to the *Tables* page.

The means in Panel *b* of the table indicate that solutions take longer for the hard problems and longer for novices than experts. Furthermore, it appears that the advantage of

Table 15.4 Data from a design with problems (*B*) nested within levels of difficulty (*A*); experience (*C*) is the between-participants factor

(a) Response times (each row contains the data for one participant)

C	A_1				A_2				A_3			
	B_{11}	B_{12}	B_{13}	B_{14}	B_{21}	B_{22}	B_{23}	B_{24}	B_{31}	B_{32}	B_{33}	B_{34}
1	4.4	5.2	5.8	4.5	5.5	5.7	4.2	4.8	5.4	4.2	5.6	5.6
1	4.4	6.0	6.1	4.3	5.7	5.3	4.0	6.4	5.5	3.7	5.2	6.7
1	5.0	5.9	6.3	4.8	5.3	6.8	4.6	6.2	6.3	5.1	4.9	5.3
1	4.4	5.0	5.5	3.4	5.2	5.2	3.5	4.9	5.8	4.6	5.4	4.5
1	4.8	5.5	5.3	5.4	4.9	5.1	4.0	5.2	5.1	4.5	5.2	5.1
C_1 Mean	4.60	5.525	5.80	4.48	5.32	5.62	4.06	5.50	5.62	4.42	5.26	5.44
2	5.0	7.8	9.8	6.0	8.4	8.7	5.9	8.4	8.9	5.1	7.9	9.5
2	3.9	6.6	8.6	5.2	6.1	8.3	4.9	6.1	7.6	3.9	7.7	9.0
2	3.9	5.6	6.5	3.6	6.5	7.2	4.1	6.2	7.6	2.8	5.6	7.1
2	6.1	7.1	8.3	5.7	6.9	9.1	5.2	6.5	9.5	6.4	7.5	8.9
2	6.0	7.8	9.1	6.6	7.9	9.9	5.9	8.0	8.8	5.4	7.0	9.2
C_2 Mean	4.98	6.98	8.46	5.42	7.16	8.64	5.20	7.04	8.48	4.72	7.14	8.74
Problem mean	4.79	6.25	7.13	4.95	6.24	7.13	4.63	6.27	7.05	4.57	6.20	7.09

(b) Condition means

	Easy (A_1)	*Medium* (A_2)	*Hard* (A_3)	*Mean*
Expert (C_1)	5.100	5.125	5.185	5.137
Novice (C_2)	6.460	7.010	7.270	6.913
Mean	5.780	6.068	6.228	6.025

the experts increases with increased problem difficulty. This is largely because problem difficulty has relatively little effect on expert performance. We next consider the ANOVA in Table 15.5 to see how to test these trends.

15.3.2 *Partitioning Sums of Squares and Degrees of Freedom*

As in our analysis of the data in Table 15.1, we have two within-participants sources of variance: *A* (problem difficulty level) and *B/A* (problems nested within difficulty level). In the present design, we have the added between-participants variable, *C* (experience), so participants now are nested within levels of *C*. The *SV* column in the ANOVA of Table 15.5 includes these four sources of variance (*C*, *S/C*, *A*, *B/A*) as well as additional interaction terms. The *AC* term is the interaction of difficulty level with experience, and SS_{AC} is calculated as in all past designs. Because all problems are present at both levels of experience, *C* also crosses with the nested problems, giving rise to the *CB/A* term.

15.3.3 *Expected Mean Squares and Quasi-F (F′) Ratios*

Once again, the expected mean squares follow the rules of thumb. For example, consider $E(MS_C)$ in Table 15.5; *S/C* and *CB/A* both contribute to this expectation because we assume that both participants (*S*) and problems (*B*) are randomly sampled from their respective

Table 15.5 ANOVA of the data of Table 15.4

SV	*df*	*SS*	*MS*	*EMS*
Between-Participants	$ac - 1 = 9$			
Experience (*C*)	$c - 1 = 1$	94.70	94.70	$\sigma_e^2 + ab\sigma_{S/C}^2 + n\sigma_{C\times B/A}^2 + nab\theta_C^2$
Participants within *C* (*S/C*)	$c(n - 1) = 8$	41.76	5.22	$\sigma_e^2 + ba\sigma_{S/C}^2$
Within-Participants	$ac(ab - 1) = 110$			
Problem difficulty (*A*)	$a - 1 = 2$	4.11	2.06	$\sigma_e^2 + nc\sigma_{B/A}^2 + b\sigma_{A\times S/C}^2 + nbc\theta_A^2$
AC	$(a - 1)(c - 1) = 2$	2.80	1.40	$\sigma_e^2 + n\sigma_{C\times B\vert A}^2 + b\sigma_{A\times S\vert C}^2 + nb\theta_{AC}^2$
AS/C	$c(n - 1)(a - 1) = 16$	5.00	.31	$\sigma_e^2 + b\sigma_{A\times S/C}^2$
Problems within *A* (*B/A*)	$a(b - 1) = 9$	111.47	12.39	$\sigma_e^2 + nc\sigma_{B/A}^2$
CB/A	$a(b - 1)(c - 1) = 9$	25.27	2.81	$\sigma_e^2 + n\sigma_{C\times B/A}^2$
SB/AC	$ac(n - 1)(b - 1) = 72$	18.70	.26	σ_e^2

Box 15.2 Quasi-*F* Test of *C* for the Data of Table 15.4 (*C* Is Between Participants, *B* Is Nested in Levels of *A*)

To test the effects of experience (*C*),

$$F' = MS_C \,/\, (MS_{S/C} + MS_{C\times B/A} - MS_{Residual})$$

Substituting values of mean squares from Table 15.5 into the preceding equation, we have

$$F' = 94.70 \,/\, (5.22 + 2.81 - .26) = 12.19$$

The numerator degrees of freedom = $c - 1$, or 1, and the denominator degrees of freedom are

$$\frac{(MS_{S/C} + MS_{C\times B/A} - MS_{Residual})^2}{MS_{S/C}^2 \,/\, df_{S/C} + MS_{C\times B/A}^2 \,/\, df_{C\times B/A} + MS_{Residual}^2 \,/\, df_{Residual}}$$
$$= (5.22 + 2.81 - .26)^2 \,/\, \left(5.22^2 \,/\, 8 + 12.81^2 \,/\, 9 + .26^2 \,/\, 72\right) = 14.09$$

$p = .004$; with $\alpha = .05$, we can conclude that more experienced problem solvers require significantly less time to solve.

populations, and they are the only essential (i.e., appearing to the left of the slash) subscripts other than *C*. The reader should apply the rules to verify the other expected mean squares in Table 15.5. Given the expected mean squares of Table 15.5, no single mean square provides an error term for *C*, *A*, or *AC*. Again, we are forced to construct quasi-*F* ratios. The *F′* test of the effect of experience (*C*) is illustrated in Box 15.2. We leave the remaining tests as an exercise.

15.4 The Replicated Latin Square Design

A disadvantage of hierarchical designs is the increased error variance inherent in presenting different items at different levels of the independent variable. A Latin square design can be used to address that variability. In a Latin square design, the assignment of lists to treatments is counterbalanced so that some of the variability associated with lists can be removed if the proper analysis is performed. In our hypothetical word naming experiment, four lists of words could be created. Each group of participants would read each list of words in a different font with the assignment of lists to fonts counterbalanced over the groups such that each list appears equally often in each font across all participants. The analysis of data from this Latin square design would allow removal of variance due to differences among the lists, increasing the efficiency of the design.

The Latin square design can also be used to counterbalance the order of treatment conditions, thereby allowing the removal of some of the variability associated with treatment order. In the example described earlier, if we had the same items in four different font conditions, responses to the four fonts might be collected in four trial blocks. Participants would be divided into four groups and the assignment of fonts to trial blocks would be counterbalanced over groups so that each font would appear equally often in each block. The advantage of such counterbalancing is that this would equate the fonts with respect to practice or fatigue effects and enable the removal of variance due to such effects in the data analysis.

We introduced the Latin square in Section 12.6, using a single square to illustrate how the square is selected randomly from the population of possible squares. We discussed its advantages and disadvantages, with particular emphasis on its potential for reducing error variance relative to other designs. You may find it helpful to review that material before considering extensions of the design.

In the most common use of the Latin square design, there are n participants in each row of the square. We refer to this as a *replicated square* because we can view the design as involving n replications of a single square. Table 15.6 provides an illustration of this design. In the table, A is the treatment variable, B is the blocking variable, and G represents groups of n participants; the groups correspond to the rows of the square and differ with respect to the assignment of treatments to blocks. The numbers in each cell are the cell means, each

Table 15.6 An example of a Latin square with cell means (scores are number of errors in a detection experiment, A is number of noise elements, and levels of B are successive blocks of 25 trials)

		Blocks				
Group	*n*	B_1	B_2	B_3	B_4	*Mean*
G_1	5	7.6 (A_2)	10 (A_4)	8.2 (A_3)	6.4 (A_1)	8.05
G_2	5	9.6 (A_4)	11 (A_2)	7 (A_1)	8.6 (A_3)	9.05
G_3	5	8.6 (A_1)	10.4 (A_3)	7 (A_2)	10.2 (A_4)	9.05
G_4	5	9.6 (A_3)	7.8 (A_1)	9 (A_4)	3.8 (A_2)	7.55
	Mean	8.85	9.8	7.8	7.25	8.425

The treatment means are:

A_1	A_2	A_3	A_4
7.45	7.35	9.2	9.7

Note: The complete data set is on the website in the file *Table 15_6LatinSqData*.

of which is based on five scores; that is, $n = 5$.[1] Blocks may be trial blocks, in which case the five participants in Group 1 are tested first in condition A_2, next in condition A_4, and so on. Or the experiment might require detection of targets of four different sizes; size might be the treatment variable, *A*, and blocks might represent quadrants of the screen in which the target could appear. As still another example, *A* might represent study times for lists of words in a memory experiment and *B* would represent the lists. Many other examples of blocking variables are possible.

We will begin our analysis of the means in Table 15.6 by first discussing the partitioning of the sums of squares. Following that, we consider possible structural models, the expected mean squares based on these, and the hypothesis tests corresponding to each model.

15.4.1 Partitioning the Sums of Squares

To make the example in Table 15.6 more concrete, assume a hypothetical experiment in which 20 participants must quickly identify a target (e.g., a digit) in a noisy display. The independent variable of interest is the number of distractor elements (e.g., letters): $A = 6$, 8, 10, or 12. A block contains the same number of distractors for all 25 trials, although the target and the distractor items change from trial to trial. The numbers of errors (incorrect responses or failures to respond within a time limit) in each block of 25 trials are the basis for the means in Table 15.6. The errors increase as the number of distractor elements increases.

We may begin the process of partitioning the variance in the design by viewing it as a mixed design with one between-participants factor of group and one within-participants factor corresponding to the number of distractors, *A*. That analysis is summarized in the left panel of Table 15.7. The reader may confirm these results by analyzing the individual scores in the website *Tables* file, *Table 15_6LatinSqData*. However, our partitioning is incomplete because it fails to recognize the blocking variable, *B*, as a component of the experimental design. Because *B* was systematically incorporated in the design, we can measure the variance associated with *B* and remove it from the *Groups* × *A* (*GA*) interaction. This is easily accomplished by conducting a second mixed-factors ANOVA in which *B* replaces *A* as the treatment variable, producing the output summarized in the middle panel of Table 15.7.

Table 15.7 Three partitions of the sums of squares based on Table 15.6

Ignores blocking factor			*Ignores treatment factor*			*Recommended analysis*		
SV	*df*	*SS*	*SV*	*df*	*SS*	*SV*	*df*	*SS*
Groups (*G*)	3	33.75	*G*	3	33.75	*G*	3	33.75
S/G	16	341.80	*S/G*	16	341.80	*S/G*	16	341.80
A	3	86.65	*B*	3	76.85	*A*	3	86.65
GA	9	129.55	*GB*	9	139.35	*B*	3	76.85
SA/G	48	533.80	*SB/G*	48	533.80	*BCR*[a]	6	52.70
						WSR[b]	48	533.80
Total	79	1,125.55	Total	79	1,125.55	Total	79	1,125.55

[a] *BCR* is the "between-cells residual"; $SS_{BCR} = SS_{GA} - SS_B = SS_{GB} - SS_A$
[b] *WSR* is the "within-participants residual"; $SS_{WSR} = SS_{SA/G} = SS_{SB/G}$

The results of the two analyses are then used to obtain the partitioning summarized in the right panel of Table 15.7. In this analysis, we are able to distinguish not three, but four within-participants sources of variability – the treatment variable, *A*; the blocking variable, *B*; a "between-cells residual" term computed as ($SS_{GA} - SS_B$) or, equivalently ($SS_{GB} - SS_A$); and a term that we will call "within-participants residual" (*WSR*), which corresponds to *SA/G* from our original analysis or, equivalently, *SB/G* from our second analysis. The *df* for the between-cells residual (*BCR*) term are computed as the difference between the *df* for the *GA* interaction and the *df* for the main effect of *B*; that is, $df_{BCR} = df_{GA} - df_B = (a - 1)^2 - (a - 1) = (a - 1)(a - 2)$. The *df* for the remaining terms are found in the usual fashion.

The right-most partitioning is the analysis we recommend for this experimental design. An astute reader may wonder why that partitioning does not explicitly contain a treatments × blocks (*AB*) interaction term. In fact, every possible combination of levels of *A* and *B* is represented in the design, so it is possible to recover information about the interaction. However, there is a confounding inherent in the design that prevents us from identifying a single term corresponding to the *AB* interaction. Groups may differ because each represents a different subset of assignments of treatments to blocks; that is, a different subset of *AB* interaction effects (see Exercise 15.8). This accounts for $(a - 1)$ of the $(a - 1)(a - 1)$ *df* associated with the interaction. The remaining *AB* interaction variability is embedded in the *BCR* source. A simple demonstration of this is to show that the sum of squares for *AB* is divided between the *G* and *BCR* sources; that is, $SS_{AB} = SS_G + SS_{BCR}$. From Table 15.6,

$$SS_{AB} = 5\times\left[(7.6-8.425)^2+(10-8.425)^2+\ldots+(3.8-8.425)^2\right]-86.65-76.85=86.45$$

But $SS_G + SS_{BCR} = 33.75 + 52.70$, which also equals 86.45. Therefore, we may view the *Groups* and *BCR* sources as each capturing some of the *AB* interaction variability. The *AB* interaction is confounded with the *Groups* and *BCR* sources.

With the partitioning of the variability established, we are now able to consider alternative models of the data, the corresponding expected mean squares, and the resulting *F* tests.

15.4.2 Expected Mean Squares and F Tests

We assume the following structural model for the replicated-square design:

$$Y_{ijkm} = \mu + \eta_{i/m} + \alpha_j + \beta_k + (\alpha\beta)_{jk} + \varepsilon_{ijkm} \tag{15.3}$$

where *i* indexes the participant within a row of the Latin square ($i = 1, 2, \ldots, n$), *j* indexes the level of *A* ($j = 1, 2, \ldots, a$), *k* indexes the level of *B* ($k = 1, 2, \ldots, a$), and *m* indexes the row within the square (the group; $m = 1, 2, \ldots, a$). The population of participant effects, the $\eta_{i/m}$, is assumed to be independently and normally distributed with mean zero and variance $\sigma^2_{S/G}$. The effects of *A*, the α_j, are assumed to be fixed. *B* may have fixed or random effects. If random, then the effects of *B*, the β_k, and the *AB* interaction effects, the $(\alpha\beta)_{jk}$, are distributed independently and normally with mean zero and respective variances σ^2_B and σ^2_{AB}.

The expected mean squares for the replicated Latin square design are presented in Table 15.8. Note that there are two panels: one containing the *EMS* when *B* is assumed to have fixed effects and another when *B* is assumed to have random effects. In both panels, the *AB* interaction variance contributes to the *Group* and *BCR* expectations, as we showed

Table 15.8 Expected mean squares and error terms for the replicated-square design

B has fixed effects			*B has random effects*		
SV	*EMS*	*Error term*	*SV*	*EMS*	*Error term*
Groups (*AB*)′	$\sigma_e^2 + a\sigma_{S/G}^2 + na\theta_{AB}^2$	*S/G*	Groups (*AB*′)	$\sigma_e^2 + a\sigma_{S/G}^2 + na\sigma_{AB}^2$	*S/G*
S/G	$\sigma_e^2 + a\sigma_{S/G}^2$		*S/G*	$\sigma_e^2 + a\sigma_{S/G}^2$	
A	$\sigma_s^2 + na\theta_A^2$	*WSR*	*A*	$\sigma_e^2 + n\sigma_{AB}^2 + na\theta_A^2$	*BCR*
B	$\sigma_e^2 + na\theta_B^2$	*WSR*	*B*	$\sigma_e^2 + na\sigma_B^2$	*WSR*
BCR (*AB*)′	$\sigma_e^2 + n\theta_{AB}^2$	*WSR*	*BCR* (*AB*′)	$\sigma_e^2 + n\sigma_{AB}^2$	*WSR*
WSR	σ_e^2		*WSR*	σ_e^2	

in discussing the partitioning of sums of squares. For this reason, we use the label $(AB)'$ as a second designation of those sources; it is this designation that is operational when we apply the rules of thumb to determine *EMS*. Note that the *AB* interaction contributes to the *A* source of variance only *when B has random effects*, as the rules of thumb dictate. This is the reason for the difference between the two panels in Table 15.8.

In the example of the detection experiment in which the number of distractor elements was varied and counterbalanced over blocks of trials, it is reasonable to assume that trial blocks have fixed effects. In that case, the expected mean squares are those in the left panel of Table 15.8 and the error term for testing both *A* (number of distractor elements) and *B* (trial blocks) is the same error term as in the mixed designs of Chapter 14 (remember that $SS_{WSR} = SS_{SA/G} = SS_{SB/G}$). This would be the case for any design in which the blocking variable is assumed to have fixed effects.

In other experiments, recall scores might be obtained for lists that have been studied for various amounts of time, or comprehensibility ratings might be obtained for lists of sentences varying in syntax, or brain activity might be recorded for sets of pictures varying in emotional content. In these examples, the blocks in Table 15.6 represent sets of stimuli. It is reasonable to assume that the stimuli have been randomly sampled from some large population of stimuli and randomly divided into several sets. Suppose each group has a different pairing of set with study time or emotional content, with the pairings arranged so that the design forms a Latin square, as in Table 15.6. Then the appropriate error term for testing *A* (e.g., study time or emotional content) is the *BCR* source.

Table 15.9 presents two sets of *F* tests and *p*-values based on the data of Table 15.6. The tests correspond to the assumption either that *B* has fixed effects or that *B* has random effects. As the *EMS* of Table 15.8 showed, when the effects of *B* are assumed to be fixed the test of the treatment variable, *A*, is the same as in the mixed design of Chapter 14. The within-participants error term is appropriate for testing both *A* and *B* main effects, as well as providing a test of the partial interaction $(AB)'$ of *A* and *B*.

When *B* is assumed to have random effects, following the rules stated in Chapter 14, σ^2_{AB} contributes to $E(MS_A)$. You will recall that in previous designs, the presence of an additional random effects factor in a design (typically a stimulus factor) has complicated the tests of some sources of variance; specifically, we have needed to construct quasi-*F* tests. Note,

Table 15.9 Two analyses of the data of Table 15.5

			B Fixed[a]		B Random[b]	
SV	df	MS	F	p	F	p
G (AB)′	3	1125.00	0.53	0.67	0.53	0.67
S/G	16	21.36				
A	3	28.88	2.60	0.06	3.29	0.10
B	3	25.62	2.30	0.09	2.30	0.09
BCR (AB)′	6	8.78	0.79	0.58		
WSR	48	11.12				

[a] and *B* are tested against *WSR*.
[b] *A* is tested against *BCR*.

however, that in the replicated Latin square design a single source of variance – *BCR* – has the appropriate *EMS* to serve as the error term to test the effect of *A*. Unless there is a large effect of *A*, this *F* test will lack power because the *df* associated with *BCR* will typically be small except when the Latin square is large. Note that despite the larger value of the *F* statistic in the random-effects case, the *p*-value is larger than in the fixed-effects test of *A*, reflecting the reduced degrees of freedom.

To increase power for the test of *A* when *B* has random effects, some researchers choose to pool the *BCR (AB)′* and *WSR* mean squares when there is strong evidence that there is no *AB* interaction ($\sigma^2_{AB} = 0$). The advantage of pooling is that it results in a large gain in error degrees of freedom. However, in the absence of convincing evidence that there is no *AB* interaction, pooling will result in a positively biased test of *A* (see Section 14.6.1).

15.4.3 Alternative Analyses of the Replicated Latin Square Design

It is not uncommon for researchers to conduct statistical analyses that are not entirely consistent with their experimental designs. The design currently under consideration is a good case in point.

One common but inappropriate approach to analyzing the replicated Latin square design is to ignore both the *Group* and blocking variables and treat it as a simple one-factor repeated-measures design. The only sources of variance distinguished are participants, *A*, and the *SA* interaction, with the effect of *A* being tested against *SA*. In terms of the recommended analysis summarized in the right panel of Table 15.7, the repeated-measures analysis corresponds to pooling the *Groups* sums of squares with the sums of squares for the *S/G* term, and – more importantly – pooling the sums of squares associated with *B* and *BCR* with the sums of squares for the *WSR* term. If *B* is assumed to have fixed effects, the pooling of the *B*, *BCR*, and *WSR* sums of squares has the unfortunate consequence of inflating the error term for the test of the treatment effect, *A*. As a result, the test of *A* will be negatively biased and Type 2 errors are likely to increase (Pollatsek & Well, 1995). In that case, an experiment designed to reduce error variance through systematic manipulation of the blocking variable actually results in lower power because the statistical analysis is incompatible with the experimental design. In contrast, if *B* is assumed to have random effects, the pooling of *B*, *BCR*, and *WSR* will underestimate the appropriate error term, and thus the test of *A* will be positively biased.

The other common statistical approach to the replicated Latin square design is to treat it as a mixed-factors design; that is, to conduct the ANOVA summarized in the left panel of Table 15.7. This analysis is an improvement over the simple repeated-measures analysis because it does distinguish the group variable. If B is assumed to have fixed effects, the resulting ANOVA produces an unbiased test of A. However, information about the effects of the blocking variable and its interaction with A are lost. Further, if B is assumed to have random effects, the test of A against SA/G will be positively biased. Once again, the lesson is that the statistical analysis should be consistent with the experimental design.

15.4.4 *Advantages and Disadvantages of the Latin Square Design*

As we discussed in Chapter 12, the major advantage of the Latin square design is that it allows the researcher to remove error variance stemming from two sources, participants and a blocking variable such as trials, or sets of stimuli. It also provides a test of the blocking variable. For example, if the assignment of treatments is counterbalanced over trials, we have the advantage over the treatments × blocks design of being able to estimate the function relating mean scores to trials.

One potential disadvantage is that there are fewer degrees of freedom associated with the error term and therefore potentially lower power than in the treatments × blocks design. This is particularly true if the blocking variable, B, is best viewed as having random effects. In that case, the between-cell residual is the appropriate error term. The BCR error term will have few degrees of freedom unless the square is large; at least 5 × 5, or preferably larger.

The Latin square design shares with other repeated-measures designs one other potential problem: The responses to treatments may depend upon the treatments preceding it. For example, suppose A in Table 15.6 represents three drugs, and B represents successive days of testing. Further suppose that mean response times under drugs 3 and 1 differ. Is this because these drugs do differentially affect performance, or is it because drug 3 was preceded twice by drug 1 whereas drug 1 was preceded twice by drug 2? In other words, is the difference due to the drugs themselves or to different *carry-over effects* from the different drugs that preceded them? One response to this problem is to increase the period between treatments; for example, test every other day. Alternatively, Cochran and Cox (1957, pp. 135–139) describe methods for calculating the sums of squares due to carry-over, and for removing that variability from the sums of squares for treatments, and Namboodiri (1972) describes several design-based solutions.

15.5 Including Between-Participants Variables in the Replicated Square Design

Consider a hypothetical experiment in which time viewing an ambiguous word in a sentence is measured as a function of whether the preceding context in the sentence supported the less frequent sense of the word (A_1), was neutral with respect to the sense of the word (A_2), or supported the more frequent sense of the word (A_3). For example, a word such as *port* usually refers to a place where ships leave or enter, but it can refer to a type of wine; the preceding sentence can predict either of those meanings or be completely neutral. Assume that we create three lists of such words, approximately equated for length and total English-language frequency. These are the blocks in Table 15.10. Finally, assume that we

Table 15.10 Cell means (of reading times in ms) for a replicated Latin square with a between-participants factor

		Lists (B)			
Reader (C)	*Row*	B_1	B_2	B_3	*Row mean*
	1	338.00 (A_2)	346.00 (A_1)	341.50 (A_3)	341.83
Good (C_1)	2	344.75 (A_3)	345.50 (A_2)	355.25 (A_1)	348.50
	3	356.75 (A_1)	347.00 (A_3)	352.00 (A_2)	351.92
	Mean	346.50	346.17	349.58	347.42
	1	359.75 (A_2)	367.25 (A_1)	360.75 (A_3)	362.58
Poor (C_2)	2	355.25 (A_3)	360.25 (A_1)	365.25 (A_2)	360.25
	3	353.25 (A_1)	351.50 (A_3)	353.25 (A_2)	352.67
	Mean	355.08	359.67	359.75	358.50
Block mean		351.29	352.92	354.67	

Context

Reader	A_1	A_2	A_3	*Mean*
Good	352.67	345.17	344.42	347.42
Poor	361.92	357.75	355.83	358.50
Mean	357.29	351.46	350.13	

Note: A_1 = context supports subordinate meaning, A_2 = context is neutral, A_3 = context supports dominant meaning; B = lists of sentences; C = reader. Each of the six *Reader* × *Row* conditions represents a group of four participants. The complete data set with columns organized both by A and by B is on the website in the file labeled *Table 15_10LatinSqData_2 data.*

wish to see whether context has different effects on the reading time for good (C_1) and poor (C_2) readers.

In general, we have an $a \times a$ Latin square, with n participants in each row of the square at each level of C; therefore, there are acn participants, each tested in a blocks. In the present example, $a = 3$, $c = 2$, $n = 4$, so that there are 24 participants and a total of 72 scores. Table 15.10 presents the design and the means for the participants in each cell of the design.[2] The means indicate that the context supporting the less common meaning (A_1) results in longer reading times. We also find that poor readers are slower than good readers, although the difference varies only slightly as a function of context. Finally, we note that reading times differ slightly over lists (the block means), but the differences are small. Significance tests are needed but we must first consider the sources of variance and their expected mean squares.

The analysis is summarized in Table 15.11 for the case where B is assumed to have random effects. As indicated in the SV column and expected mean squares of Table 15.11, both the row and between-cells residual (BCR) terms reflect possible AB interaction effects. The $C \times Row$ and $C \times BCR$ terms reflect possible ABC interaction effects. This also is indicated in the SV and EMS columns of Table 15.11. These interaction effects are again a function of the confounding inherent in incomplete block designs; that is, designs in which participants are not tested in the complete set of combinations of A and B.

Table 15.11 Analysis of the design of Table 15.10 assuming *B* has random effects

SV	*df*	*EMS*	*Error term*
C	c − 1	$\sigma_e^2 + a\sigma_{S/C\times Row}^2 + \underline{an\sigma_{BC}^2} + a^2n\theta_C^2$	quasi-F
Row (AB)′	a − 1	$\sigma_e^2 + a\sigma_{S/C\times Row}^2 + acn\sigma_{AB}^2$	S/C × Row
C × Row (ABC)′	(a − 1)(c − 1)	$\sigma_e^2 + a\sigma_{S/C\times Row}^2 + n\sigma_{ABC}^2$	S/C × Row
S/C × Row	ac(n − 1)	$\sigma_e^2 + a\sigma_{S/C\times Row}^2$	
A	a − 1	$\sigma_e^2 + \underline{nc\sigma_{AB}^2} + acn\theta_A^2$	BCR
B	a − 1	$\sigma_e^2 + acn\sigma_B^2$	WSR
AC	(a − 1)(c − 1)	$\sigma_e^2 + \underline{n\sigma_{ABC}^2} + an\theta_{AC}^2$	C × BCR
BC	(a − 1)(c − 1)	$\sigma_e^2 + an\sigma_{BC}^2$	WSR
BCR (AB)′	(a − 1)(a − 2)	$\sigma_e^2 + nc\sigma_{AB}^2$	WSR
C × BCR (ABC)′	(a − 1)(a − 2)(c − 1)	$\sigma_e^2 + n\sigma_{ABC}^2$	WSR
WSR	ac(a − 1)(n − 1)	σ_e^2	

Note: If *B* has fixed effects, the underlined components are deleted from the *EMS*, and the error terms change accordingly.

Main and interaction effects are calculated as in all preceding chapters. Recall that the term labeled "*WSR*" is simply *SA*/*C* × *Rows*, using the notation familiar from repeated-measures designs. The only terms whose calculations might pose a problem are the *BCR* (between cells residual) and its interaction with *C*. These can be viewed as differences between terms derived from standard mixed-design ANOVAs. This observation again forms the basis for using computer packages to analyze the data. Any computer program capable of handling two between-participants variables (*C*, *Row*) and one within-participants variable (*A* or *B*) can then be used to do two analyses. In one analysis, specify *A* as the within-participants variable and obtain SS_A; in the other analysis, specify *B* as the within-participants variable and obtain SS_{BxRow}. Then compute:

$$SS_{BCR} = (SS_{B\times Row} - SS_A) \text{ and } SS_{C\times BCR} = (SS_{B\times C\times Row} - SS_{AC}).$$

The results will be the same if SS_C is subtracted from $SS_{A\times Row}$ and SS_{BC} is subtracted from $SS_{AB\times Row}$. The entries in the *EMS* column in Table 15.11 were derived assuming that *B* has random effects, the most reasonable assumption when *B* represents sets of items. If *B* is a fixed-effects variable, as when *B* represents time periods, the expected mean squares and error terms are changed, as described in the note to Table 15.11.

Table 15.12 presents the numerical results for the analysis of the data set in Table 15.10. The results are clear: Good readers are faster than slow readers, context has an effect on reading times, and no interactions approach significance.

Table 15.12 ANOVA of the data in Table 15.10

SV	*df*	*SS*	*MS*	*Error term*	*F*	*p*
Reader (*C*)	1	2,211.13	2,211.13	*quasi-F*[a]	10.16	0.01
Row	2	72.33	36.17	*S/C* × *Row*	0.14	0.87
C × *Row*	2	1,204.00	602.00	*S/C* × *Row*	2.27	0.13
S/C × *Row*	18	4,776.08	265.34	*WSR*		
Context (*A*)	2	697.33	348.67	*BCR*	645.69	0.002
Blocks (*B*)	2	136.75	68.38	*WSR*	0.92	0.41
AC	2	34.33	17.17	*C* × *BCR*	3.10	0.24
BC	2	53.58	26.79	*WSR*	0.40	0.67
BCR	2	1.08	0.54	*WSR*	0.01	0.99
C × *BCR*	2	11.08	5.54	*WSR*	0.08	0.92
WSR	36	1,341.17	74.51			

[a] $F_C = MS_C / (MS_{BC} + MS_{S/C \times Row} - MS_{WSR})$ with $df_{denom} = (MS_{BC} + MS_{S/C \times Row} - MS_{WSR})^2 / (MS^2_{BC}/df_{BC} + MS^2_{S/C \times Row}/df_{S/C \times Row} - MS^2_{WSR}/df_{WSR}) = 10.70$

15.6 Summary

In this chapter, we described structural models and data analyses for two modifications of the repeated-measures designs of the preceding two chapters:

- *Hierarchical designs* that involve the nesting of stimuli within levels of the independent variable.
- *Latin square designs* that involve assigning treatments to levels of a nuisance variable so that the assignment is counterbalanced over groups of participants.

These designs have advantages and disadvantages relative to each other and to the repeated-measures and mixed designs of Chapter 14. Hierarchical designs, by nesting items within levels of the within-participants factor, *A*, avoid effects of repeating stimuli. However, differences among sets of items contribute to the error variance against which the effects of *A* are tested. Latin squares reduce error variance due to blocking factors such as trial blocks of lists of items. However, they do so at a cost of error degrees of freedom, and therefore power of the hypothesis test. The choice of design will depend on factors specific to the research such as the nature of the independent variable, the response required of the participant, the number of available participants, and the number of levels of the independent variable.

Hierarchical models are very flexible in their application and are particularly useful in dealing with designs such as those in this chapter. Many different assumptions about the effects of variables and the pattern of variances and covariances can readily be incorporated in this approach. Also, missing scores are not a problem. The price for these and other benefits is paid in both the complexity of the approach and in the decisions imposed during the analysis of data. A discussion of the models and the analyses is beyond the scope of this book, but other sources are available (e.g., Luke, 2004; Quené & van den Bergh, 2004; Raudenbush & Bryk, 2002).

Exercises

15.1 [Understanding nested terms] To understand what the nested terms represent, it may help to note that sums of squares for nested terms such as items-within-moods (*B/A* in Table 15.3) represent a pooling of sums of squares across the levels of the higher-order factor (e.g., *Mood*). Furthermore, the mean square for a source such as *B/A* is an average of the mean squares at the levels of *A*. A possible use of this understanding is that it provides an approach to the ANOVA in the absence of software that calculates nested sums of squares. Do an ANOVA of the data within each mood of Table 15.1 (*Table 15_1Nested Data* in the *Tables* page at the website), and then pool the sums of squares for *B* and *SB* and compare the result with $SS_{B/A}$ and $SS_{SB/A}$.

15.2 [Analyzing data from nested design] Assume that pairwise comparisons had been planned to test differences among the *Mood* means in Table 15.1. Perform the test comparing the sad and neutral moods (A_1 and A_3). What is the criterion *p*-value if the *FWE* rate is .05?

15.3 [*EMS* and quasi-*F*s] Fifty randomly sampled words are divided into five lists that differ with respect to meaningfulness (*M*). All 50 words are presented on a computer monitor to each of 20 participants and the time to read each word aloud is recorded. Present all *SV*, *df*, and *EMS*, and the formula for the test of *M*, including the *df*.

15.4 [*EMS* and quasi-*F*s] A researcher wishes to study the effects of viewing televised violence upon the behavior of children. Fifteen cartoon episodes are drawn from a large random sample previously rated by adult viewers. Five of these are rated as low-violence episodes, five more are medium, and five more are high in violence. Thus there are three levels of violence (*V*) with five different episodes (*E*) at each level. There are 30 participants, with equal numbers of 6-, 8-, and 10-year-olds. Viewing time is measured on each of 15 days with a different episode viewed on each day; the order of episodes is random. Present *SV*, *df*, and *EMS*, and the formula for the test of *A* and its degrees of freedom.

15.5 [Analyzing data from a nested design] In a hypothetical experiment, on a given trial 10 participants each read a sentence designed to prime a positive or negative attitude toward some personality trait. Following that, the participants read a description of an individual that is ambiguous with respect to the trait and must rate the individual on a 1 (very negative) to 7 (very positive) scale. The file *EX15_5* contains a data set with four conditions: A sentence primed either a positive (*P*) or negative (*N*) attitude and was either relevant (*R*) or irrelevant (*I*) to the trait that was rated after reading the subsequent description. There are five different randomly sampled items (prime–description pairings) in each condition, so that there are 20 ratings for each of the 10 participants. The variables are *valence* ($V = 1$ or 2, positive or negative), *relevance* ($R = 1$ or 2, relevant or irrelevant), and *items* (*Items*, five different items within each of the four conditions). Perform the proper analysis and discuss the results.

15.6 [General omega-squared] Estimate general ω^2 for *R*, *V*, and *RV* for the design of Exercise 15.5.

15.7 [*EMS* and quasi-*F*s] Twenty participants were tested in a visual detection experiment. Ten pictures of objects were presented three times, each time in a different context (C). A second factor was the English-language frequency (*F*) of the object's name; five objects were high frequency and five were low frequency.

a) State the *SV*, *df*, and *EMS*. Assume a model in which the highest-order nested term estimates only σ^2_e.
b) State the formula for the *F* ratios and error *df* for testing context (C) and frequency (*F*).

15.8 [Understanding *SS* in Latin square design] Assume the following 4 × 4 Latin square:

	B_1	B_2	B_3	B_4
G_1	A_1	A_4	A_2	A_3
G_2	A_3	A_2	A_4	A_1
G_3	A_4	A_1	A_3	A_2
G_4	A_2	A_3	A_1	A_4

A is a treatment variable and *B* is the position in the sequence of treatments. As is common in many experiments using the Latin square design, assume that each group (*G*) represents a different level of an independent variable; for example, age or level of experience. Or a different treatment might be applied to each group in the design.

a) Suppose the population effects of *A* are $\alpha_1 = -4$, $\alpha_2 = 0$, $\alpha_3 = 1$, and $\alpha_4 = 3$. Also, let the effects of *B* be $\beta_1 = -2$, $\beta_2 = 2$, $\beta_3 = 2$, and $\beta_4 = -2$. Assume $\mu = 10$ and there are five participants in each group. Add the effects of *A* and *B* to μ to create the data set; for example, the score in the left-most cell for G_4 is $\mu + \alpha_2 + \beta_1 = 10 + 0 - 2 = 8$. No other effects are present in the data.
b) Calculate the sums of squares for *A*, *B*, and *G* for these data.
c) Now add in the following set of *AB* interaction effects. Let γ_{jk} represent the interaction effect of A_j and B_k.

	B_1	B_2	B_3	B_4
G_1	$\gamma_{11} = 2$	$\gamma_{42} = 1$	$\gamma_{23} = -2$	$\gamma_{34} = -2$
G_2	$\gamma_{31} = 0$	$\gamma_{22} = -2$	$\gamma_{43} = 0$	$\gamma_{14} = -3$
G_3	$\gamma_{41} = -3$	$\gamma_{12} = -3$	$\gamma_{33} = -2$	$\gamma_{24} = 3$
G_4	$\gamma_{21} = 1$	$\gamma_{32} = 4$	$\gamma_{13} = 4$	$\gamma_{44} = 2$

What are the *A*, *B*, and *G* sums of squares for these data?

d) What is the potential problem suggested by the preceding results?

15.9 [Comparing *RM* and Latin square designs] Consider the following 4 × 4 Latin square:

$$
\begin{array}{c} \\ S_1 \\ S_2 \\ S_3 \\ S_4 \end{array}
\begin{array}{cccc} C_1 & C_2 & C_3 & C_4 \\
\end{array}
$$

$$
\begin{array}{c|cccc|}
 & C_1 & C_2 & C_3 & C_4 \\
S_1 & 25(A_1) & 16(A_4) & 24(A_2) & 18(A_3) \\
S_2 & 19(A_2) & 19(A_1) & 13(A_3) & 12(A_4) \\
S_3 & 13(A_4) & 18(A_3) & 20(A_1) & 16(A_2) \\
S_4 & 17(A_3) & 19(A_2) & 18(A_4) & 17(A_1)
\end{array}
$$

a) Calculate the SS_A, SS_C, SS_S, and $SS_{residual}$.
b) Equation 12.4 estimates the MS_{SA}, assuming we had used a participants × treatments design instead of the Latin square, and Equation 12.5 uses that estimate to obtain an estimate of the relative efficiency of the two designs. Estimate MS_{SA} from the data set above.
c) What is your estimate of the relative efficiency of the two designs?

15.10 [Analyzing data from a Latin square design] In a study of decision making, two factors were manipulated. The task either resembled one seen during a practice session or did not (experience, E), and the amount of information available was either high or low (information level, I). Each participant was tested under all four combinations of E and I with the assignment of EI combinations to four randomly sampled problems (P) counterbalanced through the use of a Latin square. Decision times were

$$\begin{array}{c} \\ S_1 \\ S_2 \\ S_3 \\ S_4 \end{array} \begin{array}{cccc} P_1 & P_2 & P_3 & P_4 \\ \end{array}$$

$$\begin{matrix} & P_1 & P_2 & P_3 & P_4 \\ S_1 & 1.4(E_1I_1) & 2.2(E_2I_2) & 1.5(E_1I_2) & 1.5(E_2I_1) \\ S_2 & 2.1(E_1I_2) & 1.5(E_1I_1) & 2.0(E_2I_1) & 2.4(E_2I_2) \\ S_3 & 2.8(E_2I_2) & 2.1(E_2I_1) & 1.4(E_1I_1) & 1.8(E_1I_2) \\ S_4 & 2.3(E_2I_1) & 2.1(E_1I_2) & 2.7(E_2I_2) & 1.6(E_1I_1) \end{matrix}.$$

Perform the ANOVA.

15.11 [Comparing two Latin square designs] A researcher wishes to investigate cognitive performance as a function of drug type (T) and dosage (D). Thirty-two participants are randomly assigned to one of two drugs, and each participant is given a different one of four dosages of the same drug on four different occasions (O). A Latin square design is used with four sequences (R, row) of dosages and eight participants in each sequence. Half of the participants in each sequence receive D_1 and half receive D_2.

a) State the SV, df, and error terms.
b) Another way to run this study would be to create a single 8 × 8 Latin square with two participants in each row (so there are still 128 total scores). The eight treatments would be all possible combinations of the drug (T) and dosage (D) levels. Write out the SV, df, and error terms for this case.
c) What are the advantages and disadvantages of the two designs?

15.12 [Comparing potential designs] A researcher is interested in gambling behavior under variations in initial stake (I), payoffs (P), and probability of winning (W). There are three levels of each of these variables. Eighty-one participants are available for this experiment. Many possible experimental designs could be used. Suggest several alternative designs and discuss their relative merits.

15.13 [Analyzing data from a Latin square design] The *Ex15_13* file in the *Exercises* page of the book's website contains a data set for a 4 × 4 Latin square design. There are 12 participants with three in each row (R) of the square. A is the treatment factor and C is the column (or position in time) factor. Assume that C has fixed effects. Perform the ANOVA.

15.14 [Consequences of inappropriate analysis] Presumably, you analyzed the *Ex15_13* data set correctly. However, some individuals might treat the data as if they were obtained from a participants × treatments design. Pool terms from your answer to Exercise 15.13 to arrive at this incorrect ANOVA. How do the F and p-values relate to those previously obtained? Explain the reason for the difference in results.

15.15 [Analyzing data from a Latin square design] The *Ex15_15* file contains data (very) loosely modeled after a study by Witvliet, Ludwig, and Vander Laan (2001). Participants in that experiment were instructed to think of an individual who had mistreated or offended them. They imagined a response to this individual as they read four scripts representing hurting someone (S_1), bearing a grudge (S_2), empathizing (S_3), and forgiving (S_4). Various physiological measures were obtained in each of several segments of time in several counterbalanced blocks. The *Ex15_15* file contains heartbeat change scores similar to averages in the Witvliet article. There are 20 cases in the file, five in each of the four rows of a Latin square. The within-participant factors are S (the script) and C (the ordinal position in the sequence of presentation). Note that two scripts have negative valences ($SN1$ and $SN2$) and two have positive valences ($SP1$ and $SP2$); the effect of valence should be considered in the analysis. Perform the analysis and summarize your conclusions.

15.16 [Analyzing data from a Latin square design] Assume that the experiment described in Exercise 15.15 was replicated with two groups of 20 participants each; $H1$ and $H2$ represent high- and low-hostility groups as measured prior to the experiment. The data are in the *Ex15_16* file. Carry out the ANOVA and state your conclusions.

Notes

1 The complete data set is on the website in the file *Table 15_6LatinSqData*. The data appear twice in the file, once organized with the within-participants variable as A and once with the within-participants variable as B.

2 The complete data set with columns organized both by A and by B is at the website in the file labeled *Table 15_10LatinSqData_2 data*.

Chapter 16

Integrated Analysis III

16.1 Overview

In this chapter, we review and extend the developments in the preceding chapters on repeated-measures designs. To provide a context, we consider data from a hypothetical experiment. The design involves one between-participants factor and two within-participants factors; one of the within-participants factors represents items nested within levels of the other within-participants factor. This is an example of the hierarchical design of Chapter 15. We have the following goals in this chapter:

- To present an integrated analysis of the example data. We explore the data; carry out the ANOVA, including quasi-*F* tests where appropriate; and estimate measures of importance.
- To review the use of the Latin square design to investigate the factors in our example.
- To review the factors relevant to the choice between randomized assignment of items to levels of within-participants variables and counterbalancing of the assignment.

16.2 Introduction to the Experiment

Fredrickson and Kahneman (1993) investigated the effects of duration of a film clip on the emotional response of individuals who witnessed the video. Are pleasant videos deemed more pleasant when longer? Are unpleasant videos deemed more unpleasant when the unpleasant events take a longer time? Our hypothetical experiment is a greatly simplified version of their study and the data are artificial, although designed to mimic the effect of duration of the videos. We assume an experiment in which only unpleasant videos of various lengths are presented. Therefore, we varied duration with different videos nested within durations. We also varied the delay of the test; one group of participants rated each video immediately after viewing it whereas a second group rated the videos from memory after viewing all the film clips.[1] In summary, we ask:

1. What is the effect of duration of an unpleasant experience upon the viewer's perception of the experience? A reasonable hypothesis is that continued exposure to an unpleasant stimulus will increase the viewer's discomfort.
2. Is the mean rating affected by the time lag between viewing and rating the videos? One possibility is that the negative effect of films viewed earlier will dissipate over time, resulting in a lower average rating of unpleasantness when ratings are made after all films have

DOI: 10.4324/9781003453550-19

been viewed. Another possibility is that the ratings of unpleasantness will be greater in the delay condition because participants will have seen multiple unpleasant videos before providing any ratings.

3. Does the effect of duration depend upon the time of testing? That is, do duration and test delay interact? The effect of duration may be more pronounced when each video is rated immediately after being viewed.

16.3 Method

16.3.1 Stimulus Materials

Thirty videos were viewed and rated for unpleasantness by a group of 10 pilot participants. From these, 15 with roughly equal distributions of unpleasantness ratings were included in the experiment. They were randomly divided into three groups of videos, and then edited so that five videos lasted approximately 30 s; five others had a 60 s duration, and the remaining five videos were 90 s long. Each video began with a title; for example, "Aftermath of Hiroshima."

16.3.2 Participants and Design

Twenty-four college-age volunteers were randomly assigned to two groups of 12 participants each; the groups differed with respect to whether each video was rated immediately after viewing or after all the videos had been presented (variable *C*). The within-participants factors were duration of the video (*A*) and specific film clips (*B/A*); five videos were nested within each of the three durations.

16.3.3 Procedure

The videos were shown in a different random order for each participant. The immediate-test group rated each video immediately after viewing. The delayed-test group viewed all 15 videos, and then were presented the video titles in the order of viewing and asked to rate each video. Ratings were on a 100-point scale, with zero being "not at all unpleasant" and 100 being "very unpleasant."

16.4 Results and Discussion

16.4.1 Exploratory Analyses

Table 16.1 presents the means for the delay × duration cells.[2] These means are also plotted, together with standard error bars, in Figure 16.1. It appears that participants find the clips more unpleasant the longer they are exposed to them; this is true of both test delay conditions, although the increase in ratings with duration is slightly greater when participants rated the videos immediately after each clip. The means are also higher in the immediate than in the delayed test condition. Finally, we note that the error bars are roughly similar in length across durations, indicating that the within-cell variances in the three durations do not differ greatly.

Table 16.1 Mean unpleasantness ratings of videos

Time of test	*Video duration (in seconds)*			
	30	60	90	Mean
Immediate	47.867	54.383	57.933	53.394
Delayed	46.367	51.367	54.300	50.678
Mean	47.117	52.875	56.117	52.036

Note: Test delay is a between-participants variable, and duration is a within-participants variable. There are 12 participants in each level of delay and each cell mean is based on ratings of five videos.

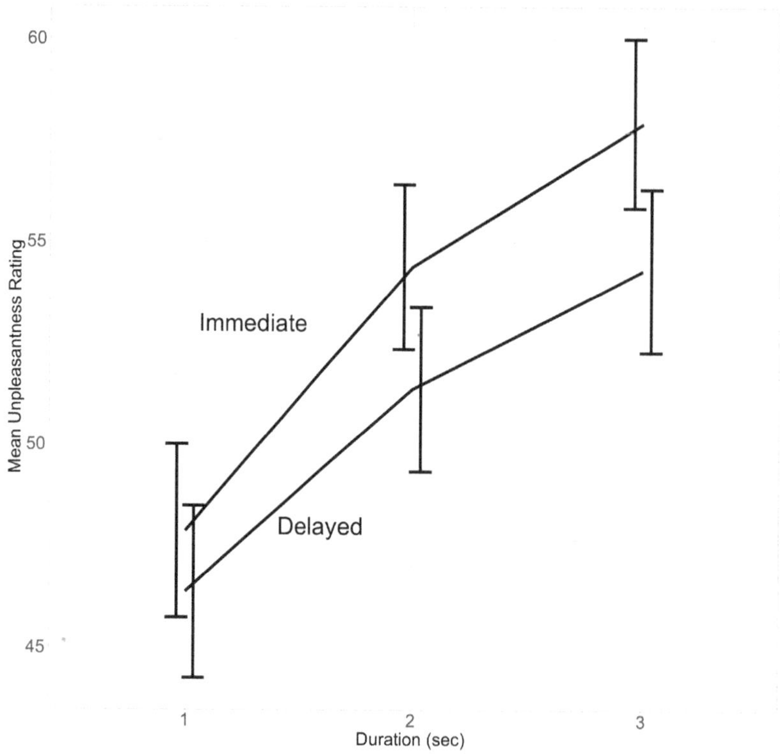

Figure 16.1 Plot of ratings of videos with standard error bars.

There are other considerations before we perform the analysis of variance. Is there evidence of any outliers that might distort the pattern of means? Are the data approximately normally distributed, as assumed by the ANOVA model? Are the variances homogeneous across items or across levels of the independent variables? To address these questions, we had SPSS save unstandardized residuals in an ANOVA of *Y* as a function of Duration, Items, and Delay. Similar options are available in other software packages. For example, in R, these residuals are stored as part of the object produced by the *aov* function; the command *aov*(data = dat, Y ~ Duration*Test*Item)$residuals will reveal them. The complete set of residuals is also available in the *IA3 Residuals* file on the *Tables* page of the book's website. Each score's residual is computed as a sum of all between- and within-participants error components associated with the score, thus the effects of the manipulated variables are removed. We plotted boxplots and *Q–Q* plots for each of the 15 videos and obtained tests of normality for each set of residuals. There were no significant departures from normality, the variances were quite similar across factor levels, and only four outliers were detected among the 360 scores, fewer than we would expect by chance.

Another issue relevant to repeated-measures analyses is the assumption of sphericity. The Greenhouse–Geisser and Huynh–Feldt epsilon adjustments for degrees of freedom were > 0.9 for both the duration and delay by duration sources. In summary, our exploration of the data suggested several trends for the effects of duration and test delay upon the mean ratings and appeared consistent with assumptions underlying the analysis. We therefore proceed to the analysis of variance to test the trends noted.

16.4.2 Tests of Hypotheses

For complex nested designs, statistical programs such as SPSS and R require detailed input that describes the crossed and/or nested relationships among the experimental factors. So, the first step in analyzing the data is to list all the sources of variance and their corresponding degrees of freedom, which provide a useful check on whether any sources of variance have been omitted. It is traditional to begin with the between-participants factors, in this case, time of test (delay, or factor *C*), which has two levels and thus $df = 1$. Of course, participants were nested within delay: *S*/*C*. There are 12 participants per test delay condition, so $df = c(n - 1) = 2(12 - 1) = 22$. These are listed in the first two rows of Table 16.2.

The design includes two within-participants factors: duration of the video (factor *A*, three levels so 2 *df*), and the five specific videos within each duration (*B*/*A*, with $a(b - 1) = 3(5 - 1) = 12$ *df*). Each duration occurs at both levels of test delay, so *A* and *C* are crossed and can interact (AC, $df = (a - 1)(c - 1) = 2$. Each participant is exposed to videos of all durations, so *S* and *A* are crossed and can interact; however, because participants are tested at only one delay, the *SA* interactions are nested within the levels of delay: *SA*/*C* ($df = c(n - 1)(a - 1) = 2(12 - 1)(3 - 1) = 44$). All videos were tested at both delays, so those factors are crossed and can interact, although the interactions are nested within the durations of the videos (*BC*/*A*, $df = a(b - 1)(c - 1) = 3(5 -1)(2 - 1) = 12$). Finally, participants see videos of every duration, and therefore those factors can interact, although the interactions occur within the combination of cells in which the participant and videos are nested: *SB*/*AC*, $df = ac(n - 1)(b - 1) = 3(2)(12 - 1)(5 -1) = 264$. These sources of variance are also listed in Table 16.2. We can assure ourselves that all sources of variance have been accounted for by summing the *df*, 1 + 22 + 2+. . .+ 264 = 359, to confirm they equal the total *df* for the experiment: $nabc - 1 = (12)(3)(5)(2) - 1 = 359$.

Next, the expected mean squares should be generated using the rules of thumb from 14.1: These are shown in Table 16.2. The appropriate error terms against which each factor is tested are derived from the *EMS*, as usual. In particular, the *EMS* terms for delay, duration, and their interaction remind us that there is no single source of variance that provides an error term against which those effects can be tested. Instead, as described in Chapters 14 and 15, quasi-*F* statistics are calculated. These involve combinations of mean squares that have the appropriate expectations for testing the effects of interest. Formulas for F'_1, the numerator mean square (e.g., MS_C) divided by its error mean square, are provided in Table 16.3; formulas for the error degrees of freedom are also presented there.

Table 16.2 Sources of variance (*SV*), degrees of freedom (*df*), and expected mean squares for the design of Table 16.1

SV	*df*	*EMS*
Delay (*C*)	$c - 1 = 1$	$\sigma_e^2 + ab\sigma_{S/C}^2 + n\sigma_{C\times B/A}^2 + nab\theta_C^2$
Participants within *C* (*S/C*)	$c(n - 1) = 22$	$\sigma_e^2 + ab\sigma_{S/C}^2$
Duration (*A*)	$a - 1 = 2$	$\sigma_e^2 + nc\sigma_{B/A}^2 + b\sigma_{A\times S/C}^2 + nbc\theta_A^2$
AC	$(a - 1)(c - 1) = 2$	$\sigma_e^2 + n\sigma_{C\times B/A}^2 + b\sigma_{A\times S/C}^2 + nb\theta_{AC}^2$
A × *S/C*	$c(n - 1)(a - 1) = 44$	$\sigma_e^2 + b\sigma_{A\times S/C}^2$
Items within *A* (*B/A*)	$a(b - 1) = 12$	$\sigma_e^2 + nc\sigma_{B/A}^2$
C × *B/A*	$a(b - 1)(c - 1) = 12$	$\sigma_e^2 + n\sigma_{C\times B/A}^2$
Residual	$ac(n - 1)(b - 1) = 264$	σ_e^2

Table 16.3 ANOVA of the rating data summarized in Table 16.1

Source	*df*	*Sums of squares*	*Mean square*	*F*	*p*
Delay (*C*)	1	664.23	664.23	1.77	0.20
Error(*C*)[a]	20.81	7,820.31	375.86		
S/C	22	8,474.31	385.20		
Duration (*A*)	2	4,986.87	2,493.34	72.33	0.00
Error(*A*)[b]	7.93	273.53	34.47		
AC	2	72.32	36.16	2.08	0.22
Error(*AC*)[c]	4.90	85.23	17.40		
A × *S/C*	44	1,176.21	26.73		
Items/duration (*B/A*)	12	473.48	39.46		
C × *B/A*	12	268.58	22.38		
Residual	264	8,372.73	31.72		

[a] $\text{Error}(C) = MS_{S/C} + MS_{C\times B/A} - MS_{Residual} = 375.863;$

$df_{\text{error}(C)} = MS^2_{\text{error}(C)} / \left[\left(MS^2_{S/C} / df_{S/C}\right) + \left(MS^2_{C\times B/A} / df_{C\times B/A}\right) + \left(MS^2_{Residual} / df_{Residual}\right)\right] = 20.81$

[b] $\text{Error}(A) = MS_{B/A} + MS_{A\times S/C} - MS_{Residual} = 34.474;$

$df_{\text{error}(A)} = MS^2_{\text{error}(A)} / \left[\left(MS^2_{B/A} / df_{B/A}\right) + \left(MS^2_{A\times S/C} / df_{A\times S/C}\right) + \left(MS^2_{Residual} / df_{Residual}\right)\right] = 7.93$

[c] $\text{Error}(AC) = MS_{A\times S/C} + MS_{C\times B/A} - MS_{Residual} = 17.399;$

$df_{\text{error}(AC)} = MS^2_{\text{error}(AC)} / \left[\left(MS^2_{C\times B/A} / df_{C\times B/A}\right) + \left(MS^2_{A\times S/C} / df_{A\times S|C}\right) + \left(MS^2_{Residual} / df_{Residual}\right)\right] = 4.90$

To run the ANOVA in SPSS, follow the steps described in Section 15.2.4 to build a structural model for the analysis that includes each of the sources of variance in Table 16.2 except the residual (*SB/AC*).[3] Once the model has been built, the analysis proceeds as in the previous chapters. The error terms used for each *F* or quasi-*F* are displayed beneath the ANOVA table in the output; these must be compared with the *EMS* to be sure the ratio has an expectation of 1 when the null hypothesis is true. For the data summarized in Table 16.1, SPSS uses appropriate error terms for the key factors of interest, namely duration (*A*), time of test (C), and their interaction (AC), but uses erroneous error terms for the tests of participants within delay (S/C), and of items within durations (B/A).

In R, we can build the structural model as described in Section 15.2.4. Alternatively, we can streamline the command with this notation: *summary*(*aov*(data = dat, Y ~ (Duration/Item)*(Test/Subject))), which indicates that participants are nested within Test (delay), and Items are nested within Duration, and that Duration and Test delay are crossed. The output includes sums of squares, *df*, and mean squares for each source of variance in Table 16.2. Then, the proper F or quasi-F can be calculated using the *EMS* in Table 16.2.

A summary of the results of this ANOVA are shown in Table 16.3, where we see that duration has a significant effect. As videos increased in length, so did viewers' feelings of discomfort as indicated by ratings of unpleasantness. The difference in mean unpleasantness ratings as a function of test delay was not significant; therefore, we cannot conclude that increased delay lowers ratings of unpleasantness in the population sampled. Furthermore, the nonsignificant AC interaction means that we cannot conclude that the delay of test affects the rate of increase of unpleasantness ratings with increased duration, despite the somewhat steeper slope in the immediate condition in Figure 16.1. It is possible that these effects are present in the sampled population, but, if so, we will require more participants and items to have sufficient power to determine this.

16.4.3 Estimating General ω^2 (Omega-Squared)

A measure of the contributions of our factors to the variability in the data is desirable and is required by most journals. The expected mean squares in Table 16.2 lead directly to one such measure. Continuing with our example of the experiment in which videos were rated for unpleasantness, we calculated estimates of general ω^2 for duration (A), delay (C), and their interaction. Although there are several different definitions of omega-squared, we defined ω^2_g as the ratio of the variance of the source of interest divided by that variance plus all variances involving participants plus error variance. The variance and omega-squared estimates are contained in Table 16.4.

From the results in Panel b of Table 16.5, it is clear that only the duration makes a more than negligible contribution to the variability in the data. The ratings were lower following

Table 16.4 Variance estimates and general ω^2 for a hierarchical design

(*a*) Variance estimates

Variance	*Estimator*	*Estimate*
Delay (C)	$[(c-1)/abcn](MS_C - MS_{error(C)})$	.80
Duration (A)	$[(a-1)/abcn](MS_A - MS_{error(A)})$	13.66
AC	$[(a-1)(c-1)/abcn](MS_{AC} - MS_{error(AC)})$	.10
S/C	$(MSs/c - MS_{Residual})/ab$	23.57
B/A	$(MS_{B/A} - MS_{Residual})/nc$	0.32
$A \times S/C$	$(MS_{A\times S/C} - MS_{Residual})/b$	0
$C \times B/A$	$(MS_{C\times B/A} - MS_{Residual})/n$	0
Error	$MS_{Residual}$	31.72

Note: Because the variance estimates for $A \times S/C$ and $C \times B/A$ were negative, they were set to zero.

(*b*) Estimates of ω^2

Source	*Estimator of* ω^2_g	*Estimate*
Delay (*C*)	$\hat{\sigma}^2_C / \left(\hat{\sigma}^2_C + \hat{\sigma}^2_e + \hat{\sigma}^2_{S/C} + \hat{\sigma}^2_{A \times S/C}\right)$	.014
Duration (*A*)	$\hat{\sigma}^2_A / \left(\hat{\sigma}^2_A + \hat{\sigma}^2_e + \hat{\sigma}^2_{S/C} + \hat{\sigma}^2_{A \times S/C}\right)$	.198
AC	$\hat{\sigma}^2_{AC} / \left(\hat{\sigma}^2_{AC} + \hat{\sigma}^2_e + \hat{\sigma}^2_{S/C} + \hat{\sigma}^2_{A \times S/C}\right)$	.002

Table 16.5 A Latin square design (*a*) and data (*b*)

(*a*)

Row	B_1	B_2	B_3
1	A_3	A_1	A_2
2	A_1	A_2	A_3
3	A_2	A_3	A_1

(*b*)

Participant	*Row*	*Delay* = C_1 *(immediate test)*						*Row mean*
		A_1	A_2	A_3	B_1	B_2	B_3	
1	1	51.0	61.8	63.4	63.4	51.0	61.8	
2	1	57.0	58.8	66.4	66.4	57.0	58.8	
3	1	53.6	55.6	53.6	53.6	53.6	55.6	55.617
4	1	41.6	48.6	56.0	56.0	41.6	48.6	
5	2	49.8	58.2	62.0	49.8	58.2	62.0	
6	2	36.6	44.8	49.6	36.6	44.8	49.6	51.450
7	2	46.4	56.4	60.6	46.4	56.4	60.6	
8	2	48.4	50.2	54.4	48.4	50.2	54.4	
9	3	54.8	62.2	63.2	62.2	63.2	54.8	
10	3	38.8	49.0	50.2	49.0	50.2	38.8	
11	3	48.0	52.6	57.8	52.6	57.8	48.0	53.117
12	3	48.4	54.4	58.0	54.4	58.0	48.4	
C_1 means		47.867	54.383	57.933	53.233	53.500	53.450	53.394
		Delay = C_2 *(delayed test)*						
13	1	57.8	62.8	61.4	61.4	57.8	62.8	
14	1	47.4	48.6	54.2	54.2	47.4	48.6	
15	1	40.4	48.2	52.2	52.2	40.4	48.2	50.700
16	1	42.2	48.0	45.2	45.2	42.2	48.0	
17	2	51.0	56.4	62.2	51.0	56.4	62.2	
18	2	37.8	43.8	47.8	37.8	43.8	47.8	50.200
19	2	46.6	53.8	56.0	46.6	53.8	56.0	
20	2	45.0	48.2	53.8	45.0	48.2	53.8	
21	3	47.0	50.8	55.6	50.8	55.6	47.0	
22	3	49.2	46.6	51.4	46.6	51.4	49.2	51.133
23	3	46.0	54.6	55.8	54.6	55.8	46.0	
24	3	46.0	54.6	56.0	54.6	56.0	46.0	
C_2 means		46.367	51.367	54.300	50.000	50.733	51.300	50.678
Overall means		47.117	52.875	56.117	51.617	52.37	52.375	52.036

Note: Scores are means of sets of five ratings for each participant, arranged by duration (*A*) and item set (*B*). *C* represents delay of test, and row refers to the assignment of durations to item sets.

a delayed test than an immediate test and this effect increased slightly with duration, but neither the F tests nor the ω^2 estimates indicated that these patterns were true of the population means. In particular, the very small ω^2 estimates indicated that little variance is attributable to the delay or delay × duration factors. These estimates of variance ratios strongly support the conclusion that – for the durations and test delays of this study – only the duration of the videos has a substantial influence on the unpleasantness ratings.

16.5 An Alternative Design: The Latin Square

16.5.1 The Experimental Design

So far, we have assumed a design in which the order of presentation of the 15 videos and their assignment to durations was randomized for each participant with the constraint that there are five stimuli shown at each duration. As discussed in Chapters 12 and 15, the Latin square design is potentially more efficient, capable of reducing error variance due to the order of testing or variability in the stimuli. With this in mind, we assume that the 15 videos are divided into three sets of five. These sets are assigned to durations in such a way that the assignment is counterbalanced over all participants. Panel *a* of Table 16.5 illustrates the design in which each level of B represents a set of five videos and the levels of A are the three durations of the study in which participants rate how unpleasant each film clip was. As in the preceding sections, we assume that the ratings were made immediately after each clip or delayed until all clips were shown; this is represented by the letter C.

Panel *b* contains the means of the ratings of the videos. These means are also in the file *Table 16_5 Lsq Data* on the *Tables* page of the book's website. They are presented twice, once organized by duration and once by item set. For example, 51 is the mean rating of five items seen by Participant 1 at the shortest duration, A_1; for Participant 1, these five items also make up the second set of videos, B_2. The advantage of having two organizations of the cell means is that the same file can be analyzed with duration as the within-participants factor, and a second time with the item set as the within-participants factor. The complete Latin square analysis can be obtained from these two preliminary analyses.

16.5.2 The Data Analysis

Table 16.6 contains three possible partitionings of the sums of squares based on the data of Table 16.5. The left-most column presents a standard mixed-design analysis with rows (R) and test delay (C) as between-participants factors and duration of the videos (A) as a within-participants factor. The middle partitioning replaces duration by set (B), referring to each set of five videos, or items. Finally, the right-most partitioning drops the terms that reflect the interactions of the three Latin square factors (A, B, and R) because there aren't enough degrees of freedom to include both the main effects and these interactions in this incomplete-block design.[4] The unanalyzed interaction effects potentially contribute to the between-cell residual (BCR) variability and to the interaction with delay (C×BCR).

Before turning to the hypothesis tests based on the right-most partitioning in Table 16.6, compare the values of the sums of squares with similarly labeled terms in Table 16.3. Note that SS_C in Table 16.3 is exactly five times larger than the same term in Table 16.6. The same is true for SS_A and SS_{AC}. The reason for this is that in Table 16.6, the sums of squares are

Table 16.6 Three partitions of the sums of squares based on Table 16.5

Ignores blocking factor			*Ignores treatment factor*			*Recommended analysis*		
SV	*df*	*SS*	*SV*	*df*	*SS*	*SV*	*df*	*SS*
Delay (*C*)	1	132.85	Delay (*C*)	1	132.85	Delay (*C*)	1	132.85
Rows (*R*)	2	65.62	Rows (*R*)	2	65.62	Rows (*R*)	2	65.62
CR	2	45.17	*CR*	2	45.17	*CR*	2	45.17
S/CR	18	1,584.07	*S/CR*	18	1,584.07	*S/CR*	18	1,584.07
Duration (*A*)	2	997.33	Set (*B*)	2	7.13	Duration (*A*)	2	997.33
AC	2	14.46	*BC*	2	3.54	*AC*	2	14.46
AR	4	18.25	*BR*	4	1,008.45	Set (*B*)	2	7.13
ACR	4	4.71	*ABR*	4	15.63	*BC*	2	3.54
WSR	36	212.28	*WSR*	36	212.28	*BCR*[a]	2	11.12
						C × *BCR*[b]	2	1.17
						WSR	36	212.28
Total	71	3,075.74	Total	71	3,075.74	Total	71	3,075.74

[a] $SS_{BCR} = SS_{AR} - SS_B = SS_{BR} - SS_A$
[b] $SS_{C \times BCR} = SS_{ACR} - SS_{BC} = SS_{BCR} - SS_{AC}$

Table 16.7 ANOVA of the data in Table 16.5

SV	*df*	*SS*	*MS*	*Error term*	*F*	*P*
Delay (*C*)	1	132.85	132.85	see *Note*	1.58	0.23
Row (*R*)	2	65.62	32.81	*S/C* × *R*	0.37	0.70
C × *R*	2	45.17	22.59	*S/C* × *R*	0.26	0.77
S/C × *R*	18	1,584.07	88.00			
Duration (*A*)	2	997.33	498.67	*BCR*	89.68	0.01
A × *C*	2	14.46	7.23	*C* × *BCR*	12.25	0.08
Set (*B*)	2	7.13	3.57	*WSR*	0.61	0.55
B × *C*	2	3.54	1.77	*WSR*	0.32	0.73
BCR	2	11.12	5.56			
C × *BCR*	2	1.17	0.59			
WSR	36	212.28	5.90			

Note: *C* is tested against $MS_{S/CR} + MS_{BC} - MS_{WSR} = 83.87$; the error *df* for testing *C* are 16.25; the formula is: $(MS_{S/CR} + MS_{BC} - MS_{WSR})^2 / \left[\left(MS^2_{S/CR} / df_{S/CR}\right) + \left(MS^2_{BC} / df_{BC}\right) + \left(MS^2_{WSR} / df_{WSR}\right)\right]$.

based on means of five items. As we said in Chapter 5, the variance of a mean is the variance of the scores divided by the number of scores. This also holds true for sums of squares.

Table 16.7 contains the tests of the Latin square data. The error terms are dictated by the expected mean squares presented earlier in Table 15.11. Those expectations indicate that the delay factor is tested by a quasi-*F* statistic. We conclude that duration of the videos influences ratings of unpleasantness, and no other effects are significant.

16.5.3 Estimates of General Omega-Squared (ω^2_g)

The expected mean squares in Table 15.11 provide the basis for estimating ω^2 for the delay and duration terms and their interaction. Because several of the variance estimates are negative, they were set to zero. Table 16.8 contains the formulas for the estimates of the

Table 16.8 Population variance estimates and partial ω^2 for a Latin square design

(*a*) ANOVA table with estimates of population variances

SV	df^a	$\hat{\sigma}^2$
Delay (*C*)	$c - 1$	$(c-1)(MS_C - MS_{S/C \times Row} - MS_{BC} + MS_{WSR}) / a^2nc = .680$
Row	$a - 1$	$(MS_{Row} - MS_{S/C \times Row}) / acn = 0$
C × *Row*	$(a-1)(c-1)$	$(MS_{C \times Row} - MS_{S/C \times Row}) / an = 0$
S/C × *Row*	$ac(n-1)$	$(MS_{S/C \times Row} - MS_{WSR}) / a = 27.48$
Duration (*A*)	$a - 1$	$(a-1)(MS_A - MS_{BCR}) / a^2nc = 13.70$
AC	$(a-1)(c-1)$	$(a-1)(c-1)(MS_{AC} - MS_{C \times BCR}) / a^2nc = .184$
Set (*B*)	$a - 1$	$(MS_B - MS_{WSR}) / acn = 0$
BC	$(a-1)(c-1)$	$(MS_{BC} - MS_{WSR}) / an = 0$
BCR	$(a-1)(a-2)$	$(MS_{BCR} - MS_{WSR}) / nc = 0$
C × *BCR*	$(a-1)(a-2)(c-1)$	$(MS_{C \times BCR} - MS_{WSR}) / n = 0$
Residual (*WSR*)	$ac(n-1)(a-1)$	$\hat{\sigma}_e^2 = MS_{WSR} = 5.90$

[a] $a = 3$, $c = 2$, and $n = 4$

(*b*) Estimates of ω_g^2

SV	*Estimator*	*Estimate*
Delay (*C*)	$\dfrac{\hat{\sigma}_C^2}{\hat{\sigma}_e^2 + \hat{\sigma}_C^2 + \hat{\sigma}_{S/CR}^2 + \hat{\sigma}_R^2 + \hat{\sigma}_{CR}^2}$	.020
Duration (*A*)	$\dfrac{\hat{\sigma}_A^2}{\hat{\sigma}_e^2 + \hat{\sigma}_A^2 + \hat{\sigma}_{S/CR}^2 + \hat{\sigma}_R^2 + \hat{\sigma}_{CR}^2}$	.291
AC	$\dfrac{\hat{\sigma}_{AC}^2}{\hat{\sigma}_e^2 + \hat{\sigma}_{AC}^2 + \hat{\sigma}_{S/CR}^2 + \hat{\sigma}_R^2 + \hat{\sigma}_{CR}^2}$	.005

population variances and for the estimates of ω^2. As in our analysis of the hierarchical design, it is again clear that only the duration of the videos has any effect on participants' ratings of unpleasantness.

16.5.4 Which Design?

Whether we consider *F* tests or measures such as ω^2, the hierarchical and Latin square designs have yielded similar conclusions, although the ω^2 values are slightly larger in Table 16.8 than in Table 16.4. The similarity of results is not surprising because the analyses were based on the same data set. If we had run two experiments, using a different design in each, there would have been some differences in the two sets of rating data. This leaves the question of which design is preferable. Although there is no single answer, the following questions should be considered before choosing a design:

1. Will items contribute considerable variance? If this is likely, counterbalancing the assignment of treatments to item sets will permit the removal of much of the item variability, increasing the efficiency of the Latin square design.

2. Will treatments and items interact? If so, the between-cell residual is likely to be inflated by that interaction variance, reducing the efficiency of the Latin square design.
3. How many factor levels will there be? Larger squares have more error degrees of freedom associated with both the *BCR* and *WSR* terms, thus increasing the effectiveness of the Latin square design. Although we used a small square for our example, generally it is wise to have squares with at least five levels of the Latin square factors.

If the Latin square is the design of choice, the data should be properly analyzed. We have found that in many, perhaps most, instances researchers have counterbalanced the assignment of factor levels and then analyzed the data as if the design were a standard repeated-measures design, or a mixed design if there was a between-participants factor. If substantial sources of variance are neglected, ratios of mean squares may not be distributed as *F* and error rates may be increased.

16.6 Summary

In this chapter, we reviewed and elaborated on the developments in Chapters 13–15 and extended the material in those chapters.

- We analyzed data from a standard mixed design and from a Latin square in which there was a between-participants factor and items were nested within levels of a within-participants factor.
- We reviewed the expected mean squares for these designs, using them to justify *F* tests and estimates of effect size in the form of partial ω^2.
- We reviewed the considerations involved in choosing between random assignment of items to levels and counterbalancing.

Exercises

16.1 a) The *EX16_1* data set on the book's website contains percentages of target detections for 20 participants for each of three days in a perception experiment. Explore the data and discuss the results of your exploration.
b) Perform the ANOVA on the data. What do you conclude?

16.2 A follow-up to the experiment in Exercise 16.1 is planned. Half of the targets (*B*) on each of the three days (*A*) will be digits and half letters. In addition, the background (*C*, distractor elements) will be either digits or letters. Present the sources of variance, *df*, and *EMS*, and indicate the appropriate error terms to test the main and interaction effects of *A*, *B*, and *C*, and their interactions. Assume that $N = 44$ and use numerical values wherever possible (*df*s, *EMS* coefficients).

16.3 A different way of designing the experiment described in Exercise 16.2 is to have half of the targets on each day be letters and half digits while maintaining a constant background type throughout the three days. However, context would be varied between groups with half of the participants seeing a background of digits and the others seeing a background of letters. Present the sources of variance, *df*, and *EMS*, and indicate the appropriate error terms to test the main and interaction effects of *A*, *B*, and *C*, and their interactions. Assume $N = 44$ and use numerical values wherever possible (*df*s, *EMS* coefficients).

16.4 a) Discuss the pros and cons of the designs in Exercises 16.2 and 16.3.
b) What effects would changes in the correlation (ρ) and the degrees-of-freedom adjustment (ε) have on the power of the tests of targets and context in each design?

16.5 Sixteen participants read 15 short texts. Although all participants were fluent in English, eight had learned it as a second language (*Group* = 1 or 2; native speaker or bilingual). Each text contained an ambiguous word whose meaning was made clear by a short disambiguating region of text following the word. Time looking at the disambiguating region was measured. In five of the texts, the ambiguous word was preceded by a context supporting its meaning ($A = 1$); in five, the contexts were neutral ($A = 2$); and in the remaining five texts ($A = 3$), the contexts implied the wrong interpretation, The data are at the book's website in the file *EX16_5*. Perform the ANOVA and discuss the results.

16.6 In studies such as that described in Exercise 16.6, scores are often missing and researchers calculate *minF′* to test hypotheses. The files *EX16_6S* and *EX16_6I* on the book's website were constructed by first randomly deleting a small number of scores from the *EX_16_5* files. The numbers in the *EX16_6S* file were calculated by averaging the remaining item scores in each level of *A* for each participant. The numbers in the *EX16_6I* file were calculated by averaging the scores in each level of *C* for each item. Calculate the *minF′* value and its degrees of freedom for the test of the *A* effect. Box 14.2 provides an example of the calculations.

16.7 Twenty air controllers were presented with a series of four simulated flight situations and their response times were recorded. The sequence of situations (*A*) was counterbalanced over four blocks of time (*B*). The data are on the website in the *EX16_7* file.
a) Ignoring block and group variability, we can treat the design as a simple 20 participants by four levels of *A* design. Present the sources of variance, *df*, sums of squares, mean squares, *F* ratios, and *p*-values, assuming this is the design.
b) Reanalyze the data taking its Latin square properties into consideration. What is the relationship between the two sets of sums of squares in the analyses in parts (a) and (b)?
c) What changes, if any, would there be in the analysis if *B* is assumed to have random effects?

16.8 The experiment described in Exercise 16.7 was replicated, but this time there were 24 participants, 12 under 60 years old ($Age = 1$) and 12 more than 60 years old ($Age = 2$). The data are in the file *EX16_8*; *Group* refers to the order of the situations, *A* again refers to the situation, and *B* refers to the position in time. Perform the ANOVA and discuss the results.

Notes

1 We don't recommend using this hypothetical design in a real experiment because it confounds time of test with number of videos viewed before providing a rating.
2 The complete data set is at the website in the file *Table 16_11 A3 data.*
3 One can also build the model in an SPSS syntax file, which then can be saved and modified for other similar analyses.
4 The Latin square is an incomplete-block design because each participant, or block, is tested only under some (actually $1 / a$) of the possible combinations of factors *A* and *B*.

Part 4

Correlation and Regression

Chapter 17

An Introduction to Correlation and Regression

17.1 Introduction to the Correlation and Regression Chapters

Although the ANOVAs we have so far discussed are widely used to determine whether independent and dependent variables are related, they have some important limitations. Standard ANOVA treats factors as qualitative categorical variables even if they are inherently quantitative and continuous, thereby ignoring potentially useful information. In ANOVA, levels of a factor are merely considered to be *different* from one another. The *ordering* of levels or *how different* the levels are from one another is not considered. Also, the neat partitioning of variability into nonoverlapping components associated with main effects and interactions that is the hallmark of ANOVA only occurs when the factors are uncorrelated. Factors are uncorrelated in designed experiments with equal numbers of scores in each cell, but generally not otherwise.

We now begin a series of chapters that develop a general framework which incorporates both categorical and quantitative variables and can deal with correlated factors:

- Chapter 17 introduces the basic ideas of correlation and regression.
- Chapter 18 develops statistical inference and power analysis for correlation and considers some alternative measures.
- Chapter 19 extends the discussion of bivariate regression and considers inference and power analysis for regression.
- Chapters 20–22 develop regression procedures when there is more than one predictor variable. This use of regression is known as *multiple regression*.
- Chapter 23 introduces the idea of coding categorical variables so that they can be used along with quantitative variables in regression, thereby developing a very general framework for statistical analysis. We then show that ANOVA is a special case of regression.
- Chapter 24 deals with analysis of covariance (ANCOVA) within the regression framework.
- Chapter 25 integrates several multiple regression analyses and focuses on interpretation.

17.2 Overview of Chapter 17

In Chapter 2 we considered how to graph and summarize distributions of single variables. We also briefly introduced scatterplots and the Pearson correlation coefficient as ways of describing and summarizing bivariate data. Procedures for characterizing bivariate data are important because we rarely study individual variables in isolation; rather, we usually are

DOI: 10.4324/9781003453550-21

interested in how variables are related to one another. For example, we may wish to know how cholesterol level varies with age or whether math skills are related to verbal skills in children. A major goal of this chapter is to discuss in more detail how to represent data in ways that allow us to see how, and how strongly, variables are related, and to present statistics that summarize important aspects of the relationship. A major focus is to expand our understanding of the correlation coefficient as a measure of the degree to which two variables have a linear (i.e., straight-line) relationship. As we will show, the value of the correlation coefficient depends on the similarity of the corresponding z scores of the variables being correlated, making it sensitive to the variances of these variables in the sample. This can lead to much confusion about how to interpret and how, or whether, to compare correlations.

Another major goal of the chapter is to develop the basics of linear regression. If two variables are systematically related, it should be possible to use information about one of the variables to predict the other. For example, knowing a father's height is useful in predicting the height of his son. Even though any single prediction may not be very accurate, if we consider many fathers and sons, on the average we can predict a son's height more accurately if we use information about the father's height than if we do not. As we would expect, the more closely the heights of fathers and sons are related, the better we can predict. We would like to develop procedures for making the best predictions possible with the information that we have available. Linear equations that use information about one variable to make optimal predictions about a second variable are referred to as *bivariate regression equations*.

In Chapter 17, we have the following goals:

- Provide examples of bivariate relationships and discuss how they may be represented with scatterplots.
- Discuss how to use "smoothers" or "fit lines" to extract systematic relationships between two variables in the face of random variability.
- Discuss the Pearson correlation coefficient as a measure of the strength of a linear relationship.
- Introduce least-squares linear regression, first with z scores and then with raw scores.
- Discuss the major differences between regression and correlation.
- Introduce the concept of regression toward the mean and provide examples of how people misunderstand, or fail to appreciate, the consequences of this phenomenon.
- Consider measures of linear fit such as the coefficient of determination and the standard error of the estimate.
- Discuss the interpretation of the correlation coefficient and factors that influence it.

17.3 Some Examples of Bivariate Relationships

Consider subtraction and multiplication accuracy scores (percent correct) for third-graders from the *Royer* data set on the arithmetic skills of grade-school children (included on the book's website). The basic statistics for the 28 third-grade students with values on both variables are given in Table 17.1. The multiplication accuracy scores tend to be lower than the subtraction scores, $t(27) = 3.09$, $p = .005$, and exhibit greater variability,[1] $z = 2.78$, $p = .023$.

Table 17.1 Descriptive statistics for the 28 third-grade students who have both subtraction (*subacc*) and multiplication accuracy scores (*multacc*; obtained using the SPSS Explore module)

Descriptives

			Statistic	*Std. error*
subacc	Mean		87.0240	2.30531
	95% confidence interval for mean	Lower bound	82.2939	
		Upper bound	91.7541	
	5% Trimmed mean		88.1999	
	Median		89.1815	
	Variance		148.805	
	Std. deviation		12.19856	
	Minimum		46.67	
	Maximum		100.00	
	Range		53.33	
	Interquartile range		15.45	
	Skewness		−1.473	.441
	Kurtosis		3.129	.858
multacc	Mean		78.4690	3.42575
	95% confidence interval for mean	Lower bound	71.4400	
		Upper bound	85.4981	
	5% Trimmed mean		79.8302	
	Median		79.4470	
	Variance		328.601	
	Std. deviation		18.12736	
	Minimum		29.41	
	Maximum		100.00	
	Range		70.59	
	Interquartile range		25.52	
	Skewness		−.835	.441
	Kurtosis		.813	.858

17.3.1 *Scatterplots*

How can we best characterize the relationship between subtraction and multiplication accuracy in this sample of third-graders? We might expect that children with higher subtraction scores will also have higher multiplication scores, because we know that some children are better at arithmetic than others. Perhaps the best way of determining whether there is a relationship between two variables is to use a *scatterplot*, in which each data point has coordinates that represent the scores for one student (see Section 2.7.1). The scatterplot for the 28 third-graders for whom we have both multiplication accuracy and subtraction accuracy scores is presented in Figure 17.1. Note that some statistical packages (in this case, R, using the *scatterplot* function in the {car} package) allow us to present histograms or box plots along the borders of the scatterplot, so we can see information about both the univariate and joint distributions in the same display.

What we see in the scatterplot is a tendency for larger multiplication scores to go together with larger subtraction scores – we say there is a *positive* relationship between the

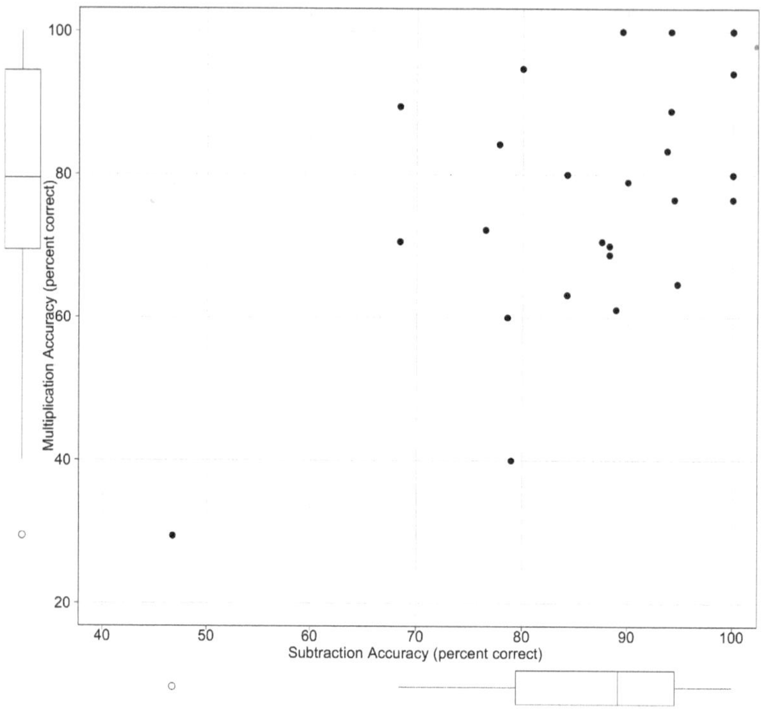

Figure 17.1 Scatterplot for subtraction and multiplication accuracy for the 28 third-grade children having both subtraction and multiplication scores, with box plots for each variable on the borders.

two variables. If larger scores on one variable tend to go together with smaller scores on the other, we have a *negative* relationship. The scatterplot in Figure 17.2 shows the mean time to answer multiplication problems plotted against multiplication accuracy. The graph shows a negative relationship, reflecting the tendency for children who are more accurate to take less time to answer. It is equally appropriate to say that the graph reflects the tendency for children who respond more quickly to have higher accuracy. Scatterplots allow us to observe the relationship between two variables but do not convey any information about why the relationship appears as it does.

We conclude this section with two additional examples of scatterplots that we will discuss later in the chapter. The first uses data obtained from an introductory college statistics class. Table 17.2 contains two scores for each of 18 students, the score on a math skills pretest taken during the first week of class and the score on the final exam. The scatterplot for the 18 data points is presented in Figure 17.3. Not surprisingly, the pretest and final scores covary: Students who score higher on the pretest tend to do better on the final exam. In Section 17.6.1, we will show how to find the equation that does the best job in predicting the 18 final exam scores from the corresponding pretest scores – that is, the equation for the line shown in Figure 17.3. That regression equation is useful both for describing the relationship between the two variables and for predicting final exam performance for students who take the pretest in future classes.

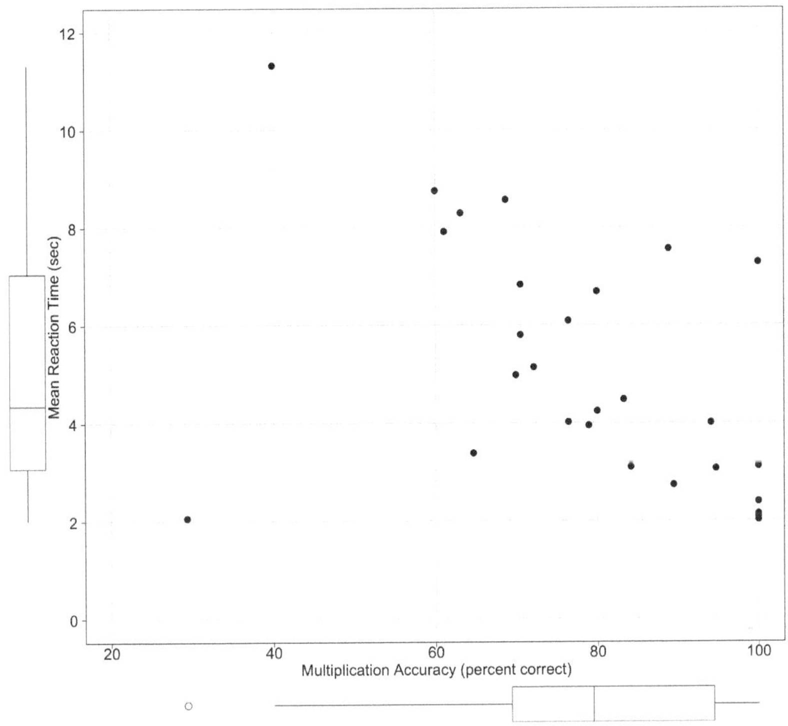

Figure 17.2 Scatterplot of response time and multiplication accuracy for third-graders in the *Royer* data set.

Table 17.2 Statistics class example data

Pretest score	*Final exam score*
X	Y
29	47
34	93
27	49
34	98
33	83
31	59
32	70
33	93
32	79
35	79
36	93
31	90
35	77
29	81
32	79
34	85
36	90
25	66

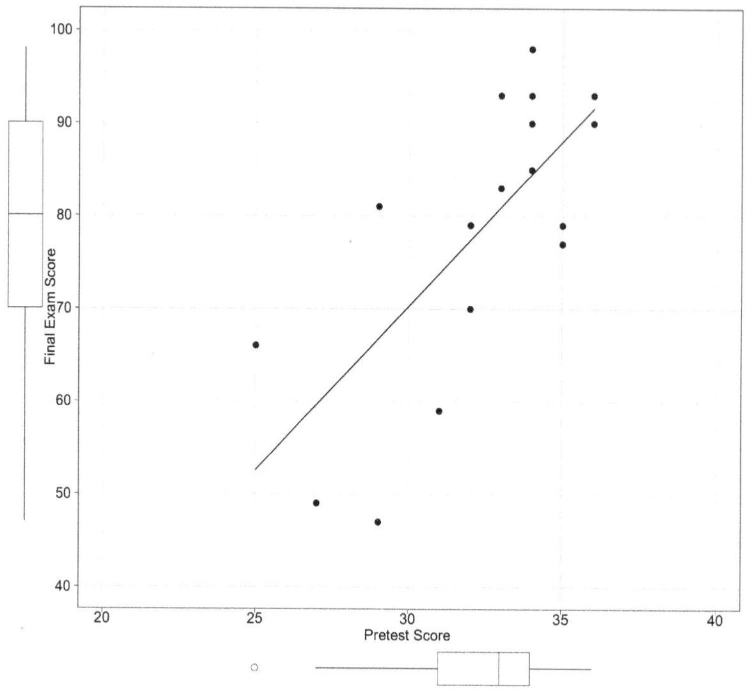

Figure 17.3 Scatterplot for pretest and final exam scores in a statistics class.

To obtain the second scatterplot, presented in Figure 17.4, we first found the mean of the four seasonal total cholesterol scores (*TC*) for each of the 431 individuals who had scores in all four seasons in the *Seasons* study conducted by researchers at the University of Massachusetts Medical School, and then plotted *TC* against age. Although there is a great deal of variability in the cholesterol scores, there seems to be a tendency for older people to have higher cholesterol scores. We used the *geom_smooth* function in the {ggplot2} package in R to include a "fit line" (these curves are also known as "smoothers") to help us determine the nature of any systematic relationship between the two variables. We consider this procedure next.

17.3.2 *Extracting the Systematic Relationship Between Two Variables*

Scatterplots can be very useful in helping us understand how, and to what extent, two variables are related. However, real data are often extremely messy. Any systematic relationship that exists between the variables may be obscured by variability due to factors such as individual differences and measurement error. In such cases we try to see through the "noise" to extract the "signal," that is, the underlying systematic relationship, if one exists. This can be difficult to do by eye, especially when there are many data points and a great deal of variability, as in the plot of cholesterol level against age in Figure 17.4.

One way of trying to get at the underlying relationship is to use some type of averaging to reduce the complexity of the display. For example, we can find the total cholesterol score

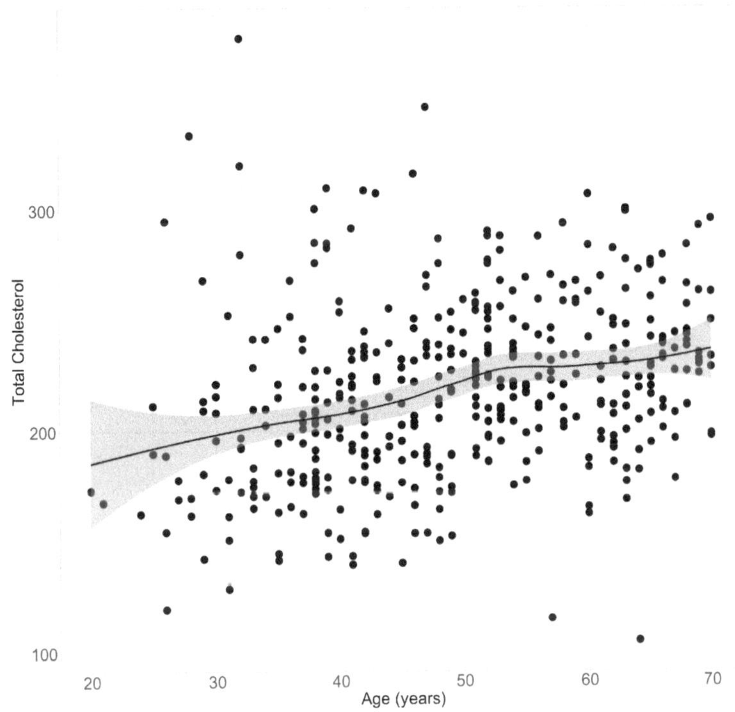

Figure 17.4 Scatterplot for total cholesterol and age with LOESS smoothing.

for each age and plot it against age, producing a plot of a "moving average" of cholesterol scores against age. Some statistical packages assist us in understanding the relationship by fitting curves called smoothers or fit lines to the data points in the scatterplot. There are many different types of smoothing functions available in statistical software, and a variety of options within each type.

An example of one kind of smoothing is provided by Figure 17.4, which displays the scatterplot for cholesterol and age with LOESS smoothing, using R. LOESS is a version of LOWESS (LOcal WEighted Scatterplot Smoothing; Cleveland, 1979; Cleveland, Devlin, & Grosse, 1988). For each value of X, LOWESS plots the Y score predicted by a procedure that gives more weight to data points near the value of X than data points that are further away (for details, see Cook & Weisberg, 1999). The resulting curve indicates a positive relationship between cholesterol and age that approximates a straight line. The gray band around the smoother indicates the 95% confidence interval for the smoother, conditional on the value of X; we will have more to say about intervals like this in Chapter 19.

In practice, we usually first look to see whether there is a systematic tendency for the variables to have a straight-line relationship because this is the simplest way that two variables can be related. Then we look further to see whether there are systematic departures from linearity. The most common numerical measures that are used to summarize the relationship between two quantitative variables are those that (1) indicate how well a straight line "captures" or summarizes the scatterplot, and (2) describe the straight line that gives the best fit.

17.4 Linear Relationships

A straight line can be represented by an equation of the form

$$Y = b_0 + b_1X$$

where b_0 and b_1 are constants because all points (X, Y) that satisfy this *linear equation* fall on a straight line. The constant b_1 is the *slope* of this line and indicates the rate of change of Y with X. We can see from the equation for a straight line that for every one-unit increase in X, Y changes by b_1 units. The constant b_0 is the *Y-intercept*, the value of Y when X is equal to zero. This equation describes a perfect linear relationship between X and Y. In fact, perfect linear relationships do not occur in behavioral data, but there are many occasions when a line provides a good approximation of the relationship between two variables.

Each of the scatterplots in Figure 17.5 contains about a dozen (X, Y) data points. If all the data points fall exactly on a straight line, we say there is a perfect linear relationship

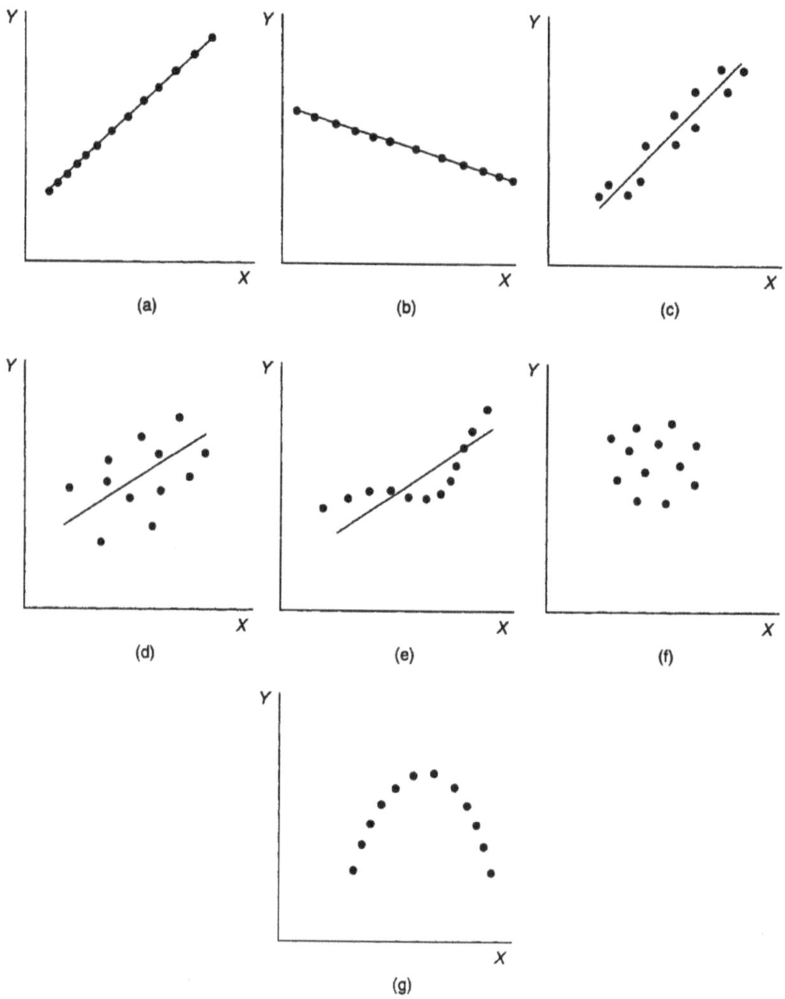

Figure 17.5 Examples of scatterplots.

between X and Y. This linear relationship is said to be positive if the slope of the line is positive; that is, if Y increases as X increases, as in Panel (a). The linear relationship is negative if Y decreases as X increases, as in Panel (b). In Panel (c), there is a systematic increase in Y as X increases. However, not all the data points fall on the straight line that seems to best capture this systematic increase, although they cluster closely around it. In this case, we say that there is a strong positive linear relationship between X and Y, but not a perfect one. In Panel (d), there is less clustering around the line, indicating a weaker linear relationship. In Panel (e), there is a linear component to the relationship between X and Y; that is, the best-fitting straight line seems to capture part of how Y and X are related. However, not only do the points fail to cluster closely around the line, but there also seems to be a systematic nonlinear component to the relationship. In Panels (f) and (g), there is no overall linear relationship between X and Y; no straight line passing through the center of either "cloud" of data points is better at characterizing the overall relationship between X and Y than a line parallel to the x-axis. In (g), however, X and Y are positively related for small values of X and negatively related for large values, whereas in (f) there does not seem to be any indication of a linear relationship for any part of the X distribution.

17.5 Introducing Correlation and Regression Using z Scores

17.5.1 Explaining the Pearson Correlation Coefficient

The Pearson correlation coefficient is a measure of the extent to which two quantitative variables are linearly related. We indicated in the previous section that the more tightly the data points are clustered around a straight line, the stronger the degree of linear relationship. However, if we plotted raw scores, the appearance of the scatterplot and the apparent degree of clustering around the best-fitting straight line would depend on the units in which X and Y are measured. This is not true for z scores, which have no units and are defined in terms of the relative standings of scores within their distributions. The correlation coefficient is a measure of similarity of the corresponding z scores of X and Y.

In Chapter 2, we indicated that each member of a set of scores $X_1, X_2, X_3, \ldots, X_n$ can be converted to a z score using

$$z_{X_i} = \frac{X_i - \bar{X}}{s_X} \tag{17.1}$$

where $\bar{X}$ is the mean of the set of scores and s_x is the standard deviation. The z score that corresponds to X_i tells us the number of standard deviations that X_i is above or below the mean of the distribution. We also showed that the mean of a complete set of z scores is zero, and that the standard deviation and variance both have a value of 1.

The Pearson correlation coefficient for two variables, X and Y, is given by

$$r_{XY} = \frac{1}{N-1}\sum_{i=1}^{N} z_{X_i} z_{Y_i} \tag{17.2}$$

The symbol r is used to denote the Pearson correlation coefficient in a sample, and ρ (the Greek letter rho) denotes the correlation in a population. The correlation coefficient is basically the average of the products of corresponding z scores (it would be exactly the average if we divided by N instead of $N - 1$ when we obtained the standard deviations of X and Y).

In Appendix 17.1, we show that if there is a perfect positive relationship between *X* and *Y*, then $z_y = z_x$. In this case, the correlation is

$$r_{XY} = \frac{1}{N-1}\sum_{i=1}^{N} z_{X_i} z_{Y_i} = \frac{1}{N-1}\sum_{i=1}^{N} z_{X_i}^2 = 1$$

This result follows from the fact that the variance of a set of *z* scores is 1 and the expression for this variance is

$$\frac{1}{N-1}\sum_{i=1}^{N}\left(z_{X_i} - \bar{z}_X\right)^2 = \frac{1}{N-1}\sum_{i=1}^{N} z_{X_i}^2$$

because $\bar{z}_X$, the mean of a set of *z* scores, is equal to 0. If there is a perfect negative relationship, $z_y = -z_x$, so that $r_{xy} = -1$. If there is no linear relationship between *Y* and *X*, there will be no systematic tendency for the products of the corresponding *z* scores, $Z_{X_i} Z_{Y_i}$, to be positive (when z_{X_i} and z_{Y_i} have the same sign) or negative (when they have opposite signs). Therefore, we would expect these products to more or less cancel out when averaged, so that r_{xy} is approximately equal to 0. In general, the value of *r* can range from −1 to +1; values of *r* close to the limits indicate strong linear relationships between *X* and *Y*, whereas values close to 0 indicate very weak linear relationships. Going back to the examples we introduced earlier in the chapter, for multiplication and subtraction accuracy for third-graders (Figure 17.1), $r = .59$; for multiplication accuracy and the time taken to answer (Figure 17.2), $r = -.49$; and for cholesterol level and age (Figure 17.4), $r = .29$.

Figure 17.6 shows the data for the final exam and pretest scores (Figure 17.3) using *z* scores rather than raw scores. In this figure, the origin (0, 0) falls at the mean of *X* and the

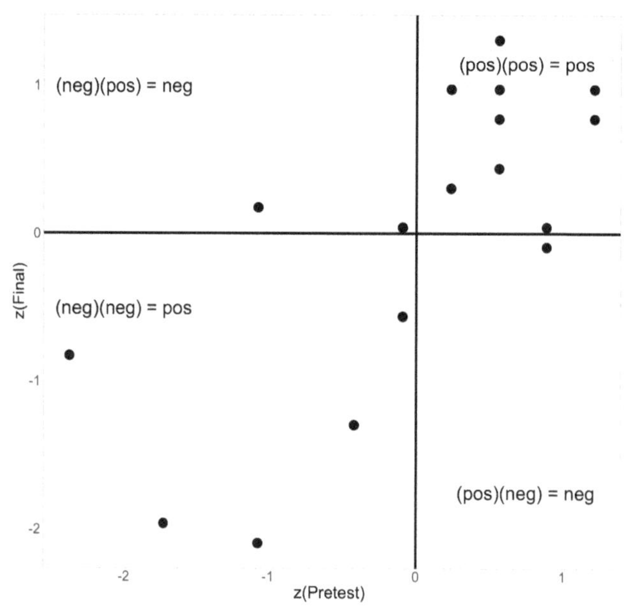

Figure 17.6 Scatterplot for z scores of the pretest and final exam scores.

mean of Y because the mean of a set of z scores is zero. Notice that most of the points fall in the upper-right and lower-left quadrants of the plot, where the signs of the z scores for X and Y agree (e.g., both positive or both negative). Points in those quadrants contribute positively to the sum in Equation 17.2, because their products are positive. In the other two quadrants (upper left, lower right), the z scores for X and Y have opposite signs (i.e., one negative and one positive); points in these quadrants subtract from the sum in Equation 17.2 because their products are negative. In all cases, points closer to the origin contribute smaller values to the sum of the cross-products, and those further from the original contribute larger values. Visualizing the data in this way may help you interpret the correlation coefficient; in this example, $r = 0.73$.

17.5.2 *The Sample Covariance*

The value of the Pearson correlation coefficient can always be obtained from Equation 17.2. However, other expressions for the Pearson correlation coefficient are often encountered. For example, if we substitute the expressions for z scores (Equation 17.1) into Equation 17.2, we get

$$r_{XY} = \frac{1}{N-1}\sum_{i=1}^{N}\left(\frac{X_i - \bar{X}}{s_X}\right)\left(\frac{Y_i - \bar{Y}}{s_Y}\right) = \frac{s_{XY}}{s_X s_Y} \tag{17.3}$$

The symbol s_{xy} in Equation 17.3 denotes the sample *covariance* of X and Y, which we can write as

$$s_{XY} = \frac{1}{N-1}\sum_{i=1}^{N}\left(X_i - \bar{X}\right)\left(Y_i - \bar{Y}\right) = r_{XY}s_X s_Y \tag{17.4}$$

The covariance is the *amount* of variance shared by X and Y.

The covariance provides information about the degree of linear relationship between X and Y. However, it is not usually employed as an index of strength of relationship because its value changes when we change the units of measurement of the variables. For example, if we measured the heights and weights of a number of people, and then found the covariance of height and weight, the numerical value of the covariance would be 12 times larger if we measured height in inches than if we measured it in feet. In contrast, the correlation coefficient would be the same in both cases. Correlation is the covariance of z scores.

17.5.3 *Using Software for Correlation and Covariance*

Software makes the calculation of correlations and covariances straightforward. In R, the *cor* and *cov* functions in the {stats} package calculate the correlation and covariance, respectively. For example, if the pretest and final exam scores from Table 17.2 are stored as variables in a data.frame called dat, then the command *cor*(dat$pretest, dat$final) returns 0.725. By default, a Pearson correlation is calculated; the *method* option can be used to select a different type of correlation (e.g., Spearman's *rho*, see Section 17.7.1). The covariance function is similar: *cov*(dat$pretest, dat$final) returns 33.3, which equals the value calculated from Equation 17.4: (.725)(3.06)(14.99) = 33.3.

In SPSS, choose *Correlate* from the *Analyze* menu, and then select *Bivariate*. Choose the variables of interest and click OK. For the data in Table 17.2, $r = 0.725$. To compute the covariance, begin as for correlation, then click on the "options" bar and tick the box for "Cross-product deviations and covariances," then click Continue and OK. Again, the result is 33.3.

17.5.4 Least-Squares Linear Regression Using z Scores

The correlation coefficient plays a major role in regression equations that predict one variable from another. Consider a set of N pairs of z scores $\left(z_{X_i}, z_{Y_i}\right)$. Suppose we wish to predict z_Y from z_X *by* a linear equation; that is, an equation of the form $\hat{z}_{Y_i} = a + bz_{X_i}$, where we indicate the predicted value by $\hat{z}_{Y_i}$, and a and b are the intercept and slope of the prediction line. How are we to determine the values of a and b that give the best predictions for the set of N paired z scores? We first must specify what we mean by "best predictions." In one sense, this is easy – we will consider "best predictions" to be those that produce the least amount of error.

But we must also specify an appropriate measure of error to minimize. Suppose for the ith prediction, our predicted value is $\hat{z}_{Y_i}$ when the actual value is Z_{Y_i}. Therefore, the ith prediction error is $z_{Y_i} - \hat{z}_{Y_i}$, and the mean error (ME) for the entire set of N predictions is given by

$$ME = \frac{1}{N}\sum_{i=1}^{N}\left(z_{Y_i} - \hat{z}_{Y_i}\right) = \frac{1}{N}\sum_{i=1}^{N}\left(z_{Y_i} - a - bz_{X_i}\right) \tag{17.5}$$

Although ME may seem like an intuitively appealing choice, it is not an appropriate measure of prediction error. For one thing, complete sets of z scores must sum to zero. Even if they did not sum to zero, positive and negative prediction errors would tend to cancel one another out, possibly producing small values of ME even when there are many large discrepancies between predicted and observed scores.

Other, more desirable candidates for a measure of error that does not allow for this type of "cancelling out" are the *mean absolute error* (MAE) and *mean squared error* (MSE), where

$$MAE = \frac{1}{N}\sum_{i=1}^{N}\left|z_{Y_i} - \hat{z}_{Y_i}\right| = \frac{1}{N}\sum_{i=1}^{N}\left|z_{Y_i} - a - bz_{X_i}\right| \tag{17.6}$$

$$MSE = \frac{1}{N}\sum_{i=1}^{N}\left(z_{Y_i} - \hat{z}_{Y_Y}\right)^2 = \frac{1}{N}\sum_{i=1}^{N}\left(z_{Y_i} - a - bz_{X_1}\right)^2 \tag{17.7}$$

Both measures are useful. The MSE is mathematically easier to work with and is the one used in introductions to regression. Regression equations obtained by minimizing the MSE are referred to as *ordinary least-squares* (OLS) regression equations. Because absolute deviations are less variable than squared deviations, the MAE is used in some developments of robust regression.

Using calculus, we can show that the *MSE* is minimized if $a = 0$ and $b = \frac{1}{N-1}\sum_{i=1}^{N} z_{X_i} z_{Y_i}$ that is, if the intercept is zero and the slope is the Pearson correlation coefficient, *r*. Therefore, the least-squares linear regression equation for predicting z_y from z_x is given by

$$\hat{z}_{Y_i} = r_{XY} z_{X_i} \tag{17.8}$$

The slope of the regression line for predicting one *z* score from another is called the *standardized regression coefficient*[2] or the *beta weight*. When we predict z_y from z_x the beta weight is the same as the correlation coefficient, r_{xy}.

If we stick with *z* scores and use the same set of data to find the least-squares linear equation that predicts z_x from z_y, we obtain the same equation with *X* and *Y* interchanged.

$$\hat{z}_{X_i} = r_{XY} z_{Y_i} \tag{17.9}$$

One of the many important differences between correlation and regression is that the Pearson correlation coefficient is *symmetric* in *X* and *Y*. That is, if we interchange *X* and *Y*, the expression for *r* is unchanged and the interpretation remains exactly the same – the correlation between *X* and *Y* is exactly the same as the correlation between *Y* and *X*. There is no such symmetry when we consider regression. Although the beta weight is r_{xy} in both Equations 17.8 and 17.9, these equations represent different regression lines.

Consider Panel (a) of Figure 17.7. Suppose that there is a moderate linear relationship between z_y and z_x and that the elliptical envelope in the diagram contains a large number of standardized data points (i.e., paired values of z_x and z_y). In the figure, the envelope is symmetrical about a straight line with a slope of 1 drawn through the origin; that is, the line with the equation $z_y = z_x$. Now consider a series of vertical strips drawn through these data points; Panel (a) shows two of these strips. Each strip can be thought of as containing data points that have roughly the same value of z_x but different values of z_y. Now imagine finding the mean value of the z_ys in each of these vertical strips – we would expect the mean to fall at about the middle of each strip. The least-squares regression line for predicting z_y from z_x is approximately the line that connects these means. As we can see from the figure, the resulting regression line, $\hat{z}_{Y_i} = r_{XY} z_{X_i}$, has a slope less than 1 unless $z_y = z_x$.

To consider what happens when we predict z_x from z_y, look at Panel (b) of Figure 17.7, which has the same data display as in Panel (a). The regression line for predicting z_x from $z_Y, \hat{z}_{X_i} = r_{XY} z_{Y_i}$ is approximately the line that passes though the means of the *horizontal* strips that are drawn through the elliptical cloud of data points. As we can see by comparing Panels (a) and (b), the two regression lines are different unless there is a perfect linear relationship. The reason for this has to do with the topic we consider next.

17.5.5 Regression Toward the Mean When Predicting z_y from z_x

From Equation 17.8 and Figure 17.7, we can see that unless there is a perfect linear relationship, the magnitude of $\hat{z}_{Y_i}$ will be smaller than that of Z_{X_i}. If there is a perfect positive linear relationship, $\hat{z}_{Y_i} = z_{X_i}$; if there is no linear relationship, $\hat{z}_{Y_i} = 0$, the mean of the z_y scores. This is because in the absence of any other information, it can be shown

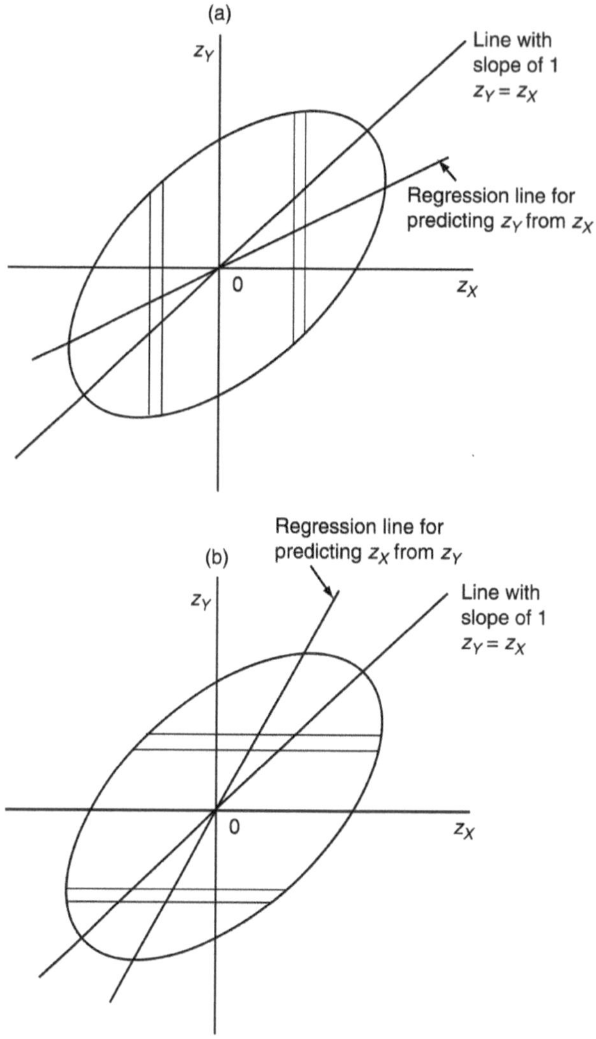

Figure 17.7 Regression lines for predicting (a) z_y from z_x and (b) z_x from z_y when there is an imperfect relationship between X and Y.

that the best least-squares predictor for a set of scores is their mean, and the mean z score is 0. Therefore, if there is a moderate positive linear relationship, $\hat{z}_{Y_i}$ should have a value somewhere between 0 and Z_{X_i}. This tendency of predicted z scores to be less "extreme" than the z scores used to predict them is called *regression toward the mean* and is a characteristic of bivariate regression whenever two variables have equal variances and do not have a perfect linear relationship. The prediction equations we have been discussing are given the name "regression equations" in recognition of this characteristic. We will consider this issue in more detail when we discuss raw-score regression in the next section.

17.6 Least-Squares Linear Regression for Raw Scores

17.6.1 *Predicting Y from X*

We have already discussed predicting z_y from z_x using the regression equation $\hat{z}_{Y_i} = r_{XY} Z_{X_i}$. However, we are often more concerned with finding the raw-score equation $\hat{Y}_i = b_0 + b_1 X_i$ that best predicts Y from X. We can do this by noting that

$$z_{X_i} = \frac{X_i - \bar{X}}{s_X} \text{ and } \hat{z}_{Y_i} = \frac{\hat{Y}_i - \bar{Y}}{s_Y}$$

and then substituting into Equation 17.8 and solving for $\hat{Y}_i$. Alternatively, we can use calculus to derive formulas for the slope and intercept that minimize the measure of prediction error, $MSE = \frac{1}{N}\sum_i \left(Y_i - \hat{Y}_i\right)^2$ (see, for example, Myers & Well, 2003). Either way, we find

$$\hat{Y}_i = \bar{Y} + r\frac{s_Y}{s_X}\left(X_i - \bar{X}\right) \tag{17.10}$$

so that the slope and intercept for the regression of Y on X are given by

$$b_1 = r\frac{s_Y}{s_X} \tag{17.11}$$

$$b_0 = \bar{Y} - b_1\bar{X} \tag{17.12}$$

We can visualize the operation of minimizing the average of the squared prediction errors to obtain the best-fitting regression line. Figure 17.8 (a) repeats the scatterplot of third-grade multiplication and subtraction accuracies (Figure 17.1), this time with a prediction line superimposed on the plot. We are predicting multiplication accuracy from subtraction scores in these children, so the prediction errors are the differences between the observed multiplication scores and the predicted values on the prediction line. Two of these errors are shown with gray arrows pointing from the observed value to their predicted values on the prediction line; the length of the arrow indicates the magnitude of the prediction error for those two points. We squared each prediction error, and the results are shown with light-gray squares. These squares show just two of the squared prediction errors that contribute to the average squared prediction error; in fact, every point has a squared error. Some squared errors are small, and some are larger. The prediction line that minimizes the average size of these squared errors (literally, the squares in Figure 17.8a) is the regression line described by Equations 17.10–17.12.

Applying Equations 17.11 and 17.12 to the third-graders' arithmetic data shown in Figure 17.1 produces the linear regression equation that best predicts multiplication accuracy (Y) from subtraction accuracy (X),

$$\hat{Y} = 1.63 + .88\,X$$

A difference of one point on the subtraction test translates into a predicted difference of 0.88 points on the multiplication exam. The prediction for the multiplication score of a

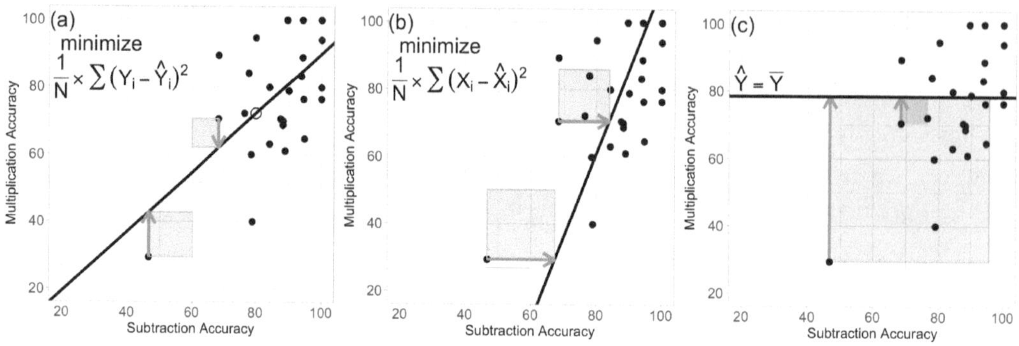

Figure 17.8 Graphical representation of (a) the regression of *Y* on *X*; (b) the regression of *X* on *Y*; and (c) ignoring *X* when predicting *Y*.

student who scored 80 on the subtraction test can be found by substituting 80 for *X* in the preceding equation. Our predicted multiplication score 1.63 + (.88)(80) = 72.03, or 72, rounding to the nearest integer; this point is marked with an open circle in Figure 17.8(a). Although this equation was developed in a data set in which we already knew both the subtraction and multiplication scores for each student, it serves as a useful way to represent how the two variables are related. Moreover, the equation can be useful in predicting the performance of students in future classes.

In contrast to what happens in correlation, *X* and *Y* play different roles in the regression of *Y* on *X*: *X* is the *predictor variable* and *Y* is the *criterion* or *dependent variable*, the variable that is predicted. In the general case where $s_x \neq s_y$, the regression equation and the correlation coefficient tell us different things about the linear relationship. The regression equation describes the straight line that is best for predicting *Y* from *X*, whereas the correlation coefficient serves as a measure of how strongly *Y* and *X* are linearly related. If we solve for *r* in Equation 17.11, we get

$$r = b_1 \frac{S_X}{S_Y} \tag{17.13}$$

From this equation we can see that the same correlation may arise from different combinations of the slope and the standard deviations of *X* and *Y*. For example, both of the combinations $b_1 = 1$, $s_x = 3$, $s_y = 5$ and $b_1 = .5$, $s_x = 6$, $s_y = 5$ correspond to *r*s of .6. Because of this, we must be extremely cautious if we wish to compare the relationships between *X* and *Y* in different groups. Two groups that have the same regression slope may have different correlations, and two groups that have the same correlation may have different slopes. If we are primarily concerned with whether the rate of change of *Y* with *X* differs in two groups, we should compare the slopes, not the correlation coefficients.

17.6.2 *Predicting X from Y*

So far, we have discussed the regression equation for predicting *Y* from *X* that is optimal in the sense that it minimizes $\Sigma(Y - \hat{Y})^2/N$ (see Panel (a) of Figure 17.8). Exactly the same reasoning can be used to find the regression equation for predicting X from Y. In this case, the index of error that is minimized is $\Sigma(X - \hat{X})^2/N$, the mean of the squared residuals when Y is used to predict X. Two of these prediction errors are indicated by the gray arrows in Panel (b) of Figure 17.8, and their squares are shown in light gray. Notice that the squared errors in Panel (b) are not the same as those in Panel (a).

The expressions that have been developed for predicting Y from X can be transformed into expressions for predicting X from Y by simply interchanging X and Y in Equation 17.13. The regression equation becomes.

$$\hat{X}_i = \bar{X} + r\frac{s_X}{s_Y}\left(Y_i - \bar{Y}\right)$$

Of course, whether it makes any sense to predict X from Y depends on the situation. We would have little reason to predict subtraction scores from multiplication scores, because subtraction is taught first.

17.6.3 Regression Toward the Mean in Predicting Y from X

We showed in Section 17.5.4 that whenever there is an imperfect linear relationship between X and Y, the values of z_y associated with any given value of z_x will, on the average, be closer to zero than z_x is. We can extend this result to raw scores by rewriting Equation 17.10 as

$$\hat{Y}_i - \bar{Y} = r\frac{s_Y}{s_X}\left(X_i - \bar{X}\right) \qquad (17.14)$$

Here we see that the best prediction for Y_i must be closer to $\bar{Y}$ than X_i is to $\bar{X}$ whenever $s_x = s_y$ and r is less than 1.

Regression toward the mean must occur for predictions whenever two variables with equal variances are not perfectly linearly related. However, the finding that predicted scores are less extreme than those used to do the predicting often leads to faulty reasoning. Because people often ignore the fact that regression effects can be produced by random variability, measurement error, or any other factors that result in less-than-perfect correlation, they often feel compelled to provide more "interesting," elaborate, and often unwarranted explanations.

We must keep the possibility of regression toward the mean in mind both when designing research and when considering many everyday life phenomena. Suppose we are interested in assessing the effects of some kind of instruction in children. Say, a group of children take a multiple-choice test, and then are given a period of instruction, followed by a second test. Usually, children who score near the top of the distribution on the first test will tend to show some decline in relative performance on a second test (just think what it takes to score at the top of the distribution), whereas children with extremely low scores on the first test will, on average, score relatively better on the second. This pattern seems to suggest that the intervening instruction is more effective for the students who scored lower on the first test than for those who scored higher. However, even if instruction was equally effective for all children, we would expect regression toward the mean if the two tests did not measure exactly the same thing or if there were random fluctuations in children's alertness and success in guessing. Therefore, if we want to assess the effects of instruction, we must include an appropriate control group and conduct an experiment.

Another illustration is the so-called sophomore slump. Individuals who perform particularly well in their first year of academics or athletics will, on average, perform relatively less well in their second year. It is usually thought that this finding requires an explanation in terms of overconfidence and poor work habits that result from early success. However, such explanations may be unnecessary because even without any interventions or distractions, an individual's performance will generally fluctuate so that there are periods of better-than-usual performance and periods of worse-than-usual performance followed by performance that is more normal for the individual. This has far-reaching implications: Fluctuations in children's behavior may cause parents to overestimate the effectiveness of interventions such as punishment. Fluctuations in the intensity of symptoms in sufferers of many chronic diseases may cause patients to overestimate the effectiveness of alternative medicine and food supplements. Fluctuations in crime rates may cause observers to overestimate the effects of changes in social policy or state laws.

Another important variation on this theme has to do with the misclassification of individuals on the basis of test results. For example, people are classified as being depressed, or as having a high cholesterol level, on the basis of tests that are not perfectly reliable. Because the test scores are subject to regression effects, we would predict that many of these classifications would change if a second test was given. Suppose being classified as having high cholesterol depends on scoring two or more standard deviations above the mean on a test. If we can assume normal distributions and that the results of a second test correlate .80 with those of the first, it can be shown that only about 43% of those individuals classified as having high cholesterol on the first test will be classified as having high cholesterol on the second (see Campbell & Kenny, 1999).

We conclude this section by noting that regression toward the mean is inevitable only if the variances of both variables are equal. As we can see in Equation 17.14, regression effects must occur for raw scores if

$$|r|\frac{s_Y}{s_X} < 1$$

However, if $s_y > s_x$, regression toward the mean need not always occur for raw scores; in fact, if $|r|\, s_y / s_x > 1$, we will have the opposite effect, which has been called egression from the mean (Ragosa, 1995). Although there are many situations in which regression effects occur for raw scores, there are others in which these effects occur for standardized, but not for raw, scores.

17.6.4 The Coefficient of Determination

The square of the correlation coefficient, r^2, is called the *coefficient of determination* and is a commonly encountered measure of strength of linear relationship. The r^2 measure is usually described as *the proportion of the variance in Y accounted for, or explained by, X* (or, because the measure is symmetric in X and Y, as the proportion of the variance in X accounted for by Y). This language sounds impressive and vaguely causal, but all it means is that r^2 is the proportion by which the squared prediction error is reduced if the regression equation is used to predict the Y scores instead of using $\bar{Y}$ to predict each of the Ys. Figure 17.8 (c) depicts this situation and shows two example squared prediction errors, but, as in Panels (a) and (b), each point has its own squared prediction error. Notice that the squared prediction error for the left-most point is much larger than in Panels (a) and (b); on average, the squared errors are smaller in Panel (a) than Panel (c), reflecting the fact that X contains information about the likely value of Y. The detailed interpretation is as follows:

1. If we do not use any information about X when predicting the corresponding value of Y, the best least-squares prediction for each Y can be shown to be $\bar{Y}$, the mean of the Y scores. If we use $\bar{Y}$ as the prediction for each of the Y scores, the sum of the squared prediction errors for the entire set of N predictions is the total variability in the Y scores, the sum of squares of Y, $SS_y = \Sigma(Y_i - \bar{Y})^2$. That is, SS_y is literally the sum of all the squared prediction errors in Figure 17.8 (c).
2. If we do use the information about X and predict using the regression equation, as in Figure 17.8 (a), the sum of the squared errors for the set of predictions is $SS_{residual} = \Sigma(Y_i - \hat{Y}_i)^2$. Substituting the expression for $\hat{Y}_i$ from Equation 17.10 and simplifying, the sum of the squared residuals can be shown to be

$$SS_{residual} = \left(1 - r^2\right)SS_Y \tag{17.15}$$

3. The *amount* by which prediction error is reduced when the regression equation is used instead of the mean is, therefore,

$$\begin{aligned} SS_{regression} &= SS_Y - SS_{residual} = SS_Y - \left(1 - r^2\right)SS_Y \\ &= r^2 SS_Y \end{aligned} \tag{17.16}$$

4. Therefore, the *proportion* by which prediction error is reduced (or the proportion of the variability in Y accounted for by the regression on X) is

$$\frac{SS_{regression}}{SS_Y} = \frac{r^2 SS_Y}{SS_Y} = r^2$$

The coefficient of determination, r^2, is therefore a measure of how well the linear regression equation fits the data. According to the Cohen (1977, 1988) guidelines, r^2 values of .01, .09, and .25 correspond to small, medium, and large linear relationships, respectively. For the arithmetic data (Figure 17.1), the correlation between the subtraction and multiplication accuracy is .594, so the coefficient of determination is $(.594)^2 = .353$. This tells us that the variability of the Y scores about the regression line is $(1 - .353) = .647$ of their variability about $\bar{Y}$. Therefore, if we use the regression equation to predict Y instead of using $\bar{Y}$, we will reduce the squared prediction error by about one-third.

We should note that r^2 has frequently been misinterpreted, and that some of these misinterpretations have resulted in inappropriate claims being made in the literature. For example, the statement has been made in a number of psychology textbooks that children achieve about 50% of their adult intelligence by age 4. There is no valid basis for this statement; it is based on a misunderstanding of r^2. Specifically, it is based on the results of several longitudinal studies that found IQ scores at age 17 to have a correlation of about .7 with IQ scores at age 4 (see Bloom, 1964). The resulting r^2 of about .5 (or 50%) provides an indication of how predictable adult IQ is from IQ at age 4, using linear regression. However, this result says nothing about the relative *amounts* of intelligence at age 4 and age 17, and therefore provides no evidence for the statement.

17.6.5 Partitioning of Variability in Regression and the Standard Error of the Estimate

Equation 17.16 indicates how the variability in the Y scores can be partitioned into two components – the variability of the predicted scores about the mean and the variability of the actual Y scores about the regression line. Rewriting Equation 17.16, we have

$$\begin{aligned} SS_Y &= SS_{regression} + SS_{residual} \\ &= r^2 SS_Y + \left(1 - r^2\right) SS_Y \qquad (17.17) \\ \Sigma\left(Y_i - \bar{Y}\right)^2 &= \Sigma\left(\hat{Y}_i - \bar{Y}\right)^2 + \Sigma\left(Y_i - \hat{Y}_i\right)^2 \end{aligned}$$

Among other things, this helps us understand an important measure of error of prediction in regression called the *standard error of the estimate*. Although r^2 is a commonly used measure of strength of relationship, measures such as the *variance of the estimate* (basically, the mean of the squared raw-score prediction errors) or its square root, the *standard error of the estimate*, provide more useful information than either r or r^2 about the accuracy of predictions based on the raw-score regression equation. The standard error of the estimate is the most commonly used measure of variability of the data points around the regression line and is typically provided as part of the regression output by software packages. For the regression of Y on X, the standard error of the estimate is defined as

$$s_{Y.X} = \sqrt{\frac{\Sigma\left(Y_i - \hat{Y}_i\right)^2}{N-2}} = \sqrt{\frac{SS_{residual}}{N-2}} = \sqrt{\frac{\left(1-r^2\right)SS_Y}{N-2}} \qquad (17.18)$$

In Equation 17.18, there are $N - 2$ *df* associated with $SS_{residual}$ because there are N data points and two restrictions; that is, two *df* are used up by estimating the intercept and slope of the regression equation. We will encounter the standard error of estimate again when we develop hypothesis testing and confidence interval estimates for regression statistics.

17.6.6 Using Software for Least-Squares Linear Regression

Least-squares linear regression is commonly available in software packages. In SPSS, select *Regression* from the *Analyze* menu, and then choose *Linear*. Move Y to the dependent variable box and X to the independent variable box, and then click OK. The unstandardized coefficients in the output correspond to the b_0 and b_1 of Equations 17.11 and 17.12, and the standardized coefficient is the slope (i.e., correlation) from Equation 17.9. Note that r, r^2, and the standard error of the estimate are also presented in the "model summary" section of the output.

In R, we can use the *lm* function in the {stats} package to fit a linear model. For example, if Y and X are two variables in a data frame called dat, we can regress Y on X using the command *lm*(Y ~ X, data = dat). The output reports the values of b_0 and b_1. We can request a summary of that linear model, like this: *summary*(*lm*(*Y* ~ X, data = dat)). In that case, the output provides additional information, such as r^2 and the standard error of the estimate, which are called "Multiple R-squared" and "Residual standard error." To calculate the $SS_{regression}$ and $SS_{residual}$, use the *anova* function: *anova*(*lm*(*Y* ~ X, data = dat)).

17.7 More About Interpreting the Pearson Correlation Coefficient

Now that we have introduced linear regression and concepts such as the standard error of the estimate, we are in a position to discuss the interpretation of the correlation coefficient in more detail. It is important to understand the characteristics of correlation coefficients and the factors that influence them because correlations seem to be the statistics of choice in some research areas. Moreover, they often serve as the raw material for more complex procedures such as factor analysis.

17.7.1 Ten Things to Remember About Correlation

1. *The Pearson correlation coefficient depends on the variabilities of* X *and* Y. Because the correlation coefficient is a measure of the similarity of the standardized values of *X* and *Y* (see Figure 17.6), s_x and s_y affect *r* because they influence the standardization of *X* and Y. From Equation 17.13, we have

   $$r = b_1 \frac{s_X}{s_Y}$$

   This means that for a given regression slope, different correlations will result from different standard deviations of *X* and *Y*. Even if the values for both the slope and standard error of estimate, $s_{y.x}$, are held constant, Equations 17.13, 17.17, and 17.18 can be used to show that different values of *r* will result for different values of s_x and s_y. For example,

   - For a given b_1 and $s_{y.x}$, *r* increases with s_x.
   - For a given $s_{y.x}$, *r* increases with b_1, s_x, and s_y.
   - For a given s_x and $s_{y.x}$, *r* increases with b_1 and s_y.

   Therefore, if we are concerned with *how* Y changes with X, we should generally look at b_1, not *r*.

2. *Because the correlation coefficient depends on* s_x and s_y, *it is a sample-specific measure.* Given a population in which *Y* and *X* are linearly related with a population correlation of ρ, the value of *r* obtained in a sample will depend on how the sample is selected. If the sample is largely selected from only part of the distribution of *X* (say, any of the regions A, B, or C in Figure 17.9), *r* will tend to be smaller than ρ because s_x will tend to be smaller than the value of σ_x in the population. Note that this dependence on the variability of *X* in the sample does not hold for the unstandardized estimate of slope, β_1.

   This type of bias for correlations is frequently referred to as the *restriction of range* problem; all other things being equal, the sample correlation will be smaller if there is less variability in X. A frequently cited example of restriction of range is the low correlation between Graduate Record Examination (GRE) scores and success in graduate school as measured by grades or faculty ratings (e.g., Dawes, 1971), which was one of several factors contributing to the recent abandonment by many universities of the

GRE scores as predictive measures. However, even if GRE scores were an excellent measure of ability, the correlation with performance would be expected to be quite low because of the restricted range – most students admitted to graduate programs have relatively high GRE scores.

Conversely, if we select a sample that overrepresents low and high values of X (i.e., is largely selected from regions A and C in Figure 17.9), r will tend to be larger than ρ. Investigators interested in the relationship between two variables will sometimes drop the middle scores for one or both of them. Although this procedure may be acceptable for determining *whether* there is a linear component to the relationship, the correlation in this type of sample should definitely not be considered to be an estimate of the value of the correlation in the whole population.

This dependence on sample variability raises the question of what comparisons of correlations from different samples actually mean. Two correlations may be different not because the samples differ in the variability about the regression line or because the slopes are different, but merely because one of the samples has a broader range of X values. Because the correlation coefficient is sample specific and generally does not provide a description of the linear relationship between X and Y, some authors (e.g., Achen, 1982; Tukey, 1969) have recommended that the correlation coefficient not be used.

We agree with Achen and Tukey that characteristics of the regression equation such as the intercept, slope, and standard error of the estimate describe a linear relationship more usefully than does r. Further, researchers should often be interested in the rate of change of Y with X (i.e., with b_1) even when they think they are interested in r. However, we do not believe that the correlation coefficient can or should be abandoned; instead, it should be used with an understanding of exactly what it does and does not measure.

3. *Size*. How large must a correlation coefficient be in order to indicate that there is a "meaningful" linear relationship? Cohen (1977, 1988) has discussed guidelines according to which rs of .10, .30, and .50 correspond to small, medium, and large effects.

Figure 17.9 Scatterplot with different regions marked to illustrate the effect of the variability of X and Y on the correlation coefficient.

Cohen arrived at these values by noting the sizes of correlations encountered in the behavioral sciences and by considering how strong a correlation would have to be before the relationship could be perceived by an observer. These values should be considered only as very rough guidelines and not as criteria for importance. In some contexts, even small correlations might be of great practical significance. We should also emphasize that unless the sample is large, the correlation may be quite different in the sample than in the population from which the sample was selected.

4. *Symmetry*. We have already noted that expressions for the correlation coefficient are *symmetric* in X and Y. The correlation between cholesterol level and age is the same as the correlation between age and cholesterol level. This is not the case for regression equations (see Figure 17.8). The equation for predicting Y from X will be the same as that for predicting X from Y only if (a) there is a perfect linear relationship between X and Y and (b) X and Y both have the same mean and standard deviation – that is, if $Y = X$.
5. *Linearity*. The Pearson correlation coefficient is a measure of strength of the *linear* relationship between X and Y. It is not a measure of relationship in general because it provides no information about whether there is a systematic nonlinear relationship between the two variables. As can be seen in Panels (e) and (g) of Figure 17.5, two variables can have a systematic curvilinear component to their relationship in addition to, or instead of, a linear one. Therefore, finding a correlation coefficient of zero does not

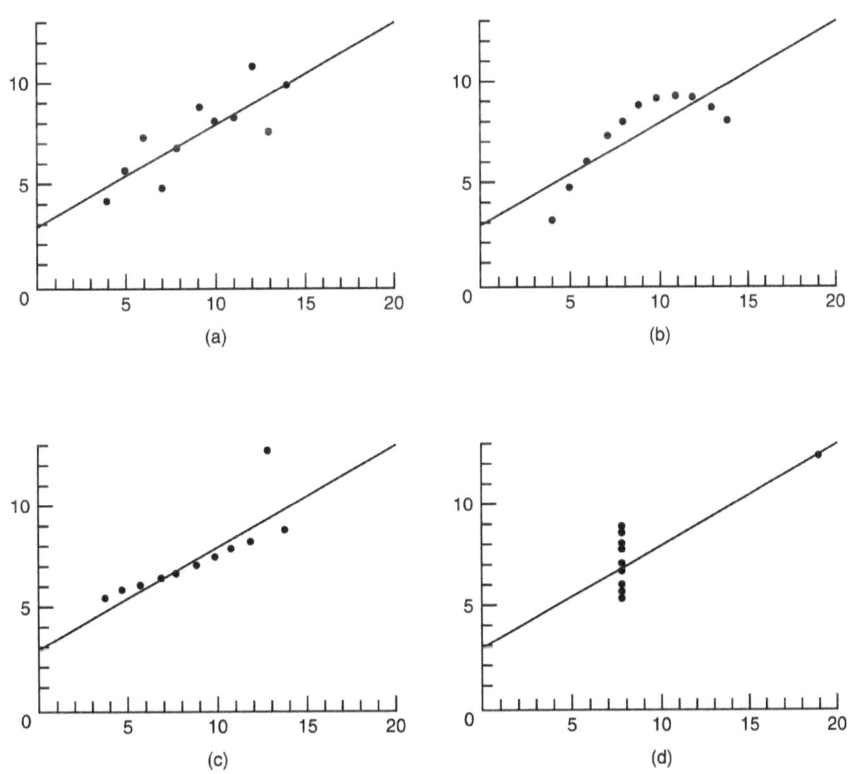

Figure 17.10 Scatterplots for four hypothetical data sets from Anscombe (1973).

necessarily mean that the variables are independent. The data points in all four panels of Figure 17.10 have identical correlations and best-fitting straight lines. For each of the four panels, $r = .82$, and the slope of the least-squares regression line that predicts Y from X is 0.5. However, whereas Panel (a) displays a moderate linear relationship with no curvilinear component, Panel (b) displays a systematic curvilinear relationship that has a linear component. It cannot be emphasized strongly enough that if you want to understand how variables are related, you must plot them and not simply rely on statistics such as the correlation coefficient or the slope of the best-fitting regression line.

6. *What happens when units of measurement are changed?* The Pearson correlation coefficient, in contrast to other statistics such as the mean, standard deviation, covariance, and the unstandardized regression coefficient, is a *dimensionless* quantity. By this we mean that r is not expressed in terms of any units; it is simply a number between −1 and +1. Remember that r is the average product of pairs of z scores, and that a z score expresses in standard deviation units the location of a score in a distribution. The values of the z scores and thus of r do not change if we change the units in which we measure X and/or Y (i.e., if we multiply each score by a constant and/or add a constant to each score, as when we change units from ounces to kilograms, or from degrees Fahrenheit to degrees centigrade). It follows that *just knowing the correlation between two variables tells us nothing about the mean or variance of either of the variables.* In contrast, changing units does affect b_1 and the raw-score variability about the best-fitting lines.

   Unfortunately, the fact that the correlation coefficient has no units is one possible reason for its popularity. When we use r or r^2 or any other standardized measure of effect size, we do not think in terms of units and therefore may not pay close attention to our measures. What does a one-point difference on a seven-point anxiety scale really mean? And does the difference between 6 and 7 on the scale mean the same thing as the difference between 3 and 4? If we do not understand our measures, we will not recognize when they are not good measures of the underlying variables that are our real concern. As Tukey (1969, p. 89) puts it,

   > Given two perfectly meaningless variables, one is reminded of their meaninglessness when a regression coefficient is given, since one wonders how to interpret its value. A correlation coefficient is less likely to bring up the unpleasant truth – we think we know what $r = -.7$ means.

7. *The correlation coefficient is not resistant to the effects of outliers; it is not robust.* The value of the correlation coefficient can be greatly influenced by the presence of extreme data points. Therefore, if there are extreme data points, it is important to identify them and, if possible, determine why they are extreme. Consider the large size of the squared prediction error for the left-most point in Figure 17.8(a). If we omit this one data point, the correlation drops from .59 to .39. There are measures of correlation that are more resistant than the Pearson coefficient because they diminish the importance of extreme scores. An example is the *Spearman rho coefficient* for which the X and Y scores are first ranked, and then the ranks are correlated. We will have more to say about such measures in the next chapter.
8. *Correlation does not imply causation.* Because correlation is a measure of strength of relationship, it is tempting to consider the correlation coefficient as a measure of the extent to which changes in X *cause* changes in Y. However, *no statistic implies causation.* A correlation between two variables means that one is useful in predicting the

other, but it does not necessarily mean that they are causally related. For example, the fact that in elementary school there is a positive correlation between shoe size and verbal ability does not mean that foot growth *causes* enhanced verbal ability, or vice versa. Rather, in this example the correlation follows from the fact that both physical and mental growth occur as children get older. It is impossible to determine causality without conducting a true experiment in which an independent variable is manipulated. However, despite repeated warnings, the tendency to lapse into causal thinking when dealing with correlation and regression is widespread. We will have much more to say about this as we continue our presentation of regression.

9. *Measurement error*. Because there is always error in measuring X and/or Y, r_{xy} underestimates the "true" correlation coefficient that would be obtained if X and Y could be measured without error. This should not be surprising. If our data are contaminated by large amounts of measurement error, they can hardly be expected to reveal strong systematic relationships.
10. *The shapes of the X and Y distributions constrain the possible values of r*. The marginal distributions of X and Y place constraints on the possible values of the correlation between X and Y. If both X and Y have identical, symmetrical distributions, it is possible that any value of r from −1 to +1 might occur, depending on how the values of X and Y are paired. However, if X and Y have distributions that are different from one another, or if one or both of the distributions are asymmetric, the full range of correlations from −1 to +1 cannot occur, no matter how the values of X and Y are paired.

    In Figure 17.11, distribution (a) is positively skewed and distribution (b) is negatively skewed. If larger scores in (a) are paired with smaller scores in (b), and vice versa, it is possible to obtain a correlation of −1; however, it is not possible to obtain a correlation of +1. For one thing, there are no scores in (b) that have positive z scores as large as those in the upper tail of (b). For another, if we attempted to pair larger scores in (a) with those in (b), it would soon become apparent that there are not enough large scores in (a) to match up with those in (b). If we did the best we could (i.e., paired off scores that had the same rank order), the scatterplot would show that we had a curvilinear relation with a correlation less than +1. Similarly, if we had two variables whose marginal distributions were both positively skewed as in (a), or were both negatively skewed as in (b), it would be possible to have a correlation of +1, but not one of −1. Because of these constraints, we should plot the univariate distributions of X and Y as well as the scatterplot when we try to understand the relationship between X and Y.

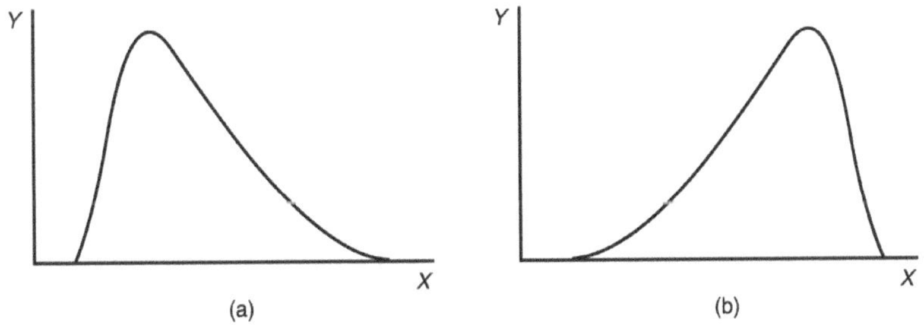

Figure 17.11 Two skewed distributions to illustrate how the distributions of X and Y constrain the possible values of the correlation coefficient.

17.7.2 Concerns About Combining Data Across Groups

When data from different groups are combined, the correlation and regression statistics for the resultant data set may not characterize the relationship between *X* and *Y* in any of the groups. The problem occurs because the aggregate statistics reflect not only the relationship between *X* and *Y* within the different groups, but also the differences among the group means. Suppose, for example, that we were interested in finding the correlation between height and weight. Because people who identify as male tend to be both taller and heavier than people who identify as female, we would expect a situation like that depicted in Panel (a) of Figure 17.12, in which the correlation would be larger for the combined group than for either male- or female-identifying individuals. This expectation is confirmed if we consider the *Seasons* data set. Here, the correlation between height and weight is .29, both for men and for women considered separately. However, if we combine the data across gender identities, the correlation is .53.

As another example, if we correlated hours worked with weekly pay, we would expect positive correlations for individuals with and without college degrees. However, college graduates tend to be paid at higher hourly rates than those without a degree, resulting in higher weekly pay for the same or fewer hours worked. This situation is like that depicted in Panel (b) of Figure 17.12, in which the correlation would now be lower in the combined group than for either group considered separately. In extreme cases, it is conceivable that *X* and *Y* could be positively correlated in each of several groups but negatively correlated when the groups were combined, as in Panel (c), or negatively correlated in each group but positively correlated when combined, as in Panel (d). The message is that we must be very cautious when combining data from meaningful subgroups.

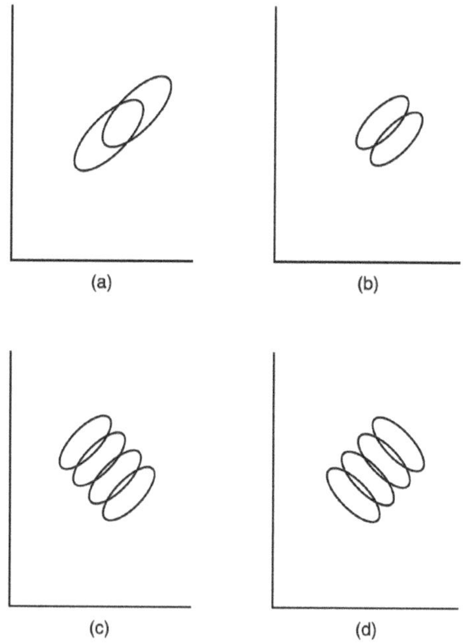

Figure 17.12 Scatterplots illustrating how combining the data from different groups can affect the correlation coefficient.

The summary statistics for the combined data set may not only fail to describe each of the constituent groups appropriately, but they may also fail to describe *any* of the groups appropriately. This is important to keep in mind, not only when dealing with correlation, but also when considering more complicated analyses that may take correlations or covariances as inputs, such as factor analysis and structural equation modelling. If the correlations do not adequately characterize the subgroups making up a data set, neither will quantities based on these correlations, such as the factors obtained in a component or factor analysis. Of course, this caution does not apply only to correlation coefficients; it applies to any statistic. For example, the regression slope would be positive for each of the groups represented in Panel (c) of Figure 17.12 but would be negative for the combined data set.

17.7.3 Ecological Correlations: Correlations Based on Rates or Averages

Researchers sometimes compute the mean of X and Y scores for each of several groups, and then correlate the averages to assess the relationship between X and Y. For example, suppose that for the *Royer* data set we compute the average subtraction accuracy score and the average multiplication accuracy score for the third-, fourth-, fifth-, and sixth-graders. We then correlate the four pairs of means and find that $r = .82$. However, if we compute the correlation of the subtraction and multiplication scores within each grade, we get rs of .59, −.18, .23, and .43, for the third through sixth grades, respectively. The point illustrated here is that a correlation based on group *averages* may not tell us anything useful about the correlations based on the *individuals* within the groups.

There are two reasons for discrepancies between correlations based on group averages and correlations based on individuals within groups. For one thing, the group means do not convey information about the within-group variability. For another, the factors that cause the variability across the groups may not be the same ones responsible for the within-group variability. For groups such as those represented in Panel (c) of Figure 17.12, we could have strong positive correlations in each of the groups, yet the correlation based on the group means could be strongly negative. Again, correlations between group averages do not communicate the relationship between X and Y between individuals within the groups.

17.8 What About Nonlinear Relationships?

So far, we have focused on describing the linear component of the relationship between two variables and on measures of its strength. Indeed, correlation coefficients and regression slopes are by far the mostly commonly reported measures of relationship for quantitative variables. This is reasonable, given that an approximately linear relationship is the simplest way that two variables can be related. However, there will certainly be situations when it is apparent from the scatterplot and the smoothers that the relationship has a nonlinear component. How are we to describe and measure the strength of this component or to describe the overall function that best seems to fit the data? We will answer this question when we discuss multiple regression in Chapter 20.

17.9 Concluding Remarks

When we explore how two quantitative variables are related in a sample of data, the first step is to plot the data and look at both the scatterplot and the univariate distributions.

Inspecting the scatterplot by eye and using smoothers or fit lines, we can try to extract, and describe, any underlying systematic relationship. We can use the Pearson correlation coefficient as a measure of the strength of linear relationship and the regression equation as a description of the straight line that allows the best prediction of one variable from the other. We can then try to determine whether there are any systematic departures from linearity, and if so, we can describe them. When we get to multiple regression, we will discuss how to fit various kinds of functions to the scatterplot.

We must also consider the possibility that the processes that determine how the variables are related may not be the same for all cases. Separate clusters of points may suggest the presence of subpopulations for which the variables are related differently or have different means. Outliers or "extreme" data should be examined closely because they may have a very large influence on statistics such as the correlation coefficient or regression slope. Extreme data points may come from participants who perform in ways that are qualitatively different from the majority of participants. Extreme data points may also arise because of errors in data collection or recording. If there are a few extreme outliers, we should examine our data records for errors, and we may wish to describe the data both when the extreme scores are included and when they are not. Statistical packages make it easy to identify outliers and to perform these analyses.

17.10 Summary

Our goal in Chapter 17 has been to give some examples of bivariate relationships and to discuss how to extract and summarize them.

- When concerned with exploring the relationship between two variables, display the data in a scatterplot. Procedures such as using smoothers or fit lines are available to help extract systematic relationships from a background of noise.
- The Pearson correlation coefficient is a number between −1 and +1 that serves as a measure of the strength of the linear relationship between two variables.
- Least-squares bivariate linear regression produces the linear equation that best predicts one variable using information about the other. We developed linear regression, first for z scores, and then for raw scores.
- When two variables have an imperfect linear relationship and have the same variance, predictions of one variable from the other must regress toward the mean.
- The coefficient of determination, r^2, is the proportion by which squared prediction error is reduced by using the regression equation, and the standard error of the estimate, $s_{y.x}$, is a measure of the amount of prediction error that remains.
- If we predict Y from X, the variability in the Y scores, SS_y, can be partitioned into two components: $SS_{regression}$, the amount by which squared prediction error has been reduced by using the regression equation, and $SS_{residual}$, the amount of squared prediction error that remains.
- There are many potential pitfalls in the interpretation of the Pearson correlation coefficient and on understanding the influences of different variables on the correlation. This may be especially true when data are combined from different groups.
- To understand how two variables are related, we should not just look at the correlations and regression slopes. *We must look at and plot the data.*

Appendix 17.1

Proof that $z_Y = \pm z_X$ when $Y = b_0 + b_1X$

We want to show that if X and Y have a perfect linear relationship, $z_y = z_x$ when the relationship is positive and $z_y = -z_x$ when it is negative. For any data point (X, Y) that falls on a straight line, we have $Y = b_0 + b_1X$. Substituting into the usual expressions for the mean and standard deviation and simplifying, we have $\bar{Y} = b_0 + b_1\bar{X}$ and $s_y = \pm b_1s_x$, with $s_y = +b_1s_x$ when b_1 is positive and $s_y = -b_1s_x$ when b_1 is negative. Therefore, if there is a perfect linear relationship between X and Y,

$$\begin{aligned} z_Y &= \frac{Y-\bar{Y}}{s_Y} \\ &= \frac{b_0 + b_1X - \left(b_0 + b_1\bar{X}\right)}{\pm b_1 s_X} \\ &= \frac{X-\bar{X}}{\pm s_X} \\ &= \pm z_X \end{aligned}$$

Exercises

17.1 [Calculating correlation, regression equations, and proportion of variance explained] Given the following data,

X	Y
1	11
2	3
3	7
4	9
5	9
6	21

a) Draw a scatterplot.
b) What is the correlation between Y and X?
c) What is the least-squares equation for the regression of Y on X?
d) What is the proportion of variance in Y accounted for by X?
e) Find the equation for the regression of X on Y.
f) What is the proportion of the variance in X accounted for by Y?

17.2 [Correlations on raw data and z scores] Given the following data for three variables X, Y, and W:

W	X	Y
12	4	7
8	6	9
4	13	3
17	12	14
18	13	16

Using software, find the correlations among *W*, *X*, and *Y*. Standardize the variables and recompute the correlations. They should be identical. Why?

17.3 [The effect of transformations on correlation] A psychologist is interested in predicting *Y* from *X* in two distinct situations. She finds the following results:

Situation 1	*Situation 2*
$b_1 = 38.41$	$b_1 = .25$
$s_y = 512.31$	$s_y = 8.44$
$s_x = 2.00$	$s_x = 23.17$

a) In which situation is the correlation between *X* and *Y* higher?
b) You are given a large number of data points (*X*, *Y*) and find that the correlation between *X* and *Y* is $r_{xy} = .70$. You now add 10 to each of the *X* scores. What happens to the correlation coefficient; that is, what is the new correlation between *Y* and the transformed *X*?
c) You have the same situation as in part (b) – except instead of adding 10 to each of the *X* scores, you multiply each of the *Y* scores by 3. Now what is the value of the correlation coefficient?
d) Now perform both operations, multiply each *Y* score by 3, and add 10 to the product. What happens to the correlation coefficient?

17.4 [Understanding influential points]

a) Using the *Royer* data set available on the book's website, find the correlation between multiplication accuracy (*multacc*) and the time taken to solve multiplication problems (*multrt*) for third-graders.
b) Generate the scatterplot for these variables.
c) The correlation coefficient is not very resistant to the influence of data points that deviate strongly from the best-fitting straight line. Here, case 64 can be shown to be the most influential data point; that is, its removal results in more change in *r* than the removal of any other data point. What is *r* if case 64 is removed?

17.5 [Interpreting correlation] For each part, indicate whether the use of the correlation coefficient is reasonable. If it is not, indicate why not.

a) The research division of the Old Southern Casket and Tobacco Corporation has just released the results of a study that they argue is inconsistent with the negative health claims made against cigarette smoking. For a large sample of heavy smokers, a substantial positive correlation was found between the total number of cigarettes smoked during a lifetime and length of life, a result they claim leads to the conclusion that cigarette smoking is beneficial to health.
b) It is found that for eighth-grade children there is a fairly strong negative correlation between the amount of television watched and school performance as measured by grades. It is claimed that this finding constitutes proof that watching television interferes with intellectual ability and has a negative effect on the ability to focus attention. Does this argument seem valid?

17.6 [The effect of transformations on correlation] A psychologist is interested in predicting Y from X in two distinct groups. She finds the following results:

Group 1	*Group 2*
$b_1 = 1$	$b_1 = 4$
$s_y = 20$	$s_y = 10$
$s_x = 10$	$s_x = 2$

a) In which situation is the correlation between X and Y higher?
b) You are given a large number of data points (X, Y) and find that the correlation between X and Y is $r_{xy} = 0.70$. You now transform X by multiplying each of the X scores by 10 and adding 3 to the each of the products. You also transform Y by multiplying each of the Y scores by 2. What is the correlation coefficient between the transformed variables?

17.7 [Relating regression slope and correlation] In a large study of income (Y) as a function of years on job (X), the data for 2,000 workers with a college degree and 2,000 without a degree are the following:

	College graduate		*No college degree*	
	Income(Y)	*Years (X)*	*Income*	*Years*
Mean	80	15	76	10
s^2	324	100	289	25
r_{xy}	.333		.235	

Note: Income is recorded in thousands of dollars.

a) Find b_{yx} (i.e., $b_{Income,\ Years}$, the regression coefficient for the regression of income on years of service) for each group separately.
b) What is your best estimate of the amount by which salary increases per year for people with and without college degrees? Is this result consistent with differences in the correlations in part (a)? Explain.
c) Using separate regression equations for each group, what salary would you predict for people with and without college degrees with 10 years of experience? With 20 years of experience?

17.8 [Understanding correlations in separate and combined groups] Using the *Seasons* data file available on the book's website, correlate height with weight, and then correlate height with weight separately for participants identified as men and as women. How might you account for the discrepancies among the three correlations?

17.9 [Interpreting correlation] For parts (a)–(c), indicate whether the use of the correlation coefficient and/or the conclusion drawn is reasonable. If it is not, indicate why not.

a) A clinical psychologist reads a description of a study in which a correlation of −.80 was obtained between a measure of anxiety and a measure of emotional stability. Deciding to verify the result, he administers the same measures of anxiety and emotional stability to a random sample of patients in a Veterans Administration (VA) hospital. The observed correlation of −.20 between measures is not significant. He concludes that he has no evidence of any relationship (at least any linear relationship) between anxiety and emotional stability.

b) Martians are tall and skinny and do not weigh very much. Jovians are shorter but weigh a lot more. Height and weight correlate pretty highly for each group, about $r = .60$. Would you expect the correlation between height and weight for a mixed group consisting of equal numbers of Martians and Jovians to be about the same, bigger, or smaller than .60? Why?

c) It is reported in the press that getting a degree from a four-year college is highly correlated with lifetime earnings; that is, it is worth several hundred thousand dollars a year in lifetime earnings.

17.10 [Correlation in separate and combined groups] Using the *Seasons* data set, verify that the correlation between cholesterol level (*TC*) and age is .506 for participants identified as women but only .148 for women 50 years of age or over and .264 for women under 50 years of age. How do you explain the discrepancy? What are the corresponding results for participants identified as men?

17.11 [Correlation in separate and combined groups] Using the data for participants identified as men in the *Seasons* data set, if we consider individual men, the correlation between *TC* and age is .062. What is this correlation if we consider levels of the variable *agegrp*; that is, if we find the means value of *TC* and age at each level of *agegrp*, and use these as our data points?

17.12 [Regression to the mean]

a) After each of two practice landings, pilot trainees discuss their performance with their instructors. The instructors find that trainees who make poor landings the first time tend to make better landings the second time, whereas trainees who make good landings the first time tend to do worse the second time. The instructors conclude that the criticism that follows poor performance tends to make pilots do better and that the praise that follows good performance tends to make them do worse. Therefore, the instructors decide to be critical of all landings, good or bad. Is this a reasonable strategy?

b) After the first examination in a course, students who scored in the bottom 25% of the distribution are given special tutoring. On the next examination, all of these students score above the average for the whole class. Can we conclude that the tutoring was effective or could the results simply be due to regression toward the mean?

c) An educational psychologist wants to see whether ability to spell has any effect on ability to read. To this end, he selects two groups of participants, a group of poor spellers and a group of good spellers; the groups are matched on mathematics performance to roughly equate overall academic skill. He now administers a reading test to both groups and finds that the good spellers do better on the average than the poor spellers. Does this mean that spelling ability affects reading ability?

d) Given what you know about regression toward the mean, explain how fluctuations in the intensity of pain and other symptoms may cause patients to overestimate the therapeutic effects of alternative medicine.

17.13 [Regression to the mean] Assume that the correlation between the adult heights of fathers and sons is .5, and further, that the mean and standard deviations of the heights of adult males are 70.0 and 3.0 inches, respectively.

a) Given the information that a father is 76 inches tall, what is the best linear prediction for the adult height of his son?

b) Given that the adult height of a son is 73 inches, what is the best linear prediction for the height of his father?
c) Given the phenomenon of regression toward the mean, why wouldn't we expect all men to have about the same height in a few more generations?

17.14 [Regression to the mean] Students are classified as gifted and gain admittance to a special, enriched program of studies if they score more than two standard deviations above the mean on an entrance exam. José scores 2.1 standard deviations above the mean (at the 98th percentile) on form A of the test and is classified as gifted. Later, he is tested on form B, a parallel form of the entrance exam that has the same mean and standard deviation as form A, and correlates .90 with it.

a) Would you expect José to be classified as gifted based on his performance on this exam? Why or why not?
b) What are the consequences of this finding?
c) Mary takes form B of the test and scores 1.89 standard deviations above the mean. She then takes form A. What is the best prediction for Mary's score on form A?

17.15 [Using real data in R] Use the *penguins* data set in the {palmerpenguins} package in R.

a) Make a scatterplot of bill length and bill depth, including a smoother.
b) Correlate bill length (*bill_length_mm*) and bill depth (*bill_depth_mm*).
c) Select the Adelie species and repeat part (b). Explain any differences in these two correlations.
d) Regress body mass on flipper length and report the resulting equation.
e) Select the Adelie species and repeat part (d). Explain and differences in your results.

Notes

1 Determined by using a test suggested by Sandik and Olsson (1982) for testing dependent variances. In this test, first the absolute deviations about the median are obtained for subtraction and for multiplication scores, and then the Wilcoxon signed-rank test is performed on these absolute deviations. For a more general discussion of tests for dependent variances, see Wilcox (1989).

2 Note that although z_y and z_x are z scores, $\hat{z}_y$ is not a z score, because, although the mean of the $\hat{z}_y$s is zero, the variance is r^2, not 1. Note also that the correlation between z_y and z_x is the same as the correlation of Y and X.

Chapter 18

More About Correlation

18.1 Overview

In Chapter 17, we developed the Pearson correlation coefficient as a descriptive statistic. In Chapter 18, we develop procedures for making inferences about correlation and we introduce other measures of correlation. The specific topics we consider are as follows:

- *Statistical inference about correlation*, including confidence intervals, significance tests, and power.
- *Correlation adjusted for the effects of other variables*, known as partial and semipartial (or part) correlation.
- *Special cases and alternative measures of correlation* for ranked data or dichotomous variables.

18.2 Inference About Correlation

So far, we have discussed the Pearson correlation coefficient as a descriptive statistic. We have not addressed the issue of what we can say about ρ_{xy}, the correlation between X and Y in a population, based on finding a correlation coefficient of r_{XY} in a sample selected from the population. Because we have focused on descriptive statistics to this point, we have not made any assumptions about the joint distribution of X and Y, although we have pointed out that certain characteristics of the distributions of X and Y can limit the range of possible values of r (see Section 17.7.1). However, we now turn to the topic of how to make inferences about correlation. This requires that we specify a model for the joint distribution of X and Y in the population.

18.2.1 A Model for Inference About Correlation

Most of the statistical tests for means discussed earlier in the book have assumed that the underlying populations of scores are normally distributed. When we discuss inference about the population correlation, we usually assume that X and Y are both random variables and that the population of (X, Y) pairs has a *bivariate normal distribution* (see Appendix 18.1 for the density function). Think of the bivariate normal distribution as the two-dimensional generalization of the univariate normal distribution.

We can graphically represent the bivariate normal distribution in several ways. In Figure 18.1, the plane defined by the X and Y axes contains all possible pairings of X

DOI: 10.4324/9781003453550-22

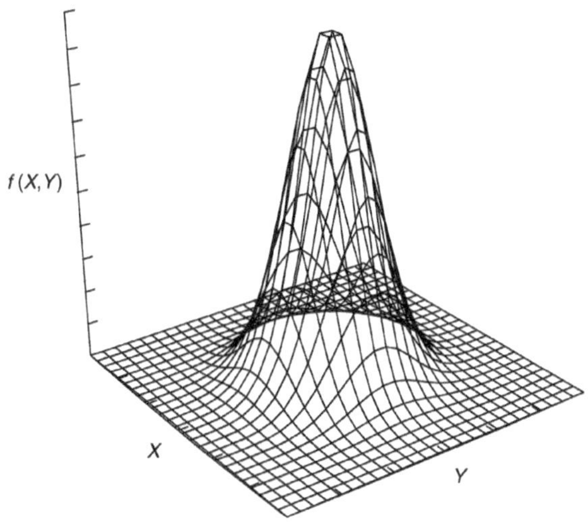

Figure 18.1 An example of a bivariate normal distribution.

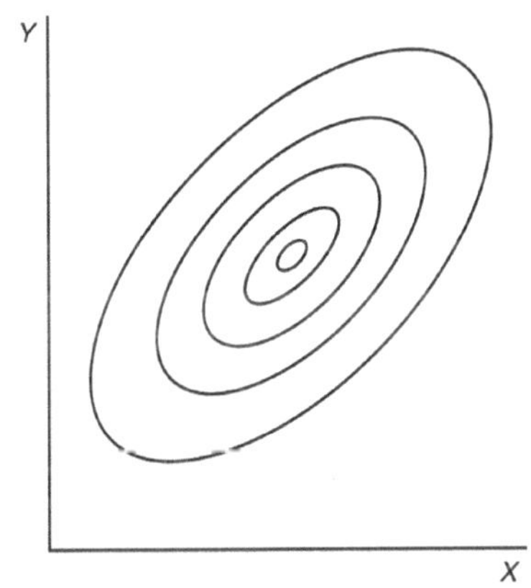

Figure 18.2 Example isodensity contours for a bivariate normal distribution.

and Y. The bivariate normal density function can be represented by a bell-shaped surface that rises above the X–Y plane. The height of the surface above the X–Y plane represents the probability density. For any value of either variable, the density function of the other variable is normally distributed.

A set of planar slices through the bivariate normal surface, with planes parallel to the X–Y plane and at different values of Z, define a family of ellipses, as shown in Figure 18.2. Each point on one of these ellipses will have the same probability density, and therefore these ellipses are called isodensity contours. These ellipses are like contour lines on a topographic map. Because of the "peaked" shape of the surface, the smaller ellipses in Figure 18.2 correspond to larger values of probability density. The more eccentric the ellipses (i.e., the less they look like circles), the larger the correlation between X and Y. A set of concentric circles corresponds to a correlation of zero.

Some characteristics of a bivariate normal distribution in X and Y are as follows:

- The marginal distributions (i.e., the distribution of one variable, collapsing over values of the other variable) of X and of Y are normal with variances σ^2_X and σ^2_Y, respectively.
- If X and Y have a correlation of zero, they are independent. That is, *given bivariate normality, the only possible systematic relationship between X and Y is a linear one.* The conditional means of Y (i.e., the means of the Y scores at particular values of X) fall on the straight line with the equation $\mu_{Y.X} = \mu_Y + \beta_{YX}(X - \mu_X)$ where $\beta_{YX} = \rho\sigma_Y/\sigma_X$ is the slope of the population regression equation that predicts Y from X. The conditional distributions of Y are normal with variance $\sigma^2_{Y.X} = \sigma^2_Y(1 - \rho^2)$. The conditional means of X fall on the straight line with the equation $\mu_{X.Y} = \mu_X + \beta_{XY}(Y - \mu_Y)$ where $\beta_{XY} = \rho\sigma_X/\sigma_Y$ is the slope for the regression of X on Y, and the conditional distributions of X are normal with variance $\sigma^2_{X.Y} = \sigma^2_X(1 - \rho^2)$.

The inferential procedures to which we now turn assume bivariate normal population distributions. If this assumption is violated, results may be biased.

18.2.2 Using the t Distribution to Test the Null Hypothesis H_0: $\rho = 0$

We can test whether there is a linear component to the relationship between X and Y in the population by considering the null hypothesis H_0: $\rho = 0$. When $\rho = 0$, the statistic

$$t = \frac{r}{\sqrt{\dfrac{1-r^2}{N-2}}} \tag{18.1}$$

has approximately a t distribution with $N - 2$ degrees of freedom. Therefore, we can use this expression as the test statistic with which to test the null hypothesis H_0: $\rho = 0$.[1] For example, the correlation between subtraction and multiplication accuracy for the 28 third-graders in the *Royer* data set is .594. The value of the t statistic with 26 df is $(.594)/\sqrt{(1-.594)^2/26} = 3.77$ We can use Appendix Table C.3, or the command 1 – *pt*(3.77, 26) in R, to find the p-value is less than .001. Therefore, we can reject the null hypothesis that H_0: $\rho = 0$.

Power calculations for tests of the null hypothesis H_0: $\rho = 0$ are readily performed with available software. G*Power 3.1 can calculate the sample size necessary to obtain any desired level of power, given a specified effect size. Suppose we want the sample size necessary to have power = .80 for rejecting the null hypothesis H_0: $\rho = 0$, using a two-tailed t test with $\alpha = .05$. Further suppose that we expect the population correlation to be "medium sized," $\rho = .30$ according to Cohen's 1988 guidelines. To use G*Power 3.1 to calculate the desired sample size, select *t tests* as the *Test family*, and then select *Correlation* from the *Tests* menu and choose the *Point biserial model*.[2] For the type of power analysis, select *A priori*: *Compute required sample size* ρ, set $r = .30$, $\alpha = .05$, power = .80, and request a two-tailed test in the *Input Parameters* boxes. Now click on *Calculate*. G*Power 3.1 indicates that to have power equal to .80, it is necessary to have sample size of $N = 82$ (see Figure 18.3).

It is important to estimate this required N so that we can decide whether it is worth using the resources necessary to achieve a reasonable level of power. If we simply went ahead

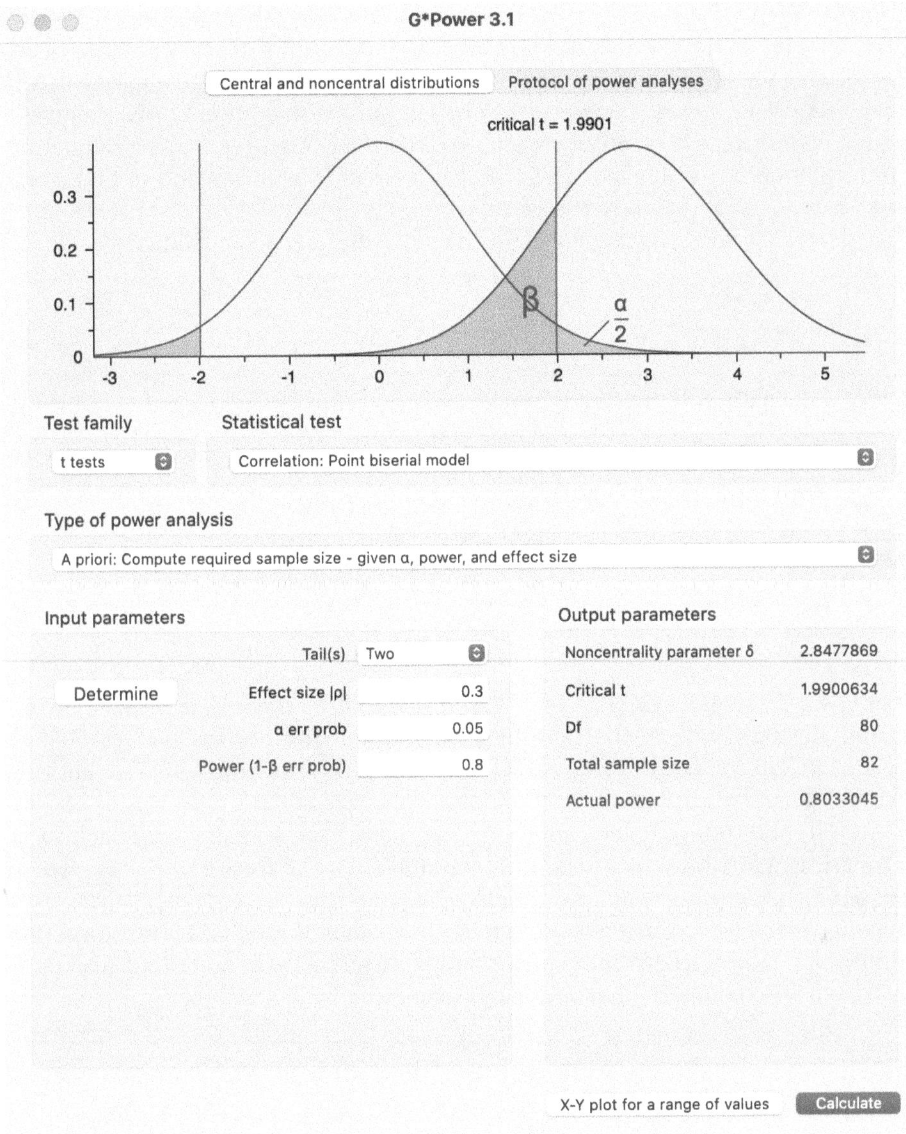

Figure 18.3 G*Power 3.1 screen for calculating the sample size necessary to have power = .80 for a two-tailed t test of H_0: ρ = 0 at α = .05 if ρ is really .30.

with a study using, say, 30 participants selected from a population in which ρ was .30, the estimated power for rejecting H_0: ρ = 0 against H_1: ρ ≠ 0 at α = .05 would only be about .38.

18.2.3 Using the Fisher Z Transformation

Although the t test for the null hypothesis ρ = 0 is quite robust with respect to the assumption of normality (see, for example, Edgell & Noon, 1984), it cannot be used to test null hypotheses about ρ other than ρ = 0, nor can it be used to develop confidence intervals for ρ. This is because the shape and the standard error of the sampling distribution of *r* depend on ρ, the very value we are trying to estimate. When ρ differs from 0, the sampling distribution of *r* is skewed, even for large sample sizes. The sampling distribution of *r* is negatively (left) skewed if ρ is positive and positively (right) skewed if ρ is negative. However, an appropriate transformation of ρ will allow us to test these other hypotheses and to find confidence intervals for ρ.

Fisher showed that if bivariate normality can be assumed, a logarithmic transformation

$$Z_r = \frac{1}{2}\log_e\left[\frac{1+r}{1-r}\right] \tag{18.2}$$

where $\log_e$ is the natural logarithm is approximately normally distributed with mean

$$Z_\rho = \frac{1}{2}\log_e\left[\frac{1+\rho}{1-\rho}\right]$$

and standard error

$$\sigma_r = \frac{1}{\sqrt{N-3}}$$

for sample sizes as small as $N = 10$. The effect of this *Fisher Z transformation*[3] is to stretch out the right tail of the sampling distribution for positive values of ρ and to stretch the left tail for negative values of ρ, to make the transformed distribution more symmetric. Thus, we can calculate confidence intervals for ρ by first using the normal distribution to find the confidence interval for Z_ρ, and then transforming back to *r*s. The transformation also allows us to test null hypotheses of the form H_0: $\rho = \rho_{hyp}$, where ρ_{hyp} is not equal to 0.[4]

Given an observed correlation *r*, we can find the value of the Fisher transformation, Z_r, using Appendix Table C.10, or the transformation menu found in most statistical packages. The $100(1 - \alpha)\%$ confidence interval for ρ is then given by

$$Z_r \pm z\sigma_r$$

where *z* is the value that cuts off the upper α/2 of the standard normal distribution; for the 95% confidence interval, $z_{.025} = 1.96$. For example, for the 28 third-graders in the *Royer* data set, the correlation between subtraction and multiplication accuracy is .594. The Fisher transformation of .594 is approximately .684. The 95% confidence interval for Z_ρ is given by

$$684 \pm (1.96)\frac{1}{\sqrt{25}}$$
$$= .684 \pm .392$$

That is, the confidence interval extends from *Z* values of .292 to 1.076. Transforming back to *r* scores using Appendix Table C.10 (or the hyperbolic tangent function of a calculator) yields 95% confidence limits of .284 and .792 for ρ. Note that because the Fisher *Z* transformation is a nonlinear function of r, the upper and lower limits of the confidence interval are not equally distant from the observed value of r.

We can reject the null hypothesis H_0: $\rho = \rho_{hyp}$ at $\alpha = .05$ for all values of ρ_{hyp} that do not fall in the confidence interval; that is, for all hypothesized values of ρ less than .284 or greater than .792. We can also test the hypothesis H_0: $\rho = \rho_{\text{hyp}}$ by using the test statistic

$$z = \frac{Z_r - Z_{\rho_{hyp}}}{\sqrt{\frac{1}{N-3}}} \tag{18.3}$$

Even though we can reject the null hypothesis that $\rho = 0$ for the example in which we found $r = .594$ in a sample of size $N = 28$, we cannot be certain about the exact value of ρ because the 95% confidence interval is quite wide. It is sobering to consider just how large confidence intervals for ρ are, even for moderately large sample sizes. Even if we had obtained the sample correlation of .594 from a sample of $N = 100$, the 95% confidence interval would extend from approximately .45 to .71 – still quite large. We strongly recommend finding confidence intervals for correlation coefficients.

We can again calculate power and the sample sizes necessary to obtain desired levels of power. For example, suppose we want to test the hypothesis H_0: $\rho = .30$ against H_1: $\rho \neq .30$ at $\alpha = .05$, with power = .80, using Equation 18.3. In G*Power 3.1, the t-based point biserial correlation test we used in Section 18.2.2 no longer applies because the null hypothesis for ρ is not zero. To enter a non-zero null hypothesis value for ρ and a specific alternative hypothesis value, we need the *Exact* test family and the *Correlation: Bivariate normal model* statistical test. Click on the *Options* button and select the radio button to *use large sample approximation (Fisher Z)*. If we assume that the true value of ρ (H_A) is .50, G*Power determines that the required sample size is $N = 140$. A similar *sample size*, 139, is found if we *use exact distribution* to calculate power from the sampling distribution of r.

It is important to note that if *a priori* power calculations are based on a correlation estimate that comes from a small sample, the confidence interval for the required sample size may be so large that the power calculation will be of little use. For example, suppose we want to find the sample size necessary to have power = .90 to reject the hypothesis H_0: ρ = .20 against H_1: $\rho \neq .20$, at $\alpha = .05$, and that we base our calculations on the estimate r = .594, found in a sample of 28 third-graders. Assuming $\rho = .594$, the sample size required to achieve power = .90 is 49 using the Fisher Z approximation. But the 95% CI for ρ extends from .284 to .792. If we use these two limits of the CI as our assumed values of ρ, we find that the required sample size is only 17 if $\rho = .792$, but it is 1,321 if $\rho = .284$!

18.2.4 Using Software for Significance Tests and Confidence Intervals of r

Statistical software can readily compute a sample correlation, as well as test the observed value against a null hypothesis and generate confidence intervals. In R, the *cor.test* function in the {stats} package performs a hypothesis test against H_0: $\rho = 0$ and reports the 95% confidence interval for r. Different confidence levels may be selected using the *conf.level* option (e.g., *conf.level* = 0.9). For example, *cor.test*(Royer_acc$multacc, Royer_acc$subacc) will reproduce the t test from Section 18.2.2 and the CI reported in Section 18.2.3. The *FisherZ* function in the {DescTools} package takes r as input and returns Z_r; *FisherZInv* reverses that process, taking Z_r as input and returning r. The upper and lower bounds of the 95%

confidence interval for Z_ρ can be obtained from the *CorCI* function in the {DescTools} package, which takes r and N as input and returns the confidence interval bounds. Optionally, different confidence levels may be selected using the *conf.level* option, and a one-tailed hypothesis test can be conducted using *CorCI* with the option *alternative* = "greater" or "less."

In SPSS, choose *Correlate* from the *Analyze* menu, and then select *Bivariate*. Move the variables of interest to the "Variables" box and click OK. The results will include a two-tailed test of significance using the t test from Section 18.2.2. For a confidence interval, click the *Confidence Interval* bar after selecting variables of interest, and the tick the box for "Estimate confidence interval of bivariate correlation parameter," set the desired confidence level, then click Continue and OK. The result will include a confidence interval based on Fisher's Z transformation.

18.2.5 What If the Assumption of Bivariate Normality Is Violated?

The model underlying statistical inference in this chapter assumes bivariate normality (see Section 18.2.1). However, we know that in real data sets, even univariate distributions rarely follow ideal normal distributions (e.g., see Micceri, 1989). There are two easily detected conditions that indicate a violation of bivariate normality. First, the bivariate normality assumption requires that both X and Y are normally distributed, so it is obviously violated if the marginal distributions of X and/or Y are not normal. Second, even if the marginal distributions are normal, the assumption will be violated if there is any systematic relationship between X and Y other than a linear one. Recall that *Q–Q* plots are useful in detecting violations of the normality assumption (see Chapter 2). In addition, tests for violations of normality and linearity are discussed in Section 19.7.

As with any violations of the assumptions of a test, our concern is that either the Type 1 error rate will be inflated, power will be reduced, or both. There have been several simulation studies to investigate the robustness of significance tests for correlation under violations of the normality assumption (e.g., Edgell & Noon, 1984; Havlicek & Peterson, 1977; Lee & Rodgers, 1998). The results show that tests of the hypothesis H_0: $\rho = 0$ are quite robust with respect to Type 1 error. Type 1 error rates are close to their nominal values, even for skewed distributions.

Although Type 1 error rate is generally well controlled when the assumption of bivariate normality is violated, the loss of power can be substantial. For example, Lee and Rodgers (1998) calculated the power for the test of H_0: $\rho = 0$ when the true ρ was .4. They compared the power when samples were drawn from a normal population to the power when samples were drawn from a highly skewed population. For sample sizes of both 30 and 60, sampling from the skewed population cut power by more than half compared to sampling from the normal population. This severe decline in power suggests that when the normality assumption is severely violated, we should consider alternatives to the usual tests.

An alternative procedure that may be useful when the normality assumption is not satisfied is *bootstrapping*. Bootstrapping is a general purpose, computationally intensive approach to inference (see, for example, Efron & Diaconis, 1983; Efron, 1988) in which the sampling distribution of any statistic can be obtained by repeatedly sampling, with replacement, from the observed sample. For example, a 95% confidence interval could be obtained for a correlation coefficient by randomly selecting a large number of samples of size N from the observed sample, and then calculating the correlation coefficients for each of these samples and ordering them. If we bootstrapped 1,000 samples, the 95% confidence interval for ρ would be defined by the 25th and 975th largest correlations; the 90% confidence interval would be defined by the 50th and 950th largest correlations.

Bootstrapping requires software. In SPSS, begin the analysis as described in Section 18.2.4, then click on the *Bootstrap* bar and tick the box to *Perform bootstrapping*. Set a desired *Number of samples*, recognizing that running more samples results in better defined confidence bounds but also takes more time to run. We recommend no fewer than 1,000 samples. For the third-graders in the *Royer* data set, a bootstrap of the 28 children with both multiplication and subtraction accuracy scores resulted in confidence bounds of .152 and .830 for 1,000 bootstrapped samples, and .165 and .823 with 10,000 samples; both intervals are wider than the CI based on the Fisher *Z* transformation discussed in Section 18.2.3 (.274 to .788).[5] Bootstrapping in R is extremely flexible and runs much more quickly than in SPSS, but it does require a modest amount of coding ability and the use of the *boot* and *boot.ci* functions in the {boot} package. An example is included in Appendix 18.2.

18.2.6 Testing Whether Independent Correlations Differ

Another natural question to ask about relationships between two variables is whether they differ in magnitude. For example, does the relationship between high school GPA and college GPA differ for people who identify as men and those who identify as women or as nonbinary? It is possible to test whether the correlation between *X* and *Y* is the same in two different populations. We can test H_0: $\rho_1 = \rho_2$, by using the test statistic

$$z = \frac{Z_{r_1} - Z_{r_2}}{\sqrt{\frac{1}{N_1 - 3} + \frac{1}{N_2 - 3}}} \tag{18.4}$$

As an example of using Equation 18.4 with real data, the correlation between subtraction and multiplication accuracy for the 28 third-graders in the *Royer* data set is .594; for the 25 fifth-graders, it is .225. To determine whether these correlations are significantly different at $\alpha = .05$, we can substitute in Equation 18.4. If we do so, we obtain

$$z = (.684 - .229) / .292 = 1.56$$

This is less than the upper critical value of $z_{crit} = 1.96$, so we do not have sufficient evidence to reject the null hypothesis that the correlations in the populations of third- and fifth-graders are equal.

Suppose that we wish to run the study again with a larger sample. We can readily do *a priori* power calculations for this test if we are willing to specify the size of the effect, $q = Z_{r_1} - Z_{r_2}$. For example, if we can assume that the two population correlations are .594 and .225, so that $q = .684 - .229 = .455$, we can use G*Power 3.1 to show that the sample size necessary to have power = .90 for a two-tailed test of H_0: $\rho_1 = \rho_2$ at $\alpha = .05$ is 105 in each group (see Figure 18.4). However, we should be cautious about using these calculations when we design research using correlations from small samples. The 95% CI for the correlation between subtraction and multiplication accuracy for third-graders extends from .284 to .792, and the 95% CI for fifth-graders extends from −.206 to .583. In this situation, we should perform several different power calculations to assess the magnitude of the task

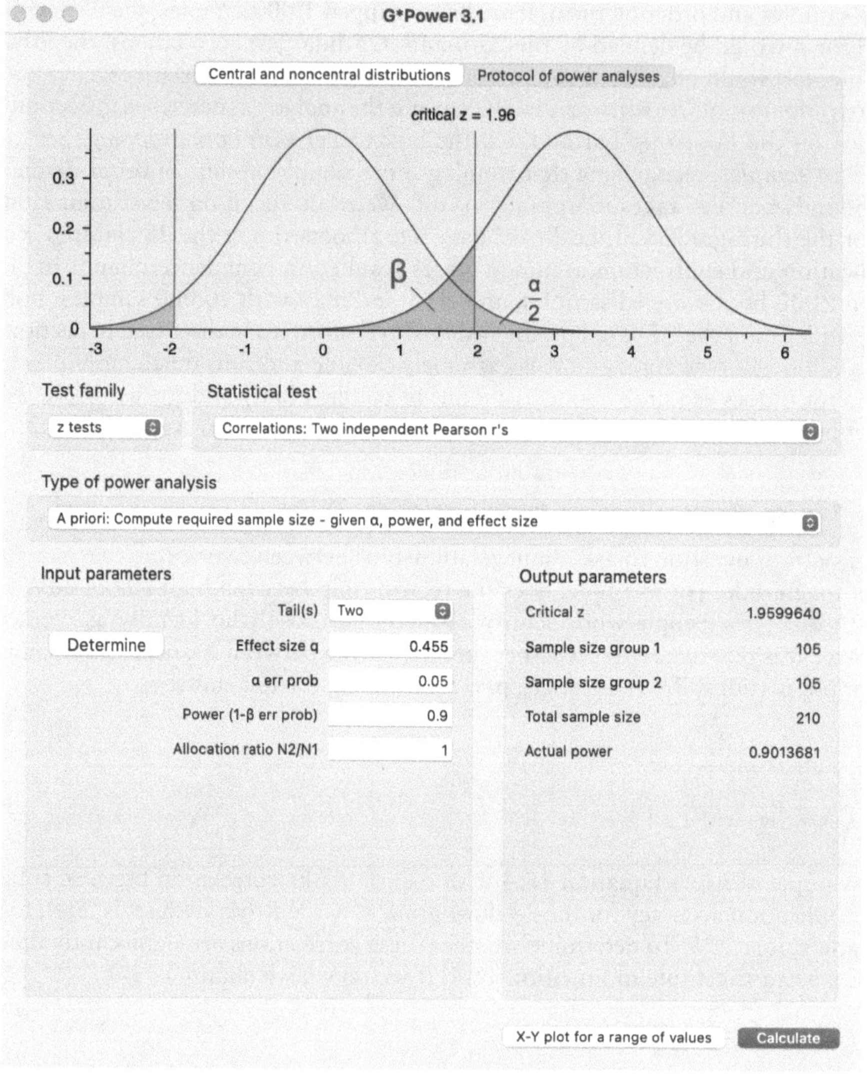

Figure 18.4 G*Power 3.1 screen for calculating the sample size necessary to have power = .90 for a two-tailed *t* test of H_0: $\rho_1 = \rho_2$ at $\alpha = .05$ if ρ_1 is really .594 and ρ_2 is really .225.

before us. As earlier, we may estimate the required sample size based on the results of a previous study. Also, we can base our calculations on Cohen's guidelines that small, medium, and large values of *q* are given by .10, .30, and .50, respectively. Finally, we may base our calculations on the smallest values of *q* that are thought to be meaningful from either a theoretical or practical perspective, if such values are available.

It is important to note that equal differences in Z_rs are equally detectable and therefore can be said to represent equal changes in linear relationship at different points along the range of possible values (see Cohen, 1977, 1988). In contrast, *equal differences between the* rs *do not correspond to equal differences in relationship at different points along their range*. For example, the .3 difference between *r*s of .8 and .5 corresponds to a *q* of .45, whereas the .3 difference between *r*s of .4 and .1 corresponds to a *q* of only .32.

As a second example of the application of Equation 18.4, in Section 17.7.2 we found that the correlations between total cholesterol level and age in the *Seasons* data set for participants who identify as women (N = 211) and as men (N = 220) were .51 and .06, respectively. Substituting into Equation 18.4, we find z = 5.15, a value well beyond any entry in Appendix Table C.2 (using R, 1 – *pnorm*(5.15) is 0 to 6 decimal places). We conclude that the population correlations are different.

Equation 18.4 also provides a basis for obtaining confidence intervals for $Z_{\rho_1} - Z_{\rho_2}$. The 95% confidence interval is given by

$$Z_{r_1} - Z_{r_2} \pm 1.96\sqrt{\frac{1}{N_1 - 3} + \frac{1}{N_2 - 3}} \tag{18.5}$$

Although Equation 18.5 can be used to find a confidence interval for , we cannot transform back to obtain a confidence interval for $\rho_1 - \rho_2$. This is because the Fisher Z transformation is not linear, so $Z_{\rho_1 - \rho_2}$ is not the same as $Z_{\rho_1} - Z_{\rho_2}$. This nonlinearity is not a problem for hypothesis testing because if $Z_{\rho 1} \neq Z_{\rho 2}$ then $\rho_1 \neq \rho_2$.

Zou (2007) provided a procedure for finding the confidence interval for the difference between two independent correlations. To find, say, the 95% confidence interval for $\rho_1 - \rho_2$, we start by finding the 95% CIs for each correlation, using the Fisher Z transform as we showed earlier. In the data for the previous example, the 95% CI for the correlation between age and *TC* for participants who identify women, ρ_1, extends from .403 to .604. We denote the lower limit of the interval by l_1 and the upper limit by u_1. We then find the 95% CI for the corresponding correlation for participants who identify as men, ρ_2, as the interval from –.072 to .192. We denote the lower limit by l_2 and the upper limit by u_2. Then, the lower and upper limits of the 95% CI for the difference between the correlations, $\rho_1 - \rho_2$, are given by

$$L = r_1 - r_2 - \sqrt{(r_1 - l_1)^2 + (u_2 - r_2)^2}$$

and

$$U = r_1 - r_2 + \sqrt{(u_1 - r_1)^2 + (r_2 - l_2)^2}$$

Substituting the values from our age and *TC* example, we find that the 95% CI for the difference between the two correlations extends from .28 to .61.

It is also possible to test the null hypothesis that more than two independent correlation coefficients are all equal. To test that J ρs are all equal, use the test statistic.

$$\chi^2_{J-1} = \sum_j (N_j - 3)Z_j^2 - \frac{\left[\sum_j (N_j - 3)Z_j\right]^2}{\sum_j (N_j - 3)} \tag{18.6}$$

where Z_j is the Fisher Z transformation of the jth correlation coefficient. For example, Table 18.1 presents the correlations between subtraction accuracy and multiplication

Table 18.1 Correlations between subtraction and multiplication accuracy for third-, fourth-, fifth-, and sixth-graders in the *Royer* data

Grade	*Correlation*	*N*	*95% CI*
Third	.594	28	.284 to .792
Fourth	−.194	28	−.528 to .194
Fifth	.225	23	−.206 to .583
Sixth	.431	26	.052 to .701

accuracy for fourth-, fifth-, and sixth-graders; these are −.194, .225, and .431, respectively. The obtained value of the test statistic is $\chi^2 = 6.899 - (10.283)^2 / 68 = 5.344$. For $\alpha = .05$ and $(J - 1) = 2\ df$, $\chi^2_{crit.05,2} = 5.99$. Therefore, we do not have sufficient evidence to reject the hypothesis that the three population correlations are identical.

18.2.7 Testing Hypotheses About Dependent Correlations

In the preceding section, we focused on comparisons of correlations of the same two variables across different groups (e.g., the correlation between subtraction and multiplication scores for third- versus fifth-graders). In this section, our concern is with evaluating multiple correlations on the same set of participants. Two questions are of interest: First, which variables are correlated? Second, which correlations differ in magnitude? We first consider the question of evaluating individual correlation coefficients.

When data on several variables have been collected from the same set of participants, the correlations may be displayed as a *correlation matrix*. If the variables are X_1, X_2, X_3, and X_4, the correlation matrix is

$$\begin{array}{c} \\ X_1 \\ X_2 \\ X_3 \\ X_4 \end{array} \begin{array}{c} \begin{array}{cccc} X_1 & X_2 & X_3 & X_4 \end{array} \\ \begin{bmatrix} 1 & r_{12} & r_{13} & r_{14} \\ r_{21} & 1 & r_{23} & r_{24} \\ r_{31} & r_{32} & 1 & r_{34} \\ r_{41} & r_{42} & r_{43} & 1 \end{bmatrix} \end{array}$$

If there are k variables, the correlation matrix consists of k^2 elements. The k elements on the major diagonal (the one that goes from the upper left to the lower right) are each equal to 1 because any variable is perfectly correlated with itself. The $k^2 - k = k(k - 1)$ off-diagonal elements can be divided into a set of $k(k - 1)/2$ correlations above the diagonal and a set of equal size below it. The matrix is symmetric because elements on opposite sides of the diagonal are identical; $r_{12} = r_{12}$ and, in general, $r_{ij} = r_{ji}$ for all i and j.

When k is large, the number of correlations will be large. For example, if $k = 20$, $k(k - 1) / 2 = 190$. If we tested each correlation for significance at $\alpha = .05$, the Type 1 error rate for the entire family of correlations would be very high. Although most software packages will dutifully provide the significance level for each correlation coefficient as though it were the only one tested, it is as necessary to control Type 1 error here as when we perform multiple t tests on the differences between means. One way of controlling Type 1 error when we have numerous significance tests is to adjust the significance levels, using the

Dunn–Bonferroni procedure described earlier. However, this adjustment will be conservative because the tests are not independent of one another.

To protect against excessive Type 1 error, Steiger (1980) has recommended the routine use of a simple test of the hypothesis that all off-diagonal elements of a correlation matrix are equal to zero. If this hypothesis cannot be rejected, tests on the individual correlations in the matrix are not likely to be meaningful unless motivated by *a priori* considerations. The hypothesis can be tested using the statistic

$$\chi^2 = (N-3)\sum_{j>}\sum_{i} Z_{ij}^2 \tag{18.7}$$

with $k(k - 1) / 2$ *df*, where Z_{ij} is the Fisher *Z*-transformed value of r_{ij} and the summation is over the $k(k -1) / 2$ squared *Z*s that correspond to the *r*s that are above (or below) the diagonal. For example, for the *Seasons* data set, if we correlate age, height, total cholesterol (*TC*), and body-mass index (*BMI*) for the 207 participants who identify as women and have scores for all four variables, the correlation matrix is

	Age	*Height*	*TC*	*BMI*
Age	1.000	−.126	0.513	0.070
Height	−.126	1.000	−.195	−.093
TC	0.513	−.195	1.000	0.152
BMI	0.070	−.093	0.152	1.000

First, we find the Fisher *Z* transformations for each of the six correlations above the main diagonal. These are then squared and summed, yielding

$$\Sigma\Sigma Z_{ij}^2 = (-.127)^2 + (.567)^2 + (.070)^2 + (-.198)^2 + (-.093)^2 + (.153)^2 = .382$$

so, from Equation 18.7, $\chi^2 = (204)(.382) = 77.93$. Because the observed value of χ^2 is much larger than the critical value $\chi^2_{.05,\,6} = 12.59$, we can reject the hypothesis that all the off-diagonal correlations are zero. Had we failed to reject the overall null hypothesis, and if we had no *a priori* knowledge about the correlations, we would conduct no further tests on these correlations. However, given a significant result, we may then test individual correlations. Further, even if the null hypothesis is not rejected, we may test some of the individual correlations on *a priori* grounds. For example, the literature suggests that there is a positive correlation between *age* and *TC*. Based on this sample, we find the 95% confidence interval for the correlation between *TC* and *age* extends from .41 to .61 and can therefore reject the null hypothesis that this correlation is zero in the population.

Let's now turn to the general question of whether a particular variable, *X*, has a different correlation with variable *Y* than with variable *Z*. For example, we may want to determine whether the correlation between *TC* and *age* is significantly different than the correlation between *TC* and *BMI*. Note that these are *dependent* rather than *independent* correlations because both are based on data from the same participants. Although this is a reasonable question to ask, it turns out that testing hypotheses about dependent correlations involves expressions that are complicated, not intuitive, and often very tedious to calculate, especially when the correlations to be tested do not have a variable in common.

Some useful sources that discuss tests of dependent correlations are Meng, Rosenthal, and Rubin (1992), Olkin and Finn (1990, 1995), and Steiger (1980). Perhaps the simplest test for two dependent correlations that have a variable in common is given by Meng et al. (1992). Given the null hypothesis $H_0 : \rho_{YX_1} = \rho_{YX_2}$, we can use the test statistic

$$z = \left(Z_{r_{YX_1}} - Z_{r_{YX_2}}\right)\sqrt{\frac{N-3}{2\left(1-r_{X_1X_2}\right)h}} \tag{18.8}$$

where

$$h = \frac{1-f\overline{r^2}}{1-\overline{r^2}}$$

and

$$f = \frac{1-r_{X_1X_2}}{2(1-\overline{r^2})}$$

and $\overline{r^2}$ is the mean of $r^2_{YX_1}$ and $r^2_{YX_2}$. To test whether the correlation between *TC* and *age* (.513) is significantly different from the correlation between *TC* and *BMI* (.152), we may substitute into Equation 18.8. We have $\overline{r^2}$= .143, f = .543, and h = 1.076, so that z = 4.18. Therefore, we may reject the null hypothesis at p = .000. Note that the test statistic given in Equation 18.8 is only appropriate if the two correlations have a common variable. If we wish to test the hypothesis H_0: $\rho_{xy} = \rho_{wq}$, the test statistic is considerably more complicated (see Steiger, 1980).

18.2.8 Using Software to Compare Correlations

The option to compare correlations has not been incorporated in the pull-down menus in SPSS. Thus, to compute inferential statistics or confidence intervals requires the development of an SPSS syntax file that takes correlations and sample size as input and computes the necessary steps in the relevant equation (e.g., Equation 18.8 for dependent correlations with a variable in common).

In R, the *concor* function in the {cocor} package is a highly flexible function that can compare independent or dependent correlations. For independent correlations, the data must be stored in two separate data frames. Suppose that the data frame *datW* contains variables named *tc* and *age* for the 211 *Seasons* participants who identify as women, and *datM* is analogous for the 220 participants who identify as men. Then, the command *cocor*(data = *list*(datW, datM), ~tc + age | tc + age) returns Fisher's Z test (Equation 18.4), z = 5.10, as well as the 95% CI for the difference based on Zou's (2007) procedure: [.27, .61]. The vertical bar in the equation ~tc + age | tc + age divides the correlation analysis into calculations for the first data set (datW, left of the bar), and for the second data set (datM, right of the bar). To run the same analysis on published correlations, use the command *cocor.indep.groups*(r_1, r_2, N_1, N_2), replacing the *r*s and *N*s with appropriate values.

To compare dependent correlations that have a variable in common, we can use the *concor* function with a single data frame. Assume that the data from the *Seasons* data set for the 207 participants who identify as women and have scores for *age*, *height*, *TC*, and *BMI* are stored in a data frame called dat. We can test the null hypothesis that the correlation of *TC* and *age* equals the correlation of *TC* and *BMI* with this command, *cocor*(data = dat, ~tc + age | tc + bmi, test = "steiger1980"), which returns $z = 4.21$, the same value (except for rounding error) as we obtained from Equation 18.8.

Finally, consider a common situation in which a researcher wants to assess the stability of a correlation in a group of individuals over time. These correlations are dependent because they are measured on the same individuals, and they include no overlapping data (because a variable measured at Time 1 will not be identical to that same variable measured at Time 2). The function *cocor.dep.groups.nonoverlap* in the {cocor} package computes the calculations recommended by Steiger (1980). This function takes a series of observed correlations in a particular order as input, followed by the sample size and any desired options (e.g., *conf.level*). More concretely, suppose that we are comparing the correlation of variables *A* and *B* on *N* individuals at times 1 and 2. The command *cocor.dep.groups.nonoverlap*(r_{A1B1}, r_{A2B2}, r_{A1A2}, r_{A1B2}, r_{A2B1}, r_{B1B2}, N) in the {cocor} package computes several closely related z tests and Zou's (2007) confidence interval.

18.2.9 Some Cautions About Tests That Compare Correlations

In Sections 18.2.6–18.2.8, we discussed how to test whether two correlations differ in size. However, given that we have previously emphasized that correlations are dependent on sample characteristics such as the variance of X and Y, we should give serious thought to exactly what we are testing when we compare two correlations. Suppose, for example, we find that the correlations between income and number of years of education are significantly different for people living in two different states. This means that there is a stronger linear relationship in one group than the other. However, the correlations may differ because the variances of income and/or years of education differ in the two groups, rather than because the rate of change of income with years of education is different. Also, a researcher who wishes to compare correlations should carefully evaluate whether constraints on the sampling of observations might have resulted in a biased estimate of ρ for either or both correlation coefficients (see Section 17.7).

Our second point is that we are often more interested in the slope of the regression line than the strength of the linear relationship. If our concern is about how much income changes with additional education and whether this differs across groups, we should compare regression slopes, not correlation coefficients. Just because the correlations are significantly different for the two groups does not necessarily mean that the regression slopes will be. Conversely, finding that the slopes are significantly different does not necessarily mean that the correlations are significantly different.

18.3 Partial and Semipartial (or Part) Correlations

In observational studies, many variables may be correlated with one another. In this section, we consider measures of the correlation between two variables that attempt to statistically control for the effects of other variables.

18.3.1 The Partial Correlation Coefficient

We noted in the last chapter that two variables may be correlated because they are both influenced directly or indirectly by other variables. For example, verbal ability is correlated with shoe size in young children because both verbal ability and shoe size increase with age. However, we may wish to ask whether there would still be a correlation even if the effects of age could somehow be *controlled* or *partialed out* of both variables. How can we find a measure of the relationship between size and verbal ability that is not contaminated by the effects of chronological age? By far the best way is to collect data from samples of children who are all approximately the same age. However, we can also attempt to control statistically for the effects of age in a sample of children of varying ages by finding a partial correlation.

If we use the notation r_{XY} = Corr (*X*, *Y*) to stand for the correlation between *X* and *Y*, then $r_{XY|W}$, the partial correlation between X and Y *with the effects of W partialed out*, is given by

$$r_{XY|W} = Corr(X|W, Y|W)$$

In this expression, $X \mid W = X - \hat{X}$, where $\hat{X}$, is the value of *X* predicted from the regression of *X* on *W*; therefore, *X*|*W* is the part of *X* that is not predictable from *W* with linear regression. Similarly, *Y*|*W* = *Y* – $\hat{Y}$ is the residual that results when *Y* is regressed on *W*. It is possible to express $r_{XY|W}$ in terms of the simple correlations between *X*, *Y*, and *W* as follows:

$$r_{XY|W} = \frac{r_{XY} - r_{XW}r_{YW}}{\sqrt{\left(1 - r_{XW}^2\right)\left(1 - r_{YW}^2\right)}} \tag{18.9}$$

Suppose *X* represents shoe size, *Y* represents verbal ability, and *W* represents age. If the correlations of both size and verbal ability with age are .7 ($r_{XW} = r_{YW} = .7$), and the correlation between shoe size and verbal ability is .5 ($r_{XY} = .5$), then $r_{size,verbal|age}$ would have a value of (.5 – .49) / (1 – .49) = .02. In other words, if we take into account the relationship between shoe size and age, and between verbal skills and age, the correlation between shoe size and verbal ability virtually disappears.

More than one variable may be partialed out of a correlation. Suppose that we wish to partial out the effects of variables *W* and *Q* from the correlation between *X* and *Y*. The partial correlation $r_{XY|WQ}$ is given by Corr(X | WQ, Y | WQ) = Corr(X – $\hat{X}$, Y – $\hat{Y}$), where $\hat{X}$ is the prediction of *X* based on a linear regression equation that contains both *W* and *Q* (we will cover this type of regression in Chapter 20), and $\hat{Y}$ is the corresponding prediction of *Y*. The same logic holds no matter how many variables there are to be partialed out; the partial correlation can always be obtained by correlating the two sets of residuals that result when *X* and *Y* are regressed on these variables. Such partial correlations are readily obtained by using statistical packages that either produce the partial correlations directly or provide the appropriate residuals that can then be correlated. If the raw data are not available but the first-order correlations are, Equation 18.10 indicates how to find the partial correlation between *X* and *Y*, partialing out *W* and *Q*. In effect, Equation 18.9 is first used to remove the effects of *Q* from r_{XY}, r_{XW}, and r_{YW}, and then is used again to remove the effects of *W*.

$$r_{XY|WQ} = \frac{r_{XY|Q} - r_{XW|Q}r_{YW|Q}}{\sqrt{\left(1 - r_{XW|Q}^2\right)\left(1 - r_{YW|Q}^2\right)}} \tag{18.10}$$

How are we to interpret the pattern of correlations that we observe when computing partial correlations? Often, the decision to calculate partial correlations is motivated by a desire to make causal statements. If $r_{XY|W}$ is about the same size as r_{XY}, it is said that W has no effect. If r_{XY} is significant but $r_{XY|W}$ is close to 0, it is said that either

1. r_{XY} is spurious, occurring because W influences both X and Y, not because X and Y influence one another; or
2. W is an intervening, or mediating, variable – perhaps W is influenced by X and in turn influences Y.

However, as we mentioned in the last chapter, the fact that two variables are correlated does not, by itself, mean that they are causally related. Causal conclusions require experiments to be run.

How are we to interpret the results of an analysis based on partial correlations? When we compute $r_{XY|W}$, what is removed from X and Y are any linear components *predictable* from W. Consider an example in which X measures parents' education, Y measures their children's performance in elementary school, and W is the number of books in the home. If $r_{XY|W}$ is considerably smaller than r_{XY} – that is, if the correlation between school performance and parent's education is much smaller when the number of books in the home is partialed out – this does not necessarily mean that simply providing the family with lots of books will have any effect on the children's school performance, nor does it mean that providing the books will reduce the influence of parental education on the child's educational performance. Partialing number of books out of the correlation between parental education and school performance not only removes any *direct effect* of books on school performance, but it also removes any components of parental education and children's school performance *predictable* from number of books. We would expect the number of books to be correlated with parental education, as well as with other important variables such as socioeconomic status and parental encouragement of children's achievement. Therefore, when the number of books is partialed out of the correlation between X and Y, any effects of these other variables predictable from number of books are removed as well.

In sum, the pattern of correlations we observe does not tell us the source of the pattern. Specifically, we must be very cautious about making causal inferences. If we have a well-developed causal theory, we can make decisions about which partial correlations make sense to compute and we can give them an interpretation within the context of the model. However, this does not work in reverse. We should not start with a bunch of partial correlations and use them to develop causal theories. The same pattern of results, say, a large r_{XY} but small $r_{XY|W}$, may be consistent with two or more very different causal models. These difficulties in interpretation also arise for other techniques in which there are multiple correlated variables and one or more of them are statistically "controlled" (e.g., analysis of covariance, multiple regression). We will be in a better position to discuss these issues further after we have considered multiple regression in Chapter 20.

18.3.2 Confidence Intervals and Significance Tests for Partial Correlation Coefficients

Confidence intervals and significance tests for partial correlation coefficients are completely analogous to those calculated in Section 18.2. The null hypothesis H_0: $\rho_{XY|W} = 0$ can be tested by using

$$t = r_{XY|W}\sqrt{\frac{N-3}{1-r_{XY|W}^2}} \text{ with } N-3\ df$$

and the more general null hypothesis H_0: $\rho_{XY|W} = \rho_{hyp}$ can be tested by using

$$z = \left(Z_r - Z_{\rho_{hyp}}\right)\sqrt{N-4}$$

where the Zs are Fisher transformations. The $100(1 - \alpha)\%$ confidence interval for ρ may be found by finding the limits for Z_ρ,

$$Z_r \pm z_{\alpha/2}\sqrt{\frac{1}{N-4}}$$

and then transforming back to thc corrclation scale.

In general, if r is the partial correlation with p variables partialed out, the expressions become

$$t = r\sqrt{\frac{N-2-p}{1-r^2}} \text{ with } N-2-p\ df$$

and

$$z = \left(Z_r - Z_{\rho_{hyp}}\right)\sqrt{N-3-p}$$

and the confidence limits on ρ can be obtained by transforming

$$Z_r \pm z_{\alpha/2}\sqrt{\frac{1}{N-3-p}}$$

18.3.3 The Semipartial (or Part) Correlation Coefficient

In addition to the partial correlation coefficient, it is possible to compute the semipartial correlation coefficient. This statistic has a useful interpretation in terms of multiple regression and will be discussed further in Chapter 20. For now, we simply indicate what it is and how to compute it.

The semipartial correlation coefficient $r_{Y(X|W)}$ is the correlation between Y and $X|W$, where $X | W = X - \hat{X}$, the residual when X is regressed on W. The coefficient may be obtained by regressing X on W, and then correlating the resulting residuals with Y, or by using Equation 18.11,

$$r_{Y(X|W)} = \frac{r_{XY} - r_{XW}r_{YW}}{\sqrt{1-r_{XW}^2}} \tag{18.11}$$

Notice that the part of X that can be predicted by W has been removed, but no adjustment has been made to Y.

18.3.4 *Using Software for Partial and Semipartial (or Part) Correlation*

Partial correlations are straightforward to calculate in SPSS. Simply select *Correlate* from the *Analyze* menu, and then choose *Partial*. Select the variables to correlate (X and Y in the equations in Section 18.3.1) and move the variable to be partialed out (W in the prior equations) to the "controlling for" box and click OK. For the 207 female-identifying participants in the *Seasons* data set with data on the *age*, *bmi*, *tc*, and *height* variables, the partial correlation of *age* and *tc* with *bmi* partialed out is 0.509.

Computing a semi-partial or part correlation in SPSS requires first regressing X on W to obtain the residuals: Select *Regression* from the *Analyze* menu, then *Linear*. Set up the regression of X on W, click the "save" bar, and tick the box to save unstandardized residuals; these will be stored in your data file with the name *RES_1* (assuming this is the first residual you have saved). Finally, compute the correlation of Y and *RES_1*, following the steps described in Section 18.2.4; the result will be the semipartial correlation, $r_{Y(X|W)}$. With the same data as above, $r_{tc(age|bmi)} = 0.504$.

In R, the {ppcor} package provides the functions we need for partial correlation (*pcor* or *pcor.test*) and for semipartial or part correlation (*spcor.test*). For partial correlation, *pcor* can take a data frame as input (i.e., *pcor*(dat)), in which case the output will be a matrix of partial correlations of each pair of variables with all the other variables partialed out, as well as a matrix of corresponding p-values. To compute a single partial correlation, *pcor.test*(X, Y, W) returns $r_{XY|W}$. For example, if the data described above are saved in a data frame called datW, then *pcor.test*(datW$age, datW$tc, dat$bmi) = $r_{age,tc|bmi}$ = .509. The function *spcor.test*(Y, X, W) returns $r_{Y(X|W)}$. For example, *spcortest*(datW$tc, datW$age, dat$bmi) = $r_{tc(age|bmi)}$ = .503. Note that any cases with missing data must be deleted for these functions to work; the *na.omit* function in the {base} package deletes rows with missing data on any variable.

18.3.5 *Constraints in Sets of Correlation Coefficients*

Given three variables X, Y, and W, there are three correlation coefficients, r_{XY}, r_{XW}, and r_{YW}. The range of possible values that can be taken on by any one of these correlations is constrained by the values taken on by the other two. As the most extreme example, if W has a perfect linear relationship with both X and Y, then X and Y must have a perfect linear relationship. That is, if $|r_{XW}| = |r_{YW}| = 1$, then $|r_{XY}| = 1$. However, what can we say about the possible values of r_{XY} if r_{XW} and r_{YW} are both equal to some other value (e.g., .7)?

Because $r_{XY|W}$ is a correlation, it must take on a value between −1 and +1. If we solve Equation 18.9 for r_{XY}, we have

$$r_{XY} = r_{XW}r_{YW} + r_{XY|W}\sqrt{\left(1-r_{XW}^2\right)\left(1-r_{W}^2\right)}$$

Therefore, the value of r_{XY} must lie between

$$r_{XW}r_{YW} - \sqrt{\left(1-r_{XW}^2\right)\left(1-r_{YW}^2\right)} \text{ and } r_{XW}r_{YW} + \sqrt{\left(1-r_{XW}^2\right)\left(1-r_{YW}^2\right)}$$

Substituting into these expressions, we see that if $r_{XW} = r_{YW} = .7$, r_{XY} must have a value between −.02 and 1.00. A strong negative correlation between X and Y is not possible if X and Y both have large positive correlations with W. Violating these constraints – either because of missing data (see later) or because of adjustments to the correlations – may cause complications if the correlations are used as inputs to multivariate procedures such as factor analysis or principal components analysis.

18.4 Missing Data in Correlation

The constraint discussed Section 18.3.5 assumes that all correlations are calculated with data from the same cases. In other words, if a case is missing data on any of X, Y, or W, the case is dropped when any of the correlations are calculated (i.e., listwise deletion). If, instead, the correlation for any two variables is calculated without regard to whether there are missing data on other variables (i.e., pairwise deletion), then different correlations may depend on different subsets of the sample. This can result in inconsistent correlations, especially if the missing data are nonrandom. We should also note that if pairwise deletion is used and the missing data are MCAR (missing completely at random; see Chapter 13), pairwise deletion produces parameter estimates that are consistent and therefore approximately unbiased in large samples. In contrast, if the missing data are MAR (missing at random) or MNAR (missing not at random), pairwise deletion can lead to seriously biased parameter estimates. The practical implication of these observations is that listwise deletion should be used when observations are missing from data sets containing three or more variables.

18.5 Other Measures of Correlation

In this section, we introduce several classes of correlation measures other than the usual Pearson product-moment correlation coefficient. We first discuss two measures used when one or both variables are dichotomous (i.e., they can take on only two possible values). Next, we discuss several measures of correlation used with ranked data.

18.5.1 The Point-Biserial and the Phi Correlation Coefficients

We are frequently confronted with dichotomous data (e.g., pass/fail, correct/error, experimental/control). Even though these are each categorical variables with two levels and there is nothing inherently quantitative about them, we can express each dichotomy as levels of a quantitative variable that may be correlated with other variables. For example, we can correlate a dichotomous variable with a continuous variable (e.g., passing or failing an individual test item with the overall test score), or correlate two dichotomous variables (e.g., did or didn't take driver training with pass/fail on a driver licensing exam). We can find the correlation by assigning any two different numbers to the categories that make up the dichotomy. Usually the two numbers 0 and 1 are used, but the size of the correlation would be the same for any pair of numbers; e.g., if we used 31 and 57 to represent the two levels instead of 0 and 1, the correlation would be the same.

When the usual Pearson r formula is applied to a data set in which one variable is continuous and the other variable takes on the values 0 and 1, the result is called the *point-biserial correlation coefficient*. There are specialized formulas for the correlation that take advantage of the fact that one of the variables is dichotomous, but they will give the same numerical result as the Pearson r applied to the same variables. The point-biserial correlation coefficient can be tested for significance by using the test statistic presented in Equation 18.1, and power for the test can be conveniently calculated by G*Power 3.1 by using the point-biserial model.

Given data from two independent groups, we could test the null hypothesis that the group population means are equal by a t test. If we do this, we get the same observed t and test of H_0: $\rho = 0$ as if we

1. formed $N = n_1 + n_2$ data points (X, Y) in which each Y score was paired with either $X = 1$ if the score belonged to group 1, or $X = 0$ if the score belonged to group 2, and then
2. found r_{XY}, and tested H_0: $\rho = 0$ by using Equation 18.1.

Solving Equation 18.1 for r yields

$$r = \sqrt{\frac{t^2}{t^2 + N - 2}} = \sqrt{\frac{t^2}{t^2 + df_{error}}} \tag{18.12}$$

Because of this relationship, some authors recommend that r be used as a measure of effect size to accompany t tests and, because $F(1, df_{error}) = t^2(df_{error})$, that r be used as the measure of effect size for F tests in which $df_1 = 1$ (see, for example, Rosenthal, 1991).

When both X and Y are dichotomous variables and we apply the Pearson r formula, the result is called the *phi coefficient* (ϕ). As is the case for the point-biserial coefficient, specialized formulas for ϕ exist, but these always give the same result as applying the Pearson r to the dichotomous data. Table 18.2 contains the calculation of the correlation between survival (survive, die) and treatment (drug, no drug), using one of the expressions for ϕ.

Table 18.2 An example of the calculation of the phi coefficient

		X: 1	X: 0
Y	1	*a*	*b*
	0	*c*	*d*

If there are a cases for which both X and Y are 1, b cases for which X is 0 and Y is 1, c cases for which X is 1 and Y is 0, and d cases for which X and Y are both 0, then the phi coefficient may be calculated as

$$\phi = \frac{ad - bc}{\sqrt{(a+b)(c+d)(a+c)(b+d)}}$$

Consider an example in which drug therapy and outcome are correlated. Assume the following contingency table:

		Outcome: Survive	Outcome: Die
Treatment	Drug	60	40
	No drug	40	60

the phi coefficient for treatment and outcome is

$$\phi = \frac{(60)(60) - (40)(40)}{\sqrt{(100)(100)(100)(100)}} = .20$$

The example also demonstrates that, depending on the context, even a small correlation can correspond to an important effect. In the example, the value of the correlation between treatment and survival is .2; that is, only $.2^2 = .04$ of the variance in survival is accounted for by the treatment. Yet this small correlation corresponds to a 20% increase in survival rate when the drug is administered, a difference that is obviously important.

The ϕ coefficient is closely related to the χ^2 test for independence, and we can show that

$$\chi_1^2 = N\phi^2 \tag{18.13}$$

The χ^2 statistic with 1 *df* can be used to test the hypothesis that *X* and *Y* are independent in the population, whereas the ϕ coefficient (or ϕ^2, which can be interpreted as the proportion of variance accounted for) can be used as a measure of the strength of the relationship between *X* and *Y*.

Note that with large enough samples, even very small effects may be statistically significant. Using Equation 18.13 and Appendix Table C.4 (or *qchisq*(.95, df = 1) in R), we find that we can reject the null hypothesis that *X* and *Y* are independent at $\alpha = .05$, if $\chi^2 = N\phi^2 > 3.84$ or $\phi^2 > 3.84 / N$. It follows that for $N = 1{,}000$, we would be able to reject the null hypothesis of independence even if ϕ^2 was only .00384, so that $\phi = .062$. In this case the "significant" correlation would only account for about one-third of 1% of the variance.

18.5.2 Correlation Coefficient for Ranked Data: Spearman rho

Sometimes we wish to obtain correlations for data that occur in the form of ranks. We may, for example, have two judges rank a set of stimuli according to some quality, and obtain the correlation between the two sets of rankings as a measure of reliability. Even if *X* and *Y* are continuous variables, we may wish to convert to ranks, either because we do not believe that equal differences in the *X* and/or *Y* scores necessarily correspond to equal differences in the underlying variable that is measured or because we desire measures that are more resistant to the effects of outliers than the usual Pearson *r*.

The special case of the Pearson *r* for ranked data is referred to as the *Spearman correlation coefficient* (r_S) or sometimes as the *Spearman rho coefficient*. Although the value of the Spearman coefficient can always be obtained by applying any of the usual Pearson *r* formulas to the ranked data, one frequently encounters a simple formula that takes advantage of the characteristics of ranks. If there are no ties, the ranks of *N* scores are the integers 1 to *N*. Therefore, the mean of a set of *N* ranks is $(N + 1) / 2$, and the variance of the ranks can be shown to be $N(N + 1) / 12$. Substituting such expressions into the Pearson *r* formula and simplifying yields

$$r_S = 1 - \frac{6\sum_i D_i^2}{N\left(N^2 - 1\right)} \tag{18.14}$$

where D_i is the difference between the *X* and *Y* ranks for the *i*th case. An example of a calculation using Equation 18.14 is given in Table 18.3.

Equation 18.14 should not be used if there are ties. Instead, when there are ties, all the scores in a group of ties are given the mean of the ranks they would have received had there been no ties. For example, if after the nine largest scores have been ranked we find that four scores are tied for 10th place, each receives the rank of 11.5 (the mean of 10, 11, 12,

Table 18.3 Calculation of r_S for a set of ranked data

X	*Rank of X*	Y	*Rank of Y*	D	D^2
81	9	20	8	1	1
59	3	16	5	−2	4
37	1	12	2	−1	1
79	8	21	9	−1	1
63	5	19	7	−2	4
72	7	17	6	1	1
42	2	9	1	1	1
61	4	14	3	1	1
83	10	25	10	0	0
70	6	15	4	2	4

$$r_S = 1 - \frac{6\Sigma_i D_i^2}{N(N^2 - 1)} = 1 - \frac{(6)(18)}{(10)(99)} = .89$$

and 13), and the next largest score receives a rank of 14. When this happens, the variances of ranks will not be the simple expressions assumed by Equation 18.14, so the Pearson *r* should be used on the ranks. Most statistical packages will perform both the ranking and the computation of the Pearson *r*. For example, in R, the *cor.test* function the {stats} package will compute Spearman's *rho* and a significance test when the option *method* = *c*("spearman") is used. In SPSS, choose *Correlate* from the *Analyze* menu, followed by *Bivariate*, then tick the box for *Spearman*.

For $N > 10$, we can test the null hypothesis that the ranks of *X* and *Y* have a correlation of zero in the population by using the test statistic given in Equation 18.1 with $N - 2$ *df*. Although this test is not appropriate for smaller samples, Zar (1972) has developed tables for the critical values of *rho* for small *N*; these have been reproduced in Siegel and Castellan (1988).

Finally, there are two different approaches to the problem of assessing agreement between two sets of ranks: the Kendall tau (τ and Goodman–Kruskal gamma (γ). Both are measures of *monotonicity*, the tendency for the underlying measures to increase or decrease together, a weaker condition than linearity. These measures are both affected by the presence of tied ranks, and Masson and Rotello (2009) showed that γ is an inconsistently biased statistic even for large samples. The interested reader is referred to Kendall (1938), Goodman and Kruskal (1954), and Siegel and Castellan (1988).

18.6 Summary

The first section of the chapter dealt with inference about correlation. Among the topics discussed were the following:

- The use of the *t* and normal distributions for testing hypotheses, and finding confidence intervals for ρ. We also discussed power, and the use of G*Power 3.1 to perform power calculations.
- The use of bootstrapping for finding confidence intervals for ρ when the assumption of bivariate normality was severely violated.

- Tests for whether the ρs in two independent populations were different, and the associated power calculations.
- Tests of hypotheses about dependent correlations; that is, those obtained from the same participants. We first discussed a test for the hypothesis that all the off-diagonal elements in a correlation matrix are equal to 0 in the population. We also discussed procedures for testing whether pairs of dependent correlations differed from one another.

The second part of the chapter dealt with partial and semipartial correlations. We showed how to calculate these correlations and that we could perform significance tests and form confidence intervals for them. However, we emphasized that the decision to find these correlations should be motivated by a well-developed causal model and that they should be interpreted within the context of the model. The final section of the chapter considered a measure of correlation based on ranked data.

Appendix 18.1
The Bivariate Normal Density Function

Both X and Y are assumed to be random variables and the density function that characterizes their joint distribution is

$$f(X,Y) = \frac{1}{2\pi\sigma_X\sigma_Y\sqrt{1-\rho_{XY}^2}}e^{-B}$$

where

$$B = \frac{1}{2\left(1-\rho_{XY}^2\right)}\left[\frac{\left(X-\mu_X\right)^2}{\sigma_X^2} + \frac{\left(Y-\mu_Y\right)^2}{\sigma_Y^2} - 2\rho_{XY}\frac{\left(X-\mu_X\right)\left(Y-\mu_Y\right)}{\sigma_X\sigma_Y}\right]$$

Although the equation for the density function may look formidable, we can think of it as a generalization of the one-dimensional normal distribution to two dimensions. In practice, we will not have to perform mathematical operations involving the density function and will use tables and software that deal with it. However, we should be aware of the characteristics of the bivariate normal distribution and of its importance in making inferences about the population correlation. We can see that the equation represents a family of bivariate normal distributions, with each member of the family defined by a combination of the parameters μ_x, μ_y, σ_x, σ_y, and ρ_{xy}.

Appendix 18.2
Bootstrapping Data in R to Compute a Sample Statistic

In R, the *boot*(data, statistic, R) function in the {boot} package bootstraps R samples (a number, not to be confused with the software name!), with replacement, from a data set. The same statistic is calculated for each sample and its value is returned as part of the output. The statistic can be any sample statistic; our example is the correlation, computed with the *cor* function in the {stats} package. The index i indicates the specific cases of X and Y to be used to calculate the sample statistic; the values of i change for every bootstrapped sample.

Here is example code to bootstrap 1,000 correlations of *X* and *Y* in a data frame called dat, saving the results in Boot.output:

```
Boot.output <- # save the results in Boot.output
  boot(data = dat, # the boot function takes input from data frame dat
    statistic = function(dat, i) { # calculates whatever function you put here
      cor(dat[i, "X"], dat[i, "Y"])}, #i indexes the cases
    R = 1000 # the number of replicates
    ) #closes the boot command line
```

The resulting values are a bootstrapped sampling distribution of r_{XY} stored in Boot.output$t. An example distribution of r_{XY} is shown in Figure A18.2, with the correlation in the original sample (r = .59) marked with a solid line and the sampled correlations at the 2.5 and 97.5 percentiles denoted with dashed lines. These boundaries were identified with the *boot.ci* function in the {boot} package, which takes the output of *boot* as its input (e.g., *boot.ci*(Boot.output)) and returns the boundaries of a 95% CI defined in several ways. To use percentiles to define the boundaries, use option *type* = "perc."

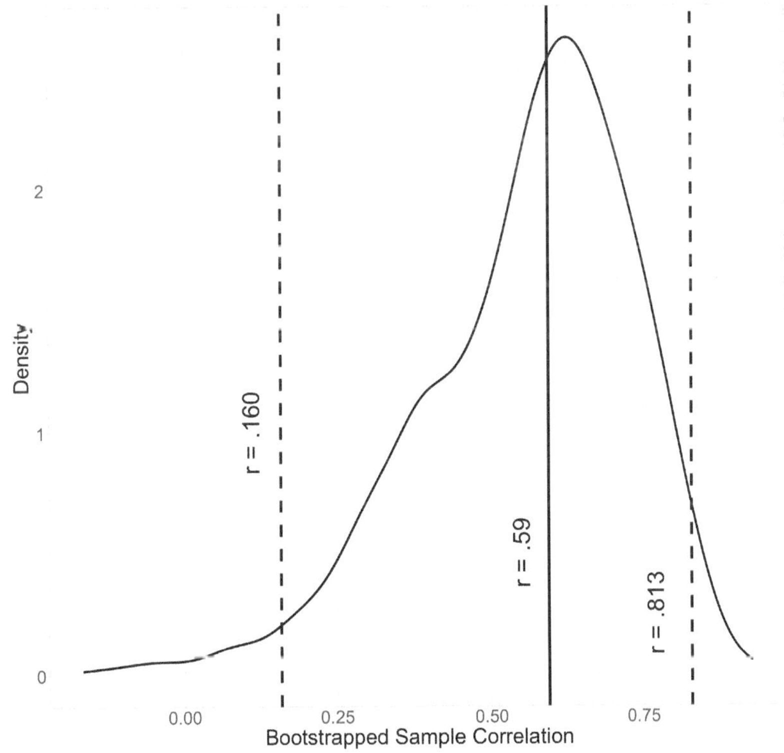

Figure A18.2 Example distribution of 1,000 bootstrapped correlations of multiplication and subtraction accuracy scores from the 28 third-graders in the *Royer_acc.xlsx* data file.

Exercises

18.1 [Hypothesis testing and power with r] In an introductory statistics course with 19 students, scores on the final examination correlate −.30 with the number of hours studied.

a) Using the t distribution, test the null hypothesis H_0: $\rho = 0$, assuming a two-tailed test with $\alpha = .05$.
b) Test the same hypothesis, using the Fisher Z transform and the normal distribution.
c) Can we conclude that studying too much interferes with test taking?
d) Find the 95% confidence interval for the population correlation coefficient, ρ. Find the 50% confidence interval.
e) Assuming that the population correlation really is −.30, use G*Power 3.1 to find the number of participants necessary to have a power of .80 for rejecting H_0: $\rho = 0$, using a two-tailed test t test with $\alpha = .05$.
f) Using the normal distribution and the test in (b), find the number of participants that would be necessary to have the power be .80 for the significance test.

18.2 [Comparing correlations] Each year, a random sample of $N = 200$ freshmen admitted to the Elite Institute of Technology (EIT) must take a standardized skills test when they first enroll. Two years ago, the correlation between the test and first-year GPA was .22. Last year, after the test had been revised, the correlation rose to .35.

a) If the two entering classes can be considered random samples of EIT freshmen, test whether the two correlations are significantly different at $\alpha = .05$.
b) Assuming that the population correlations for the two versions of the test were actually .22 and .35, use G*Power 3.1 to determine how large N in each class would have to be in order to have a power of .80 to reject the null hypothesis that the population correlations were the same (assume two-tailed test with $\alpha = .05$).
c) According to Cohen's guidelines, qs of .10, .30, and .50 correspond to small, medium, and large effect sizes. How many participants would be required to have a power of .90 for a two-tailed test of the medium-sized difference between the correlations of two groups? Use $\alpha = .05$.
d) Using Zou's procedure for finding the confidence interval for the difference between two correlations described in Section 18.2.6, find the 95% CI for the difference of the correlations obtained in part (a).

18.3 [Comparing correlations] Using the information in Exercise 17.7, test whether the correlation between income and years on the job is significantly different for people with and without a college degree. Use $\alpha = .05$, two-tailed.

18.4 [Hypothesis testing with correlations] For three independent groups, the data are as follows (use $\alpha = .05$ for any significance tests):

	G_1	G_2	G_3
n:	103	52	67
r:	.60	.45	.20

For the following, assume nondirectional alternative hypotheses with $\alpha = .05$:

a) Test H_0: $\rho_2 = 0$.
b) Test H_0: $\rho_1 = \rho_2$.
c) Test H_0: $\rho_2 = \rho_3$.
d) Find the 95% confidence interval for ρ_2.

18.5 [Testing correlations and partial correlations] A random sample of 39 students are given tests of abstract reasoning (A), quantitative reasoning (Q), and verbal skills (V). The resulting correlation matrix is

	A	Q	V
A	1.00		
Q	.30	1.00	
V	.50	.20	1.00

Note that these correlations are not independent because all the correlations are based on the same students.

a) Test the hypothesis that all the off-diagonal elements in the matrix (here the .30, .50, and .20 correlations) are equal to 0 in the population.
b) Test the hypothesis that in the population, abstract reasoning correlates equally with verbal ability and with quantitative reasoning. That is, test H_0: $\rho_{AV} = \rho_{AQ}$ against the alternative hypothesis H_1: $\rho_{AV} \neq \rho_{AQ}$.
c) Find $r_{AV|Q}$, the partial correlation between A and V with Q partialed out. Test whether it is significantly different from 0.

18.6 [Using R to test dependent correlations] Using the *Seasons* data set, assess whether the population correlation between depression (*beck_d*) and hours of exposure to weekday sunlight (*dirwdc*) is the same in the winter (1) as in the summer (3). The variables are coded for season by adding a number to the end; for example, *beck_d1* contains winter depression scores and *dirwdc3* contains summer hours of weekday sunlight exposure. Using list-wise exclusions for missing data, you should have 381 complete cases. Note that these correlations are *not independent* because they are based on data from the same individuals.

18.7 [Understanding ϕ] If we have two binary variables X and Y, we can find the correlation between them (called the "phi coefficient," ϕ), using the expression in Table 18.2.

a) What is the value of ϕ for the following 2 × 2 table?

		Item 2: Pass	Item 2: Fail	
Item 1	Pass	20	20	40
	Fail	50	10	60
		70	30	100

b) Given the marginal frequencies, what are the minimum and maximum values of ϕ that are possible?
c) Given the marginal frequencies, for what cell values would ϕ be 0?

18.8 [Correlations with sums of variables] Sometimes we will find that a correlation has been computed between some variable X and another variable T which is the sum of a number of variables including X (e.g., $T = X + Y$). Under these circumstances, we can expect a positive correlation between X and T even if X is not related to Y because X is part of T. Show that in general

$$r_{X,T-X} = \frac{r_{XT}s_T - s_X}{\sqrt{s_X^2 + s_T^2 - 2r_{XT}s_Xs_T}}$$

where $r_{X,T-X}$ is the correlation between X and the part of T not containing X.

18.9 [Correlations with difference scores] Note that the previous question has implications for the interpretation of correlations involving change or difference scores. Suppose that X refers to pretest scores and T to posttest scores. Therefore, $T - X$ refers to change scores. If we assume that $s_{pre} = s_{post}$, the equation in Exercise 18.8 reduces to

$$r_{X,T-X} = r_{pre,\,change} = \frac{r_{pre,post} - 1}{\sqrt{2\left(1 - r_{pre,post}\right)}}$$

We would not expect a perfect correlation between pre- and posttest scores for a lot of reasons, including random error. Suppose $r_{pre,post} = .70$. What do we expect for $r_{pre,change}$? Remember, because of regression toward the mean, we would expect the correlation between pretest scores and change scores to be negative.

18.10 [Correlations with sums of variables] A researcher tries to develop a new questionnaire to measure some personality trait. The instrument is made up of many items, each of which is scored numerically. The total score, T, is supposed to represent the degree to which a person has the trait. The researcher likes the instrument but thinks it will be too time-consuming to administer all of it, so she arbitrarily divides the instrument into two parts, each containing half the items. Let's refer to the score on one of the halves as X and the score on the other half as Y (so that $T = X + Y$). She finds that the correlation between X and T is high ($r_{XT} = .7$) and concludes that the correlation between the scores on the two halves of the instrument (r_{XY}) must also be pretty high, so that she can get by with using only one of the halves to measure the personality trait. If we can assume that the variances of X and Y are equal, what is r_{XY}?

18.11 [Bootstrapping to find a CI] Here are 10 students' scores on two midterm exams:

Student	1	2	3	4	5	6	7	8	9	10
Exam 1	77	79	79	87	84	80	81	77	78	78
Exam 2	78	79	80	84	84	77	82	79	74	83

a) Find the Pearson, Spearman, and Kendall correlation coefficients across exam scores.
b) Bootstrap the sample to estimate the 95% CI for Pearson r.
c) Compute the 95% CI for Pearson r and compare to your answer in (b).

18.12 [Computing correlations in real data] Using the *penguins* data from the {palmerpenguins} package in R,

a) Correlate bill length with bill depth.
b) Correlate bill length with bill depth, partialing out body mass.
c) Are there any reasons to be cautious when interpreting these correlations? If yes, make a plot to evaluate the situation.

Notes

1 Note that there are other ways to test this hypothesis. We could use the Fisher *Z* transformation along with the normal distribution (see the next section), or we could work directly with the sampling distribution of *r*. SPSS uses the *t* test described in this section to test H_0: $\rho = 0$.

2 G*Power 3.1 refers to the procedure that uses the *t* distribution to calculate power for this test as the "point-biserial model" because the procedure has a slight positive bias except for tests of the point-biserial correlation coefficient (see Section 18.4.1). For tests of the "usual" Pearson correlation coefficient for which *X* and *Y* are both quantitative continuous variables, this power calculation slightly overestimates power and underestimates the sample size required to achieve a desired level of power. This bias decreases with *N* and is negligible for $N > 30$.

3 The Fisher *Z* transformation of *r* is the inverse of the hyperbolic tangent of *r* (i.e., $Z_r = \tanh^{-1} r$). This function is available on most scientific calculators.

4 Strictly speaking, the *Z* transformation is biased by an amount $r / 2(N - 1)$ (see Pearson & Hartley, 1954, p. 29). However, this bias will be negligible unless *N* is small and *r* is large, and we ignore it here. Note also that the sample correlation *r* is a biased estimator of ρ in normal populations, and the bias can be as much as perhaps .04 under realistic conditions (see Zimmerman, Zumbo, & Williams, 2003). An approximately unbiased estimator of ρ is $\hat{\rho} = r\left[1 + \frac{1 - r^2}{2N}\right]$ but this estimator is rarely used.

5 Because bootstrapping involves random sampling, the boundaries of your bootstrapped samples on the *Royer* data may differ slightly.

Chapter 19

More About Bivariate Regression

19.1 Overview

In Chapter 19 we extend our coverage of bivariate linear regression in preparation for introducing multiple regression in Chapters 20–23. The primary goal of Chapter 19 is to discuss procedures for making inferences about bivariate regression. We also consider whether modifications of these procedures or their interpretation are required because of characteristics of the data set. The topics in Chapter 19 are as follows:

- *Inference about bivariate regression*, including confidence intervals and power calculations.
- *Inference about predictions* about values of Y at a given value of X.
- *Regression in nonexperimental research* in which the values of X are sampled from a population.
- *Consequences of measurement error in Y and in X.*
- *Unstandardized vs standardized regression coefficients.*
- *Assumptions that underlie inference.*
- *Identifying outliers and influential data points.*

We conclude the chapter by touching on several procedures for dealing with data sets that may cause difficulties for the usual, ordinary least-squares (OLS) regression.

19.2 Inference in Linear Regression

19.2.1 The Standard Fixed Linear Regression Model

Our goal in this section is to develop a basis for making inferences in situations in which the systematic relationship between two variables, X and Y, is captured by a straight line. Making inferences requires that we make certain assumptions about the population of data points. We first consider the fixed regression model in which we select certain values of X, and for each of these selected values, we randomly sample values of Y. We assume that X and Y are related in the population according to the equation

$$Y_i = \beta_0 + \beta_1 X_i + \varepsilon_i \tag{19.1}$$

where Y_i is the value of the dependent variable for the ith case, β_0 and β_1 are the Y intercept and slope of the line, X_i is the value taken on by the predictor variable for the ith case, and

DOI: 10.4324/9781003453550-23

ε_i is a random error component. We further assume that the error component, ε, is independently and normally distributed with mean 0 and variance σ^2_e; that is,

$$E(\varepsilon_i) = 0$$
$$\text{Var}(\varepsilon_i) = \sigma_e^2 \text{ for all } i \text{ (homogeneity of variance or homoscedasticity)}$$
$$\text{cov}\left(\varepsilon_i, \varepsilon_{i}'\right) = 0 \text{ except when } i = i' \text{(independence)}$$

Because we sample multiple values of Y at each value of X, there is a distribution of Y scores at each value of X. The *conditional distribution* of Y at a given value of X has a population mean, which we indicate by the symbol $\mu_{Y.X}$. Our assumptions about the population regression line imply that $\mu_{Y.X}$ lies on the straight line $\mu_{Y.X} = \beta_0 + \beta_1 X$, and that the deviation of Y from its conditional population mean is due solely to random error. If the systematic relationship between X and Y is not linear, our measure of error will include more than random variability, and the significance tests developed from this model may be biased. In this model, X is assumed to be a fixed-effect variable that is measured without error. In other words, we assume that the values of X have been selected by the researcher and that, if we replicated the study, the same values of X would be used. If these conditions are satisfied, it can be shown that b_1 and b_0, the least-squares estimators of β_1 and β_0 that we developed in Chapter 17, are both unbiased (e.g., $E(b_1) = \beta_1$) and consistent (i.e., as the sample sizes are made larger, the estimates more closely approximate the parameter values). The estimators are

$$b_1 = r\frac{s_Y}{s_X} \tag{19.2}$$

and

$$b_0 = \bar{Y} - b_1\bar{X} \tag{19.3}$$

In the next few sections, we discuss how to make statistical inferences about the slope and intercept of the regression equation and about the predictions made by the equation. In every case, we can find confidence intervals for θ, the population parameter of interest, by finding

$$\hat{\theta} \pm t_{a/2} SE(\hat{\theta}) \tag{19.4}$$

and can test hypotheses about θ by using the test statistic

$$t = \frac{\hat{\theta} - \theta_{hyp}}{SE(\hat{\theta})} \tag{19.5}$$

where $\hat{\theta}$ is the estimate of θ obtained from the sample and $SE(\hat{\theta})$ is the estimated standard error of $\hat{\theta}$. For example, the 95% confidence interval for β_1 is given by $b_1 \pm t_{.025}\, SE(b_1)$.

The general form of both Equations 19.4 and 19.5 should be familiar: They have appeared in many of the previous chapters. We are also familiar with estimating β_0 with b_0 (Equation 19.3) and β_1 with b_1 (Equation 19.2). All that remains, then, is to learn how to estimate their standard errors, $SE(b_0)$ and $SE(b_1)$; these are derived in Appendix 19.1. The results are

$$SE(b_0) = s_{Y.X}\sqrt{\frac{1}{N} + \frac{\bar{X}^2}{SS_X}} \tag{19.6}$$

and

$$SE(b_1) = \frac{s_{Y.X}}{\sqrt{SS_X}} \tag{19.7}$$

where $s_{Y.X}$ is the standard error of estimate and $SS_X = £\left(X_i - \bar{X}\right)^2$.

Although the exact forms of the expressions for the standard errors are not entirely intuitive, they have characteristics that should seem reasonable. For example,

1. We would expect that the greater the variability of the data points around the regression line, the more uncertainty there is about the location of the regression line. Therefore, the standard errors of b_0 and b_1 should increase as $s_{Y.X}$ increases.
2. We would expect more stable estimates of the regression parameters when the sample contains a wider range of X values than if the sample contains only a narrow range of Xs. Therefore, the standard errors of both b_0 and b_1 should vary *inversely* with some measure of variability in the X scores such as s_x or SS_X.
3. Because the least-squares regression line must pass though the point $(\bar{X}, \bar{Y})$, variability in the slope will affect the Y intercept less if $\bar{X}$ is close to the y-axis (i.e., if $\bar{X}$ is close to $X = 0$). Therefore, the standard error of b_0 increases as $\bar{X}$ increases.

19.2.2 A Numerical Example

The standard regression equation (Equation 19.1) has two parameters about which we might want to draw inferences: β_0 and β_1. The value of the intercept of the population regression line, β_0, is the conditional mean of Y when X has a value of zero. Estimates of β_0 are not always of interest, but we will see in Chapter 22 that there are circumstances in which they may be useful. In contrast, the slope of the population regression line, β_1, is almost always of interest. If it differs from zero, there is a linear relationship between X and Y. More importantly, the value of the slope tells us how much, on the average, Y increases (or decreases) with an increase of one unit in X.

As an example, consider the data presented in Table 19.1 and plotted in Figure 19.1. In a hypothetical visual search experiment, 20 participants each look at a computer screen and are presented with an array of letters. They are asked to indicate as quickly as they can whether a specified target letter is present in the array. Groups of five participants are assigned to array sizes of two, four, six, and eight letters, and the times (in milliseconds) to respond correctly when the target letter is present are recorded. Each participant contributes one data point. Generally, when the letter arrays are larger, it takes longer to respond to the presence of a target letter, so the data are reasonably well fit by a linear equation. Here the assumption that X is a fixed-effect variable is satisfied because the array sizes are chosen by the researcher and can be measured without error. Note that this is not the case for the statistics class example presented in Chapter 17, in which we select a sample of

Table 19.1 Data for the search experiment example

Size	*Time*
2	418
2	428
2	410
2	445
2	471
4	475
4	455
4	418
4	524
4	516
6	537
6	500
6	480
6	511
6	529
8	550
8	617
8	590
8	608
8	548

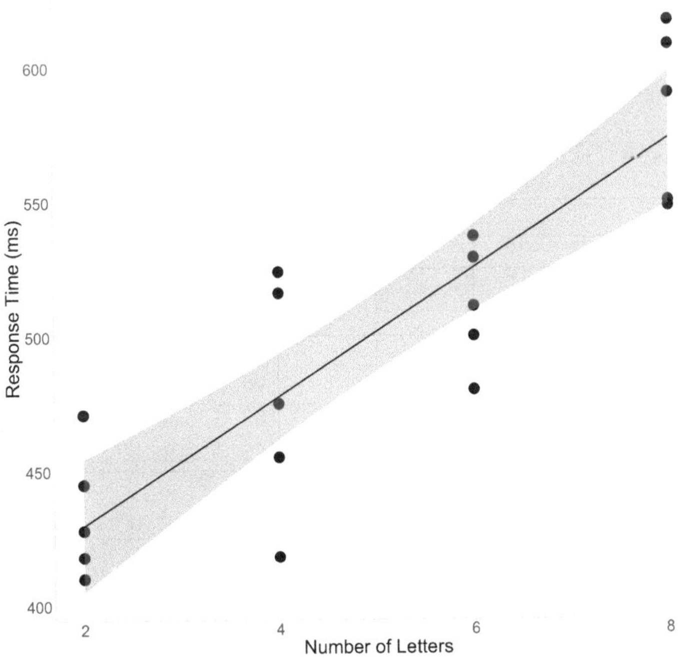

Figure 19.1 Scatterplot for the search experiment data of Table 19.1 with regression line and 95% confidence curves for the location of the linear. Note that the width of the confidence interval increases with the distance of X from $\bar{X}$.

students and obtain values of X (pretest score) and Y (final exam score) from each student, so that both X and Y are random variables.

We regressed Y (reaction time) on X (display size) using the methods described in Section 17.6.6. The results from R are shown in Figure 19.2; the SPSS output is given in Figure 19.3. Although the reports are organized differently, the analyses agree. We see that the bivariate regression equation that best predicts reaction time (Y) from array size (X) is

$$\hat{Y}_i = 381.90 + 23.92X_i$$

(In the SPSS output, these coefficients are in the column labeled "Unstandardized B.") The regression reports also display t tests for the significance of b_0 and b_1; in each case the t statistic is the ratio of the coefficient to its standard error. We see that both the slope and the intercept are significantly different from zero at $p < .001$. SPSS includes the standardized coefficients as well; these beta values are the intercept and slope from the regression of z_Y on z_X. Because the regression line must pass through the origin when the variables are standardized, the standardized intercept must be 0. The standardized slope, or beta coefficient, has the same value as the correlation coefficient in bivariate regression.[1]

In the model summary table of the SPSS output, we see the *multiple correlation coefficient*, $R_{Y.X}$ = Corr $(Y, \hat{Y})$ = .873. This is the correlation between the actual value of Y and $\hat{Y}$, the value of Y predicted from the regression equation. For bivariate regression, the multiple correlation is the absolute value of r_{XY} because the magnitude of a correlation is unchanged by a linear transformation. Note that the multiple correlation coefficient can never be negative. The R output does not include $R_{Y.X}$ but does provide its square, Multiple R-squared.

```
Call:
lm(formula = Time ~ Size, data = dat)

Residuals:
    Min      1Q  Median      3Q     Max
-59.58 -22.75   -2.16   21.24   46.42

Coefficients:
            Estimate Std. Error t value Pr(>|t|)
(Intercept)  381.900     17.227  22.169 1.62e-14 ***
Size          23.920      3.145   7.605 5.00e-07 ***
---
Signif. codes:  0 '***' 0.001 '**' 0.01 '*' 0.05 '.' 0.1 ' ' 1

Residual standard error: 31.45 on 18 degrees of freedom
Multiple R-squared:  0.7627,    Adjusted R-squared:  0.7495
F-statistic: 57.84 on 1 and 18 DF,  p-value: 5e-07
```

Figure 19.2 Regression output for the search experiment example, using the *summary* and *lm* functions in R.

Model Summary

Model	R	R Square	Adjusted R Square	Std. Error of the Estimate
1	.873[a]	.763	.749	31.45222

a. Predictors: (Constant), Size

ANOVA[a]

Model		Sum of Squares	df	Mean Square	F	Sig.
1	Regression	57216.640	1	57216.640	57.839	<.001[b]
	Residual	17806.360	18	989.242		
	Total	75023.000	19			

a. Dependent Variable: Time

b. Predictors: (Constant), Size

Coefficients[a]

Model		Unstandardized Coefficients B	Unstandardized Coefficients Std. Error	Standardized Coefficients Beta	t	Sig.
1	(Constant)	381.900	17.227		22.169	<.001
	Size	23.920	3.145	.873	7.605	<.001

a. Dependent Variable: Time

Figure 19.3 SPSS regression output for the search experiment example.

Both reports include the *standard error of the estimate* for the regression of Y on X,

$$s_{Y.X} = \sqrt{\frac{\pounds\,(Y_i - \hat{Y}_i)^2}{N-2}} = \sqrt{\frac{SS_{residual}}{N-2}} = 31.452$$

a measure of the variability around the regression line that we discussed in Chapter 17. In the R output, this value is called "Residual standard error." As we shall see, the standard error of the estimate is a component of the standard errors of most of the parameter estimates we consider in this chapter.

As we developed in Section 17.6.5, the total variability in the Y scores, $SS_Y = \Sigma(Y_i - \bar{Y})^2$, can be partitioned into two components, the variability accounted for by the bivariate regression equation,

$$SS_{regression} = \Sigma(\hat{Y} - \bar{Y})^2 = b_1^2 SS_X = r^2 SS_Y$$

and the variability not accounted for by the regression,

$$SS_{residual} = \Sigma(Y - \hat{Y})^2 = \left(1 - r^2\right) SS_Y$$

Table 19.2 Explanation of the ANOVA table in the regression output for bivariate regression

SV	*SS*	*df*	*MS*	*F*
Regression	$\Sigma(\hat{Y}_i - \bar{Y})^2 = r^2 SS_Y = b^2_1 SS_X$	1	$\frac{SS_{regression}}{1}$	$\frac{MS_{regression}}{MS_{residual}}$
Residual	$\Sigma(Y_i - \hat{Y}_i)^2 = (1 - r^2)SS_Y$	N − 2	$\frac{SS_{residual}}{N-2}$	
Total	$\Sigma(Y_i - \bar{Y})^2 = SS_Y$	N − 1		

An ANOVA table containing these terms is given in the SPSS output in Figure 19.3 and detailed in Table 19.2. In R, if you store the regression results in regr.out, then *anova*(regr.out) returns the same ANOVA table. In either case, the F is the ratio $MS_{regression}/MS_{residual}$. A significant F indicates that both r and b_1 are significantly different from zero; that is, the null hypotheses H_0: $\rho = 0$ and H_0: $\beta_1 = 0$ can both be rejected. The F can be used to test the significance of the correlation coefficient because it can be expressed as the square of the t that was presented in Chapter 17 as the test statistic for the null hypothesis H_0: $\rho = 0$. Also, the F can be written as the square of $t = b_1/SE(b_1)$, the test statistic for the null hypothesis H_0: $\beta_1 = 0$. Note that the square of the t for b_1, $(7.605)^2 = 57.836$, is, within rounding error, the same as the value of F in the ANOVA table.

We can use the standard errors shown in Equations 19.6 and 19.7 to find confidence intervals and test hypotheses about the Y intercept and slope of the regression line. For the slope of the regression equation, the 95% confidence interval for β_1 is given by

$$b_1 \pm t_{.025,18} SE(b_1) = 23.92 \pm (2.101)(3.145) = 17.31,\ 30.53$$

We can test the null hypothesis H_0: $\beta_1 = \beta_{1hyp}$ by using the test statistic

$$t = \frac{b_1 - \beta_{1_{hyp}}}{SE(b_1)} \text{ with } N-2\ df \tag{19.8}$$

If we wished to test the null hypothesis H_0: $\beta_1 = 0$, a two-tailed t test for the slope is equivalent to the F test in the ANOVA table. This may be seen by squaring the expression for the t,

$$t^2 = \frac{b_1^2}{SE(b_1)^2} = \frac{b_1^2}{s^2_{Y.X} / SS_X} = \frac{b_1^2 SS_X}{s^2_{Y.X}} = \frac{MS_{regression}}{MS_{residual}} = F$$

Suppose we wish to test the hypothesis H_0: $\beta_1 = 20$ against the alternative hypothesis H_1: $\beta_1 > 20$. The value of the test statistic is

$$t = \frac{23.92 - 20}{3.145} = 1.246$$

This result is, of course, consistent with the confidence interval limits previously calculated; those limits informed us that a two-tailed test cannot reject any hypothesized value of the slope between 17.31 and 30.53.

If we are concerned with the Y intercept, we can find confidence intervals and test hypotheses just as we did for the slope. For example, the 95% confidence interval for β_0 is

$$b_0 \pm t_{.025,18} SE(b_0) = 381.90 \pm (2.101)(17.227) = 345.71,\ 418.09$$

We know that the test of H_0: $\beta_0 = 350$ against H_1: $\beta_0 \neq 350$ at $\alpha = .05$ will not be significant because 350 lies within the 95% confidence interval for β_0.

19.2.3 Power Calculations

We can use G*Power 3.1 to perform *a priori* power tests for the hypothesis H_0: $\beta_1 = 0$ by entering information about α, the desired level of power, and the effect size,[2]

$$f^2 = \frac{R^2}{1 - R^2}$$

where R is the multiple correlation coefficient. According to the Cohen guidelines for regression, f^2 values of .02, .15, and .35 are considered "small," "medium," and "large." To run this analysis, it is helpful to view the options in the *Correlation and Regression* list of the *Tests* pull-down menu. Look for linear regression with a *fixed model*; the only options are for multiple regression, but we can reduce the number of predictors to 1. Next, consider the test of interest, which in this case focuses on the difference of R^2 from 0; selecting that option will take you to a window like the one shown in Figure 19.4. Given $f^2 = .15$, $\alpha = .05$, and desired power = .80, G*Power 3.1 indicates that a sample of size $N = 55$ is required for a "medium" effect size with one predictor variable. Similarly, sample sizes of 395 and 25 are required to obtain power = .80 for "small" and "large" effects in bivariate regression.

19.2.4 Testing Independent Slopes for Equality

We used the *Seasons* data set and regressed total cholesterol level (*TC*) on *age*, separately for the samples of 211 participants who identified as women and the 220 who identified as men and had data on both variables. The regression equations for women and men, respectively, are

$$\text{predicted } TC \text{ for women} = 131.70 + 1.71\ age\ (r = .51)$$

and

$$\text{predicted } TC \text{ for men} = 211.91 + 0.20\ age\ (r = .06)$$

In Section 18.2.5, we showed that the correlations between cholesterol level and age differed significantly for these groups, $z = 5.15$, $p = .000$. We can also test whether the regression slopes differ. As we discussed in Chapter 17, comparing slopes answers a different question than comparing correlation coefficients (see Exercise 19.5). The slope is a measure of the rate of change of Y with each unit change in X. The correlation

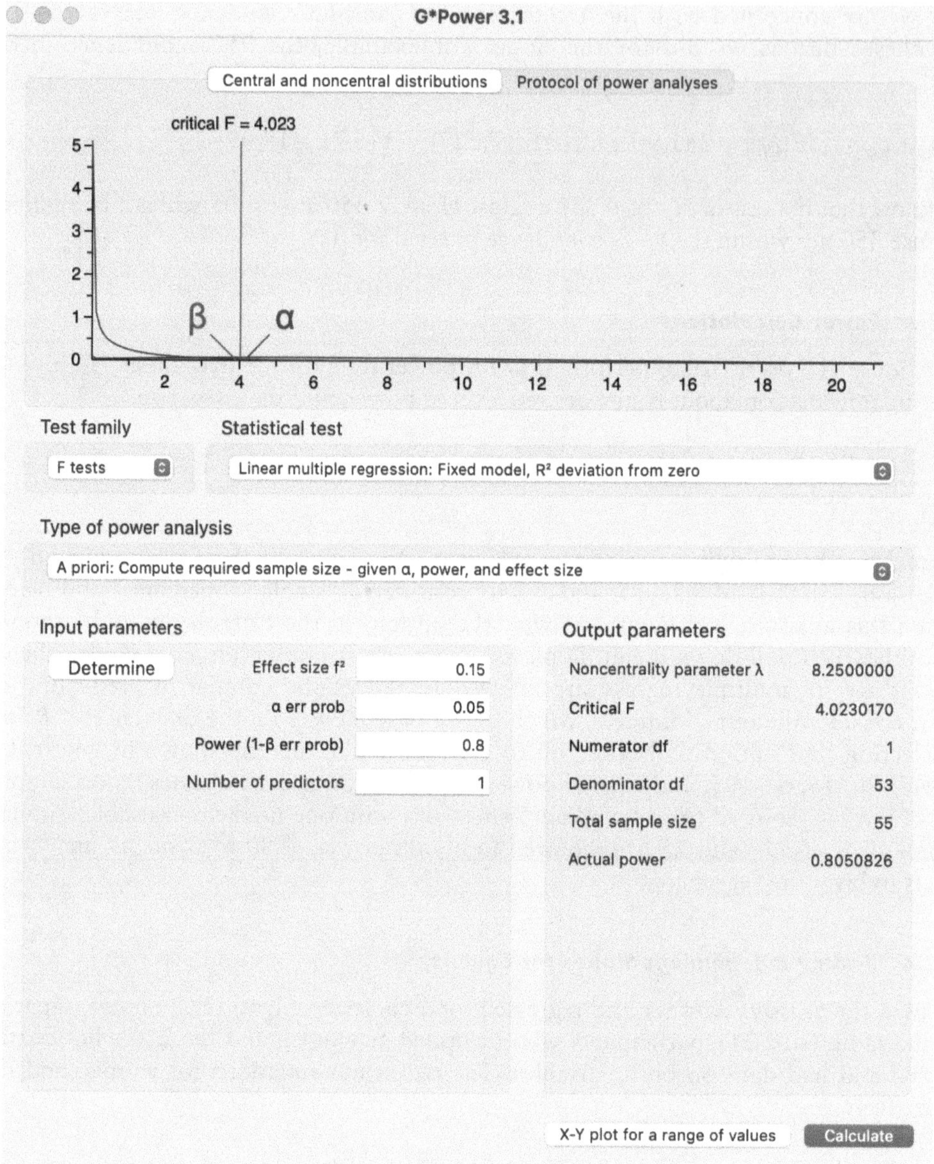

Figure 19.4 G*Power 3.1 display for *a priori* power calculations for H_0: $\beta_1 = 0$, given medium-sized effect ($f^2 = .15$, according to Cohen's guidelines), $\alpha = .05$, and power = .80.

coefficient in an index of the strength of the linear relationship between Y and X and can be thought of as a measure of the fit of the regression line. Correlation depends not only on the slope, but on the degree of scatter around the regression line as well as the variances of X and Y.

It is straightforward to compare two independent slopes in the context of multiple regression, as we will show in Chapter 22. However, even without using multiple regression, we can test the null hypothesis that two independent slopes are equal (H_0: $\beta_{11} - \beta_{12} = 0$), using the test statistic

$$t = \frac{b_{11} - b_{12}}{s_{Y.X}\sqrt{\frac{1}{SS_{X_1}} + \frac{1}{SS_{X_2}}}} \quad \text{with } N_1 + N_2 - 4 \; df \tag{19.9}$$

where $s_{Y.X}$ is the best estimate of the standard error of the estimate based on both groups (see Box 19.1). Substituting in the data for the regression of cholesterol and age for male- and female-identifying *Seasons* participants, we have

$$t = \frac{1.712 - 0.198}{36.352\sqrt{\frac{1}{28714.56} + \frac{1}{30944.92}}} = 5.08$$

so that the slopes are significantly different with $p = .000$.

Box 19.1 Testing for the Equality of Two Independent Regression Slopes

Suppose we wish to test whether the slopes are identical in populations 1 and 2; that is, H_0: $\beta_{11} - \beta_{12} = 0$. We can estimate the difference in the population slopes by $b_{11} - b_{12}$, and, because under the usual regression assumptions the *b*s can be expressed as linear combinations of the *Y*s, the ratio

$$\frac{(b_{11} - b_{12}) - (\beta_{11} - \beta_{12})}{SE(b_{11} - b_{12})}$$

is distributed as *t* with $df = N_1 + N_2 - 4$ if the null hypothesis is true. From Table 19.3 or Appendix 19.1, we know that $\text{Var}(b_1) = \sigma_e^2 / SS_X$. Further, because the groups are independent,

$$\begin{aligned} Var(b_{11} - b_{12}) &= Var(b_{11}) + Var(b_{12}) \\ &= \sigma_e^2\left(\frac{1}{SS_{X_1}} + \frac{1}{SS_{X_2}}\right) \end{aligned}$$

where SS_{X_1} and SS_{X_2} are the sums of squares of *X* in groups 1 and 2. Therefore, we can estimate the standard error of $b_{11} - b_{12}$ by $SE(b_{11} - b_{12}) = s_{Y.X}\sqrt{\frac{1}{SS_{X_1}} + \frac{1}{SS_{X_2}}}$

where the best estimate of σ^2_e is given by the weighted average of the estimates from group 1 and group 2:

$$s^2_{Y.X} = \frac{df_1 s^2_{Y.X_1} + df_2 s^2_{Y.X_2}}{df_1 + df_2} = \frac{SS_{residual_1} + SS_{residual_2}}{N_1 + N_2 - 4}$$

Combining this information, the test statistic

$$t = \frac{b_{11} - b_{12}}{s_{Y.X}\sqrt{\dfrac{1}{SS_{X_1}} + \dfrac{1}{SS_{X_2}}}} \text{ with } N_1 + N_2 - 4\ df$$

can be used to test the null hypothesis H_0: $\beta_{11} = \beta_{12}$.

Before leaving this topic, we note that the test of whether independent slopes differ depends on the assumption that the two within-group error variances are equal in the population. If this assumption is violated, the power of the test can be seriously affected, even when the sample sizes are equal. DeShon and Alexander (1996) and Overton (2001) discuss this problem and offer some possible remedies.

19.3 Using Regression to Make Predictions

In this section, we discuss inference about predictions of (1) the mean of the distribution of Y scores at a given value of X and (2) a single Y score in this distribution.

19.3.1 Obtaining a Confidence Interval for a Conditional Mean

There are occasions when it is useful to find the confidence interval for the predicted mean Y score for a given value of X. For example, a college admissions director might be interested in predicting the mean of the population of freshman grade point averages (GPA) for students with a high school GPA of 3.0; this information might inform their admissions decisions. According to the regression model, the expected value of Y at $X = X_j$ (i.e., the conditional mean) is $\mu_{Y.X_j} = \beta_0 + \beta_1 X_j$. We can show that

$$\hat{Y}_j = \hat{\mu}_{Y.X_j} = b_0 + b_1 X_j$$

is an unbiased estimator of $\mu_{Y.Xj}$, and that the estimated standard error is given by

$$SE\left(\hat{\mu}_{Y.Xj}\right) = s_{Y.X}\sqrt{h_{jj}} \tag{19.10}$$

where

$$h_{jj} = \frac{1}{N} + \frac{\left(X_j - \bar{X}\right)^2}{SS_X} \tag{19.11}$$

is the so-called *leverage* of X_j, a measure of how much of an outlier the case is with respect to the distribution of X. (See Appendix 19.2 for more detail.)

For the search experiment data, the best estimate for the conditional mean of Y at X = 4, $\mu_{Y.X} = 4$, is given by $\hat{\mu}_{Y.X=4} = 381.90 + (23.92)(4) = 477.58$, and the estimated standard error is

$$SE(\hat{\mu}_{Y.X=4}) = (31.452)\sqrt{\frac{1}{20} + \frac{(4-5)^2}{100}} = 7.70$$

Therefore, the 95% confidence interval for $\mu_{Y.X} = 4$ is

$$\hat{\mu}_{Y.X=4} \pm t_{.025,18} SE(\hat{\mu}_{Y.X=4}) = 477.58 \pm (2.101)(7.70) = 461.39,\ 493.77$$

As we can see from Equations 19.10 and 19.11, the standard error and therefore the confidence interval depend on the value of X_j. They are smallest when $X_j = \bar{X}$, and increase the more X_j deviates from $\bar{X}$. This can be seen in Figure 19.1, which is the scatterplot for the search experiment data, with the regression line and the 95% confidence interval also displayed. Hypothesis tests may be conducted in the same way as in the previous section.

19.3.2 Obtaining a Confidence Interval for a New Value of Y at X_j

We just showed how to find a confidence interval for the conditional mean of the Y scores at $X = X_j$. Now, we show how to find a confidence interval for the Y score of a *new individual* who has $X = X_j$. That is, we wish to estimate *one of the scores* from the population of scores with mean $\mu_{Y.X_j}$, where $Y_{new\,j} = \mu_{Y.X} + \varepsilon$. In our college admissions example, this situation is relevant to a decision about the admission of a particular applicant to the college. The estimate of the conditional mean, $\hat{\mu}_{Y.X_j} = b_0 + b_1 X_j$, is an unbiased estimate of $Y_{new\,j}$. Therefore, if we wish to find the confidence interval for $Y_{new\,j}$, the variability associated with $\hat{\mu}_{Y.X_j}$ contributes to error in our estimate. However, we must also consider the variability of the individual Y scores around the conditional mean. The variance of the estimate for $Y_{new\,j}$ is obtained by combining these two sources of variability, $s^2_{Y.X} + s^2_{Y.X}h_{jj} = s^2_{Y.X}(1 + h_{jj})$, so that the standard error of $Y_{new\,j}$ is given by $s_{Y.X}\sqrt{1 + h_{jj}}$ where h_{jj}, the leverage, was defined by Equation 19.11.

For the search experiment data, the predicted reaction time for a new participant with an array size of 4 is 381.90 + (23.92)(4) = 477.58, the same as the predicted conditional mean. However, the estimated standard error is

$$31.452\sqrt{1 + \frac{1}{20} + \frac{(4-5)^2}{100}} = 32.38$$

so that the 95% confidence interval for Y_{new} is 477.58 ± (2.101)(32.38) = 477.58 ± 68.03. This interval is much wider than the 95% confidence interval for the conditional mean that we previously found to be 477.58 ± 16.19: predicting an individual score is harder than predicting a mean.

Note that finding this 95% confidence interval does not allow us to conclude that 95% of the population of Y scores corresponding to X = 4 lie within the interval 477.58 ± 68.03.

As always, the correct interpretation of the confidence interval is based on what would be expected to happen during repeated sampling: Assume that:

1. We select many samples of size N, using the same values of X in each sample.
2. For each sample, we find the 95% confidence interval for the Y score of a new individual with $X = 4$.
3. For each sample we observe whether the Y score falls within the confidence interval.

Then, if the assumptions of the model are valid, the 95% confidence intervals will contain the actual scores in 95% of the samples. Table 19.3 summarizes some of the population parameters that we might wish to estimate, the sample statistics that we use as estimators, and the standard errors that can be used to form confidence intervals.

19.3.3 Using Software to Make Predictions

In R, we can use the same function, *predict* in the {stats} package, to calculate the confidence interval for a conditional mean of Y at a given value of X, and a confidence interval (often called a prediction interval) for a new single value of Y at a given value of X; the difference between the two is simply an option we set.

Begin by running the regression as described in Section 17.6.6, saving the results. We'll call that output reg.out. Then, create a new data frame – let's call it newdata – that contains

Table 19.3 Summary of bivariate regression statistics

Statistic	*Expression*	*Expected value*	*Estimated standard error for finding confidence interval*
Slope	$b_1 = \frac{\Sigma(X_i - \bar{X})(Y_i - \bar{Y})}{SS_X}$ $= r\frac{s_Y}{s_X}$	β_1	$s_{Y.X}\frac{1}{\sqrt{SS_X}}$ where $s_{Y.X} = \sqrt{\frac{SS_{residual}}{N-2}}$ is the standard error of estimate
Y intercept	$b_0 = \bar{Y} - b_1\bar{X}$	β_0	$s_{Y.X}\sqrt{\frac{1}{N} + \frac{\bar{X}^2}{SS_X}}$
Mean value of Y at X_j	$\hat{\mu}_{Y.X_j} = b_0 + b_1X_j$	$\mu_{Y.X_j}$	$s_{Y.X}\sqrt{h_{jj}}$ where $h_{jj} = \frac{1}{N} + \frac{(X_j - \bar{X})^2}{SS_X}$ is the leverage
New score at X_j	$\hat{Y}_{new\,j} = \hat{\mu}_{Y.X_j} = b_0 + b_1X_j$	$\mu_{Y.X_j}$	$s_{Y.X}\sqrt{1 + h_{jj}}$
Residual of the jth case	$e_j = Y_j - \hat{Y}_j$	0	$s_{Y.X}\sqrt{1 - h_{jj}}$

the *X* value(s) of interest. Continuing the example from Table 19.1 and Sections 19.3.1 and 19.3.2, we will make predictions from *X* = 4. In the data file, *X* is called Size, so newdata <- *data.frame*(Size = 4). Next, use the *predict* function to generate the predicted value of *Y*: *predict*(reg.out, newdata, *interval* = "confidence") will return a fit value (predicted *Y*) for each *X* in newdata, and the lower and upper bounds of the confidence interval for a conditional mean of *Y*. In our example, these are 461.39 and 493.77, equal to those calculated in Section 19.3.1. Changing the interval option to "predict," *predict*(reg.out, newdata, *interval* = "predict"), will return a fit value (predicted *Y*) for each *X* in newdata, as well as the lower and upper bounds of the confidence interval for a new individual with each *X* in newdata. In our example, the upper and lower bounds are 409.39 and 545.61, just as we calculated in Section 19.3.2.

In SPSS, begin by running the regression as described in Section 17.6.6, then click the *Save* button. In the window that pops up, tick the boxes for *Unstandardized* in the *Predicted Values* section, and for both the *Mean* and *Individual* options in the *Prediction Intervals* section. These values will be saved to your data file: *PRE_1* stores the predicted values of *Y* for each *X*, *LMCI_1* and *UMCI_1* hold the lower and upper bounds of the confidence interval for the conditional mean of Y, *LICI_1* and *UICI_1* hold the lower and upper bounds of the prediction interval for the *Y* score of a new individual with each possible value of *X*.

19.4 Regression Analysis in Nonexperimental Research

In the regression model introduced in Section 19.2.1, *X* is assumed to be fixed and measured without error. In other words, the values of *X* scores are known constants. This condition will generally only be fully satisfied when *X* is an independent variable that is manipulated in an experiment. We can consider *X* to be a fixed-effect variable in the search experiment example because the researcher selects the array sizes to be used. When a statistical inference is made, conclusions are drawn about the populations of *Y* scores at these selected levels of *X*. However, regression is commonly used with data collected in nonexperimental research in which both *X* and *Y* take on values that can vary from sample to sample and are therefore both random variables. For example, in the *Seasons* study that we have referred to throughout the book, a sample consisting of individuals between the ages of 20 and 70 was selected from the membership of a health maintenance organization and many variables were then measured. In the previous section we considered the regression of cholesterol level (*Y*) on age (*X*) from the *Seasons* data set. What should we do if both *X* and *Y* are random variables?

The usual strategy is to continue to treat *X* as though it was fixed; that is, to make our inferences conditional on the values of *X* that were obtained in the sample. In this case, we can use the same calculations for hypothesis tests and confidence intervals as for the fixed-X case that we discussed earlier. However, if we do so, we must remember that our inferences extend only to situations that have the same distribution of *X* scores as the current sample. Treating age as a fixed-effect *X* variable in the *Seasons* data set when interpreting the results of the test of equality of regression slopes for men and women in the previous section does not represent a great limitation on the inference. The age distributions for men and women are very similar and are fairly flat between the ages of 30 and 70. Moreover, age is one variable that we can usually expect to obtain without (much) error.

But suppose we wish to treat *X* as a random variable that would be allowed to vary if we repeated the study. It can be shown that the least-squares estimates for the regression coefficients will be unbiased and consistent if we can assume that the values of *Y* are drawn from a normal population with mean $\mu_{Y.X} = \beta_0 + \beta_1 X + \varepsilon$ with constant variance σ^2_e, and that X and ε, the error component, are independent.[3] The estimated standard errors calculated assuming that X is fixed will now be somewhat too small because they do not take account of the random variability in X; however, for fairly large sample sizes (say, $N > 50$), this extra uncertainty does not usually have much practical importance (Berk, 2004).

The assumption that the error component is independent of X may be violated in nonexperimental research in which uncontrolled nuisance variables may vary systematically with the variables of interest. If we regress Y on X, using data from an observational study, there may be important variables left out of the equation that are correlated with X. If so, the effects of these variables will contribute to the error component, ε, which will then be correlated with X. In this case, b_1 will not be an unbiased or a consistent estimator of β_1. We discuss this issue in more detail in Chapter 21.

19.5 Consequences of Measurement Error in Bivariate Regression

Measurement error is always a bad thing, although a certain amount is unavoidable. However, the consequences of poor measurement can be worse in standard applications of regression than in ANOVA because error in measuring the predictor variable can cause a great deal of trouble, especially when the error is systematic.

First, consider measurement error in Y, the dependent variable. Increased measurement error in Y will contribute to the error term, causing s_Y and $SS_{residual}$ to be larger. Therefore, the standard error of the estimate will be larger with the consequence that confidence intervals will be larger, significance tests will have less power, and R^2 will be smaller. Further, because the standardized regression coefficient is $b^*_1 = r = b_1 s_X / s_Y$, increasing random error in Y systematically attenuates the sample standardized coefficient. However, if the increased error in Y is random, estimates of the unstandardized regression coefficient will be unbiased.

The situation is worse when there is error in the predictor variable, X. We can see this by picturing a scatterplot for two variables that have a strong linear relationship – say, we regress height on weight. Now imagine that a random error component is added to each of the X (weight) scores. In the scatterplot, this will result in some of the X values being randomly moved to the left, and others to the right. This "smearing" of the X values will cause the slope of the regression of Y on X to become smaller. This consequence of error in the predictor variable is important, so let's consider it in more detail.

Although the usual regression model assumes that X is measured without error, the assumption is often not realistic. If we consider the situation where measurement error is present, we can express the observed value of X as consisting of two components: X_t, the "true" value (i.e., the value if there was no measurement error), and δ, an error component, so that

$$X = X_t + \delta$$

Therefore, the regression equation

$$Y = \beta_0 + \beta_1 X + \varepsilon$$

can be written as

$$Y = \beta_0 + \beta_1 X_t + (\varepsilon + \beta_1 \delta) \tag{19.12}$$

If we can assume that the measurement error in X is random and that the error component is uncorrelated with X_t, we have $E(\delta) = 0$ and $\sigma_X^2 = \sigma_{X_t}^2 + \sigma_\delta^2$. In the example of regressing height on weight, this situation might occur if the scale we used to measure weight was noisy but unbiased – say, on some trials it gives high readings and on others, low readings, but the tendency to read high or low is not related to any characteristic of the person being weighed.

If ε, δ, and X_t are all normally and independently distributed, the regression of Y on X will be linear with a slope of

$$\beta_1 = \beta_{1t}\left[\frac{\sigma_{X_t}^2}{\sigma_X^2}\right] = \beta_{1t}r_{XX} \tag{19.13}$$

where β_{1t} is the "true" slope (i.e., the slope of the equation that would be obtained by regressing Y on X_t) and $r_{XX} = \sigma_{X_t}^2/\sigma_X^2$ is the *reliability* of X (i.e., the proportion of the variability in the actual X scores that is accounted for by the true scores). Even if X_t is not normally distributed, the result holds for large samples and approximately for small samples when the reliability is high. The implication of this result is that if there is error in measuring X, the obtained slope, b_1 underestimates the magnitude of the true slope and the amount of underestimation depends directly on the amount of measurement error. If there is a great deal of measurement error, the reliability will be low, and β_1 will be much closer to zero than the true slope.

The situation is worse if we cannot assume that X_t is independent of the error component. Suppose, for example, we do not use a scale to measure weight, but instead ask each participant to provide a self-report. Such estimates of weight will probably contain systematic as well as random error because the reported weights will tend to be less than the true weights, especially for participants who are seriously overweight. Moreover, the degree of underreporting may be greater for some subgroups of participants than others. The net result is a correlation between the predictor variable and the error component, violating the assumptions made in the last section for the case in which X is a random variable. *Given this type of violation, OLS regression provides neither unbiased nor consistent estimators.* Possible remedies for dealing with correlated predictors and error components exist, but they are beyond the scope of this text (see, for example, Berk, 2004).

Although discussion of measurement sometimes gets lost in treatments of regression, it is critically important. We should always try to use the most reliable measures we can and should always be aware of the consequences of using poor measures. If we can get in trouble using measures of weight, think what can happen if we use poorly constructed assessments of psychological variables.

19.6 Unstandardized vs Standardized Regression Coefficients

When we regress Y on X, we get the regression equation

$$\hat{Y} = b_0 + b_1X$$

We can also choose to look at the standardized regression equation; that is, the equation that would result if the regression was performed using z scores,

$$\hat{z}_Y = b_1^* z_X = rz_X$$

For example, in the statistics class example discussed in Chapter 17, the raw-score regression equation for predicting final exam score (Y) from the pretest score (X) is $\hat{Y} = -36.083 + 3.546X$ and the corresponding standardized regression equation is $\hat{z}_Y = .725z_X$. The unstandardized coefficient b_1 is the change in $\hat{Y}$ (in units of Y) corresponding to a one-unit increase in X. For the statistics class example, each increase of one point on the pretest corresponds to a predicted increase of 3.546 points on the final exam. The standardized coefficient is the change in $\hat{z}_Y$ corresponding to an increase of one unit in z_X – or equivalently, b^*_1 is the change in $\hat{Y}$ in s_Y units corresponding to an increase of one s_X unit in X. For the statistics class example, this means that an increase of one standard deviation in X corresponds to an increase of .725 in $\hat{z}_Y$ (or, equivalently, an increase of .725 s_Y in $\hat{Y}$).

For any given sample, the standardized and unstandardized coefficient will have the same sign, and their significance tests will yield identical results. But does one of these coefficients better characterize the nature of the relationship? Sources such as the publication manual of the American Psychological Association state that both the unstandardized and standardized coefficients should be reported in results sections. We have no problem with this; however, unless we are specifically interested in changes stated in terms of standard deviations or in terms of relative standing, we will usually be better off working with the unstandardized coefficients for the following reasons:

1. If the scales are meaningful, one-unit changes are more understandable than changes of one standard deviation.
2. If the scales are not meaningful or if there is a great deal of measurement error, the use of unit-free measures such as R^2 and standardized coefficients is more likely to obscure this fact. Standardizing poor measures will not help. As Achen (1977, p. 806) puts it, "To replace the unmeasurable by the unmeaningful is not progress."
3. The standard deviations of X and Y in two separate samples will be different, so that standardized regression coefficients are sample specific in the same way as correlation coefficients. In fact, for bivariate regression, the standardized regression coefficients *are* the correlation coefficients and therefore should usually not be used to make comparisons across groups.
4. The standardized regression coefficient is equal to the unstandardized coefficient multiplied by s_X/s_Y. Therefore, with two separate samples, A and B, we may find that the unstandardized coefficient is larger in A than in B but that the standardized coefficient is larger in B than in A because of differences in the standard deviations (see Exercise 19.5).
5. Finally, because the standardized coefficient depends on the variances of all the variables that contribute to the error term and therefore to s_Y, whether or not they are included in the model, it is less stable than the unstandardized coefficient when variances and covariances vary across samples.

19.7 Checking for Violations of Assumptions

Because our conclusions can be seriously in error if there are severe violations of the assumptions, we next discuss how to check for violations. First, we emphasize that when we try to understand our data, we should not rely only on summary statistics such as the correlation coefficient or the slope of the regression line. It is also important to plot the data and to use the diagnostics that are usually provided by statistical software. In Figure 17.10, we presented scatterplots for four very different data sets that have identical values of N,

$b_1, b_0, SS_X, SS_{regression}, SS_{residual}, SE(b_1)$, $\bar{Y}$, b_1, b_0, SS_X, $SS_{regression}$, $SS_{residual}$, $SE(b_1)$, and r (Anscombe, 1973). As we discussed in Chapter 17, although these summary statistics strongly suggest that the relationship between Y and X is the same for all four data sets, a comparison of the four scatterplots makes clear that this is not so. One plot (a) is well fit by a straight line whereas a second (b) is curved. A third plot (c) has just one point that departs from a straight line, and a fourth would have zero X variance if one point were deleted. To understand the relationship between Y and X, we simply *must* look at the scatterplot, determine whether there are systematic deviations from linearity, determine whether there are influential data points, and check the other assumptions that underlie inference.

19.7.1 Checking Regression Assumptions Using Residuals

Valuable information about whether the assumptions of the regression model are valid may be obtained by studying residuals; that is, differences between the observed and predicted values of Y. The residuals, $e_i = Y_i - \hat{Y} = Y_i - (b_0 + b_1X_i)$, provide information about the population error components, $\varepsilon_i = Y_i - (\beta_0 - \beta_1X_i)$. Statistical software generally provides residuals and standardized residuals and allow them to be plotted in a variety of ways. If the distribution of residuals differs strongly from that assumed for the error components, the assumptions of the model may not be satisfied. Moreover, the nature of the difference can tell us which assumptions have been violated and suggest appropriate remedial measures.

Residuals cannot provide us with information about the assumption $E(\varepsilon_i) = 0$ because when least-squares regression is used, the residuals are constrained to sum to zero. The residuals can, however, provide useful information about whether there are violations of the assumptions of linearity, homogeneity of variance, normality, and independence of error. If the assumptions of linearity and homoscedasticity (homogeneity of variance) are both valid, when residuals are plotted against either X or $\hat{Y}$, the data points should lie within a horizontal band, as indicated in Panel (a) of Figure 19.5. Any other pattern suggests that the assumptions are not valid or that some kind of error has been made. For example, plots such as that in Panel (b) indicate that the relationship between Y and X is nonlinear and that the appropriate model should contain additional predictors such as X^2. Plots such as that in Panel (c), in which the residuals are more spread out for some values of X or $\hat{Y}$ than others, indicate that the variance of estimate is not constant (the homoscedasticity assumption is violated). We plot residuals against predicted values of Y rather than against Y because it can be shown that e is not correlated with $\hat{Y}$ (or, therefore, with X) but has a correlation of $\sqrt{1-r^2}$ with Y. We should also note that although plots of residuals against X and against $\hat{Y}$ provide equivalent information for bivariate regression (because Y is simply a linear function of X), this will not be the case when there is more than one predictor variable.

Another assumption of the regression model is that all relevant variables have been included in the model. If a relevant variable is omitted, the error component, ε, will consist of more than chance variability. We can determine whether an additional variable, W, belongs in the model by plotting the residuals against W. If the residual varies systematically with W, then W should be included in the model.

We used SPSS to regress total cholesterol level (TC) on *age* for women; the output for the regression is presented in Figure 19.6. The coefficients table shows that the best-fitting regression line is predicted $TC = 131.87 + 1.712$ *age*. The significant relationship between

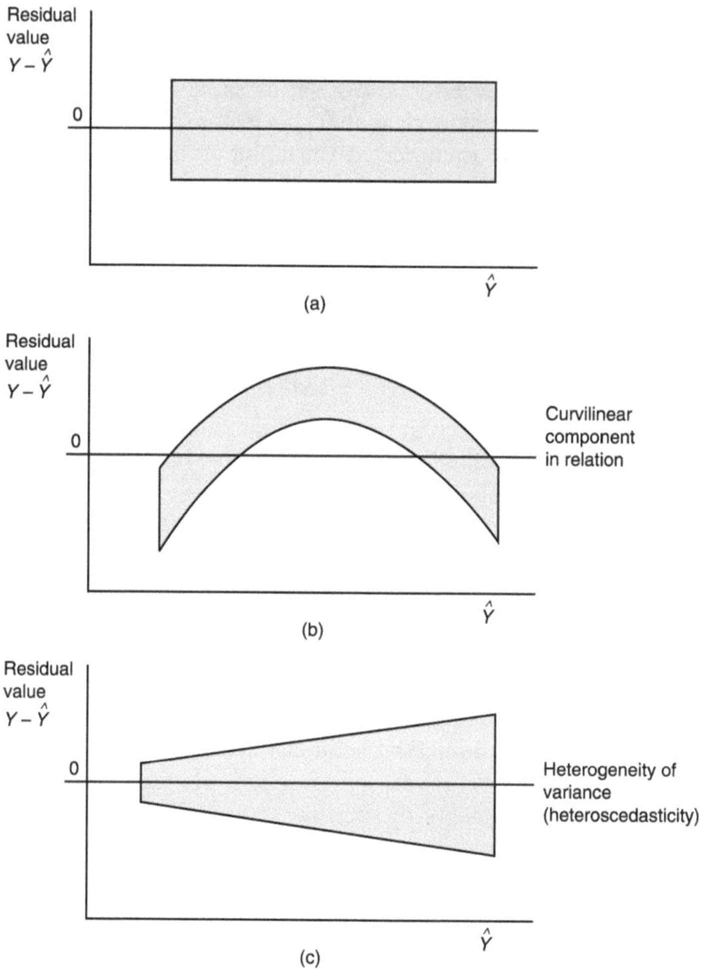

Figure 19.5 Some possible patterns for plots of residuals vs predictions in regression.

TC and *age* is demonstrated equivalently by the *t* value of 8.476 for the test of the slope, and by the *F* value of 71.838 (i.e., t^2) in the ANOVA table. We also used SPSS to plot the standardized residuals against the standardized predictions by (see Figure 19.7).[4] In Figure 19.7, the plot does not suggest any obvious curvilinearity or heteroscedasticity, although there are several outliers that we will examine more closely later.

19.7.2 An F Test for Departures from Linearity

The assumption that the conditional means of *Y* are a linear function of *X* in the population is basic to the inferential procedures that we have discussed in this chapter. Departures from linearity that are suggested by scatterplots or plots of residuals may be tested for significance by employing a procedure based on partitioning $SS_{residual}$ into two components, one

Model Summary[b]

Model	R	R Square	Adjusted R Square	Std. Error of the Estimate	Durbin-Watson
1	.506[a]	.256	.252	34.218963	2.084

a. Predictors: (Constant), age
b. Dependent Variable: tc

ANOVA[a]

Model		Sum of Squares	df	Mean Square	F	Sig.
1	Regression	84117.481	1	84117.481	71.838	<.001[b]
	Residual	244725.926	209	1170.937		
	Total	328843.407	210			

a. Dependent Variable: tc
b. Predictors: (Constant), age

Coefficients[a]

Model		Unstandardized Coefficients B	Std. Error	Standardized Coefficients Beta	t	Sig.
1	(Constant)	131.870	10.053		13.117	<.001
	age	1.712	.202	.506	8.476	<.001

a. Dependent Variable: tc

Residuals Statistics[a]

	Minimum	Maximum	Mean	Std. Deviation	N
Predicted Value	166.10110	251.67909	214.70616	20.013981	211
Residual	-114.553818	133.360168	.000000	34.137392	211
Std. Predicted Value	-2.429	1.847	.000	1.000	211
Std. Residual	-3.348	3.897	.000	.998	211

a. Dependent Variable: tc

Figure 19.6 SPSS output for the regression of *TC* on age for women.

based on systematic departures from linearity and the other based on *pure error* – that is, the variability around the conditional means of Y at the different values of X.

If the linear model is appropriate, the conditional Y means (i.e., the means of the TC values for different ages) all fall on a straight line. In this case, the variability about the straight line is the same as the variability about the means, so that $SS_{residual}$ consists only of pure error. If the linear model is not appropriate, $SS_{residual}$ consists of not only a pure error component that reflects variability about the conditional means, but also a "nonlinearity" component that reflects the extent to which the conditional means are not a perfect linear function of X. Assume that X takes on the values $X_1, X_2, \ldots, X_j, \ldots, X_a$, and that there are n_j values

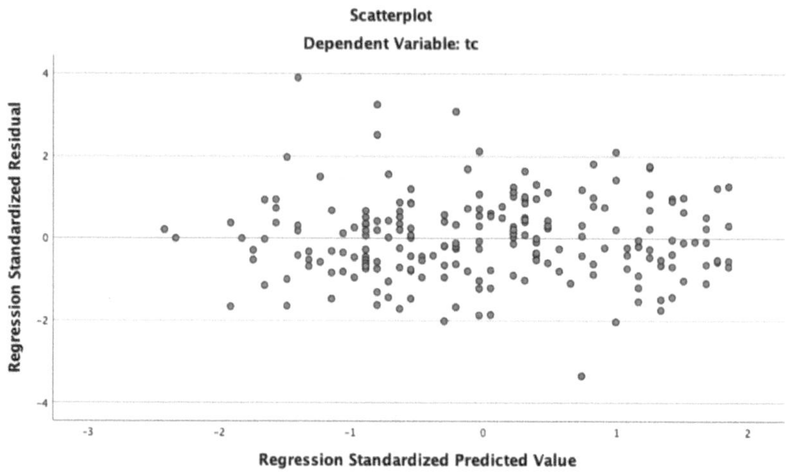

Figure 19.7 Scatterplot of standardized residuals vs standardized predictions for the regression of *TC* on age for women (from SPSS).

$Y_{1j}, Y_{2j}, \ldots, Y_{ij}, \ldots, Y_{n_j j}$ of Y at X_j. The predicted Y score at X_j is obtained from the linear equation $\hat{Y}_j = b_0 + b_1 X_j$. The identity

$$Y_{ij} - \hat{Y}_j = \left(Y_{ij} - \bar{Y}_{.j}\right) + \left(\bar{Y}_{.j} - \hat{Y}_j\right)$$

residual = pure error + nonlinearity
suggests the following partitioning of error variance:

$$\begin{aligned} \Sigma\Sigma\left(Y_{ij} - \hat{Y}_j\right)^2 &= \Sigma\Sigma\left(Y_{ij} - \bar{Y}_{.j}\right)^2 + \Sigma\Sigma\left(\bar{Y}_{.j} - \hat{Y}\right)^2 \\ SS_{residual} &= SS_{pure\ error} + SS_{nonlinearity} \end{aligned} \tag{19.14}$$

The pure error *SS* term is associated with $N - a$ *df*; there are $n_j - 1$ *df* at each of the a levels of X, and $N = \Sigma\, n_j$. The corresponding mean square, $\Sigma\Sigma(Y_{ij} - \bar{Y}_{.j})^2 / (N - a)$, estimates the variance of the scores around the conditional means of Y. The nonlinearity *SS* term is obtained by subtracting $SS_{pure\ error}$ from $SS_{residual}$. It has $a - 2$ *df* because there are a means and 2 *df* are used up in estimating the slope and intercept of the linear regression equation; equivalently, $(N - 2) - (N - a) = a - 2$. The corresponding mean square estimates a quantity that is the sum of $\sigma^2_{pure\ error}$ and a component that reflects the departure from linearity. Therefore, the linearity assumption may be tested by using

$$F = \frac{MS_{nonlinearity}}{MS_{pure\ error}} \tag{19.15}$$

with $a - 2$ and $N - a$ *df*.

Suppose we wish to test whether the relationship between *TC* and *age* departs significantly from linearity for the *Seasons* participants who identify as women. Looking at Figure 19.6,

```
> datW$age<-as.factor(datW$age)
> summary(aov(data=datW,tc~age))
             Df Sum Sq Mean Sq F value   Pr(>F)
age          46 128075    2784   2.274 8.43e-05 ***
Residuals   164 200768    1224
---
Signif. codes:  0 '***' 0.001 '**' 0.01 '*' 0.05 '.' 0.1 ' ' 1
```

Figure 19.8 Results of an ANOVA of total cholesterol (*tc*) on *age* for the *Seasons* participants who identify as women, using R.

we see from the ANOVA table of the regression output that $SS_{residual}$ = 244,725.9 with 209 *df*. Now, all we need to complete the analysis is to find $SS_{pure\ error}$, the variability of the cholesterol scores about the means of the cholesterol scores for different ages. The easiest way to do this is to perform an ANOVA in which the dependent variable is *TC* and *age* is treated as a categorical independent variable.[5] We can do this even for predictor variables such as *age* that we would not usually consider to be categorical. The test only requires that some of the values of the predictor have more than one value of *Y* associated with them so that an estimate of $SS_{pure\ error}$ may be obtained. The results of the one-way ANOVA performed with R are displayed in Figure 19.8. We see that the sum of squares term for the "Residuals" source of variance is 200,768 with 164 *df*; this is the $SS_{pure\ error}$ term in Equation 19.14. Subtracting this from $SS_{residual}$ in the regression analysis, we find $SS_{nonlinearity}$ = 244,725.9 – 200,768 = 43,957.9 with 45 *df*. Substituting into the test statistic of Equation 19.15, we have $F(45, 164) = 976.84 / 1224 = 0.80$. This result does not provide evidence of a significant departure from linearity. We can summarize the steps to test for systematic departures from linearity as follows:

1. First find $SS_{residual}$ by regressing *Y* on *X*.
2. Then find $SS_{pure\ error}$ by running an ANOVA on *Y* with *X* as the factor. In SPSS, the "within-groups" *SS* in the ANOVA is $SS_{pure\ error}$; using *aov* in R, the "Residuals" *SS* is $SS_{pure\ error}$.
3. Find $SS_{nonlinearity} = SS_{residual} - SS_{pure\ error}$ and $df_{nonlinearity} = df_{residual} - df_{pure\ error}$ and substitute into the test statistic given in Equation 19.15.

19.7.3 Normality

As indicated earlier, statistical software allows easy review of the residuals with either normal probability plots or histograms. A virtue of the normal probability plot is that if the residuals are normally distributed, the points fall on a straight line. Figure 19.9 displays both a histogram with a normal distribution smoother and a normal *Q–Q* plot for the standardized residuals of the regression of cholesterol on age for *Seasons* participants who identified as women. The distribution of residuals is slightly heavy-tailed and positively skewed. If there were large deviations from normality, we could consider transformations of the *Y* variable. Violations of the linearity and homogeneity of variance assumptions may cause the residuals to depart from normality, so that generally the linearity and homogeneity

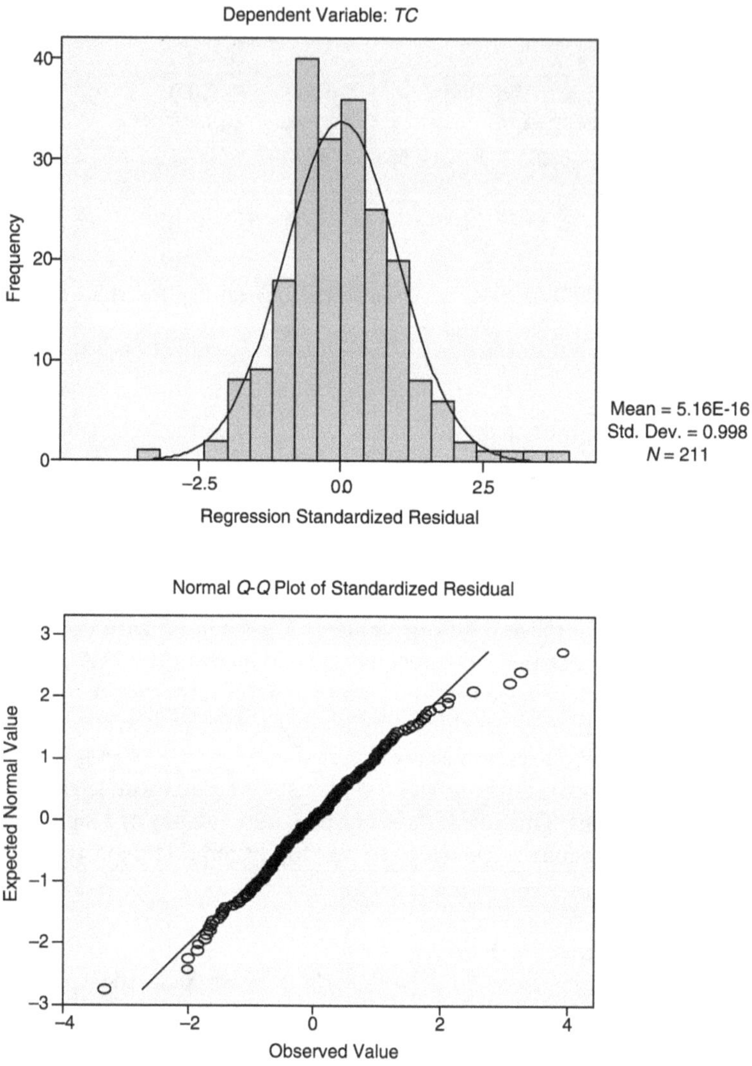

Figure 19.9 A histogram with a superimposed normal distribution and a normal Q–Q plot of the standardized residuals for the regression of cholesterol level on age for women.

of variance assumptions should be checked and addressed before looking for violations of the normality assumption.

19.7.4 Independence

We assume that the error components, the ε_i, are independent of one another. If they are positively correlated, perhaps because some important variables have been omitted from the model, standard errors calculated by the usual ordinary least-squares procedures may underestimate the true standard errors of the regression coefficients and the confidence intervals, and hypothesis tests based on them will not be appropriate. The residuals cannot be strictly independent because they are all based on the same estimates of b_0 and b_1. Nonetheless, if N is reasonably large, this unavoidable dependency will be very small so that residuals can meaningfully be examined for evidence of lack of independence. This is typically done by examining the pattern of residuals as a function of time of observation. The rationale for doing this is that data are usually collected and recorded sequentially. If the error components are independent, the residuals should not vary systematically over time. Systematic variation may reflect changes in participants, measuring devices, or surroundings. When the residuals are plotted against time or case number, the result should again look like the top panel of Figure 19.5.

It is possible for error components to exhibit different kinds of serial correlations. For example, the residual corresponding to case i may tend to be similar (or dissimilar) in size to those corresponding to case i – 1 or i – 2. Several packages print values of the Durbin–Watson D statistic that forms the basis for a test of serial correlation in adjacent residuals. The D statistic

$$D = \frac{\sum_i (e_i - e_{i-1})^2}{\sum_i e_i^2} \tag{19.16}$$

will be small when sequentially adjacent residuals are positively correlated and large when they are negatively correlated. D is approximately equal to $2(1 - r_{i,i-1})$, where $r_{i,i-1}$ is the correlation between sequentially adjacent residuals. Therefore, D can range from 0 to 4, with larger deviations in either direction from a value of 2 providing stronger evidence of serial correlation. A more detailed discussion of the test, and appropriate tables for assessing significance, may be found in Draper and Smith (1998). Under some conditions, weighted least-squares can be used to perform the regression analysis when the data are serially correlated (e.g., see Draper & Smith, 1998). In the current example, as we can see from the model summary table in the SPSS regression output in Figure 19.6, the Durbin–Watson statistic is close to 2, indicating that the serial correlation for adjacent residuals is close to zero.

To obtain the Durbin–Watson statistic in SPSS, begin the regression as described in Section 17.6.6, then click on the *Statistics* bar and tick the box for Durbin–Watson in the *Residuals* section. In R, we can use the *dwtest* function in the {lmtest} package; *dwtest*(datW$tc ~ datW$age) returns 2.084, p = .073. the same value as in SPSS.

19.8 Locating Outliers and Influential Data Points

In Chapter 2, we introduced outliers and extreme outliers as points that were more than 1.5, or 3, times the H-spread of the data. In bivariate regression, a point can be an outlier in the Y dimension if its value does not fit the general pattern established by the other data points. In other words, if its Y value, conditional on X, is unusual or extreme, then it is an outlier that may have a relatively large residual. It may also be an *influential point* if its removal would substantially change the ordinary least-squares regression outcome. If a point is an outlier in the X dimension, we say it has high *leverage*, much like an individual sitting at the extreme end of a playground teeter totter.

As we will see, outliers are not necessarily influential points, but it is always important to check for the presence of both outliers and influential data points. Once located, these

Table 19.4 The first 10 cases for the regression of cholesterol level on age for women[a] with example measures of outliers and influence

Case	*Estimate*	*Residual*	*Leverage*	*Cook*	*Student*	*Sepred*
1	222.583	.	0.005	.	.	2.532
2	205.467	.	0.006	.	.	2.596
3	215.736	−63.486	0.005	0.008	−1.871	2.359
4	239.698	−9.198	0.012	0.000	−0.270	3.774
5	224.294	8.456	0.006	0.000	0.247	2.613
6	241.410	−59.535	0.013	0.021	−1.760	3.934
7	227.717	.	0.007	.	.	2.812
8	195.198	−15.823	0.009	0.001	−0.464	3.294
9	219.159	42.716	0.005	0.004	1.253	2.414
10	226.006	.	0.006	.	.	2.707

[a] The dots indicate missing data. These come from participants who did not have cholesterol data recorded in each season.

points can be checked to determine whether they reflect different processes than the rest of the data or if they are attributable to recording or transcription errors. If so, they can be corrected or deleted from the data. If the influential points cannot be attributed to an error or failure of some sort, the appropriate way to deal with them depends on the specific research problem.

It may already be clear that our simple Chapter 2 definition of an outlier is insufficient in the context of regression. We will need a variety of measures to identify outliers and influential points. In this section, we continue to work with the regression of *TC* on *age* in the female-identified *Seasons* participants. In Table 19.4, for a small set of these participants, we have provided several of the measures we will describe in the next sections so that you may compare your own analyses to these results. Doing so will help confirm your understanding.

19.8.1 Locating Outlying Residuals and Predictors

We first deal with detecting extreme residuals, and then with extreme values of the predictor. Measures for detecting these outliers are listed in Table 19.5.

Locating Outlying Residuals. Whether a given residual is an outlier depends not only on its absolute size but also on the distribution of the other residuals. Therefore, if one is interested in locating extreme outliers, it makes sense to use some sort of standardized measure in which the raw residual is divided by something like the standard deviation. Finding that a residual has a z score of 4.50 relative to the distribution of residuals informs us more directly that it is an outlier than finding that it has an absolute value of 34.58. Although there is nothing very complicated about this basic idea, the commonly used statistical packages provide a variety of measures termed *standardized* or *Studentized residuals*. Unfortunately, different statistical software uses different names to refer to some of these measures. These naming choices will be detailed in Section 19.8.3.

We first note that the standard error for a given residual e_j is given by

$$SE(e_j) = s_{Y.X}\sqrt{1-h_{jj}} \tag{19.17}$$

where h_{jj} is the leverage of X_j (see Equation 19.11; the derivation of Equation 19.17 is in Appendix 19.2). Dividing a raw residual by its standard error results in an *internally Studentized residual* (Velleman & Welsch, 1981),

$$t_j = \frac{e_j}{s_{Y.X}\sqrt{1-h_{jj}}} \tag{19.18}$$

Although this measure seems a straightforward way to identify outlying data points, it has the important limitation that if a data point (X_j, Y_j) is far from the other data points, it may have a strong influence on the regression line (see the next section). An influential data point will pull the regression line towards itself, thereby reducing its residual, but in doing so, it will increase the residuals for many of the other data points. Because of this interdependency of the residuals, a better way to obtain an index of the extent to which a data point is an outlier is to determine its distance from a regression line that does not depend on the data point under consideration, but rather is based on the other $N - 1$ data points.

The *deleted prediction* for the *j*th case is defined as

$$\hat{Y}_j^{(-j)} = b_0^{(-j)} + b_1^{(-j)} X_j \tag{19.19}$$

where $\hat{Y}_j^{(-j)}$ is the prediction of Y from X_j, using the regression equation in which the Y intercept and slope, $b^{(-j)}_0$ and $b^{(-j)}_1$, are obtained from the $N - 1$ cases that remain when the data point (X_j, Y_j) is not included (i.e., it is "deleted" from the analysis).

The *deleted residual* for the *j*th case, $e^{(-j)}_j$, is defined as the difference between Y_j and its deleted prediction,

$$e_j^{(-j)} = Y_j - \hat{Y}_j^{(-j)} = \frac{e_j}{1-h_{jj}} \tag{19.20}$$

The ratio of the deleted residual to its standard error is called the *externally Studentized residual*, and can be expressed as

$$t_j^{(-j)} = \frac{e_j^{(-j)}}{SE\left(e_j^{(-j)}\right)} = \frac{e_j}{s_{Y.X}^{(-j)}\sqrt{1-h_{jj}}} \tag{19.21}$$

where the deleted standard error of estimate

$$s_{Y.X}^{(-j)} = \sqrt{\frac{\sum_{i \neq j}\left(Y_i - \hat{Y}_i^{(-j)}\right)^2}{N-3}} \tag{19.22}$$

is based on the $N - 1$ data points that remain after case *j* has been deleted.

We recommend the use of externally Studentized residuals to identify outlying residuals because they can be tested by the *t* statistic defined in Equation 19.21. However, as usual, when a large number of significance tests are performed, Type 1 error rate should be controlled. This can be accomplished conveniently for the family of N residuals by using the Bonferroni inequality; that is, by conducting each test at the α/N level of significance. With

Table 19.5 Measures for locating outliers

Measure	*Equation*	*Criterion*
	Measures for detecting extreme residuals	
Standardized residual	$r_j = \dfrac{e_j}{s_{Y.X}}$	
Internally Studentized residual	$t_j = \dfrac{e_j}{s_{Y.X}\sqrt{1-h_{jj}}}$	
Externally Studentized residual	$t_j^{(-j)} = \dfrac{e_j}{s_{Y.X}^{(-j)}\sqrt{1-h_{jj}}}$	$t_{.025/N}$
where	$s_{Y.X}^{(-j)} = \sqrt{\dfrac{\sum_{i \neq j}\left(Y_i - \hat{Y}_i^{(-j)}\right)^2}{N-3}}$	
	Measures for detecting outlying values of predictors	
Mahalanobis distance	$D_j = \left[\dfrac{X_j - \bar{X}}{s_X}\right]^2 = \dfrac{(N-1)\left(X_j - \bar{X}\right)^2}{SS_X}$	
Leverage	$h_{jj} = \dfrac{1}{N} + \dfrac{\left(X_j - \bar{X}\right)^2}{SS_X}$	$2(p+1)/N$
Centered leverage (SPSS)	$c_{jj} = \dfrac{\left(X_j - \bar{X}\right)^2}{SS_X} = h_{jj} - \dfrac{1}{N}$	$2\,p/N$

$\alpha = .05$ and 211 cases, the critical t is approximately 3.73. Looking at saved Studentized deleted residuals in SPSS, we can see, for example, that case 311 is an outlier with an externally Studentized residual of 4.068.

Locating Outlying Values of the Predictor. The measures considered to this point are concerned with determining the extent to which Y_j differs from its predicted value. We might also be interested in the extent to which X_j differs from the mean of the X scores. Some statistical packages provide the *Mahalanobis distance*, which for bivariate regression is just a squared z score,

$$D_j = \left[\frac{X_j - \bar{X}}{s_X}\right]^2 = \frac{(N-1)\left(X_j - \bar{X}\right)^2}{SS_X} \tag{19.23}$$

Another useful measure for identifying outliers in the X distribution that is commonly available in the regression output is the *leverage* that we discussed earlier:

$$h_{jj} = \frac{1}{N} + \frac{\left(X_j - \bar{X}\right)^2}{SS_X}$$

Leverage is closely related to the Mahalanobis distance and can be expressed in terms of it as

$$h_{jj} = \frac{1}{N} + \frac{D_j}{N-1}$$

It can be shown that the sum of the leverages for a data set is equal to $p + 1$, where p is the number of predictor variables; therefore, for bivariate regression, the h_{jj} must sum to 2 and have a mean value of $2/N$. Hoaglin and Welsch (1978) suggest that values of h_{jj} greater than $2(p + 1)/N$ should be considered large. Belsley, Kuh, and Welsch (1980) caution that this cutoff will identify too many points when there are only a few predictor variables but recommend it because it is easy to remember and use. If we use SPSS, we should note that when we ask for the leverages to be saved, what we get are *centered leverages* that are calculated from centered scores: $X_i - \bar{X}$.

$$c_{jj} = \frac{\left(X_j - \bar{X}\right)^2}{SS_X} = h_{jj} - \frac{1}{N} = \frac{D_j}{N-1} \tag{19.24}$$

For centered leverages, Hoaglin and Welsch's suggested criterion is $2p / N$.

19.8.2 Influential Points

We should look at cases that have large residuals. However, we look even more closely at cases that have an unusually large influence on the regression equation, and thereby on the predictions made using it. Note that not all outliers will have large influences. For example, outliers in Y that have leverage near 0 will have little influence on the regression function. As we shall see, cases that have large residuals and extreme values of X (high leverage) will have the greatest influence (see Equation 19.27). There are several measures that are commonly used to detect influential points. These measures are summarized in Table 19.6.

Table 19.6 Measures for detecting influential data points

Measure	*Equation*	*Criterion*
Measure of the influence of the *j*th data point on the fitted (predicted) value of Y_j		
DFFITS	$\text{DFFITS}_j = \frac{\hat{Y}_j - \hat{Y}_j^{(-j)}}{s_{Y.X}^{(-j)}\sqrt{h_{jj}}}$	$2\sqrt{(p+1)/N}$
Measure of the influence of the *j*th data point on all fitted values		
Cook's distance	$CD_j = \frac{\sum_i \left(\hat{Y}_i - \hat{Y}_i^{(-j)}\right)}{(p+1)s_{Y.X}^2}$	$F_{.50,\, p+1,\, N-p-1}$
Measure of the influence of the *j*th data point on the *k*th regression coefficient		
DFBETAS	$DFBETAS_{jk} = \frac{b_k - b_k^{(-j)}}{SE^{(-j)}(b_k)}$	$2/\sqrt{N}$

One way of assessing the influence of the jth case on the regression equation is to compare the results of the analysis when the jth case is present with the results when it is deleted. Therefore, the difference in the fitted (i.e., predicted) value, $\hat{Y}_j$ when case j is included and when it is excluded from the regression equation, $DFFIT_j = \hat{Y}_j - \hat{Y}_j^{(-j)}$, can be considered an index of the effect of the jth case.

Both $DFFIT_j$ and its standardized value,

$$DFFITS_j = \frac{\hat{Y}_j - \hat{Y}_j^{(-j)}}{s_{Y.X}^{(-j)}\sqrt{h_{jj}}} \tag{19.25}$$

where $s^{(-j)}_{Y.X}$ is as defined in Equation 19.22, can be easily calculated for each data point (see Section 19.8.3). A few criteria have been suggested for a case to be considered influential. We suggest a general cutoff of 2 and a size-adjusted cutoff of $2\sqrt{\frac{(p+1)}{N}}$ for $DFFITS$, where p is the number of predictor variables.

Cook (1977) proposed a measure that takes into consideration the effect of deleting case j on all N residuals. This measure, known as *Cook's distance*, can be expressed as

$$CD_j = \frac{\sum_i \left(\hat{Y}_i - \hat{Y}_i^{(-j)}\right)^2}{(p+1)s_{Y.X}^2} \tag{19.26}$$

where $p = 1$ for bivariate regression and, in general, p is the number of predictor variables in the regression equation. Cook and Weisberg (1982) suggest that Cook's distance should be compared to an F distribution with $p + 1$ and $N - p - 1$ df. A value of Cook's distance is considered large if it exceeds the $F(p + 1, N - p - 1)$ at the median of the distribution (i.e., where the p-value equals .5; in R, $qf(.5, p + 1, N - p - 1)$ returns the critical F). For regressions with more than five or six predictor variables, this leads to a criterion Cook's distance value of about 1; however, for bivariate regression with a sample size of about 200, as in the current example, the criterion would be about 0.70 ($qf(.5, 2, 209) = .695$).

Another useful expression for Cook's distance is

$$CD_j = \left(\frac{t_j^2}{p+1}\right)\left(\frac{h_{jj}}{1-h_{jj}}\right) \tag{19.27}$$

where t_j is the internally Studentized residual of Equation 19.18. This expression makes it clearer that the influence of a data point depends on both its residual – captured by t^2_j – and the extent to which it is an outlier – captured by its leverage, h_{jj}.

The final measures we consider here reflect differences in the regression coefficients b_0 and b_1 that result when case j is excluded from the analysis. The change in the value of the kth regression coefficient when the jth data point is deleted is given by

$$DFBETAS_{jk} = \frac{DFBETA_{jk}}{SE^{(-j)}(b_k)} \tag{19.28}$$

where $k = 0$ for the Y intercept and $k = 1$ for the slope. The standardized change is given by

$$DFBETAS_{jk} = \frac{DFBETA_{jk}}{SE^{(-j)}(b_k)} \tag{19.29}$$

where the denominator is simply the usual standard error for b_k, except that $s^{(-j)}_{Y.X}$ replaces $s_{Y.X}$. The *DFBETAS* measure has a suggested size-related cutoff of $2 / \sqrt{N}$.

Cook's distance is extremely useful for identifying influential data points that influence the fit of the regression, and the *DFBETAS* measure is of particular interest if we are concerned with the stability of the regression coefficients. Given that statistical software can readily provide these measures of influential data points, we can consider all of them.

19.8.3 Integrating Analyses of Outlying Points and Influential Points

We have now discussed several measures designed to help us detect data points that may either have been recorded erroneously or may have a disproportionate influence on our regression statistics, or both. How are we to use this information? There is no single answer; rather, we must evaluate each case on its own merits. As an example, let's look back at our regression of *TC* on age for women in Table 19.5, where case 311 seems to deserve special attention. It not only has an externally Studentized residual of 4.068, but also a *DFFITS* of .487, which is greater than the criterion of .195. It also has *DFBETAS* values of .452 and –.397 for b_0 and b_1, respectively, both of which exceed the criterion of .137. The Cook's distance for case 368 is .110. This does not exceed the criterion, but it is more than twice the size of the next largest value. When we examine case 311, we find that the data come from a participant who is relatively young (32 years old), but who has extreme *TC* (320) and *BMI* (41.1) values. There has not been any obvious error in recording the data. Although any one cholesterol reading can be in error, the cholesterol levels for this participant are over 300 in each season.

We cannot drop a data point from our analysis just because we do not like it, but we can assess whether – and by how much – our conclusions would change if the data point were excluded. If we redo the regression analysis excluding case 311, the results are much the same: The slope changes from 1.71 to 1.79, the intercept from 131.87 to 127.49, and the correlation between the observed and predicted values from .51 to .54. In the present example, dropping an extreme outlier did not affect the overall regression greatly because our analysis was based on a large number of data points. But what if the removal of this case had changed the overall analysis? After checking to make sure that our extreme points did not result from some sort of equipment or clerical error, we could present the results both with the extreme points included and without them. If our basic conclusions were supported by both sets of analyses, we could feel more confident about them. We could also collect additional data. Or we could conduct analyses using some form of *robust regression* – the term refers to procedures that are less sensitive to extreme points and some violations of assumptions.

19.8.4 Using Software to Locate Outliers and Influential Points

Outliers

In SPSS, begin the regression as described in Section 17.6.6, then click on the *Save* button, which will reveal a new window with several subsections (*Residuals*, *Influence Statistics*, *Distances*, *Predicted Values*, etc.). In the *Residuals* section, ticking the box for *Studentized* will result in SPSS computing the internally Studentized residuals and saving them in a new variable in the data file called *SRE_1*. To save externally Studentized residuals in a new variable called *SDR_1*, tick the *Studentized deleted* box. The *Standardized* option will save standardized residuals in a variable called *ZRE_1*. Several distance statistics are available in the *Distance* section of the *Save* window: *Mahalanobis* distances (calculated from centered values of *X*; saved as *Mah_1*), and *Leverage* (*Lev_1*), as well *Cook's* (*Coo_1*).

In R, begin by running the regression as described in Section 17.6.6, storing the result in an object we'll call *reg.out*. The predicted values are stored in reg.out$fitted.values and residuals are stored in reg.out$residuals, and the {stats} package includes a number of helpful functions: *rstandard*(reg.out) returns the internally Studentized residuals, and *rstudent*(reg.out) returns the externally Studentized residuals. Leverage can be calculated using the *hatvalues* function with the results of the regression as input, *hatvalues*(reg.out).

To compute Mahalanobis distance in R, take care *not* to use the *mahalanobis* function, which computes distances to the centroid of your data matrix. Instead, recall that in bivariate regression Mahalanobis distances are squared *z* scores, and that *z* scores are calculated by the *scale* function in {base} R. For data stored in a data frame called dat with an *X* variable called *X*, *scale*(dat$X)^2 will return Mahalanobis distances equal to those from Equation 19.23.

Influential Points

In SPSS, we again begin the regression as described in Section 17.6.6, then click on the *Save* button. Measures of influence are available by ticking the boxes in the *Influence Statistics* section of the window that pops up: *DfBetas* and *Standardized DfBetas* (saved as *DFB0_1*, *DFB1_1*, and *SDB0_1*, *SDB1_1*, respectively), as well as *DfFits* and *Standardized DfFits* (saved as *DFF_1* and *SDF_1*). *Cook's* distance is an option in the *Distance* section of the same window; results are saved in *COO_1*.

In R, with regression results stored in *reg.out*, the function *influence.measures*(reg.out) in the {stats} package will return *dfbetas* (standardized dfbeta values, one for intercept and one for slope), *dffits* (standardized differences in fit), *Cook's distance*, and *leverage* (called *hat*). Each of these measures is also available as its own function (*dfbetas*, *dffits*, *cooks.distance*, *hatvalues*) and takes as input the output of a regression analysis (e.g., reg.out). The unstandardized difference in betas can also be calculated with the *dfbeta* function.

19.9 Limitations of Ordinary Least-Squares Regression

Ordinary least-squares (OLS) regression, which minimizes the mean squared error of prediction, works well when its assumptions are met. When any of the assumptions described in Section 19.7 are violated, however, OLS regression can yield biased estimates and inflated standard errors. One type of violation occurs when there is heteroscedasticity in the data– when the variability in *Y* conditional on *X* is not the same for all values of *X* – which results in reduced power because the standard error of the estimate, $s_{Y.X}$, is increased. A solution may be to transform the *Y* variable; several possible transformations – and the potential pitfalls of using transformations – were discussed in Chapter 6. Another possible solution involves using a robust regression technique called weighted least-squares (WLS) regression. WLS is identical to OLS regression except that residuals are not all weighted equally. Instead, residuals based on predictor values for which there is less error variance in *Y* are weighted more heavily than residuals based on predictor values that have more error variance. The rationale is that the predictor variables associated with less prediction error are more useful, and the benefit can be smaller standard errors for b_0 and b_1.

Another problematic situation occurs when one or more data points are highly influential in the regression equation. As we discussed in Section 19.8.3, if the data are not obviously in error, the analysis should be reported both with and without the influential points. Alternatively, there are robust regression procedures that are particularly useful in situations in which there are groups of data points that collectively, but not individually, have a strong influence on the regression. For example, rather than minimizing squared prediction errors, the least absolute deviations (LAD) procedure minimizes the sum of the absolute value of the residuals, and the least median of squares (LMS) procedure minimizes the median of the squared errors. The LAD and LMS approaches are robust to outliers in the Y direction, and the LMS procedure is also robust to outliers in X (i.e., points with high leverage). These methods are beyond the scope of this book; details are available in other sources (e.g., Huynh, 1982; Draper & Smith, 1998).

A third type of challenge for OLS regression estimates stems from violations of the independence assumption: when the data set consists of *groups of related scores*, OLS regression procedures will produce biased results. For example, we would normally run a search experiment of the type discussed in Section 19.2.2 as a repeated-measures design, collecting a number of detection response times from each participant at each of the array sizes. With a repeated-measures design, we would expect that scores collected from a given participant would tend to be more similar than scores collected from different participants. If we ignored the fact that each participant provides several data points and analyzed the data in the usual way, the test of the null hypothesis, H_0: $\beta_1 = 0$, would be biased because we would expect severe violations of the independence assumption (Lorch & Myers, 1990).

Several multilevel modeling procedures, called *hierarchical linear modeling* or *multilevel regression*, have been developed that address this issue (see, for example, Goldstein, 1995; Kreft & de Leeuw, 1998; Raudenbush & Bryk, 2002). We can think of these procedures as performing regressions for each participant and weighting the regression coefficients so that those coefficients based on more variable data are given less weight when used in higher-level analyses. Although we will briefly refer to these hierarchical regression analyses again when we deal with multiple regression, detailed coverage of the relevant estimation procedures is beyond the scope of this book.

19.10 Summary

- Confidence intervals and significance tests for the population slope and intercept, as well as confidence intervals for predictions of Y scores for particular values of X, were discussed. We also considered how to test the equality of independent regression slopes.
- G*Power 3.1 was used to calculate power.
- The assumptions underlying regression were discussed. We described strategies for checking for these assumption and for dealing with their violation.
- We defined outliers and influential data points and detailed several measures that can be used to identify these points.

Appendix 19.1

The Standard Errors of b_0 and b_1

We start by showing that b_1 can be expressed as a linear combination of the Ys. From Equation 17.11,

$$b_1 = r\frac{s_Y}{s_X}$$

Substituting expressions for r (Equation 17.3), s_Y, and s_X and simplifying,

$$b_1 = \left(\frac{1}{N-1}\left(\sum\frac{(X_i-\bar{X})}{s_X}\frac{(Y_i-\bar{Y})}{s_Y}\right)\right)\frac{s_Y}{s_X} = \frac{1}{(N-1)s_X^2}\sum(X_i-\bar{X})(Y_i-\bar{Y})$$

$$= \frac{\sum(X_i-\bar{X})(Y_i-\bar{Y})}{SS_X} = \frac{\sum(X_i-\bar{X})Y_i}{SS_X} - \frac{\sum(X_i-\bar{X})\bar{Y}}{SS_X}$$

$$= \frac{\sum(X_i-\bar{X})Y_i}{SS_X} \text{ because}$$

$$\frac{\sum(X_i-\bar{X})\bar{Y}}{SS_X} = 0$$

Therefore, b_1 can be expressed as a linear combination of the Ys; that is,

$$b_1 = \sum f_i Y_i \text{ where } f_i = \frac{X_i-\bar{X}}{SS_X}$$

The variance of b_1,

$$var(b_1) = var(\sum f_i Y_i) = var(f_1Y_1 + f_2Y_2 + \ldots + f_NY_N)$$

$$= var(f_1Y_1) + var(f_2Y_2) + \ldots + var(f_NY_N), \text{ assuming independence}$$

$$= f_1^2\sigma_1^2 + f_2^2\sigma_2^2 + \ldots + f_N^2\sigma_N^2$$

$$= \sigma_e^2\left(f_1^2 + f_2^2 + \ldots + f_N^2\right), \text{ assuming homoscedasticity}$$

$$= \sigma_e^2\sum f_i^2$$

Now, because

$$f_i = \frac{X_i-\bar{X}}{SS_X}$$

$$var(b_1) = \sigma_e^2 \sum f_i^2 = \sigma_e^2 \sum \frac{(X_i - \bar{X})^2}{(SS_X)^2}$$

$$= \frac{\sigma_e^2}{SS_X} \text{ because } SS_X = \Sigma(X_i - \bar{X})^2$$

so the standard error of b_1 is

$$\frac{\sigma_e}{\sqrt{SS_X}}$$

and the **estimated standard error of b_1** is

$$SE(b_1) = \frac{s_e}{\sqrt{SS_X}}$$

To find the standard error of b_0, we start by showing that b_0 can be expressed as a linear combination of the Ys. Starting with Equation 17.12,

$$b_0 = \bar{Y} - b_1\bar{X} = \frac{1}{N}\Sigma Y_i - \bar{X}b_1 = \frac{1}{N}\Sigma Y_i - \bar{X}\Sigma f_i Y_i = \Sigma\left(\frac{1}{N} - \bar{X}f_i\right)Y_i$$

so,

$$b_0 = \Sigma g_i Y_i \text{ where } g_i = \frac{1}{N} - \frac{\bar{X}(X_i - \bar{X})}{SS_X}$$

Assuming independence and homoscedasticity, as for b_1,

$$var(b_0) = \sigma_e^2 \Sigma g_i^2$$

Substituting and simplifying, we have

$$var(b_0) = \sigma_e^2\left(\frac{1}{N} + \frac{\bar{X}^2}{SS_X}\right)$$

so that the **estimated standard error of b_0** is

$$SE(b_0) = s_e\sqrt{\frac{1}{N} + \frac{\bar{X}^2}{SS_X}}$$

Appendix 19.2

The Standard Errors of $\hat{Y}_j$ and e_j

In Appendix 19.1, we found that

$$b_1 = \Sigma f_i Y_i \text{ where } f_i = \frac{X_i - \bar{X}}{SS_X}$$

and

$$b_0 = \Sigma g_i Y_i \text{ where } g_i = \frac{1}{N} - \frac{\bar{X}(X_i - \bar{X})}{SS_X}$$

Putting these together, we see

$$\hat{Y}_j = b_0 + b_1 X_j = \Sigma g_i Y_i + X_j \Sigma f_i Y_i$$

$$= \sum_i (g_i + X_j f_i) Y_i = \sum_i \left(\frac{1}{N} - \frac{\bar{X}(X_i - \bar{X})}{SS_X} + X_j \frac{X_i - \bar{X}}{SS_X} \right) Y_i$$

$$= \sum_i \left(\frac{1}{N} + \frac{(X_j - \bar{X})(X_i - \bar{X})}{SS_X} \right) Y_i$$

so,

$$\hat{Y}_j = \sum_i h_{ij} Y_i \text{ where } h_{ij} = \frac{1}{N} + \frac{(X_j - \bar{X})(X_i - \bar{X})}{SS_X}$$

And

$$\hat{Y}_j = h_{jj} Y_j + \sum_{i \neq j} h_{ij} Y_i \text{ where } h_{jj} = \frac{1}{N} + \frac{(X_j - \bar{X})^2}{SS_X} \text{ is the leverage of case } j.$$

Then,

$$var(\hat{Y}_j) = var\left(\sum_i h_{ij} Y_i \right) = \sigma_e^2 \sum_i h_{ij}^2$$

But expanding and simplifying reveals that

$$\sum_i h_{ij}^2 = h_{jj}$$

Therefore,

$$var\left(\hat{Y}_j\right) = \sigma_e^2 h_{jj}$$

And so the standard error for the mean of Y at X_j is

$$SE\left(\hat{Y}_j\right) = s_e\sqrt{h_{jj}}$$

Finally, to derive the standard error of the residual of the jth case, SE(e_j), we begin with

$$e_j = Y_j - \hat{Y}_j = Y_j - \sum_i h_{ij} Y$$

So

$$var\left(e_j\right) = var\left(Y_j\right) + var\left(\sum_i h_{ij} Y\right) - 2cov\left(Y_j, \sum_i h_{ij} Y\right)$$

The first term on the right-hand side of the equation is equal to σ_e^2. The second term equals $\sigma_e^2 h_{jj}$. The last term is equal to $-2\sigma_e^2 h_{jj}$ because $cov\left(Y_j, Y_j\right) = \sigma_e^2$ and $cov\left(Y_j, Y_{j'}\right) = 0$ for $j \neq j'$. Therefore, $var\left(e_j\right) = \sigma_e^2(1 - h_{jj})$ and

$$SE\left(e_j\right) = s_e\sqrt{1 - h_{jj}}$$

Exercises

19.1 [Using software for regression and prediction] Use statistical software to analyze the statistics class data (file *statistics class data* on the book's website).

a) Regress final on pretest.
b) Write out the regression equation. What are the values of the standard errors of estimate for b_0 and b_1, SE(b_0), and SE(b_1)?
c) Using the regression equation, estimate the mean of the population of final exam scores with a pretest score of (i) 24; (ii) 37. Find the 95% confidence interval for each of these population means.
d) Based on your answers to (c), which estimate is more likely to be closer to the actual population value? Explain why, in terms of the leverages associated with the two pretest values.
e) Find the 95% confidence interval for the final score of an individual student with a pretest score of 24.

19.2 [Testing assumptions using residuals] The data set *EX19_2* contains response time (Y) to a target on a screen as a function of intensity level (X); the intensity levels have been coded from 1 to 5 for convenience. There are 10 participants at each value of X.

a) First, using statistical software, plot the scatter diagram, including a smoother or fit line. Then, test whether there is a linear relationship between Y and X. Save the residuals for the regression.

   i. Write out the best-fitting linear equation, using the numbers from your regression analysis. Use this equation to predict Y for each of the five X values.
   ii. Is there a significant linear relationship? Report the appropriate test statistic and *df*.

b) Now plot the residuals against the estimates. That is, produce a plot of residuals as a function of $\hat{Y}$. Include this graph with your answer. Does it suggest any problem with your analysis?
c) Fill in the following table:

SV	*df*	SS	MS	F	P
Linearity					
Lack of fit (nonlinearity)					
Pure error					

Note that if you perform an ANOVA on Y with X, treating X as a categorical independent variable, the *SS* accounted for by X is the sum of the linear and nonlinear *SS* (i.e., accounts for all the variability in the group means). The error term of the ANOVA provides an estimate of the "pure error" variability (i.e., the residual variability when all the systematic effects are partitioned out).

d) Now regress Y on both X and X^2. That is, Y should be the dependent variable and the predictor variables should be X and $XSQ = X*X$ or $XSQ = X$^2. Again, save the residuals. This estimates the parameters β_0, β_1 and β_2 for the population model

$$Y = \beta_0 + \beta_1 X + \beta_2 X^2 + \varepsilon$$

   i. Write out the equation for predicting Y with numbers taken from the output. Are β_1 and β_2 different from zero? Explain.
   ii. Does this model provide a better account of the data than the linear model? Explain.
   iii. Plot the residuals for this model against $\hat{Y}$. Do you see any problem now?

19.3 [Regression and prediction using real data]

a) Open the *Seasons* data set and select the data for women. Regress cholesterol level on age. Write out the regression equation and indicate the values of the standard error of the estimate, $SE(b_1)$, and $SE(b_0)$.
b) Using the regression equation, estimate the means of the populations of cholesterol scores for women of ages (i) 30 and (ii) 50. Find the 95% confidence interval for each of these population means.
c) Which estimate is more likely to be closer to its actual population value, that for 30- or for 50-year-old women? Explain why.

d) What is the 95% confidence interval for the cholesterol score of a randomly chosen 30-year-old woman?

19.4 [Comparing regression and ANOVA] Given the following data from a between-participants experiment in which the dependent variable is a performance measure *Y*:

Drug dosage (D)			
10	*20*	*30*	*40*
27	38	69	60
17	32	64	57
14	10	59	55
20	26	57	30
15	29	35	50

a) Regress *Y* on *D*. What is the best linear equation? Is the slope of the regression line significantly different from 0?
b) Perform an ANOVA, using *D* as the independent variable. Is the *D* effect significant? How exactly does the null hypothesis in part (b) differ from that in part (a)?

19.5 [Regression slopes vs correlations] We have previously considered data from a large study of income (*Y*) as a function of years on job (*X*); the data 2,000 workers with a college degree and 2,000 without a degree are as follows:

	College graduate		*No college degree*	
	Income (Y)	*Years (X)*	*Income*	*Years*
s^2	324	100	289	25
r_{xy}	.333		.235	

Note: Income is recorded in thousands of dollars.

In Exercise 17.7 we found that the correlation between income and years of service was significantly larger for college graduates than for people without degrees, $z = 3.38$, $p < .001$. Now, find b_{YX} (i.e., $b_{Income,\ Years}$), the regression coefficient for the regression of income on years of service for college graduates and for people without degrees.

a) What is your best estimate of the amounts by which salary increases per year for college graduates and for people without degrees?
b) Is the rate of increase significantly different for these groups?
c) Is this result consistent with differences in the correlations? Explain.

19.6 [Comparing regression and ANOVA] In a search experiment, participants are required to check for the presence of some target character in an array of characters. There are four different array sizes, $X = 2, 4, 6$, and 8. Ten participants are assigned to each array size. The time to respond for each of the 40 participants (*Y*) is recorded. The data for the four array sizes are as follows:

X_j	2	4	6	8
$\bar{Y}_{.j}$	480	520	540	540
s^2_j	360	315	324	333

Two researchers, A and B, have different views about the analysis. A uses the ANOVA design model

$$Y_{ij} = \mu + \alpha_j + \varepsilon_{ij}$$

to test the hypothesis H_0: $\mu_1 = \mu_2 = \mu_2 = \mu_4$, and B assumes the linear regression model

$$Y_{ij} = \mu_Y + \beta_1\left(X_j - \bar{X}\right) + \varepsilon_{ij}$$

and tests the hypothesis H_0: $\beta_1 = 0$.

a) Are A and B testing equivalent hypotheses? Briefly, justify your answer. If your answer is "no," are the two null hypotheses related? That is, if A's is false, should B's be true? Or if B's is false, should A's be true?
b) Use ANOVA to test H_0: $\mu_1 = \mu_2 = \mu_3 = \mu_4$; and use regression to test H_0: $\beta_1 = 0$.
c) Must SS_A always be larger than $SS_{regression}$?
d) Determine whether there is a significant departure from linearity in the data, using $\alpha = .05$.

19.7 [Testing equality of two slopes] Two groups of 40 people each participate in the kind of search experiment described in the chapter. For each group, $SS_X = 200$. For Group 1, we obtain $b_1 = 30.0$ and $s_e = 15.5$; for Group 2, $b_1 = 19.0$ and $s_e = 12.2$. Test whether the slopes for these groups differ significantly at $\alpha = .05$.

19.8 [Regression with repeated-measures data]

a) The search experiment described in Section 19.2 is rerun as a repeated-measures study; two different research assistants (RAs) each run half of the participants. In one condition, letters are used as stimulus material. Each of the 20 participants in this condition is tested at all four array sizes, and slopes are obtained for each participant by performing separate regressions. The slopes are as follows:

RA1	35	25	29	37	20	24	18	31	30	25
RA2	17	19	29	19	23	25	20	18	22	25

Find the 95% confidence interval for the difference in population slopes for the participants run by the two research assistants when letters are used.

b) In a second condition using different participants, digits (i.e., the numbers 0 through 9) are used as stimulus material. In this second condition, the slopes are

RA1	30	19	28	38	16	26	22	28	33	21
RA2	19	21	24	22	20	23	20	15	25	28

Test whether the slopes are significantly different for the participants run by the two RAs in this condition. What exactly can you conclude from the significance test?

c) From the results of both conditions, test the following, using slope as the dependent variable:

i. The interaction between RA and type of stimulus material (letters vs digit).
ii. The main effect of RA.
iii. The main effect of type of stimulus material.

19.9 [Interpreting correlation and regression] For a bivariate regression the multiple correlation coefficient R is .40. What does this tell you about the accuracy of predictions made with the regression equation?

19.10 [Regression in real data] Using the *Seasons* data set, we established in the chapter that the model assumptions were reasonably well satisfied for the regression of cholesterol level on age for participants who identify as women. We also established that the regression results were not strongly distorted by the presence of outliers and influential data points. Go through the same types of steps for the regression of cholesterol level on age for the male-identifying participants and write a brief report about what you find.

Notes

1 Unfortunately, both the standardized coefficients and the population parameters of the unstandardized coefficients are referred to as "betas." We will most often be concerned with unstandardized coefficients, and so will usually reserve the use of the β notation to refer to their population parameters. We will use b^* to refer to the sample standardized coefficient.

2 Under the assumptions described in Section 19.2.1, the noncentrality parameter is $\lambda = Nf^2$.

3 Different authors make somewhat different assumptions when dealing with random-effects predictors. Useful discussions about what happens when both X and Y are random variables may be found in Fox, 1997, pp. 113–114; Hays, 1994, pp. 637–638; Mittelhammer, Judge, and Miller, 2000, pp. 17–24 and 225–235; Neter et al., 1996, p. 85; and especially Berk, 2004, pp. 69–73.

4 In SPSS, set up the regression then click on the "Plots" bar and select *SRESID to be on the y-axis and *ZPRED to be on the x-axis; click Continue and OK. In R, if the regression output is stored in reg.out, the function *rstandard*(reg.out) in {stats} will return standardized residuals; these can be plotted against the predicted values of Y, which are in reg.out$fitted.values.

5 In R, this means *age* must be a factor. For data in a data frame called dat, we can convert the *age* variable using the *as.factor* function in the {base} package: dat$age <- *as.factor*(dat$age).

Chapter 20

Introduction to Multiple Regression

20.1 Overview

In chapters 17 and 19 we considered situations in which a dependent variable was regressed on a *single* predictor variable. Among the examples we discussed were the regressions of response time on stimulus array size, final exam score on pretest score, and cholesterol level on age. However, in most research situations, there are many relevant variables, and it is often useful to consider more than one predictor. If our goal is to generate accurate predictions, surely predictions should be better if we base them on more information. For example, in the statistics class example, we would expect to predict final exam performance better if we considered other measures of ability along with pretest score. If, on the other hand, our goal is to use regression for the much more difficult task of developing or testing an explanatory model, we may gain a better understanding of the situation if we study several variables simultaneously because dependent variables of interest are often influenced by many other variables.

In the next few chapters, we develop the basic ideas of multiple regression: regression of Y on more than one predictor variable. Following the introduction in the current chapter, we go into more detail in Chapter 21 about inference, assumptions, and power calculations. In Chapter 22 we extend our discussion of some of the issues that complicate the use of multiple regression analysis to inform theory development and consider how to add and interpret curvilinear and interaction components. In Chapter 23, we show how to incorporate qualitative categorical variables into regression, thereby providing a powerful and flexible framework within which ANOVA and analysis of covariance (ANCOVA, Chapter 24) are special cases. In the current chapter, our goals are as follows:

- *Illustrate preliminary analyses of a data set including multiple predictor variables.*
- *Introduce the basic terms and concepts of multiple regression and discuss the partitioning of variance in multiple regression.*
- *Consider measures of fit and strategies for reducing bias in fit.*
- *Illustrate multiple regression analyses of the example data set.*
- *Discuss the meaning of the regression coefficients and some of their limitations in developing explanatory models.*
- *Introduce the concept of suppression*, in which adding a predictor variable to a regression equation improves prediction even if the added variable has little or no correlation with Y.

DOI: 10.4324/9781003453550-24

20.2 An Example With Preliminary Analyses

Before introducing the concepts and analyses associated with multiple regression, we describe a data set with multiple possible predictors and illustrate preliminary analyses on the data. This data set will later be used to illustrate multiple regression analyses.

In Chapter 19, we found that cholesterol level tends to increase with age in the *Seasons* participants who identify as women. However, cholesterol level also changes systematically with other variables, such as weight. In the current section, we perform regressions involving total cholesterol level (*TC*), *age*, and body-mass index (*BMI*), which is defined as weight (in kilograms) divided by the square of height (in meters). In doing so, we use data from female-identified participants in the *Seasons* study who were 20–65 years of age when they entered the study.

If we are interested in how *TC*, *BMI*, and *age* are related, the first step is, as always, to look at the data. Figure 20.1 contains the scatterplot matrix for the three variables. The distribution of *BMI* scores, shown in the lower right panel, is positively skewed and highly peaked, and from the descriptive statistics in Panel *a* of Table 20.1, we find that both skewness and kurtosis are very large compared to their standard errors. The box plot for *BMI* scores in Figure 20.2 indicates that there are outliers, and we can see these points clearly in the scatterplots. When we plot *TC* against *BMI* in Figure 20.3 and apply a LOWESS

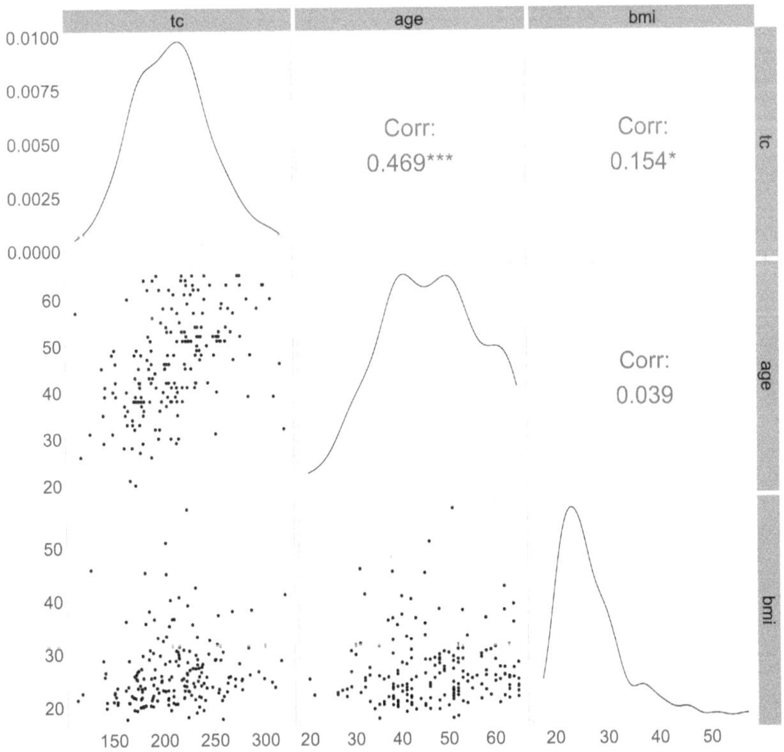

Figure 20.1 Scatterplot matrix for *TC*, *age*, and *BMI* using data from all women 65 years of age or younger and box plot for *BMI*.

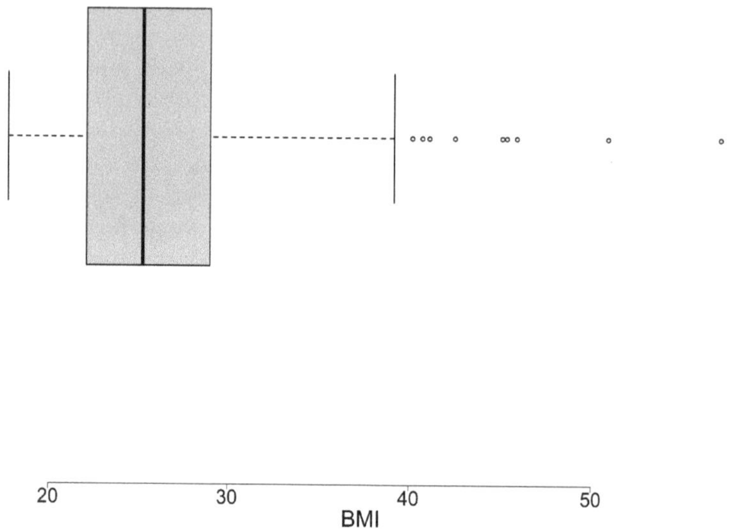

Figure 20.2 Boxplot for *BMI* using data from all female-identified *Seasons* participants aged 65 years or younger.

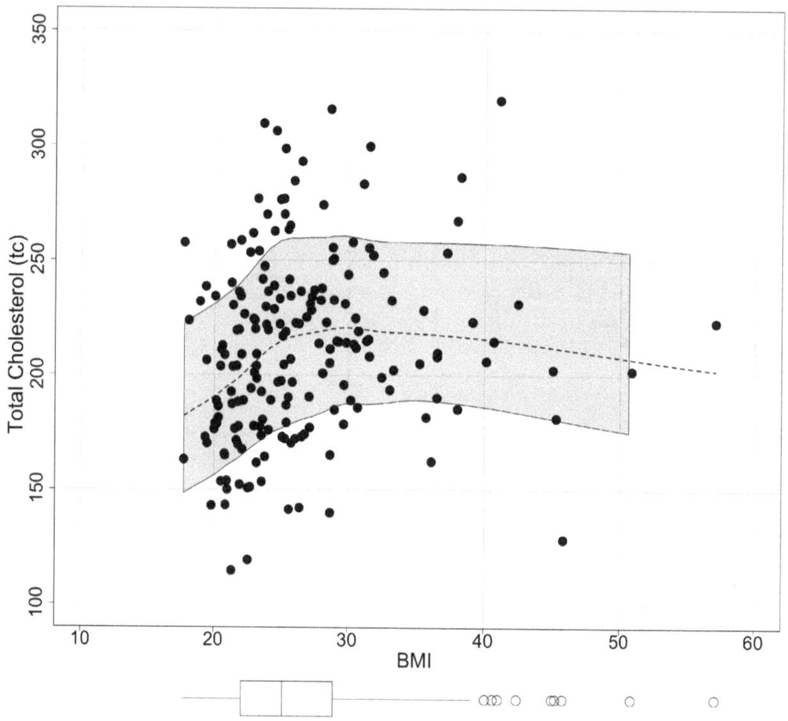

Figure 20.3 Scatterplot of *TC* vs *BMI* using LOWESS smoothing for women 65 years of age or less.

Table 20.1 Descriptive statistics for women aged 65 years or less

(a) for all cases with data on all three variables

Descriptive Statistics

	N	*Minimum*	*Maximum*	*Mean*	*Std.*	*Skewness*		*Kurtosis*	
	Statistic	*Statistic*	*Statistic*	*Statistic*	*Statistic*	*Statistic*	*Std. Error*	*Statistic*	*Std. Error*
Age	190	20.00	65.00	46.74	10.59	−.08	.18	−.74	.35
TC	190	114.88	320.00	211.61	39.61	.28	.18	−.01	.35
BMI	190	17.69	57.11	26.60	6.30	1.68	.18	3.91	.35
Valid N (listwise)	190								

(b) for cases with BMI of 40 or less

Descriptive Statistics

	N	*Minimum*	*Maximum*	*Mean*	*Std.*	*Skewness*		*Kurtosis*	
	Statistic	*Statistic*	*Statistic*	*Statistic*	*Statistic*	*Statistic*	*Std. Error*	*Statistic*	*Std. Error*
Age	181	20.00	65.00	46.93	10.63	−.12	.18	−.72	.36
TC	181	114.88	316.00	211.58	39.17	.24	.18	−.17	.36
BMI	181	17.69	39.12	25.67	4.67	.83	.18	.26	.36
Valid N (listwise)	181								

Table 20.2 Pairwise Pearson correlations (*cor*) for *TC*, *age*, and *BMI* for women aged 65 years or less with *BMI* less than or equal to 40

	Var1	*Var2*	*cor*	*p*
1	*Age*	*TC*	0.49	2.05e-12
2	*BMI*	*TC*	0.23	1.79e-03
3	*BMI*	*Age*	0.12	1.19e-01

smoother (see Chapter 17), it appears that the *BMI* outliers tend to have relatively low *TC* scores. Note from the box plot that *BMI* scores of 40 and above are outliers, and from the scatterplot we see that these outliers introduce a strong curvilinear component to the relationship between *TC* and *BMI*. Because we are primarily concerned with describing the relationships among the variables for participants who do not have extreme scores, we will exclude the data points of the nine participants whose *BMI* scores were greater than 40. As can be seen in Panel *b* of Table 20.1, when we exclude the outlying *BMI* scores, the ratios of the skewness and kurtosis measures to their standard errors are much smaller.

Information about the correlations among *TC*, *age*, and *BMI* is shown in Table 20.2, which contains R output for the 181 women aged 20 to 65 years with *BMI*s of 40 or less who have scores on all three measures. The significant correlations of *TC* with both *age*,

$r = .49$, $p = .000$, and with *BMI*, $r = .23$, $p = .002$, suggest that we might be able to predict *TC* better using information about both *age* and *BMI* than by using information about only one of these measures.

Figures 20.4 and 20.5 contain the R outputs for the regressions of *TC* (Y) on *age* (X_1), and on *BMI* (X_2), respectively. Let's consider the outputs for these analyses.

```
> tc.age<-lm(data=dat,tc~age)
> summary(tc.age)

Call:
lm(formula = tc ~ age, data = dat)

Residuals:
     Min       1Q   Median       3Q      Max
-114.959  -21.771   -0.209   21.290  112.538

Coefficients:
            Estimate Std. Error t value Pr(>|t|)
(Intercept) 126.5308    11.5379  10.967  < 2e-16 ***
age           1.8123     0.2398   7.557 2.05e-12 ***
---
Signif. codes:  0 '***' 0.001 '**' 0.01 '*' 0.05 '.' 0.1 ' ' 1

Residual standard error: 34.2 on 179 degrees of freedom
Multiple R-squared:  0.2419,    Adjusted R-squared:  0.2376
F-statistic: 57.11 on 1 and 179 DF,  p-value: 2.055e-12
```

Figure 20.4 R output from regression of *TC* vs *Age* for women 65 years of age or younger with BMI of 40 or less.

```
> tc.bmi<-lm(data=dat,tc~bmi)
> summary(tc.bmi)

Call:
lm(formula = tc ~ bmi, data = dat)

Residuals:
    Min      1Q  Median      3Q     Max
-88.294 -26.977  -3.414  23.448 102.130

Coefficients:
            Estimate Std. Error t value Pr(>|t|)
(Intercept) 161.9441    15.9131   10.18  < 2e-16 ***
bmi           1.9340     0.6101    3.17  0.00179 **
---
Signif. codes:  0 '***' 0.001 '**' 0.01 '*' 0.05 '.' 0.1 ' ' 1

Residual standard error: 38.22 on 179 degrees of freedom
Multiple R-squared:  0.05316,    Adjusted R-squared:  0.04787
F-statistic: 10.05 on 1 and 179 DF,  p-value: 0.001793
```

Figure 20.5 R output from regression of *TC* vs *BMI* for women 65 years of age or younger with BMI of 40 or less.

Looking at the Estimate columns of the coefficients panels in Figures 20.4 and 20.5, we can see that the regressions of *TC* on *age* and on *BMI* yield the equations

$$\hat{Y} = 126.531 + 1.812X_1 \tag{20.1}$$

and

$$\hat{Y} = 161.944 + 1.934X_2 \tag{20.2}$$

In Equation 20.1, the slope of 1.812 indicates that each 1-year increase in age is associated with an increase of 1.812 units in predicted *TC*. In Equation 20.2, we see that a one-unit increase in *BMI* is associated with an increase of 1.934 units in predicted *TC*.

To sum up, our application of the bivariate analyses we learned in Chapter 19 revealed several violations of assumptions. Specifically, the distribution of *BMI* was not normal because of the presence of several outliers (i.e., scores above 40). Further, the scatterplot of *TC* against *BMI* revealed a curvilinear relationship when *BMI* scores above 40 were included in the data. However, because our interest is in the population of more typical *BMI* scores, these problems were solved by eliminating the nine observations including *BMI* scores above 40. Subsequent analyses of the reduced data set showed clear positive relationships between *TC* and *BMI*, and between *TC* and *age*, *when each predictor variable is considered in isolation*. However, our goal is to understand how *age* and *BMI jointly* contribute to the prediction of *TC*. To answer this question, we apply multiple regression analyses.

20.3 The Multiple Regression Model

In bivariate regression, our concern was estimating the parameters of the linear model

$$Y_i = \beta_0 + \beta_1 X_i + \varepsilon_i$$

Given our data sample, we found values of b_0 and b_1 that minimize the mean squared error obtained by using the prediction equation

$$\hat{Y}_i = b_0 + b_1 X_i$$

We can extend this analysis to multiple linear regression by considering models in which the dependent variable, *Y*, is expressed as a linear function of several predictor variables, X_1, X_2, X_3, . . ., X_p. Although the additional predictor variables result in more complexity and require the introduction of some new concepts, many of the basic ideas underlying bivariate and multiple regression are the same.

In multiple regression, if the population model is

$$Y_i = \beta_0 + \beta_1 X_{i1} + \beta_2 X_{i2} + \ldots + \beta_p X_{ip} + \varepsilon_i$$

we can estimate the βs by finding the values of b_0, b_1, b_2, . . ., b_p in the equation

$$\hat{Y}_i = b_0 + b_1 X_{i1} + b_2 X_{i2} + \ldots + b_p X_{ip}$$

that minimize $MSE = \frac{1}{N}\sum_i \left(Y_i - \hat{Y}_i\right)^2$ for the N data points in our sample. Although the logic of minimizing mean squared prediction error is familiar from bivariate regression, there are some new complications that arise in multiple regression. A central issue stems from the potentially complex interrelationships among the predictor variables and the dependent measure. As we will see, these relationships complicate the interpretation of the regression coefficients and the partitioning of variance.

20.4 The Partitioning of Variability in Multiple Regression

20.4.1 *The Multiple Correlation Coefficient*

Our approach to analyzing the relationship between Y and a set of predictor variables, X_i, is analogous to the approach we followed in bivariate correlation and regression. In Chapter 18, we defined the correlation coefficient, r, as a measure of the degree of linear relationship between Y and X and introduced the coefficient of determination, r^2, as the proportion of the variability in one of the variables "accounted for" by the regression on the other. Both concepts have parallels when we investigate the relationship between a criterion variable, Y, and a collection of predictors, $X_1, X_2, X_3, \ldots, X_p$.

We define the multiple correlation coefficient, $R_{Y.123...p}$, as the correlation between Y and $\hat{Y}$, where

$$\hat{Y}_i = b_0 + b_1X_{i1} + b_2X_{i2} + \ldots + b_pX_{ip}$$

is the prediction of Y obtained from the multiple regression equation that contains the p predictors. If Y is perfectly predicted by the multiple regression equation, then $R = 1$. If the multiple regression equation predicts no better than the equation $Y = \bar{Y}$, then $R = 0$. When there is a single predictor variable, X, the multiple correlation coefficient reduces to $R_{Y.X} = | r_{XY} |$, the absolute value of the bivariate correlation coefficient. Although the limits for r are ± 1, R can vary only between 0 and 1. The proportion of the variability in Y accounted for by the regression on p predictor variables is $r^2_{Y\hat{Y}} = R^2_{Y.12...p}$. Therefore, we can write

$$R^2_{Y.12...p} = \frac{SS_{regression}}{SS_Y}$$

where $SS_{regression} = \sum_i \left(\hat{Y}_i - \bar{Y}\right)^2$ is the amount of variability in Y accounted for by the regression.

20.4.2 *Partitioning SS_Y into $SS_{regression}$ and $SS_{residual}$*

As was the case with bivariate regression, the variability of Y can be partitioned into a component accounted for by the regression, $SS_{regression}$, and a component not accounted for by the regression, $SS_{residual}$:

$$\Sigma\left(Y_i - \bar{Y}\right)^2 = \Sigma\left(\hat{Y}_i - \bar{Y}\right)^2 + \Sigma\left(Y_i - \hat{Y}_i\right)^2$$

$$SS_Y = SS_{regression} + SS_{residual}$$

Table 20.3 ANOVA table for multiple regression

SV	*df*	*SS*	*MS*	*F*
Regression	p	$\Sigma(\hat{Y}-\bar{Y})2$ or $R^2_{\hat{Y}.1....p}SS_Y$	$\frac{R^2_{\hat{Y}.1....p}SS_Y}{p}$	$\frac{MS_{reg}}{MS_{residual}}$
Residual	$N-1-p$	$\Sigma\left(Y_i-\hat{Y}_i\right)^2$ or $\left(1-R^2_{\hat{Y}.1....p}\right)SS_Y$	$\frac{\left(1-R^2_{\hat{Y}.1....p}\right)SS_Y}{N-1-p}$	
Total	$N-1$	$SS_Y=\Sigma\left(Y_i-\bar{Y}\right)^2$		

where $SS_{regression} = R^2SS_Y$ and $SS_{residual} = (1 - R^2)SS_Y$. It is convenient to express the partitioning of variability in an ANOVA table of the form of Table 20.3. SS_Y is associated with $N - 1$ *df* because one *df* is used to estimate the population mean. Of these $N - 1$ *df*, p are associated with the regression sum of squares because coefficients for each of the p predictors must be estimated. The remaining $N - 1 - p$ *df* are associated with the residual *SS*. Note that when there is only one predictor, $N - 1 - p = N - 2$, the result presented for bivariate regression.

Under standard assumptions that will be discussed in Chapter 21, if the p population regression coefficients $\beta_1, \beta_2, \ldots, \beta_p$ are all 0, the ratio

$$\frac{MS_{regression}}{MS_{residual}} = \frac{R^2SS_Y/p}{\left(1-R^2\right)SS_Y/(N-1-p)} = \frac{R^2/p}{\left(1-R^2\right)/(N-1-p)} \tag{20.3}$$

will be distributed as F with p and $N - 1 - p$ *df*. Therefore, the ratio of mean squares tests the null hypothesis that $\beta_1 = \beta_2 = \ldots = \beta_p = 0$. In Section 20.6, we will regress *TC* on both *age* and *BMI* score for the 181 female-identified *Seasons* participants with data on all three measures (so that $p = 2$ and $N = 181$), and we will find that $R = .522$. Substituting the square of this value into Equation 20.3, we have

$$F = \frac{.522^2/2}{\left(1-.522^2\right)/(181-1-2)} = 33.33$$

20.4.3 Partitioning $SS_{regression}$

In bivariate regression, $SS_{regression}$ corresponds to the variance in Y accounted for by the single predictor variable, X; in multiple regression, $SS_{regression}$ corresponds to the variance in Y accounted for by the entire set of p predictor variables in the regression equation. We are often interested in trying to identify the contributions of the different X variables to the prediction of Y. The pattern of correlations among the predictor variables has direct implications for the partitioning of $SS_{regression}$ into the variance components associated with each predictor. Two general patterns of relationships among the predictors may be distinguished.

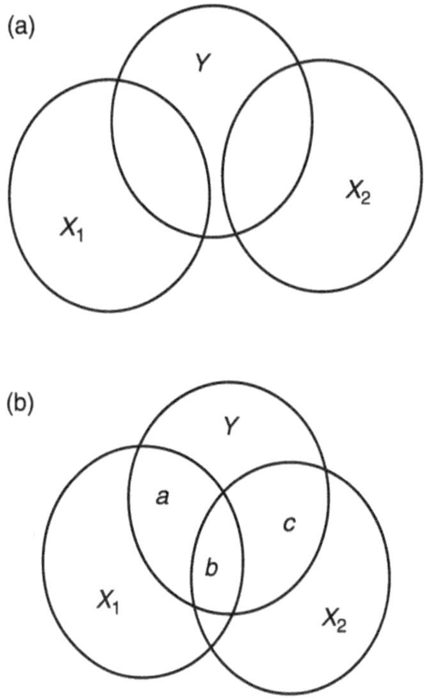

$R^2_{Y.1} = a + b$

$R^2_{Y.2} = b + c$ $\quad$ $r^2_{Y(2|1)} = c$

$R^2_{Y.12} = a + b + c$ $\quad$ $r^2_{Y(1|2)} = a$

Figure 20.6 Representations of variability in the criterion variable, Y, accounted for by uncorrelated and correlated predictor variables. (a) Uncorrelated predictors: Variabilities accounted for by X_1 and X_2 do not overlap so that $R^2_{Y.12} = r^2_{Y1} + r^2_{Y2}$ (b) Correlated predictors: Variabilities accounted for by X_1 and X_2 overlap so that $R^2_{Y.12}$ is not the sum of the r^2_{Yj}s.

If the p predictor variables in a multiple regression are mutually uncorrelated, $SS_{regression}$ can be partitioned into nonoverlapping components associated with each of the predictors. This situation typically occurs only when we have a true experiment. Panel (a) of Figure 20.6 represents this situation: X_1 and X_2 overlap with Y but not with each other, indicating that the variability in Y collectively accounted for by X_1 and X_2 is the sum of variabilities accounted for separately by X_1 and by X_2. In this situation,

$$SS_{regression} = SS_{Y.1} + SS_{Y.2}$$

where $SS_{Y.j} = r^2_{Yj}SS_Y$ and r_{Yj} is the correlation between Y and X_j. Because for two predictors, $SS_{regression} = R^2_{Y.12}SS_Y$, it follows that when the predictors are uncorrelated,

$$R^2_{Y.12} = r^2_{Y1} + r^2_{Y2} = \sum_j r^2_{Yj}$$

More generally, if p predictors account for the non-error variability in Y and are mutually uncorrelated,

$$SS_{regression} = SS_{Y.1} + SS_{Y.2} + \ldots + SS_{Y.P}$$

and

$$R^2_{Y.12\ldots p} = r^2_{Y1} + r^2_{Y2} + \ldots + r^2_{Yp} = \sum_j r^2_{Yj}$$

In words, when the predictors are uncorrelated with one another, the proportion of the variability of Y they collectively account for is the sum of the proportions of variability accounted for by the individual predictors.

However, unless we have a true experiment, predictor variables are almost always correlated with one another. The predictors usually share variability, as in Panel (b) of Figure 20.6, where the correlation between X_1 and X_2 is indicated by the overlap of their circles. Note that if we add the overlap of Y with X_1 to the overlap of Y with X_2, the area labeled (b) is added in twice. When any of p predictors are correlated, the proportion of variability in Y they account for is not the sum of the proportions associated with the individual predictors; instead, it must be adjusted for overlapping variability.

An expression for R^2 that takes the correlations between predictors into account is

$$R^2_{Y.12\ldots p} = \frac{\Sigma r_{Yj} b_j s_j}{s_Y} = \Sigma r_{Yj} b_j^* \tag{20.4}$$

where b_j is the unstandardized regression coefficient of X_j in the multiple regression equation; s_j and s_Y are the standard deviations of X_j and Y, respectively; and b^*_j is the standardized regression coefficient of X_j. For example, when we regress *TC* (Y) on *age* (X_1) and *BMI* (X_2) in Section 20.5, we will obtain the following results:

	TC	*Age*	*BMI*
b_j		1.737	1.474
r_{Yj}		0.492	0.231
s_j	39.173	10.630	4.670

so that, substituting in Equation 20.4, we have

$$R^2 = [(0.492)(1.737)(10.630) + (0.231)(1.474)(4.670)] / 39.173 = 0.272$$

In the current example, the sum of the r^2 values for the correlations between *TC* and *age* and between *TC* and *BMI*, 0.295, is not very different from the value of 0.272 obtained earlier, because the correlation between *age* and *BMI* is small (see Figure 20.1).

The increase in R^2 when X_2 is added to a regression equation that already contains X_1 is $r^2_{Y(2|1)}$, the square of the semipartial correlation coefficient introduced in Chapter 18. As we mentioned there, $r_{Y(2|1)}$ is the correlation of Y with the component of X_2 that is not predictable from X_1. In terms of Panel (b) of Figure 20.6, we may think of $r^2_{Y(2|1)}$ as the proportion of the Y circle that overlaps X_2 but not X_1. In general, the squared semipartial correlation coefficient

$r^2_{Y(p+1|12...p)}$ is the increase in R^2 that follows from adding a $(p + 1)^{st}$ predictor to a regression equation that already contains p predictors. That is,

$$r^2_{Y(p+1|12...p)} = R^2_{Y.12...p+1} - R^2_{Y.12...p} \qquad (20.5)$$

In Equation 20.5, $r_{Y(p+1|12...p)}$ is the correlation between Y and $X_{p+1} | X_1, X_2, \ldots, X_p$, where the latter term represents the residuals of the regression of X_{p+1} on $X_1, X_2, \ldots, X_p$. Applying Equation 20.5 to the *TC* data, when *age* is added to a regression equation that already contains *BMI* as a predictor, the proportion of the variance of *TC* accounted for is increased by .219. If *BMI* is added to an equation when age is already included as a predictor, the increase is .030. Note that because *age* and *BMI* are somewhat correlated and therefore account for overlapping variability in *TC*, these increases are smaller than the proportions of variance accounted for by *age* and *BMI* when each is the only predictor in the equation. In Chapter 21, we discuss "partial *F* tests" that allow us to test whether the addition of one or more variables significantly increases the variability accounted for by a regression equation.

20.4.4 The Adjusted (or Shrunken) Multiple Correlation Coefficient and Cross-Validation

When a multiple regression equation is developed from a sample of data, the multiple correlation coefficient R and its square are commonly used indices of how well the equation fits the data in the sample. These measures are also often used as estimates of how the regression equation fits the population from which the sample was obtained. However, using R or R^2 as measures of fit can be misleading because R is a *positively biased* estimator of the population coefficient. Particularly if the sample is small and there are many predictors, a regression equation that predicts well in the sample may predict poorly in the population.

With enough predictors, the regression equation *must* fit the sample well no matter how the predictors and the criterion are related in the population. Just as any two data points can be fit perfectly by a straight line, any $p + 1$ data points can be fit perfectly by a linear regression equation with p predictor variables and therefore, the resulting value of the sample R must be 1. With more than $p + 1$ data points, the R in the sample need not be 1, but will tend to be larger than R in the population if the ratio of the number of cases, N, to the number of predictors, p, is small. This bias occurs because of *capitalization on chance*: The regression equation takes advantage of chance fluctuations in scores that allow for increased predictability in the sample but not in the population. The bias in R can be reduced by working with larger samples: Although the recommended sample size depends to some extent on the nature of the research problem and the purpose of the analysis, the N/p ratio should be large – perhaps 30 or more – if the size of R is to be taken very seriously.

A common adjustment for positive bias has been provided by Wherry (1931). The proportion of variance in the population that can be accounted for by the relationship between Y and a set of X_js is denoted by ρ^2_{XY}, which can be expressed as

$$\rho^2_{XY} = 1 - \frac{\sigma^2_e}{\sigma^2_Y}$$

If we replace the population variances by their unbiased estimates, we have

$$R^2_{adjusted} = 1 - \frac{SS_{residual} / (N-1-p)}{SS_Y / (N-1)}$$

which can be rewritten as

$$R^2_{adjusted} = 1 - \left(\frac{SS_{residual}}{SS_Y}\right)\left(\frac{N-1}{N-1-p}\right)$$

But $SS_{residual} / SS_Y = 1 - R^2$. Therefore, substituting into the last equation, we have Wherry's formula,

$$R^2_{adjusted} = 1 - \left(1 - R^2\right)\left[\frac{N-1}{N-1-p}\right] \quad (20.6)$$

The adjusted (or "shrunken") squared multiple correlation coefficient is provided in the regression output of most statistical software. For the regression of *TC* on *age* and *BMI*, Equation 20.6 yields an adjusted R^2 of 1 – (1 –.272)(180 / 178) = .264. Note that if we set the adjusted R^2 equal to zero and solve Equation 20.6 for R^2, we get

$$R^2 = \frac{p}{N-1}$$

This gives the value of R^2 for which the adjusted R^2 will be 0. Using this equation, we can see that if we have 10 predictors and 50 cases, an R^2 value of .204 (or multiple R of .452) would correspond to an adjusted R of 0. In short, the unadjusted values of R and R^2 can severely overestimate the strength of the relationship between Y and a set of predictors when the N / p ratio is small!

If our interest is prediction, the Wherry adjustment may not be ideal because we are less interested in how a regression equation fits the population than in how well the equation fits *another sample* drawn from the same population. Herzberg (1969; see the discussion in Stevens, 1986) gives two adjustment equations that attempt to estimate R^2 if we want to use the regression equation developed in sample 1 to make predictions for the data of sample 2. If the predictors are random variables, as is usual in non-experimental research, the adjustment equation is

$$R^2_{adjRP} = 1 - \left(\frac{N-1}{N-p-1}\right)\left(\frac{N-2}{N-p-2}\right)\left(\frac{N+1}{N}\right)\left(1 - R^2\right) \quad (20.7)$$

If the predictors are fixed, the adjustment equation is

$$R^2_{adjFP} = 1 - \left(\frac{N-1}{N}\right)\left(\frac{N+p+1}{N-p-1}\right)\left(1 - R^2\right) \quad (20.8)$$

The shrinkage is greater if we use the Herzberg equations instead of the Wherry equation. For example, given a sample of size 100 with 10 predictors and an R^2 of .50, $R^2_{adjusted}$ = .444, whereas $R^2_{adj\ RP}$ = .374 and $R^2_{adj\ FP}$ = .383. In our current example, because of the high *N/p* ratio, there is not much shrinkage: *N* = 181, *p* = 2, and R^2 = .272, so that $R^2_{adjusted}$ = .264 and $R^2_{adj\ FP}$ = .252.

The best way of obtaining a more realistic estimate of the population *R* is to employ a procedure called *cross-validation* that avoids capitalizing on chance by developing the regression equation using one sample (called the screening or training sample) and testing it by a second sample (called the calibration or test sample). The cross-validated *R* is the correlation between (1) the predicted *Y* scores obtained when the regression equation developed in the screening sample is applied to the calibration sample and (2) the actual *Y* scores in the calibration sample. Because the regression coefficients are obtained from one sample and the value of *R* is obtained from a second sample, the cross-validated *R* cannot systematically capitalize on sampling variability.

The problem of capitalization on chance is most insidious when the variables used in the regression equation are chosen from a larger pool of possible predictors – a very common situation. Variables in the pool that are useful for predicting in the sample will be those chosen to be added to the regression equation and thus increase the multiple correlation. In this situation, chance variation in the sample may well be what determines the choice of variables for inclusion in the equation. If the *N* / *p* ratio is small and variables in the regression equation are chosen from a larger set of possible predictors, the shrinkage achieved by cross-validation can be dramatic.

We can illustrate this point with the *Seasons* data set. We selected the data from a random 25% of the participants who identified as women aged 65 years or less with *BMI* scores of 40 or less. Then, we used *TC* as the dependent variable and arbitrarily chose *height*, *BMI*1, *host*1, *anger*1, *irrit*1, *anxiety*1, *dirwdc*1, *beck_d*1, *beck_d*2, *beck_d*3, and *beck_d*4 as 11 possible predictors. Using an automated procedure called stepwise regression (to be discussed in Chapter 21), we found that the regression equation containing five predictor variables that best predicts *TC* is[1]

$$\hat{tc} = 204.43 + 10.876\ host1 - 11.216\ anxiety1 + 2.231\ beck\_d1$$

Here, the regression equation is based on 32 randomly chosen cases in the sample having data on the dependent variable and all 11 predictors, so the *N* / *p* ratio is 2.9. The *R* obtained for the sample is .42, and the adjusted *R* is .30. We then cross-validated by (1) using the same equation to predict *TC* scores for the remaining 94 cases with data on all the 11 original predictors, and then (2) finding the correlation between the predicted and the actual *TC* scores for these 94 participants. This correlation is very small, *r* = .10, so that the cross-validated R^2 is $.10^2$ = .01. The cross-validated *R* of .10 is a more realistic measure of how well the regression equation fits the population than the *R* of .42 or the adjusted *R* of .30 obtained in the subsample used to generate the equation. In other words, because of the small sample and the large number of potential predictors, the regression equation predicts very poorly except in the sample that was used to generate it.

In summary, if the *N* / *p* ratio is small, the multiple correlation coefficient for a sample will overestimate the usefulness of the regression equation in the population and in other samples. This overestimation is larger when the predictors in the equation are chosen from larger pools of potential predictors. We strongly recommend the use of cross-validation to counter the effects of capitalization on chance.

20.5 Using Software for Multiple Regression and Cross-Validation

Using software for multiple regression is similar to the bivariate regression approaches detailed in Section 17.6.6. Additional predictors are simply added to the *Independent* variable list in SPSS. In R, additional predictors are included in the regression equation in the *lm* function with a + between them: reg.out <- *lm*(data = dat, Y ~ X1 + X2); confidence intervals for the intercept and partial slope parameters can be obtained using the *confint* function in the {stats} package, which takes the output of regression as input: *confint*(reg.out).

Cross-validation involves several steps that can be done easily and quickly in R:

1. Identify a training set of data by randomly sampling a percentage of the cases and saving the sampled case numbers. The *sample* function in {base} R is a good tool to use. Assuming we have 126 cases and we want to sample 25% of them, train.index <- *sample*(*seq*(1:126), 32) will store a random sample of 32 integers from the set 1 to 126.
2. Assign those cases to a training set, train.set <- datcv[train.index,], and the complement of cases to a test set, test.set <- datcv[-train.index,].
3. Run the regression including all predictor variables, and save the results: model.cv <- *lm*(tc ~., data = train.set). At this point, you can run stepwise regression (see Chapter 21) on these predictors using the *step* function in {stats}, although that is not a necessary part of cross-validation. If you include this step, be sure to store the results: smodel.cv <- *step*(model.cv, *scope* = *list*(*lower* = *lm*(Y ~ 1), *upper* = *lm*(Y ~ .)), *direction* = "both"). The scope parameters specify the range of models under consideration, where Y ~ 1 indicates a model with an intercept only, and *Y* ~ . indicates a model with all predictors included.
4. Compute and store the $\hat{Y}$ values for the test cases (see Section 19.3.3), using the model generated with the training cases: predictions.cv <- smodel.cv %>% *predict*(test.set).
5. Finally, compute R^2 for the predicted and observed values: *cor*(predictions.cv, test.set$tc)^2. Additional cross-validation methods in R are described by De Rooij and Weeda (2020).

In SPSS, cross-validation has not been automated but it can be accomplished easily by the following steps:

1. Select a random sample of cases using the *Select Cases* option from the *Data* pull-down menu, and then clicking on the radio button for *Random sample of case*, choose a sample size.
2. Run the regression on the selected sample, which is the training set.
3. Use the *Transform* menu to *compute* a new variable, $\hat{Y}$, by typing in the *Numeric Expression* box the regression equation you just obtained. Instead of clicking OK, click on the *If...* bar at the bottom of the *Compute Variable* window and select the radio button to *Include if case satisfies condition*, then type "filter_$=0" in the box (without the quote marks), then click Continue and OK. This will compute the predicted value of *Y* only for the cases that were not in the training set that was used to develop the equation.
4. Finally, select all cases and compute the square of the correlation between the observed and newly predicted values of *Y*; this will be R^2 for the test cases.

20.6 Multiple Regression Analyses of the *TC* Data

We have identified and discussed the components of the regression equation, and we have explained how to use software to generate the analyses. We now illustrate the computations using the data from the *TC* study for which preliminary analyses were presented in Section 20.2. Figure 20.7 presents the R output for the regression of *TC* on both *age* and *BMI*.

Breaking down the analysis presented in Figure 20.7:

1. The Estimate column in the coefficients panel provides the least-squares estimates of the Y intercept (b_0) and the *unstandardized regression coefficients* (or *unstandardized partial slope coefficients*) for X_1 and X_2, b_1 and b_2.[2] These entries tell us that the best least-squares regression equation that includes both *age* and *BMI* as predictors is

$$\hat{Y} = 92.239 + 1.737X_1 + 1.474X_2 \tag{20.9}$$

The plot of Equation 20.9 is displayed in Figure 20.8, along with the data points; the residuals are denoted with lines connecting each data point with its predicted value. Instead of the regression *line* that we obtained using bivariate regression, we now have a two-dimensional regression *plane*. If *BMI* is held constant, a one-unit (i.e., 1-year) change in *age* corresponds to an increase of 1.737 units in predicted *TC*. Similarly, if *age* is held constant, a one-unit change in *BMI* corresponds to an increase of 1.474 units in

```
> tc.age.bmi<-lm(data=dat,tc~age+bmi)
> summary(tc.age.bmi)

Call:
lm(formula = tc ~ age + bmi, data = dat)

Residuals:
     Min       1Q   Median       3Q      Max
-107.790  -19.146   -2.168   20.492  114.959

Coefficients:
            Estimate Std. Error t value Pr(>|t|)
(Intercept)  92.2387    16.9214   5.451 1.65e-07 ***
age           1.7370     0.2372   7.322 8.12e-12 ***
bmi           1.4739     0.5400   2.730  0.00698 **
---
Signif. codes:  0 '***' 0.001 '**' 0.01 '*' 0.05 '.' 0.1 ' ' 1

Residual standard error: 33.6 on 178 degrees of freedom
Multiple R-squared:  0.2723,    Adjusted R-squared:  0.2642
F-statistic: 33.31 on 2 and 178 DF,  p-value: 5.154e-13
```

Figure 20.7 R output from regression of *TC* vs *Age* and *BMI* for women 65 years of age or younger with BMI of 40 or less.

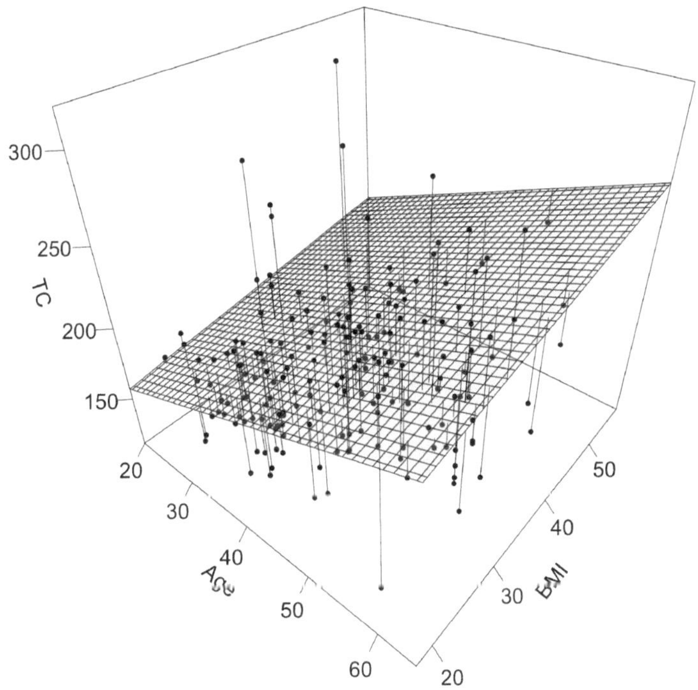

Figure 20.8 Plane for the regression of total cholesterol level (*TC*) on *age* and *BMI*, with observed points and residuals shown.

predicted *TC*. Compare these coefficients to the values we found for the bivariate regression coefficients in the preliminary analyses of Section 20.2: for *age*, the coefficient was 1.812 and for *BMI* the coefficient was 1.934. The difference between the multiple and bivariate regression coefficients is due to the fact the *age* and *BMI* are correlated ($r = .12$), so the multiple regression coefficients change when each predictor's coefficient is adjusted for the other predictor.

2. Near the bottom of the output in Figure 20.7, we see that the squared multiple correlation coefficient, *Multiple R-squared*, is .272. This means that the *TC* values predicted by Equation 20.9 have a correlation of $\sqrt{.272} = .522$ with the actual *TC* values in the sample. Multiple *R*-squared is sometimes called the *coefficient of multiple determination*, and it tells us the proportion of the variance in *TC* that is "accounted for" by the regression on *age* and *BMI*. In this example, we conclude that 27.2% of the variability in *TC* is due to variability in *age* and *BMI*. This means that by using the regression equation to predict the *Y* scores, the variability in the *Y* scores that is unaccounted for is reduced by the proportion .272. To be more specific, the residual variability that remains when the regression equation is used to predict the *Y* scores is $1 - .272 = .728$ of the residual variability that would result if $\bar{Y}$ was used to predict each of the *Y* scores (recall Figure 17.8). As we saw in Section 20.4.4, the sample-based multiple correlation coefficient is a positively biased estimator of the population coefficient. The adjusted *R* results from one type of attempt to remove the positive bias (see Equation 20.6); here, that adjustment reduces the Multiple *R*-squared from .272 to the *Adjusted R squared* of .264.

3. The R output includes the *Residual standard error*, 33.6, near the bottom of the report in Figure 20.7. That is the standard error of the estimate. As with bivariate regression, the standard error provides a measure of how well the regression equation predicts cholesterol level. The equation provides a prediction, $\hat{Y}$, for each combination of X_1 and X_2. The standard error of the estimate,[3] s_e, is the square root of the sum of the squared deviations of the actual cholesterol levels from the predicted levels, divided by the *df*, so that

$$s_e = \sqrt{\frac{\Sigma\left(Y_i - \hat{Y}_i\right)^2}{df}} = \sqrt{\frac{SS_{residual}}{df_{residual}}}$$

For the general case in which *Y* is regressed on *p* predictor variables, it can be expressed as

$$s_e = \sqrt{\frac{\left(1 - R^2_{Y.12...p}\right)SS_Y}{N - 1 - p}} \tag{20.10}$$

If the underlying model for the population is $Y_i = \beta_0 + \beta_1 X_1 + \beta_2 X_2 + \ldots + \beta_p X_p + \varepsilon_i$, the standard error of the estimate provides an estimate of the standard deviation of ε.

4. The ANOVA table from R, shown in Figure 20.9, allows us to compute the total sum of squares associated with cholesterol scores, 276,210. We also see that it can be partitioned into two components: The sum of squares accounted for by the regression, $SS_{regression}$, which equals the sum of the *SS* for *age* and for *BMI*, and the sum of squares left unaccounted for, $SS_{residual}$, where

$$SS_{regression} = \Sigma\left(\hat{Y}_i - \bar{Y}\right)^2 = R^2 SS_Y = 75{,}221$$

and

$$SS_{residual} = \Sigma\left(Y_i - \hat{Y}_i\right)^2 = (1 - R^2)SS_Y = 200{,}989$$

```
> anova(tc.age.bmi)
Analysis of Variance Table

Response: tc
           Df Sum Sq Mean Sq F value    Pr(>F)
age         1  66808   66808 59.1664 9.512e-13 ***
bmi         1   8413    8413  7.4503   0.00698 **
Residuals 178 200989    1129
---
Signif. codes:  0 ‘***’ 0.001 ‘**’ 0.01 ‘*’ 0.05 ‘.’ 0.1 ‘ ’ 1
```

Figure 20.9 ANOVA table for the regression of total cholesterol level (*TC*) on *age* and *BMI*.

```
> confint(tc.age.bmi)
                  2.5 %     97.5 %
(Intercept) 58.8463711 125.630946
age          1.2688488   2.205119
bmi          0.4083089   2.539513
```

Figure 20.10 Confidence intervals for the unstandardized regression coefficients of total cholesterol level (*TC*) on *age* and *BMI*.

Note that a test of whether the 2-predictor regression equation accounts for significant variability in total cholesterol is found by computing

$$F = MS_{regression}/MS_{residual} = (SS_{regression}/p)/(SS_{residual}/(N - p - 1) = 33.31$$

which matches the result we found in Section 20.4.2.

5. For each regression coefficient in Figure 20.7, a t statistic is formed by dividing b by its standard error. This tests the null hypothesis that the corresponding β is equal to 0. The t of 7.322 (with $p = .000$) for *age* indicates that when *BMI* is held constant, the rate of change of predicted *TC* with *age* is significantly different from zero. That is, there is a significant contribution of *age* to the predictability of *TC* over and above that provided by *BMI*. Similarly, the t of 2.730 (with $p = .007$) for *BMI* indicates there is a significant contribution of *BMI* to the predictability of *TC* over and above that provided by *age*. The significant t for b_0, the constant (i.e., intercept) of the regression equation, indicates that we can reject the null hypothesis that $\beta_0 = 0$ in the population. We calculated *confidence intervals*, using the *confint* function in the {stats} package, and the results in Figure 20.10 show the upper and lower bounds of the intervals for each coefficient. Even with a fairly large sample, the confidence intervals are quite wide. For example, the 95% confidence interval for the partial slope of predicted *TC* with *BMI* extends from 0.408 to 2.540.

20.7 The Meaning of the Regression Coefficients

We have identified and explained the components of the regression equation and have illustrated how to compute regression statistics. We now discuss the interpretation of regression coefficients in some detail, including cautions against common misinterpretations of regression statistics.

In terms of the sample regression equation, the interpretation of the unstandardized regression coefficients is straightforward.[4] Consider the regression equation

$$\hat{Y} = b_0 + b_1X_1 + b_2X_2 \tag{20.11}$$

As we have already established:

1. The intercept, b_0, is the value of $\hat{Y}$ when both X_1 and X_2 are equal to 0.
2. The regression coefficient b_1 is the rate of change of $\hat{Y}$ with X_1, holding X_2 constant. That is, if X_1 is increased by one unit and X_2 remains unchanged, the corresponding change in $\hat{Y}$ is b_1. Analogously, the regression coefficient b_2 is the rate of change of $\hat{Y}$ with X_2, holding X_1 constant.

Unfortunately, what these coefficients say about the *true model for the population* is usually much less clear. When a regression equation does a good job predicting the dependent variable, it is tempting to use the equation not only to predict outcomes, but also as an explanatory model – or at least to think of the regression coefficients in the equation as measures of the "importance" of the corresponding Xs in *influencing* (i.e., *causing* changes in) Y. However, unless the data come from a true experiment or we have a great deal of existing theory and detailed knowledge about how the data were generated, this temptation must be resisted. Although we consider the relationships between regression and explanatory models in more detail in Chapter 22, here we introduce some limitations that we must keep in mind when we are tempted to use regression equations as explanatory models and regression coefficients as measures of causal importance.

- *Regression deals with prediction, not causality.* We would expect that if changing X_1 causes important changes in Y, then X_1 should be a useful predictor of Y if we include it in a regression analysis. *However, the reverse is not necessarily true; a variable that is a very useful predictor may have no causal importance whatsoever.* The variable may be a good predictor and have a large regression coefficient because it happens to be correlated with other variables that are causally important but are not included in the equation. We should remember that X_1 may predict Y for several reasons, including the following:
  1. Some other variable (or variables) may influence both X_1 and Y. For example, shoe size is a good predictor of vocabulary size in elementary school because both shoe size and vocabulary increase with age. As another example, the number of nonfiction books in a home is a good predictor of a child's success in elementary school even if the child never reads the books. The number of books is a good indicator of the family's affluence and level of education, and these variables do influence school performance.
  2. Changes in Y may cause changes in X_1. An article in the journal *Circulation* reported that the use of diet soda drinks predicts the presence of metabolic syndrome (a constellation of metabolic risk factors) likely due to insulin resistance. This could be because the consumption of diet sodas causes the metabolic syndrome. However, an equally plausible hypothesis is that the metabolic syndrome causes people to turn to diet foods to try to control symptoms of the syndrome.
  3. Even if X_1 does have some causal influence on Y, the influence could be direct or indirect (or both). A direct effect would occur if changes in X_1 caused changes in Y. An indirect effect would occur if changes in X_1 caused changes in a third variable, X_2, and those changes in X_2 caused changes in Y. In this case we would call X_2 a *mediating variable.* A measure of the causal importance of a variable should consider both its direct and indirect effects. We will discuss mediation in Chapter 22.
- *In observational studies, the true population model is almost certainly different from the regression equation we have developed.* Suppose the true model for the population is

$$Y = \beta_0 + \beta_1 X_1 + \beta_2 X_2 + \varepsilon \tag{20.12}$$

in which the error component, ε, refers to random error. If the usual regression assumptions are satisfied (see Chapter 21), the sample coefficients b_0, b_1, and b_2 in Equation 20.11 will be unbiased estimators of the corresponding population parameters in Equation

20.12. However, just because we perform a regression that predicts the Y scores in our sample well does not necessarily mean that we have uncovered the correct population model.

We have already witnessed an example demonstrating that different data sets, possibly arising from different populations, can give rise to similar sample regression statistics. In Chapter 17 we discussed the *Anscombe* data sets in which identical bivariate regression statistics were obtained from four very different data sets (see Figure 17.10). With more predictor variables, there is far more potential for different kinds of data structures to produce similar regression equations, as well as for different combinations of predictors to account for similar amounts of variance.

Also, as we will discuss further in Chapter 22, we often work with relatively simple models that contain important variables as predictors. These models are often extremely useful for focusing our thinking and developing further research, but they rarely include all the possible relevant predictor variables. As an example, consider the research question of how student performance is influenced by reductions in class size. This research is largely inconclusive because so many variables other than class size influence student performance – examples of relevant variables include characteristics of the student, family background, and school, as well as teaching style and course content (see Exercise 22.7).

These extraneous variables may not present a severe problem if we can conduct a true *experiment* in which variables of interest are manipulated and we use randomization and matching to control other variables. If so, a model such as that expressed in Equation 20.12 may be reasonable. The error component, ε, contains the effects of variables other than those that are manipulated, but these other variables are not systematically related to the manipulated variables because of the controls present in an experiment.

However, suppose the data come from an *observational study* in which the variables of interest are systematically related to other important variables. For example, smaller classes are more likely to occur in more affluent school districts that can afford to pay their teachers better salaries and that tend to have parents with higher education levels themselves. Unless these other variables are included in the model, ε is not really a random error component and may be correlated with the variables in the model. If important variables are omitted from the model, the b_js in Equation 20.11 will generally be biased estimators of the βs in Equation 20.12.

When important variables are left out of the regression equation, the regression coefficients of variables in the equation may, in part, reflect the effects of these omitted variables. We consider this point further in the next section.

- *Both the values and the interpretations of the regression coefficients may change when other predictor variables are added to or removed from the regression equation.* Consider the following regression equations in which Y is predicted by one, two, or three predictor variables:

$$\hat{Y} = b_{Y0.1} + b_{Y1}X_1 \quad (20.13)$$

$$\hat{Y} = b_{Y0.12} + b_{Y1.2}X_1 + b_{Y2.1}X_2 \quad (20.14)$$

and

$$\hat{Y} = b_{Y0.123} + b_{Y1.23}X_1 + b_{Y2.13}X_2 + b_{Y3.12}X_3 \quad (20.15)$$

Here, we use more elaborate subscripts on the regression coefficients to specify the predictor variables in the different regression equations. For example, $b_{Y0.12}$ is the intercept for the regression equation in which Y is regressed on both X_1 and X_2, and $b_{Y1.23}$ is the coefficient of X_1 in a regression equation in which Y is regressed on X_1, X_2, and X_3.

Unless the three predictor variables are uncorrelated, the coefficients of X_1 in the three equations, b_{Y1}, $b_{Y1.2}$, and $b_{Y1.23}$, will not have the same values. We saw this in the numerical example in Sections 20.2 and 20.6, when we regressed the total cholesterol levels on *age*, *BMI*, or both. The coefficient b_{Y1} represents the rate of change of $\hat{Y}$ with X_1; the coefficient $b_{Y1.2}$ represents the rate of change of $\hat{Y}$ with X_1 if X_2 is held constant; and the coefficient $b_{Y1.23}$ represents the rate of change of $\hat{Y}$ with X_1 if both X_2 and X_3 are held constant.

This "holding constant" of the other variables is represented explicitly in the expressions for the regression coefficients. For example, $b_{Y1} = r_{Y1}\, s_Y/s_1$ whereas

$$b_{Y1.2} = \left[\frac{r_{Y1} - r_{Y2}r_{12}}{1 - r_{12}^2}\right]\frac{s_Y}{s_1} = r_{Y1|2}\,\frac{s_{Y.2}}{s_{1.2}} \tag{20.16}$$

where $r_{Y1|2}$ is the partial correlation of Y and X_1 with the effects of X_2 partialed out (see Section 18.3), $s_{Y.2}$ is the standard error of estimate when Y is regressed on X_2, and $s_{1.2}$ is the standard error of estimate when X_1 is regressed on X_2. Although the expressions for b_Y and $b_{Y1.2}$ are parallel in form, the terms in the equation for $b_{Y1.2}$ are all adjusted for X_2. Note that if $r_{12} = 0$, then $b_{Y1} = b_{Y1.2}$.

What the *b*s in Equations 20.13–20.15 have to say about the underlying model is complicated by the fact that we usually do not know the true model for the population. If X_1 and X_2 are correlated, the coefficient b_{Y1} will represent the rate of change of $\hat{Y}$ with X_1 in the sample, but it may not necessarily be a good estimator of how Y changes with X_1 in the population when X_2 is held constant. This is because if X_2 is left out of the regression equation, b_{Y1} will generally reflect the effects of *both* X_1 and X_2. If X_2 is now added to the equation, we obtain a coefficient of X_1 that represents the rate of change in $\hat{Y}$ with X_1 if X_2 is held constant.

To make this point more concrete, consider another example. Suppose we want to predict final exam performance in an introductory statistics course based on two pretests: (1) pretest 1 that measures algebra skills and (2) pretest 2 that measures abstract mathematical reasoning skills. Let's assume that people with better algebra skills tend to have better abstract reasoning skills, so the two pretests are correlated, and that performance on the final exam depends on both kinds of skills. If we regressed the final exam score only on algebra skills, we would be mistaken if we interpreted the regression coefficient as the measure of the importance of algebra skills in *influencing* the grade on the final.[5] The change in the predicted final exam score associated with a one-unit difference on pretest 1 reflects both the difference in algebra skills *and the associated difference in abstract reasoning skills*. However, if we regressed final exam score on both pretests, the coefficient of the pretest 1 variable would no longer reflect the predictive ability of abstract reasoning skills. In this case, the pretest 1 coefficient would represent the rate of change of the predicted score on the final exam with algebra skills, *holding abstract mathematical reasoning skills constant.*

20.8 Suppression Effects in Multiple Regression

One final topic deserves mention. If we compare the regression of Y on X_1 to the regression of Y on both X_1 and X_2 where the two predictors are correlated, $b_{Y1.2}$ is usually smaller than b_{Y1} because the effects of X_2 have been partialed out. However, it is possible to have *suppression effects* in which the predictive effect of X_1 is greater when X_2 is also in the equation. That is, under certain conditions, the coefficient of X_1 becomes larger when X_2 is added to the equation.

Suppose we add X_2 to a regression equation that already contains X_1. From Equation 20.5, the proportion of variance of Y accounted for by both predictors is given by

$$R^2_{Y.12} = r^2_{Y1} + r^2_{Y(2|1)} \tag{20.17}$$

where

$$r_{Y(2|1)} = \frac{r_{Y2} - r_{Y1}r_{12}}{\sqrt{1 - r^2_{12}}} \tag{20.18}$$

is the semipartial correlation between Y and the part of X_2 that is not predictable from X_1 (see Chapter 18).

In so-called *classical suppression*, $r_{Y1} > 0$, $r_{12} \neq 0$, and $r_{Y2} = 0$, so that X_2 overlaps variance with X_1 but not with Y. From Equation 20.18, the increased variability in Y accounted for by the addition of X_2 to a regression equation that contains X_1 reduces to

$$r^2_{Y(2|1)} = \frac{r^2_{Y1}r^2_{12}}{1 - r^2_{12}} \tag{20.19}$$

Also, the coefficient of X_1 when Y is regressed on both X_1 and X_2 is

$$b_{Y1.2} = \left[\frac{r_{Y1} - r_{Y2}r_{12}}{1 - r^2_{12}}\right]\frac{s_Y}{s_1} = \frac{b_{Y1}}{1 - r^2_{12}} \tag{20.20}$$

if $r_{Y2} = 0$. From Equations 20.17 and 20.19, we can see that given classical suppression, R^2 increases when X_2 is added to the regression equation even though it is uncorrelated with Y. From Equation 20.20 we see that $b_{Y1.2}$ is larger than b_{Y1} if $r_{12} \neq 0$ and increases in size as r_{12} increases. Also, given classical suppression, $b_{Y2.1} \neq 0$ even though $b_{Y2} = 0$. In what is called *net suppression*, r_{Y2} is not zero but is less than $r_{Y1}r_{12}$, so that even though X_2 now shares variance with both Y and X_1, $b_{Y1.2}$ is still larger than b_{Y1}.

Consider a hypothetical example in which suppression effects could occur: suppose, for example, that

1. Y is a non-verbal performance measure of spatial judgment ability;
2. X_1 is the score on a paper-and-pencil test of spatial ability in which the participant follows a set of complex written instructions; and
3. X_2 is a test of reading ability.

Even if reading ability has little or no correlation with spatial ability, the scores on the reading test will be correlated with the results of the group test of spatial ability because of the need to follow the written instructions. In this case, $X_1|X_2$, the score on the spatial ability test with reading ability partialed out, will be a better predictor of $Y|X_2$ than X_1 is of Y. This is because X_2 removes (or "suppresses") a source of variability in X_1 (i.e., reading ability), that is, error with respect to predicting Y.

20.9 Summary

In Chapter 20, we have developed the basic ideas of multiple regression.

- We began by presenting a data set that we used to illustrate preliminary analyses to test assumptions and gather basic information about the relationships among a dependent variable, *TC*, and two predictor variables, *age* and *BMI*, in a sample of women.
- We then stated the multiple regression equation, defined and described its components, and showed how to partition the variance in a regression study.
- We discussed cross-validation and adjusted measures of regression fit, emphasizing the point that any regression equation developed from a particular sample will always fit the sample better than it fits the population from which the sample was selected.
- We used software to illustrate how to implement regression analyses using the data set introduced at the start of the chapter.
- This was followed with a discussion of the meaning of the regression coefficients and some of their limitations in developing explanatory models. We emphasized that the machinery of regression deals with prediction, not causality.
- We noted that the coefficient of a predictor variable will generally take on different values when other predictors are added to, or removed from, the regression equation. Our discussion of suppression demonstrated that the additional predictor need not be correlated with the criterion variable.

Exercises

20.1 [Comparing regression with ANOVA and with population parameters] In a visual "search" experiment, participants are presented with a display containing an array of letters and make a response when they detect the presence of a "target letter" that was specified beforehand. Arrays can differ in the number of letters they contain and (because of differences in brightness and contrast or the presence of visual "noise") how difficult it is to identify the letters. We simulated the results of such an experiment in which number of letters and identification difficulty were varied orthogonally, using the model

$$\text{Time} = 400 + 30 \times number + 2 \times diff + \varepsilon$$

where *number* stands for the number of letters in the array (2, 4, 6, or 8), *diff* stands for identification difficulty (10 or 20 units), and ε is a number selected randomly from a normal population with mean = 0 and standard deviation = 40 to generate the 24 cases. The data can be found in the file *EX20_1* on the book's website.

a) Find the summary statistics and correlation matrix for these data.
b) Regress time on number of letters and difficulty. Are the effects of *number* and *diff* significant at $\alpha = .05$? Are these significance tests equivalent to the tests of the number of letters and difficulty main effects in a standard ANOVA? Perform an

ANOVA on time using the factors *number* and *diff* and compare the results with those that follow from the regression.

c) What are the estimates of the parameters of the model that are obtained from the regression? How do these compare with the actual parameter values ($\beta_0 = 400$, $\beta_1 = 30$, and $\beta_2 = 2$) that were used to generate the data? What are the 95% confidence intervals for β_0, β_b, and β_2? We should emphasize that in the real world, we do not know what the parameters of the model are or even the form of the model. We use the sample data to infer something about the underlying model.

20.2 [Comparing bivariate and multiple regression parameters] The file *EX20_2* contains values for three variables, Y, X_1, and X_2:

a) Verify that X_1 and X_2 are uncorrelated.
b) Verify that in this case (X_1 and X_2 uncorrelated), $R^2_{Y.12} = r^2_{Y1} = r^2_{Y2}$.
c) For this data set, what is the relationship between (i) the regression coefficient for X_1 when Y is regressed on X_1 alone and (ii) the regression coefficient for X_1 when Y is regressed on both X_1 and X_2? Is this true in general? What is the relationship between the standard errors of b_1 in (i) and (ii)?

20.3 [Deciding which predictors to include] Values for Y, X_1, and X_2 are contained in the file *EX20_3*.

a) Fit the model $Y = \beta_0 + \beta_1 X_1 + \beta_2 X_2 + \varepsilon$.
b) From the data, does it seem that both X_1 and X_2 should be included in the regression model?

20.4 [Adjusting R^2 for bias] Calculate the adjusted R^2 given the following information:

a) $R^2_{Y.1234} = .50$, $N = 10$.
b) $R^2_{Y.1234} = .50$, $N = 40$.
c) $R^2_{Y.1234} = .50$, $N = 200$.
d) $R^2_{Y.12} = .30$, $N = 40$.

20.5 [Interpreting regression fit values] Given the 15 cases in *EX20_5*,

a) Using only the first four cases, regress Y on X_1, X_2, and X_3. What is the value of $R_{Y.123}$? Comment on why $R_{Y.123}$ must take on the value that it does here.
b) Use the regression equation obtained in part (a) to predict the values of Y for each of the 15 cases.
c) Find the correlations between the scores predicted in part (b) and the actual Y values (i) for the first four cases and (ii) for the remaining 11 cases. Comment on the difference between these correlations.

20.6 [Cross-validation on real data] Using pretty much the same procedure as in Exercise 20.5, find a value for the cross-validated R for the regression of *TC* on *age* and *BMI* for women aged 65 years or younger with *BMI*s less than or equal to 40.

a) Select half the cases to serve as a training set and run the regression analysis. Report the multiple correlation coefficient, R.
b) Use the regression equation from part (a) to predict *TC* for the remaining cases, which are the test set. Calculate the correlation between the observed and predicted *TC* in the test set, and compare this value to the R you found in part (a).

c) Repeat parts (a) and (b) using 10% of the sample as the training set.
d) Comment on the different values of *R* you observe across these samples.

20.7 [Leave-one-out cross-validation using deleted predictions] Another way to find a value for the cross-validated *R* is to correlate the actual *TC* scores with the deleted predictions (SPSS calls these *adjusted predicted values*; in R, they are the difference between the externally Studentized residuals from *rstudent*, with the option type = "predictive," and the observed values). The adjusted prediction for each case is based on the regression using the remaining $N - 1$ cases, thereby limiting the damage done by capitalization on chance. Find the correlation between the adjusted prediction of *TC* and actual *TC*.

20.8 [The influence of different numbers of predictors on *R*] You add five predictor variables to a regression that already contains two predictors.

a) Is it possible for the *R* for the regression with seven predictors to be smaller than the *R* with only two predictors? Explain.
b) Is it possible for the adjusted *R* to be smaller for the regression with seven predictors? Explain.

20.9 [Multiple regression with real data] Starting with the *Seasons* file, select the data from participants identified as men with ages less than or equal to 65 and *BMI* scores less than or equal to 40. Regress *TC* on *age* and *BMI*. What is R^2? Are the regression coefficients for *age* and *BMI* significantly different from 0?

20.10 [Comparing regression coefficients when different predictors are included] Start with the analyses of Exercise 20.9. Now add height and weight as predictors. Comment on the differences in R^2 and the patterns of significant coefficients found in this analysis and the values found in Exercise 20.9. In particular, the coefficient of *BMI* in the previous exercise had the value 2.139 and was significant, $p = .004$. However, when weight and height are added as predictors the coefficient now has the value 0.221 with $p = .968$. How can these values be so different?

Notes

1 Because the cases are randomly sampled, replications of this process may yield different results.
2 SPSS calls these B coefficients to distinguish them from the *Standardized Coefficients* (Betas) that would result if the regression were performed with *z* scores.
3 A common notation is to use $s_{Y \cdot 12}$ to refer to the standard error of estimate when *Y* is regressed on X_1 and X_2, and $s_{Y.123}$ when Y is regressed on X_1, X_2, and X_3. We will use s_e when it is clear from the context what predictor variables are in the equation.
4 Here, we focus on unstandardized regression coefficients. In Chapter 22 we consider in detail the reasons we generally prefer unstandardized to standardized regression coefficients.
5 It would, however, be appropriate to interpret the regression coefficient as a measure of the importance of algebra skills as a *predictor* of final exam performance.

Chapter 21

Inference, Assumptions, and Power in Multiple Regression

21.1 Overview

In Chapter 21, we extend our discussion of multiple regression and its interpretation. Our goals are to consider the following topics:

- *The statistical model for inference.*
- *How to check for violations of the assumptions that underlie the inference model.*
- *Detection of outliers and influential points in multiple regression.*
- *Statistical inference in multiple regression*, including confidence intervals for regression coefficients, predictions, and the multiple correlation coefficient.
- *Control of Type 1 error rate.*
- *Prediction intervals in multiple regression.*
- *Power calculations in multiple regression.*
- *Automated stepwise procedures* for identifying the "best" regression equation.

21.2 Inference Models and Assumptions

As was the case for bivariate regression, the validity of our inferences in multiple regression rests upon a model and certain assumptions about the data. We again distinguish between situations in which the predictors are fixed-effect variables and situations in which they are random variables. Fixed predictors generally occur in experimental studies in which the independent variables are manipulated; Y is a random variable but the values of the Xs are selected by the researcher and are therefore considered to be fixed over replications of the experiment. Predictors are random-effects variables when they, as well as Y, are randomly sampled. Although somewhat different assumptions are made for fixed and random predictor variables, under certain conditions the procedures for testing hypotheses and forming confidence intervals are the same. However, we must remember that when the predictors are random-effects variables and we treat them as fixed, our statistical inferences are limited to situations in which the distributions of predictor variables are the same as in the current sample.

Whether X is fixed or random, we assume that the model is

$$
\begin{aligned}
Y &= \beta_0 + \beta_1 X_1 + \beta_2 X_2 + \ldots + \beta_p X_p + \varepsilon \\
&= \mu_{Y \cdot X_1 X_2 \ldots X_p} + \varepsilon
\end{aligned}
$$

DOI: 10.4324/9781003453550-25

where $\mu_{Y.X_1X_2...X_p}$ is the mean of the population of Y scores corresponding to a particular set of values for the p predictor variables. For the fixed-X situation, we assume that:

1. The values of the predictor variables are fixed and measured without error. This means that the values of the Xs are identical for each replication of the experiment.
2. The error components associated with Y are normally and independently distributed with mean 0 and variance σ^2_e.
3. No predictor variable is completely redundant; that is, no predictor variable, X_p, can be perfectly predicted from the other $p - 1$ predictors, using a linear equation. If this requirement is violated, the set of equations that must be solved to obtain the sample regression coefficients will not have a unique solution.

If the Xs are random variables, we assume 2 and 3, and further assume that the distributions of the predictor variables are independent of ε.

21.3 Testing Assumptions and Checking for Outliers and Influential Data Points

In Chapter 19 we considered how to check assumptions and to detect outliers and influential data points in bivariate regression. Here we extend the discussion to multiple regression.

Whether X is fixed or random, we assume that the underlying population model is

$$Y = \beta_0 + \beta_1 X_1 + \beta_2 X_2 + \ldots + \beta_p X_p + \varepsilon$$

and that the error components associated with each of the Y scores are normally and independently distributed with mean 0 and variance σ^2_e. As we discussed in Chapter 19, we would like to check for the assumptions of linearity, homoscedasticity, normality, and independence.

21.3.1 *Checking Assumptions by Using Residuals*

We can visually check several assumptions by using the residuals, e – the closest we can come to observing the error components, ε. We can plot histograms of e to see whether the distribution of residuals is approximately normal, and we can generate scatterplots of e against $\hat{Y}$ to determine whether there are systematic tendencies to violate assumptions (as in Figure 19.5). Under our usual assumptions, the residuals should act like normally distributed random error. If there seems to be systematic variability in the residuals, then perhaps the model has not been properly specified – perhaps additional variables should be added to the equation.

The histogram and scatterplot for the regression of *TC* on *age* and *BMI* are presented in Figure 21.1 and Figure 21.2. The standardized residuals have an approximately normal distribution. Also, although the scatterplot of e vs $\hat{Y}$ indicates that there are outliers that deserve to be looked at more closely, there are no strong suggestions of curvilinearity or heteroscedasticity.

We should also look at the *partial regression plots*. In bivariate regression, it really did not matter whether we plotted the residuals against the predicted scores or against the

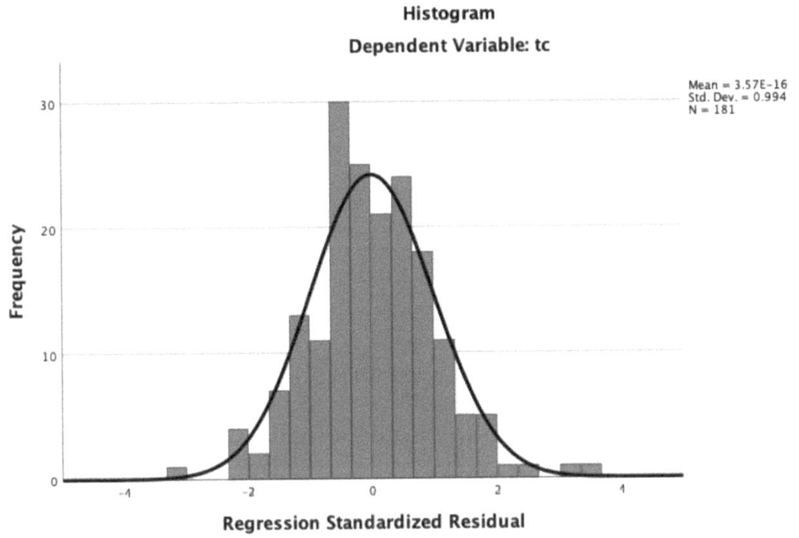

Figure 21.1 Histogram for the standardized residuals with a superimposed normal curve for the regression of *TC* on *BMI* and *age* for women aged 20–65 with *BMI* scores no larger than 40.

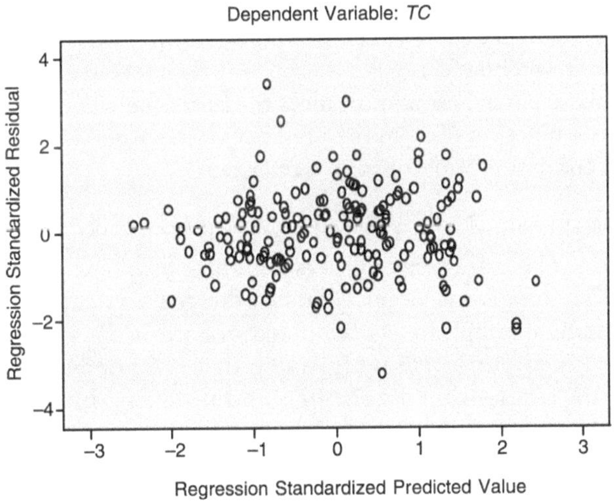

Figure 21.2 Scatterplot of the standardized residuals vs the standardized predicted values of *TC* for women aged 20–65 with *BMI* scores no larger than 40.

predictor variable, X. In that case, the predicted scores were just a linear transformation of the X scores, so both plots provide the same information. This is not the case for multiple regression because there is more than one predictor variable. For multiple regression, in addition to the plot of residuals against predicted scores, we would like to see how the residuals vary as a function of each predictor, *taking the other predictors in the equation into consideration*. Note that SPSS will readily provide these partial regression plots for each predictor in the equation if we check *Produce all partial plots* in the optional *Linear Regression: Plots* dialog box; they are also straightforward to generate using R.

Let's be more concrete in our description of partial regression plots. Suppose we have a criterion variable, Y, and three predictor variables, X_1, X_2, and X_3. We can obtain a partial regression plot for each of the predictor variables. The steps required to get the partial regression plot of Y against X_1 when X_2 and X_3 are also in the equation are the following:

1. Regress Y on X_2 and X_3, and obtain the residuals for this regression, $e_{Y.23}$.
2. Regress X_1 on X_2 and X_3, and obtain the residuals for this regression, $e_{1.23}$.
3. Now generate the scatterplot for $e_{Y \cdot 23}$ vs $e_{1 \cdot 23}$.

These partial regression plots have the following useful properties:

- If we regress $e_{Y \cdot 23}$ on $e_{1 \cdot 23}$, the slope is equal to $b_{Y1.23}$, the partial regression coefficient for X_1 in the regression equation that is obtained when Y is regressed on X_1, X_2, and X_3. The regression of Y on X_1 will not generally have the correct partial slope.
- The simple correlation between $e_{Y.23}$ and $e_{1 \cdot 23}$ is equal to the partial correlation between Y and X_1 with the other predictor variables in the equation partialed out.
- The residuals from the regression line in this plot are identical to those obtained when Y is regressed on all three predictors.
- If Y and X_1 are not linearly related, the partial regression plot will show both the linear and nonlinear components of the relation, *controlling for the other predictors*. This should help us decide whether we wish to add additional variables or other terms such as $X^2{}_1$ to the regression equation.
- We can also use these partial regression plots to determine whether the variability of the residuals varies as a function of X_j in the presence of the other predictors in the equation, thereby violating the assumption of homoscedasticity.

The partial regression plots for *TC* and *age* and for *TC* and *BMI* are displayed in Figures 21.3 and 21.4. There are some outliers and there also seems to be an increase in the variability of the *TC* residuals for larger values of the *age* residuals, suggesting a violation of the homoscedasticity assumption. As we noted in Section 19.9, this violation might be addressed with a weighted least-squares regression that is beyond the scope of this book. It is also possible that the relationship between *TC* and *BMI* may differ for younger and older participants, so that we may wish to consider whether there is an *age* × *BMI* interaction; this is a topic we will take up in Chapter 22.

21.3.2 *Testing for Departures from Linearity*

None of the residual plots show strong systematic departures from linearity. We can, however, test directly for curvilinearity in, and for interactions among, predictor variables. We take up these topics in Chapter 22.

21.3.3 *Outliers and Influential Points in Multiple Regression*

In Section 19.8.4, we introduced measures for identifying cases that might be outliers or influential points in bivariate regression; the same tools apply to multiple regression in generalized form. We can think of and use these tools in the same way as in bivariate regression. However, because more than one predictor is involved, some of these measures are

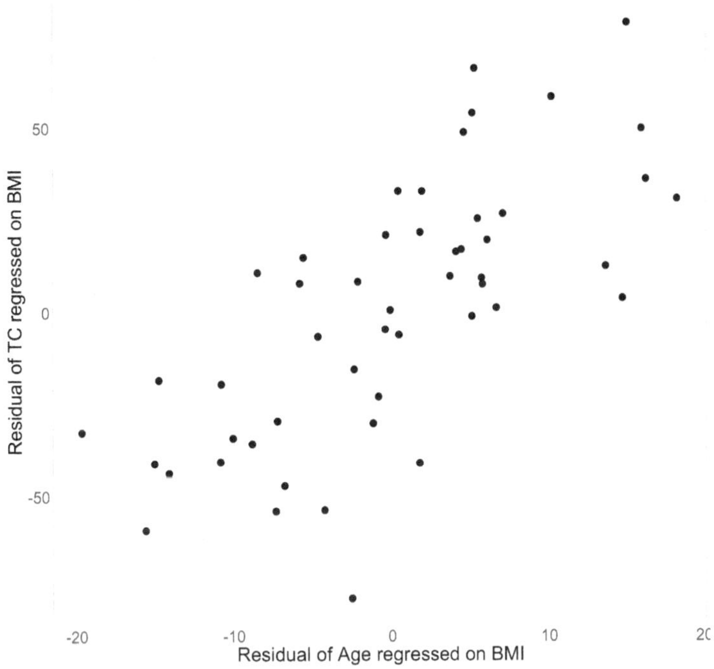

Figure 21.3 Partial regression plot for *TC* vs *age* (i.e., the plot of the residuals for the regression of *TC* on *BMI* against the residuals for the regression *of age* on *BMI*).

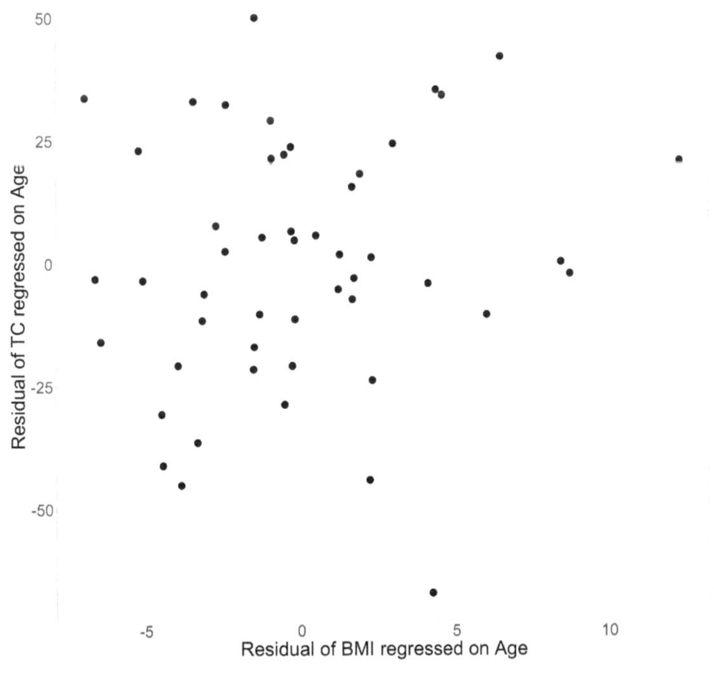

Figure 21.4 Partial regression plot for *TC* vs *BMI* (i.e., the plot of the residuals for the regression of *TC* on *age* against the residuals for the regression of *BMI on age*).

expressed in matrix notation and for this reason we will not present the equations. We can easily calculate and plot these measures, as described in Section 19.8.4, and we can readily identify extreme cases by plotting histograms of the relevant variables. Here we briefly discuss several of the more important measures to assess.

In bivariate regression, outliers in X can usually be identified by looking at the distribution of X or at the scatterplot of X vs Y. However, in multiple regression, we must depend more heavily on measures such as the leverage, h_{jj}, to identify outlying combinations of values on the predictors because multivariate outliers can occur in subtle ways. For example, the jth case may be an outlier because there are correlated deviations from the mean for several predictors; there need not be an extreme deviation for any single predictor. Moreover, even cases with large leverage may not be influential; the degree of influence is a function of the dependent variable, Y, whereas leverage is a function of the predictor variables.

As we mentioned in Chapter 19, it can be shown that the sum of the leverages is $\Sigma h_{jj} = p + 1$ where p is the number of predictor variables; therefore, the mean value of h_{jj} is $(p + 1)/N$. Hoaglin and Welsch (1978) suggest that values of h_{jj} greater than $2(p + 1)/N$ should be considered large. Belsley et al. (1980) indicate that this cutoff will identify too many data points if p is small but recommend it because it is easy to remember and use. Other guidelines mentioned by Neter, Kutner, Nachtscheim, and Wasserman (1996) are that h_{jj} values exceeding .5 indicate very high leverage whereas those between .2 and .5 indicate moderate leverage.

SPSS users should recall that SPSS reports only *centered leverages*, $h_{jj} - 1/N$, for each case. To use the cutoff criteria, we can either modify the criteria so that they apply to centered leverages or simply transform the centered leverages to regular leverages by adding $1/N$ to each of them – here we do the latter. For the regression of *TC* on *age* and *BMI*, for women aged 20–65 years with *BMI*s of ≤ 40, the largest centered leverage is .055 so that the largest h_{jj} is .055 + 1/181 = .061, a low value according to the Neter et al. (1996) guidelines, although it is larger than the Hoaglin and Welsch criterion of .033. We can see from the data set that this leverage value comes from case 569 (ID = 1068), a woman with both predictor values (*age* = 64, *BMI* = 39) near their cutoffs.

As we pointed out earlier, cases with outlying residuals do not necessarily have the greatest influence on the regression equation or the fit. We introduced Cook's distance, CD_j, in Chapter 19 as a measure of the change that would result in the overall fit of the model if the jth case was omitted. CD_j can be written as

$$CD_j = \frac{\sum_i \left(\hat{Y}_i^{(-j)} - \hat{Y}_i\right)^2}{(p+1)s_e^2}$$

where $\hat{Y}^{(-j)}_i$ is the prediction of Y_i calculated from regression coefficients obtained with the jth case deleted. A simple guideline given by Cook and Weisberg (1982) is that a Cook's distance of 1 should be considered large. However, a guideline that takes sample size and number of predictors into account is that Cook's distance values should be considered large if they exceed the cutoff $F_{.50,\, p+1,\, N^*-p-1}$. Case 569 has a Cook's distance of .029, well below the critical $F_{.50,\, 3,\, 178} = .79$. The largest Cook's distance value for the regression is .084, again well below the critical value.

Finally, although we do not discuss them here, if our major concern is with the change in the regression coefficients when a particular case is deleted, this is reflected in the *DFBETAS* values for each predictor variable; these values can be readily obtained (see Section 19.8.4) and cases with large *DFBETAS* or large Cook's distance can be checked to determine if a data coding error has occurred. If the data are not obviously in error, then our options are the same as we presented in Section 19.8: We cannot simply delete cases, but we can present the regression results with and without those cases, or we could conduct some type of robust regression.

Unfortunately, although regression diagnostics that consider the effect of deleting one point at a time work quite well when there is a single influential outlier, it is much more difficult to diagnose outliers when there are several of them. For a useful discussion of developments in the detection of multiple outliers and of robust regression, see Rousseeuw and Leroy (1987).

21.4 Testing Different Hypotheses in Multiple Regression

21.4.1 Testing the Hypothesis $\beta_1 = \beta_2 = \ldots = \beta_p = 0$

As we indicated in Section 20.4.2, if the p regression coefficients $\beta_1, \beta_2, \ldots, \beta_p$ all have the value 0 in the population, the ratio

$$\frac{MS_{regression}}{MS_{residual}} = \frac{R^2 SS_Y / p}{\left(1-R^2\right)SS_Y / (N-1-p)} = \frac{R^2 / p}{\left(1-R^2\right)/(N-1-p)}$$

will be distributed as F with p and $N - 1 - p$ df under standard assumptions. Therefore, $MS_{regression} / MS_{residual}$ can serve as the statistic to test the null hypothesis that the p regression coefficients are all zero in the population. This test asks whether we have sufficient evidence to conclude that the model

$$Y = \beta_0 + \beta_1 X_1 + \beta_2 X_2 + \ldots + \beta_p X_p + \varepsilon$$

accounts for Y in the population better than the restricted model

$$\begin{aligned} Y &= \beta_0 + \varepsilon \\ &= \mu_Y + \varepsilon \end{aligned}$$

If the restricted model is appropriate, then $\beta_1 = \beta_2 = \ldots = \beta_p = 0$, so that the best predictor for Y is $\hat{\beta}_0 = \hat{\mu}_Y = \bar{Y}$ and the multiple correlation coefficient in the population has the value 0.

Expanding our example from Chapter 20, we regressed *TC* on five predictor variables: *age* and *BMI*, as before, as well as two measures of mental health, *beck_d* and *beck_a*, and an indirect measure of self-care, *dirwdc1*, which indicates the number of hours of sunlight exposure during weekdays in the winter. We began by plotting the data to assess the assumptions and look for outliers. Figure 21.5 reveals a *beck_a* score that is about 10 points higher than any of the others, as well as some skew in the distributions of predictors but not in the distribution of *TC*. Omitting the one case with a *beck__a* score above

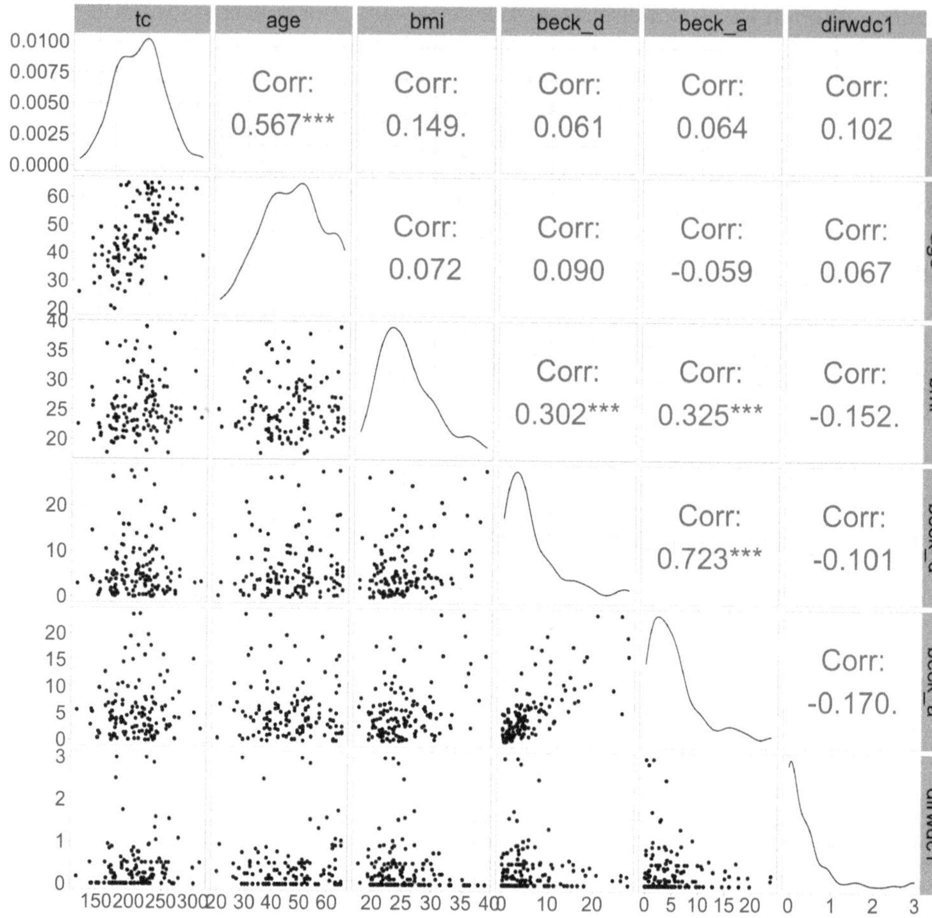

Figure 21.5 Correlation matrix of *TC*, *age*, *BMI*, *beck_d*, *beck_a*, and *dirwdc1* for women 20–65 years old with BMI <= 40, from R.

30, the regression results based on the remaining 131 cases are shown in Figure 21.6. We find

$$F(5,125) = \frac{MS_{regression}}{MS_{residual}} = \frac{\frac{60,561}{5}}{109,646/(131-1-5)} = 13.81$$

which has an associated p-value of < .001. As a result, we reject the hypothesis that the population regression coefficients are zero for all five predictors. The test assumes that $MS_{residual}$ is an estimate of the variance of the random error component. If important variables are left out of the regression equation, $MS_{residual}$ will reflect their effects as well as random error, and the test may be biased.

```
> summary(lm(data=dat,tc ~ beck_d + beck_a + age + bmi + dirwdc1))

Call:
lm(formula = tc ~ beck_d + beck_a + age + bmi + dirwdc1, data = dat)

Residuals:
    Min      1Q  Median      3Q     Max
-69.860 -15.214  -3.155  18.335 115.598

Coefficients:
            Estimate Std. Error t value Pr(>|t|)
(Intercept)  96.8351    18.2466   5.307 4.91e-07 ***
beck_d       -0.8619     0.6126  -1.407   0.1620
beck_a        1.3431     0.7692   1.746   0.0833 .
age           1.9075     0.2434   7.838 1.72e-12 ***
bmi           0.8390     0.6127   1.369   0.1734
dirwdc1       6.1568     4.6832   1.315   0.1910
---
Signif. codes:  0 '***' 0.001 '**' 0.01 '*' 0.05 '.' 0.1 ' ' 1

Residual standard error: 29.62 on 125 degrees of freedom
Multiple R-squared:  0.3558,     Adjusted R-squared:   0.33
F-statistic: 13.81 on 5 and 125 DF,  p-value: 9.812e-11

> anova(lm(data=dat,tc ~ beck_d + beck_a + age + bmi + dirwdc1))
Analysis of Variance Table

Response: tc
           Df Sum Sq Mean Sq F value    Pr(>F)
beck_d      1    625     625  0.7127    0.4002
beck_a      1    145     145  0.1653    0.6850
age         1  56950   56950 64.9249 5.307e-13 ***
bmi         1   1325    1325  1.5105    0.2214
dirwdc1     1   1516    1516  1.7283    0.1910
Residuals 125 109646     877
---
Signif. codes:  0 '***' 0.001 '**' 0.01 '*' 0.05 '.' 0.1 ' ' 1
```

Figure 21.6 Results of regression of *TC* on *age*, *BMI*, *beck_d*, *beck_a*, and *dirwdc1* for women 20–65 years old with BMI <= 40, from R.

21.4.2 *Confidence Intervals for the Squared Multiple Correlation Coefficient*

A confidence interval for ρ^2, the squared population multiple correlation coefficient, is more informative than simply stating the sample R^2 along with the results of a significance test. Using an example given by Steiger and Fouladi (1997), suppose we obtain an R^2 of .40 in a regression using five predictor variables with $N = 45$. The shrunken estimator (i.e., the adjusted R^2) is .327 and if a significance test is performed, the p is .001. However, it is more useful to know that the 95% confidence interval for ρ^2 extends from .095 to .562. This interval tells us that the range of possible values is quite wide; converting from ρ^2 to ρ, the lower limit of the 95% CI is .31 and the upper limit is .75.

An expression for the variance of R^2 is given by

$$\text{Var}\left(R^2\right) = 4\rho^2\left(1-\rho^2\right)^2\left[\frac{(N-p-1)^2}{\left(N^2-1\right)(N+3)}\right] \quad (21.1)$$

where N is the total number of cases, p is the number of predictors, and ρ^2 is the square of the population multiple correlation coefficient (Olkin & Finn, 1995). For example, if we consider the regression of *TC* on *age*, *BMI*, *beck_d*, *beck_a*, and *dirwdc1* for women aged 20–65 years with *BMI* scores of 40 or less, and take the value of R^2 from Figure 21.6 as our estimate of ρ^2, the variance of R^2 is estimated by

$$\frac{4(.3558)(1-.3558)^2(131-1-5)^2}{\left(131^2-1\right)(131+3)} = 0.004$$

We can approximate the 95% CI for ρ^2 using a normal distribution, with the following formula:

$$R^2 \pm z_{.025}SE\left(R^2\right) = .3558 \pm 1.96\sqrt{.004} = .3558 \pm .1240$$

that is, an interval that extends from approximately .23 to .48. In R, the *ci.R2* function in the {MBESS} package computes this confidence interval more precisely, taking R^2, N, and p (called K) as input: *ci.R2*(R2 = .3558, N = 131, K = 5) returns [0.199, 0.466]. The intervals are not identical in this case because the sampling distribution for R^2 is not exactly normal, an issue that is better addressed with software.

We should point out that the methods discussed in this section will not be accurate unless the assumption of multivariate normality is satisfied and the predictor variables have been specified in advance.

21.4.3 *Testing the Hypothesis* $\beta_j = \beta_{hyp}$ *and Finding Confidence Intervals for* β_j

A significant test of the overall regression equation logically implies that at least one of the predictors is linearly related to *Y*. Generally, we will want to examine how the individual *X* variables contribute to the prediction of *Y*. Under the usual assumptions, the ratio

$$\frac{b_j - \beta_j}{SE\left(b_j\right)}$$

will be distributed as t with $N - 1 - p$ *df*. Therefore, if we can estimate the standard error, we can test the hypothesis that the population intercept, β_0, or any of the population regression coefficients, β_j, are equal to any constant. In practice, the null hypothesis $\beta_j = 0$ is usually tested. Rejection of this hypothesis implies that X_j makes a significant contribution to the predictability of *Y* when it is added to the other variables in the equation. For example, in the analysis of the *TC* data summarized in Figure 21.6, the regression coefficient for *BMI* is 0.8390 and the standard error is 0.6127, so t = 0.8390 / 0.6127 = 1.369, which is not significant.

Once we have obtained the appropriate *SE*s, we can obtain confidence intervals for each parameter using

$$b_j \pm t_{\alpha/2} SE(b_j)$$

In the current example, the critical value of $t_{.025}(125)$ is 1.978 (in R, *qt*(.975, 125) = 1.978), and the standard error values from Figure 21.6 for the intercept, *age*, and *BMI* are 18.247, .2434, and .6127, respectively. Combining this information, the 95% confidence intervals for the intercept and the coefficients of *age* and *BMI* are 96.835 ± 36.093, 1.908 ± 0.481, and .839 ± 1.212. These confidence intervals match those reported in Figure 21.7.

21.4.4 *Partial F Tests: Procedures for Testing a Subset of the* β_j*s*

We can use partial F tests to determine whether adding one or more predictors to a regression equation that already contains p predictors significantly increases the predictability of Y. If we consider just one additional predictor, X_{p+1}, a test of the model

$$Y = \beta_0 + \beta_1 X_1 + \beta_2 X_2 + \ldots + \beta_p X_p + \beta_{p+1} X_{p+1} + \varepsilon$$

against the restricted model

$$Y = \beta_0 + \beta_1 X_1 + \beta_2 X_2 + \ldots + \beta_p X_p + \varepsilon$$

is equivalent to testing the hypothesis H_0: $\beta_{p+1} = 0$. This approach provides a basis for comparing any two models where the restricted model includes a subset of the predictors in the full model.

We can represent the variability in Y accounted for by regression on the variables of the restricted model as

$$SS_{Y \cdot 12 \ldots p} = R^2_{Y \cdot 12 \ldots p} SS_Y$$

```
> confint(lm(data=dat,tc ~ beck_d + beck_a + age + bmi + dirwdc1))
                 2.5 %      97.5 %
(Intercept) 60.7228764 132.9473383
beck_d      -2.0742983   0.3505851
beck_a      -0.1792437   2.8653709
age          1.4258580   2.3891864
bmi         -0.3737031   2.0516059
dirwdc1     -3.1118405  15.4255347
```

Figure 21.7 Confidence intervals for the regression coefficients of *TC* on *age*, *BMI*, *beck_d*, *beck_a*, and *dirwdc1* for women 20–65 years old with BMI <= 40.

and the variability accounted for by the larger model when the predictor X_{p+1} is added as

$$SS_{Y\cdot 12\ldots p+1} = R^2_{Y\cdot 12\ldots p+1} SS_Y$$

Therefore, the increment in variability associated with the predictor X_{p+1} is given by

$$SS_{increment} = SS_{Y\cdot p+1|12\ldots p} = \left(R^2_{Y\cdot 12\ldots p+1} - R^2_{Y\cdot 12\ldots p}\right) SS_Y$$

This increment is associated with a single *df* because only one additional regression coefficient must be estimated in the larger model. Table 21.1 presents these results in the form of an ANOVA table in which the *SS* accounted for by *Y* is first partitioned into the *SS* accounted for the larger model with $p + k$ predictors and error variance (residual); then, the *SS* for the larger model is partitioned into the *SS* for the smaller model with p predictors and the *SS* for the increment due to the remaining k predictors. The hypothesis H_0: $\beta_{p+1} = 0$ can be tested using the ratio

$$F = \frac{MS_{increment}}{MS_{residual}}$$

where the denominator is the mean square associated with the variability not accounted for by the larger model; that is, when $k = 1$,

$$MS_{residual} = \frac{\left(1 - R^2_{Y\cdot 12\ldots p+1}\right) SS_Y}{N - p - 2}$$

The numerator of the F is associated with 1 *df* if a single predictor is added ($k = 1$) and the denominator with $N - 2 - p$ *df*.

When a partial F test is used to test whether a *single* population regression coefficient is zero, the results produced are exactly equivalent to those of the t test discussed in the

Table 21.1 ANOVA table for testing the effect of adding k predictor variables to a model that already contains p predictors

SV	*df*	*SS*	*MS*
Larger model	$p + k$	$R^2_{Y.1\ldots p+k} SS_Y$	$\frac{R^2_{Y.1\ldots p+k} SS_Y}{p+k}$
Smaller model	p	$R^2_{Y.1\ldots p} SS_Y$	$\frac{R^2_{Y.1\ldots p} SS_Y}{p}$
Increment	k	$SS_{increment} = \left(R^2_{Y.1\ldots p+k} - R^2_{Y.1\ldots p}\right) SS_Y$	$SS_{increment}/k$
Residual	$(N - 1 - p - k)$	$\left(1 - R^2_{Y.1\ldots p+k}\right) SS_Y$	$\frac{\left(1 - R^2_{Y.1\ldots p+k}\right) SS_Y}{N - 1 - p - k}$

preceding section. For example, if we use this procedure to test whether hours of weekday winter sunlight exposure, *dirwdc1*, adds significantly to the prediction of *TC* over and above *age, BMI, beck_d, and beck_a*, $SS_{increment} = 1516$, $df_{increment} = 1$, and $MS_{residual} = 877$ (see Figure 21.6), so that

$$F = \frac{MS_{increment}}{MS_{residual}} = \frac{1516}{877} = 1.729$$

The *F* value obtained is the square of the *t* for the coefficient of *dirwdc1* in the summary output for the regression in Figure 21.6: $\sqrt{1.729} = 1.315$. Of course, neither test allows us to reject the null hypothesis the $\beta_{\mathrm{dirwdc1}} = 0$.

Partial *F* tests can also be used to test hypotheses which state that some subset of the β_js is equal to zero. Suppose, for example, we start with a model containing *p* predictor variables

$$Y = \beta_0 + \beta_1 X_1 + \beta_2 X_2 + \ldots + \beta_p X_p + \varepsilon$$

and we add *k* more predictor variables so that the model is now

$$Y = \beta_0 + \beta_1 X_1 + \beta_2 X_2 + \ldots + \beta_p X_p + \beta_{p+1} X_{p+1} + \ldots + \beta_{p+k} X_{p+k} + \varepsilon$$

Again, the appropriate ANOVA table is given in Table 21.1. This general approach can be used to assess the effect of adding any set of predictors to the equation and tests the hypothesis that the regression coefficients for these added predictors are all equal to zero in the population.

Let's use a partial *F* test to assess the contribution of the additional three predictors (*beck_d*, *beck_a*, and *dirwdc1*) in the five-predictor regression example we have developed in this chapter. We cannot directly compare the five predictor results to the two-predictor analysis presented in Chapter 20 because there were fewer participants with data on all five predictors ($N = 131$) than in the simpler two-predictor analysis (where $N = 181$). Fortunately, all the calculations we need are provided in the five-predictor results shown in Figure 21.6. The *SS* increment from adding *beck_d*, *beck_a*, and *dirwdc1*, is 625 + 145 + 1516 = 2,286, and *df* increment is 3, so

$$F = \frac{MS_{increment}}{MS_{residual}} = \frac{2286 / 3}{877} = 0.869$$

The increment from these three additional predictors is clearly not significant: the *p*-value is .50. This outcome means that we are not reliably better at predicting total cholesterol when these three predictors are added to our model.

21.5 Controlling Type 1 Error in Multiple Regression

The need to control overall Type 1 error for each family of hypothesis tests is emphasized in most treatments of ANOVA. However, the issue is scarcely mentioned in many treatments of multiple regression, where inflation of Type 1 error can be a very serious problem (e.g., Dar, Serlin, & Omer, 1994). Given an interesting dependent variable and a collection of possibly useful predictors, we often see attempts to assemble combinations of predictors, often without regard to theory, that just happen to result in larger increases in R^2. Because this approach can capitalize on random error in the sample (see Chapter 20), it may not produce an equation with any explanatory value or even one that is effective for predicting in samples other than the one used to generate it (see Section 20.4.4). Trying out many different regression equations and looking at the tests for R^2, as well as the tests for all the individual regression coefficients, can result in highly inflated Type 1 error for the collection of tests. If we ignore this possibility, we may add too many variables to the regression equation.

Looking at tests of individual regression coefficients is analogous to looking at tests of contrasts of means. If we have an *a priori* interest in comparing two specific means out of a larger set of means and we test this comparison at $\alpha = .05$, we can say that the Type 1 error is .05. On the other hand, if we first find the two means that differ the most and then test their difference at $\alpha = .05$, the effective Type 1 error is much larger because we had to consider all the pairwise differences to determine which one to test. Analogously, if we select a predictor variable from many candidates solely because it results in a large increase in R^2, the effective Type 1 error is much larger than the nominal significance level used for the test.

Decisions about what sets of variables should be added to the regression equation should be driven by our knowledge and theories about our research question. We should not just search through our potential predictors in a quest to find those that maximally increase R^2 because in some cases the most "effective" predictors will be those that have capitalized on chance variability in the sample. Also, some effective predictors may not be theoretically meaningful themselves, but rather gain their predictive power by being proxies for important variables that for some reason (e.g., measurement error) do not predict as well. The distinction we are trying to make here is analogous to that between dealing with planned contrasts and dealing with unexpected but large differences among means in ANOVA.

There are no fully developed procedures for controlling Type 1 error rates in multiple regression analyses. In part, this lack may be due to the complexity of the interrelationships among the predictor variables, which is a significant roadblock to estimating the error rate at all. Nonetheless, one strategy that can be helpful in reducing Type 1 errors is to treat related set of predictors as a *family* of tests, much the way we defined families of tests in ANOVA (see Section 10.5). For example, suppose we are interested in influences on classroom performance and have three sets of predictors. Say we are most interested in (1) student variables (measures of ability and motivation), then in (2) family variables (parental education, SES, support for education), and finally in (3) classroom variables (class size, teaching style). We could start by first including the set of student variables as predictors. If we determined that R^2 was significant for the set (say, at $\alpha = .05$), we could add the set of family variables and determine whether the increase in R^2 for the set was significant (again at $\alpha = .05$), and so on. In our *TC* example, we might distinguish between predictors based on mental health measures (*beck_d*, *beck_a*, *dirwdc1*) and those based on physical health (*age*, *BMI*); using partial *F* tests, we have seen that the mental health measures, as a group, did not contribute significantly to the prediction of cholesterol levels.

If the predictors can be organized into families, and if a family results in a significant R^2 for its set of predictors using the partial *F* approach of Section 21.4.4 with $\alpha = .05$, then we are justified in conducting significant tests on each of the predictors within that family with

$\alpha = .05$. In the *TC* example, we are not justified in testing any of the mental health measures on their own, because they failed to reach significance as a group. The logic of this approach extends Fisher's protected t test (Fisher, 1935) to multiple regression analyses (Cohen et al., 2003).

Fisher's protected t procedure as applied to ANOVA is as follows. First test for the significance of a categorical independent variable with a levels using α as the significance level. If the overall test is significant, then each of the $a(a - 1) / 2$ pairwise comparisons between the levels of the variable may be tested at significance level α. Tests of the individual comparisons are said to be "protected" by the significant overall test. If the overall test is not significant, we are not allowed to perform significance tests on any of the pairwise comparisons in the set (unless, presumably, we have specified them *a priori*). As applied to multiple regression, a significant R^2 or increment in R^2 for a meaningful set of predictors would allow us to look at the significance tests for each of the regression coefficients in the set at the same level of significance. If the overall test for the set of predictors is not significant, we cannot consider the tests for the individual predictors in the set, no matter how large their t statistics might be. This means that the researcher must resist the temptation to rely on the regression output provided by software, which routinely includes a value of t for each of the predictor variables. Instead, those individual test results must be reviewed only after a statistically significant family-wise assessment.

The hierarchical strategy inherent in defining families of predictor variables does have limitations. Most obviously, if no related sets of predictors are included in the regression analysis, then clearly no families of predictors can be identified. Additionally, the Fisher protected test does not completely control Type 1 error in ANOVA when the independent variable has more than three levels (Hayter, 1986), and Cohen et al. (2003) acknowledge some problems with this procedure. Nonetheless, the modified Fisher procedure at least acknowledges the issue of inflation of Type 1 error in multiple regression and offers an easily applied, if imperfect, control for it.

21.6 Inferences About the Predictions of *Y*

In some situations, the goal of a regression analysis is to make predictions. For example, a university admissions office might want to know the expected first year GPA for all students entering the university with a high school GPA of 3.0. In this case, we would want to get a confidence interval on the predicted mean university GPA for all students with a high school GPA of 3.0. Alternatively, an individual student with a high school GPA of 3.0 might want to estimate her own first year GPA.

In bivariate regression, the expected value of Y corresponding to a value X_j of X is given by $\mu_{Y.X_j} = \beta_0 + \beta_1 X_j$, which can be estimated by $\hat{Y}_j = \hat{\mu}_{Y.X_j} = b_0 + b_1 X_j$. We showed in Chapter 19 that for bivariate regression, the estimated standard error associated with the prediction of Y at $X = X_j$ is given by

$$SE\left(\hat{\mu}_{Y.X_j}\right) = s_e\sqrt{h_{jj}} \tag{21.2}$$

and that the standard error associated with an individual score is

$$SE\left(\hat{Y}_j\right) = s_e\sqrt{1 + h_{jj}} \tag{21.3}$$

where s_e is the standard error of estimate and h_{jj}, the leverage of X_j, is

$$h_{jj} = \frac{1}{N} = \frac{\left(X_j - \bar{X}\right)^2}{SS_X}$$

In multiple regression, we can find the estimated standard error for the prediction of Y associated with any combination of values of the p predictor variables in the regression equation. Because there is more than one predictor, the standard errors are most easily presented as matrix expressions. Although the expression for h_{jj} will now be more complicated because there is more than one predictor, it can be thought of and used in much the same way as in bivariate regression. If we want to find confidence intervals for the conditional means or individual scores corresponding to combinations of predictor values in our data set, we do not have to compute the leverage; we can use software to compute the confidence bounds directly or to obtain the leverage values that we can then use in Equations 21.2 and 21.3.

However, if we wish to find the standard error for the prediction of Y based on a combination of values for the p predictors, $X_{j1}, X_{j2}, \ldots, X_{jp}$, that did not occur in our sample, the estimated standard errors for $\hat{\mu}$ and $\hat{Y}$ are not directly made available in the SPSS output and must be calculated. We will not present the relevant matrix expressions or their derivation here.

For example, let's generate a confidence interval for the conditional mean of *TC* from the regression on *age* and *BMI*, assuming a new group of 40-year-old women with BMI of 30. We will use the analysis from Chapter 20 with $N = 181$, assuming the output is saved in an object called "twopred." In R, the function *predict*(twopred, data.frame(age = 40, bmi = 30), interval = "confidence") returns a predicted *TC* of 205.94 and a CI of [198.21, 213.66], see Figure 21.8. The same command, using the option *interval* = "prediction," returns the prediction interval for an individual 40-year-old woman with a BMI of 30, which includes the same predicted *TC* but a wider interval, [139.18, 272.69], because predictions for individuals are more uncertain than predictions for means. We can also predict the cholesterol level of someone whose *age* and *BMI* are not included our observed data, say a 67-year-old woman with a BMI of 45. As you can see in Figure 21.8, the resulting prediction interval

```
> newdata<-data.frame(age = c(40,67), bmi = c(30,45))
> predict(twopred, newdata, interval = "confidence")
       fit      lwr      upr
1 205.9353 198.2128 213.6579
2 274.9426 252.7608 297.1243
> predict(twopred, newdata, interval = "predict")
       fit      lwr      upr
1 205.9353 139.1760 272.6947
2 274.9426 205.0197 344.8654
```

Figure 21.8 Confidence and prediction intervals for *TC* values for new cases in which *age* = 40, *BMI* = 30, and *age* = 67, *BMI* = 45.

has a mean of 274.92 and a range of [205.02, 344.87]. Because we are making a prediction for a single new individual who falls well outside of the observed data, the interval is quite wide and should be interpreted with skepticism. Simply put, the relationship of *TC* to *age* and *BMI* is not known beyond the bounds of our data.

21.7 Power Calculations in Multiple Regression

There are several kinds of power calculations that can be performed for multiple regression, paralleling the various hypotheses that can be tested (see Section 21.4). Here, given the model

$$Y = \beta_0 + \beta_1 X_1 + \beta_2 X_2 + \ldots + \beta_j X_j + \ldots + \beta_p X_p + \varepsilon$$

we consider power for tests of the null hypotheses that

1. All p of the regression coefficients are equal to 0 in the population,
2. A particular coefficient is equal to 0, or
3. k of the p regression coefficients are equal to 0.

Most programs for calculating power require an estimate of the noncentrality parameter for the noncentral F distribution, defined as

$$\hat{\lambda} = N^* f^2 \tag{21.4}$$

where Cohen's effect size statistic for regression, f^2, is given by

$$f^2 = \frac{\Delta R^2}{1 - R^2} \tag{21.5}$$

Here, ΔR^2 is the increment in R^2 when k predictors are added to a set of $p - k$ predictors, and R^2 is the squared multiple correlation calculated for the full set of p predictors. Thus, Cohen's f^2 tells us the proportion by which the remaining unexplained variance in Y is reduced by the addition of k predictors, and the noncentrality parameter multiplies that effect size by a sample-size related variable, N^*. When the predictors are fixed-effects variables, N^* is equal to the sample size, N; when the predictors are random variables, N^* is defined by Cohen (1988) as

$$\begin{aligned} N^* &= df_{numerator} + df_{denominator} + 1 \\ &= k + (N - p - 1) + 1 \\ &= N - p + k \end{aligned} \tag{21.6}$$

Maxwell (2000, p. 436) comments that Cohen's (1988) formulation appears to provide a small adjustment for the random nature of predictors that are usually encountered in psychological research. Therefore, a convenient way to perform *a priori* power calculations is to use G*Power 3.1 to determine the N^* required to achieve the desired level of power. If the predictors are fixed, then N, the required sample size, equals N^*; if the predictors are random,[1] then from Equation 21.6, $N = N^* + p - k$. For larger values of N, the differences in estimated power obtained using N and $N - p + k$ will be small.

We now consider the three types of power calculations:

- *Case 1 – power for tests that all the regression coefficients are equal to zero in the population.* The test that all regression coefficients equal zero is equivalent to the test that the population squared multiple correlation, ρ^2, equals zero. For case 1, $p = k$, so that $N^* = N$, the required sample size, whether we have fixed- or random-effects predictors.

  As an example, suppose we want to estimate the sample size required to have power = .90 at $\alpha = .05$ for the test of the hypothesis that all population regression coefficients are zero. If we expect that $R^2 = .20$ when we regress the criterion variable on four predictors, the effect size is

$$f^2 = \frac{\Delta R^2}{1-R^2} = \frac{R^2}{1-R^2} = \frac{.20}{1-.20} = .25$$

  In G*Power 3.1, select

  1. *F tests* as the *Test family*.
  2. *Linear Multiple Regression: Fixed model, R^2 deviation from zero* as the *Statistical test*, and
  3. *A priori: Compute required sample size – given α, power, and effect size* as the *Type of power analysis*.

  Then insert the values $f^2 = .25$, *Power* = .90, $\alpha = .05$, and *Number of predictors* = 4 and click on *Calculate*. G*Power 3.1 indicates that the required sample size is 67.

- *Case 2 – power calculations for the test that a particular coefficient is zero in the population.* Here $k = 1$, so that for random predictors $N^* = N - p + 1$, so that $N = N^* + p - 1$. As an example, suppose we plan to test whether adding a predictor to a regression equation that already contains three predictors will significantly improve the predictability of the dependent variable. This is the same as testing the significance of the coefficient of the predictor in a regression that contains all four predictors. Suppose that we want to find the N required to have power = .80 if R^2 with three predictors is .25 and we expect the fourth predictor to increase R^2 to .30, so that $f^2 = .05/(1 - .30) = .071$. In G*Power 3.1 (see Figure 21.9), select

  1. *F tests* as the *Test family*,
  2. *Multiple Regression: Linear Multiple Regression: Fixed model, R^2 increase* as the *Statistical test*, and
  3. *A priori: Compute required sample size given α, power, and effect size* as the *Type of power analysis*.

  Then, insert $f^2 = .071$, $\alpha = .05$, *Power* = .80, *Number of tested predictors* = 1, and *Total number of predictors* = 4. If we click on *Calculate*, we find $N^* = 113$. If the predictors are fixed, $N = 113$. If they are random, $N = 113 + p - 1 = 116$.

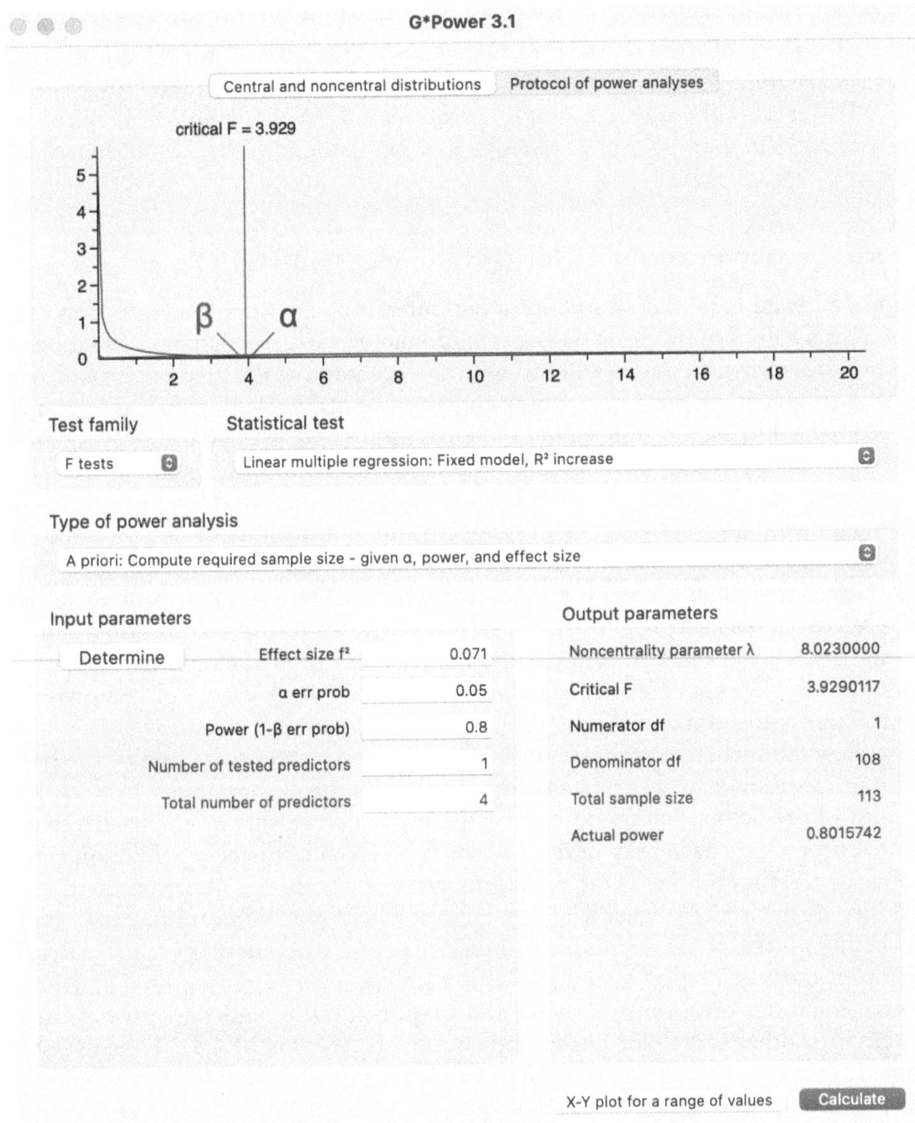

Figure 21.9 G*Power 3.1 dialog box for finding N^* necessary to have power = .80 when one predictor variable is to be added to a regression equation containing three predictors and is expected to increase R^2 from .25 to .30.

- *Case 3 – power calculations for testing whether all members of a specified subset that contains k of the p coefficients are equal to zero.* Assuming that the predictors are random variables, we again use G*Power 3.1 to find N^*, and the required sample size is $N = N^* + p - k$ if the predictors are random-effects variables.

Suppose we have a total of five predictors and want to test whether a particular subset of three predictors collectively adds significantly to the prediction over and above the contribution of the other two. If we expect that R^2 will increase from .44 to .50 when the subset of three predictors is added to the equation, then the f^2 for the additional three predictors is .06/.50 = .12. To find the sample size required to achieve power = .80, we use G*Power as in the case 2 example, except now $f^2 = .12$, *Number of tested predictors* = 3, and *Total number of predictors* = 5. N^* is found to be 95, so that $N = N^* + p - k = 95 + 5 - 3 = 97$.

21.8 Automated Procedures for Developing Prediction Equations

Sometimes we want to predict some criterion of interest by developing a regression equation that contains a subset of the potentially useful predictor variables that are available. In predicting, we are normally concerned both with the accuracy of the predictions and with the costs involved in making them. If our only concern was accuracy, then for large samples we would be inclined to use as many valid predictors as possible in the regression equation; on the other hand, concerns about costs and simplicity would motivate us to use fewer predictors. Because in many types of research most of the predictor variables are correlated with one another, including all of them in a regression equation would not only be cumbersome but would also introduce a good deal of redundancy. Several automated procedures have been developed that allow a compromise between these concerns. These procedures include *forward selection*, *backward elimination*, and *stepwise regression*; all are commonly available in statistical software. For example, in SPSS, the "Method" defaults to "Enter," which includes all variables, but can be changed to "Forward," "Backward," or "Stepwise" from the main linear regression window.

Using these automated procedures, it is often possible to select a subset of the potential predictors that account for nearly as large a proportion of the variability in Y as does the entire pool of predictors. Before describing the methods, we should emphasize that these automated procedures have been developed solely to produce the best prediction equations according to certain criteria. These equations need not be best – nor even good – in any explanatory or theoretical sense. Running an automated regression routine may be useful for predicting outcomes within a sample but is a poor way to develop theory. Recall our discussion of cross-validation techniques and why they are so important: The best-fitting regression equations are sample specific and may have little ability to predict outcomes accurately in another sample or in the population (see Section 20.4.4). Thus, we strongly recommend that cross-validation be used to validate a predictive model that is generated by any of the automated procedures discussed next.

21.8.1 Forward Selection

In the forward selection procedure, the regression equation is built up one predictor variable at a time. On the first step, the predictor that has the highest correlation (positive or negative) with the criterion variable is selected. If it fails to meet the criterion for inclusion, the procedure ends with no predictors in the equation, and the final equation is $\hat{Y}_i = \bar{Y}$. If the first predictor meets the criterion and is added to the equation, a second predictor is then selected and tested to determine whether it should be entered into the equation. The second predictor considered is the one that would result in the greatest increment in R^2 if added to the equation. If the second predictor does not meet the criterion for inclusion, the procedure terminates with only a single predictor in the equation. If it does meet the criterion, on the third step, a third predictor is selected and tested, and so on. At each step, a partial F test is performed on the selected variable, and the criterion for inclusion is stated in terms of the critical value or the significance level of the F. For example, using the five predictors and 131 cases described in Section 21.4.1, forward selection finds that the best-fitting regression model is $\widehat{TC} = 103.423 + 1.847age + .860BMI$ with $R^2 = .333$.

For procedures like forward selection, it should be clear that the usual significance levels obtained from the F distribution are not appropriate. This is because at each step several possible predictors are examined and only the one that produces the greatest increment in R^2 is tested. If only a single predictor is to be chosen from a pool of m possible predictors, the situation is analogous to choosing the largest member of a family of m contrasts and testing it for significance (see Section 21.4). As in the case of contrasts, if a single predictor is to be chosen, the Bonferroni procedure may be used to control Type 1 error; that is, to use $\alpha^* = \alpha / m$ as the criterion for significance, where α is the probability of at least one Type 1 error (see Chapter 10 for a discussion of the issue of family wise error rates and procedures for controlling them).

Wilkinson (1979) has discussed the case in which a subset of k predictors is to be chosen, where $1 < k < m$, and has provided tables of the upper 95th and 99th percentage points of the sample R^2 distribution in forward selection based on simulations (other tables and discussions of this problem can be found in Hocking, 1983; Rencher & Pun, 1980; and Wilkinson & Dallal, 1982). These tables are more conservative than the usual F tables. For example, with $N = 35$ and $\alpha = .05$, if all four members of a set of predictor variables are to be included in the regression equation, it is appropriate to use the standard F test to test R^2 for significance. When this is done, it is found that the sample R^2 must exceed .26 to reject the hypothesis that the population multiple correlation coefficient is 0. However, if the four predictors are to be selected from a larger set of 20 predictors by a forward selection procedure, according to Wilkinson's tables, the sample R^2 must exceed .51 to reject the null hypothesis. Many researchers do not seem to be aware of this problem; for a sample of 66 published papers that reported significant forward selection analyses according to the usual F tests, Wilkinson found that 19 were not significant when his tables were used.

21.8.2 Backward Elimination

As you may expect from its name, backward elimination is the opposite of forward selection. Whereas forward selection begins with no predictors in the equation and adds them to the equation one by one, backward elimination begins with all the predictors in the equation and removes them one by one until the final equation is obtained. At each step, the predictor in the equation that produces the smallest change in R^2 is tested to determine whether it should be removed from the equation. Again, the criterion for removal is generally stated in terms of the significance level of a partial F test. If the selected variable is removed, another predictor is selected and tested on the next step. The procedure terminates when a predictor that has been selected for testing does not meet the criterion for elimination; the final equation includes that predictor and all the other predictors remaining in the equation at that point.

Despite the common goal of backward elimination and forward selection procedures, and their apparent symmetry, they can result in different final equations. For example, using the five predictors and 131 cases described in Section 21.4.1, backward elimination finds that the best-fitting regression model is $\widehat{TC} = 124.00 + 1.873age$ with $R^2 = .321$. The backward elimination procedure dropped *BMI* as a predictor because doing so did not result in a significant reduction in R^2 given the other predictors in the model at that step in the process. In contrast, the forward selection procedure included *BMI* because doing so improved the fit significantly. If this variability in the final equation surprises you, you may wish to review the discussion of the meaning of regression coefficients and suppression effects in Chapter 20.

21.8.3 *Stepwise Regression*

Stepwise regression, the most popular of these automated selection procedures, is a combination of forward selection and backward elimination. The procedure is essentially the same as forward selection except that after each new predictor has been added to the regression equation, all the predictors already in the equation are re-examined to determine whether they should be removed. A partial *F* test is performed on the predictor already in the equation that produces the smallest increment in R^2. If the predictor no longer satisfies the criteria for inclusion, it is removed from the equation. Statistical software allows the user to set the significance levels (or critical *F* values) for entering or removing a variable. Continuing with the five-predictor *TC* example, stepwise regression finds that the best-fitting regression model is $\widehat{TC} = 103.423 + 1.847age + .860BMI$ with $R^2 = .333$, the same model identified using forward selection.

It is not difficult to see why it is sometimes desirable to remove a predictor that had been entered early in the analysis. For example, suppose that X_7 is highly predictable from X_4 and X_9 but is more highly correlated with *Y* than either of them. Even though X_7 may enter the equation early because of its high correlation with *Y*, it will become superfluous after X_4 and X_9 are entered. Even if X_7 contributes significantly to the predictability of *Y* by itself, it may not make a significant contribution over and above that provided by the other two variables.

It is important to emphasize that when predictor variables entered into the equation are selected from a larger pool, the significance levels printed out by stepwise processes are not "real" *p*-values. Because many researchers seem to be unaware of this fact, stepwise regression outputs are frequently misinterpreted. As Wilkinson has stated, stepwise regression routines are probably the most notorious source of "pseudo *p*-values" in the field of automated data analysis. As with forward selection, we recommend that Wilkinson's (1979) tables be used to test R^2 for significance.

We again emphasize that the sole motivation for the automated procedures described in this section is to develop useful prediction equations that include subsets of the available predictors. There is *no reason* to think that the equations they produce are reasonable explanatory models. Variables that are useful predictors need not be important components of a good theory or causal explanation of the situation. The automated procedures may include theoretically uninteresting variables in the regression equations they produce and they may fail to include the important variables. Consider, for example, a stepwise regression with several predictors that are highly correlated both with the criterion and with each other. The correlation between the criterion and the predictor included on the first step may be only marginally greater than the correlation between the criterion and the other predictors. Nonetheless, including the first predictor may prevent any of the others from being entered into the equation on subsequent steps. Even though the other predictors add significantly to the predictability of Y in the absence of the first variable, they may not do so when the first variable is in the equation. Forward regression procedures are more likely than backward elimination to exclude potentially useful predictors in situations that involve suppression (see Section 20.8). Even in situations in which both the suppressor variable and the variable whose variance is suppressed may be useful for predicting Y, it is possible that neither variable may make it into the equation unless the other variable is already there, if a forward selection procedure is used.

Because predictor variables are included in the regression equation if they are useful in the sample, stepwise procedures are extremely susceptible to capitalization on chance, especially when the sample is small. Recall that in the discussion of capitalization on chance in Chapter 20 we presented an example in which stepwise regression produced an equation that fit a sample fairly well ($R = .42$). Nonetheless, we demonstrated that the equation was of no use for predicting outside of that sample (cross-validated $R = .10$). If stepwise regression is ever to be used, it is essential to use cross-validation to evaluate the generality of the resulting regression equation.

We conclude this section by reiterating that *using these automated stepwise procedures is a very poor way to develop theory*. Automated programs may be easy to use but are a poor substitute for thinking. Even minimal knowledge about the research situation along with careful exploration of the data should lead to better preliminary models than an automated procedure. We repeat this point because we keep encountering researchers who are ignorant of, or resistant to, this advice. As Wilkinson (1998) has stated in the SYSTAT manual:

> Stepwise regression is probably the most abused computerized statistical technique ever devised. If you think you need automated stepwise regression to solve a particular problem, it is almost certain that you do not. Professional statisticians rarely use automated stepwise regression because it does not necessarily find the "best" fitting model, the "real" model, or alternative "plausible" models. Furthermore, the order in which variables enter or leave a stepwise program is usually of no theoretical significance. You are always better off thinking about why a model could generate your data and then testing that model. (p. 351).

21.9 Summary

- In Chapter 21, we discussed statistical inference in multiple regression. We also considered the models underlying inference and introduced several statistical tests and power calculations for the regression coefficients.
- We discussed how to check for violations of the assumptions that underlie the inference model and how to detect outliers and influential points in multiple regression.
- We described confidence intervals for regression coefficients and for the multiple correlation coefficient.
- We described the challenges of controlling Type 1 errors in multiple regression.
- Finally, we considered some automated procedures available in statistical packages for finding the "best" prediction equations. Because many researchers are under the impression that these automated procedures are useful in developing theory, we emphasized the point that they should not be used to develop explanatory models.

Exercises

21.1 [Bivariate and multiple regression] In an experiment designed to determine the effects of drug dosage on performance, the following data (available as data set *EX21_1*) are obtained:

Dosage in milligrams

10	*20*	*30*	*40*
6.8	10.4	10.7	8.9
2.8	6.4	14.4	12.5
5.2	13.1	15.9	12.7
4.8	8.7	10.6	7.4
	12.4		8.5
	7.2		

a) Perform an ANOVA to test the effect of dosage on performance.
b) Regress performance on dosage to determine whether there is a significant linear relationship between performance and dosage.
c) Determine whether there is a significant nonlinear relationship between performance and dosage.
d) Determine whether there is a significant quadratic component to the relationship by creating a new variable, *dosagesq*, formed by squaring dosage (you will need to compute a new variable), and then regress performance on dosage and *dosagesq*. Find the best-fitting quadratic equation.

21.2 [Stepwise regression] Data set *EX21_2* contains hypothetical data for verbal achievement of students in elementary school; potential predictor variables are *squality* and *tquality*, measures of school and teacher characteristics, and *sbackground* and *pbackground*, measures of student and parent background.

a) Do the school and teacher measures contribute to the predictability of verbal ability? Do the student and parent measures? Consider regression equations containing different combinations of predictors in arriving at your answer.
b) Perform a stepwise regression using one of the standard packages. Do you think that the regression equation identified by the stepwise regression is a reasonable explanatory model of the situation?

21.3 [Planning sample sizes] Using data from an observational study, we plan to perform a multiple regression analysis with six predictor variables. If we want to test the hypothesis that all population regression coefficients are equal to 0 at $\alpha = .05$, how many cases do we need to have a power of .80:

a) if we expect an R^2 of .30?
b) if $R^2 = .20$?
c) if $R^2 = .10$?

21.4 [Testing all regression coefficients] We perform a multiple regression with six predictors and find that $R^2 = .20$. Can we reject the null hypothesis that all the regression coefficients = 0 in the population if $N = 50$?

21.5 [Planning sample size] Suppose we wish to add a seventh predictor to a regression equation and expect it to increase R^2 from .30 to .35. How many cases are required to have power = .90?

21.6 [Planning sample size] Find the number of cases required to yield power = .80 for a test of a medium-sized effect (f^2 = .15) for a predictor to be added to a regression equation already containing five predictors.

21.7 [Interpreting the contribution of a predictor] Suppose we conduct a multiple regression analysis on data from an observational study with four predictor variables and N = 40. We find that adding variable X_4 to the other three predictors increases R^2 from .21 to .27. Can we reject the null hypothesis that the coefficient of variable X_4 has the value 0 in the population? Estimate how many cases we need to have a power of .80 to reject the null hypothesis if α = .05.

21.8 [Testing contribution of a group of predictors] Suppose we conduct a multiple regression analysis on data from an observational study with six predictor variables and N = 80. We find that adding both variables X_5 and X_6 to the first four predictors increases R^2 from .31 to .37.

a) Can we reject the null hypothesis that the coefficients of X_5 and X_6 equal zero in the population at α = .05?
b) Estimate how many cases we need to have a power of .90 to reject the null hypothesis if we were to rerun the study.

21.9 [Interpreting regression coefficients in real data] Consider the data set *tcforwomen* on the book's website. If we regress *TC* on *schoolyr*, a crude measure of years of education, we find that *schoolyr* is a significant predictor of *TC*, b = 1.42, $t(178)$ = −3.02, p = .003; more education is associated with lower cholesterol levels. This fits well with our stereotype that more educated people take better care of themselves and have better diets. However, there may be alternate explanations for this education effect. Try to think of one or more, and explore your account(s) by using the data set.

Note

1 Note that for cases 2 and 3 later, our recommended procedure for random-effects predictors may produce estimates of the required sample size that are slightly too large. This is because G*Power 3.1 bases the denominator *df* of the *F* statistic on N^* instead of N, and so the *df* will be too small by $p - k$. This can be ignored unless N is small and $p - k$ is large.

Chapter 22

Additional Topics in Multiple Regression

22.1 Overview

If our goal is to maximize prediction accuracy, we should look at measures of fit such as R or R^2 when we perform regression analyses. However, R is not a reasonable measure for assessing the usefulness of explanatory models because it is not necessarily an index of causal effects, nor is it comparable across samples. If our concern is with explanation or theory development, our goal is to understand the direct and indirect effects of the independent variables on the dependent variable of interest. Therefore, our focus should be on interpreting the regression coefficients. In Chapter 22, we discuss several issues that affect our ability to interpret regression coefficients. We introduced many of these issues in the three preceding chapters and we expand upon them here. The topics in Chapter 22 are as follows:

- *The consequences of model mis-specification*, either by omitting an important predictor or by including an extraneous variable.
- *Measurement error and missing data.*
- *Multicollinearity*, which occurs when predictor variables are completely or highly redundant.
- *Direct vs indirect effects and mediating variables.*
- *Testing for curvilinearity.*
- *Testing for interactions between predictors.*
- *Distinguishing curvilinearity from interactions.*
- *Limitations of OLS regression*, such as application to situations with dichotomous dependent variables or to designs in which variables are nested.

22.2 Specification Errors and Their Consequences

A regression equation is correctly specified if it contains the same variables as the true population model. If parameters of the population model

$$Y_i = \beta_0 + \beta_1 X_{i1} + \beta_2 X_{i2} + \ldots + \beta_p X_{ip} + \varepsilon_i$$

are estimated by the coefficients of the sample regression equation,

$$\hat{Y}_i = b_0 + b_1 X_{i1} + b_2 X_{i2} + \ldots + b_p X_{ip}$$

DOI: 10.4324/9781003453550-26

then, given the usual regression assumptions, the b_js can be shown to be unbiased estimators of the β_js. However, as we pointed out in Chapter 20, if the b_js are obtained from a regression equation that omits important predictors in the population model, or adds variables that are not in the model, they will generally be biased estimators of the β_js in the true population model.

Considering the consequences of mis-specification is important because in non-experimental research, we rarely work with "true" models. In developing explanations, we often start with crude, incomplete models, and then attempt to extend and refine them. As we conduct our analyses, questions may arise that lead us to collect more data and perform additional analyses, so that we modify our models in ways that help us better focus our thinking about the research question. Although we must be aware of their limitations, regression analyses can play a very useful role in advancing our understanding, even when the equations they produce are sample specific and cannot be thought of as representing complete, true models.

In many contexts, the most useful models are not "true" population models. Box (1979, p. 209; also see useful discussions in Berk, 2004; King, 1991) has stated, "All models are wrong, but some are useful." Problems in the social sciences are often too complex to be represented completely by any specific theory or model, especially one that can be expressed as a single regression equation. Relatively simple models with few predictors that capture the essence of a research problem may contribute more usefully to scientific discourse and/or have more direct practical applications than more cumbersome complex models even if the larger models are more correct. However, when we work with models that do not match the true underlying population model, we must be aware of how regression coefficients may be affected when predictor variables are left out of, or added to, a regression equation.

22.2.1 *Omitting Relevant Variables*

We pointed out in Chapter 20 that if an important variable in the true population model is omitted from a regression equation and if this omitted variable is correlated with one or more variables that are included in the equation, then the coefficients of the variables in the equation will reflect not only the effects of the corresponding predictors, but also the effects of the omitted variable.

Let's illustrate this point in more detail. Suppose that the true model in the population is

$$Y = \beta_{012} + \beta_{Y1 \cdot 2} X_1 + \beta_{Y2 \cdot 1} X_2 + \varepsilon$$

where we use the notation $\beta_{Y1 \cdot 2}$ to emphasize that the regression coefficient of X_1 comes from a model that includes both X_1 and X_2. If we regress Y only on X_1, thereby mis-specifying the model by omitting X_2 from the equation, we obtain the sample regression equation

$$\hat{Y} = b_{01} + b_{Y1} X_1$$

In Appendix 22.1, we show that the expected value of b_{Y1} is not $\beta_{Y1.2}$, the coefficient of X_1 in the true model, but rather $\beta_{Y1.2} + \beta_{Y2.1}\,\beta_{21} = \beta_{Y1.2} + \beta_{Y2.1}\,\rho_{12}\,\sigma_2/\sigma_1$. This expression contains $\beta_{Y1.2}$ plus an additional term that depends on $\beta_{Y2.1}$ as well as on the regression of X_2 on X_1, reflecting the fact that if X_2 is omitted from the regression, its effects on Y are partially expressed through X_1. Although things get more complicated when there are more predictors, the regression coefficients will generally be biased estimators of the population parameters if important variables are left out. The exception to this statement is when the omitted variables are *uncorrelated* with the variables that are included in the equation – we can see this earlier because the additional biasing term disappears when $\rho_{12} = 0$.

It is important to understand that it is the lack of correlation with other variables that allows us to make causal statements about independent variables based on *experiments*. In well-designed experiments, the effects of nonmanipulated variables are controlled by procedures such as randomization and matching, so that variables in the model are not meaningfully correlated with those that are omitted. In contrast, in observational studies, nuisance variables may be hopelessly confounded (i.e., correlated) with the variables of interest in ways that make it impossible to sort out causality. In this case, the best we can do is to include the most important nuisance variables in our regressions and try to control for them statistically.

The message here is that although simple models may be very useful, if these models omit important variables that are correlated with variables in the model, we may arrive at faulty conclusions (see the example in Section 20.7).

22.2.2 *Including Additional Variables*

Because of the negative consequences of omitting important variables, researchers sometimes include additional variables in their regression equations in an attempt to make sure that nothing important has been left out. Adding variables that are not correlated with variables in the equation does not bias parameter estimates of the variables in the equation. However, including these additional variables will use up degrees of freedom and reduce power. The loss of degrees of freedom will also inflate the standard errors of the relevant variables, making parameter estimates less precise.

If, on the other hand, we add predictors that are correlated with variables in the equation, the coefficients will change. For example, suppose that we are interested in the effects of parental education on children's performance in elementary school and therefore regress some measure of children's school performance on the average number of years of parental schooling. If we now add other variables to the equation such as family income, class size, teacher pay, and number of books in the home – all measures correlated with parental education – the coefficient of years of parental schooling will change because it now reflects the effects of years of parental schooling *partialing out* the effects of these other variables.

22.3 Measurement Error in Multiple Regression

As with other types of analyses, we need to have good measures of the variables when we use multiple regression. However, the consequences of measurement error differ depending on whether we are talking about the dependent variable or the predictor variables. Increased measurement error in the *dependent variable* increases the size of the error component, so that s_Y and s_e become larger and R^2 becomes smaller. This increases the size of confidence intervals and decreases the power of significance tests. However, if there are no major violations of assumptions, ordinary least-squares (OLS) estimates of the *unstandardized* regression coefficients remain unbiased. In contrast, because the *standardized* coefficient of X_j is the unstandardized coefficient multiplied by s_j/s_Y, standardized coefficients (often called "beta" weights) become smaller as measurement error in the dependent variable increases.

Now consider measurement error in the *predictor variables*. The usual regression inference model (see Chapters 19 and 21) assumes that predictor variables are fixed and measured without error. However, in many applications, values of the predictors are sampled and are measured with error. Measurement error in the predictors may result in serious problems, especially when some of the measurement error is systematic.

If the measurement error in the predictors is random, then at least the predictors may be independent of the error component for the model. The presence of random measurement error in any predictor will attenuate the estimate of the corresponding unstandardized regression coefficient, simply because the presence of noise in the predictor variable weakens its relationship to *Y*. Figure 22.1 demonstrates this effect using the statistics exam data introduced in Chapter 17: The left panel shows the regression of final exam scores on the pretest scores, assuming they are measured without error; the right panel shows the regression after the pretest scores have been "smeared" by adding a random value sampled from $N(0, 3)$ to each pretest score. R^2 drops from 0.526 when the pretest scores are error-free to 0.134 when they are measured with error.

If the predictors are uncorrelated, the presence of random measurement error in one predictor will not systematically influence the estimated regression coefficient of any other predictor. If, on the other hand, the predictors in a multiple regression are correlated, as is

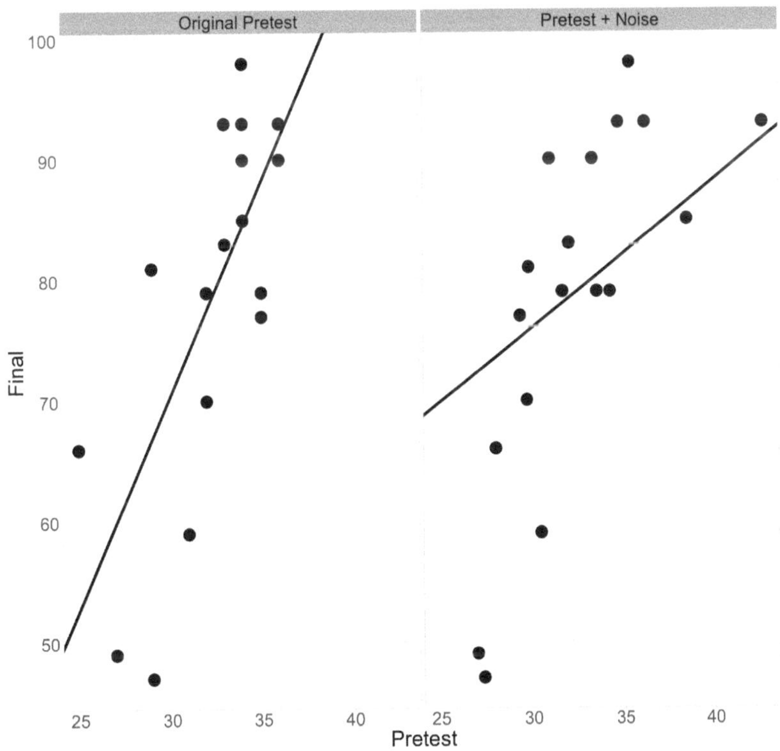

Figure 22.1 Illustration of the effect of random noise in a predictor variable. In the right panel, randomly sampled values from $N(0, 3)$ were added to the pretest scores.

usually the case, the result is more complicated. The coefficient of any predictor, X_j, is the rate of change of $\hat{Y}$ with X_j, partialing out the effects of the other predictor variables in the regression equation. When there is measurement error in predictors correlated with X_j, it is possible that too much or too little of the effects of the other variables may be partialed out. So, depending on the details, the unstandardized (b_j) and standardized (b^*_j) coefficients may be too large *or* too small.

The situation is at its worst when there is systematic error in the measurement of the predictors (e.g., when there are reporting biases such as consistently underreporting body weight or overreporting hours worked; see Chapter 19), so that there are correlations between these predictors and the error component. If this happens, the OLS estimates of the coefficients will be neither unbiased nor consistent.

Some possible remedies are available. We can try to use variables that are correlated with our predictors but not with the error component (so-called *instrumental variables*, see Berk, 2004). Or we can use structural equation modeling, a statistical technique for testing and estimating causal relationships that uses multiple indicators of underlying concepts and explicitly models error. However, detailed discussion of both procedures is beyond the scope of the current discussion. At the level of coverage in this book, the best advice is to keep the issue of measurement in mind and to use the best (cleanest) possible measures available. It should go without saying that if we use poor and error-ridden measures, we can have little confidence in the results of our statistical analyses: That is a clear example of the adage that "garbage in equals garbage out."

22.4 Missing Data in Multiple Regression

In designing research, we should use measuring instruments and data collection procedures that minimize the occurrence of missing data, especially on the dependent variable. Depending on the details, the presence of missing data will result in our analyses having reduced power and may well result in biased parameter estimates and tests. The three options offered by SPSS to deal with missing data in the regression module are (a) exclude cases listwise (i.e., listwise deletion), (b) exclude data pairwise (i.e., pairwise deletion), and (c) replace with mean (i.e., mean substitution).

If we conduct multiple regressions using a set of variables and have some missing data, software allow us to use the *listwise deletion* option; that is, each regression is conducted by using only those cases with complete data on all the variables *used in the regression*. However, this approach may lead to different regressions being performed on somewhat different subsets of the data. For example, if we use the *Seasons* data set from the website and regress *TC* on *sex*, *age*, and *BMI*, the regression is performed by using the data from the 416 participants with complete data on the four variables. If we add *Beck_d* (the mean score on the Beck Depression Scale over the four seasons) to the list of predictors, the regression is now based on the data from only the 313 cases with complete data on all five variables. If it is important to compare the results of the regressions, we may want to specify that only the data from the 313 participants with complete data on all variables be used in both analyses.

If there are modest amounts of missing data, listwise deletion can have good properties. If the missing data are MCAR (that is, missing completely at random – see the discussion of missing data in Chapter 13), the data from complete cases can be thought of as randomly selected from the original sample. If so, estimates of the regression coefficients obtained by listwise deletion will be unbiased if all the usual regression assumptions are satisfied. The standard errors, of course, will be larger because they are based on fewer observations.

It can be shown (e.g., see Allison, 2002) that if missing data on the dependent variable, Y, are *associated with* the values of Y and/or if missing data on the predictors are associated with the values of Y, then listwise regression will yield biased estimates (in this case, we can think of the missing data as an extreme case of systematic measurement error). However, as Allison points out, regression with listwise deletion is quite robust with respect to violations of the MAR assumption when the probability of missing data on Y and on the predictors does not depend on the values of Y.

Pairwise deletion is not recommended. The idea behind pairwise deletion is to use as much data as possible. Multiple regression can be performed, using the means, standard deviations, and correlations of all the variables in the regression equation. In pairwise deletion, these quantities are calculated by using whatever data are available. For example, if Y is the dependent variable and there are two predictors X_1 and X_2, we could obtain the correlation between Y and X_1 without regard to X_2, and the correlation between X_1 and X_2 without regard to Y. The problem is that if we use the pairwise deletion option in the standard statistical packages, the parameter estimates and test statistics may be biased unless the missing data are MCAR. In some cases, it may not be possible to perform certain higher-order analyses at all. Given pairwise deletion, the constraints normally expected in a correlation matrix (see Section 18.3.5) may not hold because different correlations may be obtained from substantially different subsets of cases and may be based on different numbers of observations.

Replacing missing scores on a variable by the variable mean (i.e., marginal mean imputation) is also not recommended because it is well known that this procedure produces biased estimates of variances and correlations. As mentioned in Chapter 13, there are other, more sophisticated methods of imputing missing data that can produce unbiased estimates if the missing data are MCAR or MAR. Also, when there are repeated-measures, as, for example, in the case of depression scores measured at each season of the year, hierarchical regression modeling procedures that use all the available data can be used instead (see Section 22.9).

22.5 Multicollinearity

Multicollinearity occurs when predictor variables included in a regression equation are correlated. In that case, the correlated predictors can themselves be predicted from the other predictors in the equation; they may not provide much, if any, unique information that is predictive of Y. If at least one predictor variable can be perfectly predicted from the others, then there is perfect multicollinearity. Given perfect multicollinearity, an infinite number of regression equations will fit the data equally well, so the statistical software cannot identify a solution to the regression analysis. If we have only two predictors, one way of thinking about the problem is that if the predictors are perfectly correlated, so that X_1 and X_2 account for exactly the same variance in Y, there is no basis for attributing this variability to one or the other predictor.

A useful way to think about the problem is to consider Figure 20.8, which showed the regression plane for *TC* predicted from *age* and *BMI*, and Figure 22.2. If Y is regressed on X_1 and X_2, the points ($\hat{Y}$, X_1, X_2) that satisfy the equation

$$\hat{Y} = b_0 + b_1X_1 + b_2X_2$$

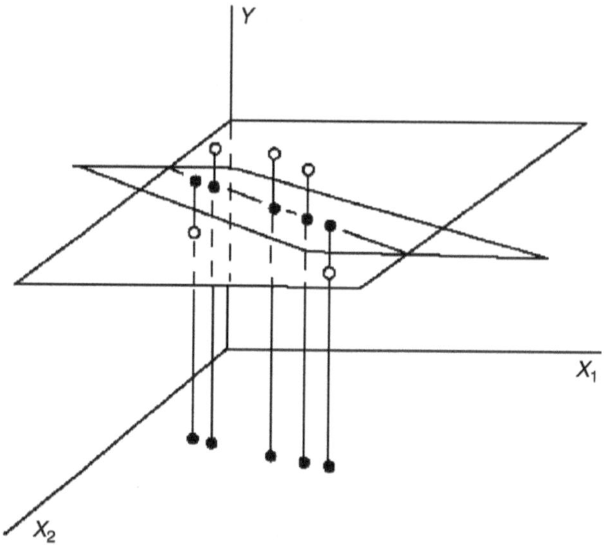

Figure 22.2 Illustration of perfect multicollinearity: The open circles represent data points (X_1, X_2, Y). The lower group of filled circles represents the (X_1, X_2) coordinates of the data points. If X_1 and X_2 are perfectly correlated, these filled circles will fall on a straight line. The upper set of filled circles represents the points (X_1, X_2, $\hat{Y}$) where the $\hat{Y}$s are the best least-squares predictions of Y corresponding to (X_1, X_2). Note that these points fall on a straight line in the three-dimensional space, and that an infinite number of planes with different values of b_0, b_1, and b_2 will contain that straight line. Therefore, it is impossible to specify a unique regression plane.

will lie along the surface of a plane in the three-dimensional space defined by axes Y, X_1, and X_2. If X_1 and X_2 are not highly correlated, as in Figure 20.8, the data points constrain the values of b_0, b_1, and b_2 that define the best-fitting regression plane. If the orientation of the plane was changed, b_0, b_1, and b_2 would take on different values, so that the predictions and therefore the fit of the regression equation would be different. However, if X_1 and X_2 are perfectly correlated, as in Figure 22.2, the points ($\hat{Y}$, X_1, X_2) will lie along a straight line in the three-dimensional space, and any of the infinite number of planes that contain the line will fit the data equally well, making it impossible to specify unique values for b_0, b_1, and b_2. Figure 22.2 shows two planes that fit equally well but which would have very different coefficient values. Box 22.1 describes a simple physical demonstration of multicollinearity, if Figure 22.2 is difficult to interpret.

It is rare to find perfect multicollinearity unless redundant variables such as subscale scores and total score, or age and year of birth, are included. However, it is not at all rare to find situations in which the predictors are highly collinear. In such cases, statistical software will perform the regression,[1] but the values of the regression coefficients may be extremely unstable. The reason is that variability in Y that is shared by the predictors will be "split up" among them. The nature of this split will depend on the details of the data and may vary widely from sample to sample, leading to dramatically increased standard errors for the regression coefficients.

Box 22.1 A Physical Demonstration of Multicollinearity and Its Consequence

1. Place three objects of different heights, such as a soda can, a wine bottle, and a bottle of hand sanitizer, on a table in a roughly triangular arrangement. Conceptually, one corner of the table defines the point ($X_1 = 0, X_2 = 0$), and thus the edges of the table define the values of X_1 and X_2 for the objects' locations. The Y value at each (X_1, X_2) point is represented by the height of each object.
2. Place a flat object like a piece of paper or a hardcover book, so that it touches the tops of your three objects. This surface is the best-fitting regression plane of Y predicted from X_1 and X_2.
3. Notice that the regression plane fits perfectly and uniquely: Any other placement of the paper creates a gap between the top of an object and its predicted value on the surface of the regression plane, which is a residual.
4. Next, rearrange the three objects so that they are in a straight line in order of their heights.
5. Try placing the paper on the objects. You will find that you can do so, but that there is no unique solution: The regression plane can be rotated around the line that connects the tops of the three objects. That is a physical demonstration of perfect multicollinearity, because the value of X_2 can be perfectly predicted from X_1, and vice versa.
6. Finally, shift the line of objects from step 4 to parallel to one side of the table (call it X_1). As you rotate the regression plane, notice that the slope of the plane along X_1 is invariant because it is determined by the relative heights of your objects. In contrast, the slope along X_2 (the perpendicular side of the table) changes sign as the plane rotates: It is sometimes positive, sometime negative, and occasionally zero. This is one possible consequence of multicollinearity: The regression coefficients are underdetermined and, under some conditions, may even change sign.

22.5.1 *Measures of Multicollinearity*

Two measures are useful for diagnosing multicollinearity problems: The *tolerance* and its reciprocal, the *variance inflation factor* (*VIF*). Tolerance is a measure of how *nonredundant* a predictor is with the other predictor variables. With only two predictor variables, the tolerance is 1 minus the square of the correlation between the two predictors. In general,

$$\textit{tolerance of the } j^{\text{th}} \textit{ predictor variable} = 1 - R_j^2.$$

where $R^2_{j.}$ is the square of the multiple correlation when X_j is regressed on all the other predictor variables in the equation. In other words, the tolerance is a measure of how well X_j can be predicted by the other predictor variable(s) in the equation.

The variance inflation factor is 1/tolerance; there is one VIF for each predictor variable. As you might expect from its name, the VIF provides an indication of the amount that the variance in each regression coefficient is increased compared to an analysis in which all the predictors are uncorrelated. We can see this by considering the standard error of the regression coefficients, which we defined in Equation 19.7 for the bivariate case:

$$SE(b_1) = \frac{s_e}{\sqrt{SS_X}}$$

where s_e is the standard error of the estimate and SS_X is the sum of squares of X. In multiple regression, the standard error of the coefficient of the jth predictor variable can be shown to be a function of its tolerance:

$$SE(b_j) = \frac{s_e}{\sqrt{SS_j}}\sqrt{\frac{1}{1 - R_{j.}^2}} = \frac{s_e}{\sqrt{SS_j}}\sqrt{\frac{1}{tolerance\ of\ X_j}} \tag{22.1}$$

where

$$SS_j = \sum_i \left(X_{ij} - \bar{X}_{.j}\right)^2$$

is the sum of squares of the jth predictor. The ratio under the right-most square root sign, 1 divided by the tolerance of X_j, is the VIF for the predictor. From Equation 22.1, we can see that as the tolerance of X_j decreases, the corresponding VIF increases, and consequently, so does the estimated *SE*. If the predictor X_j has a tolerance of 0 (i.e., if $R^2_j = 1$), this means that X_j can be perfectly predicted by a linear combination of the other predictors in the regression equation and is therefore completely redundant. Consequently, if any of the predictors has a tolerance of zero, we cannot obtain least-squares estimates of the regression coefficients because the set of equations that must be solved to find the b_js does not have a unique solution. In general, a t test of the regression coefficient b_j will have less power as the tolerance becomes smaller. This makes sense, because the more X_j is redundant with the other predictors, the less new information it provides, and the less we can expect it to contribute to the predictions of Y.

What values of tolerances or VIFs indicate a multicollinearity problem? One commonly recommended rule of thumb is that a VIF of 10 or more or, equivalently, a tolerance of .10 or less indicates reason for concern. Cohen et al. (2003) make a strong case that these cutoff values are too liberal, as the negative consequences of multicollinearity can occur at much lower values of VIF. As they note, however, the problems stemming from multicollinearity are continuous in nature, and thus there is no easy decision rule to guide us.

The demonstration in Box 22.1 shows that one possible consequence of multicollinearity is a change of sign and/or a large change in magnitude of a regression coefficient as other predictors are included in the analysis. Thus, a comparison of the bivariate regression coefficient for each predictor alone with the corresponding coefficient in the multiple regression equation can provide an additional check on potential multicollinearity issues. Of course, the correlations among the predictors also provide some clues to potential problems. In short, the assessment of multicollinearity requires checking several aspects of the data and applying sound judgment.

22.5.2 Using Software to Obtain Measures of Multicollinearity

Statistical software provides diagnostic information that can be used to determine whether there is a potentially serious multicollinearity problem. In R, assume that the regression results from the *lm* function have been stored in reg.out. Then, the *vif* function in the {car} package, *vif*(reg.out), returns a vector of variance inflation factors, one for each predictor. Tolerances are simply 1 / *vif*(reg.out). In SPSS, set up the regression and click on the *Statistics* button, then check the box for *Collinearity diagnostics*. The resulting output will include tolerance and VIF for each predictor in the model.

22.5.3 Strategies for Dealing With Multicollinearity

Several possible remedies have been suggested if we find that there is a multicollinearity problem in our data:

- *Deleting some of the predictors that are responsible for the problem.* If we have several measures of the same latent variable, this may be appropriate. However, if we have several different kinds of measures that are highly correlated, this cure may be worse than the disease. Deleting variables may result in specification errors and correlations between predictors and the error component that themselves can have very serious consequences, as we discussed earlier. A better approach may be to combine predictors, as we describe next.
- *Combining clusters of highly related predictor variables into new variables that represent common underlying factors.* Decisions about which variables to combine should be based on theoretical considerations. Procedures such as *principal components analysis* and *factor analysis* can also provide suggestions about possible underlying processes, thereby providing information about which variables might be combined. Roughly speaking, these approaches identify weighted combinations of predictor variables that, together, account for systematic variability in *Y*. By evaluating the weights of the predictors, the factors can sometimes be interpreted as reflecting latent variables such as "verbal ability" or "self-esteem." Those methods are beyond the scope of this book.
- *Centering.* This involves replacing each score by its corresponding deviation score; that is, the score minus the mean of the variable. For example, when we regressed *TC* on *age* and *BMI* using data from women aged 20–65 with *BMI* scores not exceeding 40, the mean of the age scores was 46.928. To center the *age* variable, we would replacing *age* with a new variable, *agec* = *age* – 46.928. Centering can be useful in providing more interpretable regression coefficients (e.g., see Sections 22.7.2 and 22.8.4) and will also often reduce multicollinearity.

22.6 Regression With Direct and Mediated Effects

It is possible for an independent variable to influence the dependent variable directly, or it can influence the dependent variable indirectly, through its effect on a mediating variable. Figure 22.3 provides an illustration of this distinction. In Panel (a), we have a model in which X_1 exerts a direct effect on Y. This causal influence is indicated by the arrow, and the direction of the arrow indicates the direction of causality. In Panel (b), we have a causal model that illustrates both direct and mediated effects. In this case, a change in X_1 directly affects Y but also affects Y indirectly through the mediator variable, X_2. There are many examples of the kinds of mediating influences illustrated in Panel (b): Changes in the workplace environment may influence job perception, which in turn influences productivity; or changes in parents' attitudes about education may change children's attitudes, which in turn influence academic performance. Theories specifying processes that intervene between independent and dependent variables are important in many fields of psychology. A *Web of Science* search reveals more than 54,000 citations of the seminal Baron and Kenny (1986) paper that discussed ways of testing for a mediating effect.

We must consider the possibility of mediating effects if we try to determine the "importance" of a variable. Even if we include the correct variables in the regression equation and there is a causal relationship between X_j and the outcome, Y, the regression coefficient b_j does not estimate the total influence of X_j on Y. Rather, the regression coefficient reflects the *direct effect* of X_j on Y, that is, the rate of change of Y with X_j, *holding all the other variables in the equation constant.* If changes in X_j cause changes in some of the other predictor variables, in turn causing changes in Y, we would have both direct and indirect effects. In assessing the importance of X_j, we must be concerned with both the direct and indirect effects. Given a valid causal model, structural equation modeling can be used to assess the total effect of changing a variable. However, as always, these estimates may be misleading if important variables are omitted from the model or if the model is otherwise invalid.

Again, consider the two causal models in Figure 22.3. In Panel (a), a change of one unit in X_1 causes a change of c units in Y. In Panel (b), we have a causal model that illustrates both direct and mediated effects. For this model, changing X_1 by one unit directly causes a change in Y of c' units and a change in X_2 of a units. A change of one unit in X_2 causes a change of b units in Y. Therefore, the indirect effect of X_1 on Y though the mediator X_2 is ab, so that the total effect of X_1 on Y is $c' + ab$. Baron and Kenny (1986) have discussed four steps necessary to establish that a data set is consistent with the presence of mediation. These steps, translated into our terminology, are as follows:

Step 1. Regress Y on X_1. If the regression coefficient b_{Y1} is nonzero, this is consistent with (but does not necessitate) X_1 having a causal effect on Y.

Step 2. Regress X_2 on X_1. If the regression coefficient b_{21} is nonzero, this is consistent with X_1 having a causal effect on the possible mediating variable X_2.

Steps 3, 4. Regress Y on both X_1 and X_2. The coefficient of X_1, $b_{Y1.2}$, is the rate of change of predicted Y with X_1 when X_2 is held constant, and so, *in terms of the model in Panel (b)*, $b_{Y1.2}$ is an estimate of c', the causal path from X_1 to Y. A finding that b_{Y1} is nonzero but $b_{Y1.2}$ is zero is consistent with a model in which the entire effect of X_1 on Y is mediated by the intervening variable, X_2. A finding that $b_{Y1.2}$ is nonzero, but smaller than b_{Y1}, is consistent with partial mediation by X_2.

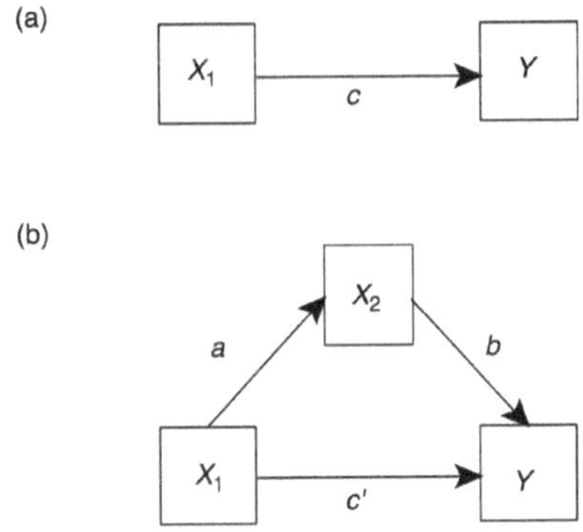

Figure 22.3 Causal models illustrating direct and indirect (mediated) effects.

22.6.1 *Some Cautions*

We emphasize that these interpretations are valid only in terms of the causal model given in Panel (b) of Figure 22.2. If our theory does not support this causal model, the best we can say is that our data are consistent with mediation but are also consistent with other causal models as well, perhaps involving reverse causality and additional variables. Also, we would like to have a statistical test for any mediation effect. One that is widely used is the Sobel test, but a limitation of the test is that it is not very powerful. MacKinnon, Lockwood, Hoffman, West, and Sheets (2002) offer a useful discussion of 14 methods for testing the statistical significance of a mediating effect. Finally, we should be aware that, as in any regression analysis, the results and interpretation may be distorted by measurement error. If, for example, the mediating variable X_2 is less reliable than X_1, the direct effect will appear larger, and the mediating effect will appear smaller than they should. If we want to pursue serious testing of a theory that involves mediation, we should use multiple indicators of the underlying latent variables and use structural equation modeling.

22.7 Testing for Curvilinearity in Regression

22.7.1 *Testing for Curvilinearity Using Continuous Predictor Variables*

When we regressed *TC* on *age* and *BMI* (see Figure 20.7), we found that there were significant effects of both predictor variables. However, it is possible that the relationship is not strictly linear and that the prediction of *TC* scores would be better if we included curvilinear components of the predictors as well. A curvilinear relationship would simply mean that the function relating the predictor variable(s) and the dependent measure is not linear over the entire range represented in our data. It might be that *Y* increases linearly up to some point, and then gradually levels off or even decreases. Or, the predicted values of *Y* may be piecewise linear, that is, linear over a range of predictor values but with different slopes at different points. If there are only two different slopes, then we may be able to describe the relationship with a quadratic (X^2) component; if there three distinct slopes (e.g., rising, falling, rising, or the reverse), then a cubic component (X^3) may be included in the model.[2]

In general, given a dependent variable Y and a predictor X, we can test for a quadratic component by regressing Y on X^2 as well as X, so that the regression equation is

$$\hat{Y} = b_0 + b_1X + b_2X^2 \tag{22.2}$$

If the regression coefficient of the squared term is significant, we can then test for the presence of a cubic component by regressing Y on X^3, as well as X and X^2, yielding the equation

$$\hat{Y} = b_0 + b_1X + b_2X^2 + b_3X^3 \tag{22.3}$$

and so on. *Note that we cannot test for the presence of a higher-level component without including the lower-level components as well.* For example, we cannot test for the presence of quadratic component by regressing only on X^2, because X^2 will usually be highly correlated with X. The regression of Y on X^2 alone would reflect both linear and quadratic effects, and we might get significance even if there was no quadratic component. If we wish to test for the presence of a quadratic component, we must partial out the linear component.

If we regress *TC* on *BMI* and BMI^2 for all women aged 65 or less in the *Seasons* data set, we find the regression equation

$$\widehat{TC} = 79.561 + 8.118\ BMI - 0.112\ BMI^2 \tag{22.4}$$

Both the coefficients of *BMI* and BMI^2 are significant; $b_1 = 8.118$, $t(187) = 3.060$, $p = .003$ and $b_2 = -0.112$, $t(187) = -2.735$, $p = .001$, respectively.

There is one bend in a quadratic function. Because the partial regression coefficient of BMI^2 is negative, the rate of change of $\widehat{TC}$ with *BMI* is negatively accelerated. That is, the amount by which $\widehat{TC}$ increases as *BMI* increases becomes less as *BMI* gets larger, and for large enough values of *BMI*, $\widehat{TC}$ starts decreasing as *BMI* increases. Had the coefficient of the quadratic component been positive, this would indicate that the rate of change of $\widehat{TC}$ with *BMI increases* as *BMI* gets larger.

A significant quadratic coefficient in a regression equation implies that the slope is different at different values of X. We may wish to determine the slope at specific values of X. For example, assuming a causal relationship between *BMI* and *TC*, a doctor might prioritize weight loss as a means of reducing cholesterol for patients with *BMI* values in a lower range where *BMI* changes more strongly predict *TC* changes and might prescribe medication for patients with higher *BMI* values that don't strongly predict *TC*. For Equation 22.2, using calculus, we can show that the slope for any value of X is equal to $b_1 + 2b_2X$. Therefore, the slope of the curve at $BMI = 25$ is $8.118 + (2)(-0.112)(25) = 2.518$, and the slope at $BMI = 35$ is $8.118 + (2)(-0.112)(35) = 0.218$. In other words, changing *BMI* from 25 to 24 has a much larger influence on the predicted value of *TC* than changing *BMI* from 35 to 34. The maximum of the curve is found at $BMI = -b_1/2b_2 = -8.118/(2)(-0.112) = 36.241$. Substituting this value into the regression equation, we obtain $\widehat{TC} \approx 227$.

22.7.2 Centering the Predictor

To make the regression coefficients more interpretable, we may also choose to subtract a constant, C, from each value of X. Ordinarily, C will be the mean of X; subtracting $\bar{X}$ from

X is referred to as *centering*. If we subtract the mean from each value of X, Equation 22.2 becomes

$$\hat{Y} = b_0^c + b_1^c(X - \bar{X}) + b_2^c(X - \bar{X})^2 \quad (22.5)$$

where the superscript "*c*" on each regression coefficient indicates that the corresponding predictor has been centered about its mean.

If we expand this equation and compare it term-by-term with Equation 22.2, we find that $b^c_2 = b_2$. That is, the regression coefficient of the squared term is the same regardless of whether the mean is subtracted from X before the regression is performed. This is a general finding. In any kind of polynomial regression, the regression coefficient of the highest-order term will be the same whether or not a constant is subtracted from X. We can also show that the slope of $\hat{Y}$ vs X is $b^c_1 + 2b^c_2(X - \bar{X})$, so that we can interpret b^c_1 as the slope at $X = \bar{X}$. Moreover, the significance test for b^c_1 allows us to determine whether the slope differs significantly from zero at $X = \bar{X}$.

The mean *BMI* score for all women in the *Seasons* data set aged 65 years or less who have cholesterol data is 26.599. If we let $BMIC = BMI - 26.599$ and regress *TC* on *BMIC* and $BMIC^2$, we find that

$$\widehat{TC} = 216.047 + 2.144\ BMIC - 0.112\ BMIC^2 \quad (22.6)$$

Both the regression coefficients for *BMIC* and $BMIC^2$ differ significantly from 0; $b^c_1 = 2.144$, $t(187) = 3.459$, $p = .001$; $b^c_2 = -0.112$, $t(187) = -2.735$, $p = .007$, respectively. Equations 22.4 and 22.6 are consistent with one another. However, in Equation 22.4, the constant in the equation and the coefficient of *BMI* represent the value of predicted *TC* and the slope of predicted *TC* vs *BMI* at $BMI = 0$ – not very useful information, considering that no *BMI* in the data set was less than 15 and no individual could ever have $BMI = 0$. In Equation 22.6, the *BMI* values have been centered, so the constant and coefficient of *BMIC* represent the values of predicted *TC* and the slope *at the mean value of BMI*. Centering the predictor also results in lower correlations among the polynomial terms. For example, the correlations between *BMI* and BMI^2, *BMI* and BMI^3, and BMI^2 and BMI^3 are .986, .943, and .985, respectively. If the *BMI* variable is centered, however, the corresponding correlations are .696, .677, and .951. Note that although the constant subtracted from each value of X is usually the mean, any other constant, C, may be subtracted, so that the regression coefficient of $X - C$ is the slope at $X = C$.

We can now go on to test whether there is a cubic component (so that the curve has two bends), by adding X^3 to a regression equation that already contains X and X^2. We would get a cubic function if, for example, the value of the response variable rose, then fell, and then leveled off or rose again as the value of the predictor increased. In principle, we can test for higher-order components by adding larger powers of X to the regression. However, making predictions about curves with two bends in them largely exhausts the sophistication of existing theories in many research areas. If we can develop meaningful predictions about functions with three or more bends, we can test them by adding higher-order polynomial components to the regression.

22.7.3 Testing for Curvilinearity Using Quantitative Categorical Variables: Trend Analysis

We have just described tests for curvilinearity when the predictor variable(s), $X_1, \ldots X_p$, are continuous variables like *BMI*. We can also use multiple regression to assess the same type of polynomial patterns – or *trends* – in Y when the predictors are ordered categorical variables, like age group, grade level, or cancer stage; in this case, we call it *trend analysis*. To do so, we regress Y first on X, then on X and X^2, then on X, X^2, and X^3, and so on, and we test the increments in variability accounted for when higher-order components are added to regressions that already contain the lower-order components.

For example, consider the addition accuracy scores for grades one to five in the *Royer* data set that are presented in Table 22.1 and plotted in Figure 22.4. *Accuracy* first increases with *grade*, and then levels off. We would expect to find both a linear trend, because the best-fitting straight line has a positive slope, and a quadratic trend, because we have a negatively accelerated curve. The steps in the analysis are illustrated in Table 22.1. If we regress *accuracy* (Y) on *grade* (X), the variability accounted for by the regression is $R^2_{Y.X}SS_Y = SS_{linear} = 4{,}690.170$. If we now regress Y on X and X^2, $SS_{quadratic}$ is the increment in $SS_{regression}$ that results when X^2 is added to a regression equation that already contains X. Similarly, if we regress Y on X, X^2, and X^3, SS_{cubic} is the increment in $SS_{regression}$ that results when X^3 is added to a regression equation that already contains X and X^2. All of these trend components have 1 *df* and can be tested against the within-groups error term, $MS_{error} = 219.318$. The ANOVA table in Panel *b* of Table 22.1 presents the results of the tests. As expected, we find significant linear and quadratic trends, $F(1, 135) = 21.385$, $p = .000$, and $F(1, 135) = 8.495$, $p = .004$, respectively. We should note that because there are five levels of *grade*, and therefore 4 *df* if we treat grade as a categorical variable, the regression of Y on X, X^2, X^3, and X^4 must account for all of the variability in the group means; that is, $R^2_{Y \cdot X,X^2,X^3,X^4} SS_Y = 6{,}626.772 = SS_{Between}$. Just as any two points can be fit perfectly by a straight line, the points for the five group means can be fit perfectly by a fourth-degree polynomial.

Table 22.1 An example of trend analysis with output from SPSS

(a) Percent addition accuracy as a function of grade for *Royer* data

Grade (X)	*1*	*2*	*3*	*4*	*5*
n	19	28	32	30	26
$\bar{Y}$	71.82	84.66	91.97	92.34	91.98
s	30.23	15.26	8.24	7.30	9.20

(b) Trend analysis for the data in *Panel* a

$SS_{linear} = R^2_{Y \cdot X}\, SS_Y = 4{,}690.170$
$SS_{quadratic} = (R^2_{Y \cdot X,X^2} - R^2_{Y \cdot X})SS_Y = 6{,}553.310 - 4{,}690.170 = 1{,}863.140$
$SS_{cubic} = (R^2_{Y \cdot X,X^2,X^3} - R^2_{Y \cdot X,X^2})SS_Y = 6{,}617.701 - 6{,}553.310 = 64.391$
$SS_{quartic} = (R^2_{Y \cdot X,X^2,X^3,X^4} - R^2_{Y \cdot X,X^2,X^3})SS_Y = 6{,}626.772 - 6{,}617.701 = 9.071$

SV	*df*	*SS*	*MS*	*F*	*P*
Grade	4	6,626.772	1,656.693	7.554	.000
linear	1	4,690.170	4,690.170	21.385	.000
quadratic	1	1,863.140	1,863.140	8.495	.004
cubic	1	64.391	64.391	0.294	.589
quartic	1	9.071	9.071	0.041	.840
Error	130	28,511.337	219.318		

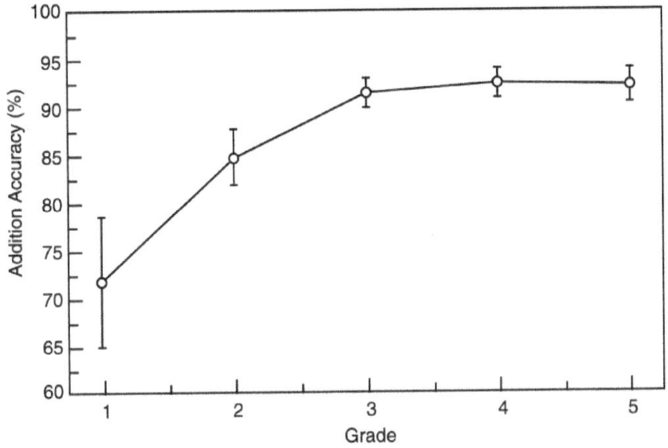

Figure 22.4 Plot of percent addition accuracy vs grade using standard errors as error bars.

22.8 Including Interaction Terms in Multiple Regression

22.8.1 *Introduction*

The regression equation $\hat{Y} = b_0 + b_1X_1 + b_2X_2$ is *additive* because the effect on $\hat{Y}$ of changing the values of either of the predictors does not depend on the value of the other predictor. This kind of model will be unrealistic whenever the relationship between Y and a predictor, X_1, is moderated by (i.e., it depends on the value taken on by) another variable X_2. In this case we would say we have an interaction between X_1 and X_2.

We dealt with interactions in detail when we discussed ANOVA. In the context of ANOVA, we have an interaction when the effect of a factor is different for different levels of a second factor. We can translate this thinking directly to multiple regression analyses. However, we can extend the concept when we work with multiple regression because we can now deal not only with categorical factors, but with a mix of categorical and quantitative variables. Therefore, it is possible to construct models in which the effect of one factor is a specified function of the levels of a second factor.

22.8.2 *Testing Interactions Between Quantitative and Dichotomous Predictors*

Consider the regression of *TC* on *age* and *sex* for participants in the *Seasons* study who are no older than 65 and have *BMI* scores no larger than 40. In this data set, *sex* is coded as a dichotomous categorical variable with participants identifying as male having the value 0 and participants identifying as female having the value 1. The regression output is provided in Figure 22.5, which provides the coefficients for the regression equation

$$\widehat{TC} = 178.009 + 0.900\ age - 8.645\ sex \tag{22.7}$$

Equation 22.7 is a misleading representation of the relationship of *TC* to *age* and *sex* because the relationship between *TC* and *age* differs for the male- and female-identifying

```
> reg.out<-lm(data=dat,tc ~ age + sex)
> summary(reg.out)

Call:
lm(formula = tc ~ age + sex, data = dat)

Residuals:
     Min       1Q   Median       3Q      Max
-130.833  -26.731   -3.774   20.873  170.704

Coefficients:
            Estimate Std. Error t value Pr(>|t|)
(Intercept) 178.0094     9.5620  18.616  < 2e-16 ***
age           0.8996     0.1905   4.723 3.33e-06 ***
sex          -8.6454     4.0456  -2.137   0.0333 *
---
Signif. codes:  0 '***' 0.001 '**' 0.01 '*' 0.05 '.' 0.1 ' ' 1

Residual standard error: 38.55 on 361 degrees of freedom
Multiple R-squared:  0.0717,    Adjusted R-squared:  0.06656
F-statistic: 13.94 on 2 and 361 DF,  p-value: 1.471e-06
```

Figure 22.5 R output for regression of *TC* on *age* and *sex*, using *Seasons* data from participants aged 65 years or less with $BMI \leq 40$.

participants. A better understanding of this relationship may be obtained by regressing *TC* on *age* separately for men and women, yielding

$$\widehat{TC} = 221.143 - 0.0005\ age \tag{22.8}$$

and

$$\widehat{TC} = 126.531 + 1.812\ age \tag{22.9}$$

These equations are plotted in Figure 22.6. It is apparent that predicted *TC* changes little with *age* for men but increases strongly with *age* for women. If we applied the test for equality of independent regression coefficients that we developed in Chapter 19, we would find that the regression slopes are significantly different.

We can easily test the interaction of age and sex by regressing *TC* on *age*, *sex*, and *age* × *sex*, a variable formed by multiplying *age* and *sex*. As you can see in Figure 22.7, this regression equation is

$$\widehat{TC} = 221.143 - 0.0005\ age - 94.612\ sex + 1.813\ Age \times Sex \tag{22.10}$$

Note that Equations 22.8–22.10 are consistent: if we substitute *sex* = 0 and *sex* = 1 into Equation 22.10, we recover Equations 22.8 and 22.9. We can also readily interpret each of the coefficients in Equation 22.10. The coefficient of *age*, –0.0005, is the rate of change

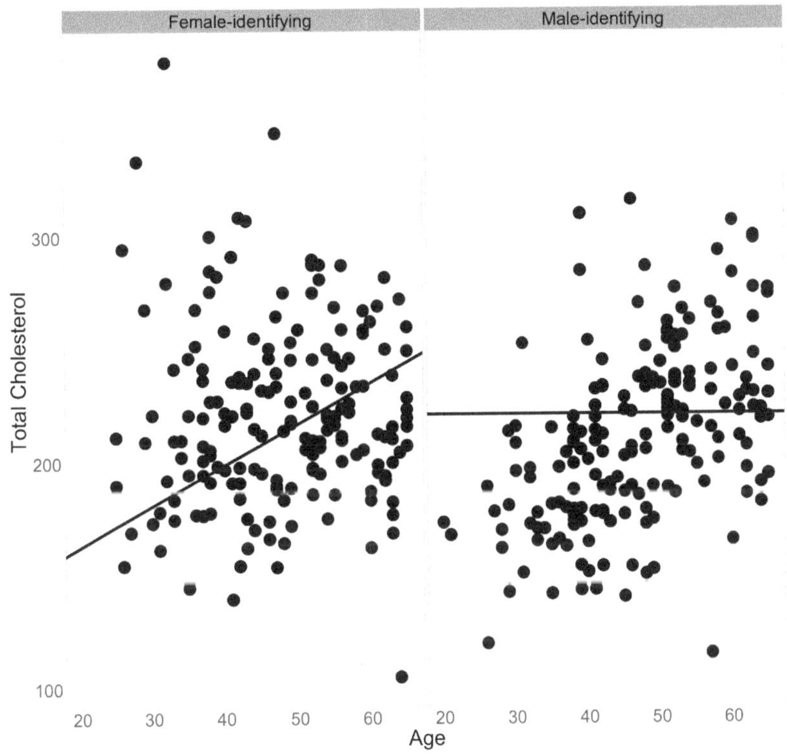

Figure 22.6 Scatterplot of *TC* versus *age* for men and women with regression functions, using data from participants aged 65 years or less with *BMI* ≤ 40.

of predicted *TC* with *age* when *sex* = 0; that is, for the male-identifying participants. The coefficient of *sex*, –94.612, is the rate of change of predicted *TC* with *sex* (i.e., the difference in predicted *TC* for men and women) for *age* = 0. We can better understand the coefficient of the product term, 1.813, by reorganizing the terms in Equation 22.10 and rewriting the equation as either

$$\widehat{TC} = 221.143 - 94.612\ sex + (-0.0005 + 1.813\ sex)\ age \tag{22.11}$$

or as

$$\widehat{TC} = 221.143 - 0.0005\ age + (-94.612 + 1.813\ age)sex \tag{22.12}$$

From these equations, we can see that the rate of change of predicted *TC* with *age* is a function of *sex*, namely, –0.0005 + 1.813 *sex*. Therefore, because in this data set the variable *sex* has only two levels, 1.813 can be interpreted as the difference between the slopes of predicted *TC* with *age* for women and men. Similarly, the change in predicted *TC* with *sex*, that is, the difference in predicted *TC* for men and women, is a linear function of *age*, –94.612 + 1.813 *age*. The coefficient of the interaction component, 1.813, can also be

```
> reg.out<-lm(data=dat,tc ~ age*sex)
> summary(reg.out)

Call:
lm(formula = tc ~ age * sex, data = dat)

Residuals:
     Min       1Q   Median       3Q      Max
-116.364  -24.485   -2.633   23.010  156.372

Coefficients:
              Estimate Std. Error t value Pr(>|t|)
(Intercept)  2.211e+02  1.277e+01  17.313  < 2e-16 ***
age         -4.573e-04  2.602e-01  -0.002    0.999
sex         -9.461e+01  1.795e+01  -5.272 2.33e-07 ***
age:sex      1.813e+00  3.693e-01   4.909 1.39e-06 ***
---
Signif. codes:  0 '***' 0.001 '**' 0.01 '*' 0.05 '.' 0.1 ' ' 1

Residual standard error: 37.37 on 360 degrees of freedom
Multiple R-squared:  0.1299,	Adjusted R-squared:  0.1227
F-statistic: 17.92 on 3 and 360 DF,  p-value: 7.328e-11
```

Figure 22.7 R output for regression of *TC* on *age*, *sex*, and the *age* × *sex* interaction using data for participants aged ≤ 65 years and *BMI* ≤ 40

interpreted as the amount by which the predicted difference in *TC* levels for women and men becomes more positive (or less negative) for each additional year of age. For example, for *age* = 60, we would predict the difference to be –94.612 + (1.813)(60) = 14.168, whereas for *age* = 30, we would predict the difference to be –94.612 + (1.813)(30) = –40.222. That is, predicted *TC* for women is about 40 units lower than for men at age 30, but about 14 units higher at age 60.

As can be seen in the R output in Figure 22.7, the coefficient of the product term, 1.813, is significant, $t(363) = 4.909$, $p = .000$. Therefore, we can reject the null hypothesis that the rate of change of predicted *TC* with *age* is the same for men and women (or equivalently, the null hypothesis that the difference in predicted *TC* for men and women is the same at each value of age).

We should emphasize that when we test the interaction between two predictors, we must include both predictors in the regression equation along with the product of the two predictors, as we did in Equation 22.10. That is, to test the interaction, we evaluate the effect of the product term after the effects of the predictors that are the constituents of the product term have been partialed out by including them in the equation (see, for example, Cohen, 1978). It is *not* appropriate to test the interaction by including only the product term in the regression equation – although this is a common mistake. For one thing, the product term may be highly correlated with its constituents. In the current example, the product term has correlations of .168 and .953 with age and sex, respectively. It is also important to note that interactions with categorical moderator variables tend to have very small effect sizes (see Aguinis, Beaty, Boik, & Pierce, 2005), so we must use large samples if it is important to investigate them.

22.8.3 *Testing the Interaction Between Two Quantitative Predictors*

In Chapter 20 we regressed *TC* on *age* and *BMI* for women aged 20–65 years and obtained the equation

$$\widehat{TC} = 92.239 + 1.737\ age + 1.474\ BMI \tag{22.13}$$

This is the equation of a two-dimensional plane in the three-dimensional space that has as its axes predicted *TC*, *age*, and *BMI* (see Figure 20.8). Because this equation is additive, if *BMI* is held constant, a 1-year increase in age corresponds to an increase of 1.737 units in predicted *TC* when *BMI* is held constant, *no matter at what value BMI is held constant*. Also, if *age* is held constant at any value, a one-unit increase in *BMI* corresponds to a 1.474-unit increase in predicted *TC*. However, it is possible that Equation 22.13 is unrealistic: The rate of change of predicted *TC* with *age* may differ for different values of *BMI*, or equivalently, the rate of change of predicted *TC* with *BMI* may differ for different values of *age*. We can investigate the possibility that *age* and *BMI* interact by creating a new variable *age* × *BMI* that is the product of *age* and *BMI*, and regressing *TC* on *age*, *BMI*, and *age* × *BMI*. As can be seen in the R output in Figure 22.8, the resulting regression equation is

$$\widehat{TC} = -70.241 + 5.209\ age + 7.844\ BMI - 0.135\ Age \times BMI \tag{22.14}$$

The coefficient of the product term is significant; $t(186) = -2.546$, $p = .012$. We can best understand the interpretation of this coefficient, −0.135, by grouping terms and rewriting Equation 22.14 as

$$\widehat{TC} = -70.241 + 5.209\ age + (7.844 - 0.135\ age)\ BMI \tag{22.15}$$

and as

$$\widehat{TC} = -70.241 + (5.209 - 0.135\ BMI)\ age + 7.844\ BMI \tag{22.16}$$

Now we can see that, according to Equations 22.15 and 22.16, the rate of change of predicted *TC* with *BMI* is a linear function of *age*, 7.844 − 0.135 *age*, and the rate of change of predicted *TC* with *age* is a linear function of *BMI*, 5.209 − 0.135 *BMI*. When we plot Equation 22.14 (see Figure 22.9), we see that because we included the product term, we no longer have a plane, but rather a curved surface on which the slope for one predictor decreases as the value of the other predictor increases.[3] The slope of predicted *TC* with *age* decreases by 0.135 for each one-unit increase in *BMI*, and the slope of *TC* with *BMI* decreases by 0.135 for each one-unit increase in *age*. For example, according to Equation 22.16, the slope of predicted *TC* with *age* for $BMI = 20$ is $5.209 - (0.135)(20) \approx 2.51$ and for $BMI = 30$ it is approximately 1.16. The slope of predicted *TC* vs *BMI* at $age = 30$ is $7.844 - (.135)(30) \approx 3.79$, and for $age = 55$ it is approximately 0.42. *BMI* is a less useful

```
> reg.out<-lm(data=dat,tc ~ age*bmi)
> summary(reg.out)

Call:
lm(formula = tc ~ age * bmi, data = dat)

Residuals:
     Min       1Q   Median       3Q      Max 
-114.395  -18.255   -0.109   20.132  116.370 

Coefficients:
             Estimate Std. Error t value Pr(>|t|)    
(Intercept) -70.2413    65.9523  -1.065 0.288312    
age           5.2086     1.3833   3.765 0.000226 ***
bmi           7.8437     2.5576   3.067 0.002503 ** 
age:bmi      -0.1354     0.0532  -2.546 0.011741 *  
---
Signif. codes:  0 '***' 0.001 '**' 0.01 '*' 0.05 '.' 0.1 ' ' 1

Residual standard error: 33.1 on 177 degrees of freedom
Multiple R-squared:  0.298,	Adjusted R-squared:  0.2861 
F-statistic: 25.05 on 3 and 177 DF,  p-value: 1.476e-13
```

Figure 22.8 R output for regression of *TC* on *age*, *BMI*, and the *age* × *BMI* interaction using *Seasons* data for female-identifying participants aged ≤ 65 years and *BMI* ≤ 40

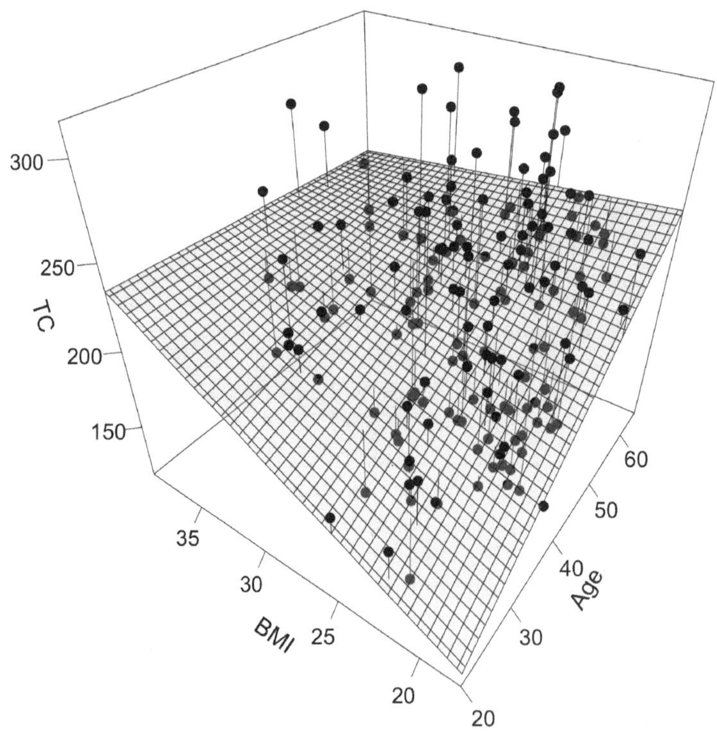

Figure 22.9 Regression surface for the regression of total cholesterol level (*TC*) on *age*, *BMI*, and the *age* × *BMI* interaction with observed points and residuals shown.

predictor of *TC* for older participants, and *age* is a less useful predictor of *TC* for participants with large *BMI* values.

How do we interpret the other regression coefficients in Equation 22.14 and their tests of significance in Figure 22.8? The coefficient of *age*, 5.209, is the slope of predicted *TC* with *age* for *BMI* = 0. The standard error, 1.383, and the corresponding *t* value of 3.765 are also conditional on the value of *BMI* = 0. Similarly, the coefficient of *BMI*, 7.844, is the slope of predicted *TC* vs *BMI* for *age* = 0, and the standard error and *t* also hold only for *age* = 0.

We are not likely to be interested in what is predicted to happen at *age* = 0 and *BMI* = 0, but we may well be interested in testing whether a coefficient is significant at a specified value of the other predictor. For example, we may wish to test whether the slope of predicted *TC* with *BMI* is significant at *age* = 20. There are several ways to conduct this test. We can find the value of the slope by substituting *age* = 20 into Equation 22.15. We can also develop expressions for the standard error of the slope at age = 20 and conduct the appropriate *t* test. However, the easiest way to perform this test of a simple effect is to transform the variable and redo the regression.

Suppose we create the variable *agem20* = *age* – 20. Then we can regress *TC* on *agem20*, *BMI*, and *agem20×BMI*, a variable obtained by multiplying *agem20* and *BMI*. The resulting regression equation is

$$\widehat{TC} = 33.931 + 5.209\ agem\,20 + 5.135\ BMI - 0.135\ agem\,20 \times BMI$$

The interpretation of the coefficients is comparable to that of Equation 22.14. Here we are most interested in the coefficient of *BMI*, 5.135, which is the slope of predicted *TC* with *BMI* at *agem20* = 0, that is, at *age* = 20. The coefficient is significantly different from 0, $t(177) = 3.765$, $p = .000$. So, we can say that the slope for *BMI* is significant at *age* = 20. Note that the coefficient of the product term is not affected by transforming *age* to *agem20* = *age* – 20. The same approach can be used to test the coefficient at any other age. Similarly, we can test the coefficients of *age* at different levels of *BMI* by transforming *BMI* and redoing the regressions.

22.8.4 *Using Centering to Reduce Multicollinearity*

When we add a product term to a regression equation that already contains the terms making up the product, we are likely to have multicollinearity. This is because a product term will generally be highly correlated with its constituents. For example, in our continuing example using women aged 20–65 with *BMI* scores no greater than 40, *age* and *BMI* have correlations of .801 and .675 with the product term *Age* × *BMI*. For the regression of *TC* on *age*, *BMI*, and *Age* × *BMI*, the variance inflation factors (see Section 22.5.1) are 35.53, 23.44, and 64.42, respectively, which suggests a possible serious problem with multicollinearity. Taking the square root of the VIFs, we find that the standard errors of the regression coefficients are at least 4.8 times larger when the *Age* × *BMI* term is included than when it is not. Because of this, it is often suggested that regressions that include product terms should be conducted with centered variables (see, for example, Jaccard, Turrisi, & Wan, 1990).

Suppose we take the suggestion and center all the variables by replacing them by their deviation scores, resulting in the centered variables *agecent* (equal to *age minus the mean of the age scores*), *BMIcent*, *TCcent*, and the product term *Agecent* × *BMIcent* (note that we obtain the product term by multiplying the centered variables, not by multiplying the variables and then centering the product). The multicollinearity problem is no longer present: *agecent* and *BMIcent* have correlations of –.007 and .009 with the interaction term, and the VIFs are 1.01, 1.01, and 1.00.

Note that the algorithms built into the standard statistical packages were able to deal with the computational challenges presented by the regression with the uncentered variables, so that the coefficient of the product term is the same, –0.135, for both analyses. However, the coefficients of the lower-order terms are different and have more meaningful interpretations when the centered variables are used. Given centering, the coefficient of X_j represents the slope of predicted Y with X_j when the values of the other predictors equal their means.

22.8.5 *Do We Have an Interaction or Curvilinearity or Both?*

Because predictor variables are usually correlated, testing for an interaction or for curvilinearity is more complicated than we have so far discussed. Suppose the true model is

$$Y = \beta_0 + \beta_1 X_1 + \beta_2 X_2 + \beta_3 X_1^2 + \varepsilon$$

Because we never know what the true model is, we may not test for curvilinearity but instead decide to test for the presence of an interaction by regressing Y on X_1, X_2, and $X_1 \times X_2$. If we have enough power, we may find that the coefficient of $X_1 \times X_2$ is significant, not because the true model contains an interaction term, but because the product term in the regression is correlated with the squared term that is in the true model. Conversely, if the true model is

$$Y = \beta_0 + \beta_1 X_1 + \beta_2 X_2 + \beta_3 X_1 X_2 + \varepsilon$$

and we test for curvilinearity by regressing Y on X_1, X_2, $X^2{}_1$, and $X^2{}_2$, if we have sufficient power, we may find that the coefficients of one or both of the squared terms are significant, not because the squared terms are present in the model, but because of their correlations with the product term that is present in the model but left out of the regression equation.

Also, omitting terms from the regression can result in finding misleading interaction and curvilinear terms. We may find, for example, a positive interaction even when the true interaction is negative and we may find a negative quadratic component even when the coefficient of the squared term in the true model is positive. Suppose that the true model is

$$Y = \beta_0 + \beta_1 X_1 + \beta_2 X_2 + \beta_3 X_1^2 + \beta_4 X_1 X_2 + \varepsilon$$

where β_3 is positive and β_4 is negative. If we decide to test for the presence of the interaction by regressing Y on X_1, X_2, and $X_1 \times X_2$, but not on $X^2{}_1$, we may find that the coefficient of the product term is positive even though it is negative in the true model because of the correlation between the product term and the $X^2{}_1$ term that was not included in the model. Similarly, if we regress on X_1X_2, and $X^2{}_1$, and omit $X_1 \times X_2$ from the equation, we may find that the coefficient of the squared term is negative, despite being positive in the true model. Therefore, Ganzach (1997) recommends that we include both the interaction and the quadratic terms in the regression when either interaction or quadratic terms are evaluated. Not everyone agrees with Ganzach's recommendation because including the squared terms may lower the power for finding significant interactions. Aiken and West (1991) and Shepperd (1991) recommend against including squared terms when evaluating interactions unless there is theoretical justification for their inclusion. We recommend exploring the situation by trying it both ways.

If we regress *TC* on *age*, *BMI*, age^2, BMI^2, and *Age* × *BMI*, the interaction term is significant, $t(175) = -2.077$, $p = .039$, even though we include the squared terms in the regression, and the squared terms are not significant (regardless of whether we center the variables). Therefore, we conclude that there is a negative interaction between *age* and *BMI*.

There is yet another complication – the differential reliability of the terms in the regression. All other things being equal, predictors that are less reliable (i.e., have more measurement error) have smaller effect sizes and are tested with less power. Further, reliability will be lower for squared and product terms than for the first-order terms (see Busemeyer & Jones, 1983; MacCallum & Mar, 1995). For the typical situation in which the predictors are at least somewhat correlated, the attenuation will be worse for the quadratic terms than for the product terms. As a result of this differential reliability, when the predictors are measured with error, interaction effects will be underestimated and quadratic terms will be underestimated even more. Therefore, if the true model contained interaction and quadratic effects of equal size, the power and observed effect size would tend to be larger for the interaction term. Because of these complications, we must be cautious about making strong conclusions from the data unless our measures are reliable and the regression equation survives cross-validation assessments.

22.9 Limitations of Ordinary Least-Squares Regression

We have mentioned alternatives to OLS regression when assumptions of the model have been violated. For example, we noted that weighted least-squares regression can be a useful alternative when the data exhibit heteroscedasticity and that principal components analysis may be useful when there is substantial multicollinearity in a data set. Two other procedures that address designs that are problematic for OLS regression are *logistic regression* and *multilevel modeling*.

OLS regression cannot be applied when the dependent variable is dichotomous and the goal is to predict that outcome.[4] For example, a researcher may wish to predict whether a criminal defendant will receive a guilty verdict based on predictors such as race, strength of the evidence, severity of the crime, prior criminal record, and so forth. One clear problem for OLS regression is the assumption that the standard error of the mean is normally distributed, because it implies the model must predict outcomes other than 1 (guilty) or 0 (not guilty). A common solution to this problem is to use *logistic regression*, which transforms the linear combination of predictor variables to a probability, p, using a function called the logistic. This transformation results in predictions that are bounded at 0 and 1 and have a smooth, "S-shaped," transition from one value to the other. The interpretation of the logistic regression coefficients is quite different than we have discussed so far, and additional new concepts must be learned to perform and interpret the results. Comprehensive introductions to the topic are available in Cohen et al. (2003) and Menard (2002). A standard reference is the book by Hosmer and Lemeshow (2001).

OLS is also not applicable when the data are hierarchically organized. It is common to encounter designs where observations are nested within other variables. One example is when participants are organized in groups, such as students who are organized into classrooms. Another common example is repeated-measures designs where multiple observations are taken on each participant in a study (i.e., observations are nested within participants).

The significance tests for standard OLS regression assume that the data points are independent of one another. That assumption is obviously violated in repeated-measures designs, and it is unlikely to be true in other nested designs. Consider the student example, where we would expect that, on the average, measures from two students in the same classroom would be more similar than measures from two students selected from different classrooms, even if the classroom means do not differ systematically from one another. The consequence of this lack of independence is that any significance tests that we perform will be positively biased. Moreover, the mean levels of the dependent and predictor variables may differ systematically across individuals in our repeated-measures design or across classrooms in the assessment of students. If so, combining data without regard to the nested structure of the data may produce regression equations that do not characterize the relationships in any of the individuals or classrooms very well (see Chapter 17).

An appropriate analysis, which takes account of the hierarchical structure in the data, is called *multilevel modeling*, *hierarchical linear modeling*, or *mixed-effects modeling*. A proper discussion of this topic is beyond the scope of the present coverage. Introductions to hierarchical regression may be found in Cohen et al. (2003), Hox (2002), and Kreft and de Leeuw (1998), and an excellent, more comprehensive treatment is given by Raudenbush and Bryk (2002).

22.10 Summary

In Chapter 22 we extended our coverage of multiple regression, largely dealing with issues related to interpreting the effects of predictor variables. We can briefly summarize much of the chapter by reviewing some cautions we must keep in mind if we seriously wish to consider the "importance" of a predictor variable. If we are concerned merely with prediction, or with describing the data, our task is relatively straightforward. However, if our goal is to associate the size of a regression coefficient with the causal importance of the corresponding predictor variable, using data that have not been collected in a true experiment, we must first understand that predictive usefulness is not the same as causal influence; that is, a variable that is a good predictor may have no causal importance whatsoever. Even if we have good reason to believe that the variable in question has a causal effect, the regression coefficient may not be a good index of its causal importance, for at least the following reasons:

- The model may be mis-specified.
- There may be measurement error, either in the predictor variable we are concerned with or in other variables in the model that are correlated with it.
- There may be nonrandomly missing data on the dependent variable or predictors.
- The predictor variables may be highly correlated.
- There may be mediating effects, so that part of the effect of the variable may be achieved by influencing other variables that, in turn, affect the dependent variable.
- There may be curvilinear effects or interactions with moderator variables that have not been considered.
- If the design has a hierarchical structure, this must be considered in the analysis or else any statistical tests may be positively biased.

These cautions suggest that it is difficult to make formal causal statements based on regression analyses using data from an observational study, and even that we should view the results of our inferential tests with some skepticism because we are unlikely to have specified the "correct" model. However, despite its limitations, regression analysis can be an extremely useful component of a program for advancing knowledge about a research problem. Existing theory can suggest that certain regressions be performed. These regressions can provide useful descriptions of the data and can suggest modifications of the theory and speculations about causal effects that can often be pursued by conducting further research with a variety of techniques and tools, including controlled experiments where possible.

Appendix 22.1

To Show That b_{Y1} Is Not an Unbiased Estimator of $\beta_{Y1.2}$ if X_2 Is Left Out of the Model

We begin by first expressing b_{Y1} as

$$b_{Y1} = \frac{\Sigma\left(X_1 - \bar{X}_1\right)(Y - \bar{Y})}{SS_1}$$

where $SS_1 = \Sigma\left(X_1 - \bar{X}_1\right)^2$ so that

$$E\left(b_{Y1}\right) = E\left(\frac{\Sigma\left(X_1 - \bar{X}_1\right)(Y - \bar{Y})}{SS_1}\right)$$

If we assume X is fixed, we can rewrite the preceding equation as

$$E\left(b_{Y1}\right) = \frac{\Sigma\left(X_1 - \bar{X}_1\right)E(Y - \bar{Y})}{SS_1}$$

Substituting the population model expressions for Y and $\bar{Y}$ and simplifying, we have

$$\begin{aligned} E\left(b_{Y1}\right) &= \frac{\beta_{Y1\cdot2}\Sigma\left(X_1 - \bar{X}_1\right)^2 + E\left[\beta_{Y2\cdot1}\Sigma\left(X_1 - \bar{X}_1\right)\left(X_2 - \bar{X}_2\right)\right]}{SS_1} \\ &= \beta_{Y1\cdot2} + \beta_{Y2\cdot1}E\left(r_{12}\sqrt{\frac{SS_2}{SS_1}}\right) \\ &= \beta_{Y1\cdot2} + \beta_{Y2\cdot1}E\left(b_{21}\right) = \beta_{Y1\cdot2} + \beta_{Y2\cdot1}\beta_{21} \end{aligned}$$

Note that because $\beta_{21} = \rho_{12}\, \sigma_2 / \sigma_1$, if $\rho_{12} = 0$, then $E(b_{Y1}) = \beta_{Y1.2}$.

Exercises

22.1 [Trend analysis] Hypothetical data from an experiment designed to determine the effects of drug dosage on performance may be found in the file *EX23_1*.

a) Perform an ANOVA to test the effect of dosage on performance.
b) Using multiple regression, perform a trend analysis (see Section 22.6.3) on the same data.

22.2 [Interpreting regression coefficients] Let's consider the distinction between standardized and unstandardized regression coefficients in a small data set. Use the file *statistics class data*. If we regress final exam score (Y) on pretest score (X), we find that the regression equation is $\hat{Y} = -36.083 + 3.546\ X$ and R is .725. The unstandardized coefficient is 3.546 and the standardized coefficient is 0.725.

a) Now suppose we want to add two additional data points that are consistent with the regression equation but that change the variances of X and Y. Say we add pretest scores of 11 and 38, along with the final exam scores predicted with the regression equation, rounding off to the nearest whole number. What are the two predicted final exam scores?
b) If we now regress final scores on pretest scores for this new data set (now with 20 cases), how should the new unstandardized regression coefficient differ from the previous one?
c) How should the new standardized regression coefficient differ?
d) How should the new value of R differ?

22.3 [Using regression to test hypotheses in real data] In the next few exercises, we use some data from the *Seasons* study to explore the relationships among the variables *age*, perceived health (*Sayhlth*), level of education (*schoolyr*), body mass index (*BMI*), depression as measured by the Beck Depression Inventory (*beck_d*), and cholesterol level (*TC*). In doing so, we demonstrate the distinction between prediction and causality and consider the limitations of our ability to answer questions with the available data.

Start by considering how level of education might be predicted by sex and age in the data set *Seasons exercises*. The amount of education that people receive has increased over recent generations, so we might expect age to be a (negative) predictor of level of education. Also, women have historically received less education than men, though in recent years such gender differences have been reduced or even reversed. Perform some appropriate regressions to explore these issues.

22.4 [Using regression to interpret real data] Next, explore how perceived health (*Sayhlth* – note that larger values on this measure correspond to poorer health) is predicted by sex and level of education. It is often stated that more educated people tend to be healthier and have longer life expectancies. Is this statement consistent with the information in the *Seasons exercises* data set? If so, why might this be the case? What kinds of additional information might be useful in investigating this issue further?

22.5 [Understanding the consequences of measurement error]

a) Start by using the *Seasons exercises* data file. To make the results comparable to those obtained in some previous analyses in the chapters, select participants aged ≤ 65 years with *BMI*s of ≤ 40. Regress *TC* on age and *BMI* for both men and women. Interpret the output.

b) Now, instead of using *BMI* as a predictor, simulate additional measurement error in *BMI* by creating a new variable *bmiplusnorm05*. This new variable is obtained by adding to each value of *BMI* a random number selected from a normal distribution with mean 0 and standard deviation 5. In R, the function *rnorm*(x, 0, 5) samples x values from $N(0, 5)$, where x is an integer value. In SPSS, compute a new variable using the *Rv.Normal* function in the *Random Numbers* function group. We have included a variable created in this fashion in the data file; feel free to create your own. Note that because we are adding random components to the values of *BMI*, each time we generate the variable, the specific values will differ. How should we expect the correlation between *TC* and *bmiplusnorm05* to differ from that between *TC* and *BMI*? How do we expect the regression of *TC* on age and *bmiplusnorm05* to differ from what was obtained in part (a)? Why?

c) Repeat part (b) except replace *BMI* by *bmiplusnorm020*, obtained by adding to *BMI* a random number selected from a normal distribution with mean 0 and standard deviation 20.

d) Now consider what should happen if we have more measurement error in the dependent variable *TC* (as opposed to the predictors). We have created a new variable *tcplusnorm040*, formed by adding to *TC* a random number selected from a normal distribution with mean 0 and standard deviation 40. How should we expect results to differ if we used this new variable as the dependent variable in the regression on *age* and *BMI* instead of using *TC*?

22.6 [Interpreting regression coefficients]

a) A researcher is interested in relating measures of parent-child attachment to measures of externalization and criticism obtained from a series of interviews. A regression of the attachment measure on both externalization and criticism yields significant t tests for the coefficients of both predictor variables. What can be concluded?

b) The researcher then decides to determine whether the joint effect of externalization and criticism is an important predictor of attachment. She creates a new variable by multiplying the externalization and criticism measures for each case. She then regresses attachment on the externalization and criticism measures as well as the new variable. The regression now shows that none of the t tests for the three predictors is significant. (i) Is this an appropriate way to assess whether the joint effect of externalization and criticism is an important predictor of attachment? Why or why not? (ii) What is the most likely reason the t tests for the coefficients of externalization and criticism are not significant in the second regression even though they were significant in the first regression described in part (a)? (iii) Considering the results of the two regressions together, what can be concluded?

22.7 [Thinking about a real world regression problem] Consider a timely research question: What is the effect of reducing class size on student performance in elementary school? If it can be established that reducing class size has a sufficiently large and enduring effect, then it may be worth investing the resources that would be required to achieve the reductions. Common sense suggests that children must learn more in smaller classes in which they can be given more individual attention. However, despite many publications on the topic, there is still vigorous debate about whether the gain in learning is worth the extra resources required to achieve it, or even whether there is any meaningful gain at all. To get a flavor of this discussion, just enter "class size debate" in any internet search engine. How can there possibly be so much disagreement about the results of class size research? Suppose we were to collect relevant data from observational studies, in which we sampled many (naturally occurring) classes of different sizes and compared the performance of students as a function of classroom size. How might the results of the study be questioned and what kinds of studies might be run that would be less subject to these criticisms?

22.8 [Using software for regression with interaction] Using the data file *Seasons exercises*, perform the analyses found in Section 22.8.2 to test the interaction of age and sex with the dependent variable *TC*. Compare with Figure 22.6. Note: To make the results comparable, select participants aged 65 years or less with *BMI* scores no larger than 40.

22.9 [Using software for regression with interaction] Perform the analyses found in Section 22.8.3 to test the interaction of *age* and *BMI* for female-identified participants. Compare with Figure 22.7.

Notes

1 When standard statistical software is used, high correlations among predictors will not generally result in much difficulty if the only goal is prediction. However, multicollinearity can result in errors in the computational algorithms employed in spreadsheets such as Excel when one attempts to use them to perform multiple regression. Centering the variables (see the third remedy later and Section 22.8.4) can reduce the correlations among predictors, and can also reduce the sizes of the numbers used in calculations, thereby reducing rounding error. Nonetheless, spreadsheets and other nonstandard statistical software should not be used to perform serious statistical analyses, especially multiple regression.

2 Simonsohn (2018) described an algorithm for fitting two lines to data patterns commonly described as "U shaped." The "two-lines test" has clear advantages over the quadratic modeling described here, but the approach does not generalize to more complicated relationships (e.g., cubic or "W shaped").

3 If the coefficient of the product term was positive, the partial slope of predicted *Y* for one predictor would become larger as the value of the other predictor increased.

4 If your goal is to estimate causal effects, see Gomila (2021) for arguments in favor of using OLS regression even with a dichotomous outcome variable.

Regression With Qualitative and Quantitative Variables

23.1 Overview

We now consider how to include categorical variables with levels that differ *qualitatively* from one another within the regression framework. Examples of such variables are biological sex, usually defined with levels female and male; diagnosed mental illness, with levels schizophrenia, depression, and anxiety disorder; and treatment condition, with levels defined by different types of therapy. Except for the brief introduction to logistic regression in the last chapter and the occasional use of dichotomous predictor variables such as sex, our development of regression to this point has focused on variables that are quantitative and are treated as though they were continuous. However, qualitative categorical variables can also be incorporated into regression analyses, providing us with a general and powerful framework within which many of the analyses that we have discussed earlier can be considered as special cases, including ANOVA and ANCOVA. Viewing ANOVA and regression within the same framework can both increase our understanding of these analyses and how they are related and allow us to deal with data from designs that cannot be handled easily by the standard ANOVA approach.

Our goals in Chapter 23 are the following:

- *Discuss how qualitative categorical variables can be coded so that they may be included within the multiple regression framework.*
- *Consider different kinds of ANOVA designs with between-participants factors within the multiple regression framework.*
- *Briefly consider how repeated-measures and mixed ANOVAs can be included within the multiple regression framework.*

23.2 One-Factor Designs

23.2.1 *Coding Qualitative Categorical Variables*

It is important to distinguish between quantitative and qualitative categorical variables in regression. Suppose we have a factor with six levels that correspond to qualitatively different treatment conditions, and code this factor by using a single variable that assigns the numbers 1–6 to the treatments. *It is not correct to use this variable as a predictor in a regression analysis*. If we did, we would be treating the factor as though it was a *quantitative* variable – such coding implies that the treatment at level 3 is three times as large as the treatment at level 1, and that the treatment at level 4 is twice as large as the one at level

DOI: 10.4324/9781003453550-27

2. This makes no sense if we have a qualitative or categorical variable. Moreover, using a single coding variable only accounts for 1 *df*, whereas in ANOVA there are 5 *df* associated with a factor that has six levels.

If the treatments are *qualitatively* different from one another, we want a coding system which specifies that the treatments are different from one another without imposing any type of ordering. As we shall see, for a factor with *a* levels, $a - 1$ predictor variables are required to do the coding. In practice, we almost never have to generate these coding variables ourselves. Rather, the statistical software does it for us in their ANOVA or GLM (*General Linear Model*) modules. Here, we describe some ways of coding categorical variables because this allows greater insight into the analyses, and in some cases, allows us to better understand the options and output provided by software.

Any qualitative categorical variable can be coded by defining one or more *dummy* (or *indicator* or *code*) *variables* that take on numerical values. These numerical values are not measures of the category levels; rather, they are best thought of as labels that collectively specify category membership. The coding is particularly simple when a categorical variable, *A*, has only two levels (e.g., pass/fail on an exam). If factor *A* has only two levels, the dummy variable *X* could take on any value at one level and any different value at the other level. For the moment, assume these values are 1 and 2. The overall test of the regression of the dependent variable *Y* on *X* would then be exactly equivalent to the ANOVA *F* test for the categorical variable, *A*.

To be more specific, if we regress *Y* on *X*, the least-squares regression line must pass through the points $(1, \bar{Y}_{.1})$ and $(2, \bar{Y}_{.2})$, as can be seen in Figure 23.1. This is because the regression line minimizes the mean squared deviations of the *Y*s, and the group means minimize the mean squared deviations in each of the groups. Because the regression line passes though the two group means, it accounts for all the variability in the group means, so $SS_{regression}$ must equal the between-group variability, SS_A. Also, $SS_{residual}$, the variability unaccounted for by the regression, must equal $SS_{S/A}$.

However, if the factor *A* has more than two levels, regression on a single dummy variable will not, in general, account for all of SS_A. Consider what happens with three levels: Panel *a* of Table 23.1 presents scores at levels A_1, A_2, and A_3 of the factor *A*. In Panel *b* of the table, we code *A* with a single dummy variable, X_1, that takes on the values 1, 2, and 3. The points that represent the group means in the space defined by X_1 and *Y* will be $(1, \bar{Y}_{.1})$, $(2, \bar{Y}_{.2})$, and $(3, \bar{Y}_{.3})$, as shown in Panel (a) of Figure 23.2. In general, these three points will not be perfectly fit by a straight line. Therefore, if *Y* is regressed on X_1, the regression will usually not account for all the variability in the group means, and $SS_{regression}$ will be less than SS_A. However, we can account for all of SS_A if we define an additional dummy variable, X_2, that is not perfectly correlated with X_1 – for example, the variable that takes on the values given in Panel *b* of Table 23.1. Now if we represent the three group means in the three-dimensional space defined by *Y*, X_1, and X_2, as in Panel (b) of Figure 23.2, it is apparent that they can be perfectly fit by a two-dimensional regression *plane*. Therefore, when *Y* is regressed on both dummy variables, all the between-group variability will be accounted for by the regression, so that $SS_{regression} = SS_A$, and the *F* test for the overall regression of *Y* on X_1 and X_2 is exactly equivalent to the ANOVA *F* test for the effect of the categorical factor, *A*.

This reasoning can be extended to factors with more levels by using additional dummy variables. In general, it will take as many dummy variables to code a factor as the number of degrees of freedom associated with it; for example, five dummy variables will be required to code a categorical variable with six levels. The only requirement on these dummy variables

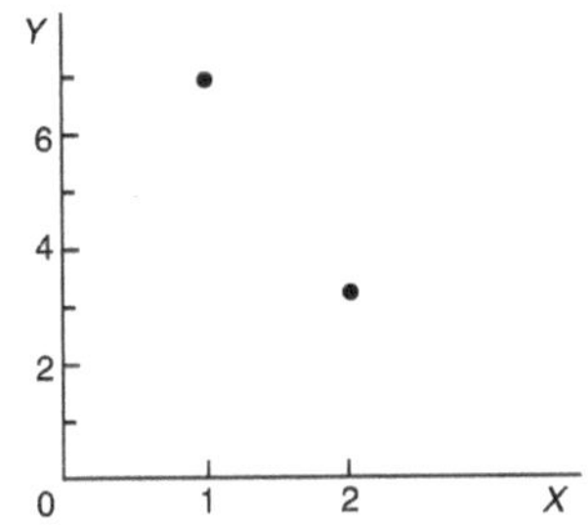

(a)

	A_1	A_2
	4	2
	8	5
	7	3
	9	
	7	
mean	7.00	3.33

Y	X
4	1
8	1
7	1
9	1
7	1
2	2
5	2
3	2

SV	*df*	*SS*	*MS*	*F*
A	1	25.21	25.21	8.10
S/A	6	18.67	3.11	

$SS_{reg} = 25.21$; $SS_{error} = 18.67$

$$F = \frac{SS_{reg}/1}{SS_{error}/(8-2)} = 8.10$$

(b)

Figure 23.1 (a) Plot of the means of A against X for the data in Panel (b). Note that the two points representing group means can be fit perfectly by a straight line. (b) Data and results of ANOVA and multiple regression.

is that they be *linearly independent*; that is, none of them may be perfectly expressed as a linear combination of the others. If a variable can be expressed as a linear combination of the other dummy variables, it is redundant, and therefore cannot contribute anything to the specification of the categories.

23.2.2 *Effect and Dummy[1] Coding*

Although the variables X_1 and X_2 in our example appropriately partition the variability in Y into components associated with *A* and *S/A*, the coefficients associated with X_1 and X_2 are not particularly meaningful. Two more common approaches to coding categorical variables, *effect coding* and *dummy coding*, produce more interpretable regression coefficients. Both coding methods allow us to specify group membership so that a regression of the dependent variable on the dummy variables will produce an analysis identical to the ANOVA. The only difference between them is that the regression coefficients produced by the two coding systems have different interpretations.

Effect Coding

In discussing ANOVA as a special case of multiple regression, we may find it useful to consider the type of dummy variable coding called *effect coding* because, as we show next,

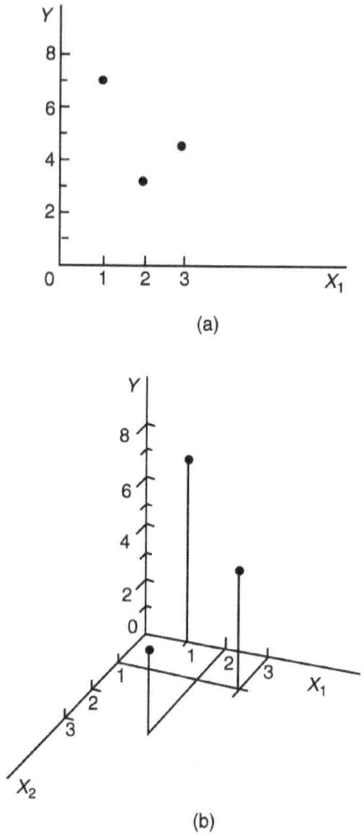

Figure 23.2 (a) Plot of the means of *A* against X_1 for the data in Table 23.1. In general, the points representing the three means cannot be fit perfectly by a straight line. (b) Plot of the means of *A* against X_1 and X_2. The points representing the three means can always be fit by a plane if X_1 and X_2 are not perfectly correlated.

it produces regression coefficients that estimate the ANOVA effects $\alpha_1, \alpha_2, \ldots, \alpha_{a-1}$, where $\alpha_j = \mu_j - \mu$. Effect coding represents group membership with dummy variables that contain 1s, 0s, and −1s, as illustrated by variables X_{E1} and X_{E2} in Panel *b* of Table 23.1. Dummy variables for effect coding are defined as follows:

$$\begin{aligned} X_{Ej} &= 1 \text{ for scores at level } A_j \\ &= -1 \text{ for scores at some arbitrary level of } A, \text{ here } A_3 \\ &= 1 \text{ for scores at level } A_j \\ &= 0 \text{ otherwise} \end{aligned}$$

Because there are three levels of *A* in Table 23.1, we require two dummy variables to account for all the between-group variability. The coding of X_{E1} and X_{E2} is presented in Panel *b* of Table 23.1. As can also be seen there, the regression of *Y* on X_{E1} and X_{E2} produces a value of $SS_{regression}$ that is equal to the SS_A obtained from the standard ANOVA, and an overall *F* statistic for the regression that is equal to the ANOVA *F*.

Table 23.1 Regression on dummy variables for a one-factor design

(a) Data and results of standard ANOVA for a one-factor design

A_1	A_2	A_3
4	2	4
8	5	5
7	3	3
9		6
7		
$\bar{Y}_{.1} = 7.00$	$\bar{Y}_{.2} = 3.33$	$\bar{Y}_{.3} = 4.50$

$$\bar{Y}_U = \frac{(7 + 3.33 + 4.50)}{3} = 4.94$$

Source	*df*	*SS*	*MS*	*F*
A	2	28.583	14.292	5.435
S/A	9	23.667	2.630	

(b) Dummy variable coding and some statistics obtained from the regressions of Y on X_1 and X_2 and on X_{E1} and X_{E2}

	Y	X_1	X_2	X_{E1}	X_{E2}	X_{D1}	X_{D2}
A_1	4	1	0	1	0	1	0
	8	1	0	1	0	1	0
	7	1	0	1	0	1	0
	9	1	0	1	0	1	0
	7	1	0	1	0	1	0
A_2	2	2	3	0	1	0	1
	5	2	3	0	1	0	1
	3	2	3	0	1	0	1
A_3	4	3	1	−1	−1	0	0
	5	3	1	−1	−1	0	0
	3	3	1	−1	−1	0	0
	6	3	1	−1	−1	0	0

The statistics for the regression of Y on X_1 and X_2, on X_{E1} and X_{E2}, or on X_{D1} and X_{D2}, are $R = .740$; $SS_{regression} = 28.583$; $SS_{residual} = 23.667$; $F = MS_{regression} / MS_{residual} = 5.435$. The regression coefficients based on effect coding are

$b_{E0} = \bar{Y}_U = 4.94, b_{E1} = \bar{Y}_{.1} - \bar{Y}_U = 2.06$, and $b_{E2} = \bar{Y}_{.2} - \bar{Y}_U = -1.61$;

And the regression coefficients based on dummy coding are

$b_{D0} = \bar{Y}_{.3} = 4.50$, $b_{D1} = \bar{Y}_{.1} - \bar{Y}_{.3} = 2.50$, *and* $b_{D2} = \bar{Y}_{.2} - \bar{Y}_{.3} = -1.17$.

We have stated that the regression coefficients resulting from effect coding correspond to the effects estimated in ANOVA. We can readily demonstrate this correspondence, first noting that the scores predicted by the regression equation for each of the groups must be the group means. For A_1, we have $\hat{Y} = \bar{Y}_{.1}$, $X_{E1} = 1$, and $X_{E2} = 0$. Substituting into the regression equation that predicts Y from X_{E1} and X_{E2}, we obtain

$$\bar{Y}_{.1} = b_{E0} + b_{E1}(1) + b_{E2}(0) = b_{E0} + b_{E1} \tag{23.1}$$

Similarly, for A_2 and A_3, respectively, we have

$$\bar{Y}_{.2} = b_{E0} + b_{E1}(0) + b_{E2}(1) = b_{E0} + b_{E2} \tag{23.2}$$

and

$$\bar{Y}_{.3} = b_{E0} + b_{E1}(-1) + b_{E2}(-1) = b_{E0} - b_{E1} - b_{E2} \tag{23.3}$$

Adding Equations 23.1–23.3, we obtain $3b_{E0} = \bar{Y}_{.1} + \bar{Y}_{.2} + \bar{Y}_{.3}$, so that

$$b_{E0} = \frac{\bar{Y}_{.1} + \bar{Y}_{.2} + \bar{Y}_{.3}}{3} = \bar{Y}_U$$

Here, $\bar{Y}_U$ is the unweighted average of the group means. If there are equal numbers of scores in each of the groups, $\bar{Y}_U$ will equal $\bar{Y}_{..}$, the grand mean of all the scores.

We can now write the coefficients of the dummy variables as deviations from the unweighted mean of the group means. Substituting $b_{E0} = \bar{Y}_U$ into Equations 23.1 and 23.2, and solving, we obtain the formulas and numerical results for b_{E1} and b_{E2} that are presented in Panel *b* of Table 23.1. For equal *n* designs, the regression coefficients correspond exactly to the estimated main effect components of the ANOVA, the $\hat{\alpha}_j$S. That is, because $\bar{Y}_{.j} = \hat{\mu}_j$, and $\bar{Y}_{..} = \hat{\mu}$,

$$b_{E1} = \bar{Y}_{.1} - \bar{Y}_{..} = \hat{\mu}_1 - \hat{\mu} = \hat{\alpha}_1$$

and

$$b_{E2} = \bar{Y}_{.2} - \bar{Y}_{..} = \hat{\mu}_2 - \hat{\mu} = \hat{\alpha}_2$$

Furthermore, because of the requirement that $\Sigma\alpha_j = 0$, $\hat{\alpha}_3$ can be found as

$$\hat{\alpha}_3 = -\hat{\alpha}_1 - \hat{\alpha}_2 = -b_{E1} - b_{E2}$$

Dummy Coding

A second way to code categorical variables is to use *dummy coding*, for which the dummy variables only take on the values 0 and 1, as illustrated by X_{D1} and X_{D2} in Panel *b* of Table 23.1. For dummy coding, the dummy variables are defined as

$$X_{Dj} = 1 \text{ for scores at level } A_j \text{, and}$$
$$= 0 \text{ otherwise}$$

Note that because we only need $a - 1$ dummy variables to code *a* groups, one group, referred to as the *reference group*, will receive 0s on all the dummy variables.

Because there are three levels of *A* in the current example, we require two dummy variables to account for all the between-group variability. Scores at A_1 receive a 1 on X_{D1} and a 0 on X_{D2}; scores at A_2 receive a 0 on X_{D1} and a 1 on X_{D2}; and scores at A_3 receive values of 0 on both X_{D1} and X_{D2}. As we can see in Panel *b* of Table 23.1, the regression of *Y* on X_{D1} and X_{D2} also produces a value of $SS_{regression}$ equal to the SS_A obtained from the standard ANOVA. Although the variability accounted for by the regression is the same whether we use dummy or effect coding, the regression coefficients are different. If we use dummy coding, the intercept, b_{D0}, takes on the value of the mean of the *reference group* (the group that has 0s on all the indicator variables). For example, in Table 23.1, the value of the intercept is equal to the mean of condition A_3. The regression coefficients for each of the dummy variables, b_{D1} and b_{D2}, take on values equal to the differences between the group coded as 1 on the dummy variable and the mean of the reference group. We can see this by noting that because the regression accounts for all the between-group variability, the prediction for each score will be its group mean.

For the scores at A_3, each dummy variable has the value 0 so that

$$\bar{Y}_{.3} = b_{D0} + b_{D1}(0) + b_{D2}(0) = b_{D0}$$

Therefore, $b_{D0} = \bar{Y}_{.3}$. For scores at A_1, we have $\hat{Y} = \bar{Y}_{.1}$. Substituting $X_{D1} = 1$, and $X_{D2} = 0$ into the equation for the regression of *Y* on X_{D1} and X_{D2}, we obtain

$$\bar{Y}_{.1} = b_{D0} + b_{D1}(1) + b_{D2}(0) = b_{D0} + b_{D1}$$

Therefore, $b_{D1} = \bar{Y}_{.1} - b_{D0} = \bar{Y}_{.1} - \bar{Y}_{.3}$. Similarly, for scores at A_2, we have

$$\bar{Y}_{.2} = b_{D0} + b_{D1}(0) + b_{D2}(1) = b_{D0} + b_{D2}$$

so that $b_{D2} = \bar{Y}_{.2} - b_{D0} = \bar{Y}_{.2} - \bar{Y}_{.3}$.

Dummy coding might be particularly useful if the design contains a control group. If so, we could let the control group serve as the reference group, and the regression coefficients would then directly equal the differences between the treatment and control means. There are many other possible ways to code categorical variables. For example, categorical variables may be coded such that regression coefficients take on the values of contrasts of possible interest. Detailed discussions of these methods may be found in sources such as Cohen et al. (2003).

No matter which coding procedure is used to code a factor *A*, when the dependent variable *Y* is regressed on the complete set of dummy variables,

$$SS_{between} = SS_A = R^2_{Y.A} SS_Y$$

where $R^2_{Y.A}$ is the square of the multiple correlation when *Y* is regressed on the $a - 1$ dummy variables that code factor *A*, and

$$SS_{within} = SS_{S/A} = \left(1 - R^2_{Y.A}\right) SS_Y$$

so that the test statistic for *A* is given by $F = \dfrac{MS_A}{MS_{S/A}} = \dfrac{R^2_{Y.A} SS_Y / (a-1)}{\left(1 - R^2_{Y.A}\right) SS_Y / (N-a)}$.

```
> X<-as.factor(c("lions","tigers","bears"))
> contrasts(X)
       lions tigers
bears      0      0
lions      1      0
tigers     0      1
> X<-relevel(X,ref = "lions")
> contrasts(X)
       bears tigers
lions      0      0
bears      1      0
tigers     0      1
```

Figure 23.3 Example of dummy codes and releveling factors in R.

23.2.3 *Using Software With Dummy Variables*

It is important to understand how categorical variables can be represented in regression because the choice of coding has direct implications for the interpretation of the intercept and the coefficients. Further, in some applications (e.g., use of the HLM software package), the value of each dummy variable must be coded in the data file. However, for the regression applications we are considering here, we rarely must code the categorical variables ourselves. For example, in SPSS, the *General Linear Model* module does not provide a choice; it simply uses dummy coding. This does not present any problem, but we should be aware what the program is doing if we try to interpret the parameter estimates in the output.

In R, we use *factors* for categorical variables. The function *as.factor*(X) in the {base} package treats the variable X as a factor; by default, the levels of X are sorted alphabetically (or numerically) and the first level of X is treated as the reference group using dummy coding. We can inspect the dummy variables that R uses in the background with the *contrasts*(X) function in the {stats} package. It reports the dummy codes for each level of X. An example is shown in Figure 23.3. If we prefer to use a different level of our factor as the reference group, we simply *relevel* the factor, setting *ref* equal to the desired level (see Figure 23.3). To use effect coding in R, we specify the contrasts to be used in the analysis. For example, the command *lm*(data = dat, $Y \sim X$, *contrasts* = *list*(X = contr.sum)) will run a regression on the data in the data frame named dat using effect coding.

23.3 Regression Analyses and Factorial ANOVA Designs

In Section 23.2, we saw how any categorical variable can be coded by a set of dummy variables and that a multiple regression analysis that uses these dummy variables as predictors provides all the information, and more, that can be obtained from a one-factor ANOVA. In Section 23.3 we extend this discussion to multi-factor ANOVA designs, first considering orthogonal designs and then the issues that arise in analyzing data from nonorthogonal or unbalanced (unequal-n) designs.

23.3.1 Orthogonal Designs

A regression analysis of a factorial design can be performed if both the factors and their interactions are coded by sets of dummy variables. Although it is generally more practical to analyze orthogonal designs with software designed to perform ANOVAs, it is useful to understand how such designs might be analyzed by regression. This will provide a deeper understanding of ANOVA as a special case of regression. In addition, greater facility with coding of categorical variables will provide the researcher with greater facility in the analysis of designs that consist of combinations of categorical and quantitative variables.

Panel *a* of Table 23.2 contains data from a 3 × 3 design with factors *A* and *C* (we use *C* rather than *B* to refer to the second factor because of the large number of *b*s already in the chapter). Panel *b* contains sets of effect dummy variables that code the design, where effect coding is used to code the categorical variables. Each set of dummy variables has as many members as the corresponding sources of variance have *df*s. *A* and *C* are coded as though each were the only factor in the design, and the set of four dummy variables that code the *AC* interaction is obtained by multiplying each dummy variable in the *A* set by each one in the *C* set. Together, the eight dummy variables code membership in the nine cells of the design. Panel *c* contains the results of an ANOVA on the data.

With effect coding, the dummy variables *within* any one of the *A*, *C*, and *AC* sets are correlated. However, if the cell frequencies are all equal, the dummy variables in any set are uncorrelated with all the dummy variables in each of the other sets; therefore, the sums of squares associated with the different sets do not overlap. Let's use the notation $R_{Y.A}$, $R_{Y.AC}$, and $R_{Y.A,AC}$ to represent the multiple correlation coefficients that result when *Y* is regressed on the sets of dummy variables that code *A*, *AC*, and both *A* and *AC*, respectively. Then, because the sets of dummy variables corresponding to *A*, *C*, and *AC* are uncorrelated, we have

$$R^2_{Y.A,C,AC} = R^2_{Y.A} + R^2_{Y.C} + R^2_{Y.AC}$$

Multiplying each of the squared correlations by SS_Y, we have

$$SS_{Between\ cell} = SS_A + SS_C + SS_{AC}$$

That is, the between-cells variability is partitioned into the main effects of *A* and *C*, and their interaction. Because we have enough coding variables to account for all the between-participants variability, $SS_{error} = SS_{residual} = (1 - R^2_{Y.A,C,AC})SS_Y$, and tests of the *A* and *C* main effects and the *AC* interaction, respectively, are provided by

$$F = \frac{R^2_{Y.A} / (a-1)}{\left(1 - R^2_{Y.A,C,AC}\right) / (N - ac)}$$

$$F = \frac{R^2_{Y.C} / (c-1)}{\left(1 - R^2_{Y.A,C,AC}\right) / (N - ac)}$$

and

$$F = \frac{R^2_{Y.AC} / (a-1)(c-1)}{\left(1 - R^2_{Y.A,C,AC}\right) / (N - ac)}$$

These test statistics have exactly the same values as the ANOVA *F*s for *A*, *C*, and *AC* that are presented in Table 23.2.

23.3.2 *Using Software for Orthogonal Designs*

With balanced orthogonal designs, the regression and ANOVA are little changed from the process described in Section 23.2.3. As for one-factor designs, dummy coding is the only

Table 23.2 Effect coding for an orthogonal 3 × 3 design

(*a*) Data

	C_1	C_2	C_3
A_1	53	88	56
	51	63	42
A_2	55	48	79
	78	42	50
A_3	79	80	69
	99	92	94

(*b*) Dummy variables formed by using effect coding

Effect:		*A*		*C*		*AC*			
	Y	X_1	X_2	X_3	X_4	X_5	X_6	X_7	X_8
A_1C_1	53	1	0	1	0	1	0	0	0
	51	1	0	1	0	1	0	0	0
A_1C_2	88	1	0	0	1	0	0	1	0
	63	1	0	0	1	0	0	1	0
A_1C_3	56	1	0	−1	−1	−1	0	−1	0
	42	1	0	−1	−1	−1	0	−1	0
A_2C_1	55	0	1	1	0	0	1	0	0
	78	0	1	1	0	0	1	0	0
A_2C_2	48	0	1	0	1	0	0	0	1
	42	0	1	0	1	0	0	0	1
A_2C_3	79	0	1	−1	−1	0	−1	0	−1
	50	0	1	−1	−1	0	−1	0	−1
A_3C_1	79	−1	−1	1	0	−1	−1	0	0
	99	−1	−1	1	0	−1	−1	0	0
A_3C_2	80	−1	−1	0	1	0	0	−1	−1
	92	−1	−1	0	1	0	0	−1	−1
A_3C_3	69	−1	−1	−1	−1	1	1	1	1
	94	−1	−1	−1	−1	1	1	1	1

(*c*) Results of ANOVA on the data in (*a*)

SV	*df*	*SS*	*MS*	*F*
A	2	2,862.333	1,431.167	7.577
C	2	64.333	32.167	0.170
AC	4	1,399.333	349.833	1.852
Error	9	1,700.000	188.889	

```
> lm(data = dat, Y ~ A*C)

Call:
lm(formula = Y ~ A * C, data = dat)

Coefficients:
(Intercept)           A2           A3           C2           C3        A2:C2        A3:C2
       52.0         14.5         37.0         23.5         -3.0        -45.0        -26.5
      A2:C3        A3:C3
        1.0         -4.5
```

Figure 23.4 Regression coefficients using dummy coding and the data from Table 23.2.

```
> lm(data = dat, Y ~ A*C,contrasts = list(A = contr.sum,C = contr.sum))

Call:
lm(formula = Y ~ A * C, data = dat, contrasts = list(A = contr.sum,
    C = contr.sum))

Coefficients:
(Intercept)           A1           A2           C1           C2        A1:C1        A2:C1
     67.667       -8.833       -9.000        1.500        1.167       -8.333        6.333
      A1:C2        A2:C2
     15.500      -14.833
```

Figure 23.5 Regression coefficients using effect coding and the data from Table 23.2.

option in SPSS. If one of the factors is continuous (i.e., *age*), then it must be entered as a covariate in the module for the *General Linear Model.*

In R, dummy coding is the default: The reference cell is set to be the first level of both factors (e.g., A_1C_1 in Table 23.2) and serves as the intercept. For the data in Table 23.2, the regression coefficients using dummy coding are shown in Figure 23.4. The mean of A_1C_1 = 52.0, the value of the intercept. The other coefficients are added to the intercept to obtain other cell means, analogous to Equations 23.1–23.3. For example, the mean of A_2C_3 is 64.5, which equals the sum of the intercept and the coefficients for A_2, C_3, and their interaction $A_2{:}C_3 = 52.0 + 14.5 - 3.0 + 1.0 = 64.5$.

To use effect coding in R, we must specify the contrasts for both factors in the analysis: *lm*(data = dat, *Y* ~ *A*C*, *contrasts* = *list*(A = contr.sum, C = contr.sum)). For the data in Table 23.2, the regression coefficients using effect coding are shown in Figure 23.5. The intercept is the grand mean of the scores, 66.67. The effect of being in cell A_1C_1 is to shift from the grand mean by the sum of the effects of A_1, C_1, and their interaction, $A_1{:}C_1$, which are given by the coefficients: $67.667 - 8.833 + 1.500 - 8.33 = 52.0$. The coefficients of cells involving the highest level of each factor (i.e., any cell involving A_3 or C_3) are found by recalling that the effects of *A* and *C* each sum to zero. This means that the effect of being in cell A_2C_3, for example, is a deviation from the grand mean equal to the effect of A_2 minus the sum of the effects of C_1 and C_2, and their interactions with A_2: $67.667 - 9.0 - 1.5 - 1.167 - 6.333 + 14.833 = 64.5$.

23.3.3 Nonorthogonal Designs

In standard ANOVA, all factors are treated as though they are qualitative categorical variables that are independent of one another (i.e., the level at which a score is located on one factor provides no information about its location on other factors). As we detailed in

Chapter 9, it is this independence or *orthogonality* that makes it possible to partition the variability in factorial designs into distinct, nonoverlapping components associated with main effects and interactions in ANOVA. However, orthogonality requires that cell frequencies be equal, and this requirement is not realistic in many research contexts. Unequal cell frequencies introduce correlations among the factors, and these correlations cause the variance components associated with the different effects to overlap. Although we discussed unequal *n* designs from an ANOVA perspective in Chapter 9 (see Section 9.8), the complexities of unequal *n* ANOVA are easier to understand within the multiple regression framework in which we expect variables to be correlated with one another.

Panel *a* of Table 23.3 contains data for a nonorthogonal design with factors *A* and *C*; Panel *b* contains the effect coding for the design. Because of the unequal cell frequencies,

Table 23.3 Effect coding for a nonorthogonal 3 × 3 design

(a) Data

	C_1	C_2	C_3
A_1	53	88	56
	51	63	42
		50	
		71	
A_2	55	48	79
	78	42	50
	39		62
A_3	79	80	69
	99	92	94
			80
			77

(b) Dummy variables formed by using effect coding

Effect:		*A*		*C*		*AC*			
	Y	X_1	X_2	X_3	X_4	X_5	X_6	X_7	X_8
A_1C_1	53	1	0	1	0	1	0	0	0
	51	1	0	1	0	1	0	0	0
A_1C_2	88	1	0	0	1	0	0	1	0
	63	1	0	0	1	0	0	1	0
	50	1	0	0	1	0	0	1	0
	71	1	0	0	1	0	0	1	0
A_1C_3	56	1	0	−1	−1	−1	0	−1	0
	42	1	0	−1	−1	−1	0	−1	0
A_2C_1	55	0	1	1	0	0	1	0	0
	78	0	1	1	0	0	1	0	0
	39	0	1	1	0	0	1	0	0
A_2C_2	48	0	1	0	1	0	0	0	1
	42	0	1	0	1	0	0	0	1
A_2C_3	79	0	1	−1	−1	0	−1	0	−1
	50	0	1	−1	−1	0	−1	0	−1
	62	0	1	−1	−1	0	−1	0	−1

(*Continued*)

Table 23.3 (Continued)

Effect:		A		C		AC			
	Y	X_1	X_2	X_3	X_4	X_5	X_6	X_7	X_8
A_3C_1	79	−1	−1	1	0	−1	−1	0	0
	99	−1	−1	1	0	−1	−1	0	0
A_3C_2	80	−1	−1	0	1	0	0	−1	−1
	92	−1	−1	0	1	0	0	−1	−1
A_3C_3	69	−1	−1	−1	−1	1	1	1	1
	94	−1	−1	−1	−1	1	1	1	1
	80	−1	−1	−1	−1	1	1	1	1
	77	−1	−1	−1	−1	1	1	1	1

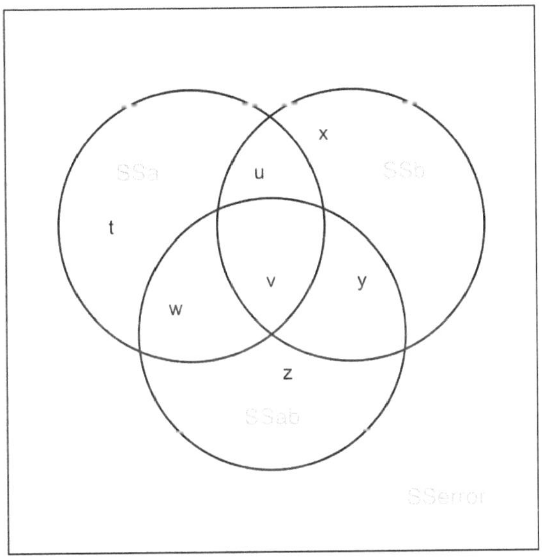

Figure 23.6 Partitioning of variability in a nonorthogonal two-factor design.

the sets of dummy variables that code the *A*, *C*, and *AC* effects are no longer uncorrelated. Therefore, in general,

$$R^2_{Y.A,C,AC} \neq R^2_{Y.A} + R^2_{Y.C} + R^2_{Y.AC}$$

because the variability associated with *A*, *C*, and *AC* overlaps, as represented by Figure 23.6. This situation creates ambiguity in interpreting effects because it is unclear how to attribute variability that is shared by two or more sources. Multiple regression analyses allow a variety of possible adjustments for this overlap, just as we saw in our discussion of ANOVA sums of squares. Next we consider three types of adjustments and make recommendations about the conditions under which each seems most appropriate.

Method 1

In considering the effect of A, we might decide to attribute to A only the variability *uniquely* associated with A. This is the variability in A that does not overlap with the other effects in the design and is represented by the area t in the upper circle of Figure 23.6. It can be obtained from

$$SS_{A|C,AC} = \left(R^2_{Y.A,C,AC} - R^2_{Y.C,AC}\right)SS_Y$$

where the subscript on the SS term on the left side of the equation is read "A adjusted for C and AC." When we adjust the sum of squares by removing the overlapping variability of all the other main or interaction effects, we obtain what are called *Type III sums of squares*. This is the default for SPSS, but not in R. You may wish to review Appendix 9.1, which describes how to obtain these different sums of squares.

Method 2

An alternative approach is to adjust the A effect only for the other main effect C, yielding the variability represented by areas t and w in the upper circle of 23.6,

$$SS_{A|C} = \left(R^2_{Y.A,C} - R^2_{Y.C}\right)SS_Y$$

where the subscript on the SS term to the left is read "A adjusted for C." In general, Method 2 adjusts any effect for all effects of the same order or lower order (e.g., main effects are adjusted for other main effects, but not for interactions). This method is known as Type II sums of squares.

Method 3

Finally, we may decide not to adjust for the contributions of the other effects at all. This yields

$$SS_A = R^2_{Y.A}SS_Y$$

In terms of Figure 23.6, the overlap of the circles is ignored; the effect of A is treated as the sum of areas t, u, v, and w. Using the terminology of Chapter 9, this is Type Ia sums of squares.

It should be clear that the three methods of analyzing effects in nonorthogonal designs will usually not produce equal estimates of effects of A. Because A will account for some variance in Y that is also accounted for by C and AC; SS_A, $SS_{A|C}$, and $SS_{A|C,AC}$ will generally not be equal. For example, for the data in Table 23.3, $SS_{A|C,AC} = 4{,}139.42$, $SS_{A|C} = 3{,}609.95$, and $SS_A = 3{,}581.08$. Depending on whether the covariations among the effects are positive or negative, the adjusted sum of squares may be smaller or larger than if there were no adjustment.[2]

The three methods of adjusting the sums of squares result in different tests of the hypotheses of interest. Table 23.4 describes the three methods of analyzing nonorthogonal factorial designs and indicates the hypotheses tested by each of them. In all three methods, interactions are adjusted for all other effects in the design, resulting in tests of the usual interaction null hypothesis

$$H_0: \mu_{jk} - \mu_{j'k} - \mu_{jk'} + \mu_{j'k'} = 0 \text{ for all } j, j', k, \text{ and } k'$$

However, the three methods produce different tests of the main effects, as would be expected given their different ways of attributing variance to A. Given this, on what basis are we to decide which approach, if any, to use? Our view is that the proper analysis depends upon what the researcher assumes to be the basis of the confounding of sources of variance in the design.

If the variables in the study are manipulated, or if the data in the cells can be viewed as samples from naturally occurring, equal-sized treatment populations, then the unequal cell sizes in the sample reflect random sampling error. That is, we may plan to have equal cell frequencies but fail to obtain them because of chance occurrences such as equipment failures or participants failing to show up. In this event, it makes sense to give each cell in the design the same weight. The corresponding overall A null hypothesis that is of interest states that the unweighted averages of the c population means at each level of A are equal. That is,

$$\mu_{1\cdot} = \mu_{2\cdot} = \ldots = \mu_{a\cdot}$$

where

$$\mu_{j.} = \frac{1}{c}\sum_k \mu_{jk}$$

is the unweighted mean of the cell means at the jth level of A. It can be shown (e.g., Myers & Well, 1995) that Method 1 tests this null hypothesis. Therefore, *Method 1 (which uses Type III sums of squares) is recommended when unequal cell frequencies occur by chance, as is usually the case in experimental designs*. This analysis is summarized in Table 23.4 and illustrated in Table 23.5.

An alternative scenario is that unequal cell sizes in a sample reflect systematic variation in the sizes of cell populations. This would often be the case in observational studies. If cell populations vary systematically in size, we may wish to test hypotheses in which the cell means are weighted according to population size. If we have reliable information about the relative sizes of the cell populations, we can use it to weight the cell means. If we do not have such information, we can use the cell frequencies as weights. In this case, the overall A null hypothesis of interest states that the weighted means of A are equal. That is,

$$\mu_{1^*} = \mu_{2^*} = \ldots = \mu_{a^*}$$

where

$$\mu_{j^*} = \frac{1}{n_{j\cdot}}\sum_k n_{jk}\mu_{jk}$$

Table 23.4 Three methods for analyzing nonorthogonal factorial designs

Method 1: Adjusting for all main and interaction effects

SV	*df*	*SS*
A	$a-1$	$SS_{A\mid C,AC}=\left(R^2_{Y.A,C,AC}-R^2_{Y.C,AC}\right)SS_Y$
C	$c-1$	$SS_{C\mid A,AC}=\left(R^2_{Y.A,C,AC}-R^2_{Y.A,AC}\right)SS_Y$
AC	$(a-1)(c-1)$	$SS_{AC\mid A,C}=\left(R^2_{Y.A,C,AC}-R^2_{Y.A,C}\right)SS_Y$
Residual	$N-ac$	$SS_{residual}=\left(1-R^2_{Y.A,C,AC}\right)SS_Y$

Hypotheses tested:

A: $\mu_{1.}=\mu_{2.}=\ldots=\mu_{a.}$ where $\mu_{j.}=\frac{1}{c}\sum_k\mu_{jk}$ is the unweighted mean of the population means for the *c* cells in the *j*th row of *A*

C: $\mu_{.1}=\mu_{.2}=\ldots=\mu_{.c}$ where $\mu_{.k}=\frac{1}{a}\sum_j\mu_{jk}$ is the unweighted mean of the population means for the *a* cells in the *k*th row of *C*

AC: $\mu_{jk}-\mu_{j'k}-\mu_{jk'}+\mu_{j'k'}=0$ for all j, k, j', k'

Usage: This method uses *Type III sums of squares* to test hypotheses about unweighted column and row means, and will usually be the method of choice when unequal cell frequencies occur by chance. This method is also known as Overall and Spiegel's (1969) method 1, Yates's (1934) weighted squares of means, and SPSS's classic regression approach.

Method 2: Adjusting for the effects of the same and lower order

SV	*df*	*SS*
A	$a-1$	$SS_{A\mid C}=\left(R^2_{Y.A,C}-R^2_{Y.C}\right)SS_Y$
C	$c-1$	$SS_{C\mid A}=\left(R^2_{Y.A,C}-R^2_{Y.A}\right)SS_Y$
AC	$(a-1)(c-1)$	$SS_{AC\mid A,C}=\left(R^2_{Y.A,C,AC}-R^2_{Y.A,C}\right)SS_Y$
Residual	$N-ac$	$SS_{residual}=\left(1-R^2_{Y.A,C,AC}\right)SS_Y$

Hypotheses tested:

$$A:\sum_k\left[n_{jk}-\frac{n_{jk}^2}{n_{.k}}\right]\mu_{jk}-\sum_{j\neq j'}\sum_k\left[\frac{n_{jk}n_{j'k}}{n_{.k}}\right]\mu_{j'k}=0 \quad \text{for } j=1,2,\ldots,a-1$$

$$C:\sum_j\left[n_{jk}-\frac{n_{jk}^2}{n_{j.}}\right]\mu_{jk}-\sum_{k\neq k'}\sum_j\left[\frac{n_{jk}n_{jk'}}{n_{j.}}\right]\mu_{jk'}=0 \quad \text{for } k=1,2,\ldots,c-1$$

AC: $\mu_{jk}-\mu_{j'k}-\mu_{jk'}+\mu_{j'k'}=0$ for all j, k, j', k'

Usage: Method 2 uses *Type II sums of squares*, which adjust an effect of interest for effects at the same or lower order, and for higher-order effects that do not include the effect of interest. If there is *no interaction*, Method 2 tests Method I's hypotheses with somewhat more power than Method 1 itself. However, if there is the possibility of an interaction, Method 2 should be avoided because it tests data-dependent hypotheses that are not useful (see the hypotheses that are tested above). This method is also known as Overall and Spiegel's method 2, Yates's fitting constants method, and SPSS's classic experimental design approach.

Method 3: Main effects not adjusted

SV	*df*	*SS*
A	$a - 1$	$SS_A = R^2_{Y.A}SS_Y$
C	$c - 1$	$SS_C = R^2_{Y.C}SS_Y$
AC	$(a - 1)(c - 1)$	$SS_{AC\|A,C} = \left(R^2_{Y.A,C,AC} - R^2_{Y.A,C}\right)SS_Y$
Residual	$N - ac$	$SS_{residual} = \left(1 - R^2_{Y.A,C,AC}\right)SS_Y$

Hypotheses tested:

A: $\mu_{1*} = \mu_{2*} = \ldots = \mu_{a*}$ where $\mu_{j*} = \frac{1}{n_{j.}}\sum_k n_{jk}\mu_{jk}$ is the weighted mean of the population means for the *c* cells in the *j*th row of *A*

C: $\mu_{*1} = \mu_{*2} = \ldots = \mu_{*c}$ where $\mu_{*k} = \frac{1}{n_{.k}}\sum_k n_{jk}\mu_{jk}$ is the weighted mean of the population means for the *a* cells in the *k*th row of C

AC: $\mu_{jk} - \mu_{j'k} - \mu_{jk'} + \mu_{j'k'} = 0$ for all *j*, *k*, *j'*, *k'*

Usage: This method tests main-effect hypotheses about the weighted row and column means. These tests may be desirable if the cell frequencies are proportional to the corresponding population sizes. This method is also known as Yates's method for proportional cell sizes.

Table 23.5 Results obtained by using the three methods with the data of Table 23.3

SV	*df*	*SS*	*MS*	*F*
Method 1				
A	2	$SS_{A\|C,AC} = 4{,}139.423$	2,069.712	11.639
C	2	$SS_{C\|A,AC} = 21.074$	10.537	0.059
AC	4	$SS_{AC\|A,C} = 1{,}103.074$	275.769	1.551
Residual	15	$(1 - R^2_{Y.A,C,AC})SS_Y = 2{,}667.833$	177.882	
Method 2				
A	2	$SS_{A\|C} = 3{,}609.950$	1,804.975	10.150
C	2	$SS_{C\|A} = 60.468$	30.234	0.170
AC	4	$SS_{AC\|A,C} = 1{,}103.074$	275.769	1.551
Residual	15	$(1 - R^2_{Y.A,C,AC})SS_Y = 2{,}667.833$	177.882	
Method 3				
A	2	$SS_A = 3{,}581.083$	1,790.542	10.089
C	2	$SS_C = 31.601$	15.800	0.089
AC	4	$SS_{AC\|A,C} = 1{,}103.074$	275.769	1.551
Residual	15	$(1 - R^2_{Y.A,C,AC})SS_Y = 2{,}667.833$	177.882	

is the weighted mean of the cell means at the *j*th level of *A*. Method 3 provides tests of main effect hypotheses based on weighted row and column means. This analysis is summarized in Table 23.4 and illustrated in Table 23.5.

Finally, under what conditions might Method 2 be used? The Method 2 approach that has been favored by some statisticians (e.g., Cramer & Appelbaum, 1980) corresponds to a hierarchical series of model tests that starts with higher-order effects. On the rationale that main effects are not very meaningful in the presence of an interaction, this approach first tests the interaction by comparing the model

$$Y_{ijk} = \mu + \alpha_j + \gamma_k + (\alpha\gamma)_{jk} + \varepsilon_{ijk}$$

against

$$Y_{ijk} = \mu + \alpha_j + \gamma_k + \varepsilon_{ijk}$$

If there is no interaction, tests of the main effects are then conducted by comparing

$$Y_{ijk} = \mu + \alpha_j + \gamma_k + \varepsilon_{ijk}$$

against

$$Y_{ijk} = \mu + \gamma_k + \varepsilon_{ijk}$$

and

$$Y_{ijk} = \mu + \alpha_j + \gamma_k + \varepsilon_{ijk}$$

against

$$Y_{ijk} = \mu + \alpha_j + \varepsilon_{ijk}$$

The Method 2 tests correspond exactly to these model comparisons. The advantage of this approach is that if there is no interaction, tests of main effects are somewhat more powerful than if Method 1 is used.

Despite the potential power advantage that Method 2 may provide over Method 1, we do not recommend using Method 2. This is because if an interaction does exist, the hypotheses that it tests depend on the pattern of cell frequencies in ways that are of little, if any, interest. For example, the null hypothesis for the effect of *A* can be shown to be

$$\sum_k \left[n_{jk} - \frac{n_{jk}^2}{n_{\cdot k}} \right] \mu_{jk} - \sum_{j \neq j'} \sum_k \left[\frac{n_{jk} n_{j'k}}{n_{\cdot k}} \right] \mu_{j'k} = 0 \text{ for } j = 1, 2, \ldots, a - 1$$

(see Carlson & Timm, 1974). Another consideration is that Method 2 is likely to produce biased tests of the main effects. If a preliminary test of the interaction does not reject the null hypothesis, tests of the main effects are conducted and based on an assumption that the interaction is zero. But remember that failure to reject a null hypothesis runs the risk of a Type 2 error. Thus, even interactions that do not approach significance can result in biased tests of the main effects (see, for example, Overall, Lee, & Hornick, 1981). Therefore, we believe that Method 2 should not be used unless there is strong *a priori* reason to assume no interaction effects, and that a test for the interaction is not significant.

The three methods considered so far are not the only possible approaches to analyzing data from a nonorthogonal design. It is possible that a logical or theoretical analysis of the research problem might dictate the order in which the sets of dummy variables are entered into the regression equation and, therefore, the nature of the adjustments. Suppose, for example, that *A* and *C* indicate levels of child and parental educational achievement, respectively. It is reasonable to assume that parental education may influence a child's educational achievement but the reverse type of influence is quite rare. In this case, it may be desirable to consider the unadjusted effects of parental educational achievement but to adjust the effects of the child's education for that of the parents. Hierarchical designs produce what are called *Type I sums of squares.*

In summary, the ability to use categorical variables in multiple regression analyses enables us to adjust sums of squares in ways that result in tests of the hypotheses of factorial ANOVA in nonorthogonal designs. We can use the Method 1 approach to test hypotheses about unweighted cell means and the Method 3 approach to test hypotheses about weighted means. Also, if we have logically or theoretically determined orderings of factors, we can perform sequential adjustments.

23.4 Testing Homogeneity of Regression Slopes Using Multiple Regression

In Chapters 19 and 22, we discussed tests for equality of slopes when one variable was regressed on another at different levels of a dichotomous (i.e., having only two levels) categorical variable. We selected participants in the *Seasons* data set aged 65 years or less with *BMI* scores no larger than 40 and tested whether the rate of change of total cholesterol level (*TC*) with *age* was the same for participants identified as men and women. Although we did not use the term there, the variable *sex* in the data set is a dummy variable that labels men by 0s and women by 1s. We showed that we could test the hypothesis that the rate of change was the same for men and women by regressing *TC* on *age*, *sex*, and an additional variable *Age* × *Sex* that was formed by multiplying *age* by *sex*. The resulting regression equation is

$$\widehat{TC} = 221.143 - 0.0005\ age - 94.612\ sex + 1.813\ Age \times Sex$$

Because the coefficient of the interaction term *age* × *sex* is significant, $t(363) = 4.909$, $p = .000$, we can reject the hypothesis that the slope of *TC* with *age* is equal for men and women, consistent with what we saw in Figure 22.5.

What if the categorical variable has more than two levels? If so, we can still test the hypothesis of homogeneity of regression slopes by determining whether there is a significant interaction. That is, we can test whether the regression of *Y* on a continuous variable, *X*, depends on the level of *A* when *A* is a categorical variable with more than two levels. However, because now more than one dummy variable is needed to code the categorical variable, the interaction term will have more than 1 *df*, and we will need to use a partial *F* test to determine whether the interaction is significant.

Consider, for example, the categorical variable education level, *EL*. We might be interested not only in its main effect on *TC*, but also in its interaction with *age*. Let *EL* = 1 correspond to individuals with a high school education or less (*schoolyr* = 1, 2, or 3), *EL* = 2 to education beyond high school but not including the bachelor's degree (*schoolyr* = 4, 5, or 6), and *EL* = 3 to at least a bachelor's degree (*schoolyr* = 7 or 8). For the moment, treat *EL* as a qualitative, categorical variable. *EL* seems to make a difference in *TC*. Using the same participants as the previous analysis, mean *TC* is 227, 217, and 210 for *EL* = 1, 2, and 3, respectively, and an ANOVA with *EL* as the independent variable is significant, $F(2, 359) = 4.908$, $p = .008$.

Now let's consider whether the effects *of EL* on *TC* depend on the age of the participants. Because *EL* has three levels, we can code it with two dummy variables – it does not matter whether we use effect, dummy, or any other kind of dummy variable coding. We can then code the interaction of *EL* with *age* by using two additional dummy variables, obtained by multiplying the two dummy variables used to code *EL* by *age*. If we have a continuous predictor variable *X* and a categorical variable *A* with *a* levels, the proportion of the variability in *Y* accounted for by a regression on *X* and the $a - 1$ dummy variables coding the categorical variable *A* is $R^2_{Y.X,A}$. If we now add the $a - 1$ dummy variables that account for the AX interaction to the regression, the proportion of variance accounted for is $R^2_{Y.X,A,AX}$. The AX interaction can be tested by the partial F

$$F = \frac{\left(R^2_{Y.X,A,AX} - R^2_{Y.X,A}\right)SS_Y / (a-1)}{\left(1 - R^2_{Y.X,A,AX}\right)SS_Y / (N-2a)}$$

The numerator of the expression corresponds to the increment in the amount of variability accounted for by the interaction and has *df* equal to the number of dummy variables needed to code the interaction. The denominator is the amount of variability not accounted for by the regression equation that contains the interaction, divided by $df = N - 1 -$ (the number of predictors in the regression equation) $= N - 1 - [1 + 2(a - 1)] = N - 2a$. The bracketed quantity, $1 + 2(a - 1)$, represents the sum of the $a - 1$ *df* for the regression of Y on A, the $a - 1$ *df* for the AX interaction, and 1 *df* for the regression of Y on X.

Using the *General Linear Model* module of SPSS to test the interaction, we would specify that EL is a categorical variable, that *age* is a continuous variable (treated as a covariate), and that we wanted to include the $EL \times$ *age* interaction in the model. Using R, we define EL as a factor and apply the effect coding method described in Section 23.3.2 and Type 3 sums of squares as detailed in Appendix 9.1. The R output for the analysis is provided in Figure 23.7. The $EL \times$ *age* interaction is not significant; $F(2, 356) = 0.641$, $p = .528$.

```
> Anova(lm(data = dat, tc ~ EL,contrasts = list(EL = contr.sum)),type = 3)
Anova Table (Type III tests)

Response: tc
             Sum Sq  Df    F value    Pr(>F)
(Intercept) 16202490   1 10415.8731 < 2.2e-16 ***
EL             15270   2     4.9083  0.007888 **
Residuals     558445 359
---
Signif. codes:  0 '***' 0.001 '**' 0.01 '*' 0.05 '.' 0.1 ' ' 1


> dat$EL <- as.factor(dat$EL)
> Anova(lm(data = dat, tc ~ EL*age,contrasts = list(EL = contr.sum)),type = 3)
Anova Table (Type III tests)

Response: tc
            Sum Sq  Df  F value    Pr(>F)
(Intercept) 498491   1 335.1596 < 2.2e-16 ***
EL            2150   2   0.7228    0.4861
age          25242   1  16.9713 4.723e-05 ***
EL:age        1905   2   0.6406    0.5276
Residuals   529488 356
---
Signif. codes:  0 '***' 0.001 '**' 0.01 '*' 0.05 '.' 0.1 ' ' 1
```

Figure 23.7 R output for the test of (a) the *EL* factor, (b) the *EL* × *age* interaction for participants with *age* ≤ 65 years and *BMI* ≤ 40. Note that here *EL* is treated as a qualitative categorical variable with three levels.

Therefore, we cannot reject the null hypothesis that the effects of *EL* do not vary with *age* (or, equivalently, that the rate of change of *TC* with *age* is the same at each level of *EL*).

The procedure can be readily extended to factorial designs. If we had *Y* and *X* scores for each cell of a 2 × 4 design with factors *A* and *C*, we could test the homogeneity of the regression slope in the eight cells of the design by performing analyses that produced partial *F* tests of the *XA*, *XC*, and *XAC* interactions.

Note that although we have been treating *EL* as a qualitative categorical variable, we could treat it as a crude quantitative variable. Although it has only three levels, they are ordered. If we regress *TC* on *EL*, treating it as a quantitative variable, we find that *EL* is significant, $b = -8.162$, $t(360) = -3.114$, $p = .002$; people with more education tend to have lower cholesterol levels. We can also test the interaction of *EL* with *age* by regressing *TC* on *age*, *EL*, and the product of *age* and *EL*. There is no evidence of an *age* × *EL* interaction, $b = -0.057$, $t(358) = -0.233$, $p = .816$.

23.5 Coding Designs With Within-Participants Factors

In a repeated-measures design, each participant is tested at every level of at least one independent variable, and participants are considered to define levels of a factor, *S*, in the design. If there are *n* participants, we can code *S* with $n - 1$ dummy variables in the same way as any other categorical variable.

Table 23.6 contains data for an *S* × *A* design with data from eight participants at four levels of the repeated-measures factor, *A*. Using effect coding, we code the eight levels of *S* with the seven dummy variables labeled in Table 23.6 as S_1–S_7, the four levels of *A* with three dummy variables, and the *SA* interaction with 21 dummy variables (SA_{11}–SA_{73}) formed by multiplying every dummy variable in the *S* set by every dummy variable in the *A* set. The sums of squares can be found from

$$SS_S = R^2_{Y.S} SS_Y$$

$$SS_A = R^2_{Y.A} SS_Y$$

and

$$SS_{SA} = R^2_{Y.SA} SS_Y$$

From our earlier coverage of repeated-measures designs, we know that for an *S* × *A* design, the appropriate test for the *A* main effect is given by $F = MS_A/MS_{SA}$.

The coding procedure can be directly extended to designs in which there are several within-participants variables, although the number of dummy variables required increases rather dramatically. If we had an *S* × *A* × *B* design with eight participants, four levels of *A* and two of *B*, coding all the main effects and interactions would require $abn - 1 = 63$ dummy variables, as many dummy variables as *df*s for each source of variance. However, if a multiple regression program was used to analyze such a design, one would only have to code the *S*, *A*, and *B* effects. Most software packages have some sort of compute or transform instruction that will create the variables needed to code the interaction effects.

Finally, the coding procedures can be extended to mixed designs that contain both within-participants and between-participants factors. Panel *a* of Table 23.7 contains a set

Table 23.6 Data and effect coding for an $S \times A$ design

Subject	A_1	A_2	A_3	A_4
1	1.4	3.2	3.2	3.0
2	2.0	2.5	3.1	5.8
3	1.4	4.2	4.1	5.6
4	2.3	4.6	4.0	5.9
5	4.7	4.8	4.4	5.9
6	3.2	5.0	6.2	5.9
7	4.0	6.8	4.5	6.5
8	5.0	6.1	6.4	6.6

		S							A			$S \times A$			
	Y	S_1	S_2	S_3	S_4	S_5	S_6	S_7	A_1	A_2	A_3	SA_{11}	SA_{21}	..	SA_{73}
A_1	1.4	1	0	0	0	0	0	0	1	0	0	1	0	...	0
	2.0	0	1	0	0	0	0	0	1	0	1	0	1	...	0
	1.4	0	0	1	0	0	0	0	1	0	0	0	0	...	0
	2.3	0	0	0	1	0	0	0	1	0	0	0	0	...	0
	4.7	0	0	0	0	1	0	0	1	0	0	0	0	...	0
	3.2	0	0	0	0	0	1	0	1	0	0	0	0	...	0
	4.0	0	0	0	0	0	0	1	1	0	0	0	0	...	0
	5.0	−1	−1	−1	−1	−1	−1	−1	1	0	0	−1	−1	...	0
A_2	3.2	1	0	0	0	0	0	0	0	1	0	0	0	...	0
	2.5	0	1	0	0	0	0	0	0	1	0	0	0	...	0
	4.2	0	0	1	0	0	0	0	0	1	0	0	0	...	0
	4.6	0	0	0	1	0	0	0	0	1	0	0	0	...	0
	4.8	0	0	0	0	1	0	0	0	1	0	0	0	...	0
	5.0	0	0	0	0	0	1	0	0	1	0	0	0	...	0
	6.8	0	0	0	0	0	0	1	0	1	0	0	0	...	0
	6.1	−1	−1	−1	−1	−1	−1	−1	0	1	0	0	0	...	0
A_3	3.2	1	0	0	0	0	0	0	0	0	1	0	0	...	0
	3.1	0	1	0	0	0	0	0	0	0	1	0	0	...	0
	4.1	0	0	1	0	0	0	0	0	0	1	0	0	...	0
	4.0	0	0	0	1	0	0	0	0	0	1	0	0	...	0
	4.4	0	0	0	0	1	0	0	0	0	1	0	0	...	0
	6.2	0	0	0	0	0	1	0	0	0	1	0	0	...	0
	4.5	0	0	0	0	0	0	1	0	0	1	0	0	...	1
	6.4	−1	−1	−1	−1	−1	−1	−1	0	0	1	0	0	...	−1
A_4	3.0	1	0	0	0	0	0	0	−1	−1	−1	−1	0	...	0
	5.8	0	1	0	0	0	0	0	−1	−1	−1	0	−1	...	0
	5.6	0	0	1	0	0	0	0	−1	−1	−1	0	0	...	0
	5.9	0	0	0	1	0	0	0	−1	−1	−1	0	0	...	0
	5.9	0	0	0	0	1	0	0	−1	−1	−1	0	0	...	0
	5.9	0	0	0	0	0	1	0	−1	−1	−1	0	0	...	0
	6.5	0	0	0	0	0	0	1	−1	−1	−1	0	0	...	−1
	6.6	−1	−1	−1	−1	−1	−1	−1	−1	−1	−1	1	1	...	1

Table 23.7 Dummy variable coding for a mixed design

(a) *Data and ANOVA table*

		C_1	C_2	C_3
	S_{11}	7	1	7
A_1	S_{21}	9	2	10
	S_{31}	7	3	8
	S_{12}	12	7	8
A_2	S_{22}	16	14	9
	S_{32}	19	11	12

SV	*df*	*SS*	*MS*	*F*
A	1	162.00	162.00	13.50
S/A	4	48.00	12.00	
C	2	85.33	42.67	17.66
AC	2	49.33	24.67	10.21
SC/A	8	19.33	2.42	

(b) Dummy variable coding for the design

		A	*S/A*				*C*		*AC*	
	Y	A_1	S/A_{11}	S/A_{12}	S/A_{21}	S/A_{22}	C_1	C_2	AC_{11}	AC_{12}
A_1S_1	7	1	1	0	0	0	1	0	1	0
	1	1	1	0	0	0	0	1	0	1
	7	1	1	0	0	0	−1	−1	−1	−1
A_1S_2	9	1	0	1	0	0	1	0	1	0
	2	1	0	1	0	0	0	1	0	1
	10	1	0	1	0	0	−1	−1	−1	−1
A_1S_3	7	1	−1	−1	0	0	1	0	1	0
	3	1	−1	−1	0	0	0	1	0	1
	8	1	−1	−1	0	0	−1	−1	−1	−1
A_2S_1	12	−1	0	0	1	0	1	0	−1	0
	7	−1	0	0	1	0	0	1	0	−1
	8	−1	0	0	1	0	−1	−1	1	1
A_2S_2	16	−1	0	0	0	1	1	0	−1	0
	14	−1	0	0	0	1	0	1	0	−1
	9	−1	0	0	0	1	−1	−1	1	1
A_2S_3	19	−1	0	0	−1	−1	1	0	−1	0
	11	−1	0	0	−1	−1	0	1	0	−1
	12	−1	0	0	−1	−1	−1	−1	1	1

of hypothetical data and the ANOVA table for a design that has one between-participants variable, *A*, and one within-participants variable, *C*, and Panel *b* presents dummy variables that code the design. The *A*, *C*, and *AC* sources of variance can be coded as in a factorial between-participants design, and *S/A* can be directly represented by coding participants separately at each level of *A*, as indicated in Table 23.7. It is not really necessary to code *SC/A* because $SS_{SC/A}$ can be obtained as a residual

$$SS_{SC/A} = SS_Y\left(1 - R^2_{Y.A,S/A,C,AC}\right)$$

However, *SC/A* could be coded by the eight dummy variables that would result from multiplying the values of variables in the C and *S/A* sets.

It should be noted that when there are different numbers of participants at each level of *A*, the standard ANOVA and GLM modules in SPSS will default to using Type III sums of squares. For example, SS_C will be obtained as $(R^2_{Y.A,C,AC} - R^2_{Y.C,AC})SS_Y$, not as $R^2_{Y.C}SS_Y$.

23.6 Summary

The two goals we had in this chapter were to discuss how categorical variables can be coded so that they can be incorporated into the multiple regression framework and to reconsider, within this framework, several of the ANOVAs that we had discussed earlier. We did not include this second goal to encourage our readers to perform ANOVAs by coding categorical variables in terms of dummy variables and then using multiple regression – although they could do so if the standard ANOVA programs were not available. Rather, we believe that considering ANOVAs from the multiple regression perspective allows us to gain a deeper understanding of these analyses.

The generality and flexibility of the multiple regression framework offer some clear advantages. The standard ANOVA approach breaks down for disproportionate-*n* designs. Thinking in terms of multiple regression – a system in which nonorthogonality is the rule rather than the exception – facilitates consideration of the kinds of adjustments that might be made. To provide appropriate analyses of nonorthogonal designs, the standard "ANOVA" programs are really multiple regression programs. We hope that this chapter provides some understanding of how these programs might work and what options they allow. Finally, the ability to include categorical and continuous variables in the same analysis not only provides a framework for better understanding ANCOVA, but also makes it clear that it is not necessary – indeed, it is wrong – to transform inherently continuous variables into categorical ones (by, for example, using median splits) in order to analyze the data (see, for example, Fitzsimons, 2008).

Exercises

23.1 [Understanding dummy and effect coding] Test scores are obtained from eight women and eight men. The data are as follows:

Gender	
Men	*Women*
27	35
18	33
16	26
27	21
24	38
30	28
32	38
26	32

a) Find the correlation between sex and test score (i.e., the point-biserial correlation coefficient). Use a dummy variable for which men are given 1s and women are given 2s. Test the correlation for significance.
b) Perform an independent-groups t test to determine whether there is a significant effect of sex.
c) How many variables are needed to code for sex? Indicate how sex could be coded using (i) effect (1, −1) and (ii) dummy (1, 0) coding.
d) Regress the dependent variable on the dummy (i.e., indicator) variables for (i) and (ii) above. Compare the significance levels with those found in (a) and (b). What are the interpretations of the regression coefficients for (i) and (ii)?

23.2 [Using different values for dummy variables] Create another dummy variable for sex, using 33 for men and −17 for women. Regress the dependent variable on this "nonsense" variable. What is the interpretation of this analysis?

23.3 [Interpreting regression coefficients with dummy variables] Given the following data from a between-participants design:

Condition		
C_1	C_2	C_3
17	11	9
33	18	12
26	14	10
27	18	8
21		14

a) How many linearly independent dummy (indicator) variables are needed to code the design?
b) Code the design, using (i) dummy coding and (ii) effect coding.
c) Regress the dependent variable on the indicator variables for (i) and (ii).
d) What are the interpretations of the regression coefficients in each case?

23.4 [Orthogonal vs nonorthogonal designs] Would the interpretations of the regression coefficients in Exercise 23.3 change if there were equal numbers of scores in each group? If so, how?

23.5 [Coding dummy variables] For the data set presented in Exercise 23.3, several coding schemes have been proposed. In each case, indicate whether regressing on the proposed set of dummy variables will result in $SS_{regression} = SS_{between}$. If a set of coding variables is not appropriate, indicate why it is not.

Condition	*Set 1*		*Set 2*		*Set 3*		*Set 4*		*Set 5*	*Set 6*		
	X_1	X_2	X_1	X_2	X_1	X_2	X_1	X_2	X_1	X_1	X_2	X_3
C_1	1	0	1	0	11	23	1	3	1	1	0	5
C_2	0	1	0	1	16	9	2	6	2	0	1	7
C_3	−1	−1	0	0	41	0	3	9	3	−1	−1	8

23.6 [The relationship between tests and hypotheses] Given the following data from a 2×3 nonorthogonal (i.e., unbalanced or unequal n) design:

	B_1	B_2	B_3
	72	49	40
	63	71	49
A_1	57	63	36
	52	48	50
	69		54
	75		
	65	56	41
	45	55	42
A_2	53	49	57
	52	52	39
	57	45	
	57		

a) Perform an ANOVA directly, using software.
b) Code the design, using effect coding (include variables for the main effects and the interaction).
c) Are the dummy variables that correspond to the different effects correlated with one another?
d) Assuming that the unequal *n*s have arisen by chance and we wish to test hypotheses about the unweighted means, perform the appropriate regression analyses and do what has to be done to test the A, B, and $A \times B$ effects. Exactly what hypotheses are tested? Compare your results with the ANOVA performed in (a).
e) Suppose you regress on just the dummy variable corresponding to the A effect, omitting the dummy variables that code for B and the $A \times B$ interaction. What hypothesis is tested by using the SS_A obtained in this analysis?

23.7 [Using regression and ANOVA on the same data] The data file *EX23_7* contains information about a dependent variable, Y, a categorical factor, A, and a predictor variable, X. We have added two dummy variables X_1 and X_2 to code the factor A, and two additional dummy variables to code the interaction between A and X, $X_1 \times X$ and $X_2 \times X$.

a) Use ANOVA to test whether the slope of the regression of Y on X is homogeneous across the levels of the A factor.
b) Perform the equivalent analysis, using the regression module and the dummy variables.

Notes

1 The term "dummy variables" is used to refer to the variables that code a categorical variable by using any of the coding procedures. The term "dummy coding" is used to refer to a specific kind of coding in which each dummy variable takes on the values 0 and 1 as described in the text.
2 This example points out that the use of overlapping circles to represent overlapping variabilities, as in Figure 23.6, is of limited utility. Here, the adjusted variabilities are larger than the unadjusted ones – something not obvious from the figure. This is because the representation does not distinguish between positive and negative covariation.

Chapter 24

ANCOVA as a Special Case of Multiple Regression

24.1 Overview

In Chapter 12, we briefly introduced analysis of covariance (ANCOVA) as one of several procedures that used information about a covariate (sometimes called a concomitant variable) to account for some of the error variance, thereby increasing the power of statistical tests on the factors of interest. Because ANCOVA is a procedure for analyzing data for a design that typically involves both categorical variables and quantitative variables, it is more naturally discussed as a type of regression rather than in terms of ANOVA. Therefore, we return to ANCOVA in this chapter to develop it more fully. Our goals in Chapter 24 are the following:

- *Discuss the rationale for ANCOVA.*
- *Introduce ANCOVA for a one-factor design as an example of multiple regression in which both quantitative and qualitative categorical variables are used as predictor variables.*
- *Present an example of the use of ANCOVA along with statistical package output.*
- *Discuss how group means can be adjusted for differences in the covariate.*
- *Consider the assumptions that underlie ANCOVA.*
- *Indicate how power calculations differ for ANOVA and ANCOVA.*
- *Consider ANCOVA for factorial designs.*
- *Briefly consider ANCOVA with more than one covariate and nonlinear ANCOVA.*

24.2 The ANCOVA Model

Treatments of ANCOVA from an analysis of variance perspective are notoriously opaque, primarily because regression is used to remove variability in the dependent variable predictable from the covariate. In Chapter 23, we showed that we can use dummy variables to code a categorical variable (say, factor A) and then use regression to perform an ANOVA. The ANOVA can be thought of a comparison between two models, a full model that contains information about group membership

$$Y_{ij} = \mu + \alpha_j + \varepsilon_{ij(ANOVA)} \tag{24.1}$$

and a restricted model that does not,

$$Y_{ij} = \mu + \varepsilon_{ij} \tag{24.2}$$

DOI: 10.4324/9781003453550-28

We can obtain $SS_{between}$ as

$$SS_{between} = SS_A = R^2_{Y.A} SS_Y$$

where $R^2_{Y.A}$ is the square of the multiple correlation coefficient that results when Y is regressed on a complete set of $a - 1$ dummy variables that code factor A. The within-group variability can be obtained as

$$SS_{within} = SS_{S/A} = \left(1 - R^2_{Y.A}\right) SS_Y$$

so that the test statistic for A is given by

$$F = \frac{MS_A}{MS_{S/A}} = \frac{R^2_{Y.A} SS_Y \,/\, (a-1)}{\left(1 - R^2_{Y.A}\right) SS_Y \,/\, (N-a)}$$

The null hypothesis tested is that the population means of Y are identical for each of the a levels of the factor A.

Performing a one-factor ANCOVA can be thought of as determining whether the categorical factor A has effects over and above those of a covariate, X, that is correlated with the dependent variable; or, equivalently, whether there are effects of A if we statistically adjust for differences in X. The hypothesis we wish to test is whether the Y population means would be identical for each of the a levels of the factor A *if each participant had the same value on* X. The ANCOVA can be thought of a comparison between two models in which Y is regressed on the covariate X:

1. A full ANCOVA model that also contains information about group membership

$$Y_{ij} = \mu + \alpha_j + \beta_{S/A}\left(X_{ij} - \bar{X}_{..}\right) + \varepsilon_{ij(ANCOVA)} \tag{24.3}$$

where $\beta_{S/A}$ equals the weighted average of the slopes that would be obtained in separate regressions for each of the groups; i.e.,

$$b_{S/A} = \frac{\sum_j b_j SS_j}{\sum_j SS_j} \tag{24.4}$$

2. A restricted model that does not include information about group membership

$$Y_{ij} = \mu + \beta_{tot}\left(X_{ij} - \bar{X}_{..}\right) + \varepsilon_{ij} \tag{24.5}$$

where β_{tot} is estimated by the slope of the regression of Y on X, ignoring group membership.

The test statistic for the ANCOVA does not contain SS_A or $SS_{S/A}$, but rather the corresponding sums of squares that have been adjusted for the covariate, called the *adjusted sums of squares*:

$$SS_{A(adj)} = \left(R^2_{Y.X,A} - R^2_{Y.X}\right) SS_Y$$

and

$$SS_{S/A(adj)} = \left(1 - R^2_{Y.X,A}\right) SS_Y$$

so that the ANCOVA test statistic is

$$F = \frac{MS_{A(adj)}}{MS_{S/A(adj)}} = \frac{\left(R^2_{Y.X,A} - R^2_{Y.X}\right) SS_Y \,/\, (a-1)}{\left(1 - R^2_{Y.X,A}\right) SS_Y \,/\, (N-a-1)}$$

where the F is distributed on $a - 1$ and $a(n - 1) - 1$ *df*; note the loss of one degree of freedom because of the estimation of $b_{S/A}$. Table 24.1 summarizes the sources of variance, and their interpretation, for both one-factor ANOVA and ANCOVA.

Table 24.1 Explanation of the terms in one-factor ANOVA and ANCOVA outputs

SV	*df*	*SS*	*Explanation*
ANOVA: Compare $Y_{ij} = \mu + \alpha_j + \varepsilon'_{ij}$ to $Y_{ij} = \mu + \varepsilon_{ij}$			
A	$a - 1$	$R^2_{Y.A} SS_Y$	"Between" variability in Y (i.e., variability accounted for by group membership)
S/A	$N - a$	$\left(1 - R^2_{Y.A}\right) SS_Y$	"Within" variability in Y (i.e., the sum of the variabilities within each of the a groups)
Total	$N - 1$	SS_Y	Total variability in Y (i.e., variability about the mean)
ANCOVA: Compare $Y_{ij} = \mu + \alpha_j + \beta_{S/A}\left(X_{ij} - \bar{X}..\right) + \varepsilon'_{ij}$ to $Y_{ij} = \mu + \beta_{tot}\left(X_{ij} - \bar{X}..\right) + \varepsilon_{ij}$			
A (adj)	$a - 1$	$\left(R^2_{Y.X,A} - R^2_{Y.X}\right) SS_Y$	Variability accounted for by the full model over and above that accounted for by regression on X without regard for group membership
X	1	$\left(R^2_{Y.X,A} - R^2_{Y.A}\right) SS_Y = SS_{reg\ (within\ groups)}$ $= SS_{reg\,(group\ 1)} + \ldots + SS_{reg\,(group\ a)}$ $= b^2_{S/A} \Sigma SS_{X_j}$	Variability accounted for by the full model over and above that accounted for by group membership alone; equivalently, the sum of the variabilities accounted for by separate regressions in each group but with a common slope
S/A (adj)	$N - a - 1$	$\left(1 - R^2_{Y.X,A}\right) SS_Y = SS_{error(group\,1)} + \ldots$ $+ SS_{error(group\ a)} = \Sigma\left(SS_{Y_j} - b^2_{S/A} SS_{X_j}\right)$	Variability left unaccounted for by the full model; equivalently, the summed residual variability in the groups; i.e., the sum of the variabilities not accounted for by the a within-group regressions using the common slope, $b_{S/A}$
Total (adj)	$N - 2$	$\left(1 - R^2_{Y.X}\right) SS_Y = SS_Y - b^2_{tot} SS_X$	Variability left unaccounted for by regression of Y on X without regard for group membership

Note: If Y is regressed on X without regard to group membership, $SS_{reg} = r^2 SS_Y = b^2_{tot} SS_X$ and $SS_{error} = SS_Y - b^2_{tot} SS_X$. Also, the common slope, $b_{S/A}$, is equal to the weighted average of the slopes, (b_js), that would be obtained in separate regressions, where the weights are the sums of squares of X for the groups.

Several points about Equations 24.3 and 24.5 should be noted. First, the slope parameters in the two models are not the same: β_{tot} is estimated by the slope of the overall regression of Y on X, ignoring group membership, whereas $\beta_{S/A}$, the *common slope* for the within-group regressions, can be shown to be estimated by the weighted average of the slopes that would be obtained in separate regressions for each of the groups, as in Equation 24.4. Therefore, we cannot perform an ANCOVA by simply performing an ANOVA on the residuals of the first regression. The second point is a matter of notation. In ANCOVA, we use μ to refer to the mean of the Y scores and $\bar{X}$ to refer to the mean of the X scores because in the usual regression inference model, the standard assumption is that X is a fixed-effect variable and Y is random.

The reason to use ANCOVA is that using regression to remove the error variance predictable from a covariate may result in a smaller error term and a reduction in bias of the group means due to the error variance. Comparing the ANOVA and ANCOVA models, we see that

$$\varepsilon_{ij(ANOVA)} = \beta\left(X_{ij} - \bar{X}_{..}\right) + \varepsilon_{ij(ANCOVA)} = \varepsilon_{pred} + \varepsilon_{res}$$

That is, the error associated with the *i*th participant in the *j*th treatment condition is made up of two components, a component predictable from the covariate (ε_{pred}) and a residual component (ε_{res}). In the ANOVA for a completely randomized design, both components contribute to the error variance, so that $\sigma^2_{\text{ANOVA}} = \sigma^2_{pred} + \sigma^2_{res}$, whereas in ANCOVA, the predictable variability is removed from the error variance, leaving only the residual component.

To illustrate the advantage of ANCOVA compared to ANOVA, consider data from a study by Myers et al. (1983; see Table 24.2 and data set *pl_with_covariate.xlsx*) in which participants learned about probability. Participants were each assigned to one of three instructional conditions based on reading different types of text: standard text (S), low-explanatory text (LE), or high-explanatory text (HE). The dependent variable was the proportion correct on a test administered after the instruction. When an ANOVA was performed, the null hypothesis that the population means for the three types of text were identical (i.e., $\mu_S = \mu_{LE} = \mu_{HE}$) could not be rejected, $F(2, 45) = 2.965$, $p = .062$. However, when a pretest measure of quantitative ability was used as a covariate in an ANCOVA, the main effect of type of text was significant, $F(2, 44) = 5.081$, $p = .010$. The data are presented in Table 24.2, and R output for the ANOVA and ANCOVA is presented in Panels (a) and (b) of Figure 24.1.

Figures 24.2 and 24.3 may be useful in understanding the potential advantage in efficiency of ANCOVA over ANOVA. Figure 24.2 schematically indicates how differences between two treatment groups may be easier to observe if the variability in Y that is predictable from X can be removed. The two clouds of data points represent two treatment groups. As can be seen from the marginal distributions plotted on the right vertical axis, the Y scores for the two groups overlap considerably. However, the distributions of the deviations of the data points from the regression lines that are superimposed on the data would show much greater separation. We should have more power if we were to test for the vertical separation of the population regression lines than if we test for differences among the marginal means.

In Figure 24.3, the restricted model for two groups is illustrated in Panel (a) and the full model is represented in Panel (b). The test of the full against the restricted model asks whether we can account for more variability by including information about group

Table 24.2 Data for the one-factor ANCOVA example: Proportion correct in the Myers, Hansen, Robson, and McCann (1983) study along with the quantitative aptitude score (*X*) for each participant

Standard		*Low explanatory*		*High explanatory*	
Y	X	Y	X	Y	X
.083	46	.083	64	.333	61
.167	41	.167	58	.333	61
.250	50	.250	51	.333	67
.250	60	.250	66	.417	52
.250	53	.250	52	.417	29
.333	74	.333	56	.500	38
.333	69	.333	68	.500	58
.333	51	.417	47	.500	60
.417	68	.417	72	.583	74
.500	49	.417	59	.583	59
.500	65	.417	61	.583	72
.500	42	.500	77	.583	48
.500	59	.583	51	.583	66
.583	69	.583	83	.667	75
.583	80	.667	93	.667	81
.750	71	.833	97	.917	90
Mean = .396	59.188	.406	65.938	.531	61.938
	$\bar{Y}_{..} = .444$	$\bar{X}_{..} = 62.355$			
$s^2 = .031$	146.029	.038	225.929	.023	239.396

```
> Anova(lm(data = dat,Y ~ text),type = 3)          # type 3 SS
Anova Table (Type III tests)

Response: Y
            Sum Sq Df  F value    Pr(>F)
(Intercept) 4.5146  1 147.3463 8.578e-16 ***
text        0.1817  2   2.9647   0.06173 .
Residuals   1.3788 45
---
Signif. codes:  0 '***' 0.001 '**' 0.01 '*' 0.05 '.' 0.1 ' ' 1
> Anova(lm(data = dat,Y ~ text + X),type = 3)        # type 3 SS
Anova Table (Type III tests)

Response: Y
             Sum Sq Df F value    Pr(>F)
(Intercept) 0.01255  1  0.6225   0.43434
text        0.20490  2  5.0812   0.01034 *
X           0.49162  1 24.3834 1.183e-05 ***
Residuals   0.88714 44
---
Signif. codes:  0 '***' 0.001 '**' 0.01 '*' 0.05 '.' 0.1 ' ' 1
```

Figure 24.1 R output for the test of (a) an ANOVA with text type as the factor, (b) an ANCOVA with text type as the factor and *X* (pretest score) as covariate.

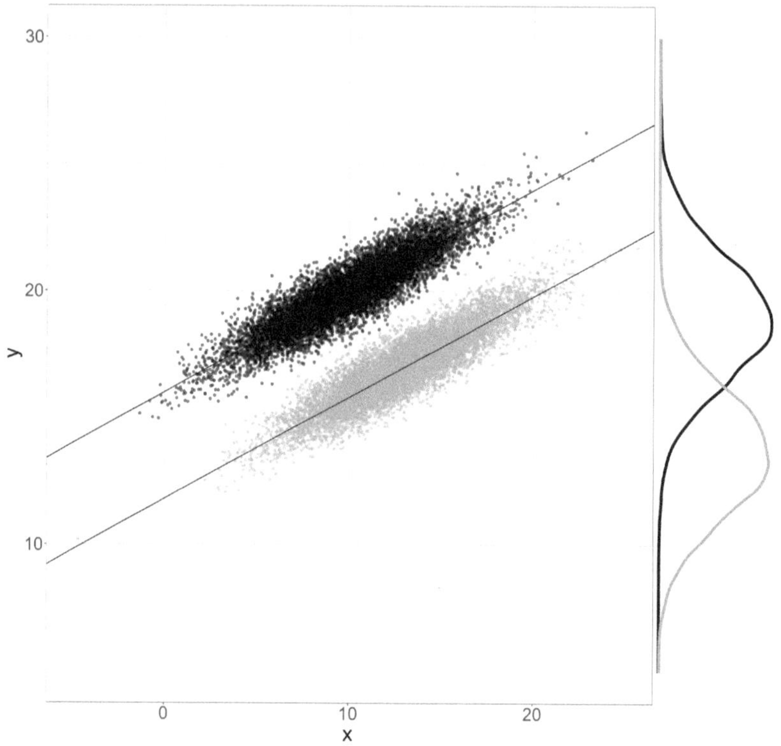

Figure 24.2 Marginal distributions of *Y* scores (right side) and the regression lines for two groups.

membership; or, equivalently, whether we can account for more variability by performing separate within-group regressions with a single, common slope than by performing a single overall regression without regard to group membership. Rejection of the null hypothesis implies that the regression lines for the different populations do not all lie on top of one another.

Using Software for ANCOVA

ANCOVA is straightforward to run using software. In SPSS, use the *General Linear Model* option in the *Analyze* pull-down menu, then select *Univariate*. For the example data in Table 24.2, move *Y* to the dependent variable box, *text* to the *Fixed Factor(s)* box, and *X* to the *Covariate(s)* box. If you wish to test the homogeneity of slopes, click on the *Model* button and create main effects for *X* and *text*, as well as their interaction. Then click Continue and OK. If you do not build a model, SPSS will test only the main effects of *X* and *text*.

In R, be sure that the treatment effect (*text*) is a factor (use *as.factor* if necessary). Then the *lm* function we are familiar with from previous chapters will perform the ANCOVA if you include the covariate in the model: *lm*(data = dat, Y ~ text + X). Store the result in ancova.out (or whatever name you prefer), and then use that as input to *Anova* with Type III sums of squares: *Anova*(ancova.out, type = 3). To test the homogeneity of slopes, you must tell R that *text* and *X* may interact: *Anova*(*lm*(data = dat, Y ~ text * X), type = 3).

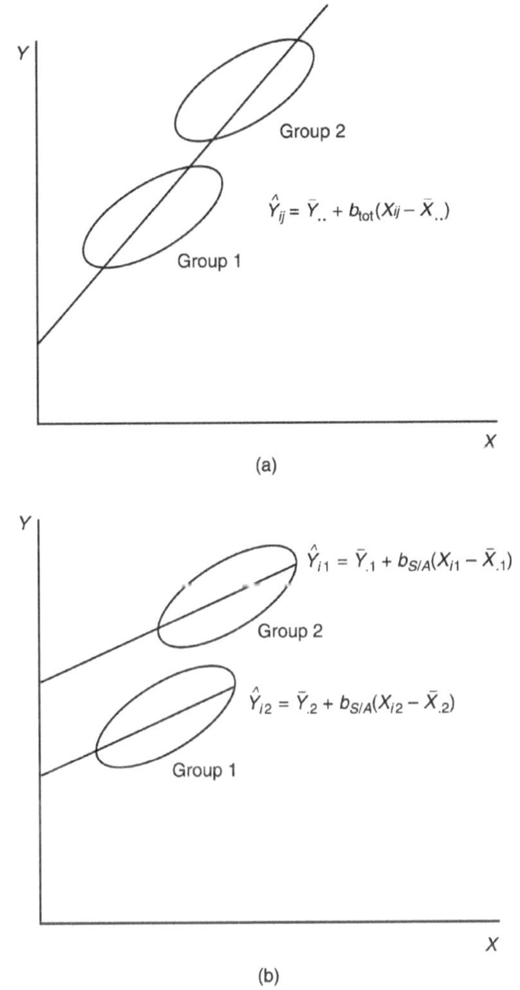

Figure 24.3 (a) Schematic representation of the regression corresponding to the reduced model of Equation 24.5. All the data are used to obtain a single regression line without regard to group membership. (b) Schematic representation of the regressions corresponding to the full ANCOVA model of Equation 24.3. Note that because the model contains only a single slope parameter, the regression lines for the groups have a common slope, $b_{S/A}$.

24.3 Adjusting the Group Means in *Y* for Differences in *X* and Testing Contrasts

In the previous section, we showed how SS_A and $SS_{S/A}$ could be adjusted for differences in the covariate. Similarly, in certain situations, it is both possible and desirable to adjust the group means for covariate differences. To do the adjustment, we predict what the group means for Y would be if the value of the covariate was held constant. We define the *adjusted mean* of the scores for group *j*, $\bar{Y}_{.j(adj)}$, as the score predicted in group *j*, using the within-group regression equation with common slope $b_{S/A}$ *if the value of the covariate is equal to the grand mean of the covariate scores*; that is, if $X_{ij} = \bar{X}_{..}$.

Because for bivariate regression of Y on X,

$$\hat{Y} = \bar{Y} + b_1(X - \bar{X})$$

the regression equation for scores in group j is

$$\hat{Y}_{ij} = \bar{Y}_{.j} + b_1\left(X_{ij} - \bar{X}_{.j}\right)$$

Substituting $b_{S/A}$ for b_1 and $X_{..}$ for X_{ij}, we have

$$\bar{Y}_{.j(adj)} = \bar{Y}_{.j} + b_{S/A}\left(\bar{X}_{..} - \bar{X}_{.j}\right)$$

or

$$\bar{Y}_{.j(adj)} = \bar{Y}_{.j} - b_{S/A}\left(\bar{X}_{.j} - \bar{X}_{..}\right)$$

The adjusted group means for the S, LE, and HE text conditions introduced in Section 24.2 are

$$\bar{Y}_{.1(adj)} = .369 - (.00732)(59.188 - 62.355) = .419$$

$$\bar{Y}_{.2(adj)} = .406 - (.00732)(65.938 - 62.355) = .380$$

and

$$\bar{Y}_{.3(adj)} = .531 - (.00732)(61.938 - 62.355) = .534$$

These three adjusted group means are the means predicted for each group for a covariate value equal to the grand mean on the covariate (i.e., 62.355 in this data set).

If the assumptions for ANCOVA are met, we can test contrasts based on the adjusted means that will generally be more powerful than those using unadjusted means. Suppose that working within the ANOVA framework, we wished to test the null hypothesis

$$H_0 : \psi = \mu_{HE} - \frac{\mu_S + \mu_{LE}}{2} = 0$$

Using the procedures described in Chapter 10, we would use the test statistic

$$t = \frac{\hat{\psi}}{s_{\hat{\psi}}} = \frac{\bar{Y}_{HE} - \dfrac{\bar{Y}_S + \bar{Y}_{LE}}{2}}{\sqrt{MS_{S/A}\Sigma\dfrac{w_j^2}{n_j}}} = \frac{.531 - \dfrac{.396 + .406}{2}}{\sqrt{(.031)(1.5/16)}} = .130/.054 = 2.411 \text{ with } 45\ df$$

and could reject the null hypothesis at $p = .019$.

An alternative approach to testing the contrast is to base it on the adjusted means (Huitema, 1980). The procedure is essentially the same as that for the unadjusted means, except that the comparison is among adjusted means and the error term contains corrections for the covariate. For a completely randomized design, the recommended test statistic is

$$t = \frac{\hat{\Psi}_{(adj)}}{s_{\hat{\Psi}(adj)}} = \frac{\bar{Y}_{HE(adj)} - \dfrac{\bar{Y}_{S(adj)} + \bar{Y}_{LE(adj)}}{2}}{\sqrt{MS_{S|A(adj)}\left(\Sigma \dfrac{w_j^2}{n_j}\right)\left(1 + \dfrac{MS_{A(X)}}{SS_{S/A(X)}}\right)}}$$

where $MS_{A(X)}$ and $SS_{S/A(X)}$ are the between-participant mean square and the within-participant sum of squares obtained when an ANOVA is with the covariate as the dependent variable and A as the treatment factor (see Figure 24.4). Substituting into the equation, we have

$$t = \frac{\hat{\Psi}_{(adj)}}{s_{\hat{\Psi}(adj)}} = \frac{.534 - \dfrac{.419 + 380}{2}}{\sqrt{(.020)(1.5/16)\left(1 + \dfrac{184.333}{9170.312}\right)}} = .135/.044 = 3.089 \text{ with } 44\ df$$

We can now reject the null hypothesis at $p = .003$. We obtained a larger value of the test statistic using the adjusted means, although we lost one error degree of freedom. In this example, the numerator of the t ratio is slightly larger for the adjusted means than for the unadjusted means. However, the main source of the difference in the two t values is attributable to the smaller error term after adjusting for the covariate (.044) than before the adjustment (.054).

The procedure we just illustrated for testing contrasts of adjusted means is appropriate when participants are assigned at random to treatments. However, in observational studies where participants are not randomly assigned to conditions, the recommended error term is different; specifically, it contains a correction that depends on the specific contrast that is tested,

$$S_{\hat{\Psi}(adj)} = \sqrt{MS_{S/A(adj)}\left(\Sigma \frac{w_j^2}{n_j} + \frac{\left(\Sigma w_j \bar{X}_{\cdot j}\right)^2}{SS_{S/A(X)}}\right)}$$

This equation follows from the expression we developed for the standard error of a predicted score when we considered bivariate regression. The implication for performing analyses is that researchers must select the error term that is appropriate to their research design. For a good discussion of these issues, see Huitema (1980).

```
> summary(aov(data=dat, X ~ text))
            Df Sum Sq Mean Sq F value Pr(>F)
text         2    369   184.3   0.905  0.412
Residuals   45   9170   203.8
```

Figure 24.4 R results of ANOVA with X as the dependent measures and *text* as the treatment factor.

As always, when several contrasts are tested, we should control Type 1 error across the set of comparisons. The procedures for controlling familywise error (*FWE*) rate were discussed in Chapter 10. If there are several planned contrasts, we can use the Dunn–Bonferroni method. For post hoc contrasts, if we wish to use the Scheffé test, the *t* statistics obtained earlier can be referred to the criterion $\sqrt{(a-1)F_{FWE,\,a-1,\,N-a-1}}$. For the Tukey post hoc test of pairwise differences, the same test statistics can be used with weights +1 and −1. If the covariate is a fixed-effect variable, the test statistic can be referred to $q_{FWE,\,a,\,df_{error}} / \sqrt{2}$, where q is a critical value of the Studentized range statistic that we used with the Tukey test in Chapter 10. There is one new consideration in the current context: if the covariate is a random variable, as is usually the case, Bryant and Paulson (1976) have shown that q should be replaced by $Q_{FWE,\,a,\,c,\,df_{error}}$, a value of the generalized Studentized range statistic in which c is the number of covariates. Tables of the generalized Studentized range statistic are available in Huitema (1980) and Kirk (1995) or using the *qtukey* function in R (see Section 10.7.1).

Using Software to Adjust Means and Test Contrasts

To adjust the means for the covariate in R, use the *effect* function in the {effects} package. It takes as input the results of an ANCOVA and information on which means are to be adjusted: *effect*("text", ancova.out) returns the adjusted means for each level of the text factor. In SPSS, set up the ANCOVA as described in Section 24.2, then click on the *EM Means* bar and move *text* to the box labeled *Display Means for*, the click Continue and OK. The adjusted means will appear as *Estimated Marginal Means* and a footnote to the table will inform you of the value on the covariate that was used for the adjustments (i.e., $X = 62.35$).

To compute contrasts in R, we can set up a matrix of contrasts as described in Box 10.1, saving the result in a matrix called *mat*. Then include the option *contrasts* = *list*(text = mat) in the *lm* function described in Section 24.2: *summary*(*lm*(data = dat, Y ~ text + X, *contrasts* = *list*(text = mat))) provides the contrast estimates using adjusted means, their standard errors, and significance tests.

In SPSS, the *text* conditions are ordered alphabetically (HE, LE, S), so we should be able compute the contrast using the Helmert contrasts that are available from the *Contrasts* button within the ANCOVA. However, doing so produces an erroneous value of the contrast, suggesting a bug in the software. This serves as a clear reminder that it is always a good idea to check software results against known values to confirm that the calculations are as expected. As an example, we confirmed that the Helmert contrasts in SPSS do accurately reproduce the contrast with unadjusted values from an ANOVA on these data; only the ANCOVA contrasts are faulty.

24.4 Assumptions and Interpretation in ANCOVA

When ANCOVA is used instead of ANOVA, increases in power may be achieved at the cost of greater complexity and more assumptions. The standard assumptions for ANCOVA break down into two groups. As in ordinary ANOVA, some assumptions are necessary for the ratio of adjusted mean squares to be distributed as *F*. However, unless certain additional assumptions are made, the ANCOVA *F* may test a different null hypothesis than ordinary ANOVA, and the adjusted means may be biased estimates of the population means. As with most other regression-related procedures, given that we have access to sophisticated statistical software, the computations are the least of our worries. As always, the greatest challenges lie in interpretation. We now discuss the assumptions underlying ANCOVA and the consequences of violating them.

24.4.1 Normality and Homogeneity of Variance

In ANCOVA, it is assumed that the conditional distributions of Y at different values of X are normal and have equal variances, just as in regression. In general, the consequences of violating assumption are similar to those for ANOVA, with the exception that they depend to some extent on the distribution of the covariate (see Huitema, 1980, for a more detailed discussion of these assumptions). ANCOVA is unlikely to be severely biased by violations of the normality and homogeneity of variance assumptions provided there are equal numbers of participants in each group and the covariate itself is approximately normally distributed.

24.4.2 Linearity

As we have so far discussed it, ANCOVA adjusts for differences in the covariate by removing the variability accounted for by a linear regression on X. If there is a systematic nonlinear component to the relationship between X and Y, the use of linear regression will not remove all the variability in Y potentially accounted for by X. The effect of moderate nonlinearity is a slight negative bias in the ANCOVA F test. However, strongly nonlinear relationships can result in severely biased F tests if linear ANCOVA is used (Atiqullah, 1964). Consider, for example, a situation in which there is a quadratic component to the relationship between Y and the covariate, X. If the curvilinearity is ignored and the X^2 that should be included in the model is omitted, it will contribute to the error component, thereby resulting in bias.

If the nature of the nonlinearities can be specified, transformations of Y or polynomial ANCOVA (see Section 24.7.3) may be used. It is recommended that the linearity assumption be checked as a preliminary step in using ANCOVA. Plotting the scatter diagrams for each group offers a quick check and a significance test for nonlinearity has previously been discussed.

24.4.3 Assumption of Homogeneity of Regression Slopes

In ANCOVA we adjust for differences in the covariate by using regression. In doing so, for each group we use a common slope, $b_{S/A}$, the pooled within-group regression coefficient, which is essentially an average of the slopes that would be obtained in separate within-groups regressions. This kind of adjustment makes the most sense if we can assume that the population slopes are the same in each of the groups. If the population group slopes differ, using any kind of "average" slope adjustment will not be appropriate for at least some of the groups. An analogy can be made between an adjusted A main effect in an ANCOVA in the presence of heterogeneous slopes and an A main effect in an ANOVA in the presence of an interaction between A and a second factor, B. If there is a large interaction, particularly if the curves cross, the F test of A may not adequately reflect the A effect at any level of B. Similarly, if the group regression coefficients vary, the effect of A is different for different values of X. Nonetheless, in ANOVA it is still useful to look at main effects even in the presence of interactions, and also to look at simple effects at different levels of the independent variables. Analogous procedures have been developed for ANCOVA.

It may help to describe the situation by using diagrams. When Y is adjusted for the effects of the covariate, X, treatment effects are interpreted in terms of vertical separation of the group regression lines instead of differences in the marginal group means. Suppose that the lines in each panel of Figure 24.6 represent regression lines obtained separately for two treatment groups. Parallel lines, illustrated in Panel (a), indicate that the treatment effects are the same for each value of the covariate. In this case, we can identify the treatment effects with the vertical distance between the lines at any value of X. On the other hand, nonparallel lines, illustrated in Panels (b) and (c), indicate that the treatment effects are not the same for each value of the covariate. In Panel (b), the two lines intersect at $X = \bar{X}_{..}$, but differ considerably for high and low values of X. In Panel (c), there is a separation between the two lines at $X = \bar{X}_{..}$, but the separation is larger for large X and smaller for small X. For Panels (b) and (c), it does not make sense to consider the average separation of the two lines; rather, it is of more interest to determine for what values of X, if any, the separations are significant.

If the slopes of the population regression lines are reasonably homogeneous as evidenced by roughly parallel lines in the data set, we may proceed with the standard ANCOVA and be confident of conclusions regarding differences between adjusted group means. In practice, researchers usually first test the homogeneity of regression slope by testing the interaction of the group factor with the covariate (see Section 23.4). The interaction test can be accomplished with software, assuming that *text* is the factor and X is the covariate and including the *text* × X interaction in the model (see Figure 24.5 for R output from this analysis). If the interaction test is not significant, as in these data, then the ANCOVA is justified. Many textbooks recommend abandoning ANCOVA if the test is significant. However, the strict dichotomy between equal and unequal slopes is an oversimplification of the problem of comparing regression lines. Small differences in slope may not seriously affect the validity of the ANCOVA and, for large samples, homogeneity may be rejected even when the slope differences are small (and conversely, for small samples, the test for homogeneity may not have much power). Therefore, it may be worth proceeding cautiously with ANCOVA even in the presence of modest heterogeneity.

If the group slopes are sufficiently different to cause concern about interpreting an effect of groups in the ANCOVA, it is appropriate to analyze the interaction of the treatment with

```
> Anova(lm(data = dat,Y ~ text * X),type=3)
Anova Table (Type III tests)

Response: Y
             Sum Sq Df F value  Pr(>F)
(Intercept) 0.00290  1  0.1395 0.71069
text        0.04765  2  1.1463 0.32755
X           0.13529  1  6.5091 0.01446 *
text:X      0.01416  2  0.3407 0.71325
Residuals   0.87298 42
---
Signif. codes:  0 '***' 0.001 '**' 0.01 '*' 0.05 '.' 0.1 ' ' 1
```

Figure 24.5 R output on the test of homogeneity of slope of Y on X at the different levels of *text*.

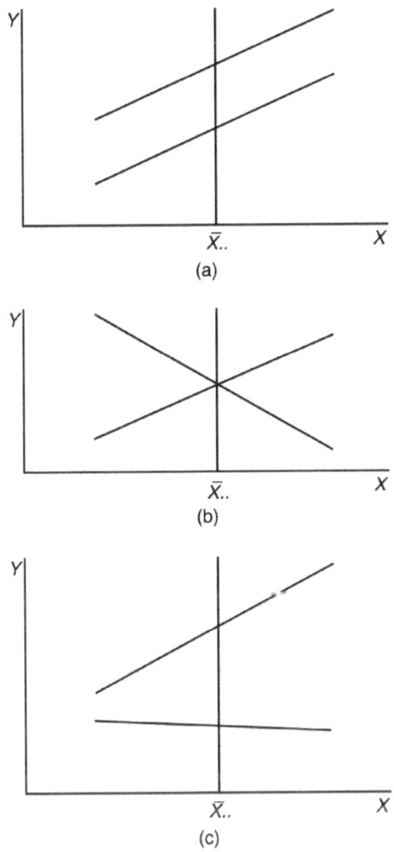

Figure 24.6 Examples of regressions with homogeneous and heterogeneous slopes.

the covariate. Thus, in situations like those depicted in Panels (b) and (c) of Figure 24.6, we would be interested in determining how the groups differ at different values of X. Johnson and Neyman (1936) have developed a procedure for establishing regions of significance on the covariate. The Johnson–Neyman technique and related procedures are described in sources such as Huitema (1980), Hunka and Leighton (1997), and Ragosa (1980). Ragosa (1980) has also developed procedures for testing the average separation between regression lines and the separation at particular values of the covariate that are appropriate both when the lines have equal slopes and when they do not. Although these procedures are potentially useful, we do not describe them here because they have not, to the best of our knowledge, been implemented in any of the standard statistical packages. However, a good description is presented in Maxwell and Delaney (2004).

24.4.4 The Assumption That the Treatment and Covariate Are Independent

Many researchers use ANCOVA in observational studies "to adjust for initial group differences in the covariate" in the mistaken belief that the statistical procedure effectively eliminates any confounding of the covariate and the grouping variable. Thus, we emphasize that *ANCOVA should not be used to analyze designs in which the covariate varies systematically with the treatment.* If the treatment influences the covariate or is otherwise systematically related to it, performing an ANCOVA will not simply reduce error variance; rather, it may adjust between-group differences in ways that are difficult to understand and that may lead to biased tests.

It was appropriate to use ANCOVA to test the effect of text condition for the Myers et al. (1983) experiment because participants were randomly assigned to the text conditions and there was no way that the treatment could influence the covariate. However, suppose everything else was kept the same except that participants were given the quantitative reasoning test *after* the experiment instead of before they read the text material. In this case, we should not use the quantitative reasoning test as a covariate because material given in the text conditions might affect the participant's quantitative reasoning score. If so, and we went ahead and used it as a covariate, the regression adjustment built into the ANCOVA would not only remove some of the error variance in the dependent variable, but *it would also remove some of the effect of the treatment.*

To illustrate another kind of problem, suppose that rather than randomly assigning participants to text conditions, the different text materials were presented to intact groups. For example, say the *LE* text was given to a class of psychology majors and the *S* and *HE* texts were given to classes of math and fine arts majors, thereby confounding text condition with major. We would say that we had a *nonequivalent-groups design.* As we noted at the start of this subsection, some researchers seem to believe that performing an ANCOVA can appropriately adjust the groups for their pre-existing differences. However, even though the groups may differ on the covariate, the confounding cannot be magically removed by performing an ANCOVA; you cannot salvage a bad experimental design with statistical analysis. The underlying groups may differ on many variables, and some of these differences are not likely to be fully predictable from a covariate. If treatments are applied to intact groups that differ from one another, ANCOVA presents the same kinds of difficulties that are always associated with interpreting the results of observational studies. Whenever the covariate varies systematically across conditions it becomes correlated with other variables that differ across groups, including the treatment itself. Performing an ANCOVA tends to adjust the effects of all these variables, but to different degrees. In some cases, adjusting for some kinds of differences might exacerbate others.

Consider another example in which the independent variable and covariate might be correlated. Suppose an experiment is conducted to evaluate three different teaching programs. Students are given material to study on their own, and then are tested. Suppose the mean test score for participants in Program 1 is higher than the means for Programs 2 and 3 and an ANOVA performed on the test scores is significant, suggesting that the three programs are not all equally effective. However, students assigned to Program 1 are observed to spend more time working with the material than students assigned to the other programs. If an ANCOVA using "study time" as the covariate reveals no significant differences, are we allowed to conclude that the three programs would be equally effective if study time was held constant? This interpretation is not necessarily correct. Statistically controlling for study time is not the same as experimentally controlling or manipulating it, so causal statements are not justified. We simply do not know from these data what would happen if study time was actually held constant. The materials used in Program 1 may be more understandable and interesting to work with than the materials used in the other programs. These qualities may be the cause of both the superior test performance and the greater study time. Using study time as a covariate will tend to remove the effects of any variables correlated with study time, including the characteristics of the program that are responsible for the superior performance. It is therefore entirely possible that Program 1 would produce superior performance even if study time was equated.

A related point addresses another approach that is sometimes used in the situation just described; namely, omitting observations in an attempt to equate groups. "Controlling" study time by throwing out data is not appropriate. Suppose that the mean daily study time is 40 minutes for the first group and 25 minutes for the other two groups. What if we analyze only the performance scores for students in the three groups who have comparable study times, say, 30–35 minutes? This is a poor strategy because in selecting students who have comparable study times, we may be selecting students who differ widely in other important characteristics. There is no reason to think that students in the first group whose study times are below that group's average are comparable in ability and motivation to students who have above-average study times in the other two groups. If we are interested in the effects of the program and of study time, there is no substitute for conducting a true experiment in which both variables are manipulated.

The assumption of independence of treatment and covariate will be met in a completely randomized design when covariates are measured before treatments have been administered. However, as already noted, measuring covariates during or after treatment can result in a relationship between the treatment and covariate even in a completely randomized design. In the case of nonequivalent-group designs, it is likely that the treatment and covariate will be correlated. In either of these situations, an ANOVA should be performed on the covariate to evaluate whether the assumption of independence of treatment and covariate has been met. If the *F* test demonstrates differences in the covariate across groups, the ANCOVA *F*s and adjusted means on the dependent variable will almost certainly be biased. Unfortunately, a nonsignificant ANOVA does not assure there will be no bias, but any bias is likely to be small.

24.4.5 Assumption That the Covariate Is Fixed and Measured Without Error

The standard model for making statistical inferences about regression assumes that the variable used to predict is a fixed-effect variable that is measured without error. In ANCOVA, this translates into the assumption that the covariate has these properties. However, the assumption that the covariate is a fixed-effect variable can generally be violated without serious consequences so long as we realize that if X is random, the statistical inference extends only to situations in which the distributions of X scores are comparable to those used in the current study.

If we randomly assign participants to treatments or assign them to treatments entirely based on their scores on the covariate, increased random error in both the dependent variable and covariate will result in reduced power but the results of the ANCOVA will not be biased. On the other hand, if we administer the different treatments to nonequivalent groups, measurement error in the covariate can result in increased bias and greater difficulties in interpretation. If the mean covariate value varies across groups and if the covariate is measured with error, the expected values of the adjusted *Y*s may differ even if the treatment has no effect. Figure 24.7 illustrates the problem for two groups in which the true scores of X and Y are perfectly correlated and there are no treatment effects. In Panel (a), X is measured without error and both group equations have the same slope and intercept. Therefore, the adjusted means for these groups must be the same and an ANCOVA would correctly reveal that there are no treatment effects. Panel (b) represents the same situation, except that now X is measured with error. The effect of the measurement error will be to "spread out"

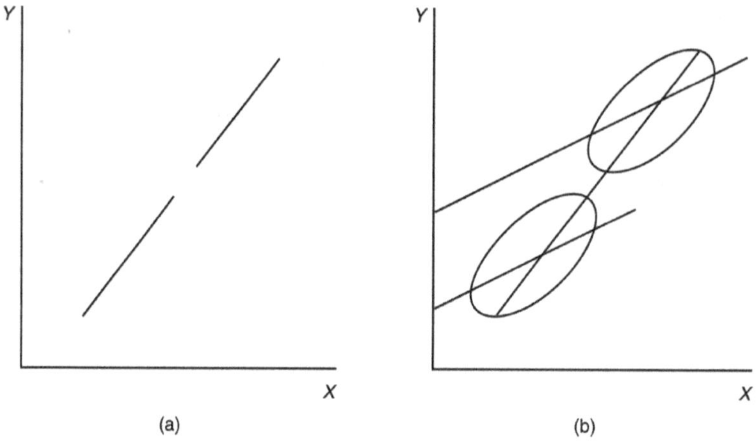

Figure 24.7 Effect of measurement error on regression slopes. Both samples are from populations in which the true scores of *Y* and *X* are perfectly correlated and no treatment is applied. *X* is measured without error in Panel (a) and with error in Panel (b).

the values of *X* and reduce the slopes for both groups (see Figure 22.1). As can be seen in Panel (b), the adjusted *Y* mean will now be larger for the group that has the larger values of *X*, even though there is no treatment effect. Because groups do not differ systematically from one another in randomized designs, measurement error in *X* will result in reduced power but will not cause the adjusted means to be biased. However, in nonequivalent-group designs, the measurement error in *X* may well introduce bias.

24.5 Using the Covariate to Assign Participants to Groups

If covariate values are available before an experiment is conducted, they can be used to assign participants to conditions. For example, if the independent variable has *a* levels, the *a* participants with the highest scores on the covariate can be randomly distributed across the treatment conditions, then the participants with the next highest *a* scores, and so on, much like a treatments × blocks design. Compared to randomly assigning participants to treatments without regard to the covariate values, this procedure will generally result in variances that are more similar across groups with respect to both *Y* and *X*. Maxwell et al. (1984), using a simulation study, found that both ANCOVA and treatments × blocks ANOVAs were more powerful when assignment was made using the covariate than when it was not. Moreover, they found that the ANCOVA analysis was more powerful than the treatments × blocks ANOVA for several different values of ρ.

24.6 Estimating Power in ANCOVA

When we discussed power for one-factor ANOVA, we considered the Cohen effect-size statistic

$$f_{ANOVA} = \sqrt{\frac{\sigma_A^2}{\sigma_e^2}}$$

According to Cohen's rough guidelines, large, medium, and small effect sizes in ANOVA correspond to f values of .40, .25, and .10, respectively. If we use these values in G*Power 3.1 for ANOVA and choose

1. *F tests* as the *Test family*;
2. *ANOVA, fixed effects, omnibus, one-way* as the *Statistical test*;
3. *A priori: Compute required sample size – given α, power and effect size* as the *Type of power analysis*;

then insert *Number of groups* = 3 (as in the current example), α = .05, and desired power = .80, we find that the required numbers of participants per group are 22, 53, and 323, respectively.

The major change when we perform power calculations for ANCOVA is that the error variance, σ^2, is adjusted for differences in the covariate and should therefore be smaller; that is, $\sigma^2_{ANCOVA} = \sigma^2_{ANOVA}(1 - \rho^2)$, where ρ is the within-groups correlation between the dependent variable and covariate. Therefore, for ANCOVA, the quantity that should be used in place of f_{ANOVA} in G*Power 3.1 is

$$f_{ANCOVA} = f_{ANOVA}\sqrt{\frac{1}{1-\rho^2}} \tag{24.6}$$

For large values of ρ, f_{ANCOVA} can be much larger than f_{ANOVA}; therefore, power may be considerably larger for an ANCOVA than for the corresponding ANOVA, and the required sample sizes will be smaller.

In the current example, the correlation between proportion correct on the test and pretest score is approximately .6 in each of the text groups. If we substitute ρ = .6 into Equation 24.6, we see that the ANOVA f values are each multiplied by $\sqrt{1/\left(1-.6^2\right)} = 1.25$ when we perform ANCOVA power calculations. For large, medium, and small ANOVA effect sizes, the f values used in ANCOVA power calculations become .50, .3125, and .124. Using G*Power 3.1, we find that the required numbers of participants in each group for α = .05 and power = .80 are 14, 34, and 210 for large, medium, and small effect sizes, considerably smaller than the group sizes of 22, 52, and 323 found earlier for ANOVA. (The same estimated sample sizes are found in G*Power using the ANCOVA f values whether the *Statistical test* is ANCOVA or ANOVA.) If the within-group correlations between the covariate and the dependent variable are smaller, say, $r = .3$, the power advantage for ANCOVA is reduced. The ANOVA f values are now multiplied by a smaller number, $\sqrt{1/\left(1-.3^2\right)} =$ 1.048, so that the ANCOVA sample sizes required to achieve power = .80 are 20, 48, and 294, respectively.

24.7 Extensions of ANCOVA

24.7.1 Factorial ANCOVA

The multiple regression approach to ANCOVA can be readily extended to factorial designs. For example, if we consider a two-factor design described in the previous section, the ANCOVA tests for A, C, and AC are provided by

$$F_A = \frac{\left(R^2_{Y.X,A,C,AC} - R^2_{Y.X,C,AC}\right)SS_Y / df_A}{\left(1 - R^2_{Y.X,A,C,AC}\right)SS_Y / (N - 2 - df_A - df_C - df_{AC})}$$

$$F_C = \frac{\left(R^2_{Y.X,A,C,AC} - R^2_{Y.X,A,AC}\right)SS_Y / df_C}{\left(1 - R^2_{Y.X,A,C,AC}\right)SS_Y / (N - 2 - df_A - df_C - df_{AC})}$$

and

$$F_{AC} = \frac{\left(R^2_{Y.X,A,C,AC} - R^2_{Y.X,A,C}\right)SS_Y / df_{AC}}{\left(1 - R^2_{Y.X,A,C,AC}\right)SS_Y / (N - 2 - df_A - df_C - df_{AC})}$$

The statistical packages will perform the appropriate analyses if we specify *A* and *C* as factors, and *X* as the covariate.

24.7.2 Using More Than One Covariate

A researcher may wish to adjust for several sources of unwanted variability by using several covariates. For example, suppose we have a one-factor design and information about two covariates, *X* and *W*, that are each linearly related to *Y*. Performing an ANCOVA that uses both covariates tests whether *A* has significant effects over and above both *X* and *W*. The appropriate test statistic is the partial *F*:

$$F = \frac{\left(R^2_{Y.X,W,A} - R^2_{Y.X,W}\right)SS_Y / df_A}{\left(1 - R^2_{Y.X,W,A}\right)SS_Y / (N - 3 - df_A)}$$

Note that the denominator of this equation has one less *df* because of the additional covariate. The adjusted means are the scores predicted by the regression equation for each group if $X = \bar{X}_{..}$ and $W = \bar{W}_{..}$. Homogeneity of regression slopes can be tested by using the partial *F* for the interactions between *A* and the covariates,

$$F = \frac{\left(R^2_{Y.X,W,A,AX,AW} - R^2_{Y.X,W,A}\right)SS_Y / 2df_A}{\left(1 - R^2_{Y.X,W,A,AX,AW}\right)SS_Y / (N - 3 - 3df_A)}$$

24.7.3 Nonlinear ANCOVA

The relationships between the dependent variable and potential covariates are not always linear. For example, according to the Yerkes–Dodson law, we would expect a curvilinear relationship between measures of motivation and performance. If we use standard ANCOVA procedures when there is substantial nonlinearity, the ANCOVA *F* tests may have little power and the adjusted means may be biased estimates of the treatment means. Therefore, it is a good idea to check for severe violations of nonlinearity by plotting scatter diagrams for each group. Also, significance tests for nonlinearity are available (see Chapter 19).

If the relationship between Y and X is nonlinear but monotonic (i.e., Y increases or decreases with X but not in a linear fashion), it may be worth checking to see whether there is a simple transformation of X such as log X or X raised to some power, for which the relationship between Y and the transformed X is approximately linear. If such a transformation can be found, the transformed value of X can be used as the covariate in a standard ANCOVA.

If the relationship between Y and X is not monotonic, a simple transformation will not achieve linearity. However, in this case it may be worthwhile to use polynomial ANCOVA, in which the ANCOVA model contains linear and higher-order polynomial components. For quadratic ANCOVA, it is assumed that the relationship between Y and X is of the form

$$Y = b_0 + b_1 X + b_2 X^2$$

For cubic ANCOVA, the polynomial function contains an X^3 term; and so forth. Quadratic ANCOVA is conducted by including both X and X^2 as covariates, so that for the one-factor design, the ANCOVA test for A becomes

$$F = \frac{\left(R^2_{Y.X,X^2,A} - R^2_{Y.X,X^2}\right) SS_Y \,/\, df_A}{\left(1 - R^2_{Y.X,X^2,A}\right) SS_Y \,/\, (N - 3 - df_A)}$$

Higher-order polynomial ANCOVA can be performed by adding X^3, X^4, and so on, as covariates. However, it is important to keep in mind that although more complex models will fit better, using more covariates results in fewer error *dfs*. Therefore, one should be careful not to use more complex models or more covariates than are necessary.

Finally, it should be noted that the powers of X (X, X^2, etc.) are highly correlated and using them in the same multiple regression will result in multicollinearity (see Chapter 22) that may cause computational difficulties and parameter instability. These problems can generally be avoided by centering the covariates; that is, by using deviation scores. For example, $x = (X - \bar{X})$ and $x^2 = (X - \bar{X})^2$ may be used instead of X and X^2 in the regression.

24.8 Summary

ANCOVA is a procedure in which a covariate can be used to account for some of the variability in the dependent variable, thereby achieving greater power for statistical tests at the cost of more assumptions and greater complexity. It is best understood as a special case of multiple regression. In Chapter 24, we:

- Developed the underlying logic of ANCOVA from a multiple regression perspective in which sources of variance are adjusted for differences on the covariate.
- Presented an example of ANCOVA and compared it with the corresponding ANOVA.
- Showed how group means could be adjusted for differences on the covariate and indicated how contrasts could be performed by using them.
- Discussed the major assumptions that underlie ANCOVA and considered some of the major issues involved in interpreting the results appropriately. In particular, we discussed the challenges posed by heterogeneity of regression slopes, by lack of independence between the covariate and the treatment, and by having measurement error in the covariate.

- Distinguished three designs in which ANCOVA is often used: (a) designs in which participants are randomly assigned to treatments, (b) designs in which participants are assigned based on their score on the covariate, and (c) designs in which treatments are assigned to pre-existing non-equivalent groups. ANCOVA is an appropriate statistical procedure for analyzing designs (a) and (b), but its use with design (c) is problematic despite the fact that it is commonly used in that situation.
- Considered power calculations in ANCOVA.
- Briefly considered extensions to factorial designs, to the use of more than one covariate, and to nonlinear ANCOVA.

Exercises

24.1 [Comparing ANOVA and ANCOVA] Consider an experiment conducted to test the effectiveness of three software packages for teaching problem-solving skills to seventh-graders. Thirty-six seventh-graders are randomly selected and assigned to the software packages with the restriction that 12 children work with each of the packages. The levels of the independent variable (P) are the software packages the children worked with; the dependent variable (Y) is the score obtained on a problem-solving achievement test administered after the children have worked with a package for six months. There are also scores (X) obtained on a problem-solving pretest administered before the children were assigned to work with the software packages. The data are given in data set *EX24_1*.xlsx.

a) Perform an ANOVA using P as the independent variable.
b) Perform an ANCOVA using P as the independent variable and X as the covariate.
c) Test whether the homogeneity of regression slope assumption is satisfied.

24.2 [Interpreting adjusted means] Eighteen participants are assigned randomly to three treatment conditions, A_1, A_2, and A_3. After the treatment is applied, values of Y, the dependent variable, are obtained. However, before the treatment is applied, values of X, a variable closely related to Y, are recorded. The data are in data set *EX24_2.xlsx*.

a) Perform an ANOVA on Y.
b) Perform an ANOVA on X.
c) Test for homogeneity of regression in the three groups.
d) Perform an ANCOVA on Y, using X as the covariate.
e) How do the hypotheses tested by the ANOVA and the ANCOVA differ?
f) What are the adjusted means for the three treatment groups?
g) What is the interpretation of the adjusted means?

24.3 [Deciding when to use ANCOVA] Discuss whether it is appropriate to perform ANCOVAs using X as the covariate in each of the following cases:

a) Measures of job satisfaction (Y) and performance evaluation by supervisors (X) are obtained for eight randomly sampled workers in each of the four departments of a company. The researchers desire to test whether job satisfaction is the same in each of the departments. The data are as follows:

D_1		D_2		D_3		D_4	
X	Y	X	Y	X	Y	X	Y
1.4	1.0	3.2	3.0	6.2	7.3	5.8	5.6
2.0	2.7	6.8	5.5	3.1	4.0	6.6	7.2
3.2	3.9	5.0	5.6	3.2	4.9	6.5	6.1
1.4	1.0	2.5	3.2	4.0	6.9	5.9	7.1
2.3	4.0	6.1	4.2	4.5	2.1	5.9	5.4
4.0	3.4	4.8	4.2	6.4	5.6	3.0	4.0
5.0	3.7	4.6	3.7	4.4	6.0	5.9	5.6
4.7	2.3	4.2	3.8	4.1	4.6	5.6	5.8

b) Thirty children are each randomly assigned to one of three remedial math skills training programs. Before entering the programs, each child takes a standardized pretest (X). At the end of six months, a standardized achievement test (Y) is given to each of the children. The researchers wish to determine whether the training programs are all equally effective. The data are as follows:

A_1		A_2		A_3	
X	Y	X	Y	X	Y
29	61	39	79	41	78
37	73	34	66	36	66
26	54	35	76	29	56
32	63	39	84	33	61
31	62	35	73	42	70
37	76	27	75	35	65
33	72	35	66	32	59
39	80	29	85	42	80
33	73	34	62	39	65
36	72	26	79	36	64

24.4 [Regression residuals vs ANCOVA] If, for the data of problem 24.2, all the Y scores are regressed on all the X scores, the regression equation obtained is

$$\hat{Y} = 9.974 + 1.816\,X$$

The residuals for this regression are as follows (and are given in data set *EX24_4.xlsx*):

A_1	A_2	A_3
−5.771	2.045	2.128
−6.138	−0.771	1.412
−2.689	−0.872	6.780
−1.404	−1.670	3.780
−3.771	0.862	5.311
−3.955	1.045	3.678

Perform an ANOVA on these residuals. Is the ANOVA on the residuals equivalent to the ANCOVA of part (d) of Exercise 24.2? Why or why not?

24.5 With the data from Exercise 24.2, use some form of dummy coding to code *A* and then perform the ANCOVA using the regression module. (The data set with dummy coding can be found in *EX24_5.xlsx*).

24.6 [ANCOVA with a two-way design] Perform an ANCOVA for the following two-factor, between-participants design:

	A_1		A_2	
	X	Y	X	Y
	24.4	15.9	22.5	24.2
	22.3	15.7	12.5	19.7
	23.3	19.2	14.2	19.2
B_1	15.8	13.4	18.6	17.9
	22.6	18.0	15.2	24.4
	24.9	22.5	23.2	28.0
	20.9	15.1	20.9	19.9
	19.6	13.7	18.1	28.2
	23.9	12.8	18.1	18.1
B_2	26.2	25.5	11.5	13.5
	18.8	17.0	22.4	19.3
	24.0	25.3	30.2	35.1

Chapter 25

Integrated Analysis IV

25.1 Overview

In the first part of this chapter, we discuss multiple regression analyses to explore the relationships between physical activity and depression, using real data collected as part of the *Seasons* study. We find that higher levels of leisure-time physical activity predict lower scores on the Beck Depression Inventory (*BDI*) for participants who identify as women, whereas higher levels of work- and household-related physical activity do not, when the three measures are included in a regression. However, there are several possible reasons for this pattern of results that cannot easily be sorted out without conducting an experiment in which physical activity is manipulated. Therefore, in the last part of the chapter, we discuss an experiment that might be designed to resolve the ambiguity, as well as the analysis of a hypothetical data set that might result.

25.2 Introduction to the Study

Lin, Halgin, Well, and Ockene (2008) have reviewed some of the literature discussing the psychological benefits of physical activity. For example, the International Society of Sport Psychology (ISSP, 1992) has published a position statement in which it concludes that physical activity is related to decreases in mild to moderate depression, neuroticism, anxiety, and stress. Most studies that have examined the effect of physical activity on depression scores have considered leisure-time exercise such as walking, running, and cycling, and have found them to be associated with lower levels of depression.

For most people, leisure-related exercise comprises a relatively small part of their total physical activity. Friedenreich, Courneya, and Bryant (1998) found that a sample of 115 women were, on average, physically active approximately 55 hours per week. However, less than 10% of this time was spent on leisure-time exercise (about 4 hours); the rest was divided between occupational (22 hours) and household (28 hours) activity. Here, we use data collected as part of the *Seasons* study to assess the possible beneficial effects of leisure-time, occupational, and household physical activity. The relevant data are contained in the *Seasons physical activity* data set on the book's website.

25.3 Method

25.3.1 Participants

The *Seasons* study[1] was a large project primarily directed at assessing seasonal variation in blood lipid levels. A detailed account of the methodology may be found in Merriam,

DOI: 10.4324/9781003453550-29

Ockene, Hebert, Rosal, and Matthews (1999), and a statement of the lipid results is presented in Ockene et al. (2004). As a part of the study, data were collected on physical characteristics such as height and weight, diet, activity, psychological measures, education, marital and employment status, health, work environment, and exposure to sunlight, as well as blood lipid characteristics. There were two sets of eligibility criteria. The first required that participants be between the ages of 20 and 70 years, have a telephone, be literate in English, and plan to remain in the area for a year. There was also a second set of criteria dealing with the participant's health record. Participants could not be taking medication to lower lipids, be participating in a cholesterol- or weight-lowering diet, have a history of cancer within the past five years, or suffer from a psychiatric disease or other condition that would limit their participation in the study. Participants were recruited from the membership of the Fallon Healthcare System, a large health maintenance organization serving central Massachusetts. Of the 5,300 potential participants contacted by telephone, 1,254 met the first set of requirements and made an appointment. Of these, 641 participants (331 who identified as men and 310 who identified as women) kept their initial appointment, met all the eligibility requirements, and completed the baseline questionnaires. Of these, 285 men and 243 women were employed and able to provide data on occupational activity.

25.3.2 Measures Used in the Analyses

Several measures were taken on each participant.

Physical Activity

In each of 15 telephone interviews spread through their five quarters of participation in the study, participants were asked to recall the amount of time they spent in each of three categories of activity (leisure, occupational, and household) and at each of four intensities (light, moderate, vigorous, and very vigorous) during the previous 24-hour period. Estimates of physical activity energy expenditure were determined, using standard metabolic equivalent (MET) values. MET is a measure of the rate at which adults burn calories and is defined as the ratio of the activity metabolic rate to the resting metabolic rate: 1 MET is approximately 1 kcal per kilogram of body weight per hour and is roughly equivalent to the energy cost of sitting quietly. The weighted sums of different activity intensities were used: light activity = 1.5 METs; moderate = 4.0 METs; vigorous = 6.0 METs; and very vigorous = 8.0 METs. Activity scores in both hours of activity and the corresponding MET scores were averaged over the interview sessions yielding the variables *tot_hours_leisure* and *tot_MET_leisure*; *tot_hours_occup* and *tot_MET_occup*; and *tot_hours_house* and *tot_MET_house*. For our purposes, the ordering of the activities in terms of intensity is more important than the details of how their MET values were estimated.

Depression

Depression was measured by the Beck Depression Index (*BDI*; Beck & Steer, 1987), which has been found to be an effective screening instrument for detecting depression in community populations. This is a self-report questionnaire on which participants rate symptoms and attitudes related to depression. Twenty-one items are each scored from 0 to 3 so that the *BDI* score can range from 0 to 63. Scores on items that received no response were imputed from the scores on the other items, yielding an overall *BDI* summary score.

Body Mass Index (BMI)

BMI scores were calculated for each participant from height and weight scores. *BMI* is defined as weight in kilograms divided by the square of height in meters. Current CDC guidelines for *BMI* are underweight = less than 18.5; normal weight = 18.5–24.9; overweight = 25.0–30.0; and obese = more than 30.0.

Education

Participants reported the highest level of education they had received on an 8-point scale: 1 = no high school; 2 = some high school, 3 = high school diploma; 4 = vocational or trade school; 5 = some college; 6 = associate's degree; 7 = bachelor's degree; and 8 = graduate degree.

Employment

Participants reported their employment status: 1 = full time; 2 = part time; and 3 = not employed.

Perceived Health

Several measures of current health were collected. Here, we use a general self-report in which participants reported their health as 1 = excellent; 2 = very good; 3 = good; 4 = fair; and 5 = poor. Therefore, larger scores on the health measure correspond to poorer self-perceived health.

25.4 Procedure

Data collection took place between December 1994 and March 1998. Each participant was followed for a period of 15 months, including a baseline quarter and four follow-up quarters. Demographic information was collected only at baseline. Depression scores and measures of physical activity were collected each quarter. Information on physical activity was obtained in three, 24-hour-recall phone interviews during each of the quarters of data collection, a total of 15 interviews per participant. For the present purposes, depression and activity were each averaged across the five quarters to provide a single score on each measure for each participant.

25.5 Results and Discussion

For the analyses in this chapter, we used data from participants who were employed full time (employed = 1; 252 identified as men and 184 as women) or part time (employed = 2; 33 men and 59 women). The data are contained in the file *Seasons physical activity* that can be found on the book's website. The *BDI* score distribution is highly skewed for both men and women; skewness = 1.441 and 1.271, respectively. We therefore used a square-root transformation of the $BDI = \sqrt{BDI + 1}$, reducing the skewness for the transformed variable to 0.668 for men and 0.512 for women. Although the patterns of significant results were very similar for the transformed and untransformed measure of depression, we usually present results for the transformed *BDI* scores.

It is well established that female-identifying participants score higher on the *BDI* than male-identifying participants (e.g., Piccinelli & Wilkinson, 2000). Some of this difference occurs because women are more willing to report symptoms of depression than men. However, even accounting for this response bias, women are considered to have higher levels of depression than men. We therefore performed regressions separately for men and women. A different strategy would be to include gender identity as a predictor in the regression analysis.

Table 25.1 displays the means and standard errors for age, education, *BMI*, self-perceived health, depression scores, and activity values (in MET units) for the 285 men and 243 women who were employed full or part time at the time of the study (i.e., employed = 1 or 2). Using Welch's independent-groups t tests because of the unequal variances for men and women on some variables, we found significant gender differences for all these measures except for age and self-reported health. Men had significantly higher values for the education measures, for *BMI*, and for both leisure and occupational activity. Women had significantly higher values for both the raw and transformed depression measures and for household activity.

Following Lin et al. (2008), who also used data from the *Seasons* study, we regressed the transformed *BDI* score on the leisure-time, occupational, and household activity measures. The regression coefficients for the analysis with transformed *BDI* as the dependent variable are displayed in Table 25.2. The only significant coefficient was that of leisure activity for women, $t(239) = -2.281$, $p = .023$, suggesting that leisure-time physical activity is a predictor of lower depression scores for women, controlling for other kinds of physical activity, whereas household and occupational physical activity are not.

However, we must consider why it is that some individuals engage in more leisure-time physical activity than others. It is possible that healthier and fitter individuals tend to spend

Table 25.1 Means and standard errors for age, education, *BMI*, perceived health, depression scores, and activity values (in MET units) for men and women who were employed. Independent-group t tests (Welch's t) were computed to test differences between men and women.

Variable	*Men (n = 285)*	*Women (n = 243)*	*t′ value*	*df′*	*p*
Age	45.649 (0.697)	45.181 (0.683)	0.480	524.155	.632
Education	5.782 (0.109)	5.370 (0.126)	2.465	500.677	.014
BMI	27.834 (0.263)	26.381 (0.378)	3.226	443.898	.002
Health	2.333 (0.050)	2.198 (0.052)	1.882	516.906	.060
BDI	5.492 (0.300)	6.793 (0.367)	−2.743	486.991	.006
Transformed *BDI*	2.376 (0.055)	2.615 (0.063)	−2.873	502.553	.004
Leisure activity	2.022 (0.121)	1.642 (0.114)	2.279	524.872	.023
Occupational activity	7.054 (0.431)	3.818 (0.256)	6.458	452.253	.000
Household activity	3.939 (0.188)	4.467 (0.179)	−2.034	524.471	.042

Table 25.2 Results of the regression of the transformed depression scores on leisure-time, occupational, and household physical activity

Predictor	*Men*			*Women*		
	b (SE)	*β*	*p*	*b (SE)*	*β*	*p*
Leisure physical activity (MET)	−.025 (.028)	−.056	.369	−.083 (.036)	−.150	.023
Occupational physical activity (MET)	.007 (.008)	.054	.385	.026 (.106)	.105	.111
Household physical activity (MET)	.006 (.018)	.021	.724	−.027 (.023)	−.077	.239

Table 25.3 Results of the regression of the leisure-time physical activity score on *BMI*, perceived health, and the occupational and household activity scores for men and women

Predictor	*Men*			*Women*		
	b (SE)	*β*	*p*	*b (SE)*	*β*	*p*
BMI	−.038 (.027)	−.083	.164	−.034 (.019)	−.113	.067
Health[2]	−.265 (.145)	−.109	.068	−.421 (.135)	−.191	.002
Occupational physical activity (MET)	−.073 (.016)	−.259	.000	−.094 (.027)	−.211	.001
Household physical activity (MET)	−.077 (.037)	−.119	.037	−.126 (.039)	−.196	.002

more energy in leisure-time physical activities. It also seems possible that these individuals will tend to have fewer symptoms of depression. That is, variations in *BMI* and health may contribute to both depression scores and measures of leisure-time activity. In addition, individuals who expend more energy on occupational and household activities will have less energy left for leisure-time activities; these same individuals may feel less in control of their lives and thus may experience higher levels of depression. These statements are supported by an exploration of the data that shows small but significant negative correlations between leisure-time physical activity and poorness of health ($r = -.217, p = .001$), *BMI* ($r = -.170, p = .008$), occupational activity ($r = -.187, p = .003$), and household activity ($r = -.168, p = .009$) for women, and much the same pattern for men.

To explore the relationships between leisure activity and the other two activities (occupational and household), as well as *BMI* and health, we regressed the leisure-time activity score on the other four measures. The results are displayed in Table 25.3. For women, the regression coefficients were all negative and highly significant for the health and physical activity measures. In other words, poorness of health, occupational activity, and household physical activity were all negative predictors of leisure-time physical activity, controlling for the other predictors in the regression equation. A similar pattern was found for men, except that the regression coefficient for health was not significant.

Given the evidence of relationships among *BMI*, health, and the three types of activity, we reanalyzed the transformed *BDI* score on *BMI*, health, and the three activity measures. As can be seen in Table 25.4, rated poorness of health is a highly significant predictor of depression for both men and women, as is *BMI* for women. Controlling for *BMI* and self-perceived health, we now find that none of the physical activity measures were significant predictors of depression.

Another possible predictor of interest is education level. Although we have only a crude measure of education (see the description in the Materials section), it is significantly positively correlated with leisure-time activity and significantly negatively correlated with *BMI*, poorness of health, depression, and occupational physical activity, for both groups of participants. If we add education to the list of predictors in Table 25.4, its regression coefficient is not significant and the pattern of significance for the other predictors is not changed for men or for women. However, if we regress transformed *BDI* on education and the three measures of physical activity, we see in Table 25.5 that education is a significant negative predictor of depression for both men and women, but that, controlling for education level, none of the regression coefficients for the activity measures are significant.

So, what can we conclude about leisure-time activity and depression from this study? When considered along with the other activity measures, leisure-time physical activity is a significant predictor of lower *BDI* scores for women. However, our other analyses suggest that the predictive power of leisure-time exercise may come about because of its relationships with health, education, *BMI*, and the other measures of activity. Leisure-time physical activity may indeed have therapeutic effects with respect to depression, but it is extremely difficult to disentangle these from the effects of other variables we have considered in this observational study. If we want definitive answers to questions about the benefits of leisure exercise, we should design an experiment in which participants are assigned to conditions that correspond to different amounts of leisure activity and the effects of variables such as education, *BMI*, and perceived health are controlled by randomization or by using blocking and including them as independent variables in a multi-factor design.

Table 25.4 Results of the regression of the transformed *BDI* score on *BMI*, perceived health, and the three types of physical activity scores for men and women

Predictor	*Men*			*Women*		
	b (SE)	*β*	*p*	*b (SE)*	*β*	*p*
BMI	.003 (.012)	.77	.777	.027 (.010)	.160	.010
Health	.408 (.065)	.369	.000	.315 (.076)	.259	.000
Leisure physical activity (MET)	−.001 (.027)	−.002	.979	−.035 (.036)	−.064	.322
Occupational physical activity (MET)	.008 (.007)	.064	.270	.027 (.015)	.110	.082
Household physical activity (MET)	.006 (.016)	.020	.725	−.027 (.022)	−.075	.230

Table 25.5 Results of the regression of the transformed *BDI* score on education level and the three types of physical activity scores for men and women

Predictor	*Men*			*Women*		
	b (SE)	*β*	*p*	*b (SE)*	*β*	*p*
Education	−.099 (.034)	−.197	.004	−069 (.032)	−.138	.034
Leisure physical activity (MET)	−.015 (.028)	−.032	.601	−.070 (.036)	−.126	.058
Occupational physical activity (MET)	−.004 (.009)	−.034	.622	.021 (.016)	.087	.188
Household physical activity (MET)	.004 (.017)	.012	.839	−.027 (.023)	−.078	.232

25.6 A Hypothetical Experimental Test of the Effects of Leisure Activity on Depression

To provide a test of the hypothesis that amount of leisure physical activity affects depression, consider a hypothetical experiment in which adult men and women perform different amounts of exercise each day. Thus, the experiment is a 2 × 2 (*Gender identity* × *Activity Level*) between-participants design. Random assignment to activity condition is used to control for other factors evaluated in the preceding study (e.g., household activity, occupational activity, *BMI*). However, it is also decided to collect information on participants' perceived health and education levels because both variables were found to correlate with depression scores in the *Seasons* study (see Tables 25.4 and 25.5). Scores on these two variables will be used as covariates to improve the power of the statistical test of the effect of leisure activity on depression scores.

To decide on the number of participants to include in the study, *a priori* power analyses were conducted (see Section 24.6). Because we will again conduct separate analyses on the data for the men and women, the relevant design has one factor (activity, two levels) with two covariates (health and education). To convert Cohen's conventional f values for small, medium, and large effects of .1, .25, and .40 to the corresponding values for the ANCOVA, ρ^2 was estimated by regressing the transformed *BDI* scores on the health and education scores in the *Seasons* data set; R^2 was .156 for men and .120 for women. (The procedure of regressing the dependent variable on both covariates and using R^2 to estimate ρ^2 is an extension of the procedures presented in Section 24.7.) To be conservative in estimating the required number of participants, the value of .120 was used. The resulting conversion of Cohen's values produced f values of .106, .2665, and .4264 for small, medium, and large effects, respectively. Using G*Power 3.1, the *N*s required for power = .80 are 702, 114, and 46, respectively. Because the planned procedure is expensive in both time and resources, it was decided that 114 men and 114 women would be recruited for the experiment.

Participants were solicited from the same lists of HMO clients used in the *Seasons* study. At the start of the experiment, all participants completed a questionnaire in which they reported how much education they had completed and their perceived level of health, as in the *Seasons* study. They then received instructions about the nature of the activity condition to which they were assigned. One condition involved engagement in regular, vigorous exercise; the other, engagement in an alternative activity with minimal physical exercise. To ensure compliance with condition assignment, every participant agreed to report to the campus gym at a scheduled time every weekday to engage in the appropriate amount of exercise. Participants in the exercise condition followed a regimen of a strenuous, one-hour workout on each weekday (rate of energy expenditure estimated at 6 METs). Participants assigned to the control condition participated in one hour of group discussion per day (estimated at 1 MET). All participants participated in the experiment for six months.[3] At the end of the six-month period, all participants took the *BDI*. Note that the difference in the energy demands of the two activities is relatively large compared to the observed variability in exercise found for the correlational study. This is an important advantage of the experiment; namely, it permits a more definitive comparison of exercise levels than occurs naturally. This, in turn, will mean a more powerful test of the effects of exercise than was possible in the observational study.

Hypothetical data from the experiment are contained in the *Activity experiment* data set on the book's website. The file includes the following variables: activity condition (0 = discussion; 1 = exercise); gender identity (coded as sex = 0 for male; sex = 1 for female); education; health; and *BDI_transform*. The means for the education, health, and *BDI* variables are presented in Table 25.6, broken down by gender. Note that the number of observations is fewer than the 114 men and 114 women who started the study because some participants failed to complete the procedure. However, attrition was low and no clear evidence that the condition influenced the number of participants who declined to finish the experiment, so there were no concerns in proceeding with the analyses. Note that the values in Table 25.6 are like those found in the observational study (see Table 25.1).

Scatterplots were generated separately for the men and women relating the *BDI* scores to each of the covariates, health, and education. There was no evidence of nonlinear relationships between *BDI* and either covariate. Preliminary regression analyses were conducted on the data for men and for women to test whether the group slopes were homogeneous. In these analyses, *BDI* was the criterion variable and gender, education, health, activity, and the interactions of activity with health and with education were included. None of the interactions involving the covariates were significant, so the assumption of homogeneity of group slopes was satisfied.

Separate ANCOVAs were conducted on the data for the men and women, using R. In each analysis, the dependent variable was specified as *BDI_transform*, the factor was activity, and the covariates were education and health. The results of the ANCOVAs are summarized in Figure 25.1 (Panels a and b), along with ANOVAs on the same data (Panels c and d). The key result in both analyses is that depression scores were lower for participants in the vigorous exercise condition than for participants in the minimal exercise condition. For the men, the adjusted means in the exercise and control conditions were 2.077 and 2.569, respectively; for the women, the adjusted means were 2.429 and 2.856, respectively. In fact, the effects of exercise were quite similar for men and women; an ANCOVA that included the data from both sexes showed main effects of both activity and sex (women's *BDI* scores were higher than men's), but no evidence of a significant interaction. Finally, we note that ANOVAs on the data showed the same effects (see Figure 25.1(c) and (d)), although with somewhat larger error terms.

In sum, because participants were assigned at random to these two conditions and the activity variable was manipulated rather than observed, we can make a stronger case from our hypothetical experiment that leisure exercise decreases depression than we could from the observational study.

Table 25.6 Means and standard errors for education, health, and *BMI* (transformed) for men and women. Independent-group *t* tests (Welch's *t*) were computed to test differences between men and women.

Variable	*Men* (n = 110)	*Women* (n = 111)	t' *value*	df'	p
Education	5.35 (0.189)	4.98 (0.200)	1.357	218.437	.176
Health	2.509 (0.077)	2.270 (0.079)	2.164	218.942	.032
Transformed *BDI*	2.319 (0.082)	2.648 (0.104)	−2.492	207.770	.014

```
> Anova(lm(data = datM, BDI_transform ~ health + activity + education),type=3)
Anova Table (Type III tests)

Response: BDI_transform
            Sum Sq  Df F value    Pr(>F)
(Intercept) 13.819   1 27.3986 8.429e-07 ***
health      15.676   1 31.0788 1.903e-07 ***
activity     6.588   1 13.0616 0.0004628 ***
education    1.102   1  2.1853 0.1423032
Residuals   53.465 106
---
Signif. codes:  0 '***' 0.001 '**' 0.01 '*' 0.05 '.' 0.1 ' ' 1
```

```
> Anova(lm(data = datW, BDI_transform ~ health + activity + education),type=3)
Anova Table (Type III tests)

Response: BDI_transform
             Sum Sq  Df F value    Pr(>F)
(Intercept)  36.759   1 36.0197 2.709e-08 ***
health        8.921   1  8.7420  0.003827 **
activity      4.897   1  4.7987  0.030652 *
education     4.288   1  4.2016  0.042831 *
Residuals   109.195 107
---
Signif. codes:  0 '***' 0.001 '**' 0.01 '*' 0.05 '.' 0.1 ' ' 1
```

```
> Anova(lm(data = datM, BDI_transform ~ activity),type=3)
Anova Table (Type III tests)

Response: BDI_transform
            Sum Sq  Df F value    Pr(>F)
(Intercept) 365.68   1 552.609 < 2.2e-16 ***
activity      8.54   1  12.908 0.0004943 ***
Residuals    71.47 108
---
Signif. codes:  0 '***' 0.001 '**' 0.01 '*' 0.05 '.' 0.1 ' ' 1
```

```
> Anova(lm(data = datW, BDI_transform ~ activity),type=3)
Anova Table (Type III tests)

Response: BDI_transform
            Sum Sq  Df  F value  Pr(>F)
(Intercept) 478.05   1 417.1795 < 2e-16 ***
activity      7.20   1   6.2863 0.01364 *
Residuals   124.90 109
---
Signif. codes:  0 '***' 0.001 '**' 0.01 '*' 0.05 '.' 0.1 ' ' 1
```

Figure 25.1 ANCOVA results from R for the Activity experiment for participants identifying as men (Panel a) and as women (Panel b), and the ANOVA results for men (Panel c) and women (Panel d).

25.7 Summary and More Discussion

In the first, observational study, we used several regression analyses to point out the difficulty of assessing the influence of different kinds of physical activity. The main question was whether leisure-time physical activity reduced symptoms of depression. We were unable to reach a definitive answer to this question because we could not separate the effects of leisure activity from the effects of other variables. Possible reasons why leisure-time physical activity may be a proxy for other measures include the following (as well as many more): people with higher *BMI* values tend to be more depressed and they also exercise less; people with less education tend to be more depressed and also tend to work in jobs that involve more physical activity, so they have less energy left to engage in leisure-related exercise; and so on.

The most direct way to assess the effect of leisure-time physical activity on depression is to conduct an experiment in which participants are randomly assigned to different exercise conditions. That is what we did in the second, hypothetical study. In addition, we included health and education measures as covariates because both variables correlated significantly with depression in the observational study. Their inclusion in the design theoretically permitted a more powerful test of the relationship between leisure activity and depression. In fact, it turned out that the covariates were significantly, but not strongly, related to the dependent variable, so there was a negligible effect on power. A significant relationship between leisure activity and depression was found, suggesting that exercise tends to reduce depression. In the experiment, there is less ambiguity in interpreting the statistical relationship between the two measures because the random assignment of participants to conditions eliminated serious confounding of leisure activity with other potential influences on depression.

Although a major goal of Section 25.6 was to consider how data from a follow-up experiment might be analyzed, it is useful to consider some design concerns that might limit the generality of the conclusions from such an experiment. We should be under no illusions that a single, simple experiment of the sort described previously would settle the issue once and for all. Some issues that would impair our ability to arrive at definitive conclusions from our simple experiment include the following.

Issues Regarding the Selection of Participants

The choice of participants used in the experiment is critical. If participants are selected from the HMO lists used in the *Seasons* study and are contacted by telephone and asked to participate in the experiment, on what basis will they decide whether to participate? Participants will self-select, and this will be influenced by what they are told about the experiment. If the study is presented to potential participants as a study on depression, rather than a study of blood lipids and exercise, this may influence whether participants with higher depression scores will decide to volunteer. Relatedly, when potential participants hear about the time commitment, many will decide against participation because of either lack of motivation to devote that much time or because other commitments (e.g., their jobs) make it onerous to find the required time.

Another issue for these potential participants, given their range of health and age, is that they may not all be capable of participating in a vigorous exercise condition. In any event, the effects of vigorous exercise might be different for different kinds of participants.

If we did not have access to the HMO patient pool and instead selected volunteers from university classes, participants would be young adults with more similar age, educational level, and health backgrounds. The results of the experiment would be useful but would not readily generalize to the population at large.

Issues Regarding the Choice of Experimental and Control Conditions

We must be careful about the control conditions that are used. In the experiment we described earlier, there were two conditions, a vigorous exercise condition and a minimal exercise condition in which participants participated in a group discussion. It could be argued that the second condition is not an appropriate control. We might wish to have several exercise conditions: low, moderate, and vigorous. Part of the motivation is that although the effect of exercise on depression is not a completely settled issue in the scientific literature, it is presented as a fact on the internet – insert "exercise and depression" into a search engine and read what comes up. Because of this, we must assume that many participants will have the expectation that an exercise program reduces depression. The consequence for us is that we must contend with placebo effects – the expectation would be that exercise should help and this expectation might have positive effects on depression. Therefore, we might be better off using several different activity conditions. Also, we might want to separate the "social" effects of getting out and going to the gym from the effects of the exercise itself. Therefore, we might want to add conditions in which participants engaged in some exercise program at home. We might also want to look at the potentially different beneficial effects of leisure exercise for participants who already engaged in different amounts of occupational or household activity. Given enough participants, time, and resources, we could be successful in sorting this all out. But it would not be simple.

We should also note that it would have been possible to design a more powerful experiment than was presented earlier. In the *Seasons* study, the *BDI* was administered quarterly, and we can see that correlations for *BDI* scores obtained roughly six months apart are approximately .7 to .8 for both men and women. Therefore, we could have used a pretest-posttest design in which we administered the *BDI* both at the beginning and at the end of the experiment and used the pretest score as the covariate in the ANCOVA.

Finally, we should note that in many research contexts, it is not possible to conduct an experiment. We may be unable to manipulate the variables of interest, or it may be unethical to do so. In this case, the best we can do is to conduct an observational study and try to control for possible moderator variables statistically.

Exercises

25.1 Using the data in the file *Seasons physical activity*, reproduce the analyses reported in Tables 25.1–25.5. Remember that these analyses were based only on participants who were employed either full or part time (*employed* = 1 or *employed* = 2).

25.2 Given the type of observational study described in Chapter 25, estimate the number of cases required to yield power = .80 for a test of a medium-sized effect ($f^2 = .15$) of leisure-time exercise on depression in a regression in which there are three additional predictors: age and measures of occupational and household exercise. This type of *a priori* power calculation is described in Chapter 21.

25.3 For the regression of depression on leisure-time, occupational, and household physical activity, the effect of leisure-time activity was significant for women, $b = -.083$, $p = .023$; see Table 25.2. Do what is necessary to determine whether this is a small, medium, or large effect ΔR^2 according to the Cohen guidelines. Remember that for multiple regression, $f^2 = \frac{\Delta R^2}{1 - R^2}$. The guidelines, as well as how to obtain the f^2, were discussed when we considered power calculations in Chapter 21.

25.4 Let's explore some of the variables that are associated with depression, using the data set *phys_exercise_problems*.

a) Generate scatterplots of depression vs *BMI* for both men and women, and then use something like LOWESS smoothers or fit lines to explore the relationship between the two variables.

b) There seems to be a suggestion that, at least for men, there is a curvilinear relationship between depression and *BMI*. Test whether there is a significant curvilinear relationship and briefly discuss how it might be interpreted.

25.5 Considering only employed participants, 252 men and 184 women were employed full time and 33 men and 59 women were employed part time. Can we reject the hypothesis that employment status and gender identity are independent?

25.6 Are the effects of leisure-time physical activity on depression the same for women employed full and part time? If there are differences, speculate about why they may be present.

25.7 Explore the relationship between scores on the depression measure and our crude measure of education.

25.8 Explore the relationship between our measure of perceived health and the measures of physical activity.

Notes

1 The *Seasons* study was funded by grant R01-HL52745 from the National Heart, Lung, and Blood Institute, Bethesda, MD, USA.

2 Higher scores on the health variable correspond to worse self-rated health.

3 Note that if this were a real experiment, we would be concerned about a high attrition rate due to the substantial time commitment required of participants.

Part 5

Epilogue

Chapter 26

Some Final Thoughts, Suggestions, and Cautions

In this book, we have discussed how to describe and make inferences from data, with special emphasis on applications of ANOVA and the regression framework for statistical analysis. We use this final chapter to highlight 22 comments, cautions, and common errors that should be kept in mind when reading, conducting, and reporting research.

26.1 Designing the Research

1. *The generality of results depends on how the sample is selected.* The use of inferential statistics allows us to make statements about a population based on a sample randomly selected from it. However, although there are formal procedures for selecting random samples from a population of interest, these are rarely used in most research contexts. There are several practical constraints that result in the use of nonrandom samples.

   One constraint is that matters of convenience often influence participant recruitment. In psychological research, many researchers take advantage of the availability of students enrolled in introductory psychology classes. Studies in cognitive, educational, social, and even clinical psychology heavily tap this source of participants. This practice limits the generalizability of study findings, although the exact nature of the limits is often unclear and is likely to depend on the nature of the phenomena under investigation. Findings from a study of basic perceptual processes can probably be generalized to young adults regardless of educational level, ethnicity, socioeconomic status, and a host of other variables. On the other hand, findings from a study of reasoning are more likely to be restricted in generality, perhaps only applying to young adults with similar educational backgrounds.

   Also, participants are usually self-selected. They choose whether to sign up for an experiment at a particular time during the semester, or whether to respond to a phone call or email requesting their participation. In addition, many research studies have eligibility criteria that further constrain sampling. For example, the *Seasons* study was conducted through a medical school and had access to the members of a large HMO. Even so, of the 5,300 potential participants initially contacted by telephone, only 641 agreed to participate, kept their initial appointments, met all the eligibility requirements (see Chapter 25), and completed the baseline questionnaires. These considerations mean that findings from a study may generalize only to those who reasonably match the characteristics of the participants who self-select and meet eligibility criteria.

   Related to issues of self-selection are attrition considerations. Participant attrition is common in some types of research. Animal research involving surgical procedures may

DOI: 10.4324/9781003453550-31

result in deaths and longitudinal research with humans must deal with participants who drop out of the study, because either participants lose motivation or researchers are unable to keep up-to-date contact information. In other types of research, the challenges of an experimental procedure may be daunting for some participants, causing them to drop out, and this may occur more frequently in some experimental conditions than in others.

As an example of how attrition may bias results, consider that when we used the statistics class data set, the analyses that predicted final exam score from pretest score did not take account of the students who took the pretest, enrolled in the course, but dropped the class before taking the final exam. It is unlikely that the students who drop a course are a random sample of the students enrolled in a course, and we could look at their records to see whether we could have predicted that they would not complete the course. Quite likely, had they been required to stay in the course and take the final exam, they would have tended to do less well than non-dropouts.

We must keep in mind how the nature of the final sample limits the generality of our findings. When writing up results, we must describe our populations and samples as clearly as possible. If control or comparison groups are part of the design, we must specify exactly how they are defined. In short, we should always attend carefully to factors that may limit our ability to generalize our findings. We should also try to design our research materials and data collection procedures in ways that minimize the occurrence of missing data and participant attrition.

2. *Sophisticated statistical analyses may not mean much unless measures are valid and reliable.* Most statistics books say a great deal about statistical procedures but very little about measurement. We have commented throughout the book that researchers must try to make sure that their instruments are measuring the correct thing (i.e., they are valid) and do so with relatively little measurement error (i.e., they are reliable). Failure to do so may seriously compromise the results of any subsequent statistical analyses, no matter how sophisticated they may be. The consequences of poor reliability on a measure may be worse than just low power. For example, in multiple regression, poor reliability on a predictor may result in distorted regression coefficients for *other* predictors correlated with it. Also, measurement error may be systematic, as when participants with high values on some emotionally charged measures tend to underreport them.
3. *The research design is important.* If our goal is to uncover the causal influence of some treatment, we should try, if possible, to design an appropriate experiment and use matching and randomization to control important nuisance variables. If we can perform an experiment, there are some designs that are more efficient than others, although often at the cost of more complexity and more assumptions (see Chapter 12). If the nature of the research problem precludes the manipulation of important independent variables and we are left with an observational study, then it will be difficult, if not impossible, to make valid causal statements – no matter how sophisticated our analyses may be. We can try to identify uncontrolled nuisance variables that may moderate the effects of possible causal variables, and then measure them and try to adjust for their effects by modifying the design or the analyses. We can also generate a regression model or a more complicated "causal" structural equation model and determine whether the model is consistent with the obtained data. This may be a very useful step in the research process, but we must remember that testing such a model cannot, by

itself, tell us whether the model is correct. We must keep in mind that there may be other plausible models that are also consistent with the data.

4. *Conduct* a priori *power calculations.* We have emphasized how statistical software such as G*Power 3.1 can be used to determine the number of participants necessary to achieve a specified level of power for finding significant results, given a specified effect size. If we fail to perform these calculations and perform a study with too few participants, the result will be an expenditure of effort and resources without much chance of finding significant results. Also, if we base our calculations on a measure of effect size obtained from a small pilot study, we must remember that the point estimate for the effect size is likely to be larger than the population effect size, and that the confidence interval for this measure may be also be large, resulting in large confidence intervals on the estimated required number of participants. We can estimate these limits by finding the required N at the upper and lower ends of the confidence interval for the effect size.

26.2 The Initial Analyses

5. *Always explore the data.* Before conducting any statistical tests, always generate a thorough description of the data and consider it carefully. Statistical software provide many graphical display options that allow us to summarize and display data. We should check for the presence of extreme outliers and data points that may have undue influence on the results. We must make sure that results do not occur because of anomalies that result from malfunctions in procedure or computer glitches during the data analysis. If we have outliers that cannot be explained by equipment failures or transcription errors, in many cases we can fall back on robust procedures, and/or we can determine the sensitivity of our findings to these questionable data points by determining whether we would come to the same conclusions regardless of whether these points are included. We should also check to see if there are patterns of missing data that might bias our results.
6. *Correlations without scatterplots can be misleading.* The Pearson correlation coefficient is a very commonly used statistic that is a measure of the extent that two variables are linearly related. However, there may be a nonlinear component to the relationship between two variables in addition to, or instead of, a linear one. It is possible that there is an important systematic relationship between two variables even though their correlation is very small. Correlations are also very sensitive to the presence of outliers. If it is important to discuss correlations, we must look at the relevant scatterplots. Moreover, we must take note of the level of analysis: Combining data across different groups can grossly distort the correlation that may be present within each group.
7. *Do not artificially categorize inherently continuous independent variables.* Because some researchers are more comfortable with t or χ^2 tests and ANOVA than with regression, we often see the use of median splits (i.e., placing all the values below the median into one category and all values above the median in another) to turn inherently continuous predictor variables into dichotomous ones. This procedure throws away useful information and categorizes observations in an arbitrary way. For example, the smallest value and the value just below the median are placed in the same category even though they may be very different, but the value just below the median and the value just above it are placed in different categories even though they may be very similar. Dichotomizing tends to reduce statistical power (cf. Irwin & McClelland, 2001)

and can also create spurious significant effects if the predictor variables are correlated (cf. Maxwell & Delaney, 1993). In the most egregious version of this kind of error, researchers start with data from an observational study, categorize several inherently continuous correlated variables, perform a multi-factor ANOVA using these variables as factors, and then discuss the results as though they had conducted an experiment. This approach is unacceptable because it pretends that continuous variables are categorical and that correlated variables are independent. More detailed criticisms can be found in Fitzsimons (2008) and Pedhazur (1997).

8. *Be careful about combining data from different subgroups.* If we analyze the data from distinct subgroups without explicitly taking note of this in our model, then variability associated with differences among the groups will contribute to the error term and result in negatively biased tests. Moreover, the statistics for the combined data set may not match those for the data sets of any of the constituent groups. As an extreme example, we indicated in Chapter 17 that the correlation and regression coefficient could be positive for a combined data set, even though both of these statistics were negative in each of the constituent data sets. In ANOVA, a common error is to include factors in an experimental design (e.g., counterbalancing variables) but then to ignore them in the statistical analysis.
9. *Pay attention to assumptions.* The validity of many statistical test results depends on assumptions made about the population. If the assumptions are grossly violated, then the true p-values for these tests may differ greatly from their nominal values. We have presented procedures for checking major assumptions and have discussed the consequences of violating them. We should use the capabilities of our statistical software to examine residuals as well as to look at summary measures such as kurtosis and skewness. Even when Type 1 error rates are not badly distorted, power may be lost when the data do not conform to the assumptions underlying the method of analysis. In several chapters, we have discussed alternative procedures including data transformations, nonparametric tests, and tests that compensate for differences in variances, or that are resistant to the effects of outliers.

26.3 Interpreting the Results

10. *Interpret p-values correctly.* Remember that a p-value for a statistical test is the conditional probability $p(\text{data} \mid H_0 \text{ true})$. If $p = .01$, this means that if the null hypothesis is true, the probability of obtaining data at least as inconsistent with the null hypothesis as was obtained in the current sample is .01. This does *not* mean that the probability that the null hypothesis is true is .01. Neither the opposite conditional probability $p(H_0 \text{ true} \mid \text{data})$ nor $p(H_0 \text{ true})$ means the same thing as $p(\text{data} \mid H_0 \text{ true})$.
11. *Interpret confidence intervals correctly.* Confidence intervals are often misinterpreted. If a researcher computes that the confidence interval for a population mean extends from, say, 5 to 10, this does *not* mean that the probability is .95 that the population mean lies between 5 and 10. Nor does it mean that in repeated samples 95% of the sample means will fall between 5 and 10. If repeated samples are taken from a population, the limits on the confidence intervals will vary because the sample means will vary; however, 95% of the intervals will contain the value of the population mean being estimated so we say we are 95% *confident* that the single interval we compute contains the population mean.

12. *Remember that there are two kinds of significance test errors.* The probability of Type 1 error is $\alpha = p(\text{reject } H_0 \mid H_0 \text{ true})$; the probability of Type 2 error is $\beta = p(\text{fail to reject } H_0 \mid H_0 \text{ false})$. The smaller we set α, the larger β will be. We must be aware of this trade-off when we choose procedures to limit Type 1 error. The cost of guarding against false effects is that we may miss important effects that are present.
13. *The finding of nonsignificant results does not necessarily mean that the null hypothesis is true.* Suppose we fail to reject the null hypothesis because the obtained p-value is not smaller than the specified significance level (i.e., $p > \alpha$). We cannot on this basis assume that the null hypothesis is true, but merely that we lack sufficient evidence to reject it. In some cases, the goal of a study is to argue that some null hypothesis is approximately true. We cannot make this case merely by finding nonsignificant results. This is because it is all too easy to find nonsignificant results by conducting a poor study with poor measures and little power that tells us almost nothing about the population. If we want to offer convincing evidence that the null hypothesis is approximately true, we must base our argument on confidence intervals for measures of effect size or turn to a Bayesian analysis.
14. *The finding of significant results does not necessarily mean that the null hypothesis is false.* By chance, we may have sampled a set of data that is relatively unlikely given the null hypothesis, resulting in a Type I error. To lower the probability of a Type I error, we can set a small α, and we can run studies that are well-powered for our planned tests. We should run replications that are well-powered, and we must stay humble about our findings: A significant result is simply a statement of probability, not a declaration of truth in the world.
15. *Distinguish between statistical significance and effect size.* Researchers commonly look at the p-values provided by statistical tests to make comments about importance. However, the p-value depends not only on the size of the effect, but also on the size of the sample. Large sample sizes can result in highly significant results even when effect sizes are too small to be of any practical or theoretical importance. For example, Cohen (1990) discussed a *New York Times* article with the headline "Children's Height Linked to Test Scores" that reported a study involving nearly 14,000 children that showed a "definite" link between age- and sex-adjusted height with intelligence and achievement scores, even after controlling for a host of other factors such as birth order and physical maturity. Before ordering growth hormone, we should note that with a sample as large as 14,000, we would get $p < .001$ with a correlation as low as $r = .0278$. Using regression to predict IQ and height with a correlation of this size, we find that a predicted change in IQ from 100 to 130 would require about a 14-foot change in height, and a predicted four-inch change in height would require a change in IQ of about 230 points. We must consider not only significance levels, but also effect sizes.
16. *Remember that measures of effect size are also statistics.* The APA task force recommends that measures of effect size always be presented for primary results, and more journals are now requiring that effect size measures be included in results sections. However, for any given study, the estimated effect size in a sample is just a statistic that has a sampling distribution of its own. Therefore, we should present confidence intervals when reporting effect size measures (e.g., see Chapter 6).
17. *Note that groups with significant differences and groups without significant differences are not necessarily significantly different from one another.* Suppose there is a significant main effect of condition (factor C) for group A_1 but the C effect is not significant

for group A_2. We cannot on this basis say that the effects of C are different for groups A_1 and A_2. If we want to reject the null hypothesis that both groups have the same C effect, we must directly test that the C effect is significantly different for A_1 and A_2; that is, we must test the $A \times C$ *interaction*. Also, note that even if there is a significant C effect at A_1 but no C effect whatsoever at A_2, this interaction may not be significant. This is because interactions (i.e., differences of differences) have larger standard errors than differences. This issue is discussed in more detail in Chapter 10.

18. *Performing multiple statistical tests can increase the overall Type 1 error rate to unacceptable levels.* As we perform more statistical tests on the same data set, we are increasingly likely to commit Type 1 errors – that is, we are more likely to find spurious significant results. Procedures for controlling Type 1 error are well developed for certain kinds of tests, whereas they are virtually nonexistent for others. In Chapter 10 we discussed procedures for controlling Type 1 error for families of contrasts. We distinguished between planned (i.e., theory-driven) and unplanned contrasts and discussed methods for controlling Type 1 error in both cases. However, there are other areas where such control is less commonly addressed. For multi-factor ANOVAs, common practice has been to test each main effect and interaction at the same significance level, usually .05, thereby resulting in a great potential inflation of Type 1 error in design with many factors. Finally, we note that the control of Type 1 error is rarely explicitly addressed in multiple regression even though these procedures are extremely vulnerable to Type 1 error inflation, especially when predictors added to the equation have been selected from much larger pools of variables. In the regression chapters we distinguish between theory-driven analyses and what might be called *ad hoc* curve fitting (i.e., the assembling of collections of predictors that just happen to account for large values of R^2). We also discussed Steiger's (1980) recommendation to perform an omnibus test of all off-diagonal elements in a correlation matrix before looking at the individual correlations (Chapter 18). This is useful because with, say, 15 variables, there are (15)(14)/2 = 105 pairwise correlations. There is nothing wrong with looking at your data very carefully and in different ways – as the saying goes, if you torture your data long enough, they will confess! The problem is that if you perform a very large number of tests without taking Type 1 error into account, they may confess to almost anything.
19. *Note that even a nonsignificant result can add to the cumulative evidence for an effect.* Suppose a researcher runs a study with 60 participants and finds an unexpected effect that is fairly large and statistically significant. Further, suppose that the effect is of theoretical importance and was not predicted, so a second researcher decides to repeat the study with 40 participants in an attempt to confirm its existence. If the effect is again significant, then there has been a successful replication and we have added confidence that the effect exists. But suppose the effect in the second study is not significant; say the estimated effect is in the same direction, but only about two-thirds the size of the effect found in the original study. How do we explain the "discrepancy" between the two studies? In fact, there is no discrepancy to explain. We would expect sampling variability in the estimate of the effect size and, in any event, would expect less significant results with the second study because it was run with fewer participants. It is true that we now have two different estimates for the effect size, and that our best estimate of effect size is some average of the two. However, the fact that both studies found the effect to be in the same direction means that the case for the existence of the effect is stronger than if we did not have the findings for the second study, despite the fact that

the results were not significant. It would take an effect size of zero or in the opposite direction in the second study to weaken the case for a significant effect. The field of meta-analysis (see, for example, Hunter & Schmidt, 1990; Rosenthal, 1991, 1995) has been developed to assess cumulative results across studies.

20. *Consider whether to compare correlations or unstandardized regression coefficients when looking at differences in bivariate relationships.* The unstandardized regression coefficient, b_1, is the slope of the linear equation that best predicts Y from X. The correlation coefficient, r, is an index of the strength of the linear relationship between Y and X. These are related but different concepts, and we have argued that the slope is often the information that is most relevant to the researcher. Moreover, it is often difficult to interpret correlation coefficients. No matter how it is expressed, the Pearson correlation coefficient is a measure that depends on the similarity of corresponding z scores. Because of this, the correlation coefficient and related measures such as the multiple correlation coefficient and standardized regression coefficients do not have units and are sample specific; that is, the values of the correlations depend not only on the slope, but also on characteristics of the samples such as their variances in ways that unstandardized regression coefficients do not. Therefore, correlations in two samples may differ because the variance of X differs, even if the rate of change of Y with X in the two samples is the same.
21. *Automated stepwise regression is not an appropriate basis for explanation.* Automated procedures that select variables to be included in a model such as stepwise regression should not be used to develop theory. Stepwise regression does not necessarily produce the best, good, or even plausible explanatory models. The p-values produced by stepwise regression programs are not "real" p-values because they do not take note of the fact that variables entered into the equation are selected from larger pools of possible predictor variables simply because they result in the largest increase in the multiple correlation coefficient. The order in which variables are added to a regression equation by an automated procedure is usually of no theoretical significance. Stepwise regression will capitalize on chance variation, especially in small samples. In Chapter 20 we considered an example of stepwise regression in which the multiple correlation coefficient was .42, but the cross-validated R was only .10. Even apart from the use of automated procedures, we have argued that attempts to maximize R^2 are not the best way to develop explanatory models.
22. *Regression equations generally are not explanatory causal models and regression coefficients are generally not measures of causal importance.* It is possible that a theory-driven causal model can be expressed as a set of regression equations, and we can perform these regressions to determine whether the model is consistent with the data. However, we cannot assemble a collection of predictors that just happen to account for a respectable proportion of the variability of an interesting outcome variable and thereby assume that we have a useful or plausible explanatory model. Among the points we mentioned in Chapters 20 and 22 were (a) the machinery of regression deals with prediction, not causation, and a variable may be a good predictor without being causally important because the effects of variables that are not included in the equation are "absorbed" by predictors in the equation that are correlated with them; (b) because of this, both the values and interpretation of regression coefficients change when other correlated variables are entered into the equation; and (c) the regression coefficients do not take account of effects mediated by other variables in the equation because the

interpretation of a regression coefficient is the rate of change of the outcome variable with the predictor, *holding all the other predictors constant*. It is also important to note that although changing one variable while holding constant several others is easy enough to understand in terms of the sample regression equation, it may not correspond to any plausible manipulation in the real world. Despite their limitations, regression analyses can be a very useful component of a program to advance knowledge and theory about a research problem by providing useful descriptions of the data and by suggesting questions and causal speculations that can be pursued by a variety of research techniques.

In earlier chapters, we have talked about data as our window on the world. Pursuing the metaphor, this window is almost always dirty, and the view it provides is easily distorted. In this book, we have discussed issues in research design, as well as how to describe and interpret data in ways that allow us best to cope with random noise and reduce bias. We have also tried to point out many of the pitfalls that should be avoided. We urge you to insist on good research design, remembering two key points: (1) garbage in implies garbage out, and (2) you cannot salvage a bad design with good analysis.

Appendices

Appendix A

Notation and Summation Operations

We must have a common language to talk about the derivations and computational formulas that relate to experimental designs and statistical summaries. Such a language exists in the notational system presented here. If you try to master it now, your efforts will be amply repaid. We begin with a few simple rules that are then applied to some elementary statistical quantities.

A.1 A Single Group of Scores

A.1.1 Some Basic Rules

In a group of scores like Y_1, Y_2, Y_3, Y_4, . . ., Y_n, the subscript has no purpose except to distinguish among the individual scores. The quantity n is the total number of scores in the group. Suppose that $n = 5$ and we want to show that all five scores are to be added together. We could write

$$Y_1 + Y_2 + Y_3 + Y_4 + Y_5$$

or, more briefly,

$$Y_1 + Y_2 + \ldots + Y_5$$

Still more briefly, we write

$$\sum_{i=1}^{5} Y_i$$

This expression is read "sum the values of Y for all i from 1 to 5." In general, $i = 1, 2, \ldots,$ n (that is, i takes on the values of 1 to n), and the summation of a group of n scores is indicated by

$$\sum_{i=1}^{n} Y_i$$

The quantity i is the *index*, and 1 and n are the *limits* of summation.[1] When the context of the presentation is clear, the index and limits are often dropped. Thus, we may often indicate by ΣY that a group of scores are to be summed. Three rules for summation follow.

RULE 1. The sum of a constant times a variable equals the constant times the sum of the variable; or

$$\Sigma CY = C\Sigma Y$$

The term C is a constant in the sense that its value does not change as a function of i; the value of Y depends on i, and Y is therefore a variable relative to i. The rule is easily proved.

$$\begin{aligned}\Sigma CY &= CY_1 + CY_2 + CY_3 + \ldots + CY_n \\ &= C(Y_1 + Y_2 + Y_3 + \ldots + Y_n) \\ &= C\Sigma Y\end{aligned}$$

RULE 2. The sum of a constant equals n times the constant, where n equals the number of quantities summed; or

$$\Sigma C = C + C + \ldots + C = nC$$

RULE 3. The summation sign follows the distributive law on quantities within parentheses.

EXAMPLE 1.

$$\sum_i^n (X_i - Y_i) = \sum_i^n X_i - \sum_i^n Y_i$$

Proof.

$$\begin{aligned}\Sigma(X - Y) &= (X_1 - Y_1) + (X_2 - Y_2) + \ldots + (X_n - Y_n) \\ &= (X_1 + X_2 + \ldots + X_n) - (Y_1 + Y_2 + \ldots + Y_n) \\ &= \Sigma X - \Sigma Y\end{aligned}$$

EXAMPLE 2.

$$\Sigma(X - Y)^2 = \Sigma X^2 + \Sigma Y^2 - 2\Sigma XY$$

Proof.

$$\begin{aligned}\Sigma(X - Y)^2 &= (X_1 - Y_1)^2 + \ldots + (X_n - Y_n)^2 \\ &= \left(X_1^2 + Y_1^2 - 2X_1Y_1\right) + \left(X_2^2 + Y_2^2 - 2X_2Y_2\right) + \ldots + \left(X_n^2 + Y_n^2 - 2X_nY_n\right) \\ &= \left(X_1^2 + X_2^2 + \ldots + X_n^2\right) + \left(Y_1^2 + Y_2^2 + \ldots + Y_n^2\right) - 2(X_1Y_1 + X_2Y_2 + \ldots + X_nY_n) \\ &= \Sigma X^2 + \Sigma Y^2 - 2\Sigma XY\end{aligned}$$

A.1.2 *Applying the Summation Rules*

We can apply the rules of summation to prove the properties of means and variances stated in Chapter 2 (Section 2.3.1). Throughout this section it should be clear that we are summing over i from 1 to n even though the index and limits are not explicitly presented in each expression.

Properties of the Mean

1. $\Sigma(Y - \bar{Y}) = 0$; the sum of all deviations of scores about their mean is zero. Applying Rule 3, we get

$$\Sigma(Y - \bar{Y}) = \Sigma Y - \Sigma \bar{Y}$$

However, $\bar{Y}$ is a constant; its value is the same regardless of the value of the index of summation. Therefore, applying Rule 2, we rewrite the last equation as

$$\Sigma(Y - \bar{Y}) = \Sigma Y - n\bar{Y}$$

Because $\bar{Y} = \Sigma Y/n$, we can rewrite this as

$$\Sigma(Y - \bar{Y}) = \Sigma Y - n\bar{Y} = \Sigma Y - n\left(\frac{\Sigma Y}{n}\right) = \Sigma Y - \Sigma Y = 0$$

2. $\Sigma(Y + k)/n = \bar{Y} + k$; if a constant is added to all scores, the mean is increased by that constant. Applying Rule 3 gives

$$\frac{\Sigma(Y + k)}{n} = \frac{\Sigma Y + \Sigma k}{n} = \frac{\Sigma Y}{n} + \frac{\Sigma k}{n}$$

Applying Rule 2 and noting that $\Sigma\, Y/n = \bar{Y}$, we have

$$\frac{\Sigma(Y + k)}{n} = \bar{Y} + \frac{nk}{n} = \bar{Y} + k$$

3. $\Sigma kY/n = k\bar{Y}$; if all scores are multiplied by a constant, the mean is multiplied by that constant. Applying Rule 1, we have

$$\frac{\Sigma kY}{n} = \frac{k\Sigma Y}{n} = k\bar{Y}$$

4. $\Sigma(Y - \bar{Y})^2$ is a minimum. Assume that there is some value $\bar{Y} + d$ such that the sum of squared deviations of all scores about it is smaller than the sum about any other value. This sum of squared distances is $\Sigma[Y - (\bar{Y} + d)]^2$. Expanding in accord with Rule 3, we have

$$\Sigma[Y - (\bar{Y} + d)]^2 = \Sigma[(Y - \bar{Y}) - d]^2 = \Sigma(Y - \bar{Y})^2 + \Sigma d^2 - 2\Sigma d(Y - \bar{Y})$$

Applying Rule 1, we rewrite the right-most term as

$$2\Sigma d(Y - \bar{Y}) = 2d\Sigma(Y - \bar{Y}) = (2d)(0)$$

because $\Sigma(Y - \bar{Y}) = 0$. Applying Rule 2, we have

$$\Sigma d^2 = nd^2$$

Therefore,

$$\Sigma[Y - (\bar{Y} + d)]^2 = \Sigma(Y - \bar{Y})^2 + nd^2$$

which is as small as possible when $d = 0$; that is, when we sum the squared deviations of scores about their mean.

Properties of the Variance

1. Adding a constant to all scores leaves the variance unchanged. If a constant k is added to all scores the new variance is

$$\hat{\sigma}^2_{Y+k} = \frac{\Sigma[(Y + k) - (\overline{Y + k})]^2}{n - 1} = \frac{\Sigma(Y - \bar{Y})^2}{n - 1} = \hat{\sigma}^2_Y$$

2. Multiplying all scores by a constant k is equivalent to multiplying the variance by k^2 and the standard deviation by k. We have

$$\hat{\sigma}^2_{kY} = \frac{\Sigma(kY - \overline{kY})^2}{n - 1} = \frac{\Sigma k^2 (Y - \bar{Y})^2}{n - 1}$$

By Rule 1 this becomes

$$\hat{\sigma}^2_{kY} = \frac{k^2 \Sigma(Y - \bar{Y})^2}{n - 1} = k^2 \hat{\sigma}^2_Y$$

z Scores

The properties proven allow us to show that the mean of a set of z scores is zero and its variance is 1. Recall that

$$z = \frac{Y - \bar{Y}}{\hat{\sigma}_Y}$$

To obtain the average of a set of n z scores, we sum them and divide by n, keeping in mind that $\Sigma(Y - \bar{Y}) = 0$. Then

$$\frac{\Sigma z}{n} = \frac{\Sigma(Y - \bar{Y})}{n\hat{\sigma}_Y} = \frac{(0)}{n\hat{\sigma}_Y} = 0$$

To prove that the variance (and therefore the standard deviation) of the z scores is 1, expand the formula for z as

$$z = \left(\frac{1}{\hat{\sigma}_Y}\right)Y - \left(\frac{1}{\hat{\sigma}_Y}\right)\bar{Y}$$

Note that $1/\hat{\sigma}_Y$ is a constant with respect to the index of summation i. Because adding (or subtracting) a constant from a variable does not change its variance (see the first property of the variance), the variance of z is the same as the variance of $(1/\hat{\sigma}_Y)$. But, from the second property of a variance, we know that the variance of a constant $(1/\hat{\sigma}_Y)$ times a variable (Y) is the squared constant times the variance of the variable. That is,

$$\hat{\sigma}_z^2 = \left(\frac{1}{\hat{\sigma}_Y}\right)^2 \hat{\sigma}_Y^2 = 1$$

A.1.3 Raw-Score Formulas

The summation rules can be applied to obtain raw-score formulas for quantities such as the variance and covariance. These raw-score or *computational* formulas contain sums of scores, squared scores, and cross-products rather than sums of squared differences and cross-products of difference scores. This allows them to minimize rounding error and makes them convenient to use with simple hand calculators that do not have variance and correlation keys.

The numerator of the expression for the variance of Y is $SS_Y = \Sigma(Y_i - \bar{Y})^2$. To get the raw-score formula for SS_Y, expand the quantity within the summation sign. Thus

$$\Sigma(Y - \bar{Y})^2 = \Sigma\left(Y^2 + \bar{Y}^2 - 2Y\bar{Y}\right)$$

Applying Rule 3, we have

$$\Sigma(Y - \bar{Y})^2 = \Sigma Y^2 + \Sigma\bar{Y}^2 - \Sigma 2Y\bar{Y}$$

Noting that $\bar{Y}^2$ is a constant and applying Rule 2, we have

$$\Sigma(Y - \bar{Y})^2 = \Sigma Y^2 + n\bar{Y}^2 - \Sigma 2Y\bar{Y}$$

The quantity $2\bar{Y}$ is a constant and, by Rule 1, can be placed before the summation sign. Thus,

$$\Sigma(Y - \bar{Y})^2 = \Sigma Y^2 + n\bar{Y}^2 - 2\bar{Y}\Sigma Y$$

Now replace $\bar{Y}$ by $\Sigma Y/n$ to get

$$\Sigma(Y - \bar{Y})^2 = \Sigma Y^2 + \frac{n(\Sigma Y)^2}{n^2} - 2\left(\frac{\Sigma Y}{n}\right)\Sigma Y$$

Simplifying, we have

$$\Sigma(Y - \bar{Y})^2 = \Sigma Y^2 - \frac{(\Sigma Y)^2}{n} \tag{A.1}$$

Dividing the right-hand side of Equation A.1 by $n - 1$ gives the raw-score formula for $\hat{\sigma}_Y^2$.

We can find the raw-score formula for the covariance of X and Y,

$$\hat{\sigma}_{XY} = \frac{\Sigma(X-\bar{X})(Y-\bar{Y})}{n-1}$$

by noting that Equation A.1 could be rewritten as

$$\Sigma(Y-\bar{Y})^2 = \Sigma(Y-\bar{Y})(Y-\bar{Y}) = \Sigma YY - \frac{(\Sigma Y)(\Sigma Y)}{n}$$

By analogy, the numerator of $\hat{\sigma}_{XY}$ has the raw-score formula

$$\Sigma(X-\bar{X})(Y-\bar{Y}) = \Sigma XY - \frac{(\Sigma X)(\Sigma Y)}{n}$$

Dividing by $n - 1$ yields the raw-score formula for $\hat{\sigma}_{XY}$

A.2 Several Groups of Scores

The simplest possible experimental design involves several groups of scores. Thus, one might have a groups of n participants each, which differ in the amount of reward they receive for their performance on some learning task. When recording the data, there would be a column for each level of amount of reward – that is, for each experimental group. The scores for a group could be written in order within the appropriate column. In referring to a score, we should designate it by its position in the column (or experimental group) and by the position of the column. Table A.1 illustrates this procedure. Note that the first subscript refers to the position in the group (row), the second to the position of the group (column). Thus Y_{22} is the second score in group 2, and in general, Y_{ij} is the ith score in the jth group.

Suppose we want to refer to the mean of a single column. The term used previously, $\bar{Y}$, is obviously inadequate because it does not designate the row or column that we want. Even $\bar{Y}_1$ is not clear, as it might just as easily refer to the mean of the first row as to the mean of the first column.[2] The appropriate designation is $\bar{Y}_{.1} = (1/n)\Sigma_i^n Y_{i1}$; the dot represents the

Table A.1 A two-dimensional matrix of scores

	Groups					
	Y_{11}	Y_{12}	...	Y_{1j}	...	Y_{1a}
	Y_{21}	Y_{22}	...	Y_{2j}	...	Y_{2a}
	.	.		.		.
	.	.		.		.
	.	.		.		.
Participants	Y_{i1}	Y_{i2}	...	Y_{ij}	...	Y_{ia}
	.	.		.		.
	.	.		.		.
	.	.		.		.
	Y_{n1}	Y_{n2}	...	Y_{nj}	...	Y_{na}

summation over i, the index that ordinarily appears in that position. Similarly, the mean of row i would be designated by $\bar{Y}_{i.} = (1/a)\ \Sigma_j^a Y_{ij}$; summation is over the index j. The mean of all an scores would be designated $\bar{Y}.. = (1/an)\Sigma\Sigma Y_{ij}$, or merely $\bar{Y}$.

Some examples using the double summation $(\Sigma_i \Sigma_j)$ may be helpful. Suppose we have

$$\sum_{j=1}^{a}\sum_{i=1}^{n} Y_{ij}^2$$

This is an instruction to set i and j initially at 1; the resulting score Y_{11} is then squared. Holding j at 1, we step i from 1 to n, squaring each score thus obtained and adding it to those previously squared. When n scores have been squared and summed, we reset the index i at 1 and step j to 2; the squaring and summing is then carried out for all Y_{i2}. The process continues until all an scores have been squared and summed. The process just described can be represented by

$$\left(Y_{11}^2 + Y_{21}^2 + \ldots + Y_{na}^2\right)$$

If we have

$$\sum_{j=1}^{a}\left(\sum_{i=1}^{n} Y_{ij}\right)^2$$

the notation indicates that a sum of n scores is to be squared. We again set j at 1, and after adding together all the Y_{i1}, square the total. The index j is then stepped to 2 and i is reset at 1; we get another sum of n scores, which is squared and added to the previous squared sum. We again continue until all an scores have been accounted for. The process can be represented by

$$\left(Y_{11} + Y_{21} + \ldots + Y_{n1}\right)^2 + \ldots + \left(Y_{1a} + Y_{2a} + \ldots + Y_{na}\right)^2$$

A third possibility is

$$\left(\sum_{j=1}^{a}\sum_{i=1}^{n} Y_{ij}\right)^2$$

which indicates that the squaring operation is carried out once on the total of an scores; we then have

$$\left[\left(Y_{11} + Y_{21} + \ldots + Y_{n1}\right) + \ldots + \left(Y_{1a} + Y_{2a} + \ldots + Y_{na}\right)\right]^2$$

Note that the indices within the parentheses show how many scores are to be summed prior to squaring, and the indices outside the parentheses show how many squared totals are to be summed. When no parentheses appear, as in $\Sigma\Sigma Y^2$, we treat the notation as if it were $\Sigma\Sigma(Y^2)$. When no indices appear outside the parentheses, it is understood that we are dealing with a single squared term, as in $(\Sigma\Sigma Y)^2$. When several indices appear together, whether inside or outside the parentheses, the product of their upper limits tells us the number of terms involved. Thus, $(\Sigma_{j=1}^{a}\Sigma_{i=1}^{n} Y)^2$ indicates that an scores are summed before the squaring.

Table A.2 Some sample data

	Group 1	*Group 2*	*Group 3*
	4	1	6
	1	7	4
	3	2	5
	2	4	4
	–	–	–
$\Sigma_i Y_{ij} =$	10	14	19
$\Sigma_i Y_{ij}^2 =$	30	70	93

Our three illustrations of the double summation can be further clarified if we use some numbers. Let us use the three groups of four scores each shown in Table A.2.
Now,

$$\sum_j \sum_i Y_{ij}^2 = 30 + 70 + 93 = 193$$

and

$$\sum_j \left(\sum_i Y_{ij} \right)^2 = (10)^2 + (14)^2 + (19)^2 = 657$$

and

$$\left(\sum_j \sum_i Y_{ij} \right)^2 = (10 + 14 + 19)^2 = 1849$$

As another example of how to use double summation, we might derive a raw-score formula for the average group variance, often referred to as the *within-group mean square.* This is the sum of the group variances divided by a, the number of groups, or

$$\frac{1}{a}\left[\frac{\sum_{i=1}^{n} \left(Y_{i1} - \bar{Y}_{.1} \right)^2}{n-1} + \ldots + \frac{\sum_{i=1}^{n} \left(Y_{ia} - \bar{Y}_{.a} \right)^2}{n-1} \right]$$

More briefly, this average is indicated by

$$\frac{1}{a(n-1)} \sum_j^a \sum_i^n \left(Y_{ij} - \bar{Y}_{.j} \right)^2$$

Now, expanding the numerator (or "sums of squares") of this quantity, we get

$$\sum_{j=1}^{a}\sum_{i=1}^{n}\left(Y_{ij}-\bar{Y}_{.j}\right)^2=\sum_{j=1}^{a}\sum_{i=1}^{n}\left(Y_{ij}^2+\bar{Y}_{.j}^2-2Y_{ij}\bar{Y}_{.j}\right)$$

Using the distributive property, we "multiply through" by Σ_i, noting that $\bar{Y}_{.j}$ varies only with j; it is constant when i is the index of summation. Terms are also rearranged so that sums are pre-multiplied by constants:

$$\sum_j\sum_i\left(Y_{ij}-\bar{Y}_{.j}\right)^2=\sum_j\left(\sum_i Y_{ij}^2+n\bar{Y}_{.j}^2-2\bar{Y}_{.j}\sum_i Y_{ij}\right)$$

Note that $\Sigma_i\bar{Y}_{.j}=n\bar{Y}_{.j}$. Although $\bar{Y}_{.j}$ is a variable relative to the index j, it is a constant relative to i, the index over which we are currently summing; therefore, Rule 2 applies.

Substituting raw-score formulas for the group means gives

$$\sum_j\sum_i\left(Y_{ij}-\bar{Y}_{.j}\right)^2=\sum_j\left[\sum_i Y_{ij}^2+n\frac{\left(\Sigma_i Y_{ij}\right)^2}{n^2}-2\left(\frac{\Sigma_i Y_{ij}}{n}\right)\sum_i Y_{ij}\right]$$

Simplifying gives

$$\sum_j\sum_i\left(Y_{ij}-\bar{Y}_{.j}\right)^2=\sum_j\left[\sum_i Y_{ij}^2-\frac{\left(\Sigma_i Y_{ij}\right)^2}{n}\right]$$

which can also be written

$$\sum_j\sum_i Y_{ij}^2-\frac{\Sigma_j\left(\Sigma_i Y_{ij}\right)^2}{n}$$

To simplify notation, we can use T (for "total") to replace ΣY. The sum of scores, for example, for group j is

$$T_{.j}=\sum_i Y_{ij}$$

and the raw-score expression just derived can be rewritten as

$$\sum_j\sum_i Y_{ij}^2-\frac{\Sigma_j T_{.j}^2}{n}$$

Notes

1 To conserve space, when we wish to indicate an index of summation in a line of text or a fraction, we will often write it as a subscript. The expression $\Sigma_i Y_i$ should be considered equivalent to ΣY_i.

2 In the example design in Table A.1, the mean of the first row would not be a quantity of interest, because we stipulated that the order within each column was arbitrary. There are designs, however, for which row means are as interesting and important as column means.

Appendix B

Expected Values and Their Applications

The view of a population parameter as the expected value of a statistic is inherent in most inferential procedures. Furthermore, many important results are derived by taking expectations of statistics. The following discussion introduces these ideas. We begin by defining an expected value, and we then present some rules for working with expectations. We then apply these rules to derive some results that were presented earlier in this book.

B.1 Definitions and Basic Rules

We repeat the earlier definitions of expected values (see Chapter 5) for convenience in dealing with the other material in this appendix. The expected value of a random variable, Y, may be viewed as a weighted average of all possible values Y can take. The weights are probabilities, $p(y)$, when Y is discretely distributed and densities, $f(y)$, when Y is continuously distributed. In the discrete case,

$$E(Y) = \sum_{y} yp(y)$$

and in the continuous case,

$$E(Y) = \int_{y} yf(y)dy$$

$E(Y)$ is read as "the expected value of Y" or "the expectation of Y." The y under the summation and integral signs is meant to remind us that the sum or integral is over all possible values of Y.

The symbol E is often referred to as an *expectation operator*, meaning that it is an instruction to sum or integrate the variable indicated. The expectation operator follows a set of rules like those presented in Appendix A for the summation operator. The most important of these rules are presented next.

RULE 1. *The expectation of a constant times a variable equals the constant times the sum of the variable:*

$$E(CY) = CE(Y)$$

This may be seen by writing

$$E(CY) = \Sigma(Cy)p(y) = C\Sigma yp(y) = CE(Y)$$

RULE 2. *The expectation of a constant is the constant:*

$$E(C) = C$$

If several events have the same numerical value C, their average value will equal C.

RULE 3. *E follows the distributive law and acts like a multiplier.* For example,

$$E(X + Y) = E(X) + E(Y)$$

To prove this, begin with the definition of $E(X + Y)$:

$$E(X + Y) = \sum_x \sum_y (x + y)p(x, y)$$

where the expression on the right indicates that each possible value of $X + Y$ is multiplied by its joint probability, and these products are then summed. Distributing this expression, we obtain

$$\begin{aligned} E(X + Y) &= \sum_x \sum_y xp(x, y) + \sum_x \sum_y yp(x, y) \\ &= \sum_x x\left[\sum_y p(x, y)\right] + \sum_y y\left[\sum_x p(x, y)\right] \\ &= \sum_x xp(x) + \sum_y yp(y) = E(X) + E(Y) \end{aligned}$$

A special case of this expression occurs when one variable is replaced by a constant; then

$$E(Y + C) = E(Y) + E(C) = E(Y) + C$$

This equation allows us to assert that

$$E(Y - \mu) = 0$$

because

$$E(Y - \mu) = E(Y) - \mu = \mu - \mu = 0$$

Another application of Rule 3 is

$$E(X + Y)^2 = E(X)^2 + E(Y)^2 + 2E(XY)$$

This leads to a proof of the statement in Chapter 2 that the variance of Y, $E(Y - \mu)^2$, equals $E(Y^2) - \mu^2$:

$$\begin{aligned} E(Y-\mu)^2 &= E\left(Y^2\right)+E(\mu)^2-2E(Y\mu) \\ &= E\left(Y^2\right)+\mu^2-2\mu E(Y), \quad \text{because } \mu \text{ is a constant} \\ &= E(Y)^2+\mu^2-2\mu^2, \quad \text{because } \mu \text{ and } E(Y) \text{ are the same entity} \\ &= E\left(Y^2\right)-\mu^2 \end{aligned}$$

RULE 4. *IF X and Y are independently distributed, then* $E(XY) = E(X)E(Y)$. To prove this, we again begin with the definition of an expectation:

$$\begin{aligned} E(XY) &= \sum_x\sum_y xyp(x,y) \\ &= \sum_x\sum_y xyp(x)p(y) \end{aligned}$$

because the joint probability $p(x,y) = p(x)p(y)$ if X and Y are independently distributed. Rearranging terms gives

$$E(XY) = [\Sigma xp(x)][\Sigma yp(y)] = E(X)E(Y)$$

A useful implication of this is that $E(X-\bar{X})(Y=\bar{Y})=0$ if X and Y are independent. This follows because $E(X-\bar{X})(Y-\bar{Y})$ then must equal $[E(X-\bar{X})][E(Y-\bar{Y})]=0\times 0$. Therefore, if X and Y are independent, their covariance (and consequently ρ) must equal zero.

B.2 Applications to Estimation

We can now show that $\bar{Y}$ is an unbiased estimate of μ; that is, $E(\bar{Y}) = E(Y)$ or μ. We have

$$\begin{aligned} E(\bar{Y}) &= E\left(\frac{\Sigma Y}{n}\right) = \frac{1}{n}E(\Sigma Y) \quad \text{by Rule 1} \\ &= \frac{1}{n}\Sigma E(Y) \\ &= \frac{1}{n}(n)E(Y) = E(Y) \end{aligned}$$

We next show that $\hat{\sigma}^2$ is an unbiased estimator of σ^2; that is, $E(\hat{\sigma}^2)=\sigma^2$. Begin by considering the sum of squares, the numerator of $\hat{\sigma}^2$:

$$\begin{aligned} E[\Sigma(Y-\bar{Y})^2] &= E\Sigma[(Y-\mu)-(\bar{Y}-\mu)]^2 \\ &= E[\Sigma(Y-\mu)^2+\Sigma(\bar{Y}-\mu)^2-2(\bar{Y}-\mu)\Sigma(Y-\mu)] \\ &= E[\Sigma(Y-\mu)^2+n(\bar{Y}-\mu)^2-2n(\bar{Y}-\mu)^2] \\ &= E[\Sigma(Y-\mu)^2-n(\bar{Y}-\mu)^2] \\ &= \Sigma E(Y-\mu)^2-nE(\bar{Y}-\mu)^2 \quad \text{by Rule 3} \end{aligned}$$

The average squared deviation of a quantity from its average is a variance; that is, $E(Y - \mu)^2 = \sigma^2$ and $E(\bar{Y} - \mu)^2 = \sigma^2/n$. Therefore,

$$E[\Sigma(Y-\bar{Y})^2] = n\sigma^2 - \frac{n\sigma^2}{n}$$
$$= (n-1)\sigma^2$$

Therefore,

$$E\left(\frac{\Sigma(Y-\bar{Y})^2}{n-1}\right) = E[\hat{\sigma}^2] = \sigma^2$$

B.3 The Mean and Variance of the Binomial Distribution

Consider a series of n Bernoulli trials and let $X = 1$ or 0, depending upon whether the trial outcome was a success or failure; $p(X = 1) = p$ and $p(X = 0) = q$. The total number of successes in the n trials is $Y = \Sigma X$. We want to derive expressions for $E(Y)$ and Var(Y), the mean and variance of the binomial distribution. We have

$$E(Y) = E(\Sigma X) = \Sigma E(X)$$
$$= \Sigma x p(x) \text{ by definition of an expected value}$$
$$= \Sigma[(1)(p) + (0)(q)] = \Sigma p = np$$

We derive the expression for the variance of the binomial distribution in a similar manner:

$$\text{Var}(Y) = \text{Var}(\Sigma X)$$

The variance of a sum of independent variables is the sum of their variances (see Section 2.5.2); therefore,

$$\text{Var}(Y) = \text{Var}(\Sigma X) = \Sigma\text{Var}(X)$$

The variance of X is $E[X - E(X)]^2 = E(X^2) - [E(X)]^2$; see the development under Rule 3, immediately preceding Rule 4. We showed above that $E(X) = p$, and

$$E(X^2) = (1^2)(p) + (0^2)(q) = p$$

by definition of an expected value. Therefore, $Var(X) = E(X^2) - [E(X)]^2 = p - p^2 = p(1-p) = pq$. Finally, we have

$$Var(Y) = \sum Var(X) = \sum pq = npq$$

Appendix C

Statistical Tables

Table C.1 The binomial probability: $p(y, n, p)$

	y	.05	.10	.15	.20	.25	.30	.35	.40	.45	.50
$n = 4$											
	0	.8145	.6561	.5220	.4096	.3164	.2401	.1785	.1296	.0915	.0625
	1	.1715	.2916	.3685	.4096	.4219	.4116	.3845	.3456	.2995	.2500
	2	.0135	.0486	.0975	.1536	.2109	.2646	.3105	.3456	.3675	.3750
	3	.0005	.0036	.0115	.0256	.0469	.0756	.1115	.1536	.2005	.2500
	4	.0000	.0001	.0005	.0016	.0039	.0081	.0150	.0256	.0410	.0625
$n = 5$											
	0	.7738	.5905	.4437	.3277	.2373	.1681	.1160	.0778	.0503	.0313
	1	.2036	.3281	.3915	.4096	.3955	.3601	.3124	.2592	.2059	.1563
	2	.0214	.0729	.1382	.2048	.2637	.3087	.3364	.3456	.3369	.3125
	3	.0011	.0081	.0244	.0512	.0879	.1323	.1811	.2304	.2757	.3125
	4	.0000	.0005	.0022	.0064	.0146	.0284	.0488	.0768	.1128	.1563
	5	.0000	.0000	.0001	.0003	.0010	.0024	.0053	.0102	.0185	.0313
$n = 6$											
	0	.7351	.5314	.3771	.2621	.1780	.1176	.0754	.0467	.0277	.0156
	1	.2321	.3543	.3993	.3932	.3560	.3025	.2437	.1866	.1359	.0938
	2	.0305	.0984	.1762	.2458	.2966	.3241	.3280	.3110	.2780	.2344
	3	.0021	.0146	.0415	.0819	.1318	.1852	.2355	.2765	.3032	.3125
	4	.0001	.0012	.0055	.0154	.0330	.0595	.0951	.1382	.1861	.2344
	5	.0000	.0001	.0004	.0015	.0044	.0102	.0205	.0369	.0609	.0938
	6	.0000	.0000	.0000	.0001	.0002	.0007	.0018	.0041	.0083	.0156
$n = 7$											
	0	.6983	.4783	.3206	.2097	.1335	.0824	.0490	.0280	.0152	.0078
	1	.2573	.3720	.3960	.3670	.3115	.2471	.1848	.1306	.0872	.0547
	2	.0406	.1240	.2097	.2753	.3115	.3177	.2985	.2613	.2140	.1641
	3	.0036	.0230	.0617	.1147	.1730	.2269	.2679	.2903	.2918	.2734
	4	.0002	.0026	.0109	.0287	.0577	.0972	.1442	.1935	.2388	.2734
	5	.0000	.0002	.0012	.0043	.0115	.0250	.0466	.0774	.1172	.1641
	6	.0000	.0000	.0001	.0004	.0013	.0036	.0084	.0172	.0320	.0547
	7	.0000	.0000	.0000	.0000	.0001	.0002	.0006	.0016	.0037	.0078
$n = 8$											
	0	.6634	.4305	.2725	.1678	.1001	.0576	.0319	.0168	.0084	.0039
	1	.2793	.3826	.3847	.3355	.2670	.1977	.1373	.0896	.0548	.0313
	2	.0515	.1488	.2376	.2936	.3115	.2965	.2587	.2090	.1569	.1094
	3	.0054	.0331	.0839	.1468	.2076	.2541	.2786	.2787	.2568	.2188
	4	.0004	.0046	.0185	.0459	.0865	.1361	.1875	.2322	.2627	.2734
	5	.0000	.0004	.0026	.0092	.0231	.0467	.0808	.1239	.1719	.2188
	6	.0000	.0000	.0002	.0011	.0038	.0100	.0217	.0413	.0703	.1094
	7	.0000	.0000	.0000	.0001	.0004	.0012	.0033	.0079	.0164	.0313
	8	.0000	.0000	.0000	.0000	.0000	.0001	.0002	.0007	.0017	.0039
$n = 9$											
	0	.6302	.3874	.2316	.1342	.0751	.0404	.0207	.0101	.0046	.0020
	1	.2985	.3874	.3679	.3020	.2253	.1556	.1004	.0605	.0339	.0176
	2	.0629	.1722	.2597	.3020	.3003	.2668	.2162	.1612	.1110	.0703
	3	.0077	.0446	.1069	.1762	.2336	.2668	.2716	.2508	.2119	.1641
	4	.0006	.0074	.0283	.0661	.1168	.1715	.2194	.2508	.2600	.2461
	5	.0000	.0008	.0050	.0165	.0389	.0735	.1181	.1672	.2128	.2461
	6	.0000	.0001	.0006	.0028	.0087	.0210	.0424	.0743	.1160	.1641
	7	.0000	.0000	.0000	.0003	.0012	.0039	.0098	.0212	.0407	.0703
	8	.0000	.0000	.0000	.0000	.0001	.0004	.0013	.0035	.0083	.0176
	9	.0000	.0000	.0000	.0000	.0000	.0000	.0001	.0003	.0008	.0020

Table C.1 (Continued)

	y	p .05	.10	.15	.20	.25	.30	.35	.40	.45	.50
$n = 10$											
	0	.5987	.3487	.1969	.1074	.0563	.0282	.0135	.0060	.0025	.0010
	1	.3151	.3874	.3474	.2684	.1877	.1211	.0725	.0403	.0207	.0098
	2	.0746	.1937	.2759	.3020	.2816	.2335	.1757	.1209	.0763	.0439
	3	.0105	.0574	.1298	.2013	.2503	.2668	.2522	.2150	.1665	.1172
	4	.0010	.0112	.0401	.0881	.1460	.2001	.2377	.2508	.2384	.2051
	5	.0001	.0015	.0085	.0264	.0584	.1029	.1536	.2007	.2340	.2461
	6	.0000	.0001	.0012	.0055	.0162	.0368	.0689	.1115	.1596	.2051
	7	.0000	.0000	.0001	.0008	.0031	.0090	.0212	.0425	.0746	.1172
	8	.0000	.0000	.0000	.0001	.0004	.0014	.0043	.0106	.0229	.0439
	9	.0000	.0000	.0000	.0000	.0000	.0001	.0005	.0016	.0042	.0098
	10	.0000	.0000	.0000	.0000	.0000	.0000	.0000	.0001	.0003	.0010
$n = 11$											
	0	.5688	.3138	.1673	.0859	.0422	.0198	.0088	.0036	.0014	.0005
	1	.3293	.3835	.3248	.2362	.1549	.0932	.0518	.0266	.0125	.0054
	2	.0867	.2131	.2866	.2953	.2581	.1998	.1395	.0887	.0513	.0269
	3	.0137	.0710	.1517	.2215	.2581	.2568	.2254	.1774	.1259	.0806
	4	.0014	.0158	.0536	.1107	.1721	.2201	.2428	.2365	.2060	.1611
	5	.0001	.0025	.0132	.0388	.0803	.1321	.1830	.2207	.2360	.2256
	6	.0000	.0003	.0023	.0097	.0268	.0566	.0985	.1471	.1931	.2256
	7	.0000	.0000	.0003	.0017	.0064	.0173	.0379	.0701	.1128	.1611
	8	.0000	.0000	.0000	.0002	.0011	.0037	.0102	.0234	.0462	.0806
	9	.0000	.0000	.0000	.0000	.0001	.0005	.0018	.0052	.0126	.0269
	10	.0000	.0000	.0000	.0000	.0000	.0000	.0002	.0007	.0021	.0054
	11	.0000	.0000	.0000	.0000	.0000	.0000	.0000	.0000	.0002	.0005
$n = 12$											
	0	.5404	.2824	.1422	.0687	.0317	.0138	.0057	.0022	.0008	.0002
	1	.3413	.3766	.3012	.2062	.1267	.0712	.0368	.0174	.0075	.0029
	2	.0988	.2301	.2924	.2835	.2323	.1678	.1088	.0639	.0339	.0161
	3	.0173	.0852	.1720	.2362	.2581	.2397	.1954	.1419	.0923	.0537
	4	.0021	.0213	.0683	.1329	.1936	.2311	.2367	.2128	.1700	.1208
	5	.0002	.0038	.0193	.0532	.1032	.1585	.2039	.2270	.2225	.1934
	6	.0000	.0005	.0040	.0155	.0401	.0792	.1281	.1766	.2124	.2256
	7	.0000	.0000	.0006	.0033	.0115	.0291	.0591	.1009	.1489	.1934
	8	.0000	.0000	.0001	.0005	.0024	.0078	.0199	.0420	.0762	.1208
	9	.0000	.0000	.0000	.0001	.0004	.0015	.0048	.0125	.0277	.0537
	10	.0000	.0000	.0000	.0000	.0000	.0002	.0008	.0025	.0068	.0161
	11	.0000	.0000	.0000	.0000	.0000	.0000	.0001	.0003	.0010	.0029
	12	.0000	.0000	.0000	.0000	.0000	.0000	.0000	.0000	.0001	.0002
$n = 13$											
	0	.5133	.2542	.1209	.0550	.0238	.0097	.0037	.0013	.0004	.0001
	1	.3512	.3672	.2774	.1787	.1029	.0540	.0259	.0113	.0045	.0016
	2	.1109	.2448	.2937	.2680	.2059	.1388	.0836	.0453	.0220	.0095
	3	.0214	.0997	.1900	.2457	.2517	.2181	.1651	.1107	.0660	.0349
	4	.0028	.0277	.0838	.1535	.2097	.2337	.2222	.1845	.1350	.0873
	5	.0003	.0055	.0266	.0691	.1258	.1803	.2154	.2214	.1989	.1571
	6	.0000	.0008	.0063	.0230	.0559	.1030	.1546	.1968	.2169	.2095
	7	.0000	.0001	.0011	.0058	.0186	.0442	.0833	.1312	.1775	.2095
	8	.0000	.0000	.0001	.0011	.0047	.0142	.0336	.0656	.1089	.1571

Table C.1 (Continued)

		p									
	y	*.05*	*.10*	*.15*	*.20*	*.25*	*.30*	*.35*	*.40*	*.45*	*.50*
	9	.0000	.0000	.0000	.0001	.0009	.0034	.0101	.0243	.0495	.0873
	10	.0000	.0000	.0000	.0000	.0001	.0006	.0022	.0065	.0162	.0349
	11	.0000	.0000	.0000	.0000	.0000	.0001	.0003	.0012	.0036	.0095
	12	.0000	.0000	.0000	.0000	.0000	.0000	.0000	.0001	.0005	.0016
	13	.0000	.0000	.0000	.0000	.0000	.0000	.0000	.0000	.0000	.0001
n = 14											
	0	.4877	.2288	.1028	.0440	.0178	.0068	.0024	.0008	.0002	.0001
	1	.3593	.3559	.2539	.1539	.0832	.0407	.0181	.0073	.0027	.0009
	2	.1229	.2570	.2912	.2501	.1802	.1134	.0634	.0317	.0141	.0056
	3	.0259	.1142	.2056	.2501	.2402	.1943	.1366	.0845	.0462	.0222
	4	.0037	.0349	.0998	.1720	.2202	.2290	.2022	.1549	.1040	.0611
	5	.0004	.0078	.0352	.0860	.1468	.1963	.2178	.2066	.1701	.1222
	6	.0000	.0013	.0093	.0322	.0734	.1262	.1759	.2066	.2088	.1833
	7	.0000	.0002	.0019	.0092	.0280	.0618	.1082	.1574	.1952	.2095
	8	.0000	.0000	.0003	.0200	.0082	.0232	.0510	.0918	.1398	.1833
	9	.0000	.0000	.0000	.0003	.0018	.0066	.0183	.0408	.0762	.1222
	10	.0000	.0000	.0000	.0000	.0003	.0014	.0019	.0136	.0312	.0611
	11	.0000	.0000	.0000	.0000	.0000	.0002	.0010	.0033	.0093	.0222
	12	.0000	.0000	.0000	.0000	.0000	.0000	.0001	.0005	.0019	.0056
	13	.0000	.0000	.0000	.0000	.0000	.0000	.0000	.0001	.0002	.0009
	14	.0000	.0000	.0000	.0000	.0000	.0000	.0000	.0000	.0000	.0001
n = 15											
	0	.4633	.2059	.0874	.0352	.0134	.0047	.0016	.0005	.0001	.0000
	1	.3658	.3432	.2312	.1319	.0668	.0305	.0126	.0047	.0016	.0005
	2	.1348	.2669	.2856	.2309	.1559	.0916	.0476	.0219	.0090	.0032
	3	.0307	.1285	.2184	.2501	.2252	.1700	.1110	.0634	.0318	.0139
	4	.0049	.0428	.1156	.1876	.2252	.2186	.1792	.1268	.0780	.0417
	5	.0006	.0105	.0449	.1032	.1651	.2061	.2123	.1859	.1404	.0916
	6	.0000	.0019	.0132	.0430	.0917	.1472	.1906	.2066	.1914	.1527
	7	.0000	.0003	.0030	.0138	.0393	.0811	.1319	.1771	.2013	.1964
	8	.0000	.0000	.0005	.0035	.0131	.0348	.0710	.1181	.1647	.1964
	9	.0000	.0000	.0001	.0007	.0034	.0116	.0298	.0612	.1048	.1527
	10	.0000	.0000	.0000	.0001	.0007	.0030	.0096	.0245	.0515	.0916
	11	.0000	.0000	.0000	.0000	.0001	.0006	.0024	.0074	.0191	.0417
	12	.0000	.0000	.0000	.0000	.0000	.0001	.0004	.0016	.0052	.0139
	13	.0000	.0000	.0000	.0000	.0000	.0000	.0001	.0003	.0010	.0032
	14	.0000	.0000	.0000	.0000	.0000	.0000	.0000	.0000	.0001	.0005
	15	.0000	.0000	.0000	.0000	.0000	.0000	.0000	.0000	.0000	.0000
n = 16											
	0	.4401	.1853	.0743	.0281	.0100	.0033	.0010	.0003	.0001	.0000
	1	.3706	.3294	.2097	.1126	.0535	.0228	.0087	.0030	.0009	.0002
	2	.1463	.2745	.2775	.2111	.1336	.0732	.0353	.0150	.0056	.0018
	3	.0359	.1423	.2285	.2463	.2079	.1465	.0888	.0468	.0215	.0085
	4	.0061	.0514	.1311	.2001	.2252	.2040	.1553	.1014	.0572	.0278
	5	.0008	.0137	.0555	.1201	.1802	.2099	.2008	.1623	.1123	.0667
	6	.0001	.0028	.0180	.0550	.1101	.1649	.1982	.1983	.1684	.1222
	7	.0000	.0004	.0045	.0197	.0524	.1010	.1524	.1889	.1969	.1746
	8	.0000	.0001	.0009	.0055	.0197	.0487	.0923	.1417	.1812	.1964
	9	.0000	.0000	.0001	.0012	.0058	.0185	.0442	.0840	.1318	.1746

Table C.1 (Continued)

	y	.05	.10	.15	.20	.25	.30	.35	.40	.45	.50
	10	.0000	.0000	.0000	.0002	.0014	.0056	.0167	.0392	.0755	.1222
	11	.0000	.0000	.0000	.0000	.0002	.0013	.0049	.0142	.0337	.0667
	12	.0000	.0000	.0000	.0000	.0000	.0002	.0011	.0040	.0115	.0278
	13	.0000	.0000	.0000	.0000	.0000	.0000	.0002	.0008	.0029	.0085
	14	.0000	.0000	.0000	.0000	.0000	.0000	.0000	.0001	.0005	.0018
	15	.0000	.0000	.0000	.0000	.0000	.0000	.0000	.0000	.0001	.0002
	16	.0000	.0000	.0000	.0000	.0000	.0000	.0000	.0000	.0000	.0000
n = 17											
	0	.4181	.1668	.0631	.0225	.0075	.0023	.0007	.0002	.0000	.0000
	1	.3741	.3150	.1893	.0957	.0426	.0169	.0060	.0019	.0005	.0001
	2	.1575	.2800	.2673	.1914	.1136	.0581	.0260	.0102	.0035	.0010
	3	.0415	.1556	.2359	.2393	.1893	.1245	.0701	.0341	.0144	.0052
	4	.0076	.0605	.1457	.2093	.2209	.1868	.1320	.0796	.0411	.0182
	5	.0010	.0175	.0668	.1361	.1914	.2081	.1849	.1379	.0875	.0472
	6	.0001	.0039	.0236	.0680	.1276	.1784	.1991	.1839	.1432	.0944
	7	.0000	.0007	.0065	.0267	.0668	.1201	.1685	.1927	.1841	.1484
	8	.0000	.0001	.0014	.0084	.0279	.0644	.1134	.1606	.1883	.1855
	9	.0000	.0000	.0003	.0021	.0093	.0276	.0611	.1070	.1540	.1855
	10	.0000	.0000	.0000	.0004	.0025	.0095	.0263	.0571	.1008	.1484
	11	.0000	.0000	.0000	.0001	.0005	.0026	.0090	.0242	.0525	.0944
	12	.0000	.0000	.0000	.0000	.0001	.0006	.0024	.0081	.0215	.0472
	13	.0000	.0000	.0000	.0000	.0000	.0001	.0005	.0021	.0068	.0182
	14	.0000	.0000	.0000	.0000	.0000	.0000	.0001	.0004	.0016	.0052
	15	.0000	.0000	.0000	.0000	.0000	.0000	.0000	.0001	.0003	.0010
	16	.0000	.0000	.0000	.0000	.0000	.0000	.0000	.0000	.0000	.0001
	17	.0000	.0000	.0000	.0000	.0000	.0000	.0000	.0000	.0000	.0000
n = 18											
	0	.3972	.1501	.0536	.0180	.0056	.0016	.0004	.0001	.0000	.0000
	1	.3763	.3002	.1704	.0811	.0338	.0126	.0042	.0012	.0003	.0001
	2	.1683	.2835	.2556	.1723	.0958	.0458	.0190	.0069	.0022	.0006
	3	.0473	.1680	.2406	.2297	.1704	.1046	.0547	.0246	.0095	.0031
	4	.0093	.0700	.1592	.2153	.2130	.1681	.1104	.0614	.0291	.0117
	5	.0014	.0218	.0787	.1507	.1988	.2017	.1664	.1146	.0666	.0327
	6	.0002	.0052	.0301	.0816	.1436	.1873	.1941	.1655	.1181	.0708
	7	.0000	.0010	.0091	.0350	.0820	.1376	.1792	.1892	.1657	.1214
	8	.0000	.0002	.0022	.0120	.0376	.0811	.1327	.1734	.1864	.1669
	9	.0000	.0000	.0004	.0033	.0139	.0386	.0794	.1284	.1694	.1855
	10	.0000	.0000	.0001	.0008	.0042	.0149	.0385	.0771	.1248	.1669
	11	.0000	.0000	.0000	.0001	.0010	.0046	.0151	.0374	.0742	.1214
	12	.0000	.0000	.0000	.0000	.0002	.0012	.0047	.0145	.0354	.0708
	13	.0000	.0000	.0000	.0000	.0000	.0002	.0012	.0045	.0134	.0327
	14	.0000	.0000	.0000	.0000	.0000	.0000	.0002	.0011	.0039	.0117
	15	.0000	.0000	.0000	.0000	.0000	.0000	.0000	.0002	.0009	.0031
	16	.0000	.0000	.0000	.0000	.0000	.0000	.0000	.0000	.0001	.0006
	17	.0000	.0000	.0000	.0000	.0000	.0000	.0000	.0000	.0000	.0001
	18	.0000	.0000	.0000	.0000	.0000	.0000	.0000	.0000	.0000	.0000
n = 19											
	0	.3774	.1351	.0456	.0144	.0042	.0011	.0003	.0001	.0000	.0000
	1	.3774	.2852	.1529	.0685	.0268	.0093	.0029	.0008	.0002	.0000

Table C.1 (Continued)

		p									
	y	*.05*	*.10*	*.15*	*.20*	*.25*	*.30*	*.35*	*.40*	*.45*	*.50*
	2	.1787	.2852	.2428	.1540	.0803	.0358	.0138	.0046	.0013	.0003
	3	.0533	.1796	.2428	.2182	.1517	.0869	.0422	.0175	.0062	.0018
	4	.0112	.0798	.1714	.2182	.2023	.1491	.0909	.0467	.0203	.0074
	5	.0018	.0266	.0907	.1636	.2023	.1916	.1468	.0933	.0497	.0222
	6	.0002	.0069	.0374	.0955	.1574	.1916	.1844	.1451	.0949	.0518
	7	.0000	.0014	.0122	.0443	.0974	.1525	.1844	.1797	.1443	.0961
	8	.0000	.0002	.0032	.0166	.0487	.0981	.1489	.1797	.1771	.1442
	9	.0000	.0000	.0007	.0051	.0198	.0514	.0980	.1464	.1771	.1762
	10	.0000	.0000	.0001	.0013	.0066	.0220	.0528	.0976	.1449	.1762
	11	.0000	.0000	.0000	.0003	.0018	.0077	.0233	.0532	.0970	.1442
	12	.0000	.0000	.0000	.0000	.0004	.0022	.0083	.0237	.0529	.0961
	13	.0000	.0000	.0000	.0000	.0001	.0005	.0024	.0085	.0233	.0518
	14	.0000	.0000	.0000	.0000	.0000	.0001	.0006	.0024	.0082	.0222
	15	.0000	.0000	.0000	.0000	.0000	.0000	.0001	.0005	.0022	.0074
	16	.0000	.0000	.0000	.0000	.0000	.0000	.0000	.0001	.0005	.0018
	17	.0000	.0000	.0000	.0000	.0000	.0000	.0000	.0000	.0001	.0003
	18	.0000	.0000	.0000	.0000	.0000	.0000	.0000	.0000	.0000	.0000
	19	.0000	.0000	.0000	.0000	.0000	.0000	.0000	.0000	.0000	.0000
$n = 20$											
	0	.3585	.1216	.0388	.0115	.0032	.0008	.0002	.0000	.0000	.0000
	1	.3774	.2702	.1368	.0576	.0211	.0068	.0020	.0005	.0001	.0000
	2	.1887	.2852	.2293	.1369	.0669	.0278	.0100	.0031	.0008	.0002
	3	.0596	.1901	.2428	.2054	.1339	.0716	.0323	.0123	.0040	.0011
	4	.0133	.0898	.1821	.2182	.1897	.1304	.0738	.0350	.0139	.0046
	5	.0022	.0319	.1028	.1746	.2023	.1789	.1272	.0746	.0365	.0148
	6	.0003	.0089	.0454	.1091	.1686	.1916	.1712	.1244	.0746	.0370
	7	.0000	.0020	.0160	.0545	.1124	.1643	.1844	.1659	.1221	.0739
	8	.0000	.0004	.0046	.0222	.0609	.1144	.1614	.1797	.1623	.1201
	9	.0000	.0001	.0011	.0074	.0271	.0654	.1158	.1597	.1771	.1602
	10	.0000	.0000	.0002	.0020	.0099	.0308	.0686	.1171	.1593	.1762
	11	.0000	.0000	.0000	.0005	.0030	.0120	.0336	.0710	.1185	.1602
	12	.0000	.0000	.0000	.0001	.0008	.0039	.0136	.0355	.0727	.1201
	13	.0000	.0000	.0000	.0000	.0002	.0010	.0045	.0146	.0366	.0739
	14	.0000	.0000	.0000	.0000	.0000	.0002	.0012	.0049	.0150	.0370
	15	.0000	.0000	.0000	.0000	.0000	.0000	.0003	.0013	.0049	.0148
	16	.0000	.0000	.0000	.0000	.0000	.0000	.0000	.0003	.0013	.0046
	17	.0000	.0000	.0000	.0000	.0000	.0000	.0000	.0000	.0002	.0011
	18	.0000	.0000	.0000	.0000	.0000	.0000	.0000	.0000	.0000	.0002
	19	.0000	.0000	.0000	.0000	.0000	.0000	.0000	.0000	.0000	.0000
	20	.0000	.0000	.0000	.0000	.0000	.0000	.0000	.0000	.0000	.0000

Table C.2 The standardized normal distribution

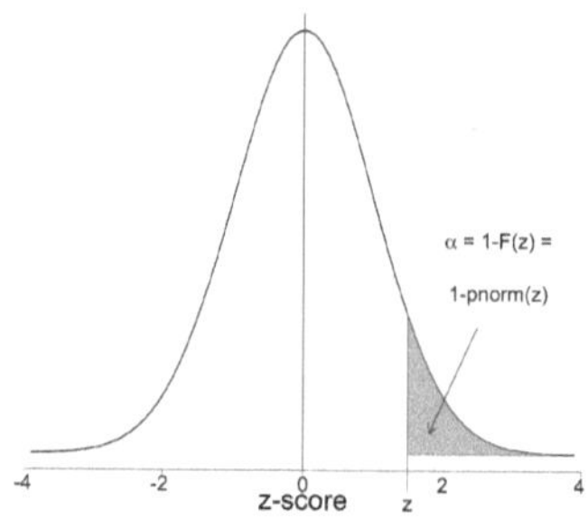

z	α	z	α	z	α	z	α	z	α
0.00	.5000	0.33	.3707	0.66	.2546	0.99	.1611	1.32	.0934
0.01	.4960	0.34	.3669	0.67	.2514	1.00	.1587	1.33	.0918
0.02	.4920	0.35	.3632	0.68	.2483	1.01	.1562	1.34	.0901
0.03	.4880	0.36	.3594	0.69	.2451	1.02	.1539	1.35	.0885
0.04	.4840	0.37	.3557	0.70	.2420	1.03	.1515	1.36	.0869
0.05	.4801	0.38	.3520	0.71	.2389	1.04	.1492	1.37	.0853
0.06	.4761	0.39	.3483	0.72	.2358	1.05	.1469	1.38	.0838
0.07	.4721	0.40	.3446	0.73	.2327	1.06	.1446	1.39	.0823
0.08	.4681	0.41	.3409	0.74	.2296	1.07	.1423	1.40	.0808
0.09	.4641	0.42	.3372	0.75	.2266	1.08	.1401	1.41	.0793
0.10	.4602	0.33	.3336	0.76	.2236	1.09	.1379	1.42	.0778
0.11	.4562	0.44	.3300	0.77	.2206	1.10	.1357	1.43	.0764
0.12	.4522	0.45	.3264	0.78	.2177	1.11	.1335	1.44	.0749
0.13	.4483	0.46	.3228	0.79	.2148	1.12	.1314	1.45	.0735
0.14	.4443	0.47	.3192	0.80	.2119	1.13	.1292	1.46	.0721
0.15	.4404	0.48	.3156	0.81	.2090	1.14	.1271	1.47	.0708
0.16	.4364	0.49	.3121	0.82	.2061	1.15	.1251	1.48	.0694
0.17	.4325	0.50	.3085	0.83	.2033	1.16	.1230	1.49	.0681
0.18	.4286	0.51	.3050	0.84	.2005	1.17	.1210	1.50	.0668
0.19	.4247	0.52	.3015	0.85	.1977	1.18	.1190	1.51	.0655
0.20	.4207	0.53	.2981	0.86	.1949	1.19	.1170	1.52	.0643
0.21	.4168	0.54	.2946	0.87	.1922	1.20	.1151	1.53	.0630
0.22	.4129	0.55	.2912	0.88	.1894	1.31	.1131	1.54	.0618
0.23	.4090	0.56	.2877	0.89	.1867	1 22	.1112	1.55	.0606
0.24	.4052	0.57	.2843	0.90	.1841	1.23	.1093	1.56	.0594
0.25	.4013	0.58	.2810	0.91	.1814	1.24	.1075	1.57	.0582
0.26	.3974	0.59	.2776	0.92	.1788	1.25	.1056	1.58	.0571
0.27	.3936	0.60	.2743	0.93	.1762	1.26	.1038	1.59	.0559
0.28	.3897	0.61	.2709	0.94	.1736	1.27	.1020	1.60	.0548
0.29	.3859	0.62	.2676	0.95	.1711	1.28	.1003	1.61	.0537
0.30	.3821	0.63	.2643	0.96	.1685	1.29	.0985	1.62	.0526
0.31	.3783	0.64	.2611	0.97	.1660	1.30	.0968	1.63	.0516
0.32	.3745	0.65	.2578	0.98	.1635	1.31	.0951	1.64	.0505

Table C.2 (Continued)

z	α	z	α	z	α	z	α	z	α
1.65	.0495	1.98	.0239	2.31	.0104	2.64	.0041	2.97	.0015
1.66	.0485	1.99	.0233	2.32	.0102	2.65	.0040	2.98	.0014
1.67	.0475	2.00	.0228	2.33	.0099	2.66	.0039	2.99	.0014
1.68	.0465	2.01	.0222	2.34	.0096	2.67	.0038	3.00	.0013
1.69	.0455	2.02	.0217	2.35	.0094	2.68	.0037	3.01	.0013
1.70	.0446	2.03	.0212	2.36	.0091	2.69	.0036	3.02	.0013
1.71	.0436	2.04	.0207	2.37	.0089	3.0	.0035	3.03	.0012
1.72	.0427	2.05	.0202	2.38	.0087	2.71	.0034	3.04	.0012
1.73	.0418	2.06	.0197	2.39	.0084	2.72	.0033	3.05	.0011
1.74	.0409	2.07	.0192	2.40	.0082	2.73	.0032	3.06	.0011
1.75	.0401	2.08	.0188	2.41	.0080	2.74	.0031	3.07	.0011
1.76	.0392	2.09	.0183	2.42	.0078	2.75	.0030	3.08	.0010
1.77	.0384	2.10	.0179	2.43	.0075	2.76	.0029	3.09	.0010
1.78	.0375	2.11	.0174	2.44	.0073	2.77	.0028	3.10	.0010
1.79	.0367	2.12	.0170	2.45	.0071	2.78	.0027	3.11	.0009
1.80	.0359	2.13	.0166	2.46	.0069	2.79	.0026	3.12	.0009
1.81	.0351	2.14	.0162	2.47	.0068	2.80	.0026	3.13	.0009
1.82	.0344	2.15	.0158	2.48	.0066	2.81	.0025	3.14	.0008
1.83	.0336	2.16	.0154	2.49	.0064	2.82	.0024	3.15	.0008
1.84	.0329	2.17	.0150	2.50	.0062	2.83	.0023	3.16	.0008
1.85	.0322	2.18	.0146	2.51	.0060	2.84	.0023	3.17	.0008
1.86	.0314	2.19	.0143	2.52	.0059	2.85	.0022	3.18	.0007
1.87	.0307	2.20	.0139	2.53	.0057	2.86	.0021	3.19	.0007
1.88	.0301	2.21	.0136	2.54	.0055	2.87	.0021	3.20	.0007
1.89	.0294	2.22	.0132	2.55	.0054	2.88	.0020	3.21	.0007
1.90	.0287	2.23	.0129	2.56	.0052	2.89	.0019	3.22	.0006
1.91	.0281	2.24	.0125	2.57	.0051	0.00	.0019	3.23	.0006
1.92	.0274	2.25	.0122	2.58	.0049	2.91	.0018	3.24	.0006
1.93	.0268	2.26	.0119	2.59	.0048	2.92	.0018	3.25	.0006
1.94	.0262	2.27	.0116	2.60	.0047	2.93	.0017		
1.95	.0256	2.28	.0113	2.61	.0045	2.94	.0016		
1.96	.0250	2.29	.0110	2.62	.0044	2.95	.0016		
1.97	.0244	2.30	.0107	2.63	.0043	2.96	.0015		

Source: Adapted from Table 1 in Pearson, E. S. and Hartley, H. O. (1958). *Biometrika Tables for Statisticians*, Vol. 1, 2nd ed. Cambridge University Press: Cambridge.

Table C.3 Percentage points of the *t* distribution

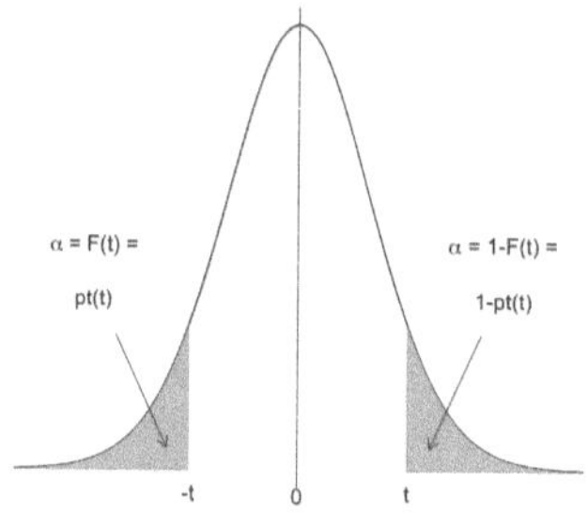

	Level of significance for a one-tailed test									
	0.4	*0.25*	*0.1*	*0.05*	*0.025*	*0.01*	*0.005*	*0.0025*	*0.001*	*0.0005*
	Level of significance for a two-tailed test									
df	*0.8*	*0.5*	*0.2*	*0.1*	*0.05*	*0.02*	*0.01*	*0.005*	*0.002*	*0.001*
1	0.325	1.000	3.078	6.314	12.706	31.821	63.657	127.32	318.31	636.62
2	.289	0.816	1.886	2.920	4.303	6.965	9.925	14.089	22.327	31.599
3	.277	.765	1.638	2.353	3.182	4.541	5.841	7.453	10.215	12.924
4	.271	.741	1.533	2.132	2.776	3.747	4.604	5.598	7.173	8.610
5	0.267	0.727	1.476	2.015	2.571	3.365	4.032	4.773	5.893	6.869
6	.265	.718	1.440	1.943	2.447	3.143	3.707	4.317	5.208	5.959
7	.263	.711	1.415	1.895	2.365	2.998	3.499	4.029	4.785	5.408
8	.262	.706	1.397	1.860	2.306	2.896	3.355	3.833	4.501	5.041
9	.261	.703	1.383	1.833	2.262	2.821	3.250	3.690	4.297	4.781
10	0.260	0.700	1.372	1.812	2.228	2.764	3.169	3.581	4.144	4.587
11	.260	.697	1.363	1.796	2.201	2.718	3.106	3.497	4.025	4.437
12	.259	.695	1.356	1.782	2.179	2.681	3.055	3.428	3.930	4.318
13	.259	.694	1.350	1.771	2.160	2.650	3.012	3.372	3.852	4.221
14	.258	.692	1.345	1.761	2.145	2.624	2.977	3.326	3.787	4.140
15	0.258	0.691	1.341	1.753	2.131	2.602	2.947	3.286	3.733	4.073
16	.258	.690	1.337	1.746	2.120	2.583	2.921	3.252	3.686	4.015
17	.257	.689	1.333	1.740	2.110	2.567	2.898	3.222	3.646	3.965
18	.257	.688	1.330	1.734	2.101	2.552	2.878	3.197	3.610	3.922
19	.257	.688	1.328	1.729	2.093	2.539	2.861	3.174	3.579	3.883
20	0.257	0.687	1.325	1.725	2.086	2.528	2.845	3.153	3.552	3.850
21	.257	.686	1.323	1.721	2.080	2.518	2.831	3.135	3.527	3.819
22	.256	.686	1.321	1.717	2.074	2.508	2.819	3.119	3.505	3.792
23	.256	.685	1.319	1.714	2.069	2.500	2.807	3.104	3.485	3.768
24	.256	.685	1.318	1.711	2.064	2.492	2.797	3.091	3.467	3.745
25	0.256	0.684	1.316	1.708	2.060	2.485	2.787	3.078	3.450	3.725
26	.256	.684	1.315	1.706	2.056	2.479	2.779	3.067	3.435	3.707
27	.256	.684	1.314	1.703	2.052	2.473	2.771	3.057	3.421	3.690
28	.256	.683	1.313	1.701	2.048	2.467	2.763	3.047	3.408	3.674
29	.256	.683	1.311	1.699	2.045	2.462	2.756	3.038	3.396	3.659
30	0.256	0.683	1.310	1.697	2.042	2.457	2.750	3.030	3.385	3.646
40	.255	.681	1.303	1.684	2.021	2.423	2.704	2.971	3.307	3.551
60	.254	.679	1.296	1.671	2.000	2.390	2.660	2.915	3.232	3.460
120	.254	.677	1.289	1.658	1.980	2.358	2.617	2.860	3.160	3.373
∞	.253	.674	1.282	1.645	1.960	2.326	2.576	2.807	3.090	3.291

Source: Adapted from Table .2 in Pearson, E. S. and Hartley. H. O. (1958). *Biometrika Tables for Statisticians*, Vol. 1, 2nd ed. Cambridge University Press: Cambridge.

Table C.4 Percentage points of the chi-square distribution

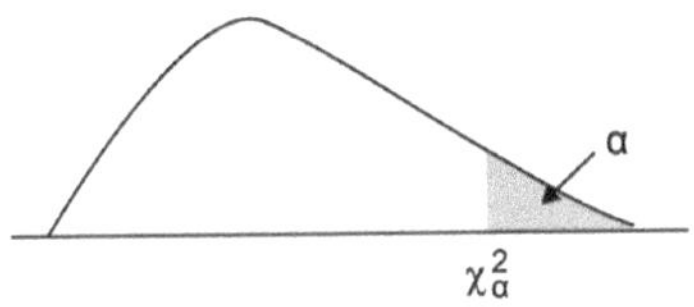

df \ α	0.995	0.9950	0.975	0.9950	0.990	0.750	0.500
1	392704.10^{-10}	157088.10^{-9}	982069.10^{-9}	393214.10^{-8}	0.0157908	0.1015308	0.454937
2	0.0100251	0.0201007	0.0506356	0.102587	0.210720	0.575364	1.38629
3	0.0717212	0.114832	0.215795	0.351846	0.584375	1.212534	2.36597
4	0.206990	0.297110	0.484419	0.710721	1.063623	1.92255	3.35670
5	0.411740	0.554300	0.831211	1.145476	1.61031	2.67460	4.35146
6	0.675727	0.872085	1.237347	1.63539	2.20413	3.45460	5.34812
7	0.989265	1.239043	1.68987	2.16735	2.83311	4.25485	6.34581
8	1.344419	1.646482	2.17973	2.73264	3.48954	5.07064	7.34412
9	1.734926	2.087912	2.70039	3.32511	4.16816	5.89883	8.34283
10	2.15585	2.55821	3.24697	3.94030	4.86518	6.73720	9.34182
11	2.60321	3.05347	3.81575	4.57481	5.57779	7.58412	10.3410
12	3.07382	3.57056	4.40379	5.22603	6.30380	8.43842	11.3403
13	3.56503	4.10691	5.00874	5.89186	7.04150	9.29906	12.3398
14	4.07468	4.66043	5.62872	6.57063	7.78953	10.1653	13.3393
15	4.60094	5.22935	6.26214	7.26094	8.54675	11.0365	14.3389
16	5.14224	5.81221	6.90766	7.96164	9.31223	11.9122	15.3385
17	5.69724	6.40776	7.56418	8.67176	10.0852	12.7919	16.3381
18	6.26481	7.01491	8.23075	9.39046	10.8649	13.6753	17.3379
19	6.84398	7.63273	8.90655	10.1170	11.6509	14.5620	18.3376
20	7.43386	8.26040	9.59083	10.8508	12.4426	15.4518	19.3374
21	8.03366	8.89720	10.28293	11.5913	13.2396	16.3444	20.3372
22	8.64272	9.54249	10.9823	12.3380	14.0415	17.2396	21.3370
23	9.26042	10.19567	11.6885	13.0905	14.8479	18.1373	22.3369
24	9.88623	10.8564	12.4011	13.8484	15.6587	19.0372	23.3367
25	10.5197	11.5240	13.1197	14.6114	16.4734	19.9393	24.3366
26	11.1603	12.1981	13.8439	15.3791	17.2919	20.8434	25.3364
27	11.8076	12.8786	14.5733	16.1513	18.1138	21.7494	26.3363
28	12.4613	13.5648	15.3079	16.9279	18.9392	22.6572	27.3363
29	13.1211	14.2565	16.0471	17.7083	19.7677	23.5666	28.3362
30	13.7867	14.9535	16.7908	18.4926	20.5992	24.4776	29.3360
40	20.7065	22.1643	24.4331	26.5093	29.0505	33.6603	39.3354
50	27.9907	29.7067	32.3574	34.7642	37.6886	42.9421	49.3349
60	35.5346	37.4848	40.4817	43.1879	46.4589	52.2938	59.3347
70	43.2752	45.4418	48.7576	51.7393	55.3290	61.6983	69.3344
80	51.1720	53.5400	57.1532	60.3915	64.2778	71.1445	79.3343
90	59.1963	61.7541	65.6466	69.1260	73.2912	80.6247	89.3342
100	67.3276	70.0648	74.2219	77.9295	82.3581	90.1332	99.3341
z	−2.5758	−2.3263	−1.9600	−1.6449	−1.2816	−0.6745	0.0000

Table C.4 (Continued)

df \ α	0.250	0.100	0.050	0.025	0.010	0.005	0.001
1	1.32330	2.70554	3.84146	5.02389	6.63490	7.87944	10.828
2	2.77259	4.60517	5.99147	7.37776	9.21034	10.5966	13.816
3	4.10835	6.25139	7.81473	9.34840	11.3449	12.8381	16.266
4	5.38527	7.77944	9.48773	11.1433	13.2767	14.8602	18.467
5	6.62568	9.23635	11.0705	12.8325	15.0863	16.7496	20.515
6	7.84080	10.6446	12.5916	14.4494	16.8119	18.5476	22.458
7	9.03715	12.0170	14.0671	16.0128	18.4753	20.2777	24.322
8	10.2188	13.3616	15.5073	17.5346	20.0902	21.9550	26.125
9	11.3887	14.6837	16.9190	19.0228	21.6660	23.5893	27.877
10	12.5489	15.9871	18.3070	20.4831	23.2093	20.1882	29.588
11	13.7007	17.2750	19.6751	21.9200	24.7250	26.7569	31.264
12	14.8454	18.5494	21.0261	23.3367	26.2170	28.2995	32.909
13	15.9839	19.8119	22.3621	24.7356	27.6883	29.8194	34.528
14	17.1170	21.0642	23.6848	26.1190	29.1413	31.3193	36.123
15	18.2451	22.3072	24.9958	27.4884	30.5779	32.8013	37.697
16	19.3688	23.5418	26.2962	28.8454	31.9999	34.2672	39.252
17	20.4887	24.7690	27.5871	30.1910	33.4087	35.7185	40.790
18	21.6049	25.9894	28.8693	31.5264	34.8053	37.1564	42.312
19	22.7178	27.2036	30.1435	32.8523	36.1908	38.5822	43.820
20	23.8277	28.4120	31.4104	34.1696	37.5662	39.9968	45.315
21	24.9348	29.6151	32.6705	35.4789	38.9321	41.4010	46.797
22	26.0393	30.8133	33.9244	36.7807	40.2894	42.7956	48.268
23	27.1413	32.0069	35.1725	38.0757	41.6384	44.1813	49.728
24	28.2412	33.1963	36.4151	39.3641	42.9798	45.5585	51.179
25	29.3389	34.3816	37.6525	40.6465	44.3141	46.9278	52.620
26	30.4345	35.5631	38.8852	41.9232	45.6417	48.2899	54.052
27	31.5284	36.7412	40.1133	43.1944	46.9630	49.6449	55.476
28	32.6205	37.9159	41.3372	44.4607	48.2782	50.9933	56.892
29	33.7109	39.0875	42.5569	45.7222	49.5879	52.3356	58.302
30	34.7998	40.2560	43.7729	46.9792	50.8922	53.6720	59.703
40	45.6160	51.8050	55.7585	59.3417	63.6907	66.7659	73.402
50	56.3336	63.1671	67.5048	71.4202	76.1539	79.4900	86.661
60	66.9814	74.3970	79.0819	83.2976	88.3794	91.9517	99.607
70	77.5766	85.5271	90.5312	95.0231	100.425	104.215	112.317
80	88.1303	96.5782	101.879	106.629	112.329	112.3291	124.839
90	98.6499	107.565	113.145	118.136	124.116	128.299	137.208
100	109.141	118.498	124.342	129.561	135.807	140.169	140.169
z	+0.6745	+1.2816	+1.6449	+1.9600	+2.3263	+2.5758	+3.0902

For *df* > 100 take

$$\chi^2 = df\left(1 - \frac{2}{9df} + z\sqrt{\frac{2}{9df}}\right)^3 \quad \text{or} \quad \chi^2 = \frac{1}{2}\left(z + \sqrt{2df - 1}\right)^2$$

according to the degree of accuracy required. *z* is the standardized normal deviate corresponding to α and is shown in the bottom line of the table.

Source: Adapted from Table 8 in Pearson, E. S. and Hartley, H. O. (1958). *Biometrika Tables for Statisticians*, Vol. 1, 2nd ed. Cambridge University Press: Cambridge.

Table C.5 Upper percentage points of the *F* distribution

df_2 \ df_1	α	1	2	3	4	5	6	8	12	24	∞
1	.001	405284	500000	540379	562500	576405	585937	598144	610667	623497	636619
	.005	16211	20000	21615	22500	23056	23437	23925	24426	24940	25465
	.01	4052	4999	5403	5625	5764	5359	5981	6106	6234	6366
	.025	647.79	799.50	864.16	899.58	921.85	937.11	956.66	976.71	997.25	1018.30
	.05	161.45	199.50	215.71	224.58	230.16	233.99	238.88	243.91	249.05	254.32
	.10	39.86	49.50	53.59	55.83	57.24	58.20	59.44	60.70	62.00	63.33
	.25	5.83	7.50	8.20	8.58	8.82	8.98	9.19	9.41	9.63	9.85
2	.001	998.5	999.0	999.2	999.2	999.3	999.3	999.4	999.4	999.5	999.5
	.005	198.50	199.00	199.17	199.25	199.30	199.33	199.37	199.42	199.46	199.51
	.01	98.49	99.00	99.17	99.25	99.30	99.33	99.36	99.42	99.46	99.50
	.025	38.51	39.00	39.17	39.25	39.30	39.33	39.37	39.42	39.46	39.50
	.05	18.51	19.00	19.16	19.25	19.30	19.33	19.37	19.41	19.45	19.50
	.10	8.53	9.00	9.16	9.24	9.29	9.33	9.37	9.41	9.45	9.49
	.25	2.56	3.00	3.15	3.23	3.28	3.31	3.35	3.39	3.44	3.48
3	.001	167.5	148.5	141.1	137.1	134.6	132.8	130.6	128.3	125.9	123.5
	.005	55.55	49.80	47.47	46.20	45.39	44.84	44.13	43.39	42.62	41.83
	.01	34.12	30.81	29.46	28.71	28.24	27.91	27.49	27.05	26.60	26.12
	.025	17.44	16.04	15.44	15.10	14.89	14.74	14.54	14.34	14.12	13.90
	.05	10.13	9.55	9.28	9.12	9.01	8.94	8.84	8.74	8.64	8.53
	.10	5.54	5.46	5.39	5.34	5.31	5.28	5.25	5.22	5.18	5.13
	.25	2.02	2.28	2.36	2.39	2.41	2.42	2.44	2.45	2.46	2.47
4	.001	74.14	61.25	56.18	53.44	51.71	50.53	49.00	47.41	45.77	44.05
	.005	31.33	26.28	24.26	23.16	22.46	21.98	21.35	20.71	20.03	19.33
	.01	21.20	18.00	16.69	15.98	15.52	15.21	14.80	14.37	13.93	13.46
	.025	12.22	10.65	9.98	9.60	9.36	9.20	8.98	8.75	8.51	8.26
	.05	7.71	6.94	6.59	6.39	6.26	6.16	6.04	5.91	5.77	5.63
	.10	4.54	4.32	4.19	4.11	4.05	4.01	3.95	3.90	3.83	3.76
	.25	1.81	2.00	2.05	2.06	2.07	2.08	2.08	2.08	2.08	2.08
5	.001	47.04	36.61	33.20	31.09	29.75	28.84	27.64	26.42	25.14	23.78
	.005	22.79	18.31	16.53	15.56	14.94	14.51	13.96	13.38	12.78	12.14
	.01	16.26	13.27	12.06	11.39	10.97	10.67	10.29	9.89	9.47	9.02
	.025	10.01	8.43	7.76	7.39	7.15	6.98	6.76	6.52	6.28	6.02
	.05	6.61	5.79	5.41	5.19	5.05	4.95	4.82	4.68	4.53	4.36
	.10	4.06	3.78	3.62	3.52	3.45	3.40	3.34	3.27	3.19	3.10
	.25	1.70	1.85	1.89	1.89	1.89	1.89	1.89	1.89	1.88	1.87

Table C.5 (Continued)

df_2 \ df_1	α	1	2	3	4	5	6	8	12	24	∞
6	.001	35.51	27.00	23.70	21.90	20.81	20.03	19.03	17.99	16.89	15.75
	.005	18.64	14.54	12.92	12.03	11.46	11.07	10.57	10.03	9.47	8.88
	.01	13.74	10.92	9.78	9.15	8.75	8.47	8.10	7.72	7.31	6.88
	.025	8.81	7.26	6.60	6.23	5.99	5.82	5.60	5.37	5.12	4.85
	.05	5.99	5.14	4.76	4.53	4.39	4.28	4.15	4.00	3.84	3.67
	.10	3.78	3.46	3.29	3.18	3.11	3.05	2.98	2.90	2.82	2.72
	.25	1.62	1.76	1.78	1.79	1.79	1.78	1.78	1.77	1.75	1.74
7	.001	29.22	21.69	18.77	17.19	16.21	15.52	14.63	13.71	12.73	11.69
	.005	16.24	12.40	10.88	10.05	9.52	9.16	8.68	8.18	7.65	7.08
	.01	12.25	9.55	8.45	7.85	7.46	7.19	6.84	6.47	6.07	5.65
	.025	8.07	6.54	5.89	5.52	5.29	5.12	4.90	4.67	4.42	4.14
	.05	5.59	4.74	4.35	4.12	3.97	3.87	3.73	3.57	3.41	3.23
	.10	3.59	3.26	3.07	2.96	2.88	2.83	2.75	2.67	2.58	2.47
	.25	1.57	1.70	1.72	1.72	1.71	1.71	1.70	1.68	1.67	1.65
8	.001	25.42	18.49	15.83	14.39	13.49	12.86	12.04	11.19	10.30	9.34
	.005	14.69	11.04	9.60	8.81	8.30	7.95	7.50	7.01	6.50	5.95
	.01	11.26	8.65	7.59	7.01	6.63	6.37	6.03	5.67	5.28	4.86
	.025	7.57	6.06	5.42	5.05	4.82	4.65	4.43	4.20	3.95	3.67
	.05	5.32	4.46	4.07	3.84	3.69	3.58	3.44	3.28	3.12	2.93
	.10	3.46	3.11	2.92	2.81	2.73	2.67	2.59	2.50	2.40	2.29
	.25	1.54	1.66	1.67	1.66	1.66	1.65	1.64	1.62	1.60	1.58
9	.001	22.86	16.39	13.90	12.56	11.71	11.13	10.37	9.57	8.72	7.81
	.005	13.61	10.11	8.72	7.96	7.47	7.13	6.69	6.23	5.73	5.19
	.01	10.56	8.02	6.99	6.42	6.06	5.80	5.47	5.11	4.73	4.31
	.025	7.21	5.71	5.08	4.72	4.48	4.32	4.10	3.87	3.61	3.33
	.05	5.12	4.26	3.86	3.63	3.48	3.37	3.23	3.07	2.90	2.71
	.10	3.36	3.01	2.81	2.69	2.61	2.55	2.47	2.38	2.28	2.16
	.25	1.51	1.62	1.63	1.63	1.62	1.61	1.60	1.58	1.56	1.53
10	.001	21.04	14.91	12.55	11.28	10.48	9.92	9.20	8.45	7.64	6.76
	.005	12.83	9.43	8.08	7.34	6.87	6.54	6.12	5.66	5.17	4.64
	.01	10.04	7.56	6.55	5.99	5.64	5.39	5.06	4.71	4.33	3.91
	.025	6.94	5.46	4.83	4.47	4.24	4.07	3.85	3.62	3.37	3.08
	.05	4.96	4.10	3.71	3.48	3.33	3.22	3.07	2.91	2.74	2.54
	.10	3.28	2.92	2.73	2.61	2.52	2.46	2.38	2.28	2.18	2.06
	.25	1.49	1.60	1.60	1.60	1.59	1.58	1.56	1.54	1.52	1.48

11	.001	19.69	13.81	11.56	10.35	9.58	9.05	8.35	7.63	6.85	6.00
	.005	12.23	8.91	7.60	6.88	6.42	6.10	5.68	5.24	4.76	4.23
	.01	9.65	7.20	6.22	5.67	5.32	5.07	4.74	4.40	4.02	3.60
	.025	6.72	5.26	4.63	4.28	4.04	3.88	3.66	3.43	3.17	2.88
	.05	4.84	3.98	3.59	3.36	3.20	3.09	2.95	2.79	2.61	2.40
	.10	3.23	2.86	2.66	2.54	2.45	2.39	2.30	2.21	2.10	1.97
	.25	1.46	1.58	1.58	1.58	1.56	1.55	1.54	1.51	1.49	1.45
12	.001	18.64	12.97	10.80	9.63	8.89	8.38	7.71	7.00	6.25	5.42
	.005	11.75	8.51	7.23	6.52	6.07	5.76	5.35	4.91	4.43	3.90
	.01	9.33	6.93	5.95	5.41	5.06	4.82	4.50	4.16	3.78	3.36
	.025	6.55	5.10	4.47	4.12	3.89	3.73	3.51	3.28	3.02	2.72
	.05	4.75	3.88	3.49	3.26	3.11	3.00	2.85	2.69	2.50	2.30
	.10	3.18	2.81	2.61	2.48	2.39	2.33	2.24	2.15	2.04	1.90
	.25	1.46	1.56	1.56	1.55	1.54	1.53	1.51	1.49	1.46	1.42
13	.001	17.81	12.31	10.21	9.07	8.35	7.86	7.21	6.52	5.78	4.97
	.005	11.37	8.19	6.93	6.23	5.79	5.48	5.08	4.64	4.17	3.65
	.01	9.07	6.70	5.74	5.20	4.86	4.62	4.30	3.96	3.59	3.16
	.025	6.41	4.97	4.35	4.00	3.77	3.60	3.39	3.15	2.89	2.60
	.05	4.67	3.80	3.41	3.18	3.02	2.92	2.77	2.60	2.42	2.21
	.10	3.14	2.76	2.56	2.43	2.35	2.28	2.20	2.10	1.98	1.85
	.25	1.45	1.55	1.55	1.53	1.52	1.51	1.49	1.47	1.44	1.40
14	.001	17.14	11.78	9.73	8.62	7.92	7.43	6.80	6.13	5.41	4.60
	.005	11.06	7.92	6.68	6.00	5.56	5.26	4.86	4.43	3.96	3.44
	.01	8.86	6.51	5.56	5.03	4.69	4.46	4.14	3.80	3.43	3.00
	.025	6.30	4.86	4.24	3.89	3.66	3.50	3.29	3.05	2.79	2.49
	.05	4.60	3.74	3.34	3.11	2.96	2.85	2.70	2.53	2.35	2.13
	.10	3.10	2.73	2.52	2.39	2.31	2.24	2.15	2.05	1.94	1.80
	.25	1.44	1.53	1.53	1.52	1.51	1.50	1.48	1.45	1.42	1.38
15	.001	16.59	11.34	9.34	8.25	7.57	7.09	6.47	5.81	5.10	4.31
	.005	10.80	7.70	6.48	5.80	5.37	5.07	4.67	4.25	3.79	3.26
	.01	8.68	6.36	5.42	4.89	4.56	4.32	4.00	3.67	3.29	2.87
	.025	6.20	4.77	4.15	3.80	3.58	3.41	3.20	2.96	2.70	2.40
	.05	4.54	3.68	3.29	3.06	2.90	2.79	2.64	2.48	2.29	2.07
	.10	3.07	2.70	2.49	2.36	2.27	2.21	2.12	2.02	1.90	1.76
	.25	1.43	1.52	1.52	1.51	1.49	1.48	1.46	1.44	1.41	1.36

Table C.5 (Continued)

df_2 \ df_1	α	1	2	3	4	5	6	8	12	24	∞
16	.001	16.12	10.97	9.00	7.94	7.27	6.81	6.19	5.55	4.85	4.06
	.005	10.58	7.51	6.30	5.64	5.21	4.91	4.52	4.10	3.64	3.11
	.01	8.53	6.23	5.29	4.77	4.44	4.20	3.89	3.55	3.18	2.75
	.025	6.12	4.69	4.08	3.73	3.50	3.34	3.12	2.89	2.63	2.32
	.05	4.49	3.63	3.24	3.01	2.85	2.74	2.59	2.42	2.24	2.01
	.10	3.05	2.67	2.46	2.33	2.24	2.18	2.09	1.99	1.87	1.72
	.25	1.42	1.51	1.51	1.50	1.48	1.47	1.45	1.43	1.39	1.34
17	.001	15.72	10.66	8.73	7.68	7.02	6.56	5.96	5.32	4.63	3.85
	.005	10.38	7.35	6.16	5.50	5.07	4.78	4.39	3.97	3.51	2.98
	.01	8.40	6.11	5.18	4.67	4.34	4.10	3.79	3.45	3.08	2.65
	.025	6.04	4.62	4.01	3.66	3.44	3.28	3.06	2.82	2.56	2.25
	.05	4.45	3.59	3.20	2.96	2.81	2.70	2.55	2.38	2.19	1.96
	.10	3.03	2.64	2.44	2.31	2.22	2.15	2.06	1.96	1.84	1.69
	.25	1.42	1.51	1.51	1.49	1.47	1.46	1.44	1.41	1.38	1.33
18	.001	15.38	10.39	8.49	7.46	6.81	6.35	5.76	5.13	4.45	3.67
	.005	10.22	7.21	6.03	5.37	4.96	4.66	4.28	3.86	3.40	2.87
	.01	8.28	6.01	5.09	4.58	4.25	4.01	3.71	3.37	3.00	2.57
	.025	5.98	4.56	3.95	3.61	3.38	3.22	3.01	2.77	2.50	2.19
	.05	4.41	3.55	3.16	2.93	2.77	2.66	2.51	2.34	2.15	1.92
	.10	3.01	2.62	2.42	2.29	2.20	2.13	2.04	1.93	1.81	1.66
	.25	1.41	1.50	1.49	1.48	1.46	1.45	1.43	1.40	1.37	1.32
19	.001	15.08	10.16	8.28	7.26	6.61	6.18	5.59	4.97	4.29	3.52
	.005	10.07	7.09	5.92	5.27	4.85	4.56	4.18	3.76	3.31	2.78
	.01	8.18	5.93	5.01	4.50	4.17	3.94	3.63	3.30	2.92	2.49
	.025	5.92	4.51	3.90	3.56	3.33	3.17	2.96	2.72	2.45	2.13
	.05	4.38	3.52	3.13	2.90	2.74	2.63	2.48	2.31	2.11	1.88
	.10	2.99	2.61	2.40	2.27	2.18	2.11	2.02	1.91	1.79	1.63
	.25	1.41	1.50	1.49	1.48	1.46	1.44	1.42	1.40	1.36	1.31
20	.001	14.82	9.95	8.10	7.10	6.46	6.02	5.44	4.82	4.15	3.38
	.005	9.94	6.99	5.82	5.17	4.76	4.47	4.09	3.68	3.22	2.69
	.01	8.10	5.85	4.94	4.43	4.10	3.87	3.56	3.23	2.86	2.42
	.025	5.87	4.46	3.86	3.51	3.29	3.13	2.91	2.68	2.41	2.09
	.05	4.35	3.49	3.10	2.87	2.71	2.60	2.45	2.28	2.08	1.84
	.10	2.97	2.59	2.38	2.25	2.16	2.09	2.00	1.89	1.77	1.61
	.25	1.40	1.49	1.48	1.47	1.45	1.44	1.42	1.39	1.35	1.29

21	.001	14.59	9.77	7.94	6.95	6.32	5.88	5.31	4.70	4.03	3.26
	.005	9.83	6.89	5.73	5.09	4.68	4.39	4.01	3.60	3.15	2.61
	.01	8.02	5.78	4.87	4.37	4.04	3.81	3.51	3.17	2.80	2.36
	.025	5.83	4.42	3.82	3.48	3.25	3.09	2.87	2.64	2.37	2.04
	.05	4.32	3.47	3.07	2.84	2.68	2.57	2 42	2.25	2.05	1.81
	.10	2.96	2.57	2.36	2.23	2.14	2.08	1.98	1.88	1.75	1.59
	.25	1.40	1.49	1.48	1.46	1.44	1.43	1.41	1.83	1.34	1.29
22	.001	14.38	9.61	7.80	6.81	6.19	5.76	5.19	4.58	3.92	3.15
	.005	9.73	6.81	5.65	5.02	4.61	4.32	3.94	3.54	3.08	2.55
	.01	7.94	5.72	4.82	4.31	3.99	3.76	3.45	3.12	2.75	2.31
	.025	5.79	4.38	3.78	3.44	3.22	3.05	2.84	2.60	2.33	2.00
	.05	4.30	3.44	3.05	2.82	2.66	2.55	2.40	2.23	2.03	1.78
	.10	2.95	2.56	2.35	2.22	2.13	2.06	1.97	1.86	1.73	1.57
	.25	1.40	1.48	1.47	1.46	1.44	1.42	1.40	1.37	1.33	1.28
23	.001	14.19	9.47	7.67	6.69	6.08	5.65	5.09	4.48	3.82	3.05
	.005	9.63	6.73	5.58	4.95	4.54	4.26	3.88	3.47	3.02	2.48
	.01	7.88	5.66	4.76	4.26	3.94	3.71	3.41	3.07	2.70	2.26
	.025	5.75	4.35	3.75	3.41	3.18	3.02	2.81	2.57	2.30	1.97
	.05	4.28	3.42	3.03	2.80	2.64	2.53	2.38	2.20	2.00	1.76
	.10	2.94	2.55	2.34	2.21	2.11	2.05	1.95	1.84	1.72	1.55
	.25	1.39	1.47	1.47	1.45	1.43	1.41	1.40	1.37	1.33	1.27
24	.001	14.03	9.34	7.55	6.59	5.98	5.55	4.99	4.39	3.74	2.97
	.005	9.55	6.66	5.52	4.89	4.49	4.20	3.83	3.42	2.97	2.43
	.01	7.82	5.61	4.72	4.22	3.90	3.67	3.36	3.03	2.66	2.21
	.025	5.72	4.32	3.72	3.38	3.15	2.99	2.78	2.54	2.27	1.94
	.05	4.26	3.40	3.01	2.78	2.62	2.51	2.36	2.18	1.98	1.73
	.10	2.93	2.54	2.33	2.19	2.10	2.04	1.94	1.83	1.70	1.53
	.25	1.39	1.47	1.46	1.44	1.43	1.41	1.39	1.36	1.32	1.26
25	.001	13.88	9.22	7.45	6.49	5.88	5.46	4.91	4.31	3.66	2.89
	.005	9.48	6.60	5.46	4.84	4.43	4.15	3.78	3.37	2.92	2.38
	.01	7.77	5.57	4.68	4.18	3.86	3.63	3.32	2.99	2.62	2.17
	.025	5.69	4.29	3.69	3.35	3.13	2.97	2.75	2.51	2.24	2.91
	.05	4.24	3.38	2.99	2.76	2.60	2.49	2.34	2.16	1.96	1.71
	.10	2.92	2.53	2 32	2.18	2.09	2.02	1.93	1.82	1.69	1.52
	.25	1.39	1.47	1.46	1.44	1.42	1.41	1.39	1.36	1.32	1.25

Table C.5 (Continued)

df_2 \ df_1	α	1	2	3	4	5	6	8	12	24	∞
26	.001	13.74	9.12	7.36	6.41	5.80	5.38	4.83	4.24	3.59	2.82
	.005	9.41	6.54	5.41	4.79	4.38	4.10	3.73	3.33	2.87	2.33
	.01	7.72	5.53	4.64	4.14	3.82	3.59	3.29	2.96	2.58	2.13
	.025	5.66	4.27	3.67	3.33	3.10	2.94	2.73	2.49	2.22	1.88
	.05	4.22	3.37	2.98	2.74	2.59	2.47	2.32	2.15	1.95	1.69
	.10	2.91	2.52	2.31	2.17	2.08	2.01	1.92	1.81	1.68	1.50
	.25	1.38	1.46	1.45	1.44	1.42	1.41	1.38	1.35	1.31	1.25
27	.001	13.61	9.02	7.27	6.33	5.73	5.31	4.76	4.17	3.52	2.75
	.005	9.34	6.49	5.36	4.74	4.34	4.06	3.69	3.28	2.83	2.29
	.01	7.68	5.49	4.60	4.11	3.78	3.56	3.26	2.93	2.55	2.10
	.025	5.63	4.24	3.65	3.31	3.08	2.92	2.71	2.47	2.19	1.85
	.05	4.21	3.35	2.96	2.73	2.57	2.46	2.30	2.13	1.93	1.67
	.10	2.90	2.51	2.30	2.17	2.07	2.00	1.91	1.80	1.67	1.49
	.25	1.38	1.46	1.45	1.43	1.42	1.40	1.38	1.35	1.31	1.24
28	.001	13.50	8.93	7.19	6.25	5.66	5.24	4.69	4.11	3.46	2.70
	.005	9.28	6.44	5.32	4.70	4.30	4.02	3.65	3.25	2.79	2.25
	.01	7.64	5.45	4.57	4.07	3.75	3.53	3.23	2.90	2.52	2.06
	.025	5.61	4.22	3.63	3.29	2.06	2.90	2.69	2.45	2.17	1.83
	.05	4.20	3.34	2.95	2.71	2.56	2.44	2.29	2.12	1.91	1.65
	.10	2.89	2.50	2.29	2.16	2.06	2.00	1.90	1.79	1.66	1.48
	.25	1.38	1.46	1.45	1.43	1.41	1.40	1.38	1.34	1.30	1.24
29	.001	13.39	8.85	7.12	6.49	5.59	5.18	4.64	4.05	3.41	2.64
	.005	9.23	6.40	5.28	4.66	4.26	3.98	3.61	3.21	2.76	2.21
	.01	7.60	5.42	4.54	4.04	3.73	3.50	3.20	2.87	2.49	2.03
	.025	5.59	4.20	3.61	3.27	3.04	2.88	2.67	2.43	2.15	1.81
	.05	4.18	3.33	2.93	2.70	2.54	2.43	2.28	2.10	1.90	1.64
	.10	2.89	2.50	2.28	2.15	2.06	1.99	1.89	1.78	1.65	1.47
	.25	1.38	1.45	1.45	1.43	1.41	1.40	1.37	1.34	1.30	1.23
	.005	9.18	6.35	5.24	4.62	4.23	3.95	3.58	3.18	2.73	2.18
30	.001	13.29	8.77	7.05	6.12	5.53	5.12	4.58	4.00	3.36	2.59
	.01	7.56	5.39	4.51	4.02	3.70	3.47	3.17	2.84	2.47	2.01
	.025	5.57	4.18	3.59	3.25	3.03	2.87	2.65	2.41	2.14	1.79
	.05	4.17	3.32	2.92	2.69	2.53	2.42	2.27	2.09	1.89	1.62
	.10	2.88	2.49	2.28	2.14	2.05	1.98	1.88	1.77	1.64	1.46
	.25	1.38	1.45	1.44	1.42	1.41	1.39	1.37	1.34	1.29	1.23

40	.001	12.61	8.25	6.60	5.70	5.13	4.73	4.21	3.64	3.01	2.23
	.005	8.83	6.07	4.98	4.37	3.99	3.71	3.35	2.95	2.50	1.93
	.01	7.31	5.18	4.31	3.83	3.51	3.29	2.99	2.66	2.29	1.80
	.025	5.42	4.05	3.46	3.13	2.90	2.74	2.53	2.29	2.01	1.64
	.05	4.08	3.23	2.84	2.61	2.45	2.34	2.18	2.00	1.79	1.51
	.10	2.84	2.44	2.23	2.09	2.00	1.93	1.83	1.71	1.57	1.38
	.25	1.36	1.44	1.42	1.41	1.39	1.37	1.35	1.31	1.27	1.19
60	.001	11.97	7.76	6.17	5.31	4.76	4.37	3.87	3.31	2.69	1.90
	.005	8.49	5.80	4.73	4.14	3.76	3.49	3.13	2.74	2.29	1.69
	.01	7.08	4.98	4.13	3.65	3.34	3.12	2.82	2.50	2.12	1.60
	.025	5.29	3.93	3.34	3.01	2.79	2.63	2.41	2.17	1.88	1.48
	.05	4.00	3.15	2.76	2 52	2.37	2.25	2.10	1.92	1.70	1.39
	.10	2.79	2.39	2.18	2.04	1.95	1.87	1.77	1.66	1.51	1.29
	.25	1.35	1.42	1.41	1.39	1.37	1.35	1.32	1.29	1.24	1.15
120	.001	11.38	7.31	5.79	4.95	4.42	4.04	3.55	3.02	2.40	1.56
	.005	8.18	5.54	4.50	3.92	3.55	3.28	2.93	2.54	2.09	1.43
	.01	6.85	4.79	3.95	3.48	3.17	2.96	2.66	2.34	1.95	1.38
	.025	5.15	3.80	3.23	2.89	2.67	2.52	2.30	2.05	1.76	1.31
	.05	3.92	3.07	2.68	2.45	2.29	2.17	2.02	1.83	1.61	1.25
	.10	2.75	2.35	2.13	1.99	1.90	1.82	1.72	1.60	1.45	1.19
	.25	1.34	1.40	1.39	1.37	1.35	1.33	1.30	1.26	1.21	1.10
∞	.001	10.83	6.91	5.42	4.62	4.10	3.74	3.27	2.74	2.13	1.00
	.005	7.88	5.30	4.28	3.72	3.35	3.09	2.74	2.36	1.90	1.00
	.01	6.64	4.60	3.78	3.32	3.02	2.80	2.51	2.18	1.79	1.00
	.025	5.02	3.69	3.12	2.79	2.57	2.41	2.19	1.94	1.64	1.00
	.05	3.84	2.99	2.60	2.37	2.21	2.09	1.94	1.75	1.52	1.00
	.10	2.71	2.30	2.08	1.94	1.85	1.77	1.67	1.55	1.38	1.00
	.25	1.32	1.39	1.37	1.35	1.33	1.31	1.28	1.24	1.18	1.00

Source: Adapted from Table .8 in Pearson, E. S. and Hartley, H. O. (1958). *Biometrika Tables for Statisticians*, Vol. 1, 2nd ed. Cambridge University Press: Cambridge.

Table C.6 Critical values of the Bonferroni *t* statistic (note that the tabled values are two-tailed)

		Number of contrasts (K)									
df	*FWE*	*2*	*3*	*4*	*5*	*6*	*7*	*8*	*9*	*10*	*15*
3	.01	7.453	8.575	9.465	10.215	10.869	11.453	11.984	12.471	12.924	14.819
	.05	4.177	4.857	5.392	5.841	6.232	6.580	6.895	7.185	7.453	8.575
	.10	3.182	3.740	4.177	4.541	4.857	5.138	5.392	5.625	5.841	6.741
4	.01	5.598	6.254	6.758	7.173	7.529	7.841	8.122	8.376	8.610	9.568
	.05	3.495	3.961	4.315	4.604	4.851	5.068	5.261	5.437	5.598	6.254
	.10	2.776	3.186	3.495	3.747	3.961	4.148	4.315	4.466	4.604	5.167
5	.01	4.773	5.247	5.604	5.893	6.138	6.352	6.541	6.713	6.869	7.499
	.05	3.163	3.534	3.810	4.032	4.219	4.382	4.526	4.655	4.773	5.247
	.10	2.571	2.912	3.163	3.365	3.534	3.681	3.810	3.926	4.032	4.456
6	.01	4.317	4.698	4.981	5.208	5.398	5.563	5.709	5.840	5.959	6.434
	.05	2.969	3.287	3.521	3.707	3.863	3.997	4.115	4.221	4.317	4.698
	.10	2.447	2.749	2.969	3.143	3.287	3.412	3.521	3.619	3.707	4.058
7	.01	4.029	4.355	4.595	4.785	4.944	5.082	5.202	5.310	5.408	5.795
	.05	2.841	3.128	3.335	3.499	3.636	3.753	3.855	3.947	4.029	4.355
	.10	2.365	2.642	2.841	2.998	3.128	3.238	3.335	3.422	3.499	3.806
8	.01	3.833	4.122	4.334	4.501	4.640	4.759	4.864	4.957	5.041	5.374
	.05	2.752	3.016	3.206	3.355	3.479	3.584	3.677	3.759	3.833	4.122
	.10	2.306	2.566	2.752	2.896	3.016	3.117	3.206	3.285	3.355	3.632
9	.01	3.690	3.954	4.146	4.297	4.422	4.529	4.622	4.706	4.781	5.076
	.05	2.685	2.933	3.111	3.250	3.364	3.462	3.547	3.622	3.690	3.954
	.10	2.262	2.510	2.685	2.821	2.933	3.028	3.111	3.184	3.250	3.505
10	.01	3.581	3.827	4.005	4.144	4.259	4.357	4.442	4.518	4.587	4.855
	.05	2.634	2.870	3.038	3.169	3.277	3.368	3.448	3.518	3.581	3.827
	.10	2.228	2.466	2.634	2.764	2.870	2.960	3.038	3.107	3.169	3.409
11	.01	3.497	3.728	3.895	4.025	4.132	4.223	4.303	4.373	4.437	4.685
	.05	2.593	2.820	2.981	3.106	3.208	3.295	3.370	3.437	3.497	3.728
	.10	2.201	2.431	2.593	2.718	2.820	2.906	2.981	3.047	3.106	3.334
12	.01	3.428	3.649	3.807	3.930	4.031	4.117	4.192	4.258	4.318	4.550
	.05	2.560	2.779	2.934	3.055	3.153	3.236	3.308	3.371	3.428	3.649
	.10	2.179	2.403	2.560	2.681	2.779	2.863	2.934	2.998	3.055	3.273
13	.01	3.372	3.584	3.735	3.852	3.948	4.030	4.101	4.164	4.221	4.440
	.05	2.533	2.746	2.896	3.012	3.107	3.187	3.256	3.318	3.372	3.584
	.10	2.160	2.380	2.533	2.650	2.746	2.827	2.896	2.957	3.012	3.223
14	.01	3.326	3.530	3.675	3.787	3.880	3.958	4.026	4.086	4.140	4.349
	.05	2.510	2.718	2.864	2.977	3.069	3.146	3.214	3.273	3.326	3.530
	.10	2.145	2.360	2.510	2.624	2.718	2.796	2.864	2.924	2.977	3.181
15	.01	3.286	3.484	3.624	34.773	3.822	3.897	3.963	4.021	4.073	4.273
	.05	2.490	2.694	2.837	2.947	3.036	3.112	3.177	3.235	3.286	3.484
	.10	2.131	2.343	2.490	2.602	2.694	2.770	2.837	2.895	2.947	3.146
16	.01	3.252	3.444	3.581	3.686	3.773	3.846	3.909	3.965	4.015	4.208
	.05	2.473	2.673	2.813	2.921	3.008	3.082	3.146	3.202	3.252	3.444
	.10	2.120	2.328	2.473	2.583	2.673	2.748	2.813	2.870	2.921	3.115
17	.01	3.222	3.410	3.543	3.646	3.730	3.801	3.862	3.917	3.965	4.152
	.05	2.458	2.655	2.793	2.898	2.984	3.056	3.119	3.173	3.222	3.410
	.10	2.110	2.316	2.458	2.567	2.655	2.729	2.793	2.848	2.898	3.089
18	.01	3.197	3.380	3.510	3.610	3.692	3.762	3.822	3.874	3.922	4.104
	.05	2.445	2.639	2.775	2.878	2.963	3.034	3.095	3.149	3.197	3.380
	.10	2.101	2.304	2.445	2.552	2.639	2.712	2.775	2.829	2.878	3.065
19	.01	3.174	3.354	3.481	3.579	3.660	3.727	3.786	3.837	3.883	4.061
	.05	2.433	2.625	2.759	2.861	2.944	3.014	3.074	3.127	3.174	3.354
	.10	2.093	2.294	2.433	2.539	2.625	2.697	2.759	2.813	2.861	3.045

Table C.6 (Continued)

		Number of contrasts (K)									
df	*FWE*	*2*	*3*	*4*	*5*	*6*	*7*	*8*	*9*	*10*	*15*
20	.01	3.153	3.331	3.455	3.552	3.630	3.697	3.754	3.804	3.850	4.023
	.05	2.423	2.613	2.744	2.845	2.927	2.996	3.055	3.107	3.153	3.331
	.10	2.086	2.285	2.423	2.528	2.613	2.683	2.744	2.798	2.845	3.026
25	.01	3.078	3.244	3.361	3.450	3.523	3.584	3.637	3.684	3.725	3.884
	.05	2.385	2.566	2.692	2.787	2.865	2.930	2.986	3.035	3.078	3.244
	.10	2.060	2.252	2.385	2.485	2.566	2.634	2.692	2.742	2.787	2.959
30	.01	3.030	3.189	3.300	3.385	3.454	3.513	3.563	3.607	3.646	3.796
	.05	2.360	2.536	2.657	2.750	2.825	2.887	2.941	2.988	3.030	3.189
	.10	2.042	2.231	2.360	2.457	2.536	2.601	2.657	2.706	2.750	2.915
35	.01	2.996	3.150	3.258	3.340	3.407	3.463	3.511	3.553	3.591	3.735
	.05	2.342	2.515	2.633	2.724	2.797	2.857	2.910	2.955	2.996	3.150
	.10	2.030	2.215	2.342	2.438	2.515	2.579	2.633	2.681	2.724	2.885
40	.01	2.971	3.122	3.227	3.307	3.372	3.426	3.473	3.514	3.551	3.691
	.05	2.329	2.499	2.616	2.704	2.776	2.836	2.887	2.931	2.971	3.122
	.10	2.021	2.204	2.329	2.423	2.499	2.562	2.616	2.663	2.704	2.862
60	.01	2.915	3.057	3.156	3.232	3.293	3.344	3.388	3.426	3.460	3.590
	.05	2.299	2.463	2.575	2.660	2.729	2.785	2.834	2.877	2.915	3.057
	.10	2.000	2.178	2.299	2.390	2.463	2.524	2.575	2.620	2.660	2.811
120	.01	2.860	2.995	3.088	3.160	3.217	3.265	3.306	3.342	3.373	3.494
	.05	2.270	2.428	2.536	2.617	2.683	2.737	2.783	2.824	2.860	2.995
	.10	1.980	2.153	2.270	2.358	2.428	2.486	2.536	2.579	2.617	2.761

Table C.7 Distribution of Dunnett's *d* statistic for comparing treatment means with a control (note that the tabled values are two-tailed)

df for MS_{error}	*FWE*	*Number of means (including control)* 2	3	4	5	6	7	8	9	10
	.10	1.94	2.34	2.56	2.71	2.83	2.92	3.00	3.07	3.12
	.05	2.45	2.86	3.18	3.41	3.60	3.75	3.88	4.00	4.11
6	.02	3.14	3.61	3.88	4.07	4.21	4.33	4.43	4.51	4.59
	.01	3.71	4.22	4.60	4.88	5.11	5.30	5.47	5.61	5.74
	.10	1.89	2.27	2.48	2.62	2.73	2.82	2.89	2.95	3.01
	.05	2.36	2.75	3.04	3.24	3.41	3.54	3.66	3.76	3.86
7	.02	3.00	3.42	3.66	3.83	3.96	4.07	4.15	4.23	4.30
	.01	3.50	3.95	4.28	4.52	4.17	4.87	5.01	5.13	5.24
	.10	1.86	2.22	2.42	2.55	2.66	2.74	2.81	2.87	2.92
	.05	2.31	2.67	2.94	3.13	3.28	3.40	3.51	3.60	3.68
8	.02	2.90	3.29	3.51	3.67	3.79	3.88	3.96	4.03	4.09
	.01	3.36	3.77	4.06	4.27	4.44	4.58	4.70	4.81	4.90
	.10	1.83	2.18	2.37	2.50	2.60	2.68	2.75	2.81	2.86
	.05	2.26	2.61	2.86	3.04	3.18	3.29	3.39	3.48	3.55
9	.02	2.82	3.19	3.40	3.55	3.66	3.75	3.82	3.89	3.94
	.01	3.25	3.63	3.90	4.09	4.24	4.37	4.48	4.57	4.65
	.10	1.81	2.15	2.34	2.47	2.56	2.64	2.70	2.76	2.81
	.05	2.23	2.57	2.81	2.97	3.11	3.21	3.31	3.39	3.46
10	.02	2.76	3.11	3.31	3.45	3.56	3.64	3.71	3.78	3.83
	.01	3.17	3.53	3.78	3.95	4.10	4.21	4.31	4.40	4.47
	.10	1.80	2.13	2.31	2.44	2.53	2.60	2.67	2.72	2.77
	.05	2.20	2.53	2.76	2.92	3.05	3.15	3.24	3.31	3.38
11	.02	2.72	3.06	3.25	3.38	3.48	3.56	3.63	3.69	3.74
	.01	3.11	3.45	3.68	3.85	3.98	4.09	4.18	4.26	4.33
	.10	1.78	2.11	2.29	2.41	2.50	2.58	2.64	2.69	2.74
	.05	2.18	2.50	2.72	2.88	3.00	3.10	3.18	3.25	3.32
12	.02	2.68	3.01	3.19	3.32	3.42	3.50	3.56	3.62	3.67
	.01	3.05	3.39	3.61	3.76	3.89	3.99	4.08	4.15	4.22
	.10	1.77	2.09	2.27	2.39	2.48	2.55	2.61	2.66	2.71
	.05	2.16	2.48	2.69	2.84	2.96	3.06	3.14	3.21	3.27
13	.02	2.65	2.97	3.15	3.27	3.37	3.44	3.51	3.56	3.61
	.01	3.01	3.33	3.54	3.69	3.81	3.91	3.99	4.06	4.13
	.10	1.76	2.08	2.25	2.37	2.46	2.53	2.59	2.64	2.69
	.05	2.14	2.46	2.67	2.81	2.93	3.02	3.10	3.17	3.23
14	.02	2.62	2.94	3.11	3.23	3.32	3.40	3.46	3.51	3.56
	.01	2.98	3.29	3.49	3.64	3.75	3.84	3.92	3.99	4.05
	.10	1.75	2.06	2.23	2.34	2.43	2.50	2.56	2.61	2.65
	.05	2.12	2.42	2.63	2.77	2.88	2.96	3.04	3.10	3.16
16	.02	2.58	2.88	3.05	3.17	3.26	3.33	3.39	3.44	3.48
	.01	2.92	3.22	3.41	3.55	3.65	3.74	3.82	3.88	3.93
	.10	1.73	2.04	2.21	2.32	2.41	2.48	2.53	2.58	2.62
	.05	2.10	2.40	2.59	2.73	2.84	2.92	2.99	3.05	3.11
18	.02	2.55	2.84	3.01	3.12	3.21	3.27	3.33	3.38	3.42
	.01	2.88	3.17	3.35	3.48	3.58	3.67	3.74	3.80	3.85
	.10	1.72	2.03	2.19	2.30	2.39	2.46	2.51	2.56	2.60
	.05	2.09	2.38	2.57	2.70	2.81	2.89	2.96	3.02	3.07
20	.02	2.53	2.81	2.97	3.08	3.17	3.23	3.29	3.34	3.38
	.01	2.85	3.13	3.31	3.43	3.53	3.61	3.67	3.73	3.78
	.10	1.71	2.01	2.17	2.28	2.36	2.43	2.48	2.53	2.57
	.05	2.06	2.35	2.53	2.66	2.76	2.84	2.91	2.96	3.01

Table C.7 (Continued)

df for MS_{error}	*FWE*	*Number of means (including control)*								
		2	*3*	*4*	*5*	*6*	*7*	*8*	*9*	*10*
24	.02	2.49	2.77	2.92	3.03	3.11	3.17	3.22	3.27	3.31
	.01	2.80	3.07	3.24	3.36	3.45	3.52	3.58	3.64	3.69
	.10	1.70	1.99	2.15	2.25	2.33	2.40	2.45	2.50	2.54
	.05	2.04	2.32	2.50	2.62	2.72	2.79	2.86	2.91	2.96
30	.02	2.46	2.72	2.87	2.97	3.05	3.11	3.16	3.21	3.24
	.01	2.75	3.01	3.17	3.28	3.37	3.44	3.50	3.55	3.59
	.10	1.68	1.97	2.13	2.23	2.31	2.37	2.42	2.47	2.51
	.05	2.02	2.29	2.47	2.58	2.67	2.75	2.81	2.86	2.90
40	.02	2.42	2.68	2.82	2.92	2.99	3.05	3.10	3.14	3.18
	.01	2.70	2.95	3.10	3.21	3.29	3.36	3.41	3.46	3.50
	.10	1.67	1.95	2.10	2.21	2.28	2.35	2.39	2.44	2.48
	.05	2.00	2.27	2.43	2.55	2.63	2.70	2.76	2.81	2.85
60	.02	2.39	2.64	2.78	2.87	2.94	3.00	3.04	3.08	3.12
	.01	2.66	2.90	3.04	3.14	3.22	3.28	3.33	3.38	3.42
	.10	1.66	1.93	2.08	2.18	2.26	2.32	2.37	2.41	2.45
	.05	1.98	2.24	2.40	2.51	2.59	2.66	2.71	2.76	2.80
120	.02	2.36	2.60	2.73	2.82	2.89	2.94	2.99	3.03	3.06
	.01	2.62	2.84	2.98	3.08	3.15	3.21	3.25	3.30	3.33
	.10	1.64	1.92	2.06	2.16	2.23	2.29	2.34	2.38	2.42
	.05	1.96	2.21	2.37	2.47	2.55	2.62	2.67	2.71	2.75
∞	.02	2.33	2.56	2.68	2.77	2.84	2.89	2.93	2.97	3.00
	.01	2.58	2.79	2.92	3.01	3.08	3.14	3.18	3 22	3.25

Source: Adapted from tables in Dunnett, C. W. (1955). A multiple comparison procedure for comparing several treatments with a control. *Journal of the American Statistical Association*, *50*, 1096–1121, and from Dunnett, C. W. (1964). New tables for multiple comparisons with a control. *Biometrics*, *20*, 482–491.

Table C.8 Critical values of the Studentized range distribution

		Number of ordered means						
Error df	*FWE*	*2*	*3*	*4*	*5*	*6*	*7*	*8*
2	.01	14.04	19.02	22.29	26.72	26.63	28.20	29.53
	.05	6.09	8.33	9.80	10.88	11.74	12.44	13.03
	.10	4.13	5.73	6.77	7.54	8.14	8.63	9.05
3	.01	8.26	10.62	12.17	13.33	14.24	15.00	15.64
	.05	4.50	5.91	6.83	7.50	8.04	8.48	8.85
	.10	3.33	4.47	5.20	5.74	6.16	6.51	6.81
4	.01	6.51	8.12	9.17	9.96	10.58	11.10	11.55
	.05	3.93	5.04	5.76	6.29	6.71	7.05	7.35
	.10	3.02	3.98	4.59	5.04	5.39	5.68	5.93
5	.01	5.70	6.98	7.80	8.42	8.91	9.32	9.67
	.05	3.64	4.60	5.22	5.67	6.03	6.33	6.58
	.10	2.85	3.72	4.26	4.66	4.98	5.24	5.46
6	.01	5.24	6.33	7.03	7.56	7.97	8.32	8.61
	.05	3.46	4.34	4.90	5.31	5.63	5.90	6.12
	.10	2.75	3.56	4.07	4.44	4.73	4.97	5.17
7	.01	4.95	5.92	6.54	7.01	7.37	7.68	7.94
	.05	3.34	4.17	4.68	5.06	5.36	5.61	5.82
	.10	2.68	3.45	3.93	4.28	4.56	4.78	4.97
8	.01	4.75	5.64	6.20	6.63	6.96	7.24	7.47
	.05	3.26	4.04	4.53	4.89	5.17	5.40	5.60
	.10	2.63	3.37	3.83	4.17	4.43	4.65	4.83
9	.01	4.60	5.43	5.96	6.35	6.66	6.92	7.13
	.05	3.20	3.95	4.42	4.76	5.02	5.24	5.43
	.10	2.59	3.32	3.76	4.08	4.34	4.55	4.72
10	.01	4.48	5.27	5.77	6.14	6.43	6.67	6.88
	.05	3.15	3.88	4.33	4.65	4.91	5.12	5.31
	.10	2.56	3.27	3.70	4.02	4.26	4.47	4.64
11	.01	4.39	5.15	5.62	5.97	6.25	6.48	6.67
	.05	3.11	3.82	4.26	4.57	4.82	5.03	5.20
	.10	2.54	3.23	3.66	3.97	4.21	4.40	4.57
12	.01	4.32	5.05	5.50	5.84	6.10	6.32	6.51
	.05	3.08	3.77	4.20	4.51	4.75	4.95	5.12
	.10	2.52	3.20	3.62	3.92	4.16	4.35	4.51
13	.01	4.26	4.96	5.40	5.73	5.98	6.19	6.37
	.05	3.06	3.74	4.15	4.45	4.69	4.89	5.05
	.10	2.51	3.18	3.59	3.89	4.12	4.31	4.46
14	.01	4.21	4.90	5.32	5.63	5.88	6.09	6.26
	.05	3.03	3.70	4.11	4.41	4.64	4.83	4.99
	.10	2.49	3.16	3.56	3.85	4.06	4.27	4.42
15	.01	4.17	4.84	5.25	5.56	5.80	5.99	6.16
	.05	3.01	3.67	4.08	4.37	4.60	4.78	4.94
	.10	2.48	3.14	3.54	3.83	4.05	4.24	4.39
16	.01	4.13	4.79	5.19	5.49	5.72	5.92	6.08
	.05	3.00	3.65	4.05	4.33	4.56	4.74	4.90
	.10	2.47	3.12	3.52	3.80	4.03	4.21	4.36
17	.01	4.10	4.74	5.14	5.43	5.66	5.85	6.01
	.05	2.98	3.63	4.02	4.30	4.52	4.71	4.86
	.10	2.46	3.11	3.50	3.78	4.00	4.18	4.33
18	.01	4.07	4.70	5.09	5.38	5.60	5.79	5.94
	.05	2.97	3.61	4.00	4.28	4.50	4.67	4.82
	.10	2.45	3.10	3.49	3.77	3.98	4.16	4.31

Table C.8 (Continued)

Error df	FWE	Number of ordered means 2	3	4	5	6	7	8
19	.01	4.05	4.67	5.05	5.33	5.55	5.74	5.89
	.05	2.96	3.59	3.98	4.25	4.47	4.65	4.79
	.10	2.45	3.09	3.47	3.75	3.97	4.14	4.29
20	.01	4.02	4.64	5.02	5.29	5.51	5.69	5.84
	.05	2.95	3.58	3.96	4.23	4.45	4.62	4.77
	.10	2.44	3.08	3.46	3.74	3.95	4.12	4.27
24	.01	3.96	4.55	4.91	5.17	5.37	5.54	5.69
	.05	2.92	3.53	3.90	4.17	4.37	4.54	4.68
	.10	2.42	3.05	3.42	3.69	3.90	4.07	4.21
30	.01	3.89	4.46	4.80	5.05	5.24	5.40	5.54
	.05	2.89	3.49	3.85	4.10	4.30	4.46	4.60
	.10	2.40	3.02	3.39	3.65	3.85	4.02	4.16
40	.01	3.83	4.37	4.70	4.93	5.11	5.27	5.39
	.05	2.86	3.44	3.79	4.04	4.23	4.39	4.52
	.10	2.38	2.99	3.35	3.61	3.80	3.96	4.10
60	.01	3.76	4.28	4.60	4.82	4.99	5.13	5.25
	.05	2.83	3.10	3.71	3.98	1.16	1.31	1.11
	.10	2.36	2.96	3.31	3.56	3.76	3.91	4.04
120	.01	3.70	4.20	4.50	4.71	4.87	5.01	5.12
	.05	2.80	3.36	3.69	3.92	4.10	4.24	4.36
	.10	2.34	2.93	3.28	3.52	3.71	3.86	3.99
∞	.01	3.64	4.12	4.40	4.60	4.76	4.88	4.99
	.05	2.77	3.31	3.63	3.86	4.03	4.17	4.29
	.10	2.33	2.90	3.24	3.48	3.66	3.81	3.93

Error df	FWE	Number of ordered means 9	10	11	12	13	14	15
2	.01	30.68	31.69	32.59	33.40	34.13	34.81	35.43
	.05	13.54	13.99	14.39	14.75	15.08	15.38	15.65
	.10	9.41	9.73	10.01	10.26	10.49	10.70	10.89
3	.01	16.20	16.69	17.13	17.53	17.89	18.22	18.52
	.05	9.18	9.46	9.72	9.95	10.15	10.35	10.53
	.10	7.06	7.29	7.49	7.67	7.83	7.98	8.12
4	.01	11.93	12.27	12.57	12.84	13.09	13.32	13.53
	.05	7.60	7.83	8.03	8.21	8.37	8.53	8.66
	.10	6.14	6.33	6.50	6.65	6.78	6.91	7.03
5	.01	9.97	10.24	10.48	10.70	10.89	11.08	11.24
	.05	6.80	7.00	7.17	7.32	7.47	7.60	7.72
	.10	5.65	5.82	5.97	6.10	6.22	6.34	6.44
6	.01	8.87	9.10	9.30	9.48	9.65	9.81	9.95
	.05	6.32	6.49	6.65	6.79	6.92	7.03	7.14
	.10	5.34	5.50	5.64	5.76	5.88	5.98	6.08
7	.01	8.17	8.37	8.55	8.71	8.86	9.00	9.12
	.05	6.00	6.16	6.30	6.43	6.55	6.66	6.76
	.10	5.14	5.28	5.41	5.53	5.64	5.74	5.83
8	.01	7.68	7.86	8.03	8.18	8.31	8.44	8.55
	.05	5.77	5.92	6.05	6.18	6.29	6.39	6.48
	.10	4.99	5.13	5.25	5.36	5.46	5.56	5.64
9	.01	7.33	7.50	7.65	7.78	7.91	8.03	8.13
	.05	5.60	5.74	5.87	5.98	6.09	6.19	6.28
	.10	4.87	5.01	5.13	5.23	5.33	5.42	5.51

Table C.8 (Continued)

		Number of ordered means						
Error df	*FWE*	*9*	*10*	*11*	*12*	*13*	*14*	*15*
10	.01	7.06	7.21	7.36	7.49	7.60	7.71	7.81
	.05	5.46	5.60	5.72	5.83	5.94	6.03	6.11
	.10	4.78	4.91	5.03	5.13	5.23	5.32	5.40
11	.01	6.84	6.99	7.13	7.25	7.36	7.47	7.56
	.05	5.35	5.49	5.61	5.71	5.81	5.90	5.98
	.10	4.71	4.84	4.95	5.05	5.15	5.23	5.31
12	.01	6.67	6.81	6.94	7.06	7.17	7.27	7.36
	.05	5.27	5.40	5.51	5.62	5.71	5.80	5.88
	.10	4.65	4.78	4.89	4.99	5.08	5.16	5.24
13	.01	6.53	6.67	6.79	6.90	7.01	7.10	7.19
	.05	5.19	5.32	5.43	5.53	5.63	5.71	5.79
	.10	4.60	4.72	4.83	4.93	5.02	5.10	5.18
14	.01	6.41	6.54	6.66	6.77	6.87	6.96	7.05
	.05	5.13	5.25	5.36	5.46	5.55	5.64	5.71
	.10	4.56	4.68	4.79	4.88	4.97	5.05	5.12
15	.01	6.31	6.44	6.56	6.76	6.76	6.85	6.93
	.05	5.08	5.20	5.31	5.40	5.49	5.57	5.65
	.10	4.52	4.64	4.75	4.84	4.93	5.01	5.08
16	.01	6.22	6.35	6.46	6.56	6.66	6.74	6.82
	.05	5.03	5.15	5.26	5.35	5.44	5.52	5.59
	.10	4.49	4.61	4.71	4.81	4.89	4.97	5.04
17	.01	6.15	6.27	6.38	6.48	6.57	6.66	6.73
	.05	4.99	5.11	5.21	5.31	5.39	5.47	5.54
	.10	4.46	4.58	4.68	4.77	4.86	4.94	5.01
18	.01	6.08	6.20	6.31	6.41	6.50	6.58	6.66
	.05	4.96	5.07	5.17	5.27	5.35	5.43	5.50
	.10	4.44	4.55	4.66	4.75	4.83	4.91	4.98
19	.01	6.02	6.14	6.25	6.34	6.43	6.51	6.59
	.05	4.92	5.04	5.14	5.23	5.32	5.39	5.46
	.10	4.42	4.53	4.63	4.72	4.80	4.88	4.95
20	.01	5.97	6.09	6.19	6.29	6.37	6.45	6.52
	.05	4.90	5.01	5.11	5.20	5.28	5.36	5.43
	.10	4.40	4.51	4.61	4.70	4.78	4.86	4.92
24	.01	5.81	5.92	6.02	6.11	6.19	6.26	6.33
	.05	4.81	4.92	5.01	5.10	5.18	5.25	5.32
	.10	4.34	4.45	4.45	4.62	4.71	4.78	4.85
30	.01	5.65	5.76	5.85	5.93	6.01	6.08	6.14
	.05	4.72	4.82	4.92	5.00	5.08	5.15	5.21
	.10	4.28	4.38	4.47	4.56	4.64	4.71	4.77
40	.01	5.50	5.60	5.69	5.76	5.84	5.90	5.96
	.05	4.64	4.74	4.82	4.90	4.98	5.04	5.11
	.10	4.22	4.32	4.41	4.49	4.56	4.63	4.70
60	.01	5.36	5.45	5.53	5.60	5.67	5.73	5.79
	.05	4.55	4.65	4.73	4.81	4.88	4.94	5.00
	.10	4.16	4.25	4.34	4.42	4.49	4.56	4.62
120	.01	5.21	5.30	5.38	5.44	5.51	5.56	5.61
	.05	4.47	4.56	4.64	4.71	4.78	4.84	4.90
	.10	4.10	4.19	4.28	4.35	4.42	4.49	4.54
∞	.01	5.08	5.16	5.23	5.29	5.35	5.40	5.45
	.05	4.39	4.47	4.55	4.62	4.69	4.74	4.80
	.10	4.04	4.13	4.21	4.29	4.35	4.41	4.47

Table C.8 (Continued)

		Number of ordered means						
Error df	*FWE*	*16*	*17*	*18*	*19*	*20*	*30*	*40*
2	.01	36.00	36.53	37.03	37.50	37.95	41.32	43.61
	.05	15.91	16.14	16.37	16.57	16.77	18.27	19.28
	.10	11.07	11.24	11.39	11.54	11.68	12.73	13.44
3	.01	18.81	19.07	19.32	19.55	19.77	21.44	22.59
	.05	10.69	10.84	10.98	11.11	11.24	12.21	12.87
	.10	8.25	8.37	8.48	8.58	8.68	9.44	9.95
4	.01	13.73	13.91	14.08	14.24	14.40	15.57	16.37
	.05	8.79	8.91	9.03	9.13	9.23	10.00	10.53
	.10	7.13	7.23	7.33	7.41	7.50	8.14	8.57
5	.01	11.40	11.55	11.68	12.81	11.93	12.87	13.52
	.05	7.83	7.93	8.03	8.12	8.21	8.88	9.33
	.10	6.54	6.63	6.71	6.79	6.86	7.44	7.83
6	.01	10.08	10.21	10.32	10.43	10.54	11.34	11.90
	.05	7.24	7.34	7.43	7.51	7.59	8.19	8.60
	.10	6.16	6.25	6.33	6.40	6.47	7.00	7.36
7	.01	9.24	9.35	9.46	9.55	9.65	10.36	10.85
	.05	6.85	6.94	7.02	7.10	7.17	7.73	8.11
	.10	5.91	5.99	6.06	6.13	6.20	6.70	7.04
8	.01	8.66	8.76	8.85	8.94	9.03	9.68	10.13
	.05	6.57	6.65	6.73	6.80	6.87	7.40	7.76
	.10	5.72	5.80	5.87	5.94	6.00	6.48	6.80
9	.01	8.23	8.33	8.41	8.49	8.57	9.18	9.59
	.05	6.36	6.44	6.51	6.58	6.64	7.15	7.49
	.10	5.58	5.66	5.72	5.79	5.85	6.31	6.62
10	.01	7.91	7.99	8.08	8.15	8.23	8.79	9.19
	.05	6.19	6.27	6.34	6.41	6.47	6.95	7.28
	.10	5.47	5.54	5.61	5.67	5.73	6.17	6.48
11	.01	7.65	7.73	7.81	7.88	7.95	8.49	8.86
	.05	6.06	6.13	6.20	6.27	6.33	6.79	7.11
	.10	5.38	5.45	5.51	5.57	5.63	6.07	6.36
12	.01	7.44	7.52	7.59	7.67	7.73	8.25	8.60
	.05	5.95	6.02	6.09	6.15	6.21	6.66	6.97
	.10	5.31	5.37	5.44	5.50	5.55	5.98	6.27
13	.01	7.27	7.35	7.42	7.49	7.55	8.04	8.39
	.05	5.86	5.93	6.00	6.06	6.11	6.55	6.85
	.10	5.25	5.31	5.37	5.43	5.48	5.90	6.19
14	.01	7.13	7.20	7.27	7.33	7.40	7.87	8.20
	.05	5.79	5.85	5.92	5.97	6.03	6.46	6.75
	.10	5.19	5.26	5.32	5.37	5.43	5.84	6.12
15	.01	7.00	7.07	7.14	7.20	7.26	7.73	8.05
	.05	5.72	5.79	5.85	5.90	5.96	6.38	6.67
	.10	5.15	5.21	5.27	5.32	5.38	5.78	6.06
16	.01	6.90	6.97	7.03	7.09	7.15	7.60	7.92
	.05	5.66	5.73	5.79	5.84	5.90	6.31	6.59
	.10	5.11	5.17	5.23	5.28	5.33	5.73	6.00
17	.01	6.81	6.87	6.94	7.00	7.05	7.49	7.80
	.05	5.61	5.68	5.73	5.79	5.84	6.25	6.53
	.10	5.07	5.13	5.19	5.24	5.30	5.69	5.96
18	.01	6.73	6.79	6.85	6.91	6.97	7.40	7.70
	.05	5.57	5.63	5.69	5.74	5.79	6.20	6.47
	.10	5.04	5.10	5.16	5.21	5.26	5.65	5.92

Table C.8 (Continued)

Error df	*FWE*	*Number of ordered means* 16	17	18	19	20	30	40
19	.01	6.65	6.72	6.78	6.84	6.89	7.31	7.61
	.05	5.53	5.59	5.65	5.70	5.75	6.15	6.42
	.10	5.01	5.07	5.13	5.18	5.23	5.62	5.88
20	.01	6.59	6.65	6.71	6.77	6.82	7.24	7.52
	.05	5.49	5.55	5.61	5.66	5.71	6.10	6.37
	.10	4.99	5.05	5.10	5.16	5.21	5.59	5.85
24	.01	6.39	6.45	6.51	6.56	6.61	7.00	7.27
	.05	5.38	5.44	5.49	5.55	5.59	5.97	6.23
	.10	4.91	4.97	5.02	5.07	5.12	5.49	5.74
30	.01	6.20	6.26	6.31	6.36	6.41	6.77	7.02
	.05	5.27	5.33	5.38	5.43	5.48	5.83	6.08
	.10	4.83	4.89	4.94	4.99	5.03	5.39	5.64
40	.01	6.02	6.07	6.12	6.17	6.21	6.55	6.78
	.05	5.16	5.22	5.27	5.31	5.36	5.70	5.93
	.10	4.75	4.81	4.86	4.91	4.95	5.29	5.53
60	.01	5.84	5.89	5.93	5.97	6.02	6.33	6.55
	.05	5.06	5.11	5.15	5.20	5.24	5.57	5.79
	.10	4.68	4.73	4.78	4.82	4.86	5.20	5.42
120	.01	5.66	5.71	5.75	5.79	5.83	6.12	6.32
	.05	4.95	5.00	5.04	5.09	5.13	5.43	5.64
	.10	4.60	4.65	4.69	4.74	4.78	5.10	5.31
∞	.01	5.49	5.54	5.57	5.61	5.65	5.91	6.09
	.05	4.85	4.89	4.93	4.97	5.01	5.30	5.50
	.10	4.52	4.57	4.61	4.65	4.69	5.00	5.20

Source: Adapted from Table II.2 in *The Probability Integrals of the Range and of the Studentized Range*. prepared by H. L. Harter. D. S. Clemm, and E. H. Guthrie. The original tables are published in WADC Tech. Rep. 58–484, Vol. 2, 1959, Wright Air Development Center.

Table C.9 Critical values for the Wilcoxon Signed-Rank test

		Number of pairs									
One-tailed	*Two-tailed*	*5*	*6*	*7*	*8*	*9*	*10*	*11*	*12*	*13*	*14*
.05	.10	0	2	3	5	8	10	13	17	21	25
.025	.05		0	2	3	5	8	10	13	17	21
.01	.02			0	1	3	5	7	9	12	15
.005	.01				0	1	3	5	7	9	12
		15	16	17	18	19	20	21	22	23	24
.05	.10	30	35	41	47	53	60	67	75	83	91
.025	.025	25	29	34	40	46	52	58	65	73	81
.01	.02	19	23	27	32	37	43	49	55	62	69
.005	.01	15	19	23	27	32	37	42	48	54	61
		25	26	27	28	29	30	31	32	33	34
.05	.10	100	110	119	130	140	151	163	175	187	200
.025	.05	89	98	107	116	126	137	147	159	170	182
.01	.02	76	84	92	101	110	120	130	140	151	162
.005	.01	68	75	83	91	100	109	118	128	138	148
		35	36	37	38	39	40	41	42	43	44
.05	.10	213	227	241	256	271	286	302	319	336	353
.025	.05	195	208	221	235	249	264	279	294	310	327
.01	.02	173	185	198	211	224	238	252	266	281	296
.005	.01	159	171	182	194	207	220	233	247	261	276
		45	46	47	48	49	50				
.05	.10	371	389	407	426	446	466				
.025	.05	343	361	378	396	415	434				
.01	.02	312	328	345	362	379	397				
.005	.01	291	307	322	339	355	373				

Table C.10 Transformation of *r* to *Z*

r	*Z*	*r*	*Z*	*r*	*Z*	*r*	*Z*	*r*	*Z*
0.000	0.000	0.200	0.203	0.400	0.424	0.600	0.693	0.800	1.099
0.005	0.005	0.205	0.208	0.405	0.430	0.605	0.701	0.805	1.113
0.010	0.010	0.210	0.213	0.410	0.436	0.610	0.709	0.810	1.127
0.015	0.015	0.215	0.218	0.415	0.442	0.615	0.717	0.815	1.142
0.020	0.020	0.220	0.224	0.420	0.448	0.620	0.725	0.820	1.157
0.025	0.025	0.225	0.229	0.425	0.454	0.625	0.733	0.825	1.172
0.030	0.030	0.230	0.234	0.430	0.460	0.630	0.741	0.830	1.188
0.035	0.035	0.235	0.239	0.435	0.466	0.635	0.750	0.835	1.204
0.040	0.040	0.240	0.245	0.440	0.472	0.640	0.758	0.840	1.221
0.045	0.045	0.245	0.250	0.445	0.478	0.645	0.767	0.845	1.238
0.050	0.050	0.250	0.255	0.450	0.485	0.650	0.775	0.850	1.256
0.055	0.055	0.255	0.261	0.455	0.491	0.655	0.784	0.855	1.274
0.060	0.060	0.260	0.266	0.460	0.497	0.660	0.793	0.860	1.293
0.065	0.065	0.265	0.271	0.465	0.504	0.665	0.802	0.865	1.313
0.070	0.070	0.270	0.277	0.470	0.510	0.670	0.811	0.870	1.333
0.075	0.075	0.275	0.282	0.475	0.517	0.675	0.820	0.875	1.354
0.080	0.080	0.280	0.288	0.480	0.523	0.680	0.829	0.880	1.376
0.085	0.085	0.285	0.293	0.485	0.530	0.685	0.838	0.885	1.398
0.090	0.090	0.290	0.299	0.490	0.536	0.690	0.848	0.890	1.422
0.095	0.095	0.295	0.304	0.495	0.543	0.695	0.858	0.895	1.447
0.100	0.100	0.300	0.310	0.500	0.549	0.700	0.867	0.900	1.472
0.105	0.105	0.305	0.315	0.505	0.556	0.705	0.877	0.905	1.499
0.110	0.110	0.310	0.321	0.510	0.563	0.710	0.887	0.910	1.528
0.115	0.116	0.315	0.326	0.515	0.570	0.715	0.897	0.915	1.557
0.120	0.121	0.320	0.332	0.520	0.576	0.720	0.908	0.920	1.589
0.125	0.126	0.325	0.337	0.525	0.583	0.725	0.918	0.925	1.623
0.130	0.131	0.330	0.343	0.530	0.590	0.730	0.929	0.930	1.658
0.135	0.136	0.335	0.348	0.535	0.597	0.735	0.940	0.935	1.697
0.140	0.141	0.340	0.354	0.540	0.604	0.740	0.950	0.940	1.738
0.145	0.146	0.345	0.360	0.345	0.611	0.745	0.962	0.945	1.783
0.150	0.151	0.350	0.365	0.550	0.618	0.750	0.973	0.950	1.832
0.155	0.156	0.355	0.371	0.555	0.626	0.755	0.984	0.955	1.886
0.160	0.161	0.360	0.377	0.560	0.633	0.760	0.996	0.960	1.946
0.165	0.167	0.365	0.383	0.565	0.640	0.765	1.008	0.965	2.014
0.170	0.172	0.370	0.388	0.570	0.648	0.770	1.020	0.970	2.092
0.175	0.177	0.375	0.394	0.575	0.655	0.775	1.033	0.975	2.185
0.180	0.182	0.380	0.400	0.580	0.662	0.780	1.045	0.980	2.298
0.185	0.187	0.385	0.406	0.585	0.670	0.785	1.058	0.985	2.443
0.190	0.192	0.390	0.412	0.590	0.678	0.790	1.071	0.990	2.647
0.195	0.198	0.395	0.418	0.595	0.685	0.795	1.085	0.995	2.994

Answers to Selected Exercises

Chapter 1

1.1 (a) The study is an experiment. Participants were randomly assigned to each condition.

(b) The sample might best be characterized as a random sample of (presumably) overweight volunteers. The volunteer aspect is important because it's not clear that the effects of the diets would be the same if used by individuals less motivated to lose weight.

(c) One alternative approach is to observe individuals who have selected a particular diet. This might be done by advertising for individuals who are on the diets of interest, requesting they volunteer as participants in a study and then simply assessing their weight. This not as attractive an option as the experiment for several reasons. First, it's not an experiment, so conclusions about the causal role of diet in weight are not possible. Second, the researcher wouldn't have baseline information on weight and health prior to the start of the diet. Some other concerns are that the targeted diets may not be equally well represented in the volunteer sample, and individuals who chose different diets may differ systematically with respect to other factors that could their weight.

1.3 (a) There are several factors other than long hours in day care that may increase aggressive behavior. Two possibilities are that (1) parents may be more likely to place more aggressive children in day care to provide a socializing experience or for other reasons; (2) parents who are more stressed may foster aggression in their children and may place those children in day care for longer hours to reduce their stress or to allow them to work more/earn more money.

(b) Ideally, a measure of aggressive behavior prior to the first enrollment in day care should be obtained. If variation in aggressive behavior as a function of hours/day in day care manifests itself only after the first placement in day care, there is evidence that time in day care has an effect. (Of course, this doesn't reveal whether the relation is due to the day care environment or separation from parents and/or siblings.) If children who were more aggressive after placement in day care were also more aggressive prior to first placement, there is an indication that factors such as those listed in part (a) are involved. Other measures such as measures of parental stress and rated quality of the home environment may also prove useful. Using the regression procedures referred to in this chapter, and developed in Chapters 19 and 20, we can investigate whether such measures predict variability in aggressive behavior beyond that predicted by hours in day care.

(c) An experiment could be performed by randomly assigning children to day care or to home care. (In this case, the home care assignment might be called a "wait list control.") To further reduce the influence of nuisance variables, care might be taken to equate the groups with respect to such factors as parents' working hours, parents' rating of their stress, and how many adult family members are in the child's home. The advantage of the experimental approach is that random assignment ensures that there will be no systematic bias due to uncontrolled factors and several factors may be controlled by equating the groups with respect to them. The disadvantage is that the approach is both unethical and impractical. It is unethical because if one treatment does cause problems, the researcher has subjected half of the children to an experience that may have a negative effect on the child's future. The experiment is impractical because it is doubtful that parents would allow a researcher to decide the preschool experience of their children.

1.5 Lower education levels can result in lower income, more repetitive jobs, and poorer health. All these consequences may, in turn, increase depression. Thus, there is some credibility to the idea that the difference in depression scores between high school educated and college educated individuals is affected by education. However, further examination suggests other possibilities. For example, if the high school educated group came from families with lower income, they may have needed to work to supplement the family income rather than focusing on school; these pressures could increase the risk of depression. It is also possible that other important factors in the family environment mediated differences in both education and depression scores. From the data available on the website, we might try to determine whether there are systematic age differences between the two groups and, if so, whether depression is related to age within each educational level. If so, the difference in depression scores might be the result of differences in age (or other nuisance variables).

Chapter 2

2.1 (a) 30.56; (b) 33.5; (c) 239,121; (d) 16,311; (e) 9.54; (f) lower hinge = 23.5, upper hinge = 37.5; (g) Using R:

```
Y<-c(21, 40, 34, 34, 16, 37, 21, 38, 32, 11, 34, 38, 26, 27, 33, 47)
mean(Y)
#[1] 30.5625
median(Y)
#[1] 33.5
sum(Y)^2
#[1] 239121
sum(Y^2)
#[1] 16311
sd(Y)
#[1] 9.542667
fivenum(Y)
#[1] 11.0 23.5 33.5 37.5 47.0
fivenum(Y)[2] #lower hinge
```

```
#[1] 23.5
fivenum(Y)[4] #upper hinge
#[1] 37.5
skewness(Y)
#[1] -0.4716718
kurtosis(Y)
#[1] 2.555271
```

2.3 (a) If $\bar{Y} = 47$ with six scores, then $\Sigma Y = 6*47 = 282$. The given scores sum to 225, so the new score must be 282 – 225 = 57. (**b**) The original $\bar{Y} = 45$. Adding a score equal to the mean has no impact on the variance because its deviation around the mean is 0.

2.5 (a) 92; (b) 232; (c) 900; (d) 315; (e) 4,605; (f) *var*(X) = 13, *var*(Y = 29.8, *var*(X+Y) = 14.3.

2.7 (a) 286.533, or 287 to the precision of the data; (**b**) 245.84; (c) 2979.6.

2.9 Standardizing shifts all means to 0 and standard deviations to 1. The z score distributions have the same shape as the raw data, and skew and kurtosis values are unchanged. The medians and ranges are not necessarily in the same order (e.g., now $\tilde{x} < \tilde{y}$). Standard scores are normally distributed only if the raw scores are.

2.11 (a) The line graph is preferable in that it more clearly reveals differences among age groups.

(b) Two aspects of the graph are notable. First, the younger age groups (*agegrp* = 1 and 2) have higher mean Beck anxiety scores than the older groups. Second, this is particularly pronounced in the winter season; although three of the four groups are most anxious then, this is markedly so for the youngest group.

(c) Considering the influence of outliers does not change our conclusions. Median trends over seasons within each age group show a trend similar to that for the means, though the differences among age groups are not quite as large when the median is viewed instead of the mean.

2.13 (a) The 2022 average salaries are reasonably symmetric, perhaps slightly left skewed, with no outliers. The mean ($5,356,148) and median ($5,475,877) are similar. The Q–Q plot indicates the data are normally distributed.

(b) In 2007, the American League (*AL*) had higher average salaries than the National League (*NL*). A side-by-side boxplot shows that the median and mean are both higher in the *AL*, so it's not just one team paying more and pulling up the average. There's also more variability in average *salary* in the *AL* than the *NL*.

(c) The average *salary* in 2007 is linearly and positively relative to the average salary in 2022, $r = 0.665$, with some scatter in the data. The relationship is stronger within each league, $r = 0.755$ for *AL* and $r = 0.708$ for *NL*, because league accounts for some of the variability in salary.

(d) The overall mean salary in 1986 is $543,269. Inflation-adjusted, that would imply a mean salary in 2022 of $1,570,047, which almost exactly equals the *minimum* mean *salary* in 2022.

(e) Team *rank* and mean *salary* in 2022 are almost perfectly negatively correlated, $r = -0.987$, which is easy to see in a scatterplot. The negative relationship occurs because the lowest *rank* is the best performing team, so higher *salary* and higher performance are strongly related.

Chapter 3

3.1 (a) .2; (b) $.2^3 = .008$; (c) (.8)(.2)(.8) = 0.128; (d) 3! / (1! 2!) * $(.2^1)(.8^2)$ = 0.384; (e) 1 – *p*(none correct) = 1 – $.8^3$ = 0.488; (f) 3! / (2! 1!)*$(.2^2)(.8^1)$ = 0.096; (g) $(.8^4)(.2)$ = 0.08192.

3.3 (a) (.7)(.6) = 0.42; (b) (.3)(.4) = 0.12; (c) (.7)(.4) = 0.28; (d) (.3)(.6) = 0.18; (e) 1 – *p*(neither A nor B hits target) = 1 – (.3)(.4) = 0.88; (f) 1 – *p*(all 4 throws miss) = 1 – (.3)(.3)(.4)(.4) = 0.9856; (g) 1 – *p*(all 4 throws hit target) = 1 – $(.7^2)(.6^2)$ = 0.8236.

3.5 (a)

Test results	HIV	No HIV	Total
Positive	997	1,485	2,482
Negative	3	97,515	97,518
Total	1,000	99,000	100,000

(b) *p*(infected | positive) = 997/2482 = 0.402; (c) *p*(not infected | negative) = 97,515/97,518 = 0.9999.

3.7 (a) 1 – *p*(positive | dementia) = 1–0.17 = 0.83; (b) 1 – *p*(positive | no dementia) = 1 – 0.008 = 0.992; (c & d) Solve these by making a table of outcomes:

Test results	dementia	No dementia	Total
Positive	340	64	404
Negative	1,660	7,936	9,596
Total	2,000	8,000	10,000

Then we can see that *p*(dementia | positive) = 340/404 = 0.842 and *p*(dementia | negative) = 1,660 / 9,596 = 0.173.

Or, use Bayes' Rule: *p*(A | B) = *p*(B | A)**p*(A) / *p*(B).

So, *p*(dementia | positive) = *p*(positive | dementia)**p*(dementia) / *p*(positive) = *p*(positive | dementia)**p*(dementia) / [*p*(positive | dementia)**p*(dementia) + *p*(positive | no dementia)**p*(no dementia)] = 0.17 * (0.2) / (.17*.2 + .008*.8) = 0.84; *p*(dementia | negative) = *p*(negative | dementia)**p*(dementia) / *p*(negative) = 0.83*0.2/0.9596 = 0.173.

3.9 (a) There's only one way to a 12, which is by getting two 6s. So, (1/6)*(1/6) = 0.028; (b) There are several ways to roll a 7: <1,6>, <6,1>, <2,5>, <5,2>,<3,4>, and <4,3>, and each has the same probability (1/6)(1/6) = 0.028. Overall, the probability of rolling a 7 is 6*0.028 = 0.168.

3.11 (a) First be sure of the ordering of the color and cut labels. One way to do this is to use the unique commands, like so: *unique*(diamonds$color) and *unique*(diamonds$cut) show that the levels are ordered well for our question.

```
> unique(diamonds$color)
[1] E I J H F G D
Levels: D < E < F < G < H < I < J
> unique(diamonds$cut)
[1] Ideal Premium Good Very Good Fair
```

Levels: Fair < Good < Very Good < Premium < Ideal
Next, use the commands *diamonds %>%*
group_by(Color = color <= "F,"Cut = cut >= "Very Good") %>%
summarise(n=n())
to obtain this table of counts:
Color Cut n
<lgl> <lgl> <int>
1 FALSE FALSE 3313
2 FALSE TRUE 24513
3 TRUE FALSE 3203
4 TRUE TRUE 22911

(b) *p*(better color | better cut) = 22,911 /47,424 = 0.483
p(better color | worse cut) = 3,203 / 6,516 = 0.492 These are similar probabilities but not identical, so better color is not independent of cut.

Chapter 4

4.1 **(a)** *p*(reject | H_0 true) = 0.05; **(b)** *p*(fail to reject | H_0 false) = β = 1 – power = 0.20; **(c)** Use Bayes' Rule: *p*(H_0 true | fail to reject) = *p*(fail to reject | H_0 true)**p*(H_0 true) / [*p*(fail to reject | H_0 true)**p*(H_0 true) + *p*(fail to reject | H_0 false)**p*(H_0 false)] = (0.95*0.30)/ [(0.95*0.30) + (0.20*0.70)] = 0.285/(0.285+0.14) = 0.67; **(d)** *p*(reject) = *p*(reject | H_0 true)**p*(H_0 true) + *p*(reject | H_0 false)**p*(H_0 false) = (.05)(.30)+(.80)(.70) = 0.575.

4.3 Let C equal the critical region. Then, **(a)** C > 10; **(b)** C = 5; **(c)** C < 2; **(d)** C $\leq$ 3 or C $\geq$ 17.

4.5 **(a)** (i) H_0: π = 0.25, H_1: π > 0.25, (ii) n = 5, (iii) reject if *y* (number of matches) > 3; **(b)** (i) H_0: π = 0.4, H_1: $\pi \neq$ 0.4, (ii) n = 15, (iii) reject if y < 2 or y > 10.

4.7 In R, the *binom.test* function in the {stats} package requires the number of successes, the number of trials, the null hypothesis value and the chosen α. We can obtain the number of successes by summing the values in each column (separately!) and the count by obtaining the length of each column. Using a two-tailed test in each case, the commands are *binom.test*(x = *sum*(EX4_7$Y1), n = *length*(EX4_7$Y1), p = .5, alternative = *c*("two.sided")) and *binom.test*(x = *sum*(EX4_7$Y2),n = *length*(EX4_7$Y2), p = .5, alternative = *c*("two.sided")). We find that we can reject the null hypothesis for *Y2*, *p* = 0.04139, but not for *Y1*, *p* = 0.2632.

4.9 **(a)** H_0: π = 0.5, H_1: $\pi \neq$ 0.5. π is the probability that the imagery procedure is better than the rote memorization approach. *N* = 12 because we have 12 students and it's a within-participant design. We set α = .05 and define a rejection region. Using R, *qbinom*(p = .025, size = 12, prob = .5, lower.tail = FALSE) returns 9, meaning the upper tail rejection region is *Y* > 9. For the lower-tail portion, *qbinom*(p = .025, size = 12, prob = .5, lower.tail = TRUE) returns 3. So, *Y* < 3 or *Y* > 9 defines the rejection region. We observe *Y* = 9 so we do not have enough evidence to reject the null hypothesis. Again, the sample size is too small to provide clarity. **(b)** We could use G*Power or R. In R, *pbinom*(3, size = 12, prob = .9, lower.tail = TRUE) + *pbinom*(9, size = 12, prob = .9, lower.tail = FALSE) returns 0.889 the total area in the rejection region assuming H_A: π = 0.9. Of course, G*Power provides the same value.

Chapter 5

5.1 (a) $p(Y > 115 \mid \mu = 100, \sigma = 15) = p(z > (115–100)/15) = p(z > 1) = 0.1587$.
(b) (i) $p(Y > 130) = p(z > 2) = 0.0228$, (ii) $p(85 < Y < 145) = p(-1 < z < 3) = p(z<3) - p(z < -1) = 0.84$, (iii) $p(70 < Y < 80) = p(-2 < z < -1.33) = 0.0690$.
(c) We need the 10th and 90th percentile values, which are +/– the same z score due to the symmetry of the normal distribution. Use Table C.2 or R. *qnorm*(.1) = –1.28 = z score at 10th percentile. We can convert this to Y with algebra. Or we can get the equivalent Y value directly: *qnorm*(.1, 100, 15) = 80.78. The same logic finds the upper value, $Y = 119.22$.
(d) *qnorm*(.75, 100, 15)= 110.12 or use Table C.2 and algebra.
(e) Because we are asking about a mean, the corresponding measure of variability in the z score calculation is the *SEM*. $p(\bar{Y} > 105) = p(z > (105–100) / (15/\sqrt{10})) = p(z > 1.054) = 0.146$.

5.3 (a) (i) $p(X < 25) = p(z < (25–30)/20) = 0.4013$, (ii) $p(X > 60) = 0.0668$, (iii) $p(15 < x < 40) = 0.4648$.
(b) $p(X > \mu_Y) = p(X > 20) = 0.6915$.
(c) (i) $\mu_W = \mu_X + \mu_Y = 30 + 20 = 50$, (ii) $\sigma^2_W = \sigma^2_X + \sigma^2_Y = 20^2 + 16^2 = 656$, (iii) $p(X > 35) = p(z > (35 - 50)/\sqrt{656}) = p(z > -0.5857) = 0.7209$.
(d) We first need to calculate her X and Y scores, from which we know her W. To find X, for example, identify the value of X at the 85th percentile of $N(30,20)$ distribution using Table C.2 and algebra or R: *qnorm*(.85, 30, 20) = 50.73. Similar logic finds that her Y value is 11.61, so her W score = $X + Y = 62.34$. Lastly, find the percentile of 62.34 in the W distribution, again using Table C.2 or R: *pnorm*(62.34, 50, *sqrt*(656)) = 0.685. Her total score is at the 68.5th percentile.

5.5 (a) (i) $p(Y < .6) = 0.6$, (ii) $0.6*0.6 = 0.36$, (iii) 0.6^{20}.
(b) The Central Limit Theorem says the sampling distribution of the mean should be approximately normal, with $\mu_{\bar{Y}} = 0.5$ and $\sigma_{\bar{Y}} = \frac{\sigma_Y}{\sqrt{n}} = \frac{1/\sqrt{12}}{\sqrt{20}} = 0.0645$.
(c) $p(\bar{Y} < .6) = p(z < (.6 - .5) / .0645) = p(z < 1.55) = 0.9395$.
(d) We used the Central Limit Theorem to conclude that the sampling of the mean is normal with the mean and standard deviation given in (b). The same approach does not apply to (a)(iii) because that question was about a set of individual scores, not a sampling distribution of the mean, and the population distribution is known to be uniform, not normal, in form.

5.7 (a) H_0: $\mu = 52.8$, H_1: $\mu > 52.8$. Reject if $z > 1.645$.
(b) $z = (56–52.8)/(10.5/\sqrt{50}) = 2.155$. Because 2.155 exceeds the critical z of 1.645, reject the null hypothesis and conclude that authoritarianism scores have increased.
(c) Start by translating the critical z score to the equivalent observed score: $1.645 = (x–52.8)/1.485$, so $x = 55.243$. Now, find the area above that x under H_A, $\mu = 57$, which you can get from its z score: $z=(55.243–57)/1.485$ and Table C.2 or in R using *pnorm*(55.243, 57, 1.485, lower.tail = FALSE) = 0.8816. Power is just about 0.88.
(d) 56 +/– 1.96(1.485) = [53.09, 58.91].
(e) The CI in part (d) either does or does not contain the true mean. If we repeated the process of sampling 50 students' scores many times and computed the 95% CI for each sample, then 95% of the intervals would actually contain the population mean.

5.9 (a) $\mu_{T-C} = \mu_T - \mu_C = 0.5\sigma$, var(T – C) = $\sigma_T^2 + \sigma_C^2 - 2\rho_{TC}\sigma_T\sigma_C = 2\sigma^2$ because T and C are independent.

(b) $p(\text{T} - \text{C} > 0) = p(z > \frac{0-0.5\sigma}{\sqrt{2\sigma^2}}) = p(z > \frac{-0.5\sigma}{\sigma\sqrt{2}}) = p(z > -0.35) = 0.6368$; (c) (i) $p(z > -1.06) = 0.8554$, (ii) $p(\bar{T} - \bar{C} > 0) = p(z > \frac{0-0.2\sigma}{\frac{\sigma}{3}\sqrt{2}}) = p(z > -0.42) = 0.6628$.

The reduction in effect size (difference between the means) causes a reduction in the power of a significance test of that difference. Similarly, if n were increased, then the standard error would be smaller and the power would increase.

5.11 (a) $\sigma / \sqrt{n} = 13.6 / \sqrt{225} = 0.9067$.

(b) 95% CI is estimate +/– $z_{.025}$*(std error of the estimate), or 2 +/– 1.96 (0.9067) = [0.2229, 3.777].

(c) zero falls outside the 95% CI and so we can reject the null hypothesis of no difference with $\alpha = .05$. Or, $z = (2-0)/0.9067 = 2.2058$, which exceeds the critical value $z = 1.96$ when $\alpha = .05$. Either way, we reject the null hypothesis.

5.13 (a) $s_{\overline{diff}} = \sqrt{\frac{s_1^2 + s_2^2 - 2r_{12}s_1s_{12}}{n}} = \sqrt{\frac{40.794^2 + 40.113^2 - 2(0.855)(40.794)(40.113)}{220}}$ = 1.467.

(b) (224.059–218.818) +/– 1.96(1.467) = [2.366, 8.116].

(c) Zero is not in the CI, so we can reject the null hypothesis of no difference with a = .05. Or we can compute $z = ((224.059-218.818)-0)/1.467 = 3.57$. Because $z=3.57$ exceeds the critical value of 1.96, we reject the null hypothesis.

5.15 (a) $\text{SE}_{\text{diff}} = \sqrt{\left(\frac{42.0^2}{112}\right) + \left(\frac{33.9^2}{98}\right)} = 5.24$.

(b) (229–215) +/– 1.96(5.24) = [3.73, 24.27] Because zero is not in the 95% CI, we can reject the null hypothesis of no difference in total cholesterol as a function of education level (HS graduate or less vs HS graduate+ some college).

5.17 (a) H_0: $\mu = 200$, H_1: $\mu > 200$.

(b) H_A: $\mu = 206$; (c) We want to detect a TC of 206, which is a z score of (206–200)/(30/sqrt(n)) = sqrt(n)/5. To have power of 0.8, this z-score needs to be above the critical z of 1.645 (one tailed test with $\alpha = .05$). How much higher? The area above the critical $z = 1.645$ under H_A has to be 0.8, which occurs when $z_A = -0.8416$ (use *qnorm*(.2, lower.tail = FALSE) or Table C.2 to get this value). That means *sqrt*(n)/5 = 1.645 + 0.8416 = 2.4866, so n = 154.58 or 155. Drawing a picture of the criteria and distributions may be helpful on this problem.

Chapter 6

6.1 (a) 90% CI is estimate +/– $t_{crit}(df = \text{n}-1)*SE_{estimate}$. Here, $t_{crit}(8) = 1.860$ and the mean score is 47.889 so the CI is 47.889 +/– 1.860(4.996) = [38.599, 57.179].

(b) The normal population mean is 60, which falls above the 90% CI in (a). This allows us to conclude that the protein-deficient infants have lower-than-typical motor skills on this test, with $\alpha = 0.10$ and this small sample. We should be cautious not to overinterpret this result.

(c) The new mean is 50.889, and the new 90% CI is 50.889 +/– 1.860(4.455) = [42.605, 59.173]. The assumed population mean of 60 is still outside the 90% CI so we can conclude that these infants still differ from the normal population with $\alpha = .10$. Using $\alpha = 0.05$ and a one-tailed test, $t(8) = 2.045$, $p = .0375 < \alpha$ so we can conclude that their motor skills are still below normal on this test.

(d) $\bar{Y}_2 - \bar{Y}_1 = 3$ and $SEM = 3.741$ / sqrt(9) = 1.247; $t(8) = 2.405$, $p = .0214 < \alpha$, so we can conclude that the infants' motor skills improved from Y_1 (lower scores) to Y_2 (higher scores).

(e) By hand, or using the code provided by Fitts (2020) for R, we can calculate $d_z = 3 \,/\, \sqrt{\dfrac{224.61+178.61}{2}}$ = 3/14.199 = 0.211, a small difference by Cohen's guidelines.

6.3 (a) $t(df = 15) = 2.0 / (5.6/sqrt(16)) = 2/1.4 = 1.429$. With $\alpha = .05$, $t_{crit}(df{=}15) = 2.131$, so we cannot reject the null hypothesis.

(b) $d_z = 2.0 / 5.6 = 0.357$.

(c) Using the observed d_z as the effect size, G*Power shows we need $N = 50$ to have power of at least .80.

(d) This part asks you to calculate post hoc power six times (two distributional assumptions × three sample sizes). Using G*Power, the results are tabulated here:

Sample Size	*t distribution*	*Standard normal distribution*
16	0.389	0.415
36	0.676	0.691
50	0.801	0.811

Note that, from Chapter 5 (Section 5.6.1), the ncp for the standard normal is the observed effect (2.0) / SEmean and that SEmean varies with sample size. For N = 16, the ncp = 1.429 from part (a). For $N = 36$, $ncp = 2.0/(5.6/sqrt(36)) = 2.144$, and for $N = 50$, $ncp = 2/(5.6/sqrt(50)) = 2.525$. The results using the t and the z are fairly similar with for small samples, and they become more similar as sample size increase because the t distribution because more like the standard normal with greater degrees of freedom.

6.5 (a) $t = (30.2 - 27.0) / (s_{pooled}{*}sqrt(1/n_1 + 1/n_2)) = 3.2 / (sqrt(20{*}8/30 + 10{*}30/30){*}sqrt(1/21 + 1/11)) = 2.196$ where $t_{crit}(30) = 2.042$, so we can reject the null hypothesis with $\alpha = .05$.

(b) $t' = (30.2 - 27.0) / sqrt(8/21 + 30/11) = 1.815$, $df' = \dfrac{\left(\dfrac{8}{21}+\dfrac{30}{11}\right)^2}{\dfrac{8^2}{21^2 * 20}+\dfrac{30^2}{11^2 * 10}} = 12.86.$

In R, we can calculate $qt(.975$, df = 12.86) = 2.16, the critical value of t. Our observed t' < the critical value so we cannot reject the null hypothesis of no difference in the means.

(c) The difference between the two test results occurs because the group with the larger sample size has the smaller variance, so the pooled estimate of a common variance (calculated in part (a)) underestimates the true variance and creates a positive bias in the test results. That is, there is a greater than .05 chance of a Type I error (see Table 6.3). Welch's t corrects that bias.

6.7 (a) We can calculate by hand or use R's *t.test* function to find the CIs, being sure to use the paired=TRUE option. See the "code for Chapter 6 Exercises Solutions.R" for details. Using R, we find the *Lab* CI is [10.512, 20.738] and for the *Natural* condition the CI is [9.286, 33.714]. By hand, we have $df = 7 = n - 1$ in each condition, so we use Table C.3 or R's *qt* function to find $t_{crit} = 2.365$. The CI is $\bar{Y}_{diff}$ +/– $2.365 * \frac{s_{diff}}{\sqrt{n}}$. In the *Lab* condition, that's 15.625 +/– $2.365(6.12/\sqrt{7})$ = [10.512, 20.738]. In the *Natural* condition, that's 21.5 +/– $2.365(14.6/\sqrt{7})$ = [9.286, 33.714]. Zero isn't in either CI, so in both conditions we can reject the null hypothesis that there is no difference from *Day 1* to *Day 2* with $\alpha = .05$, two-tailed. Indeed, *Day 1* scores are higher than *Day 2* scores in both conditions.

(b) This question compares across conditions so it's an independent groups t. We can use R's *t.test* function again, being sure to use the paired = FALSE option (which is the default). Doing so reveals $t(12.793) = 1.756$, $p = .103$, so we can't reject the null hypothesis that the two conditions have the same performance on Day 2. By hand, we use Welch's $t' = \frac{\bar{Y}_1 - \bar{Y}_2}{\sqrt{\frac{s_1^2}{n_1} + \frac{s_2^2}{n_2}}} = \frac{33.375 - 25.25}{\sqrt{\frac{112}{8} + \frac{59.4}{8}}} = 8.125 / 4.63 = 1.755$. df can be calculated from Equation 6.17; (c) H_0: $\mu_{Lab,1} - \mu_{Lab,2} = \mu_{Nat,1} - \mu_{Nat,2}$ and H_1: $\mu_{Lab,1} - \mu_{Lab,2} \neq \mu_{Nat,1} - \mu_{Nat,2}$ which is equivalent to H_0: $\mu_{Lab,diff} = \mu_{Nat,diff}$ and H_1: $\mu_{Lab,diff} \neq \mu_{Nat,diff}$. $t' = \frac{\bar{Y}_1 - \bar{Y}_2}{\sqrt{\frac{s_1^2}{n_1} + \frac{s_2^2}{n_2}}} = \frac{15.6 - 21.5}{\sqrt{\frac{6.12^2}{8} + \frac{14.6^2}{8}}} =$ $-5.9/5.597 = -1.054$, which is not significant. Using R's *t.test* function we get $p = 0.32$ and $df = 9.38$.

6.9 (a) For *Sayhlth* = 2, $d_z = 4.56/20.3 = 0.225$; for *Sayhlth* = 4, $d_z = 11.1 / 19.4 = 0.572$.

(b) The within-condition variances contribute to the denominator of both d and d_z. However, there is an adjustment for the covariance of the conditions in the denominator of d_z. The impact of that covariance depends on its magnitude and sign. (c) For the *Sayhlth* = 4 condition, we need 27 participants. For the *Sayhlth* = 2 condition, the effect size estimate is much smaller and as a consequence, we need 157 participants to achieve power of .80.

6.11 (a) $d = .8 / \sqrt{\frac{(2.25 + 1.44)}{2}} = 0.59$.

(b) G*Power tells us that we need 37 per group, or 74 participants total with a one-tailed test.

(c) We can ask G*Power to "Determine" what the effect size would be under these conditions. Click "Determine" and then enter the condition means and standard deviations, as well as the assumed correlation, and then click "calculate and transfer to main window" followed by "Calculate" in the main window. You'll see that the effect size estimate is slightly smaller than in part (a) (0.58 rather than 0.59) and the sample size needed is only 20 under these assumptions.

6.13 (a) Because the sample sizes are equal across conditions in each experiment, we can use Equation 6.8 to calculated $s_{pooled} = \sqrt{\frac{(22+27)}{2}} = 4.95$. Then $d = 3.3 / 4.95 = 0.67$ in each experiment. Next, using $t = \frac{\bar{Y}_1 - \bar{Y}_1}{s_{pooled}\sqrt{2/n}} = \frac{3.3}{4.95\sqrt{2/n}}$ we can find t for each experiment, and the $pt(t,df)$ function in R or Table C.3 to find the p-value (remember to calculate for both tails) and G*Power to compute post hoc power. The effect size, d, is constant but the p-value decreases as df increase. *Post hoc* power is inversely related to the p-values: smaller p, higher power and larger observed t. These numbers don't provide unique information: p is the probability of rejecting the null hypothesis when it is true, and power is the probability of rejecting the null hypothesis when it is false. Both depend on sample size. Notice, too, that power is low even with 50 participants.

Experiment	*1*	*2*	*3*	*4*
N	4	9	16	25
df	6	16	30	48
t	0.943	1.414	1.886	2.357
d	0.67	0.67	0.67	0.67
p	0.382	0.177	0.069	0.023
Post hoc power	.127	.267	0.450	0.641

(b) Neither p-values nor post hoc power are good measures of the importance of an effect. Both are conditional probabilities that depend on sample size. The estimated effect size, d, provides an indication of what the size of the effect might be in the population, but it cannot tell us whether that effect is important. Small effects can be important, and large effects can be trivial.

Chapter 7

7.1 (a) $\mu = (-1)(.25)+(0)(.5)+(1)(.25) = 0$; $\sigma^2 = (.25)(-1-0)^2 + (.5)(0-0)^2 + (.25)(1-0)^2 = 0.5$. (b) Drawing two scores:

Possible scores	*Mean*	*Probability*
(−1,−1)	−1	$.25^2 = 0.0625$
(−1,0) or (0,−1)	−0.5	$2*(.25(.5)) = 0.25$
(0,0) or (−1,1) or (1, −1)	0	$.5^2+.25^2+.25^2 = 0.375$
(1,0) or (0,1)	0.5	$.25(.5)+.25(.5) = 0.25$
(1,1)	1	$.25^2 = 0.0625$

Notice that the probabilities sum to 1, as they must if we've accounted for all possible events in this sample space.

(c) $E(\bar{x}) = 0$ and $\sigma^2_{\bar{x}} = .0625(-1-0)^2 + .25(-.5-0)^2 + .375(0-0)^2 + .25(.5-0)^2 + .0625(1-0)^2 = 0.25$, which is half the variance of the population from part (a). Notice that $\sigma^2_{\bar{x}} = 0.25 = \frac{\sigma^2_x}{n} = 0.5/2 = 0.25$.

7.3 (a) On each trial there are four options. If the participant is guessing at random, the probability of choosing correctly on one trial is 0.25.

(b) We compute the probability of k successes out of n trials, where $k = 0, 1, \ldots, 6$, $n = 6$, and $p = p(\text{success}) = 0.25$. $P(y = k|n,p) = p^k(1 - p)^{n-k}$, so for $k = 0$, prob = $0.25^0\ (1-.25)^6 = 1*0.178$. In R, the *pbinom(k,n,p)* function will help (remember to subtract off the probability that $y<k$), or use Table C.1.

Number Correct	0	1	2	3	4	5	6
Probability	0.178	0.356	0.297	0.132	0.033	0.004	0.000

(c) There are 100 participants in the study, so the expected number of participants with each possible number correct is simply 100 $*p(k)$. For example, we expected 132 participants to have exactly three correct responses.

(d) Each participant sees six trials with 0.25 chance of being correct on each. So the expected number correct is 6 * 0.25 = 1.5.

(e) $p(y > 1.5) = p(y = 2) + p(y = 3) + \ldots + p(y = 6) = 0.297 + 0.132 + 0.033 + 0.004 + 0.000 = 0.466$.

7.5 (a) For *Group 1*, the critical value of t with .025 area above it is *qt*(.975, *df*=11) = 2.201 and the $SE_{\bar{x}} = SD_x$ / *sqrt*(12) = 6.6/*sqrt*(12) = 1.91. Then the 95% CI for the mean is 10.7 +/– 2.201*1.91 or [6.5, 14.9]. For *Group 1*, 0 is not in the interval so we can conclude that the *Change* scores are significantly greater than 0 in that group with $\alpha = .05$. Similarly for *Group* 2, the 95% CI for the mean is 5.42 +/– 2.571*2.66 or [–1.42, 12.3]. In *Group* 2, which is much smaller, 0 is in the CI and so we cannot reject the null hypothesis that the change scores have an average of 0.

(b) $t = \dfrac{\overline{change_1} - \overline{change_2}}{\sqrt{\dfrac{s_1^2}{n_1} + \dfrac{s_2^2}{n_2}}} = 5.28 / 3.27 = 1.61$. Use Equation 6.17 to calculate df, or use *t.test* in R. We find $df = 10.24$ and the p-value is 0.069 for the one-tailed alternative hypothesis that the mean change is "greater" than 0, which is not significant. We conclude that the difference in the change scores as a function of group is not significant. Note that this is true even though one of the groups has a mean change that is itself significantly different than zero while the other does not. Testing groups independently is not the same as comparing them in a single test.

7.7 (a) The variance of the difference is $\sigma^2_1 + \sigma^2_2 - 2\sigma_1\sigma_2\rho_{12}$. Substituting with known values, we get variance = 12.5 + 12.5 – 2(12.5)(.6) = 10, so $\sigma_{diff} = 3.16$.

(b) $d_z = 1.52/3.16 = 0.48$.

(c) G*Power reveals that we need a sample of 29 for a correlated-scores design under these assumptions.

7.9 (a) We can use the *YuenTTest*(data = dat, RT ~ Group, trim = .2) function in {DescTools} in R. Doing so, we find a nonsignificant difference: trimmed $t(26.9) = 1.61$, $p = .1196$.

(b) The test in part (a) provides the 95% CI for the trimmed data: [–44.53, 366.6]. This interval is slightly narrower than the 95% CI for the untrimmed data: [–24.94, 404.6], and it's centered on a smaller difference in means (161 vs 189.8). Both intervals include zero, though, so neither allows us to reject the null hypothesis.

(c) The *wilcox.test* in R computes the Mann–Whitney U statistic we need, 390.5, and then we divide U by the number of comparisons made (the number of participants in *Group 1* * the number of participants in *Group 2*), which gives $f = 0.65$. This means that 65% of participants in *Group 1* had a higher rank *RT* than the participants in *Group 2*. The result is also not significant, $p = .075$.

Chapter 8

8.1 (a) The variances will be multiplied by 100^2. The F ratio will not change because both the numerator and denominator increase by the same factor.
(b) The variance is increased by the square of the constant.
(c) Adding a constant to all scores will not change the mean squares or the F ratios.
(d) Because the spread of the group means changes, the MS_A also changes. However, adding the same constant to all scores in a group will not change the within-group variance and therefore MS_A is unaffected.

8.3 (a) Using the *aov* function in R, we get $F(1,8) = 0.832$, $p = .388$.
(b) Using *t.test* in R with the equal-variance assumption, we get $t(8) = -.912$, $p = .388$. Recall that R orders conditions alphabetically, and that doesn't affect anything except the sign of the t statistic. Squaring the obtained t, -0.912^2 we get the obtained $F = 0.832$. We will see this relationship again: Student's $t^2(df = \text{m}) = F(1, \text{m})$.

8.5 (a) Histograms, box plots, and *Q–Q* plots indicate that the data are right-skewed (long tail for high scores). The Shapiro–Wilk tests of normality (*shapiro.test* in R) result in p-values < 0.03 for each group, indicative of nonnormal data. Here's a plot of the data with condition value jittered on the x-axis.

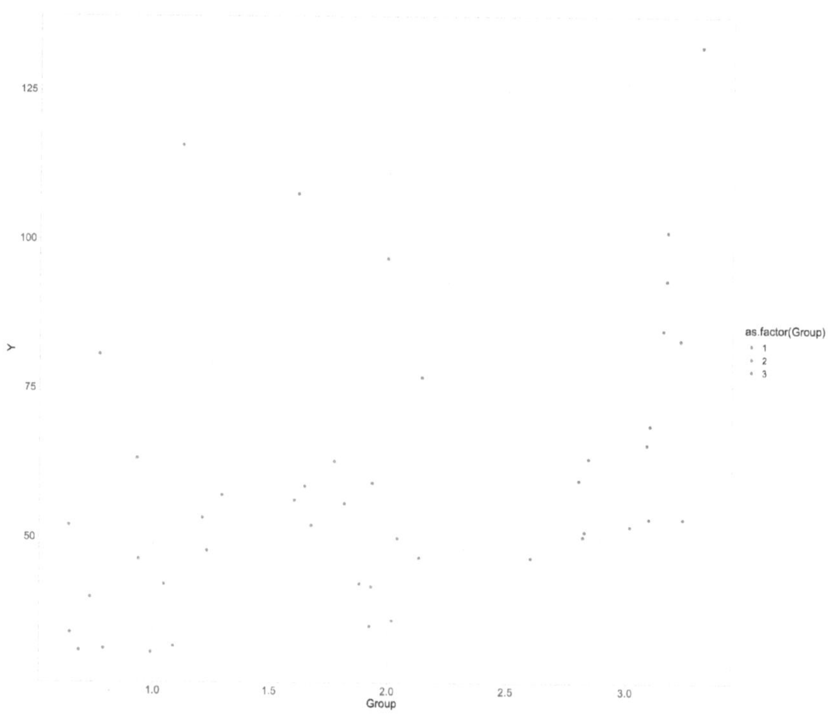

The standard deviations are similar as is skew. So we expect that the Type 1 error rate is probably not inflated, especially considering we have equal sample sizes. However, we might have higher power using a rank-based test.

(b) Using the *aov* function in R, we find that:

```
#                     Df   Sum Sq   Mean Sq   Fvalue   Pr(>F)
#as.factor(Group)     2    3056     1528      2.967    0.0623 .
#Residuals            42   21631    515
# – -
# Signif. codes: 0 '***' 0.001 '**' 0.01 '*' 0.05 '.' 0.1 ' ' 1
```

So the standard ANOVA doesn't allow us to reject the null hypothesis.
Next, running the Kruskal–Wallis *H* test using the *kruskal.test* function in R, we find:
Kruskal–Wallis chi-squared = 8.2155, df = 2, p-value = 0.01644. (The mean ranks for the 3 groups are 16.20, 22.87, and 29.93). Here, we can reject the null. The difference in results likely reflects the fact that the transformation to ranks reduces the impact of the long tails on within-group variability (i.e., error variance or $MS_{S/A}$), thus increasing power (as predicted in part (a)).

8.7 (a) For Table 1, $F(2,27) = 80/5 = 16$ and using Equation 8.16, $\omega^2 = (2)(15)/[(2)(15)+30] = 0.5$. For Table 2, $F(2,12) = 42.5/5 = 8.5$ and $\omega^2 = (2)(7.5)/[(2)(7.5)+15] = 0.5$.

(b) Changing the sample size won't generally change the estimate of ω^2 unless the error mean square changes. The major change we see is in the *F* ratio, which occurs because the contribution of the treatment effects, θ^2, to the numerator is multiplied by a larger *n*.

(c) $\eta^2 = SS_A / SS_{total} = 2(80)/(2(80)+27(5)) = 0.54$ for Table 1 and 0.59 for Table 2 via analogous calculation. With ω^2 constant at 0.5, increasing the sample size decreased η^2.

(d) (i) If $F = 1$, then $MS_A = MS_{S/A}$ and $\omega^2 = 0$ (see Equation 8.14). Alternatively, see Equation 8.16. (ii) $\eta^2 = SS_A / SS_{total} = (a-1)MS_A / [(a-1)MS_A + a(n-1)MS_{S/A}]$. If $F = 1$, this reduces to $(a-1)/(an-1)$.

(e) *F* reflects the sample size as well as treatment effects and error variances. As a result, reliance of *F* (or the associated *p*-value) may cause very small effects to appear important if *n* is large. In Chapter 6 (Table 6.2), we saw a case in which the data set with the smaller standardized effect size resulted in the larger value of *t*, and in the present example we saw that ω^2 might be invariant even as *F* varied with *n*. As we also saw in the present example, η^2 is also affected by sample size. If we want to compare effects across similar experiments differing in size, ω^2 is the better measure to use.

8.9 We begin by calculating the effect size if the theory is correct. $\sigma_A^2 = \Sigma(\mu_{.j} - \mu_{..})^2 / a$ where the μ_j = 10, 14, 18. Assuming equal *n* (per problem 8.8), the grand mean is 14, and so $\sigma_A^2 = 10.67$. Then Equation 8.17 shows that $f = sqrt(\sigma_A^2 / \sigma_e^2) = sqrt(10.67 / 30) = 0.596$. G*Power shows we need a total sample of 33 participants under these assumptions, 11 per group.

8.11 (a) Note that participant ID = 635 doesn't have any information about employment status, so we should exclude that individual from our analyses. Boxplots show right skew in each condition. Mean and median depression scores are numerically higher for the unemployed (3) and employed part-time (2) groups compared to the full-time employed condition (1). The Brown–Forsythe test (*leveneTest*(. . ., center = median)) finds evidence for heterogeneous variances: $F(2,323) = 3.27$, $p = .039$.

The variance is largest in the unemployed group (3), about double the variance in the group that's employed full time (1, the largest group by a factor of almost 4). This means that the *F* test for means would be positively biased: The largest group has the smallest variance, so $MS_{S/A}$ will be underestimated, increasing *F*.

(b) The ANOVA is:

For *Beck_D*

	Df	Sum Sq	Mean Sq	F value	Pr(>F)
as.factor(employed)	2	173	86.68	3	0.0512 .
Residuals	323	9333	28.90		

We conclude that there is no significant effect of employment status on mean depression level. Although the raw data yield a marginally significant *F*, that is due to the positive bias in that test as described in part (a). The large full-time sample with small variance compared to the part-time and unemployed groups results in the underestimation of the within-cell variance.

(c) Welch's *F* test (*welch.test*) concludes that there is no evidence in these data for a difference in depression as a function of employment status, $F(2,82) = 2.29$, $p = .107$. That's consistent with our earlier conclusion based on analysis of the transformed data.

(d) Equation 8.18 says $f = sqrt((2)(2)/326) = 0.11$, a small effect by Cohen's guidelines. We lack evidence that employment status influenced mean depression level, at least in these data.

Chapter 9

9.1 (a)

Source	Df	Sum Sq	Mean Sq	F value	Pr(>F)
A	3	1118.8	372.9	7.177	0.00287 **
B	1	77.0	77.0	1.483	0.24100
A:B	3	1003.5	334.5	6.438	0.00458 **
Residuals	16	831.3	52.0		

(b) Each main effect is a row or column mean minus the grand mean.

	A1	A2	A3	A4	$\bar{Y}_{..k}$	β_k
B1	27.333	20.000	35.667	29.333	28.063	−1.792
B2	38.000	20.333	20.000	48.333	31.667	1.792
$\bar{Y}_{.j.}$	32.667	20.167	27.833	38.833	29.875	
α_j	2.792	−9.708	−2.042	8.958		

The interactions are the cell means minus grand mean minus the row and column main effects:

	A1	A2	A3	A4
B1	−3.542	1.625	9.625	−7.708
B2	3.542	−1.625	−9.625	7.708

Notice that the sum of the main effects, and sums of the row and column interaction effects are all zero.

(c) $SS_A = nb\sum_j \alpha_j^2 = (3)(2)(2.792^2 + (-9.708)^2 + (-2.042)^2 + 8.958^2) = 1118.736$

$SS_B = na\sum_k \beta_k^2 = (3)(4)(-1.792^2 + 1.792^2) = 77.070$

$SS_{AB} = n\sum_j\sum_k \alpha\beta_{jk}^2 = 3((-3.542)^2 + \ldots + (7.708)^2) = 1003.458$. These are all the same as in part (a) except for minor differences due to rounding.

9.3 (a) While we don't normally calculate ANOVAs with a calculator, it is possible to do so, especially if your calculator has a variance function. Alternatively, use R to do some of the calculations. Start with *df*: for *A*, $(a - 1) = 1$; for *B*, $(b - 1) = 2$; for interaction, $(a - 1)(b - 1) = 2$, $df_{total} = n - 1 = 59$, and $df_{S/AB} = (n - 1)ab = 9(2)(3) = 54$. The mean of $A1 = 4.47$, for $A2 = 3.77$; bn*variance of those means equals MS_A. The means of *B* are: 3.45, 3.95, 4.95; an*variance of those means equals MS_B. MS_e = the average variance across the cells = $(2.75 + \ldots + 3.75)/6 = 3.5$; multiply by the df_e to get SS_e. For the interaction *MS*, we know $SS_{cells} = (ab - 1)n^{*}var(\text{cell means}) = 88.28$, and $SS_{AB} = SS_{cells} - SS_A - SS_B = 88.28 - 7.35 - 23.4 = 57.5$. Divide by *df* to get the *MS*.

Source	df	SS	MS	F	p
A	1	7.35	7.35	2.1	0.153
B	2	23.4	11.7	3.3	0.044
AB	2	57.5	28.8	8.2	<.001
S/AB	54	189.0	3.5		

(b) $SS_{A/B3} = (10)((6.5 - 4.95)^2 + (3.4 - 4.95)^2) = 48.05$ and $df_{A/B3} = 1$; $MS_{S/A/B3} = (5.50 + 3.75)/2 = 3.625$; $F(1,18) = 48.05 / 3.625 = 13.26$ so $p = 0.002$.

(c) $SS_{B/A2} = 10((4.3 - 3.77)^2 + (3.6 - 3.77)^2 + (3.4 - 3.77)^2) = 4.467$ with $df = 2$, so $MS_{B/A2} = 2.23$. $MS_{S/B/A2} = (1.75 + 2.25 + 3.75)/3 = 2.58$ so $F(2, 27) = 2.23/2.58 < 1$ and clearly not significant.

(d) The cell variances range from 1.75 to 5.50, a more than three-fold difference. Given that wide range, it makes sense to use only the contributing cells' variances when computing the MS_e for the simple effects. Otherwise we would risk over- or underestimating the error variance, and therefore the *F* statistic would be too small or too large.

9.5 (a) *summary*(*aov*(data = dat, *Y* ~ *A***B*)) gives:

```
#              Df  Sum Sq  Mean Sq  F value  Pr(>F)
#A              2  972     486.0    3.097    0.05248.
#B              3  3689    1229.7   7.836    0.00017 ***
#A:B            6  2237    372.8    2.375    0.03996 *
#Residuals 60   9416    156.9
```

(b) Both *A* and *B* are extrinsic (manipulated) variables so $\eta_g^2 = \frac{SS_B}{SS_B + SS_{S/AB}} = \frac{3689}{3689 + 9416} = 0.28$; (c) *A* is an intrinsic (observed) variable so $\eta_g^2 = \frac{SS_B}{SS_B + SS_A + SS_{AB} + SS_{S/AB}} = \frac{3689}{3689 + 972 + 2237 + 9416} = 0.23$; in R, using *eta_squared*(p95anova, generalized="*A*") where *p95anova* is the saved output of the *aov* in part (a), confirms that $\eta_g^2 = 0.23$.

9.7 (a) The variances differ by almost seven-fold across conditions and appear correlated with the condition means. The data also do not appear to be normally distributed: There are outliers in several conditions and the data appear right skewed in most cells. The cell sizes are equal.

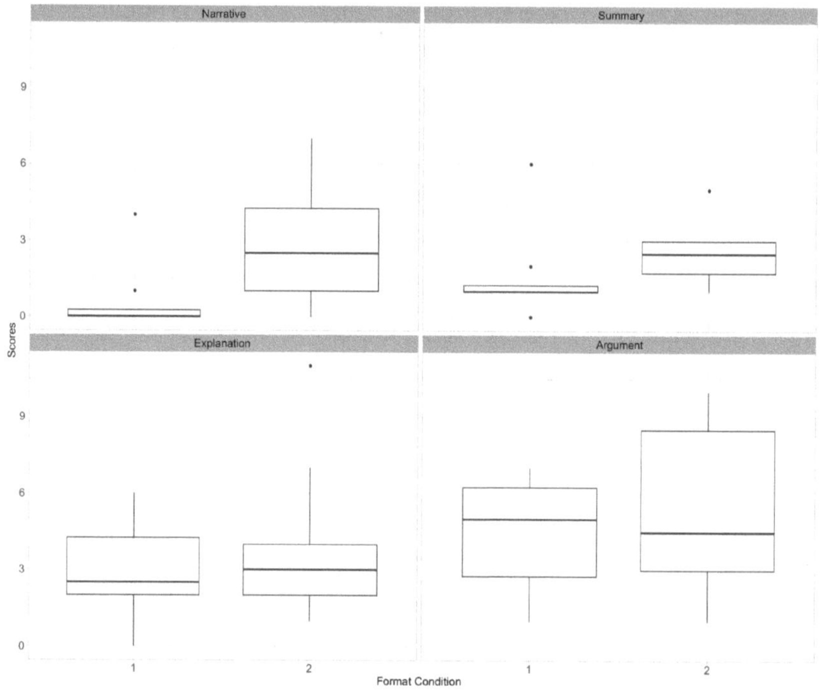

(b) Using R, *summary*(*aov*(data = dat, causal ~ instruct*format)) gives the ANOVA table:

#	Df	Sum Sq	Mean Sq	F value	Pr(>F)
#instruct	3	106.67	35.56	6.369	0.000861 ***
#format	1	26.27	26.27	4.705	0.034338 *
#instruct:format	3	5.05	1.68	0.301	0.824280
#Residuals	56	312.62	5.58		

(c) Again using R, the transformation can be done right in the command line: *summary*(*aov*(data = dat, log(causal + 1) ~ instruct*format))

	Df	Sum Sq	Mean Sq	F value	Pr(>F)
instruct	3	8.059	2.6862	8.408	0.000105 ***
format	1	2.512	2.5119	7.863	0.006923 **
instruct:format	3	1.387	0.4625	1.448	0.238711
Residuals	56	17.891	0.3195		

The statistical conclusions are the same: There are main effects of both instruction and format condition and no interaction. The *p*-values are smaller now. The boxplots show outliers and some right skew, but the variances are more similar though still not great (3:1 ratio or so).

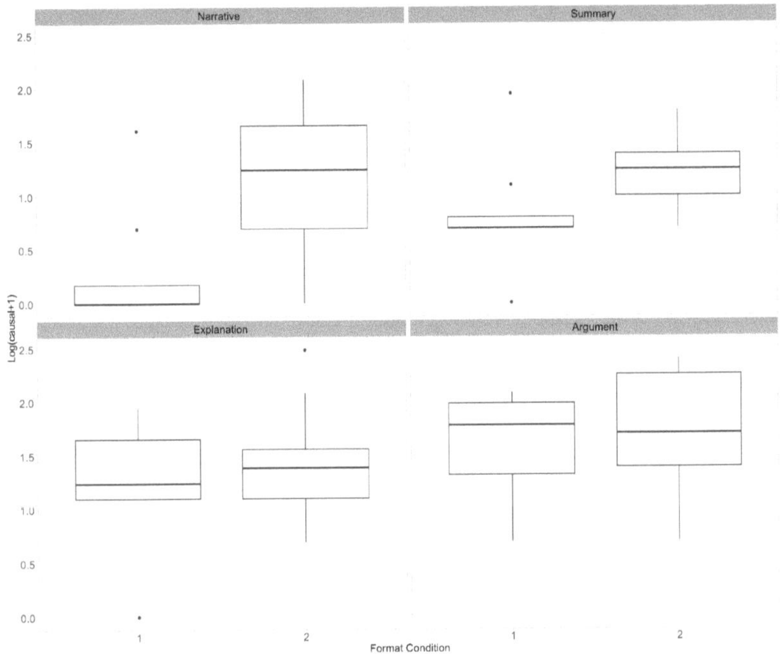

9.9 (a) The numerator is (63 – 67) – (37 – 54) = 13, or equivalently, (63 – 37) – (67 – 54) = 13.
(b) We need the variance of the linear combination in the numerator. From Appendix 5.1 (Equation 5.18), we know that the variance of a linear combination (L) is a weighted sum of the component variances and weighted covariances. The covariances are 0 in this case because the scores in each cell are independent. Assuming homogeneity of variances, we can use the MS_e = 150 for the variance estimates and n = 10, so the $var(L)$ = 4(150)/10, and the denominator of the t is the square root of that variance, or 7.75; (c) $t(36)$ = 13/7.75 = 1.68, and in R the area above 1.68 is 1 – pt(1.68,36) = 0.051 so the two-tailed p-value is 0.102. The corresponding F for the interaction is $F(1, 36) = 1.68^2 = 2.82$. We can use 1 – pf(2.82, 1, 36) = 0.102 to get the p-value.

9.11 (a) In R, *summary*(*aov*(data = dat, Y ~ A*E)) provides the ANOVA table:

	Df	Sum Sq	Mean Sq	F value	Pr(>F)
A	2	194.9	97.43	3.903	0.0341 *
E	1	112.1	112.13	4.491	0.0446 *
A:E	2	77.3	38.63	1.547	0.2333
Residuals	24	599.2	24.97		

The *EMS* are as follows:

Source	Df	EMS
A	2	$\sigma_e^2 + 10\theta_A^2$
E	1	$\sigma_e^2 + 15\theta_E^2$
AE	2	$\sigma_e^2 + 5\theta_{AE}^2$
S/AE	24	σ_e^2

(b) $\hat{\omega}_g^2(A) = \hat{\sigma}_A^2 / \left(\hat{\sigma}_A^2 + \hat{\sigma}_e^2\right)$ and $\hat{\sigma}_A^2 = df_A\ (MS_A - MS_{S/AE})/N = 2(97.43–24.97)/30 =$ 4.83 and $\hat{\sigma}_e^2 = 24.97$, so $\hat{\omega}_g^2(A) = \frac{4.83}{4.83 + 24.97} = 0.16.$

(c) If we ignore E, variance due to E and its interactions appear in MS_e. The ANOVA table (which can be confirmed using *summary*(*aov*(data = dat, Y ~ A)) in R) and the associated *EMS* are as follows:

Source	Df	SS	MS	F	P	EMS
A	2	194.9	97.43	3.34	0.051	$\sigma_e^2 + 10\theta_A^2$
S/A	27	788.6	29.21			σ_e^2

(d) $\hat{\sigma}_A^2 = 2(97.43 - 29.21)/30 = 4.55$ and $\hat{\omega}^2(A) = \frac{4.55}{4.55 + 29.21} = 0.13.$

(e) Neglecting a variable pools it and its interaction sums of squares with the error sum of squares. Because E contributed more than random error to $MS_{S/A}$, the p-value in the test of A was inflated. For the same reason, we underestimated A's contribution to the population variance, as evidenced by the reduction in our estimate of $\omega_g^2(A)$. All factors should be included in the initial analysis and only neglected if there is strong evidence that they contribute nothing beyond error variance in the population.

9.13 (a) Gender identity (G) is an intrinsic variable. The lecture (L) and movie (M) content are manipulated (extrinsic) variables. $\eta_g^2(L) = \frac{SS_L}{SS_L + SS_G + SS_{LG} + SS_{MG} + SS_{MLG} + SS_{S/MLG}} = \frac{SS_L}{SS_{Total} - SS_M - SS_{ML}}$. Because L, G, M and their interactions all have 1 df, their SS equal their MS. The $SS_{S/MLG} = 72\ MS_{S/MLG} = 72(.065) = 4.68$. Summing all the SS gives $SS_{total} = 6.291$ and thus $\eta_g^2(L) = \frac{.273}{6.291 - .445 - .341} = 0.050$. For $\eta_g^2(G) = \frac{SS_G}{SS_G + SS_{LG} + SS_{MG} + SS_{MLG} + SS_{S/MLG}} = .245/(.245+.076+.089+.142+4.68) =$ 0.047.

(b) $\hat{\omega}_g^2(M) = \frac{\hat{\sigma}_M^2}{\hat{\sigma}_M^2 + \hat{\sigma}_G^2 + \hat{\sigma}_{MG}^2 + \hat{\sigma}_{LG}^2 + \hat{\sigma}_{LMG}^2 + \hat{\sigma}_e^2}$ We need to estimate the population variances except for $MS_{S/MLG}$. For any effect (or interaction), the corresponding variance estimate is $(df_{effect} / N)*(MS_{effect} - MS_{S/MLG})$. We know that $N = 80$ and all the non-error $df = 1$. Let's multiply all of the variances by 100 – that doesn't change their ratio – and summarize the results in a table:

M	G	LG	MG	MLG	S/MLG
0.475	0.225	0.014	0.030	0.096	6.500

Then we substitute these variance estimates into the equation for omega-squared and find $\hat{\omega}_g^2(M) = 0.065$.

9.15 (a) $N = 96$ and there are $(abc) = 24$ cells in the design, which gives $n = 4$. $F_A = 2.84 = 56.8 / MS_{S/ABC}$; $MS_{S/ABC} = 20$.

(b) $\hat{\sigma}^2_{effect} = (df_{effect} / N)*(MS_{effect} - MS_e) = (3 / 96)(56.8 - 20) = 1.15$; $f^2 = 1.15 / 20 = 0.058$; (c) From G*Power 3.1 assuming $f = .25$ and $N = 24*8 = 192$, we find power = 0.63.

9.17 (a) From the given data we can compute the cell means and the grand mean = 5.

	B1	B2	*mean*
A1	10	10	10
A2	2	2	2
mean	3.6	7.33	5

$SS_{cells} = 2(10 - 5)^2 + 4(10 - 5)^2 + 8(2 - 5)^2 + 2(2 - 5)^2 = 240$.

(b) $SS_A = 6(10 - 5)^2 + 10(2 - 5)^2 = 240$; $SS_B = 10(3.6 - 5)^2 + 6(7.333 - 5)^2 = 52.27$. Then $SS_{AB} = 240 - 240 - 52.57 = -52.27$, an impossibility! This occurs because the A and B SS are correlated because of the unequal sample sizes over cells.

(c) $\hat{\alpha}_1 = (10 - 5) = 5$ $\hat{\alpha}_2 = (2 - 5) = -3$; $6(5) + 10(-3) = 0$, as expected.

(d)

Adjusting for alpha effects	B1	B2
A1	5	5
A2	5	5

Now, $SS_A = 6(5 - 5)^2 + 10(5 - 5)^2 = 0$; $SS_B = 10(5 - 5)^2 + 6(5 - 5)^2 = 0$. $SS_{cells} = 2(5 - 5)^2 + 4(5 - 5)^2 + 8(5 - 5)^2 + 2(5 - 5)^2 = 0$, so $SS_{AB} = 0$. In this toy example, what appears to be effects of B and AB are all because of A. Once A's effects are removed, the other effects disappear because they are perfectly correlated with A's effects. In real data, the SS aren't likely to go to zero if you adjust for, say, the effect of A, but the SS calculations as computed in part (a) will show misleading or nonsensical results. We will cover this again in Chapter 23.

9.19 (a) This is a nested design: Participants and groups are nested within a teaching method. We will cover nesting in detail in Chapter 15.

Source	df	MS	*Error term*	F	p
M	1	3627	*G/M*	8.64	.009
X	1	128	*G/MX*	4.00	.061
MX	1	123	*G/MS*	3.84	.066
G/M	18	420			
GX/M	18	32			
error	40	30			

(b) The reanalysis assumes that $\sigma_{GX/M} = 0$.

(c) The new error *MS* is [(18)(32) + (40)(30)] / (18 + 40) = 30.621

Source	df	MS	EMS
M	1	3627	$\sigma_e^2 + 4\sigma_{G/M}^2 + 40\theta_M^2$
X	1	128	$\sigma_e^2 + 40\theta_X^2$
MX	1	123	$\sigma_e^2 + 20\theta_{MX}^2$
G/M	18	420	$\sigma_e^2 + 4\sigma_{G/M}^2$
Pooled error	58	30.62	σ_{es}^2

The new error *MS* is [(18)(32) + (40)(30)] / (18 + 40) = 30.621.

(d) F_M (1,18) = 3627 / 420 = 8.64 is unchanged from part (a); F_X (1,58) = 128 / 30.62 = 4.18, p = .045 (now significant); F_{MX} (1,58) = 123 / 30.62 = 4.02, p = .050 (now significant).

(e) The *F* ratio for a test of *GX/M* is approximately 1, indicating that this potential source of variability does not contribute to the variance in the population.

Chapter 10

10.1 (a) H_0: $.5(\mu_{F1}+\mu_{F2})-\mu_c \geq 0$; H_1: $.5(\mu_{F1}+\mu_{F2})-\mu_c < 0$. To test H_0, calculate $t = \dfrac{\hat{\psi}}{\sqrt{MS_{S/A}\sum w_j^2/n}} = \dfrac{.5(14.6+14.9)-13.8}{\sqrt{4(.5^2+.5^2+1^2)/20}}$ = 1.734. The MS_e is based on five cells with 20 scores each, or 5(20 − 1) = 95 *df*. 1 − *pt*(1.734, 95) = 0.043 = *p*-value. Reject the null.

(b) H_0: $.5(\mu_{I1}+\mu_{I2})-\mu_c = 0$; H_1: $.5(\mu_{I1}+\mu_{I2})-\mu_c \neq 0$. $t(95)$ =(.5(11.8 + 11.7) − 13.8) / .548 = −3.74 which is clearly significant, $p < .0002$.

(c) H_0: $.5(\mu_{F1}+\mu_{F2})-.5(\mu_{I1}+\mu_{I2})=0$; H_1: $.5(\mu_{F1}+\mu_{F2})-.5(\mu_{I1}+\mu_{I2})\neq 0$.

$t(95)$ = ((14.6 + 14.9) / 2 − (11.8 + 11.7) / 2) / $\sqrt{\dfrac{4(.5^2+.5^2+(-.5)^2+(-.5)^2)}{20}}$ = 3 / .447 = 6.71. Reject the null.

10.3 (a) t = (9 − 2) / sqrt(4(2 / 5)) = 5.534, df = 5*(5 − 1) = 20, K = 4, *FWE* = 0.05 two-tailed. (i) Use *qt*(1 − (.05 / 4) / 2, df = 20) to find the critical value of 2.74 for the Dunn–Bonferroni method. (ii) For Dunnett's test, the critical value of d = 2.70 can be found in Table C.8. We can reject the null hypothesis in both cases.

(b) Use Tukey's *HSD* and find the critical value of *q*: *qtukey*(.95, nmeans = 5, df = 20) returns 4.23. Equation 10.18 says we reject the null whenever $t > q$ / sqrt(2), so here reject if 4.534 > 4.23 / *sqrt*(2) = 2.99; the difference is significant.

(c) The critical values, smallest to largest, are 2.70, 2.74, 2.99 for the Dunnett, Dunn–Bonferroni, and Tukey *HSD* tests. Those indicate the relative power of the tests; Dunnett has the highest power because the critical value is the lowest. Tukey's *HSD* test has the lowest power but allows us to test a larger family of comparisons.

(d) The CIs are estimate +/− critical value * *sqrt*(MS_e(2 / *n*)) = 7 +/− critical value * *sqrt*(4(2 / 5)); what varies is the critical value (2.70, 2.74, 2.99). The CIs are Dunnett

[3.58, 10.42], Dunn–Bonferroni [3.53, 10.47], Tukey *HSD* [3.22, 10.78]. Narrower intervals provide a more powerful test (all else being equal, a narrower test is less likely to include zero): Dunnett's test is most powerful and has the narrowest CI.

10.5 (a) $K = 5$, $df = 5(10 - 1) = 45$. Divide $FWE = .05$ over 5 two-tailed tests, leaving $EC = .01$, with .005 in each tail. So *qt*(.995, 45) = 2.70 is the critical value. t_{obs} = (10.4 – .25(8.6 + 9.5 + 9.2 + 8.0)) / *sqrt*(4*(.25 / 10 + 1 / 10)) = 2.23. $t_{obs} < t_{crit}$, so we cannot reject the null.

(b) S = +/– *sqrt*(4**F*(4, 45)) = +/– *sqrt*(4*2.58) = +/– 3.21. $t_{obs} < S$, so we cannot reject the null.

(c) The CIs are estimate +/– critical value * *sqrt*(MS_e(2 / n)) = 1.575 +/– critical value * *sqrt*(4(1.25 / 10)) = 1.575 +/– critical value * .707. The critical value depends on the type of error correction applied. For Dunn–Bonferroni in part (a), the CI is 1.575 +/– 2.70*.707 = [–0.33, 3.48]; for Scheffé in part (b), it's 1.575 +/– 3.21*.707 = [–0.70, 3.85]. The Dunn–Bonferroni is narrower because the family of tests is smaller. Both intervals include 0 so the null hypothesis cannot be rejected.

(d) Use Tukey's *HSD*, *qtukey*(.95, nmeans = 5, df = 45) = 4.02, so t_{crit} = +/– 4.02/ *sqrt*(2) = 2.84. t_{obs} = (9.5 – 8.6)/*sqrt*(4(2/10)) = 1.006. $t_{obs} < t_{crit}$ so we cannot reject the null hypothesis.

(e) t_{obs} = (.5(8.6 + 9.5) – (1 / 3)(9.2 + 8.0 + 10.4)) / *sqrt*(4 *((.5 + .3333) / 10)) = –0.26. We planned this single contrast, so we can use the standard t test with t_{crit} (45) = +/– 2.014. We cannot reject the null.

10.7 (a) The difference in means is 3.23; $s_{\hat{\psi}} = \sqrt{17.45\left(\frac{1}{19}+\frac{1}{33}\right)} = 1.203$. t_{obs} = 3.23/1.203 = 2.68. (i) *qtukey*(.95, nmeans = 4, df = 124) / *sqrt*(2) = 2.604 = t_{crit} for the Tukey–Kramer method. $t_{obs} > t_{crit}$ so we can reject H_0. (ii) There are "4 choose 2" = 6 different pairs of means to test, so $EC = FWE / 6$ = .05/6 = .008 two tailed, or .004 in each tail. The critical value of t is *qt*(.996, 124) = 2.70. Using the Dunn–Bonferroni correction, we fail to reject the null. (iii) The CI is estimate +/– t * $SE_{estimate}$. Here, 3.23 +/– critical value * 1.203. The critical value depends on the correction for multiple comparisons: 2.604 for T–K and 2.70 for D–B. The CIs are: T–K [0.10, 6.36]; D-B [–.02, 6.48]. T–K is slightly narrower.

(b) (i) Use Equations 6.16 and 6.17, or the *games_howell_test* function in {rstatix} for data in dat: *games_howell_test*(dat, *mean_d* ~ *schoolyr*, detailed = TRUE). We find t_{obs} = 2.28, adjusted df = 21.6 and adjusted-p = 0.133. The CI is [–0.702, 7.16], which includes 0 so we cannot reject H_0. When we took heterogeneity of variance into account, we lost the advantage of using a smaller *SE* that came from the large sample having the small variance. Because the n and variance both vary across *schoolyr* levels, the Games–Howell test is the most appropriate. (ii) [–0.702, 7.16].

10.9 (a) Using Brown–Forsythe test (Levene's test with the median), $F(2, 323) = 3.27$, p = .039, so the variances differ significantly and the assumption of homogeneity of variance was not appropriate. The smallest variance is for the largest group, so the MS_e is underestimated and the F is positively biased in our approach to 10.8.

(b) Use Welch's t applied to the contrast, weighted by n. $\psi = 46\bar{Y}_2 + 60\bar{Y}_3 - 106\bar{Y}_1$ = 162.42 and $SE_{\psi} = \sqrt{\sum s_j^2 \frac{w_j^2}{n_j}}$ = *sqrt*((106² (21.8) / 220) + (46² (39.8) / 46) + (60² (46.8) / 60) = 75.84, so t' = 162.42/75.84 = 2.14. Calculate df from Equation 10.10, finding 154.85. In R, 1– *pt*(2.14, df = 154.85) = .017. We can reject H_0.

10.11 (a)

	A1	A2	A3	mean
B1	20.3	23.8	30.4	24.8
B2	24.4	24.5	23.7	24.2
B3	28.1	25.3	22.8	25.4
mean	24.3	24.5	25.6	24.8

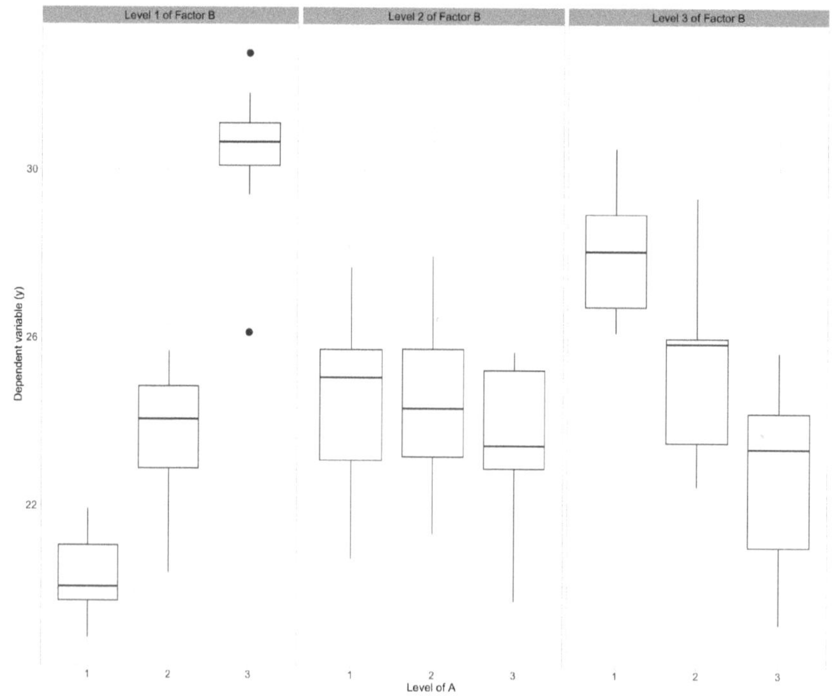

There is an apparent interaction of A and B: for $B1$, Y increases from $A1$ to $A3$ quite dramatically, whereas for $B3$, Y decreases from $A1$ to $A3$. Marginal means of A increase slightly from $A1$ to $A3$, perhaps indicating a main effect of A. Marginal means of B are similar for levels 1 and 2 and increase slightly for level 3, perhaps hinting at a main effect of B.

(b)

#	Df	Sum Sq	Mean Sq	F value	Pr(>F)
#a	2	30.6	15.29	4.469	0.0144 *
#b	2	20.3	10.15	2.968	0.0570.
#a:b	4	645.7	161.43	47.188	<2e-16 ***
#Residuals	81	277.1	3.42		

(c) Tukey *HSD* results show that $A3$ differs from $A1$ and from $A2$; this is the only significant difference.

Tukey multiple comparisons of means
95% family-wise confidence level
Fit: *aov*(formula = y ~ a * b, data = dat)
$a

	diff	lwr	upr	p adj
2-1	0.2739333	−0.86627153	1.414138	0.8345455
3-1	1.3505000	0.21029514	2.490705	0.0161003
3-2	1.0765667	−0.06363819	2.216772	0.0683020

10.13 (a) We see equal sample sizes in every cell, and equal variance.

```
# A tibble: 6 × 5
# Groups: dose [3]
#dose supp n avg.len var.len
#<ord> <fct> <int> <dbl> <dbl>
#1 0.5  OJ   10   13.2  19.9
#2 0.5  VC   10   7.98  7.54
#3 1    OJ   10   22.7  15.3
#4 1    VC   10   16.8  6.33
#5 2    OJ   10   26.1  7.05
#6 2    VC   10   26.1  23.0
#Levene's Test for Homogeneity of Variance (center = median)
#       Df F value Pr(>F)
#group 5 1.7086 0.1484 #no evidence of heterogeneity of variance
#       54
```

The boxplot shows a couple outliers. Tooth (cell) length increases with dose and is generally higher when delivered via OJ rather than ascorbic acid, except at the highest dose.

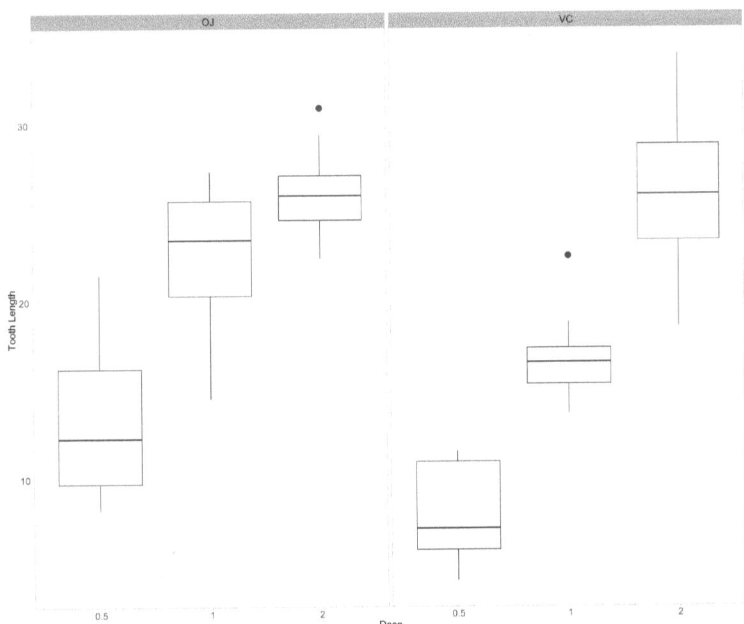

The ANOVA shows significant effects of *dose*, of delivery method (*supp*), and their interaction.

```
summary(aov(data = dat, len ~ dose*supp)) #equal n, equal var
#             Df   Sum Sq   Mean Sq   F value   Pr(>F)
# dose         2   2426.4   1213.2    92.000    < 2e-16 ***
# supp         1   205.4    205.4     15.572    0.000231 ***
# dose:supp    2   108.3    54.2      4.107     0.021860 *
# Residuals   54   712.1    13.2
# - -
# Signif. codes: 0 '***' 0.001 '**' 0.01 '*' 0.05 '.' 0.1 ' ' 1.
```

(b) $\binom{6}{2} = \frac{6!}{2!(6-2)!} = 15$ if we consider the six cell means alone. We could also compare delivery method, collapsing over dose, and there are $\binom{3}{2} = \frac{3!}{2!(3-2)!} = 3$ more comparisons of dose level ignore delivery method. That's a total of 19 comparisons. Using a Bonferroni correction, we divide α by the number of comparisons. But each factor is its own family, so we set $\alpha_{EC} = .05$ for delivery method, $\alpha_{EC} = .05 / 3 = .0167$ for dose, and $\alpha_{EC} = .05 / 15 = .0033$ for all other comparisons.

(c) Given equal n and equal variance, we use Tukey's *HSD* to make all pairwise comparisons while limiting *FWE*. We find all three levels of *dose* differ significantly, with adjusted p-values < .001. We also find OJ differs from VC, adjusted $p = .0002$. There are also quite a few differences among the individual cells that are significant.

```
Tukey multiple comparisons of means
95% family-wise confidence level

Fit: aov(formula = len ~ dose * supp, data = dat)

$dose
         diff     lwr         upr         p adj
1-0.5    9.130    6.362488    11.897512   0.0e+00
2-0.5    15.495   12.727488   18.262512   0.0e+00
2-1      6.365    3.597488    9.132512    2.7e-06

$supp
diff          lwr          upr          p adj
VC-OJ -3.7    -5.579828    -1.820172    0.0002312

$`dose:supp`
                  diff      lwr          upr           p adj
1:OJ-0.5:OJ       9.47      4.671876     14.2681238    0.0000046
2:OJ-0.5:OJ       12.83     8.031876     17.6281238    0.0000000
0.5:VC-0.5:OJ     -5.25     -10.048124   -0.4518762    0.0242521
1:VC-0.5:OJ       3.54      -1.258124    8.3381238     0.2640208
2:VC-0.5:OJ       12.91     8.111876     17.7081238    0.0000000
2:OJ-1:OJ         3.36      -1.438124    8.1581238     0.3187361
0.5:VC-1:OJ       -14.72    -19.518124   -9.9218762    0.0000000
1:VC-1:OJ         -5.93     -10.728124   -1.1318762    0.0073930
2:VC-1:OJ         3.44      -1.358124    8.2381238     0.2936430
0.5:VC-2:OJ       -18.08    -22.878124   -13.2818762   0.0000000
1:VC-2:OJ         -9.29     -14.088124   -4.4918762    0.0000069
2:VC-2:OJ         0.08      -4.718124    4.8781238     1.0000000
```

1:VC-0.5:VC	8.79	3.991876	13.5881238	0.0000210
2:VC-0.5:VC	18.16	13.361876	22.9581238	0.0000000
2:VC-1:VC	9.37	4.571876	14.1681238	0.0000058.

(d) (i) Calling doses .5, 1, and 2 Lo, Med, & High (L,M,H):

$$H_0: \left(\frac{\mu_L+\mu_M}{2}\right)_{OJ} - \left(\frac{\mu_L+\mu_M}{2}\right)_{VC} = \mu_{H,OJ} - \mu_{H,VC} \text{ or equivalently,}$$

H_0: $.5\mu_{L,OJ} + .5\mu_{M,OJ} - \mu_{H,OJ} - .5\mu_{L,VC} - .5\mu_{M,VC} + \mu_{H,VC} = 0$. H_1 is directional: >.

(ii) This is a planned contrast in an equal-variance situation so we use a standard t test. (You could also use Welch's t, for the reasons presented in Chapter 6.)

$\psi = .5(13.2 + 22.7 - 7.98 - 16.8) - (26.1 - 26.1) = 5.56$. SE $= \sqrt{MS_e \sum \frac{w_j^2}{n_j}} =$

$sqrt(13.2(6/10)) = 2.81$. $t(54) = 5.56 / 2.81 = 4.05$ and $t_{crit} = 1.67$, one-tailed, $\alpha = .05$. We conclude that the delivery method does matter more for the two lower doses than for the highest dose.

(e) No, testing a planned contrast does not require a significant main effect or interaction from an omnibus ANOVA.

10.15 (a) $SS_A = 10((24 - 18)^2 + (16 - 18)^2 + (14 - 18)^2) = 560$.

(b) (i) $SS_1 = 10(24 - 16)^2 / (1^2 + (-1)^2) = 320$; (ii) $SS_2 = 10{*}(.5(24) + .5(16) - 14)^2 / (.5^2 + .5^2 + (-1)^2) = 240$; (iii) $SS_3 = 10(24 - 14)^2 / (1^2 + (-1)^2) = 500$. Because ψ_1 and ψ_2 are orthogonal – $(1)(.5) + (-1)(.5) = 0$ – the total of their SS equals SS_A: $320 + 240 = 560$.

(c) (i) $SS_1 = 10(20 - 20)^2 / (1^2 + (-1)^2) = 0$; (ii) $SS_2 = 10{*}(.5(20) + .5(20) - 14)^2 / (.5^2 + .5^2 + (-1)^2) = 240$; (iii) $SS_3 = 10(20 - 14)^2 / (1^2 + (-1)^2) = 180$ and SS_A is now 240. The first two contrasts are still orthogonal; zeroing one of them out has no effect on the SS of the other (although they still must sum to SS_A). On the other hand, the third contrast is not orthogonal to either of the first two. That means they share variance, so changing one of the contrasts necessarily influences the other and so the SS_3 differs from part (b). Think of the SS as pieces of a pie. Two independent slices don't affect one another; removing one doesn't change the other (though it changes their total). For two overlapping – non-independent/non-orthogonal – pieces, though, you can't remove one without affecting the other.

10.17 (a) $\widehat{\psi_s} = \hat{\psi} / s_{pooled} = \sum w_j \bar{Y}_{.j} / \sqrt{MS_{S/A}} = 6/30 = 0.2$.

(b) $\hat{\psi} = (8 / 18)(24) + (10 / 18)(16) - 14 = 5.556$; $\widehat{\psi_s} = \frac{5.556}{30} = .185$. Note that this standardized contrast estimate is calculated on the original measurement scale; do not multiply by 18 to create integer weights.

Chapter 11

11.1 The boxplots show reasonable symmetry but several outliers. Shapiro–Wilk tests are marginal, indicating that the deviations from normality are not huge. We could run a rank-based ANOVA or we could trim the data. There are only 20 scores in each group, so trimming 20% from each tail is quite costly. Numerical summaries and visual review indicate that the variances are not too different across conditions; we don't see a large effect of group membership on Y. We proceed with a standard ANOVA, finding $F(2,57) = 1.526$, $p = .226$, for the effect of *Group*.

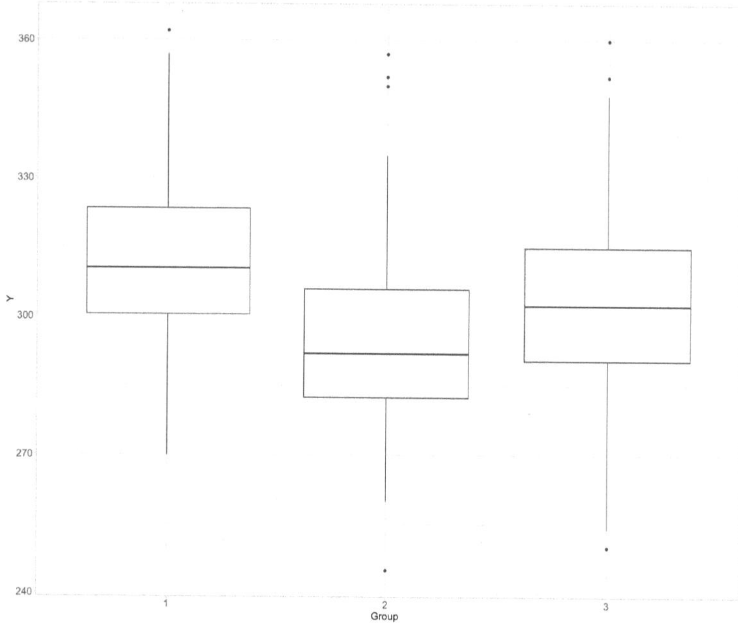

11.3 In this data set, we have equal *n* (20) in each group, widely different variances (3:1 ratio), a couple of outliers and right skew in group 3. The Shapiro–Wilk test indicates nonnormality, and the variance increases with the mean. Looks like a transformation could help, so we use a spread level plot to find the recommended power is –0.4, which is similar to 1 / *sqrt*(*Y*). We try that, as well as log(*Y*) because that often helps with right skewed data. Boxplots indicate that log(*Y*) works reasonably well, so we run the ANOVA and find a significant effect of *Group*: $F(2, 57) = 3.51$, $p = .037$.

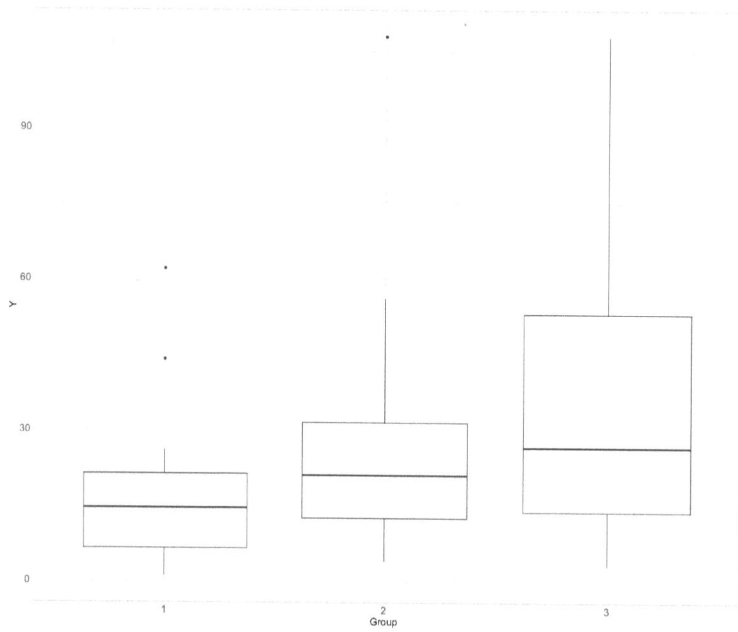

11.5 (a) The variances differ markedly so we need Welch's t. Using Equations 10.8–10.10, we compute $t' = 4.1 / 1.598 = 2.565$ and $df' = 13.49$. In R we use the *qt* function to find the critical t for a two-tailed test when there are four planned contrasts and $FWE = .05$: $qt((1 - .0125 / 2), 13.49) = 2.88$. Because $2.565 < 2.88$, we cannot reject the null hypothesis.

(b) Again we use Welch's t with the Dunn–Bonferroni adjusted to p. $df' = 6.19$ and $t' = 2.718$. There are five comparisons to the control, so each gets $EC = .01$, $qt((1 - .01 / 2), 6.19) = t_{crit}$ of 3.66. We cannot reject the null, $t' = 2.718 < 3.66 = t_{crit}$.

11.7 (a) Let Y = *younger*, D = *older*, T = *Text*, and W = *Web*. H_0: $(\mu_{YT} - \mu_{YW}) - (\mu_{DT} - \mu_{DW}) = 0$. Weights: 1, −1, −1, 1. N = 20/cell. $SS_{FA} = 20((72.50 - 83.25) - (72.25 - 89.0))^2 / 4 = 180$, the same as reported in Table 9.12.

(b) Let S = *Summary* and A = *Argument*. Then H_0: $[(\mu_{YTS} - \mu_{YWS}) - (\mu_{DTS} - \mu_{DWS})] - [(\mu_{YTA} - \mu_{YWA}) - (\mu_{DTA} - \mu_{DWA})] = 0$. The $SS_{FAI} = 10^* ([(71.25 - 76.50) - (70.5 - 88.25)] - [(73.75 - 90) - (74 - 89.75)])^2 / 8 = 211.25$.

11.9 (a) The difference in means is 4.8 and the cell variances are similar (far less than the 4:1 rule), so we will use Tukey's *HSD*. $s_{\bar{Y}} = \sqrt{MS_{S/AB}/nb}$ = *sqrt*(110/20) = 2.345.

In R, we find the critical q = *qtukey*(.95, nmeans = 5, df = 90) = 3.94. The CI is 4.8 +/− (3.94)(2.345) = [−4.439, 14.039] and we cannot reject the null hypothesis of no difference.

(b) H_0:$(\mu_{LL} + \mu_L)/2 - (\mu_{CC} + \mu_C)/2 = 0$. Inserting values into Equation 10.5, we get $t(90) = 8.85/2.345 = 3.774$;

(c) (i) The criterion is $t(90) = 1.99$; $3.774 > 1.99$ so we reject the null. (ii) Now $EC = FWE / 3 = .0167$ and the critical $t = 2.46$. (iii) The Scheffé procedure is needed. S = sqrt(4 $F_{.05,4,90}$) = *sqrt*(4*2.47) = 3.145; reject the null if $|t_{obs}| > S$.

(d) (i) Let R = *retirees*; S = *students*. H_0: $[(\mu_{LLR} + \mu_{LR})/2 - \mu_{MR}] - [(\mu_{LLS} + \mu_{LS})/2 - \mu_{MS}] = 0$; (ii) psi = (((16.1 + 14.7) / 2 − 13.5) − ((18.5 + 16.9) / 2 − 8.5)) = −7.3 and s_{psi} = *sqrt*(110*((4)(.25 / 10) + (2)(1 / 10))) = 5.74. $t_{crit}(90) = 1.99$ so the 95% CI is −7.3 +/− (1.99)(5.74) = [−18.72, 4.12]. Zero is in the CI so the null hypothesis cannot be rejected.

Chapter 12

12.1 (a) *summary*(*aov*(data = dat, $Y \sim A$)) #completely randomized design

#	Df	Sum Sq	Mean Sq	F value	Pr(>F)
#A	2	861	430.6	2.431	0.097 .
#Residuals	57	10097	177.1		

If blocking contributes variance (i.e., if either $MS_{AB} \neq 0$ or $MS_B \neq 0$), then that variance is reflected in a larger $MS_{S/A}$ when the data are treated as if they came from a completely randomized design. This negatively biases the F test for A.

(b) *summary*(*aov*(data = dat, $Y \sim A^*B$)) #treatments × blocks design

	Df	Sum Sq	Mean Sq	F value	Pr(>F)
A	2	861	430.6	3.411	0.04119 *
B	3	3269	1089.7	8.633	0.00011 ***
A:B	6	768	128.1	1.014	0.42724
Residuals	48	6059	126.2		

We see a significant effect of blocking, and the marginally significant effect of A that we saw in (a) is now significant because the variance due to blocking has been removed from the error variance.

12.3 (a) $f = sqrt((2 / 60)\ (430.6 - 126.2)) / sqrt(126.2) = 0.284$.

(b) $f = sqrt((2 / 60)\ (430.6 - 126.2)) / sqrt(175.37) = 0.241$.

(c) Using G*Power, power of the (one-way) randomized design is 0.35; power of the (main effects and interactions) treatments × blocks design is 0.468.

(d) To obtain power = .468 with f = .241 and three groups in a one-way randomized design, we need N = 84 or 28/group.

(e) *A priori* calculation with f = .284, numerator df = 2, 3 groups, we get N = 123. Taking N * 1.381 = 169.9 or 170 participants needed in the completely randomized design for equivalent power. We get a similar sample size (N = 171) if we enter f = .241 in a one-way design seeking power = .8. The relative efficiency of the two designs also estimates the ratio of sample sizes needed to achieve the same power.

12.5 (a) In R, $cor(X, Y)$ shows that r = .0845. $(1 - r^2)$ = .993 and .993*10958.933 = 10882.22, which is within rounding error of the SS total in the ANCOVA. Also, the ratio of SS_{total} in the two analyses is .993 = 10879.757 / 10957.933.

(b) $r_{S/A}^2 = (-749 + 2010.75 + 5.8)^2 / ((1736 + 3591.75 + 1874.2)(2624.55 + 2418.95 + 5053.2)) = .022$; $(1-.022)*(10096.7) = 9874.6$, which is within rounding error of the $SS_{S/A}$ in the ANCOVA.

12.7 (a) $\sigma_A^2 = \left(\frac{2}{12}\right)(32.25 - 3.92) = 4.72$; $\sigma_S^2 = \left(\frac{3}{12}\right)(18.33 - 3.92) = 3.60$; $\sigma_W^2 = 3.60 + 3.92 = 7.52$; $f = \sqrt{4.72 / 7.52} = 0.79$.

(b) power = .965.

(c) power= .53, which is lower than the repeated-measures design because subject variability appears in the error variance.

(d) The repeated-measures design would require N = 19, assuming the correlation from (b) and with three measurements each. The between-subjects completely randomized design would require N = 207.

(e) As the correlation decreases, less subject-variance is removed from the error term in a repeated-measures design, meaning a larger sample is required for the same power. With a correlation of 0, N = 69 to have power = .9 in the RM design. That means that 3 × 69 = 207 independent observations are required for that power level, the same as in the completely randomized design in part (d).

12.9 (a) $estMSE_{RM} = ((4)(7.13) + 32.96) / 5 = 12.30$.

(b) $Adjusted_{RM} = (13 / 15)(19 / 17)(12.30 / 7.13) = 1.67$. The Latin Square design is more efficient in this case, and will require fewer participants to achieve the same power.

Chapter 13

13.1 **(a, b)**

```
summary(aov(data = dat, Y ~ A + Error(S/A)))
##Error: S
```

#	Df	Sum Sq	Mean Sq	F value	Pr(>F)
#Residuals	3	70.67	23.56		

#Error: S:A

#	Df	Sum Sq	Mean Sq	F value	Pr(>F)
#A	2	6.500	3.250	2.854	0.135
#Residuals	6	6.833	1.139		

EMS

S $\sigma_e^2 + 3\sigma_S^2$

A $\sigma_e^2 + 4\theta_A^2$

S:A σ_e^2.

(c) $\hat{\sigma}_A^2 = \frac{a-1}{a}\hat{\theta}_A^2 = \frac{a-1}{a}\frac{(MS_A - MS_{SA})}{n} = \frac{2}{3}\frac{2.111}{4} = 0.352;$

$\hat{\sigma}_S^2 = \frac{(MS_S - MS_{SA})}{a} = 7.472;\ \hat{\sigma}_e^2 = 1.139;$

Then $\hat{\omega}_p^2(A) = \frac{\hat{\sigma}_A^2}{\hat{\sigma}_A^2 + \hat{\sigma}_S^2 + \hat{\sigma}_e^2} = \frac{0.352}{0.352 + 7.472 + 1.139} = 0.039$

(d) No. The participants show different interaction effects.

	A_1		A_2		A_3		
	Y_{i1}	$\eta\alpha_{i1}$	Y_{i2}	$\eta\alpha_{i2}$	Y_{i3}	$\eta\alpha_{i3}$	μ_i
S_1	12	−1.45	14	1.05	15	0.30	13.7
S_2	9	0.25	8	−0.25	10	0	9
S_3	10	−0.05	9	−0.55	12	0.70	10.3
S_4	8	1.25	6	−0.25	7	−1.00	7
μ_j	9.75		9.25		11.0		10

13.3 (a) *summary*(*aov*(data = LongForm, value ~ FactorA + Error(id/FactorA)))
#Error: id

#	Df	Sum Sq	Mean Sq	F value	Pr(>F)
#Residuals	5	105.6	21.12		

#
#Error: id:FactorA

#	Df	Sum Sq	Mean Sq	F value	Pr(>F)
#FactorA	2	216.55	108.28	21.93	0.000221 ***
# Residuals	10	49.38	4.94		

$MS_{SA} - MS_e = 4.94$
Mean $A1$ – Mean $A2$ = −1.33; $t(10) = -1.33/sqrt(4.94*(1/6 + 1/6)) = -1.036$. With $df = 10$, $p = 0.16$.

(b) var(A1–A2) = 1.10, $t(5) = -1.33/\text{sqrt}(1.10/6) = -3.106$. With $df = 5$, $p = .013$.

(c) The variances in the three levels of *A* are 2.94, 2.61, and 25.4. The *SE* based on the ANOVA MS_e is inflated by the variance of group 3, a condition that is not involved in the difference of interest. The test from part (b) is preferred for that reason.

13.5 (a) *summary*(*aov*(data=dat,Y ~ A+Error(S/A)))

```
#Error: S
#            Df    Sum Sq    Mean Sq    F value    Pr(>F)
#Residuals   7     59.09     8.441
#
#Error: S:A
#            Df    Sum Sq    Mean Sq    F value    Pr(>F)
#A           3     3.763     1.2542     6.35       0.00311 **
# Residuals  21    4.148     0.1975
```

1–*pf*(6.35, *df*1 = 3*(1/3), *df*2 = 21*(1/3)) = p = .04 assuming the lower-bound epsilon adjustment. In both cases, we can reject the null hypothesis that the μ_j are all equal.

(b) Using the Dunn–Bonferroni method (Tukey<s *HSD* method assumes independently distributed means), if the *FWE* = .05, α = (.05/6) = .008 (two-tailed). We want the value of t (7) *df* that cuts off .004 in each tail: *qt*(.004, *df* = 7, lower.tail = FALSE) = 3.67 The means are 3.275 and 4.175 and the standard deviation of the difference is .804. Therefore, the .95 simultaneous confidence interval for this difference is (4.175 – 3.275) + (3.67)(.804/ *sqrt*(8)) = [–0.14, 1.94].

13.7 (a) H_0: $\psi = 0$; H_1 $\psi \neq 0$. From a wide format data frame, *dat*2 <– *dat*2 %>% *group_by*(S) %>% *mutate*(psi = A1–*sum*(A2,A3,A4)/3) Then run a t test on psi: *t.test*(dat2$psi) to find $t(4) = 3.107$, $p = .036$. We cannot reject the null hypothesis.

(b) The *t.test* results include the CI: [.305, 5.43].

13.9 (a) The values of n are as follows:

		Correlation		
		.4	.6	.8
	.6	51	35	19
Epsilon	.8	42	29	15
	1	36	25	13

(b) As epsilon increases, a larger fraction of the degrees of freedom is operative so that a smaller n is required to maintain power = .9. With respect to the correlation, recall that the variance of difference scores decreases as the correlation increases. The error mean square is a function of the average variance of difference scores (see Exercise 13.2) and therefore will decrease as r increases. As the error mean square decreases, a smaller n is required to maintain a constant level of power.

13.11 *cochran_qtest*(dat,Smoke ~Period|S)

```
# A tibble: 1 × 6
#   .y.    n     statistic  df     p       method
#   *  <chr> <int> <dbl> <dbl> <dbl> <chr>
#1 Smoke  20    6.53       2      0.0381  Cochran's Q test
```

Q(2) = 6.53, p = .0381.

Chapter 14

14.1 (a) Fixed effects: P; Random effects: W, O

Source	EMS
W	$\sigma_e^2 + 4\sigma_{WO}^2 + 20\sigma_W^2$
P	$\sigma_e^2 + 10\sigma_{PO}^2 + 5\sigma_{WP}^2 + \sigma_{WPO}^2 + 50\theta_P^2$
O	$\sigma_e^2 + 4\sigma_{WO}^2 + 40\sigma_O^2$
WP	$\sigma_e^2 + \sigma_{WPO}^2 + 5\sigma_{WP}^2$
WO	$\sigma_e^2 + 4\sigma_{WO}^2$
PO	$\sigma_e^2 + \sigma_{WPO}^2 + 10\sigma_{PO}^2$
WPO	$\sigma_e^2 + \sigma_{WPO}^2$

(b) $$F'_1 = \frac{MS_P}{(MS_{PO} + MS_{WP} - MS_{WPO})} = \frac{2610}{640 + 330 - 320} =$$

$$4.015, \textit{the } \text{denominator } \textit{df are df}'_{error} = \frac{(MS_{PO} + MS_{WP} - MS_{WPO})^2}{\frac{MS^2_{PO}}{(p-1)(o-1)} + \frac{MS^2_{WP}}{(w-1)(p-1)} + \frac{(MS^2_{WPO})}{(w-1)(p-1)(o-1)}} = \frac{650^2}{\frac{640^2}{12} + \frac{330^2}{27} + \frac{330^2}{108}} = 10.8$$

so $p = 0.37$. We reject the null hypothesis that the programs have no effect.

(c) Testing WP against WPO, $F_{27,108} = MS_{SP} / MS_{WPO} = 1.03$, $p = .44$. Assuming that $\sigma^2_{WP} = 0$, P can now be tested against MS_{PO}. The result is $F_{3,12} = 4.078$, $p < .05$. Again, we reject the null.

14.3 (a)
```
summary(aov_car(data=dat,Y~ A*B + Error(S/A*B),anova_table=list(correction="none")))
#Univariate Type III Repeated-Measures ANOVA Assuming Sphericity
#
#              Sum Sq num Df Error SS den Df F value    Pr(>F)
#(Intercept) 811.20       1  36.467      4 88.9799 0.0007042 ***
# A           112.13       1  14.867      4 30.1704 0.0053535 **
# B             9.80       2   2.533      8 15.4737 0.0017801 **
# A:B           4.07       2  18.933      8  0.8592 0.4591948
```

Note that the error term for A is SA, for B it's SB, and for AB it's SAB. The R output shows the SS_{error} for each term; divide by the appropriate df (denDf) to find the MS for each error term.

(b)
```
> summary(aov_car(data = dat, meanY~A + Error(S/A)))

Univariate Type III Repeated-Measures ANOVA Assuming Sphericity

             Sum Sq num Df Error SS den Df F value    Pr(>F)
(Intercept) 270.400      1  12.1556      4   88.98 0.0007042 ***
A            37.378      1   4.9556      4   30.17 0.0053535 **
```

Notice that the *SS* for *A* and for *SA* (error) are one-third their values from part (a) because here each "score" is an average of three scores in the original data.

(c)

Source	EMS
S	$\sigma_e^2 + a\sigma_{SB}^2 + ab\sigma_S^2$
A	$\sigma_e^2 + \sigma_{SAB}^2 + b\sigma_{SA}^2 + n\sigma_{AB}^2 + bn\theta_A^2$
B	$\sigma_e^2 + a\sigma_{SB}^2 + na\sigma_B^2$
SA	$\sigma_e^2 + \sigma_{SAB}^2 + b\sigma_{SA}^2$
SB	$\sigma_e^2 + a\sigma_{SB}^2$
AB	$\sigma_e^2 + \sigma_{SAB}^2 + n\sigma_{AB}^2$
SAB	$\sigma_e^2 + \sigma_{SAB}^2$

(d) When *B* is random, we need a quasi-*F* test for *A*. $F_1' = \dfrac{MS_A}{MS_{SA} + MS_{AB} - MS_{SAB}}$ =33.136. From Equation 14.8, $df_{error} = 1.84$, so $p = .035$.

(e) Averaging over levels of *B* and analyzing the data as if there were only *a* scores per participant ignores the *AB* and *SAB* variability that contribute to the *A* mean square if *B* has random effects. Thus, that procedure often leads to an inflated Type 1 error rate.

14.5 We need to estimate population variances under both assumptions:

$\hat{\sigma}^2$	A, B fixed	B random
S	$(MS_S - MS_{SAB})/ab = 16.000$	$(MS_S - MS_{SAB})/ab = 16.000$
A	$(a-1)(MS_A - MS_{SA})/abn = .533$	$(a-1)[MS_A - (MS_{SA} + MS_{AB} - MS_{SAB})]/abn = .08$
B	$(b-1)(MS_B - MS_{SB})/abn = 1.04$	$(MS_B - MS_{SB})/an = 5.20$
SA	$(MS_{SA} - MS_{SAB})/b = 2.80$	$(MS_{SA} - MS_{SAB})/b = 2.80$
SB	$(MS_{SB} - MS_{SAB})/a = 7.33$	$(MS_{SB} - MS_{SAB})/a = 7.33$
AB	$(a-1)(b-1)(MS_{AB} - MS_{SAB})/abn = 1.813$	$MS_{AB} - MS_{SAB})/n = 27.20$
SAB	$MS_{SAB} = 10.00$	$MS_{SAB} = 10.00$

(a) $\hat{\omega}_g^2(A) = \hat{\sigma}_A^2 / \left(\hat{\sigma}_A^2 + \hat{\sigma}_S^2 + \hat{\sigma}_{SA}^2 + \hat{\sigma}_{SB}^2 + \hat{\sigma}_e^2\right) = .015$.

(b) Same equation but the variance estimates differ. $\hat{\omega}_g^2(A) = .002$.

(c) Now *B* and *AB* contribute to the denominator, $\hat{\omega}_g^2(A) = .001$.

14.7 (a)

```
> summary(aov_ez(data=dat,"S","Y",between=c("A"),within=c("B")))
Contrasts set to contr.sum for the following variables: A

Univariate Type III Repeated-Measures ANOVA Assuming Sphericity

              Sum Sq num Df Error SS den Df  F value    Pr(>F)
(Intercept) 8800.2      1   66.222      4 531.5570 2.097e-05 ***
A            490.9      1   66.222      4  29.6510  0.005524 **
B            447.4      2   16.444      8 108.8378 1.579e-06 ***
A:B           14.8      2   16.444      8   3.5946  0.076952 .
```

(b)
```
> summary(aov_ez(data=dat,"S","Y",between=c("A"),within=c("B")))
Contrasts set to contr.sum for the following variables: A

Univariate Type III Repeated-Measures ANOVA Assuming Sphericity

             Sum Sq num Df Error SS den Df  F value    Pr(>F)
(Intercept) 8800.2      1   66.222      4 531.5570 2.097e-05 ***
A            490.9      1   66.222      4  29.6510  0.005524 **
B            447.4      2   16.444      8 108.8378 1.579e-06 ***
A:B           14.8      2   16.444      8   3.5946  0.076952 .
```

$SS(A)$ and $MS(A)$ are one-third the size here as in part (a) because we have only one "score" per cell rather than three. $F(A)$ is identical.

(c) Filter the cases using $A == 1$ and, separately, using $A == 2$. Then run an repeated-measures ANOVA with factor B on each data set; the result will have $SS(SB/A)$ as the error term for B. $SS_{SB/A1} = 3.111$; $SS_{SB/A2} = 13.333$, sum them to get 16.444, the $SS_{SB/A}$ from part (a). Average the MS in each case (3.111/4; 13/333/4) and find the mean. It equals 2.056, $MS_{SB/A}$ from part (a).

14.9 (a)

Source	df	EMS
A	1	$\sigma_e^2 + \sigma_{SB/A}^2 + 3\sigma_{S/A}^2 + 3\sigma_{AB}^2 + 9\sigma_A^2$
S/A	4	$\sigma_e^2 + \sigma_{SB/A}^2 + 3\sigma_{S/A}^2$
B	2	$\sigma_e^2 + \sigma_{SB/A}^2 + 6\theta_B^2$
AB	2	$\sigma_e^2 + \sigma_{SB/A}^2 + 3\sigma_{AB}^2$
SB/A	8	$\sigma_e^2 + \sigma_{SB/A}^2$

(b) To test A, use a quasi-F. $F' = MS_A \,/\, (MS_{S/A} + MS_{AB} - MS_{SB/A}) = 480.899/(16556+7389-2.056) = 22.426$. Error $df = 4.97$, $p = .005$.

14.11 (a)

Source	df	EMS
Between-participants	71	
X (school type)	1	$\sigma_e^2 + 2\sigma_{S/XA}^2 + 72\theta_X^2$
A (age)	2	$\sigma_e^2 + 2\sigma_{S/XA}^2 + 48\theta_A^2$
XA	2	$\sigma_e^2 + 2\sigma_{S/XA}^2 + 24\theta_{XA}^2$
S/XA	66	$\sigma_e^2 + 2\sigma_{S/XA}^2$
Within-participants	72	
T (task)	1	$\sigma_e^2 + \sigma_{ST/XA}^2 + 72\theta_T^2$
XT	1	$\sigma_e^2 + \sigma_{ST/XA}^2 + 36\theta_{XT}^2$
AT	2	$\sigma_e^2 + \sigma_{ST/XA}^2 + 24\theta_{AT}^2$
XAT	2	$\sigma_e^2 + \sigma_{ST/XA}^2 + 12\theta_{XAT}^2$
ST/XA	66	$\sigma_e^2 + \sigma_{ST/XA}^2$

S/XA is the error terms for the between-participants effects; *ST/XA* is the error term for the within-participants effects.

(b) The most conservative approach is to base the error term only on those scores involved in these tests of simple effects. For (i) calculate the variances of the T_1 scores within each *AX* combination and average them. $df = ax(n - 1) = (3)(2)(11) = 66$. For (ii) average the T_1 variances for each of the three home-schooled groups, $df = a(n - 1) = (3)(11) = 33$. For (iii) calculate the *ST* mean square within each of the three home-schooled groups and average them. $df = a(n - 1)(t - 1) = 3(11)(1) = 33$.

14.13 (a)

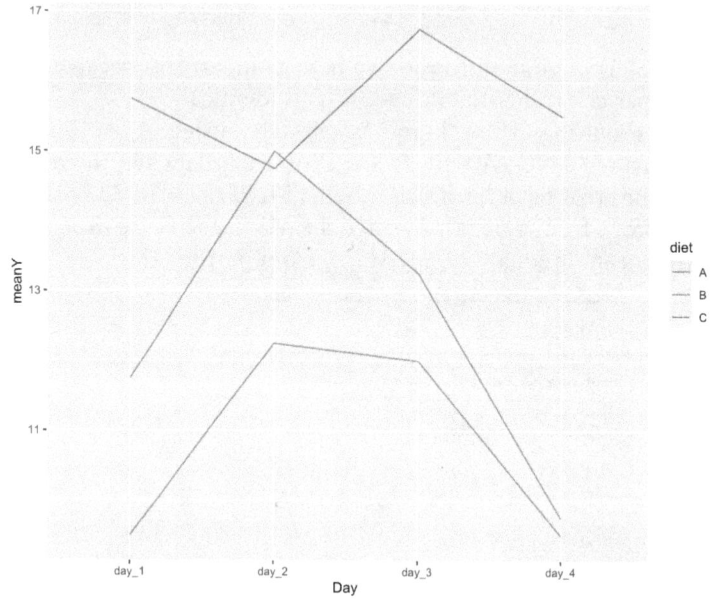

(b) As expected based on the graph, there is a main effect of *Diet*: $F(2,9) = 13.637$, $p = .002$. We'd expect that, at least, *Diet A* and *Diet C* differ. Other differences may also be significant. Looking at the within-participant effects of *Day*, the GG epsilon = .661 so we use the adjusted *p*-values and *df*. There is a main effect of *Day* ($F(1.984,17.856) = 12.923$, $p < .001$) and the noted interaction of *Day* and *Diet* ($F(3.968,17.856) = 5.103$, $p = .006$).

14.15 (a) $F_a = 80.083/8.750 = 9.152$; with 1 and 6 *df*, $p = .023$. $F_{ab} - 108/11.611 = 9.301$; with 1 and 6 *df*, *p*=. 023.

(b) We now form quasi-*F* ratios to test *A* and *AB*. To test the *A* source, $F' = MS_a / (MS_{s/a} + MS_{ac} - MS_{sc/a}) = 80.083/(8.750 + 21.271 - 2.271) = 2.885$. The denominator *df* are $df_2 = \dfrac{(MS_{S/A} + MS_{AC} + MS_{SC/A})^2}{\left(\dfrac{MS_{S/A}{}^2}{df_{S/A}}\right) + \left(\dfrac{MS_{AC}{}^2}{df_{AC}}\right) + \left(\dfrac{MS_{SC/A}{}^2}{df_{SC/A}}\right)} = 3.22$; $p = .18$. When we consider the variance due to the random factor *C*, the *p*-value is increased. If we wish to generalize to the sampled population of items, the result is no longer significant.

14.17 (a) If we consider this a two-way design with a difference score as a dependent measure, then the *F* for DV**Format* in problem 14.16(b) equals the *F* for *Format* alone in a univariate test. Both are 11.75. Then we use Equation 8.19 for Cohen's $f = sqrt((1 / 64)*10.747) = 0.41$, a large effect under Cohen's guidelines. **(b)** Subtracting *IVT–SVT* in each format, we get –3.75 for *text* and 8.281 for *web*. The interaction contrast is their difference: psi = 8.281 – (–3.75) = 12.031. To obtain a CI, we have psi +/– s_{psi} t_{crit} = 12.031 +/– s_{psi} 2.00. s_{psi} = *sqrt*((2/4*n*)MS_{error}), where 4*n* is the number of scores on which each mean is based, and the MS_{error} comes from the between-participants analysis of the differences scores: s_{psi} = *sqrt*((2 / 32)*197.154) = 3.510. The 95% CI is 12.031 +/– (2.00)(3.510) = [5.03, 19.03], which doesn't include 0.

(c) Select the data for one format at a time and run an ANOVA on the difference scores. (i) In the *text* condition f = *sqrt*((3 / 32)(.110) = .102, a small effect. (ii) $F < 1$ in the *web* condition, so our estimate of Cohen's f is 0. In either condition, instructions have little effect on the difference between the two dependent measures, relative to the error variance.

(d) Using G*Power, we enter f = .1, numerator *df* = 3, *groups* = 4, α = .05 and power = .8. The result is N = 1095, or 274/*instruction* condition.

Chapter 15

15.1 Select the cases where A = 1 (or 2 or 3) and then run a between-participants ANOVA with B as a fixed factor.

Source	SS	df	MS
B/A_1	1,808,925.00	4	452,231.00
B/A_2	623,512.15	4	155,878.04
B/A_3	695,324.35	4	173,831.09

Summing the *SS*, we get 3,127,761.5, which matches the $SS_{B/A}$ shown in Table 15.3. Averaging the *MS*, we get 260,646.71, which matches the $MS_{B/A}$ shown in Table 15.3.

15.3

Source	df	EMS
S	19	$\sigma_e^2 + 50\sigma_S^2$
M	4	$\sigma_e^2 + 20\sigma_{Items/M}^2 + 10\sigma_{SM}^2 + 200\theta_M^2$
Items/M	45	$\sigma_e^2 + 20\sigma_{Items/M}^2$
SM	76	$\sigma_e^2 + 10\sigma_{SM}^2$
SxItems/M	855	σ_e^2

The main effect of *M* is tested with a quasi-$F = MS_M / (MS_{Items/M} + MS_{SM} - MS_{SxItems/M})$. Numerator df = 4. Denominator $df = (MS_{Items/M} + MS_{SM} - MS_{SxItems/M})^2 / (MS^2_{Items/M}/45 + MS^2_{SM}/76 - MS^2_{SxItems/M}/855)$.

15.5

```
> aov_out<-summary(aov(data=dat,Y~id*R*V+Items %in% (R*V)))
> aov_out
             Df Sum Sq Mean Sq F value   Pr(>F)
id            9  14.91    1.66   1.661  0.10371
R             1   0.41    0.41   0.406  0.52495
V             1  34.44   34.44  34.541 2.78e-08 ***
id:R          9   9.34    1.04   1.041  0.41033
id:V          9  10.70    1.19   1.193  0.30397
R:V           1   8.41    8.41   8.428  0.00428 **
id:R:V        9  11.35    1.26   1.264  0.26143
R:V:Items    16  40.00    2.50   2.507  0.00206 **
Residuals   144 143.60    1.00
```

Participants (= S = *id*) and *Items* are random-effects variables. Note that the error terms for R, V, and the RV interaction are incorrect in the R output. We need quasi-Fs. For R, the error term is $MS_{RS} + MS_{Items/RV} - MS_{SItems/RV} = 2.541$ where $S = id$; for V it's $MS_{VS} + MS_{Items/RV} - MS_{SItems/RV} = 2.692$; for RV it is $MS_{RVS} + MS_{Items/RV} - MS_{SItems/RV} = 2.763$. The error df follow from Equation 14.7: for R, 12.482; for V, 13.086; for RV, 13.301. So $F_R = 0.159$; $F_V = 12.794$ (p= .003); $F_{RV} = 3.042$ (p= .104).

15.7 (a)

Source	df	EMS
C	2	$\sigma_e^2 + 10\sigma_{SC}^2 + 20\sigma_{CI/F}^2 + 200\theta_C^2$
F	1	$\sigma_e^2 + 15\sigma_{SF}^2 + 60\sigma_{I/F}^2 + 3\sigma_{SI/F}^2 + 300\theta_F^2$
CF	2	$\sigma_e^2 + 5\sigma_{SCF}^2 + 20\sigma_{CI/F}^2 + 100\theta_{CF}^2$
S	19	$\sigma_e^2 + 3\sigma_{SI/F}^2 + 30\sigma_S^2$
SC	38	$\sigma_e^2 + 10\sigma_{SC}^2$
SF	19	$\sigma_e^2 + 3\sigma_{SI/F}^2 + 15\sigma_{SF}^2$
SCF	38	$\sigma_e^2 + 5\sigma_{SCF}^2$
I/F	8	$\sigma_e^2 + 3\sigma_{SI/F}^2 + 60\sigma_{I/F}^2$
CI/F	16	$\sigma_e^2 + 20\sigma_{CI/F}^2$
SI/F	152	$\sigma_e^2 + 3\sigma_{SI/F}^2$
SCI/F	304	σ_e^2

(b) The error term for C is $MS_{SC} + MS_{CI/F} - MS_{SCI/F}$. The error for F is $MS_{SF} + MS_{I/F} - MS_{SI/F}$. The df for the error terms follow from Equation 14.7.

15.9 (a) $SS_A = 79.5$, $SS_C = 166.5$, $SS_S = 56$, and $SS_{resid} = 29$.

(b) $estMS_{SA} = \left[(a-1)MS_{resid} + MS_C\right]/a = \left[(4-1)29 + \frac{166.5}{3}\right]/4 = 17.5.$

(c) $RE_{LS\ toRM} = \left[\frac{df_{LS}+1}{df_{LS}+3}\right]\left[\frac{df_{RM}+3}{df_{RM}+1}\right]\left[\frac{MS_{SA}}{MS_{resid}}\right] = \left[\frac{6+1}{6+3}\right]\left[\frac{9+3}{9+1}\right]\left[\frac{17.5}{4.833}\right] = 3.042$

The RM design is estimated to need about three times as many participants as the LS design to achieve the same power.

15.11 (a)

Source	df	Error term
Drug Type (T)	1	S/TR
Row (R)	3	S/TR
TR	3	S/TR
S/TR	24	WCR
Occasions (O)	3	WCR
Dosages (D)	3	WCR
TO	3	WCR
TD	3	WCR
Between Cells Residual (BCR)	6	WCR
T × BCR	6	WCR
Within Cells Residual (WCR)	72	

(b)

Source	df	Error term
Row (R)	7	S/R
S/R	8	WCR
Occasions (O)	7	WCR
Drug Type (T)	1	WCR
Dosages (D)	3	WCR
TD	3	WCR
Between Cells Residual (BCR)	42	WCR
Within Cells Residual (WCR)	56	

(c) Design (b) requires fewer participants and has a simpler analysis. However, it may not be practical to run each participant on eight occasions (compared to four in design (a)). There are also questions about whether either drug type or dosing should be manipulated within-participant.

15.13 Ignoring *A*:

```
> summary(aov(data=LongFormC, Y ~ C *r + Error(id/C))) #ignores A

Error: id
          Df Sum Sq Mean Sq F value Pr(>F)
r          3  415.1   138.5   1.252  0.351
Residuals  8  885.0   110.6

Error: id:C
          Df Sum Sq Mean Sq F value  Pr(>F)
C          3  701.7  233.92  18.693 1.8e-06 ***
C:r        9  453.4   50.38   4.026 0.00302 **
Residuals 24  300.3   12.51
```

Ignoring *C*:

```
> summary(aov(data=LongFormA, Y ~ A *r + Error(id/A))) #ignores C

Error: id
          Df Sum Sq Mean Sq F value Pr(>F)
r          3  415.4   138.5   1.252  0.354
Residuals  8  885.0   110.6

Error: id:A
          Df Sum Sq Mean Sq F value   Pr(>F)
A          3  384.2  128.08  10.235 0.000158 ***
A:r        9  770.9   85.66   6.845 7.73e-05 ***
Residuals 24  300.3   12.51
```

Putting the analyses together, remembering $SS_{BCR} = SS_{CR} - SS_A = SS_{AR} - SS_C$ and *A* and *C* are tested against *WSR*:

Source	df	SS	MS	F	p
R	3	415.4	138.5	1.25	.354
S/R	8	885.0	110.6		
A	3	384.2	128.08	10.235	.0002
C	3	701.7	233.92	18.693	<.001
BCR	6	69.2	11.53		
WSR	24	300.3	12.51		

15.15

```
> summary(aov(data=LongFormC, Y ~ C *row + Error(id/C))) #ignores Script

Error: id
          Df Sum Sq Mean Sq F value Pr(>F)
row        3  10.38   3.458   1.795  0.189
Residuals 16  30.82   1.927

Error: id:C
          Df Sum Sq Mean Sq F value   Pr(>F)
C          3   5.95   1.984   7.555 0.000308 ***
C:row      9  44.24   4.916  18.719 6.87e-13 ***
Residuals 48  12.61   0.263

> summary(aov(data=LongFormT, Y ~ Script *row + Error(id/Script)))

Error: id
          Df Sum Sq Mean Sq F value Pr(>F)
row        3  10.38   3.458   1.795  0.189
Residuals 16  30.82   1.927

Error: id:Script
           Df Sum Sq Mean Sq F value   Pr(>F)
Script      3  42.46  14.154  53.894 2.13e-15 ***
Script:row  9   7.74   0.859   3.273  0.00352 **
Residuals  48  12.61   0.263

> summary(aov(data=LongFormT, Y~V*Script + Error(id/Script)))

Error: id
          Df Sum Sq Mean Sq F value Pr(>F)
Residuals 19   41.2   2.168

Error: id:Script
          Df Sum Sq Mean Sq F value   Pr(>F)
V          1  40.54   40.54 113.603 3.45e-15 ***
Script     2   1.92    0.96   2.691   0.0764 .
Residuals 57  20.34    0.36
```

Putting the analyses together, remembering $SS_{BCR} = SS_{CR} - SS_T = SS_{TR} - SS_C$ and *T*, *V*, *T/V*, and *C* are tested against *WSR*:

Source	*df*	*SS*	*MS*	*F*	*p*
Row	3	10.38	3.458	1.80	.189
S/R	16	30.82	1.927		
Script (T)	3	42.46	14.154	53.894	<.001
Valence (V)	1	40.45	40.45	154.15	.000
T/V	2	1.92	.96	3.65	.035
C	3	5.95	1.984	7.55	.000
BCR	6	1.783	.297		
WSR	48	12.606	.263		

We conclude that heartbeat change scores are affected by the script, and that the variability is affected mostly by the valence (positive vs negative).

Chapter 16

16.1 (a)

```
> outliers<- LongForm %>%
+   group_by(Day) %>%
+   identify_outliers(RT) #two outliers, both in Day 3
> outliers
# A tibble: 2 × 5
  Day      RT    id is.outlier is.extreme
  <chr> <dbl> <int> <lgl>      <lgl>
1 3        17     4 TRUE       FALSE
2 3        97    10 TRUE       FALSE
```

The data are reasonably symmetric and lacking outliers except for *Day 3*, which has an outlier in each direction. *Q–Q* plots are consistent with normality, except for those two outliers.

(b) No effects are significant. Notice that the *df* are reduced due to nonsphericity.

```
> anova_test(data=LongForm,RT ~ Day + Error(id/Day))
ANOVA Table (type III tests)

$ANOVA
  Effect DFn DFd     F     p p<.05   ges
1    Day   2  38 2.819 0.072       0.047

$`Mauchly's Test for Sphericity`
  Effect     W     p p<.05
1    Day 0.749 0.074

$`Sphericity Corrections`
  Effect   GGe     DF[GG] p[GG] p[GG]<.05   HFe      DF[HF] p[HF] p[HF]<.05
1    Day 0.799 1.6, 30.37 0.086           0.861 1.72, 32.72 0.081
```

16.3

Source	*df*	*EMS*	*Error term*
Context (C)	1	$\sigma_e^2 + 6\sigma_{S/C}^2 + 132\theta_C^2$	*S/C*
S/C	42	$\sigma_e^2 + 6\sigma_{S/C}^2$	
Days (A)	2	$\sigma_e^2 + 2\sigma_{SA/C}^2 + 88\theta_A^2$	*SA/C*
AC	2	$\sigma_e^2 + 2\sigma_{SA/C}^2 + 44\theta_{AC}^2$	*SA/C*
SA/C	84	$\sigma_e^2 + 2\sigma_{SA/C}^2$	
Targets(B)	1	$\sigma_e^2 + 3\sigma_{SB/C}^2 + 132\theta_B^2$	*SB/C*
BC	1	$\sigma_e^2 + 3\sigma_{SB/C}^2 + 66\theta_{BC}^2$	*SB/C*

16.3

Source	df	EMS	Error term
SB/C	42	$\sigma_e^2 + 3\sigma_{SB/C}^2$	
AB	2	$\sigma_e^2 + 44\theta_{AB}^2$	Residual
ABC	2	$\sigma_e^2 + 22\theta_{ABC}^2$	Residual
Residual (SAB/C)	84	σ_e^2	

16.5

```
> summary(aov(data=LongForm,Y~(GROUP/id)*(A/B)))
                 Df Sum Sq Mean Sq
GROUP             1   1910    1910
A                 2  15202    7601
GROUP:id         14   6858     490
A:B              12   6705     559
GROUP:A           2    519     259
GROUP:A:B        12   7614     634
GROUP:id:A       28   5819     208
GROUP:id:A:B    168  71774     427
```

A = Ambiguity (fixed); *B* = Items (random); *GROUP* = language group (fixed); *id* = subjects (random)

Source	EMS	Error term
GROUP	$\sigma_e^2 + ab\sigma_{S/Group}^2 + n\sigma_{BxGroup/A}^2 + nab\theta_{Group}^2$	$MS_{S/Group} + MS_{BxGroup/A} - M_{residual}$
S/GROUP	$\sigma_e^2 + ab\sigma_{S/Group}^2$	
A	$\sigma_e^2 + ng\sigma_{B/A}^2 + b\sigma_{SA/Group}^2 + nbg\theta_A^2$	$MS_{Group/A} + MS_{SA/Group} - M_{residual}$
B/A	$\sigma_e^2 + ng\sigma_{B/A}^2$	
AxGROUP	$\sigma_e^2 + n\sigma_{BxGroup/A}^2 + b\sigma_{SA/Group}^2 + nb\theta_{AxGroup}^2$	$MS_{SA/Group} + MS_{BxGroup/A} - M_{residual}$
BxGROUP/A	$\sigma_e^2 + n\sigma_{BxGroup/A}^2$	
SA/GROUP	$\sigma_e^2 + b\sigma_{SA/Group}^2$	
Residual	σ_e^2	

Df for error terms follow from Equation 14.7: for *Group* = 9.39, for *A* = 2.348, for *AxGroup* = 4.762

Source	df	MS	F	p
GROUP	1	1910	2.739	.13
S/Group	14	490		
A	2	7601	22.399	.03
B/A	12	559		
AxGROUP	2	259	0.625	.57
BxGROUP/A	12	634		
SA/GROUP	28	208		
Residual	168	427		

16.7 (a)
```
> summary(aov(data=LongFormA,Y~id*A+Error(id/A)))

Error: id
   Df Sum Sq Mean Sq
id 19  928.7   48.88

Error: id:A
      Df Sum Sq Mean Sq
A      3    633  211.10
id:A 57   4757   83.46
```

(b)
```
> summary(aov(data=LongFormB, Y ~ B *Group + Error(id/B))) #ignores A

Error: id
          Df Sum Sq Mean Sq F value Pr(>F)
Group      3  155.7   51.90   1.074  0.388
Residuals 16  773.0   48.31

Error: id:B
          Df Sum Sq Mean Sq F value Pr(>F)
B          3    158   52.60   0.637  0.595
B:Group    9   1270  141.08   1.709  0.113
Residuals 48   3963   82.56
```

Putting these analyses together, $BCR = SS_{GA} - SS_B = SS_{GB} - SS_A = 1270 - 633 = 636$. Groups are tested against *S/G*. Situation (*A*) is tested against *WSR*; in (a), *SA* is a pool of *B*, *BCR*, and *WSR*.

Source	*df*	*SS*	*MS*	*F*	*p*
Group (G)	3	155.7	51.9	1.074	.388
S/G	16	773.0	48.31		
Situation (A)	3	633	211.1	2.557	.066
Blocks (B)	3	158	52.6	0.637	.595
BCR	6	636	106	1.281	.284
WSR	48	3963	85.56		

Although in this example, the *B* and *BCR* terms contribute relatively little variability and therefore there is little change in the significance level, if either contributed more variability, the error term in part (a) would be inflated. Consequently, the test of *A* against *SA* would yield a ratio not distributed as *F*, and there would be negative bias in the test.

(c) If *B* represented a factor having random effects, *BCR* would be the appropriate error term for *A*, assuming the presence of *AB* interaction effects in the population.

Chapter 17

17.1 (**b**) 0.620. (**c**) $\hat{Y} = 3.00 + 2.00X$. (**d**) 0.3846. (**e**) $\hat{X} = 1.58 + .19Y$. (**f**) 0.3846.

17.3 (a) r is bigger in Situation 2. Using Equation 17.11, s_x / s_y is much bigger in Situation 2. **(b, c, d)** The correlation is unchanged in each case.

17.5 (a) The longer you live, the more time you have to smoke cigarettes. If we are concerned about the influence of smoking on longevity, we should look at the rate of cigarette smoking (cigarettes/day) not the total number.

(b) The data do not allow us to make a causal statement. It could be that less able or less motivated students spent less time on schoolwork and therefore had more time to watch TV. We cannot conclude that the TV watching caused the poor performance.

17.7 (a) For college grads, $b_1 = .599$, for employees without college degrees, $b_1 = .799$.

(b) For each extra year of employment, college grads' salaries are increased .599 and nongrads' salaries are increased .799. The difference is smaller for college grads even though their correlation of income and experience is higher, in part because of much larger variability in years of experience.

(c) To predict income from years of experience, we can use $\hat{Y} = Y + b_1(X - \bar{X})$. For 10 years of experience, we predict for people without college degrees, $\hat{Y} = 76 + .799(10 - 10) = 76.0$ and for college grads, $\hat{Y} = 80 + .599(10 - 15) = 77.0$. For 20 years of experience, the predictions are 84.0 for non-degree employees and 83.0 for college grads.

17.9 (a) The patients in the VA hospital may constitute a restricted sample. For example, the anxiety scores may tend to be quite high. The sample should not be used to make inferences about the general population.

(b) The correlation between height and weight for the mixed group would be expected to be lower than .60. If the differences in mean height and weight for Martians and Jovians were great enough, the correlation might even be negative.

(c) This is a classic case of inferring causation from correlation. People who graduate from college may indeed make more money, but it is not obvious how much of their financial success can be attributed directly to graduating from college. Graduates may be smarter and more motivated and organized than nongraduates and therefore more successful. Graduates may also be more likely to come from wealthier and more stable families.

17.11 0.394.

17.13 (a) The best prediction is 73 inches = .5(3 / 3)(76–70).

(b) The best prediction for the height of the father is 71.5 inches. Again, there is regression toward the mean: 71.5 = .5(3/3)(73–70).

(c) Even though there is regression toward the mean for the predicted score whenever the absolute value of the correlation is less than 1, this does not imply the distribution of height will become less variable over time. Whenever there is less than perfect prediction, the actual scores can be thought of as being distributed around the prediction. If these distributions are summed, the original distribution should be reproduced.

17.15 (b) −0.229.

(c) 0.386. Within each species of penguin, the correlation is positive. It is negative overall (ignoring species) because Gentoos tend to have smaller bill depth but greater length than the Adelies. This is an example of the pattern in Figure 17.12c.

(d) Overall, predicted body mass = −5872.09 + 50.15*flipper length.

(e) For the Adelie penguins, predicted body mass = −2508.09 + 32.69*flipper length. The slope is shallower for the Adelies because the range of flipper lengths is narrower than overall. Try plotting all three groups to see the effect.

Chapter 18

18.1 (a) $t(17) = -1.30$, p = 0.212, cannot reject null.
(b) $z = -1.24$, $p = 0.216$, cannot reject null.
(c) No, even if the correlation had been significant, we could not conclude that studying interferes with test performance. More likely, students having difficulty may study more, but still perform more poorly.
(d) *CorCI*(−.3,19) returns [−.664, .179]. *CorCI*(−.3,19, *conf.level* = .5) returns [−.445, −.140]
(e) To use G*Power 3.1 to calculate the number of subjects required to have power = .80 for the test in (a) if $\rho = -.30$, we select *t* tests as the *Test family*; *Correlation: Point biserial model* as the *Statistical test*; and *A priori . . .* as the *Type of power analysis*. The effect size is | r | = .30. G*Power 3.1 indicates that we need N = 82 to get power = .80.
(f) If we intend to use the Fisher *Z* and perform the test in (b), to calculate the required sample size, we must select *Exact* as the *Test family*, *Correlation: Bivariate normal model* and *A priori . . .* as the *Type of power analysis*. Again, the effect size is .30. The population ρ (for the null hypothesis) is 0. Now we must click on the *Options* tab and select *Use large sample approximation (Fisher Z)*. The obtained result is N = 85.

18.3 *cocor.indep.groups*(.333, .235, 2000, 2000) returns $Z = 3.372$, $p = .0007$. We reject the null hypothesis and conclude that the correlation is significantly greater with a college degree than without.

18.5 (a) Equation 18.7, using Fisher *Z* to transform the correlations, equals 15.79, $p = .0014$.
(b) *cocor.dep.groups.overlap*(.3, .5, .2, 39) compares $r = .3$ and $r = .5$ using many different tests. Steiger's (1980) test using average correlations returns $z = -1.0846$, *p*-value = 0.2781, and we cannot reject the null. (The same conclusion is reached for every test.)
(c) Using Equation 18.9: (.5 − .3*.2)/*sqrt*((1 − .3^2)(1 − .2^2)) = 0.47 = r_{AVIQ}. $t(36) = 3.20$, $p = .003$.

18.7 (a) $\Phi = -.36$.
(b) $\varphi_{max} = .56$, $\varphi_{min} = -.80$.
(c) To obtain $\varphi = 0$, we need independence, so that *p*(pass item 1 and pass item 2) = *p*(pass 1)*p*(pass 2) = .4 × .7 = .28. The frequency of the *pass-pass* cell would then be $Np = 28$. The other frequencies may be filled in so as to preserve the marginals.

18.9 Substituting into the equation, we obtain $r_{pre,\ change} = -.39$.

18.11 (a) Pearson = .653, Spearman = .597, Kendall = .471.
(b) Bootstrapped 95% CI = [.0438, .9444] with 2,000 samples (your endpoints may differ slightly).
(c) *CorCI*(.6527,10) = [.039, .909]. The CIs are quite similar, although the bootstrapped CI is slightly wider than the computed CI.

Chapter 19

19.1 (a)

```
> summary(lm(dat$final~dat$pretest))

Call:
lm(formula = dat$final ~ dat$pretest)

Residuals:
    Min      1Q  Median      3Q     Max
-19.764  -8.633   1.503   7.753  14.236

Coefficients:
            Estimate Std. Error t value Pr(>|t|)
(Intercept) -36.0832    27.2951  -1.322 0.204770
dat$pretest   3.5465     0.8421   4.212 0.000662 ***
---
Signif. codes:  0 '***' 0.001 '**' 0.01 '*' 0.05 '.' 0.1 ' ' 1

Residual standard error: 10.64 on 16 degrees of freedom
Multiple R-squared:  0.5258,    Adjusted R-squared:  0.4961
F-statistic: 17.74 on 1 and 16 DF,  p-value: 0.0006621
```

(b) $\widehat{Final} = -36.08 + 3.55 pretest$

$SE(b_0) = 27.30$, $SE(b_1) = .84$.

(c)

```
> new<-data.frame(pretest = c(24,37))
> predict(regout, new, interval="confidence")
       fit      lwr       upr
1 49.03202 33.32850  64.73555
2 95.13609 85.17051 105.10168
```

CI for *pretest* = 24, [33.328, 64.736]; for *pretest* = 37, [85.171, 105.102].

(d) The estimate at *pretest* = 37 is likely to be more accurate. Because it is closer to the mean of the *pretest* scores (i.e., 32.3), it has a smaller leverage and thus a smaller standard error.

(e)

```
> predict(regout, new, interval="prediction")
       fit      lwr      upr
1 49.03202 21.55106 76.51299
```

Prediction interval at *pretest* = 24, [21.55, 76.51].

19.3 (a) $TC = 131.87 + 1.712\ age$, $s_e = 34.22$, $SE(b_1) = .202$, and $SE(b_0) = 10.05$.

(b) For *age* = 30, CI is [174.54, 191.89]; for *age* = 50, CI is [212.76, 222.14].

(c) The interval is narrower (and hence the estimate is more likely to be closer to the population parameter) for 50-year-old women because 50 is closer to the mean *age* (43.898) than 30, and hence the standard error is smaller.

(d) Prediction interval is [115.20, 251.23].

19.5 (a) Using $b_1 = r\ SY / SX$, we find there is a slope of .599 for college graduates (*CG*) and .799 for people without college degrees (*ND*). Each additional year of service corresponds to about an additional \$600 (.599 × \$1,000) for college grads and about \$800 for people without a college degree.

(b) We can test whether the slope difference is significant using the test statistic (see Box 19.1) $t = \frac{b_{CG} - b_{ND}}{SE(b_{CG} - b_{ND})} = \frac{b_{CG} - b_{ND}}{s_e\sqrt{\frac{1}{SS_{X_{CG}}} + \frac{1}{SS_{X_{ND}}}}}$ where $s_e^2 = \frac{SS_{resid}}{N_{CG} + N_{ND} - 4} = \frac{(1 - r_{CG}^2)SS_{Y_{CG}} + (1 - r_{ND}^2)SS_{Y_{ND}}}{3996} = \frac{575855.86 + 545806.95}{3996} =$ 280.70 so that s_e =16.75. Thus, $t(3996) = -2.38$, $p < .02$. The salary increment per year of experience for non-degreed employees is significantly *greater* than for college grads.

(c) So here we have a situation in which the correlation is significantly larger for college grads than people without degrees even though the slope is significantly larger for non-degree employees than for college grads. The reason for this apparent paradox is that the college grads have greater variability in their years of service. Here, the unstandardized regression coefficient tells one story and the standardized regression coefficient (here the correlation coefficient) tells another. It is important to understand the difference.

19.7 We can test whether the slope difference is significant using the test statistic (see Box 19.1) $t = \frac{b_{11} - b_{12}}{SE(b_{11} - b_{12})} = \frac{b_{11} - b_{12}}{s_e\sqrt{\frac{1}{SS_{X_1}} + \frac{1}{SS_{X_2}}}} = \frac{30.0 - 20.0}{s_e\sqrt{\frac{1}{200} + \frac{1}{200}}}$ and s_e^2 is the weighted average of $s_{e1}^2 = 15.5^2$ and $s_{e2}^2 = 12.2^2$ or 194.545, so $s_e = 13.95$. Substituting into the equation for t, we get $t(76) = 7.17$ and we can reject the null hypothesis that the slopes are equal.

19.9 The R of .40 tells us that in the sample, the variability of the actual Y scores about the regression line is only 84% of their variability about the line $\hat{Y} = \bar{Y}$; i.e., the proportion of the variability accounted for by the regression is $.40^2 = .16$.

Chapter 20

20.1 (a)

```
        var1   var2  cor        p
1     Number   Time 0.76 1.91e-05
2 Difficulty   Time 0.34 1.05e-01
3 Difficulty Number 0.00 1.00e+00

  Number Difficulty meanTime sdTime
   <dbl>      <dbl>    <dbl>  <dbl>
1      2         10     493.   10.5
2      2         20     532    35.7
3      4         10     546.   63.0
4      4         20     583.   40.5
5      6         10     584.   30.4
6      6         20     647.   44.7
7      8         10     626.   70.5
8      8         20     670    40.0
```

(b) *Time* = 402.375 + 22.825 *Number* + 4.575 *Difficulty*. For *Number*, $t(21) = 6.192$, $p < .001$; for *Difficulty*, $t(21) = 2.775$, $p = .011$. In the ANOVA, *Number* and *Difficulty* are both significant and their interaction is not. The results of the regression are not equivalent to that of an ANOVA. The regression treats the predictors as quantitative variables and tests whether the rate of change of time with one of the variables is different from 0 in the population, holding the other variable constant. In the ANOVA, the test of the *Number* main effect addresses the question of whether the population means for the different levels of number are all the same.

(c) The estimates are 402.375, 22.825, and 4.575 for β_0, β_1, and β_2, and the CIs are shown as follows. The true values of the population parameters each fall within the 95% CI.

```
> confint(lm(data=dat,Time ~ Number+Difficulty))
                 2.5 %     97.5 %
(Intercept) 335.987917 468.762083
Number       15.159280  30.490720
Difficulty    1.146786   8.003214
```

20.3 (a) Regressing Y on $X1$ and $X2$ yields the regression equation $\widehat{Y} = -16.294 + 9.196X1 + 9.941X2$.

(b) In the initial regression, the coefficients of $X1$ and $X2$ both differ significantly from 0, $t(14) = 2.420$, $p = .030$ and $t(14) = 3.378$, $p = .005$, respectively. Therefore, both variables should be included. $X1$ and $X2$ are correlated with one another, $r = 0.39$, but the correlation is not significant ($p = .125$).

20.5 (a) $R^2 = 1$ because there are four parameters (intercept + three partial slopes) and four observations, so there are no remaining *df*; the model is saturated.

(b)
```
> predict(lm(data=dat2,Y~X1+X2+X3),dat)
         1          2          3          4          5          6          7
 0.1500000  0.0900000 -3.3600000 -1.3400000 -2.2468104  3.8562583  2.3668006
         8          9         10         11         12         13         14
 3.8305393 -4.8775844 -5.4127538 -5.0672577 -5.7813418 -0.3176991  1.7162598
        15
 1.9269231
```

The predictions for the first four cases are identical (to seven decimal places) to the observed Y values in those cases.

(c) If the predicted values are stored in p and the original data in *dat*, then *cor*(p[1:4], dat$Y[1:4]) will give the correlation of observed and predicted Y values for the first four cases, which is 1.0, and *cor*(p[5:15], dat$Y[5:15]) will give the correlation for the remaining 11 cases. This second correlation is only 0.244, much lower because the N/p ratio is so small (4/4 = 1 in this example).

20.7 With the selected data stored in a data frame called datcv,

```
> model.jack <- lm(tc ~., data = datcv)
> delpred<-datcv$tc-rstandard(model.jack,type="predictive")
> cor.test(delpred,datcv$tc)

        Pearson's product-moment correlation

data:  delpred and datcv$tc
t = 7.6727, df = 179, p-value = 1.045e-12
alternative hypothesis: true correlation is not equal to 0
95 percent confidence interval:
 0.3791371 0.5998178
sample estimates:
      cor
0.497484
```

20.9 $TC = 161.9 + .010\ age + 2.14\ bmi$, $R^2 = .046$ and only *bmi* is a significant predictor: $t(180) - 2.948$, $p - .004$. Adjusted R-squared: 0.03547.

Chapter 21

21.1 (a)

```
> summary(aov(data = dat, performance ~ dosage))
            Df Sum Sq Mean Sq F value  Pr(>F)
dosage       3 132.31   44.10    7.29 0.00305 **
Residuals   15  90.74    6.05
```

(b)

```
> summary(lm(data = dat, performance ~ dosage))

Call:
lm(formula = performance ~ dosage, data = dat)

Residuals:
    Min      1Q  Median      3Q     Max
-4.4778 -2.1458  0.1278  1.3250  5.6750

Coefficients:
            Estimate Std. Error t value Pr(>|t|)
(Intercept)  5.26667    1.77641   2.965  0.00868 **
dosage       0.16528    0.06453   2.561  0.02023 *
---
Signif. codes:  0 '***' 0.001 '**' 0.01 '*' 0.05 '.' 0.1 ' ' 1

Residual standard error: 3.077 on 17 degrees of freedom
Multiple R-squared:  0.2785,    Adjusted R-squared:  0.236
F-statistic: 6.561 on 1 and 17 DF,  p-value: 0.02023

> anova(lm(data = dat, performance ~ dosage))
Analysis of Variance Table

Response: performance
          Df Sum Sq Mean Sq F value  Pr(>F)
dosage     1  62.11  62.110  6.5607 0.02023 *
Residuals 17 160.94   9.467
```

(c) Find $SS_{pure\ error} = SS_{error}$ from ANOVA = 90.74. Find $SS_{nonlinearity} = SS_{pure\ error} - SS_{residual} =$ 160.94 – 90.74 = 70.2. $df_{nonlinearity} = df_{pure\ error} - df_{residual} = 17 - 15 = 2$. So, $F(2,15) = MS_{nonlinearity} / MS_{pure\ error} = (70.2/2)/(90.74/15) = 5.8$. $1 - pf(5.8, 2, 15) = p = .0136$. There is a significant linear effect of *dosage*.

(d) Both *dosage* and *dosesq* have significant partial slopes, so there are significant linear and quadratic (or as least curved) components: performance = –4.71 + 1.12 *doseage* – .02 *dosesq*.

21.3 In each case, calculate $f^2 = \dfrac{\Delta R^2}{1-R^2}$ and then use G*Power.

(a) $f^2 = .3/.7 = .429$ so $N = 39$.

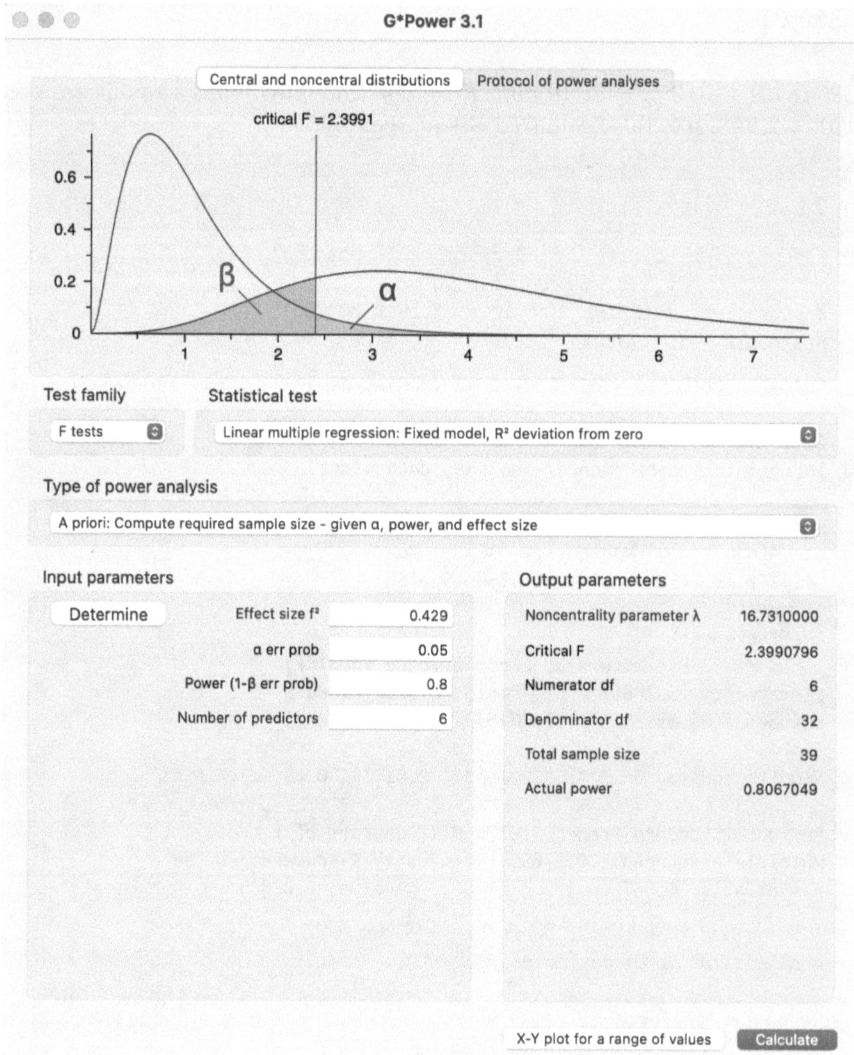

(b) $N = 62$.

(c) $N = 130$.

21.5 Calculate $f^2 = \dfrac{\Delta R^2}{1 - R^2}$ = .05/(1 – .35) = 0.77 and then use G*Power to find N^* = 139. If the predictors are random, $N = N^* + p - 1 = 139 + 7 - 1 = 145$.

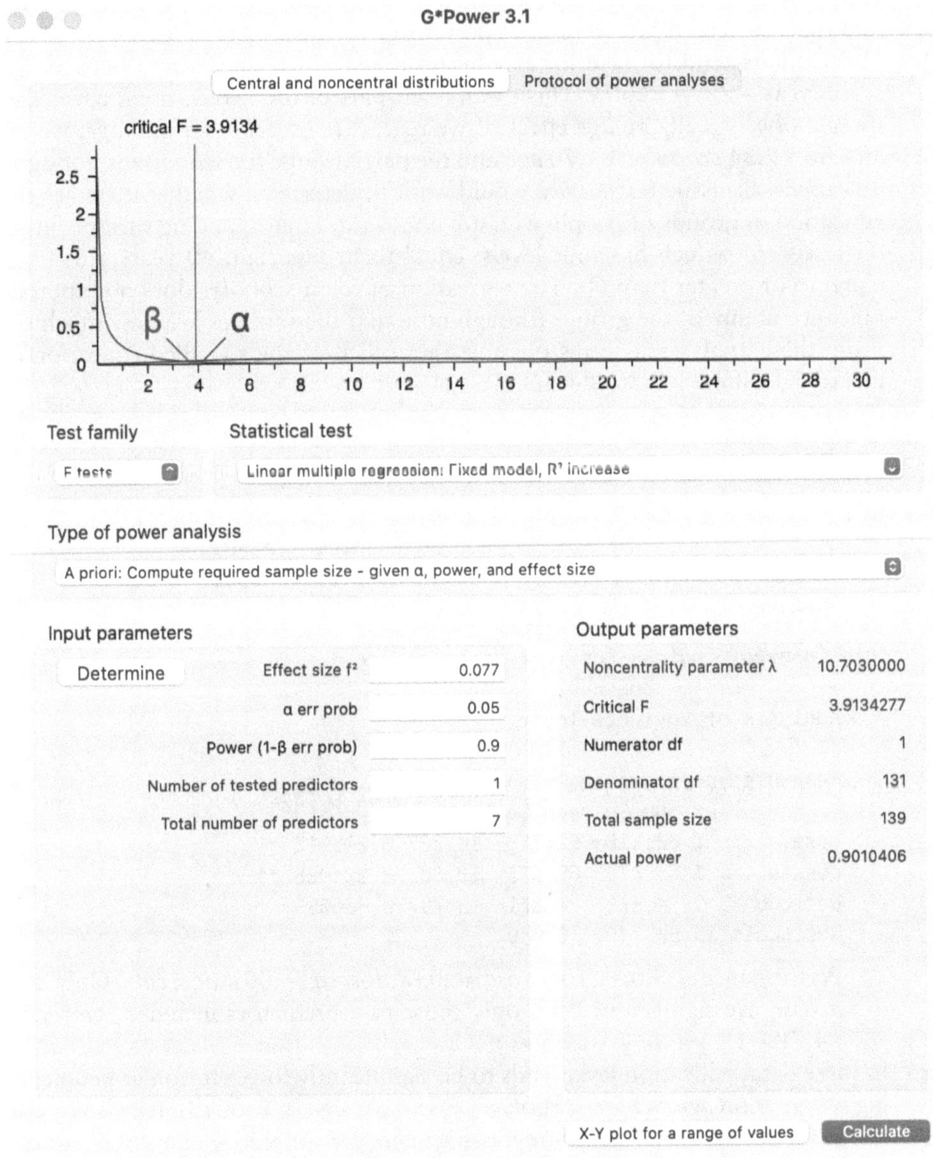

21.7 If including X_4 in the regression equation results in R^2 increasing from .21 to .27, then $\Delta R^2 = .06$ and $f^2 = \dfrac{\Delta R^2}{1 - R^2}$ = .06 / (1 – .27) = .082. To test whether the increment in predictability provided by the inclusion of X_4 is significant, use the partial-F ratio:

$$F = \frac{\left(R^2_{Y.1234} - R^2_{Y.123}\right)/1}{\left(1 - R^2_{Y.1234}\right)/(N-p-1)} = \frac{.06}{(1-.27)/35} = 2.877.$$ The critical value of $F(1,35)$ with $\alpha = .05$ is 4.121, so the coefficient of X_4 is not significant. Using G*Power, we find that we need $N^* = 98$ to have power of .8 to test X_4 so $N = 98 + 4 - 1 = 101$.

21.9 Older adults tend to have lower education levels; the correlation between *age* and *schoolyr* is −.22, $p = .003$. This means that part of the "education" effect measured by *schoolyr* is really an *age* effect. If we regress *tc* on *schoolyr* and *age*, we find *tc* = 144.06 – 2.34 *schoolyr* + 1.70 *age* and the partial slope for *schoolyr* is not significant. To pursue this issue further, we would want to determine whether there are effects of education in groups of people who are about the same age. One variable in the data set is *agegrp*, which has four levels: equal to or less than 40 years, 40–49, 50–59, equal to or greater than 60. The regression of *tc* on *schoolyr* does not approach significance in any of the groups (though note that the samples are now much smaller). From these analyses it seems possible that much of the so-called education effect is due to age differences in education.

Chapter 22

22.1 (a)
```
> summary(aov(data = dat, performance ~ as.factor(dose) ))
                Df Sum Sq Mean Sq F value  Pr(>F)
as.factor(dose)  3 132.31   44.10    7.29 0.00305 **
Residuals       15  90.74    6.05
```

(b)
```
> anova(lm(data = dat, performance ~ dose + dosesq +
+                dosecub))
Analysis of Variance Table

Response: performance
          Df Sum Sq Mean Sq F value   Pr(>F)
dose       1 62.110  62.110  10.267 0.005912 **
dosesq     1 65.375  65.375  10.807 0.004986 **
dosecub    1  4.821   4.821   0.797 0.386085
Residuals 15 90.740   6.049
```

Performance = 3.0 – .105 *dose* + .037 *dosesq* – .001 *dosecub*. Only *dose* and *dosesq* are significant. With only those two predictors included, *performance* = −4.71 + 1.12 *dose* – .02 *dosesq*.

22.3 In these data, education level tends to be significantly lower for older people: regressing *schoolyr* on *age*, we see *schoolyr* = 6.37 – .02 *age* ($p = .0013$). It's also significantly lower for women than men – regressing *schoolyr* on *sex*, where *sex* = 1 indicates a female identity, we see *schoolyr* = 5.69 – .545 *sex*, so education level is lower for female participants ($p < .001$). The interaction of *age* and *sex* is not significant as a regression predictor ($p = .20$).

22.5 (a) For men, *tc* = 16.19 + .010 *age* + 2.14 *bmi*, where the coefficient for *age* is not significant ($p = .97$) and the coefficient for *bmi* is ($p = .004$); $R^2 = .046$. For women, *tc* = 92.24 + 1.74 *age* + 1.47 *bmi*, where the coefficients for *age* ($p < .001$) and *bmi* are significant ($p = .007$); $R^2 = .272$.

(b) We would expect the correlations of *tc* with *bmiplusnorm05* to be smaller than those with *bmi*, although given that values are randomly chosen from the normal distribution, this does not have to happen. In one random sample, for women, the correlations with *tc* are .23 for *bmi*, .21 for *bmiplusnorm05*; for men, they are .21 and .08. Of course, different random samples will result in different values.

The larger the additional measurement error, the less useful the variables are as predictors, and the smaller their regression coefficients. Again for a particular random sample, the results are: for men, *tc* = 203.22 + .05 *age* + .57 *bmiplus05*, where the coefficients are not significant. For women, *tc* = 106.19 + 1.74 *age* + .93 *bmiplus05*, where the coefficients for *age* ($p < .001$) and *bmiplus05* are significant ($p = .012$); $R^2 = .268$.

(c) This time, for a random sample of noise values, for women, the correlations with *tc* are .23 for *bmi*, – .13 for *bmiplusnorm05*; for men, they are .21 and .00. Of course, different random samples will result in different values.

The measurement error is larger than in part (b), rendering the variables less useful as predictors. Again for a particular random sample, the results are: for men, *tc* = 221.2 + 00 *age* + .00 *bmiplus020*, where the coefficients are not significant. For women, *tc* = 132.5 + 1.79 *age* – .19 *bmiplus020*, where the coefficients for *age* ($p < .001$) is significant but *bmiplus020* is not ($p = .119$); $R^2 = .252$.

(d) Increased measurement error in the dependent variable increases the size of the error component, so that s_Y and s_e become larger and R^2 becomes smaller. This increases the size of confidence intervals and decreases the power of significance tests. However, if there are no major violations of assumptions, OLS estimates of the unstandardized regression coefficients remain unbiased. In contrast, because the standardized coefficient of X_j is the unstandardized coefficient multiplied by s_j / s_Y, standardized coefficients (the so-called beta weights) become systematically smaller as measurement error in the dependent variable increases. All of these statements are consistent with the regression results:

For men, *tcplusnorm040* = 164.6 + .06 *age* + 2.07 *bmi*, where, as in part (a), the coefficient for *age* is not significant ($p = .87$) and the coefficient for *bmi* is ($p = .04$); $R^2 = .024$

For women, *tc* = 143.18 + 1.26 *age* + .55 *bmi*, where the coefficients for *age* ($p < .001$) is significant but *bmi* is not ($p = .51$); $R^2 = .07$. Here, we see reduced power and larger *SE*s (thus wider CIs) for the coefficients.

We should emphasize that we have randomly chosen the random components to add to *tc*, so that if we were to create the variable *tcplusnorm040* again, the values would be different, but the two variables should have the same general characteristics.

22.7 Looking at the literature closely, we can see why there is disagreement about the conclusions. Assessing the influence of class size is an extremely complex research problem, largely because there are many variables other than class size that are associated with differences in student performance. Some of these variables reflect student characteristics such as ability and motivation. Others characterize the student's family and background; for example, socioeconomic level (SES) and family size, as well as parental education and support for education. Still others include age and racial status, given that some studies have found class size reductions to be more beneficial for young and for minoritized students. There are also classroom variables such the subject matter that is taught and the teaching styles that are used. All of these can be thought of as potential moderator variables; that is, the relationship between class size and performance may be different for different levels of these variables. Therefore, unless we can somehow control for these variables or incorporate them into our analyses, they will complicate the task of assessing the effect of class size.

Given sufficient resources, we could perform an experiment in which we manipulate some of the most important variables and try to control the others by random assignment of both students and teachers to class sizes. However, suppose that we could only perform observational studies in which we measured student performance in pre-existing classes of different sizes. Any associations between class size and performance measures will be confounded with the effects of the other important variables mentioned earlier. If we found that performance was better for smaller class sizes, this might be because, for example, smaller class sizes tend to be found in schools located in more affluent neighborhoods that have, on average, more educated and supportive families and better teachers. In this case, the best we can do is to try to measure those variables that seem most important and include them in our analyses, keeping in mind that statistical control is not the same thing as experimental control. In fact, only two major studies on this topic have been performed using random assignment of students and teachers, and both have been criticized on other methodological grounds.

We can also ask about the nature of the functional relationship between class size and student performance. Do equal reductions in class size result in equal amounts of improvement? For example, are the benefits of reducing class size from 25 to 20 equal to those achieved by reducing class size from 15 to 10? Or is the relationship between class size and student performance best described as some sort of curvilinear, or even as a discontinuous function? To investigate this question, we would have to add curvilinear components to our regression equations.

22.9

```
> summary(lm(data = dat, tc ~ bmi*age))

Call:
lm(formula = tc ~ bmi * age, data = dat)

Residuals:
     Min       1Q   Median       3Q      Max
-114.395  -18.255   -0.109   20.132  116.370

Coefficients:
             Estimate Std. Error t value Pr(>|t|)
(Intercept) -70.2413    65.9523  -1.065 0.288312
bmi           7.8437     2.5576   3.067 0.002503 **
age           5.2086     1.3833   3.765 0.000226 ***
bmi:age      -0.1354     0.0532  -2.546 0.011741 *
---
Signif. codes:  0 '***' 0.001 '**' 0.01 '*' 0.05 '.' 0.1 ' ' 1

Residual standard error: 33.1 on 177 degrees of freedom
  (6 observations deleted due to missingness)
Multiple R-squared:  0.298,	Adjusted R-squared:  0.2861
F-statistic: 25.05 on 3 and 177 DF,  p-value: 1.476e-13
```

Chapter 23

23.1 (a) $r = .509, p = .044$.

(b) $t'\ (13.9) = -2.21, p = .044$.

(c) We need one dummy variable because there are only two levels of *Gender* in this data set. (i) *men* = 1, *women* = –1 (or the reverse); (ii) *men* = 0, *women* = 1 (or the reverse).

(d) (i) Effect coding:

```
> summary(lm(data = dat,Y ~ Gender))

Call:
lm(formula = Y ~ Gender, data = dat)

Residuals:
    Min      1Q  Median      3Q     Max
-10.375  -3.875   1.312   3.969   7.000

Coefficients:
            Estimate Std. Error t value Pr(>|t|)
(Intercept)   28.188      1.442   19.55 1.47e-11 ***
Gender        -3.188      1.442   -2.21   0.0442 *
---
Signif. codes:  0 '***' 0.001 '**' 0.01 '*' 0.05 '.' 0.1 ' ' 1

Residual standard error: 5.769 on 14 degrees of freedom
Multiple R-squared:  0.2587,    Adjusted R-squared:  0.2057
F-statistic: 4.885 on 1 and 14 DF,  p-value: 0.04424
```

(ii) Dummy Coding:

```
> summary(lm(data = dat,Y ~ Gender))

Call:
lm(formula = Y ~ Gender, data = dat)

Residuals:
    Min      1Q  Median      3Q     Max
-10.375  -3.875   1.312   3.969   7.000

Coefficients:
            Estimate Std. Error t value Pr(>|t|)
(Intercept)   31.375      2.040   15.38 3.64e-10 ***
Gender        -6.375      2.884   -2.21   0.0442 *
---
Signif. codes:  0 '***' 0.001 '**' 0.01 '*' 0.05 '.' 0.1 ' ' 1

Residual standard error: 5.769 on 14 degrees of freedom
Multiple R-squared:  0.2587,    Adjusted R-squared:  0.2057
F-statistic: 4.885 on 1 and 14 DF,  p-value: 0.04424
```

Note that the ANOVA tables are the same for both analyses; in both cases the *Gender* variable accounts for all the variability in the means. However, the slope coefficients are not the same. The coefficient of EFFECT in the first analysis, −3.188, indicates that the mean for *men* is 3.188 units less than the average of the means for *men* and *women*, whereas the coefficient of DUMMY in the second analysis, −6.375, indicates that the mean of the scores for *men* is 6.375 units less than the mean of the scores for *women*.

23.3 (a) To code three levels of one variable, we need two dummy variables.

(b) In R, we don't need to actually generate these variables, but it's helpful to consider what they are doing:

	Effect Coding		*Dummy Coding*	
Y	E1	E2	D1	D2
17	1	0	1	0
33	1	0	1	0
26	1	0	1	0
27	1	0	1	0
21	1	0	1	0
11	0	1	0	1
18	0	1	0	1
14	0	1	0	1
18	0	1	0	1
9	−1	−1	0	0
12	−1	−1	0	0
10	−1	−1	0	0
8	−1	−1	0	0
14	−1	−1	0	0

(c)
```
> summary(lm(data = dat, Y ~ C,contrasts = list(C = contr.sum)))

Call:
lm(formula = Y ~ C, data = dat, contrasts = list(C = contr.sum))

Residuals:
   Min      1Q Median      3Q     Max
-7.800 -2.350  0.300  2.612  8.200

Coefficients:
            Estimate Std. Error t value Pr(>|t|)
(Intercept)   16.883      1.165  14.491 1.64e-08 ***
C1             7.917      1.616   4.900 0.000472 ***
C2            -1.633      1.710  -0.955 0.359985
---
Signif. codes:  0 '***' 0.001 '**' 0.01 '*' 0.05 '.' 0.1 ' ' 1

Residual standard error: 4.335 on 11 degrees of freedom
Multiple R-squared:  0.716,     Adjusted R-squared:  0.6644
F-statistic: 13.87 on 2 and 11 DF,  p-value: 0.0009845
```

```
> dat$C <- relevel(dat$C,ref = 3)
> summary(lm(data = dat, Y ~ C))

Call:
lm(formula = Y ~ C, data = dat)

Residuals:
   Min     1Q Median     3Q    Max
-7.800 -2.350  0.300  2.612  8.200

Coefficients:
            Estimate Std. Error t value Pr(>|t|)
(Intercept)   10.600      1.939   5.467 0.000196 ***
C1            14.200      2.742   5.179 0.000304 ***
C2             4.650      2.908   1.599 0.138149
---
Signif. codes:  0 '***' 0.001 '**' 0.01 '*' 0.05 '.' 0.1 ' ' 1

Residual standard error: 4.335 on 11 degrees of freedom
Multiple R-squared:  0.716,     Adjusted R-squared:  0.6644
F-statistic: 13.87 on 2 and 11 DF,  p-value: 0.0009845
```

The ANOVA tables are identical for these analyses.

(d) The interpretation of the coefficients depends on the coding scheme. For effect coding, b_0 = the mean of the group means of Y; b_1= mean of Y for group 1 – b_0; b_2 = mean of Y for group 2 – b_0. For dummy coding, the reference group was set to group 3, b_0 = the mean Y of group 3; b_1 = mean of group 1 – b_0; b_2 = mean of group 2 – b_0.

23.5 Set 1 is appropriate – effect coding.

Set 2 is appropriate – dummy coding.

Set 3 is appropriate because using the proposed dummy variables will result in $SS_{regression}$ = $SS_{between}$, although the coefficients will not be interpretable.

Set 4 is not appropriate – there are two coding variables, but values on X_2 are just the values on X_1 multiplied by the constant 3. X_1 and X_2 are perfectly correlated so we effectively only have one dummy variable.

Set 5 is not appropriate. There is only one dummy variable. $SS_{regression}$ will equal $SS_{between}$ only if the condition means are perfectly correlated with the values of X_1.

Set 6 is not appropriate because there are three dummy variables, but only two degrees of freedom for the effect of the condition variable. Regressing Y on any two of X_1, X_2, and X_3 will yield $SS_{regression}$ = $SS_{between}$.

23.7 (a)

```
> Anova(lm(data = dat, Y ~ A*X,contrasts =
+            list(A = contr.sum)),type=3)
Anova Table (Type III tests)

Response: Y
             Sum Sq Df F value    Pr(>F)
(Intercept) 0.00079  1  0.0382    0.8459
A           0.04765  2  1.1463    0.3275
X           0.48042  1 23.1136 1.983e-05 ***
A:X         0.01416  2  0.3407    0.7133
Residuals   0.87298 42
```

(b) Regressing on all four dummy variables and X we get $R^2 = .4406$:

```
> summary(lm(data = dat, Y ~ X+X1+X2+X3+X4),type=3)

Call:
lm(formula = Y ~ X + X1 + X2 + X3 + X4, data = dat)

Residuals:
      Min        1Q    Median        3Q       Max
-0.306634 -0.087095  0.000237  0.071985  0.304853

Coefficients:
             Estimate Std. Error t value Pr(>|t|)
(Intercept)  0.171282   0.153311   1.117   0.2703
X            0.005811   0.002406   2.415   0.0202 *
X1          -0.240689   0.240925  -0.999   0.3235
X2          -0.330506   0.226868  -1.457   0.1526
X3           0.002048   0.003909   0.524   0.6030
X4           0.002765   0.003453   0.801   0.4277
---
Signif. codes:  0 ‘***’ 0.001 ‘**’ 0.01 ‘*’ 0.05 ‘.’ 0.1 ‘ ’ 1

Residual standard error: 0.1442 on 42 degrees of freedom
Multiple R-squared:  0.4406,     Adjusted R-squared:  0.374
F-statistic: 6.615 on 5 and 42 DF,  p-value: 0.0001268
```

Regressing on the just X, X_1 and X_2, we get $R^2 = .4315$:

```
> summary(lm(data = dat, Y ~ X+X1+X2),type=3)

Call:
lm(formula = Y ~ X + X1 + X2, data = dat)

Residuals:
     Min       1Q   Median       3Q      Max
-0.30906 -0.09016 -0.00258  0.08140  0.28612

Coefficients:
             Estimate Std. Error t value Pr(>|t|)
(Intercept)  0.077687   0.098462   0.789  0.43434
X            0.007322   0.001483   4.938 1.18e-05 ***
X1          -0.115302   0.050368  -2.289  0.02692 *
X2          -0.154225   0.050551  -3.051  0.00386 **
---
Signif. codes:  0 ‘***’ 0.001 ‘**’ 0.01 ‘*’ 0.05 ‘.’ 0.1 ‘ ’ 1

Residual standard error: 0.142 on 44 degrees of freedom
Multiple R-squared:  0.4315,     Adjusted R-squared:  0.3927
F-statistic: 11.13 on 3 and 44 DF,  p-value: 1.462e-05
```

So $F = \dfrac{(.4406 - .4315)SS_Y / 2}{(1 - .4406)SS_Y / 42} = .342$ which is within rounding error of the .340 we got for the AX interaction in (a).

Chapter 24

24.1 (a)
```
> summary(aov(data = dat, Y ~ P))
            Df Sum Sq Mean Sq F value Pr(>F)
P            2    806   403.1   2.221  0.124
Residuals   33   5989   181.5
```

The effect of P is nonsignificant.

(b)
```
> Anova(lm(data = dat, Y ~ P + X),type=3) #ANCOVA
Anova Table (Type III tests)

Response: Y
            Sum Sq Df F value    Pr(>F)
(Intercept) 2116.2  1 31.9596 2.968e-06 ***
P            539.5  2  4.0739   0.02653 *
X           3869.7  1 58.4431 1.034e-08 ***
Residuals   2118.8 32
```

Now, the effect of P is significant.

(c)
```
> Anova(lm(data = dat,Y ~ P * X),type=3) #test regression slope homogeneity
Anova Table (Type III tests)

Response: Y
             Sum Sq Df F value    Pr(>F)
(Intercept) 1006.54  1 15.4347 0.0004643 ***
P             75.46  2  0.5786 0.5668236
X            816.01  1 12.5130 0.0013371 **
P:X          162.46  2  1.2456 0.3022268
Residuals   1956.38 30
```

The interaction term is not significant, so we do not reject the hypothesis of homogeneity of slopes.

24.3 (a) No, it is not appropriate to use ANCOVA here. We have a nonequivalent-groups design because the workers for whom we have satisfaction scores have not been randomly assigned to the four *departments*. Moreover, an ANOVA with X as the dependent variable yields a significant effect of *department*, $F(3, 28) = 5.602$, $p = .004$.

(b) No, it is not appropriate to use ANCOVA here. The data violate the assumption of homogeneity of slope. The test of heterogeneity of slope indicates that there is a significant interaction between the covariate X and the factor A, $F(2, 24) = 7.137$, $p = .004$

24.5 Regressing Y on X,

```
> anova(lm(data=dat,Y~X))
Analysis of Variance Table

Response: Y
          Df Sum Sq Mean Sq F value    Pr(>F)
X          1 610.23  610.23  42.689 6.887e-06 ***
Residuals 16 228.71   14.29
```

Regressing Y on X, and the dummy variables coding for A,

```
> anova(lm(data = dat, Y ~ X+X1+X2))
Analysis of Variance Table

Response: Y
          Df Sum Sq Mean Sq F value    Pr(>F)
X          1 610.23  610.23 186.159 1.771e-09 ***
X1         1 140.77  140.77  42.945 1.284e-05 ***
X2         1  42.05   42.05  12.828 0.003007 **
Residuals 14  45.89    3.28
```

From the first analysis, we see that $R^2_{Y.X}SS_Y = 610.23$. And from the second analysis, we have $R^2_{Y.X,A}SS_Y = 610.23 + 140.77 + 42.05 = 793.05$ so that

$$SS_{A(adj)} = (R^2_{Y.X,A} - R^2_{Y.X})SS_Y = 793.05 - 610.23 = 182.82$$

And

$$SS_{S/A(adj)} = \left(1 - R^2_{Y.X,A}\right)SS_Y = 45.89$$

Thus, the ANCOVA test statistic is

$$F(2,14) = \frac{MS_{A(adj)}}{MS_{S/A(adj)}} = \frac{SS_{A(adj)} / (a-1)}{SS_{S/A(adj)} / (N-a-1)} = 27.886$$

This is the same as the F we calculated in 24.2(d).

Chapter 25

25.3 $f^2 = \dfrac{\Delta R^2}{1-R^2} = \dfrac{.020}{1-.044} = .021$ = because the regression of depression (*SqrtBDIplus1*) on occupational and household physical activity for women yields $R^2 = .024$. Adding leisure-time physical activity increases R^2 to .044. Cohen's guidelines are .02, .15. and .35 for small, medium, and large effects, so this effect size is medium to medium-large.

25.5 It looks like a higher fraction of the people *employed* full time (status = 1) are men, whereas more of the part-time employees are female.

```
# A tibble: 4 x 3
# Groups:   sex [2]
    sex employed `n()`
  <dbl>    <dbl> <int>
1     0        1   252
2     0        2    33
3     1        1   184
4     1        2    59
```

We could do a chi-squared test of independence, finding that employment status and *sex* are not independent:

```
> chisq_test(dat$employed,dat$sex)
# A tibble: 1 x 6
      n statistic        p    df method          p.signif
* <int>     <dbl>    <dbl> <int> <chr>           <chr>
1   528      13.8 0.000199     1 Chi-square test ***
```

Or, we can use a correlation. Testing the correlation does not usually test independence, but it does when both variables are dichotomous.

```
> cor.test(dat$employed,dat$sex)

        Pearson's product-moment correlation

data:  dat$employed and dat$sex
t = 3.8819, df = 526, p-value = 0.0001168
alternative hypothesis: true correlation is not equal to 0
95 percent confidence interval:
 0.08273337 0.24867745
sample estimates:
     cor
0.166887
```

25.7 Our crude measure of education is a significant predictor of depression score for both men and women. As can be seen in Table 25.5, when education is included as a predictor along with the three measures of physical activity, education remains a significant predictor of depression score but the activity measures are not (although leisure activity does not miss by much). Clearly, level of education is related to scores on the depression scale.

References

Achen, C. H. (1977). Measuring representation: Perils of the correlation coefficient. *American Journal of Political Science*, *41*, 805–815.
Achen, C. H. (1982). *Interpreting and using regression*. Beverly Hills, CA: Sage.
Aguinis, H., Beaty, J. C., Boik, R. J., & Pierce, C. A. (2005). Effect size and power in assessing the moderating effects of categorical variables using multiple regression: A 30-year review. *Journal of Applied Psychology*, *90*, 94–107.
Aiken, L. S., & West, S. G. (1991). *Multiple regression: Testing and interpreting interactions*. Newbury Park, CA: Sage.
Akritas, M. G. (1990). The rank transform method in some two-factor designs. *Journal of the American Statistical Association*, *85*, 73–78.
Alexander, R. A., & Govern, D. M. (1994). A new and simpler approach to ANOVA under variance heterogeneity. *Journal of Educational Statistics*, *19*, 91–101.
Algina, J., Keselman, H. J., & Penfield, R. D. (2005a). An alternative to Cohen's standardized mean difference effect size. *Psychological Methods*, *10*, 317–328.
Algina, J., Keselman, H. J., & Penfield, R. D. (2006). Confidence interval coverage for Cohen's effect size statistic. *Educational and Psychological Measurement*, *66*, 945–960.
Allison, P. D. (2002). *Missing data*. Newbury Park, CA: Sage.
American Psychological Association (2020). *Publication manual of the American Psychological Association* (7th ed.). https://doi.org/10.1037/0000165-000
Anderson, L. R., & Ager, J. W. (1978). Analysis of variance in small group research. *Personality and Social Psychology Bulletin*, *4*, 341–345.
Anscombe, F. J. (1973). Graphs in statistical analysis. *American Statistician*, *27*, 17–21.
Atiqullah, M. (1964). The robustness of the covariance analysis of a one-way classification. *Biometrika*, *51*, 365–373.
Baayen, R. H., Davidson, D. J., & Bates, D. M. (2008). Mixed-effects modeling with crossed random effects for participants and items. *Journal of Memory and Language*, *59*, 390–412.
Balanda, K. P., & MacGillivray, H. L. (1988). Kurtosis: A critical review. *American Statistician*, *42*, 111–119.
Baron, R. M., & Kenny, D. A. (1986). The moderator-mediator variable distinction in social psychological research: Conceptual, strategic, and statistical considerations. *Journal of Personality and Social Psychology*, *51*, 1173–1182.
Bartlett, M. S. (1937). Properties of sufficiency and statistical tests. *Proceedings of the Royal Society*, *160*, 268–282.
Beck, A. T., & Steer, R. A. (1987). *Beck depression inventory: Manual*. San Antonio, TX: Psychological Corporation and Harcourt Brace Jovanovich.
Belsley, D. A., Kuh, E., & Welsch, R. E. (1980). *Regression diagnostics*. New York: Wiley.
Benjamini, Y., & Hochberg, Y. (1995). Controlling the false discovery rate: A practical and powerful approach to multiple testing. *Journal of the Royal Statistical Society, Series B*, *57*, 289–300.
Berk, R. A. (2004). *Regression analysis: A constructive critique*. Thousand Oaks, CA: Sage.
Bevan, M. F., Denton, J. Q., & Myers, J. L. (1974). The robustness of the *F* test to violations of continuity and form of treatment populations. *British Journal of Mathematical and Statistical Psychology*, *27*, 199–204.

Blair, R. C., & Higgins, J. J. (1985). A comparison of the paired samples *t* test to that of Wilcoxon's signed-rank test under various population shapes. *Psychological Bulletin, 97*, 119–128.

Bless, H., Bohner, G., Schwarz, N., & Strack, F. (1990). Mood and persuasion: A cognitive response analysis. *Personality and Social Psychology Bulletin, 16*, 331–345.

Bloom, B. S. (1964). *Stability and change in human characteristics.* New York: Wiley.

Boik, R. J. (1981). *A priori* tests in repeated-measures designs: Effects of nonsphericity. *Psychometrika, 46*, 241–255.

Boneau, C. A. (1962). A comparison of the power of the Uand *t* tests. *Psychological Review, 69*, 246–256.

Box, G. E. P. (1953). Nonnormality and tests on variances. *Biometrika, 40*, 318–335.

Box, G. E. P. (1954). Some theorems on quadratic forms in the study of analysis of variance problems: Effect of inequality of variance in the one-way classification. *Annals of Mathematical Statistics, 25*, 290–302.

Box, G. E. P. (1979). Robustness in the strategy of scientific model building. In R. L. Launer & G. N. Wilkinson (Eds.), *Robustness in statistics* (pp. 201–236). New York: Academic Press.

Bradley, J. V. (1968). *Distribution-free statistical tests.* Englewood Cliffs, NJ: Prentice-Hall.

Brown, M. B., & Forsythe, A. B. (1974a). Robust tests for the equality of variances. *Journal of the American Statistical Association, 69*, 364–367.

Brown, M. B., & Forsythe, A. B. (1974b). The ANOVA and multiple comparisons for data with heterogeneous variances. *Biometrics, 30*, 719–724.

Bryant, J. L., & Paulson, A. S. (1976). An extension of Tukey's method of multiple comparisons to experimental designs with random concomitant variables. *Biometrika, 63*, 631–638.

Busemeyer, J. R., & Jones, L. E. (1983). Analysis of multiplicative combination rules when the causal variables are measured with error. *Psychological Bulletin, 93*, 549–562.

Campbell, D. T., & Kenny, D. A. (1999). *A primer on regression artifacts.* New York: Guilford.

Campbell, D. T., & Stanley, J. C. (1963). *Experimental and quasi-experimental designs for research.* Boston: Houghton-Mifflin.

Carlson, J. E., & Timm, N. H. (1974). Analysis of nonorthogonal fixed-effect designs. *Psychological Bulletin, 81*, 563–570.

Clark, H. H. (1973). The language-as-fixed-effect fallacy: A critique of language statistics in psychological research. *Journal of Verbal Learning and Verbal Behavior, 12*, 335–359.

Cleveland, W. S. (1979). Robust locally weighted regression and smoothing scatterplots. *Journal of the American Statistical Association, 78*, 158–161.

Cleveland, W. S., Devlin, S. J., & Grosse, E. H. (1988). Regression by local fitting: Methods, properties, and computational algorithms. *Journal of Econometrics, 37*, 87–114.

Clinch, J. J., & Keselman, H. J. (1982). Parametric alternatives to the analysis of variance. *Journal of Educational Statistics, 7*, 207–214.

Cobb, J. A., & Hops, H. (1973). Effects of academic survival skill training on low achieving first graders. *Journal of Educational Research, 67*, 108–113.

Cochran, W. G. (1941). The distribution of the largest of a set of estimated variances as a fraction of their total. *Eugenics, 11*, 47–52.

Cochran, W. G. (1950). The comparison of percentages in matched samples. *Biometrika, 37*, 256–266.

Cochran, W. G., & Cox, G. M. (1957). *Experimental designs* (2nd ed.). New York: Wiley.

Cohen, J. (1962). The statistical power of abnormal-social psychological research. *Journal of Abnormal and Social Psychology, 65*, 145–153.

Cohen, J. (1973). Eta-squared and partial eta-squared in fixed factor ANOVA designs. *Educational and Psychological Measurement, 33*, 107–112.

Cohen, J. (1977). *Statistical power analysis for the behavioral sciences* (rev. ed.). New York: Academic Press.

Cohen, J. (1978). Partialled products are interactions and partialled powers are curve components. *Psychological Bulletin, 85*, 858–866.

Cohen, J. (1988). *Statistical power analysis for the behavioral sciences* (3rd ed.). New York: Academic Press.

Cohen, J. (1990). Things I have learned (so far). *American Psychologist, 45*, 1304–1312.

Cohen, J., Cohen, P., West, S. G., & Aiken, L. S. (2003). *Applied multiple regression/correlation analysis for the behavioral sciences* (3rd ed.). Mahwah, NJ: Lawrence Erlbaum Associates, Inc.

Conover, W. J., & Iman, R. L. (1981). Rank transformations as a bridge between parametric and nonparametric statistics. *American Statistician, 35*, 124–129.

Cook, R. D. (1977). Detection of influential observations in linear regression. *Technometrics, 19*, 15–18.

Cook, R. D., & Weisberg, S. (1982). *Residuals and influence in regression.* New York: Chapman & Hall.

Cook, R. D., & Weisberg, S. (1999). *Applied regression including computing and graphics.* New York: Wiley.

Coombs, W. T., Algina, J., & Oltman, D. O. (1996). Univariate and multivariate omnibus hypothesis tests selected to control Type I error rates when population variances are not necessarily equal. *Review of Educational Research, 66*, 137–179.

Cramer, E. M., & Appelbaum, M. I. (1980). Nonorthogonal analysis of variance – once again. *Psychological Bulletin, 87*, 51–57.

Crespi, L. P. (1944). Amount of reinforcement and level of performance. *Psychological Review, 51*, 341–357.

Cribbie, R. A., Fiksenbaum, L., Keselman, H. J., & Wilcox, R. R. (2012). Effect of non-normality on test statistics for one-way independent groups designs. *British Journal of Mathematical and Statistical Psychology, 65*, 56–73. http://doi.org/10.1111/j.2044-8317.2011.02014.x

Cumming, G. (2013). Cohen's *d* needs to be readily interpretable: Comment on Shieh (2013). *Behavioral Research, 45*, 968–971. http://doi.org/10.3758/s13428-013-0392-4

Cumming, G., & Finch, S. (2001). A primer on the understanding, use, and calculation of confidence intervals that are based on central and noncentral distributions. *Educational and Psychological Measurement, 61*, 532–574.

Dalton, S., & Overall, J. E. (1977). Nonrandom assignment in ANCOVA: The alternative ranks design. *Journal of Experimental Education, 46*, 58–62.

Dar, R., Serlin, R. C., & Omer, H. (1994). Misuse of statistical tests in three decades of psychotherapy research. *Journal of Consulting and Clinical Psychology, 62*, 75–82.

Davenport, J. M., & Webster, J. T. (1973). A comparison of some approximate *F* tests. *Technometrics, 15*, 779–789.

Dawes, R. M. (1971). A case study of graduate admissions: Application of three principles of human decision making. *American Psychologist, 26*, 180–188.

DeCarlo, L. T. (1997). On the meaning and uses of kurtosis. *Psychological Methods, 2*, 292–307.

Delacre, M., Lakens, D., & Leys, C. (2017). Why psychologists should by default use Welch's *t*-test instead of Student's *t*-test. *International Review of Social Psychology, 30*, 92–101. https://doi.org/10.5334/irsp.82

Delacre, M., Leys, C., Mora, Y. L., & Lakens, D. (2019). Taking parametric assumptions seriously: Arguments for the use of Welch's *F*-test instead of the classical *F*-test in one-way ANOVA. *International Review of Social Psychology, 32*, 1–12. http://doi.org/10/5334/irsp.198

De Rooij, M., & Weeda, W. (2020). Cross-validation: A method every psychologist should know. *Advances in Methods and Practices in Psychological Science, 3*, 248–263. http://doi.org/10.1177/2515245919898466

DeShon, R. P., & Alexander, R. A. (1996). Alternative procedures for testing regression slope homogeneity when group error variances are unequal. *Psychological Methods, 1*, 261–277.

Donaldson, T. S. (1968). Robustness of the *F* test to errors of both kinds and the correlation between the numerator and denominator of the *F* ratio. *Journal of the American Statistical Association, 63*, 660–676.

Downing, S. M., & Haladnya, T. M. (2006). *Handbook of test development.* Mahwah, NJ: Lawrence Erlbaum Associates, Inc.

Draper, N. R., & Smith, H. (1998). *Applied regression analysis* (3rd ed.). New York: Wiley.

Duncan, D. B. (1955). Multiple range and multiple *F* tests. *Biometrics, 11*, 1–42.

Dunn, O. J. (1961). Multiple comparisons among means. *Journal of the American Statistical Association, 56*, 52–64.

Dunnett, C. W. (1955). A multiple comparison procedure for combining several treatments with a control. *Journal of the American Statistical Association, 50*, 1096–1121.

Dunnett, C. W. (1964). New tables for multiple comparisons with a control. *Biometrics, 20*, 482–491.

Dunnett, C. W. (1980). Pairwise multiple comparisons in the unequal variance case. *Journal of the American Statistical Association, 75*, 796–800.

Edgell, S. E., & Noon, S. N. (1984). Effect of the violation of normality on the *t* test of the correlation coefficient. *Psychological Bulletin, 95*, 576–583.

Efron, B. (1988). Bootstrap confidence intervals: Good or bad? *Psychological Bulletin, 104*, 293–296.

Efron, B., & Diaconis, B. (1983). Computer-intensive methods in statistics. *Scientific American, 248*, 115–130.

Faul, F., Erdfelder, E., Buchner, A., & Lang, A.-G. (2009). Statistical power analyses using G*Power 3.1: Tests for correlation and regression analyses. *Behavior Research Methods, 41*, 1149–1160.

Faul, F., Erdfelder, E., Lang, A. G., & Buchner, A. (2007). G*Power 3: A flexible statistical power analysis program for the social, behavioral, and biomedical sciences. *Behavior Research Methods, 39*, 175–191.

Feldt, L. S. (1958). A comparison of the precision of three experimental designs employing a concomitant variable. *Psychometrika, 23*, 335–353.

Fenz, W., & Epstein, S. (1967). Gradients of physiological arousal of experienced and novice parachutists as a function of an approaching jump. *Psychosomatic Medicine, 29*, 33–51.

Field, A. P., & Wilcox, R. R. (2017). Robust statistical methods: A primer for clinical psychology and experimental psychopathology researchers. *Behaviour Research and Therapy, 98*, 19–38. http://doi.org/10.1016/j.brat.2017.05.013

Fisher, R. A. (1935). *The design of experiments*. Edinburgh, Scotland: Oliver & Boyd.

Fitts, D. A. (2020). Commentary on "A review of effect sizes and their confidence intervals, Part I: The Cohen's *d* family": The degrees of freedom for paired samples designs. *The Quantitative Methods for Psychology, 16*, 281–294. http://doi.org/10.20982/tqmp.16.4.p281

Fitzsimons, G. J. (2008). Editorial: Death to dichotomizing. *Journal of Consumer Research, 35*, 5–8.

Forster, K. I., & Dickinson, R. G. (1976). More on the language-as-fixed-effect fallacy: Monte Carlo estimates of error rates for F_1, F_2, F', and *min* F'. *Journal of Verbal Learning and Verbal Behavior, 15*, 135–142.

Fox, J. (1997). *Applied regression analysis, linear models, and related methods*. Thousand Oaks, CA: Sage.

Fredrickson, B. L., & Kahneman, D. (1993). Duration neglect in retrospective evaluations of affective episodes. *Journal of Personality and Social Psychology, 65*, 45–55.

Friedenreich, C. M., Courneya, K. S., & Bryant, H. E. (1998). The lifetime total physical activity questionnaire: Development and reliability. *Medicine and Science in Sports and Exercise, 30*, 266–274.

Friedman, M. (1937). The use of ranks to avoid the assumption of normality implicit in the analysis of variance. *Journal of the American Statistical Association, 32*, 675–701.

Funder, D. C., & Ozer, D. J. (2019). Evaluating effect size in psychological research: Sense and nonsense. *Advances in Methods and Practices in Psychological Science, 2*, 156–168. http://doi.org/10.1177/2515245919847202

Games, P. A. (1973). Type IV errors revisited. *Psychological Bulletin, 80*, 304–307.

Games, P. A., & Howell, J. F. (1976). Pairwise multiple comparison procedures with unequal n's and/or variances: A Monte Carlo study. *Journal of Educational Statistics, 1*, 113–125.

Games, P. A., Keselman, H. J., & Clinch, J. J. (1979). Tests for homogeneity of variance in factorial designs. *Psychological Bulletin, 86*, 978–984.

Games, P. A., Keselman, H. J., & Rogan, J. C. (1981). Simultaneous pairwise multiple comparison procedures when sample sizes are unequal. *Psychological Bulletin, 90*, 594–598.

Ganzach, Y. (1997). Misleading interaction and curvilinear terms. *Psychological Methods, 2*, 235–247.

Gelman, A., & Loken, E. (2014). The statistical crisis in science. *American Scientist, 102*, 460–465.

Gibbons, J. D. (1993). *Nonparametric statistics: An introduction*. Newbury, CA: Sage.

Glass, G. V. (1976). Primary, secondary, and meta-analysis of research. *Educational Researcher, 5*, 3–8.

Goldstein, H. (1995). *Multilevel statistical models*. London: Edward Arnold.

Gomila, R. (2021). Logistic or linear? Estimating causal effects of experimental treatments on binary outcomes using regression analysis. *Journal of Experimental Psychology: General, 150*, 700–709. http://doi.org/10.1037/xge0000920

Gomila, R., & Clark, C. S. (2022). Missing data in experiments: Challenges and solutions. *Psychological Methods, 27*, 143–155. http://doi.org/10.1037/met0000361

Goodman, L. A., & Kruskal, W. H. (1954). Measures of association for cross classifications. *Journal of the American Statistical Association, 49*, 732–764. http://doi.org/10.2307/2281536

Goulet-Pelletier, J.-C., & Cousineau, D. (2018). A review of effect sizes and their confidence intervals, Part I: The Cohen's *d* family. *The Quantitative Methods for Psychology, 14*, 242–265. http://doi.org/10.20982/tqmp.14.4.p242

Greenhouse, S. W., & Geisser, S. (1959). On methods in the analysis of profile data. *Psychometrika, 55*, 431–433.

Greenwald, A. G. (1993). Consequences of prejudice against the null hypothesis. In G. Keren & C. Lewis (Eds.), *A handbook for data analysis in the behavioral sciences: Methodological issues* (pp. 419–448). Hillsdale, NJ: Lawrence Erlbaum Associates, Inc.

Grissom, R. J. (2000). Heterogeneity of variance in clinical data. *Journal of Consulting and Clinical Psychology, 68*, 155–165.

Grissom, R. J., & Kim, J. J. (2001). Review of assumptions and problems in the appropriate conceptualization of effect size. *Psychological Methods, 6*, 135–146.

Harlow, L. L. (1997). Significance testing introduction and overview. In L. L. Harlow, S. A. Mulaik, & J. H. Steiger (Eds.), *What if there were no significance tests?* (pp. 1–17). Mahwah, NJ: Lawrence Erlbaum Associates, Inc.

Harris, R. J. (2001). *A primer of multivariate statistics* (3rd ed.). Mahwah, NJ: Lawrence Erlbaum Associates, Inc.

Harter, H. L., Clemm, D. S., & Guthrie, E. H. (1959). *The probability integrals of the range and of the Studentized range* (WADC Tech. Rep. 58–484). Wright-Patterson Air Force Base. Ohio: Wright Air Development Center.

Hartley, H. O. (1950). The maximum *F* -ratio as a short-cut test for heterogeneity of variance. *Biometrika, 37*, 308–312.

Havlicek, L. L., & Peterson, N. L. (1977). Effect of the violations of assumptions upon significance levels for the Pearsonr. *Psychological Bulletin, 84*, 373–377.

Hays, W. L. (1981). *Statistics* (3rd ed.). New York: Holt, Rinehart & Winston.

Hays, W. L. (1994). *Statistics* (5th ed.). New York: Holt, Rinehart & Winston.

Hayter, A. J. (1986). The maximum familywise error rate of Fisher's least significant difference test. *Journal of the American Statistical Association, 81*, 1000–1004.

Hedges, L. V., & Olkin, I. (1985). *Statistical methods for meta-analysis*. New York: Academic Press.

Herzberg, P. A. (1969). The parameters of cross validation. *Psychometrika* (Monograph supplement, No. 16).

Hill, M., & Dixon, W. J. (1982). Robustness in real life: A study of clinical laboratory data. *Biometrics, 38*, 377–396.

Hillebrand, D. K. (1986). *Statistical thinking for behavioral scientists*. Boston: Duxbury.

Hoaglin, D. C., Mosteller, F., & Tukey, J. W. (1983). *Understanding robust and exploratory data analysis*. New York: Wiley.

Hoaglin, D. C., Mosteller, F., & Tukey, J. W. (1985). *Exploring data tables, trends and shapes*. New York: Wiley.

Hoaglin, D. C., Mosteller, F., & Tukey, J. W. (1991). *Fundamentals of exploratory analysis of variance*. New York: Wiley.

Hoaglin, D. C., & Welsch, R. (1978). The hat matrix in regression and ANOVA. *American Statistician, 32*, 17–22.

Hochberg, Y. (1988). A sharper Bonferroni procedure for multiple tests of significance. *Biometrika, 75*, 800–803.

Hocking, R. R. (1983). Developments in linear regression methodology: 1959–1982. *Technometrics, 25*, 219–245.

Hoenig, J. M., & Heisey, D. M. (2001). The abuse of power: The pervasive fallacy of power calculations in data analysis. *American Statistician, 55*, 19–24.

Hogg, R. V. (1974). Adaptive robust procedures: A partial review and some suggestions for future applications and theory. *Journal of the American Statistical Association, 69*, 909–927.

Hogg, R. V., Fisher, D. M., & Randles, R. K. (1975). A two-sample adaptive distribution-free test. *Journal of the American Statistical Association, 70*, 656–667.

Hollander, M., & Wolfe, D. A. (1999). *Nonparametric statistical methods* (2nd ed.). Hoboken, NJ: Wiley.

Holm, S. (1979). A simple sequentially rejective multiple test procedure. *Scandinavian Journal of Statistics, 6*, 65–70.

Hommel, G. (1988). A stepwise rejective multiple test procedure based on a modified Bonferroni test. *Biometrika, 75*, 383–386.
Hora, S. C., & Conover, W. J. (1984). The *F* statistic in the two-way layout with rank-score transformed data. *Journal of the American Statistical Association, 79*, 668–673.
Hora, S. C., & Iman, R. L. (1988). Asymptotic relative efficiencies of the rank-transformation procedure in randomized complete-block designs. *Journal of the American Statistical Association, 83*, 462–470.
Horst, A. M., Hill, A. P., & Gorman, K. B. (2020). *Palmerpenguins: Palmer Archipelago (Antarctica) penguin data*. R package version 0.1.0. Retrieved from https://allisonhorst.github.io/palmerpenguins/. http://doi.org/10.5281/zenodo.3960218
Hosmer, D. W., & Lemeshow, S. (2001). *Applied logistic regression* (2nd ed.). New York: Wiley.
Hox, J. (2002). *Multilevel analysis: Techniques and applications*. Mahwah, NJ: Lawrence Erlbaum Associates, Inc.
Hsu, T. C., & Feldt, L. S. (1969). The effect of limitations on the number of criterion score values on the significance of the *F* test. *American Educational Research Journal, 6*, 515–527.
Hudson, J. D., & Krutchkoff, R. C. (1968). A Monte Carlo investigation of the size and power of tests employing Satterthwaite's synthetic mean squares. *Biometrika, 55*, 431–433.
Huitema, B. E. (1980). *The analysis of covariance and alternatives*. New York: Wiley.
Hunka, S., & Leighton, J. (1997). Defining Johnson–Neyman regions of significance in the three-covariate ANCOVA using mathematica. *Journal of Educational and Behavioral Statistics, 22*, 361–387.
Hunter, J. E., & Schmidt, F. L. (1990). *Methods of meta-analysis: Correcting error and bias in research studies*. Newbury Park, CA: Sage.
Huynh, H. (1982). A comparison of four approaches to robust regression. *Psychological Bulletin, 92*, 505–512.
Huynh, H., & Feldt, L. S. (1970). Conditions under which mean square ratios in repeated measurements designs have exact F-distributions. *Journal of the American Statistical Association, 65*(332), 1582–1589. https://doi.org/10.1080/01621459.1970.10481187
Huynh, H., & Feldt, L. S. (1976). Estimation of the Box correction for degrees of freedom from sample data in randomized block and split-plot designs. *Journal of Educational Statistics, 1*, 69–82.
Hyndman, R. J., & Fan, Y. (1996). Sample quantiles in statistical packages. *American Statistician, 50*, 361–365. http://doi.org/10.2307/2684934
Iman, R. L., & Conover, W. J. (1981). Rank transformations as a bridge between parametric and nonparametric statistics. *American Statistician, 35*, 124–129.
Iman, R. L., Hora, S. C., & Conover, W. J. (1984). Comparison of asymptotically distribution-free procedures for the analysis of complete blocks. *Journal of the American Statistical Association, 79*, 674–685.
Irwin, J. R., & McClelland, G. H. (2001). Misleading heuristics and moderated regression models. *Journal of Marketing Research, 38*, 100–109.
ISSP (International Society of Sport Psychology). (1992). Physical activity and psychological benefits: A position statement. *The Sport Psychologist, 6*, 199–203.
Jaccard, J., Turrisi, R., & Wan, C. K. (1990). *Interaction effects in multiple regression*. Newbury Park, CA: Sage.
James, G. S. (1951). The comparison of several groups of observations when the ratios of the population variances are unknown. *Biometrika, 38*, 324–329.
James, G. S. (1954). Tests of linear hypotheses in univariate and multivariate analysis when the ratios of the population variances are unknown. *Biometrika, 41*, 19–43.
Janky, D. G. (2000). Sometimes pooling for analysis of variance hypothesis tests: A review and study of a split-plot model. *American Statistician, 54*, 269–279.
Johnson, P. O., & Neyman, J. (1936). Tests of certain linear hypotheses and their application to some educational problems. *Statistical Research Memoirs, 1*, 57–93.
Kelley, K. (2005). The effects of nonnormal distributions on confidence intervals around the standardized mean difference: Bootstrap and parametric confidence intervals. *Educational and Psychological Measurement, 65*, 51–69.
Kendall, M. (1938). A new measure of rank correlation. *Biometrika, 30*, 81–89. http://doi.org/10.1093/biomet/30.1-2.81

Kepner, J. L., & Robinson, D. H. (1988). Nonparametric methods for detecting treatment effects in repeated-measures designs. *Journal of the American Statistical Association, 83*, 456–461.

Keren, G. (1993). Between-or within-participants design: A methodological dilemma. In G. Keren & C. Lewis (Eds.), *A handbook for data analysis in the behavioral sciences: Methodological issues* (pp. 257–272). Hillsdale, NJ: Lawrence Erlbaum Associates, Inc.

Keselman, H. J., Othman, A. R., Wilcox, R. R., & Fradette, K. (2004). The new and improved two-sample *t* test. *Psychological Science, 15*, 47–51.

Keselman, H. J., Rogan, J. C., Mendoza, J. L., & Breen, L. J. (1980). Testing the validity conditions of repeated-measures *F* tests. *Psychological Bulletin, 87*, 479–481.

Keselman, H. J., Wilcox, R. R., Othman, A. R., & Fradette, K. (2002). Trimming, transforming statistics, and bootstrapping: Circumventing the biasing effects of heteroscedasticity and nonnormality. *Journal of Modern Applied Statistical Methods, 1*, 288–309.

Keuls, M. (1952). The use of the Studentized range in connection with an analysis of variance. *Euphytica, 1*, 112–122.

King, G. (1991). Truth is stranger than prediction, more questionable than inference. *American Journal of Political Science, 35*, 1047–1053.

Kirk, R. E. (1995). *Experimental design: Procedures for the behavioral sciences* (3rd ed.). Belmont, CA: Brooks/Cole.

Kirk, R. E. (1996). Practical significance: A concept whose time has come. *Educational and Psychological Measurement, 56*, 746–759.

Kraemer, H. C. (2005). A simple effect size indicator for two-group comparisons? A comment on *r* equivalent. *Psychological Methods, 10*, 413–419.

Kraemer, H. C., & Thiemann, S. (1987). *How many participants? Statistical power analysis in research*. Beverly Hills, CA: Sage.

Kramer, C. Y. (1956). Extension of multiple range tests to group means with unequal numbers of replications. *Biometrics, 12*, 307–310.

Kreft, I., & de Leeuw, J. (1998). *Introducing multilevel modeling*. London: Sage.

Kroes, A. D. A., & Finley, J. R. (2023). Demystifying omega squared: Practical guidance for effect size in common analysis of variance designs. *Psychological Methods*, Advance online publication. http://doi.org/10.1037/met0000581

Kruskal, W. H., & Wallis, W. A. (1952). Use of ranks in one-criterion variance analysis. *Journal of the American Statistical Association, 47*, 583–621.

Lakens, D. (2013). Calculating and reporting effect sizes to facilitate cumulative science: A practical primer for *t* tests and anovas. *Frontiers in Psychology, 4*, 863–875. http://doi.org/10.3389/fpsyg.2013.00863

Lee, W.-C., & Rodgers, J. L. (1998). Bootstrapping correlation coefficients using univariate and bivariate sampling. *Psychological Methods, 3*, 91–103.

Lehmann, E. L. (1975). *Nonparametrics*. San Francisco: Holden-Day.

Lenth, R. V. (2001). Some practical guidelines for effective sample-size determination. *American Statistician, 55*, 187–193.

Levene, H. (1960). Robust tests for equality of variances. In I. Olkin (Ed.), *Contributions to probability and statistics: Essays in honor of Harold Hotelling* (pp. 278–292). Stanford, CA: Stanford University Press.

Lin, L., Halgin, R. P., Well, A. D., & Ockene, I. (2008). The relationship between depression and occupational, household, and leisure-time physical activity. *Journal of Clinical and Sport Psychology, 2*, 95–107.

Lindquist, E. F. (1953). *Design and analysis of experiments in education and psychology*. Boston: Houghton-Mifflin.

Little, R. J. A., & Rubin, D. B. (1987). *Statistical analysis with missing data*. New York: Wiley.

Lix, L. M., & Keselman, H. J. (1998). To trim or not to trim: Tests of location equality under heteroscedasticity and nonnormality. *Educational and Psychological Measurement, 58*, 409–429.

Lix, L. M., Keselman, J. C., & Keselman, H. J. (1996). Consequences of assumption violations revisited: A quantitative review of alternatives to the one-way analysis of variance *F* test. *Review of Educational Research, 66*, 579–620.

Loftus, G. R., (1978). On interpretation of interactions. *Memory & Cognition*, 6, 312–319.

Lorch, R. F., & Myers, J. L. (1990). Regression analyses of repeated-measures data: A comparison of three different methods. *Journal of Experimental Psychology: Learning, Memory, and Cognition, 16*, 149–157.

Luke, D. A. (2004). *Multilevel modeling*. Thousand Oaks, CA: Sage.

Lunney, G. H. (1970). Using analysis of variance with a dichotomous variable: An empirical study. *Journal of Educational Measurement, 7*, 263–269.

MacCallum, R. C., & Mar, C. M. (1995). Distinguishing between moderator and quadratic effects in multiple regression. *Psychological Bulletin, 118*, 405–421.

MacKinnon, D. P., Lockwood, C. M., Hoffman, J. M., West, S. G., & Sheets, V. (2002). A comparison of models to test mediation and other intervening variable effects. *Psychological Methods, 7*, 83–104.

Marascuilo, L. A., & Levin, J. R. (1972). Type IV errors and interactions. *Psychological Bulletin, 78*, 368–374.

Marascuilo, L. A., & Levin, J. R. (1973). Type IV errors and games. *Psychological Bulletin, 80*, 308–309.

Martin, P., & Bateson, P. (2007). *Measuring behaviour: An introductory guide*. Cambridge: Cambridge University Press.

Masson, M. E. J., & Rotello, C. M. (2009). Sources of bias in the Goodman–Kruskal gamma coefficient measure of association: Implications for studies of metacognitive processes. *Journal of Experimental Psychology: Learning, Memory, and Cognition, 35*, 509–527.

Mauchly, J. W. (1940). Significance test for sphericity of a normal n-variate distribution. *Annals of Mathematical Statistics, 11*, 204–209.

Maxwell, S. E. (1980). Pairwise multiple comparisons in repeated-measures designs. *Journal of Educational Statistics, 5*, 269–287.

Maxwell, S. E. (2000). Sample size and multiple regression analysis. *Psychological Methods, 5*, 434–458.

Maxwell, S. E., & Bray, J. H. (1986). Robustness of the quasi *F* statistic to violations of sphericity. *Psychological Bulletin, 99*, 416–421.

Maxwell, S. E., Camp, C. J., & Arvey, R. D. (1981). Measures of strength of association. *Journal of Applied Psychology, 66*, 525–534.

Maxwell, S. E., & Delaney, H. D. (1993). Bivariate median splits and spurious statistical significance. *Psychological Bulletin, 113*, 181–190.

Maxwell, S. E., & Delaney, H. D. (2004). *Designing experiments and analyzing data: A model comparison perspective* (2nd ed.). Mahwah, NJ: Lawrence Erlbaum Associates, Inc.

Maxwell, S. E., Delaney, H. D., & Dill, C. A. (1984). Another look at ANCOVA versus blocking. *Psychological Bulletin, 95*, 136–147.

McGraw, L., & Wong, S. (1992). The common language effect size statistic. *Psychological Bulletin, 111*, 361–365.

Mead, R., Bancroft, T. A., & Han, C. (1975). Power of analysis of variance test procedures for incompletely specified fixed models. *Annals of Statistics, 3*, 797–808.

Menard, S. (2002). *Applied logistic regression* (2nd ed.). Thousand Oaks, CA: Sage.

Meng, X.-L., Rosenthal, R., & Rubin, D. B. (1992). Comparing correlated correlation coefficients. *Psychological Bulletin, 111*, 172–175.

Merriam, P. A., Ockene, I. S., Hebert, J. R., Rosal, M. C., & Matthews, C. E. (1999). Seasonal variation of blood cholesterol levels. *Journal of Biological Rhythms, 14*, 330–330.

Micceri, T. (1989). The unicorn, the normal curve, and other improbable creatures. *Psychological Bulletin, 105*, 156–166.

Miller, R. G. Jr. (1981). *Simultaneous statistical inference* (2nd ed.). New York: Springer-Verlag.

Mittelhammer, R. C., Judge, G. C., & Miller, D. J. (2000). *Econometric foundations*. Cambridge: Cambridge University Press.

Morrison, D. F. (2004). *Multivariate statistical methods*. Pacific Grove, CA: Duxbury Press.

Morrow, L. M., & Young, J. (1997). A family literacy program connecting school and home: Effects on attitude, motivation, and literacy achievement. *Journal of Educational Psychology, 89*, 736–742.

Murray, J. E., Yong, E., & Rhodes, G. (2000). Revisiting the perception of upside-down faces. *Psychological Science, 11*, 492–496.

Myers, J. L. (1979). *Fundamentals of experimental design* (3rd ed.). Boston: Allyn & Bacon.

Myers, J. L., DiCecco, J. V., & Lorch, R. F. (1981). Group dynamics and individual performances: Pseudogroup and quasi-*F* analyses. *Journal of Personality and Social Psychology, 40*, 86–98.

Myers, J. L., DiCecco, J. V., White, J. B., & Borden, V. M. (1982). Repeated measurements on dichotomous variables: *Q* and *F* tests. *Psychological Bulletin, 92*, 517–525.

Myers, J. L., Hansen, R. S., Robson, R. R., & McCann, J. (1983). The role of explanation in learning elementary probability. *Journal of Educational Psychology, 75*, 374–381.

Myers, J. L., Pezdek, K., & Coulson, D. (1973). Effects of prose organization on free recall. *Journal of Educational Psychology, 65*, 313–320.

Myers, J. L., & Well, A. D. (1995). *Research design and statistical analysis*. Hillsdale, NJ: Lawrence Erlbaum Associates, Inc.

Myers, J. L., & Well, A. D. (2003). *Research design and statistical analysis* (2nd ed.). Mahwah, NJ: Lawrence Erlbaum Associates, Inc.

Namboodiri, N. K. (1972). Experimental designs in which each participant is used repeatedly. *Psychological Bulletin, 77*, 54–64.

Neter, J., Kutner, M. H., Nachtscheim, C. J., & Wasserman, W. (1996). *Applied linear statistical models* (4th ed.). Boston: WCB McGraw-Hill.

Newman, D. (1939). The distribution of range in samples from a normal population, expressed in terms of independent estimate of a standard deviation. *Biometrika, 31*, 20–30.

Nordstokke, D. W., & Zumbo, B. D. (2007). A cautionary tale about Levene's Tests for equal variances. *Journal of Educational Research & Policy Studies*, 7, 1–14.

Nunnally, J., & Berstein, I. (1994). *Psychometric theory* (3rd ed.). New York: McGraw-Hill.

Ockene, I. S., Chiriboga, D. E., Stanek, E. J., Harmatz, M. G., Nicolosi, R., Saperia, G., . . ., Hebert, J. R. (2004). Seasonal variation in serum cholesterol levels. *Archives of Internal Medicine, 164*, 863–870.

Okada, K. (2013). Is omega squared less biased? A comparison of three major effect size indices in one-way ANOVA. *Behaviormetrika, 40*, 129–147.

Okada, K., & Hoshino, T. (2017). Researchers' choice of the number and range of levels in experiments affects the resultant variance-accounted-for effect size. *Psychonomic Bulletin & Review, 24*, 607–616. http://doi.org/10.3758/s13423-016-1128-0

Olejnik, S., & Algina, J. (2000). Measures of effect size for comparative studies: Applications, interpretations, and limitations. *Contemporary Educational Psychology, 25*, 241–286.

Olejnik, S., & Algina, J. (2003). Generalized eta and omega squared statistics: Measures of effect size for some common research designs. *Psychological Methods, 8*, 434–447.

Olkin, I., & Finn, J. (1990). Testing correlated correlations. *Psychological Bulletin, 108*, 330–333.

Olkin, I., & Finn, J. (1995). Correlations redux. *Psychological Bulletin, 118*, 155–164.

Oshima, T. C., & Algina, J. (1992). Type I error rates for James' second order test and Wilcox's H_m test under heteroscedasticity and non-normality. *British Journal of Mathematical and Statistical Psychology, 45*, 225–263.

Overall, J. E., Lee, D. M., & Hornick, C. W. (1981). Comparison of two strategies for analysis of variance in nonorthogonal designs. *Psychological Bulletin, 90*, 367–375.

Overall, J. E., & Spiegel, D. K. (1969). Concerning least squares analysis of experimental data. *Psychological Bulletin, 72*, 311–322.

Overton, R. C. (2001). Moderated multiple regression for interactions involving categorical variables: A statistical control for heterogeneous variance across two groups. *Psychological Methods, 6*, 218–233.

Pearson, E. S., & Hartley, H. (1954). *Biometrika tables for statisticians*. London: Cambridge University Press.

Pedhazur, E. J. (1997). *Multiple regression in behavioral research: Explanation and prediction*. Fort Worth, TX: Harcourt Brace.

Peritz, E. (1970). *A note on multiple comparisons*. Unpublished manuscript, Hebrew University, Jerusalem, Israel.

Perlmutter, J., & Myers, J. L. (1973). A comparison of two procedures for testing multiple contrasts. *Psychological Bulletin, 79*, 181–184.

Piccinelli, M., & Wilkinson, G. (2000). Gender differences in depression: A critical review. *British Journal of Psychiatry, 177*, 486–492.

Pollatsek, A., & Well, A. D. (1995). On the use of counterbalanced designs in cognitive research: A suggestion for a better and more powerful analysis. *Journal of Experimental Psychology: Learning, Memory, and Cognition, 21*, 783–794.

Quené, H., & van den Bergh, H. (2004). On multi-level modeling of data from repeated-measures designs: A tutorial. *Speech Communication, 43*, 103–121.

Ragosa, D. (1980). Comparing nonparallel regression lines. *Psychological Bulletin, 88*, 307–321.

Ragosa, D. (1995). Myths and methods: "Myths about longitudinal research," plus supplemental questions. In M. Gottman (Ed.), *The analysis of change* (pp. 3–66). Mahwah, NJ: Lawrence Erlbaum Associates, Inc.

Räkkönen, K., Matthews, K. A., Flory, J. D., Owens, J. F., & Gump, B. B. (1999). Effects of optimism, pessimism, and trait anxiety on ambulatory blood pressure and mood during everyday life. *Journal of Personality and Social Psychology, 76*, 104–113.

Ramsey, P. H. (1978). Power differences between pairwise multiple comparisons. *Journal of the American Statistical Association, 73*, 479–485.

Ramsey, P. H. (1981). Power of univariate pairwise multiple comparison procedures. *Psychological Bulletin, 90*, 352–366.

Raudenbush, S. W., & Bryk, A. S. (2002). *Hierarchical linear models: Applications and data analysis methods*. Thousand Oaks, CA: Sage.

Razali, N., & Wah, Y. B. (2011). Power comparisons of shapiro–wilk, kolmogorov–smirnov, lilliefors and anderson–darling tests. *Journal of Statistical Modeling and Analytics*, 2, 21–33.

Rencher, A. C., & Pun, F. C. (1980). Inflation of R-squared in best subset regression. *Technometrics, 22*, 49–54.

Rogan, J. C., Keselman, H. J., & Mendoza, J. L. (1979). Analysis of repeated measurements. *British Journal of Mathematical and Statistical Psychology, 32*, 269–286.

Rom, D. M. (1990). A sequentially rejective test procedure based on a modified Bonferroni inequality. *Biometrika, 77*, 663–665.

Rosenberger, J. L., & Gasko, M. (1983). Comparing location estimators: Trimmed means, medians, and trimeans. In D. C. Hoaglin, F. Mosteller, & J. F. Tukey (Eds.), *Understanding robust and exploratory data analysis* (pp. 297–328). New York: Wiley.

Rosenthal, R. (1991). *Meta-analytic procedures for social research* (rev. ed.). Newbury Park, CA: Sage.

Rosenthal, R. (1995). Writing meta-analytic reviews. *Psychological Bulletin, 118*, 183–192.

Rosenthal, R., & Rubin, D. B. (1994). The counternull value of an effect size: A new statistic. *Psychological Science, 5*, 329–334.

Rosenthal, R., & Rubin, D. B. (2003). r equivalent: A simple effect size indicator. *Psychological Methods, 8*, 492–496.

Rouanet, H., & Lepine, D. (1970). Comparisons between treatments in a repeated-measurement design: ANOVA and multivariate methods. *British Journal of Mathematical and Statistical Psychology, 23*, 147–163.

Rousseeuw, J. R., & Leroy, A. M. (1987). *Robust regression and outlier detection*. New York: Wiley.

Roy, S. N., & Bose, R. C. (1953). Simultaneous confidence interval estimation. *Annals of Mathematical Statistics, 39*, 405–422.

Royer, J. M., Tronsky, L. M., & Chan, Y. (1999). Math-fact retrieval as the cognitive mechanism underlying gender differences in math test performance. *Contemporary Educational Psychology, 24*, 181–266.

Sandik, L., & Olsson, B. (1982). A nearly distribution-free test for comparing dispersion in paired samples. *Biometrika, 69*, 484–485.

Satterthwaite, F. E. (1946). An approximate distribution of variance components. *Biometrics Bulletin*, 2, 110–114.

Schafer, J. L. (1997). *Analysis of incomplete multivariate data*. London: Chapman & Hall.

Schafer, J. L. (1999). Multiple imputation: A primer. *Statistical Methods in Medical Research, 8*, 3–15.

Scheffé, H. (1959). *The analysis of variance*. New York: Wiley.

Schnorr, J. A., Lipkin, S. G., & Myers, J. L. (1966). Level of risk in probability learning: Within-and between-participants designs. *Journal of Experimental Psychology, 72*, 497–500.

Schrier, A. M. (1958). Comparison of two methods of investigating the effect of amount of reward on performance. *Journal of Comparative and Physiological Psychology, 51*, 725–731.

Seaman, M. A., Levin, J. R., & Serlin, R. C. (1991). New developments in pairwise multiple comparisons: Some powerful and practicable procedures. *Psychological Bulletin, 110*, 577–586.

Shaffer, J. P. (1979). Comparison of means: An Ftest followed by a modified multiple range procedure. *Journal of Educational Statistics, 4*, 14–23.

Shaffer, J. P. (1986). Modified sequentially rejective multiple test procedures. *Journal of the American Statistical Association, 81*, 826–831.

Shaffer, J. P. (1995). Multiple hypothesis testing. *Annual Review of Psychology, 46*, 561–584.

Shapiro, S. S., & Wilk, M. B. (1965). An analysis of variance test for normality (complete samples). *Biometrika, 52*, 591–611.

Shepperd, J. A. (1991). Cautions in assessing spurious "moderator effects." *Psychological Bulletin, 110*, 315–317.

Shieh, G. (2013). Confidence intervals and sample size calculations for the standardized mean difference effect size between two normal populations under heteroscedasticity. *Behavior Research Methods, 45*, 955–967. https://doi.org/10.3758/s13428-012-0228-7

Šidák, Z. (1967). Rectangular confidence regions for the means of multivariate normal distributions. *Journal of the American Statistical Association, 62*, 626–633.

Siegel, S., & Castellan, N. J. (1988). *Nonparametric statistics for the behavioral sciences* (2nd ed.). New York: McGraw-Hill.

Simmons, J., Nelson, L., & Simonsohn, U. (2011). False-positive psychology: Undisclosed flexibility in data collection and analysis allow presenting anything as significant. *Psychological Science, 22*, 1359–1366.

Simonsohn, U. (2018). Two lines: A valid alternative to the invalid testing of U-shaped relationships with quadratic regressions. *Advances in Methods and Practices in Psychological Science, 1*, 538–555. http://doi.org/10.1177/2515245918805755 Correction in: Simonsohn, U. (2019). Corrigendum: Two lines: A valid alternative to the invalid testing of U-shaped relationships with quadratic regressions. *Advances in Methods and Practices in Psychological Science*, 2, 410–411. http://doi.org/10.1177/2515245919894972

Steiger, J. H. (1980). Tests for comparing elements of a correlation matrix. *Psychological Bulletin, 87*, 245–251.

Steiger, J. H., & Fouladi, R. T. (1997). Noncentrality interval estimation and the evaluation of statistical models. In L. L. Harlow, L. L. Mulaik, & J. H. Steiger (Eds.), *What if there were no significance tests?* (pp. 221–257). Mahwah, NJ: Lawrence Erlbaum Associates, Inc.

Stevens, J. (1986). *Applied multivariate statistics for the social sciences*. Hillsdale, NJ: Lawrence Erlbaum Associates, Inc.

Stolberg, A. L. (2001). Understanding school- and community-based groups for children and adolescents. *PsycCRITIQUES, 46*, 55–56.

"Student" [Gosset, W. S.] (1927). Errors of routine analysis. *Biometrika, 19*, 151–164.

Tomarken, A. J., & Serlin, R. C. (1986). Comparison of ANOVA alternatives under variance heterogeneity and specific noncentrality structures. *Psychological Bulletin, 99*, 90–99.

Toothaker, L. E. (1993). *Multiple comparison procedures*. Newbury Park, CA: Sage.

Tukey, J. W. (1953). *The problem of multiple comparisons*. Unpublished manuscript, Princeton University.

Tukey, J. W. (1969). Analyzing data: Sanctification or detective work? *American Psychologist, 24*, 83–91.

Tukey, J. W. (1977). *Exploratory data analysis*. Reading, MA: Addison-Wesley.

Tukey, J. W. (1991). The philosophy of multiple comparisons. *Statistical Science, 6*, 100–116.

Tukey, J. W., & McLaughlin, D. H. (1963). Less vulnerable confidence and significance procedures for location based on a single sample: Trimming/Winsorization. I. *Sankhyā: Indian Journal of Statistics, Series A, 25*, 331–352.

Vargha, A., & Delaney, H. D. (1998). The Kruskal–Wallis test and stochastic homogeneity. *Journal of Educational and Behavioral Statistics, 23*, 170–192.

Velleman, P., & Welsch, R. (1981). Efficient computing of regression diagnostics. *American Statistician, 35*, 234–242.

Wagenmakers, E. J., Krypotos, A.-M., Criss, A. H., & Iverson, G. (2012). On the interpretation of removable interactions: A survey of the field 33 years after Loftus. *Memory & Cognition, 40*, 145–160. http://doi.org/10.3758/s13421-011-0158-0

Welch, B. L. (1938). The significance of the difference between two means when the population variances are unequal. *Biometrika, 25*, 350–362.

Welch, B. L. (1947). The generalization of Student's problem when several different population variances are involved. *Biometrika, 34*, 28–35.

Welch, B. L. (1951). On the comparison of several mean values: An alternative approach. *Biometrika, 38*, 330–336.

Welsch, R. E. (1977). Stepwise multiple comparison procedures. *Journal of the American Statistical Association, 72*, 566–575.

Wherry, R. J. (1931). A new formula for predicting the shrinkage of the coefficient of multiple correlation. *Annals of Mathematical Statistics, 2*, 440–457.

Wickham, H. (2016). *ggplot2: Elegant graphics for data analysis*. New York: Springer-Verlag.

Wickam, H., Francois, R., Henry, L., Muller, K., & Vaughan, D. (2023). *dplyr: A Grammar of Data Manipulation*. R package version 1.1.4, https://github.com/tidyverse/dplyr, https://dplyr.tidyverse.org.

Wilcox, R. R. (1987). New designs in analysis of variance. *Annual Review of Psychology, 38*, 29–60.

Wilcox, R. R. (1989). Comparing the variances of dependent groups. *Psychometrika, 54*, 305–315.

Wilcox, R. R. (1997). *Introduction to robust estimation and hypothesis testing*. San Diego, CA: Academic Press.

Wilcox, R. R., Peterson, T. J., & McNitt-Gray, J. L. (2018). Data analyses when sample sizes are small: Modern advances for dealing with outliers, skewed distributions, and heteroscedasticity. *Journal of Applied Biomechanics, 34*, 258–261. http://doi.org/10.1123/jab.2017-0269

Wiley, J., & Voss, J. F. (1999). Constructing arguments from multiple sources: Tasks that promote understanding and not just memory for texts. *Journal of Educational Psychology, 91*, 301–311.

Wilkinson, L. (1979). Tests of significance in stepwise regression. *Psychological Bulletin, 86*, 168–174.

Wilkinson, L. (1998). *SYSTAT 8 statistics manual*. Chicago, IL: SPSS, Inc.

Wilkinson, L., & Dallal, G. E. (1982). Tests of significance in forward selection regression with an *F*-to-enter stopping rule. *Technometrics, 24*, 25–28.

Wilkinson, L., & the Task Force on Statistical Inference (1999). Statistical methods in psychology journals: Guidelines and explanations. *American Psychologist, 54*, 594–604.

Williams, E. J. (1949). Experimental designs balanced for the estimation of residual effects of treatments. *Australian Journal of Scientific Research, Series A: Physical Sciences, 2*, 149–168.

Witvliet, C. V., Ludwig, T. E., & Vander Laan, K. L. (2001). Granting forgiveness or harboring grudges: Implications for emotion, physiology, and health. *Psychological Science, 12*, 117–123.

Wright, D. B., & Herrington, J. A. (2011). Problematic standard errors and confidence intervals for skewness and kurtosis. *Behavior Research, 43*, 8–17. http://doi.org/10.3758/s13428-010-0044-x

Yates, F. (1934). The analysis of multiple classifications with unequal numbers in the different classes. *Journal of the American Statistical Association, 29*, 57–66.

Yuen, K. K. (1974). The two-sample trimmed *t* for unequal population variances. *Biometrika, 61*, 165–170.

Zar, J. H. (1972). Significance testing of the Spearman rank correlation coefficient. *Journal of the American Statistical Association, 67*, 578–580.

Zeaman, D. (1949). Response latency as a function of the amount of reinforcement. *Journal of Experimental Psychology, 9*, 466–483.

Zimmerman, D. W. (2004). A note on preliminary tests of equality of variances. *British Journal of Mathematical and Statistical Psychology, 57*, 173–181. http://doi.org/10.1348/000711004849222

Zimmerman, D. W., & Zumbo, B. D. (1993). The relative power of parametric and nonparametric statistical methods. In G. Keren & C. Lewis (Eds.), *A handbook for data analysis in the behavioral sciences: Methodological issues* (pp. 481–517). Hillsdale, NJ: Lawrence Erlbaum Associates, Inc.

Zimmerman, D. W., Zumbo, B. D., & Williams, R. H. (2003). Bias in estimation and hypothesis testing of correlation. *Psicológica, 24*, 133–158.

Zou, G. Y. (2007). Toward using confidence intervals to compare correlations. *Psychological Methods, 12*, 399–413.

Zwick, R. (1993). Pairwise multiple comparison procedures for one-way analysis of variance designs. In G. Keren & C. Lewis (Eds.), *A handbook for data analysis in the behavioral sciences: Statistical issues* (pp. 43–71). Hillsdale, NJ: Lawrence Erlbaum Associates, Inc.

Index

Note: Page numbers in *italics* indicate a figure and page numbers in **bold** indicate a table on the corresponding page.

FSC
www.fsc.org
MIX
Papier | Fördert gute Waldnutzung
FSC® C083411

Kolibri 360 GmbH
Ferdinand-Jühlke-Straße 7
99095 Erfurt, Deutschland
produktsicherheit@kolibri360.de